UNIVERSITY OF NOTTINGHAM

10 0717182 X

W█████████████AWN

ROM THE LIBRARY

KU-279-885

Tyn Myint-U
Lokenath Debnath

Linear Partial
Differential Equations
for Scientists and Engineers

GEORGE GREEN LIBRARY OF
SCIENCE AND ENGINEERING

Fourth Edition

Birkhäuser
Boston • Basel • Berlin

Tyn Myint-U
5 Sue Terrace
Westport, CT 06880
USA

Lokenath Debnath
Department of Mathematics
University of Texas-Pan American
1201 W. University Drive
Edinburgh, TX 78539
USA

Cover design by Alex Gerasev.

Mathematics Subject Classification (2000): 00A06, 00A69, 34B05, 34B24, 34B27, 34G20, 35-01, 35-02, 35A15, 35A22, 35A25, 35C05, 35C15, 35Dxx, 35E05, 35E15, 35Fxx, 35F05, 35F10, 35F15, 35F20, 35F25, 35G10, 35G20, 35G25, 35J05, 35J10, 35J20, 35K05, 35K10, 35K15, 35K55, 35K60, 35L05, 35L10, 35L15, 35L20, 35L25, 35L30, 35L60, 35L65, 35L67, 35L70, 35Q30, 35Q35, 35Q40, 35Q51, 35Q53, 35Q55, 35Q58, 35Q60, 35Q80, 42A38, 44A10, 44A35 49J40, 58E30, 58E50, 65L15, 65M25, 65M30, 65R10, 70H05, 70H20, 70H25, 70H30, 76Bxx, 76B15, 76B25, 76D05, 76D33, 76E30, 76M30, 76R50, 78M30, 81Q05

Library of Congress Control Number: 2006935807

ISBN-10: 0-8176-4393-1   e-ISBN-10: 0-8176-4560-8
ISBN-13: 978-0-8176-4393-5   e-ISBN-13: 978-0-8176-4560-1

Printed on acid-free paper.

©2007 Birkhäuser Boston   *Birkhäuser*

All rights reserved. This work may not be translated or copied in whole or in part without the written permission of the publisher (Birkhäuser Boston, c/o Springer Science+Business Media LLC, 233 Spring Street, New York, NY 10013, USA) and the author, except for brief excerpts in connection with reviews or scholarly analysis. Use in connection with any form of information storage and retrieval, electronic adaptation, computer software, or by similar or dissimilar methodology now known or hereafter developed is forbidden.

The use in this publication of trade names, trademarks, service marks and similar terms, even if they are not identified as such, is not to be taken as an expression of opinion as to whether or not they are subject to proprietary rights.   ICOT17182x

9 8 7 6 5 4 3 2 1

www.birkhauser.com   (SB)

To the Memory of

U and Mrs. Hla Din

U and Mrs. Thant

Tyn Myint-U

In Loving Memory of

My Mother and Father

Lokenath Debnath

"True Laws of Nature cannot be linear."

"The search for truth is more precious than its possession."

"Everything should be made as simple as possible, but not a bit simpler."

Albert Einstein

"No human investigation can be called real science if it cannot be demonstrated mathematically."

Leonardo Da Vinci

"First causes are not known to us, but they are subjected to simple and constant laws that can be studied by observation and whose study is the goal of Natural Philosophy ... Heat penetrates, as does gravity, all the substances of the universe; its rays occupy all regions of space. The aim of our work is to expose the mathematical laws that this element follows ... The differential equations for the propagation of heat express the most general conditions and reduce physical questions to problems in pure Analysis that is properly the object of the theory."

James Clerk Maxwell

"One of the properties inherent in mathematics is that any real progress is accompanied by the discovery and development of new methods and simplifications of previous procedures ... The unified character of mathematics lies in its very nature; indeed, mathematics is the foundation of all exact natural sciences."

David Hilbert

" ... partial differential equations are the basis of all physical theorems. In the theory of sound in gases, liquid and solids, in the investigations of elasticity, in optics, everywhere partial differential equations formulate basic laws of nature which can be checked against experiments."

Bernhard Riemann

"The effective numerical treatment of partial differential equations is not a handicraft, but an art."

Folklore

"The advantage of the principle of least action is that in one and the same equation it relates the quantities that are immediately relevant not only to mechanics but also to electrodynamics and thermodynamics; these quantities are space, time and potential."

Max Planck

"The thorough study of nature is the most ground for mathematical discoveries."

"The equations for the flow of heat as well as those for the oscillations of acoustic bodies and of fluids belong to an area of analysis which has recently been opened, and which is worth examining in the greatest detail."

Joseph Fourier

"Of all the mathematical disciplines, the theory of differential equation is the most important. All branches of physics pose problems which can be reduced to the integration of differential equations. More generally, the way of explaining all natural phenomena which depend on time is given by the theory of differential equations."

Sophus Lie

"Differential equations form the basis for the scientific view of the world."

V.I. Arnold

"What we know is not much. What we do not know is immense."

"The algebraic analysis soon makes us forget the main object [of our research] by focusing our attention on abstract combinations and it is only at the end that we return to the original objective. But in abandoning one-self to the operations of analysis, one is led to the generality of this method and the inestimable advantage of transforming the reasoning by mechanical procedures to results often inaccessible by geometry ... No other language has the capacity for the elegance that arises from a long sequence of expressions linked one to the other and all stemming from one fundamental idea."

"It is India that gave us the ingenious method of expressing all numbers by ten symbols, each symbol receiving a value of position, as well as an absolute value. We shall appreciate the grandeur of the achievement when we remember that it escaped the genius of Archimedes and Appolonius."

P.S. Laplace

"The mathematician's best work is art, a high perfect art, as daring as the most secret dreams of imagination, clear and limpid. Mathematical genius and artistic genius touch one another."

Gòsta Mittag-Leffler

Contents

Preface to the Fourth Edition

"A teacher can never truly teach unless he is still learning himself. A lamp can never light another lamp unless it continues to burn its own flame. The teacher who has come to the end of his subject, who has no living traffic with his knowledge but merely repeats his lessons to his students, can only load their minds; he cannot quicken them."

Rabindranath Tagore
An Indian Poet
1913 Nobel Prize Winner for Literature

The previous three editions of our book were very well received and used as a senior undergraduate or graduate-level text and research reference in the United States and abroad for many years. We received many comments and suggestions from many students, faculty and researchers around the world. These comments and criticisms have been very helpful, beneficial, and encouraging. This fourth edition is the result of the input.

Another reason for adding this fourth edition to the literature is the fact that there have been major discoveries of new ideas, results and methods for the solution of linear and nonlinear partial differential equations in the second half of the twentieth century. It is becoming even more desirable for mathematicians, scientists and engineers to pursue study and research on these topics. So what has changed, and will continue to change is the nature of the topics that are of interest in mathematics, applied mathematics, physics and engineering, the evolution of books such is this one is a history of these shifting concerns.

This new and revised edition preserves the basic content and style of the third edition published in 1989. As with the previous editions, this book has been revised primarily as a comprehensive text for senior undergraduates or beginning graduate students and a research reference for professionals in mathematics, science and engineering, and other applied sciences. The main goal of the book is to develop required analytical skills on the part of the reader, rather than to focus on the importance of more abstract formulation, with full mathematical rigor. Indeed, our major emphasis is to provide an accessible working knowledge of the analytical and numerical methods with proofs required in mathematics, applied mathematics, physics, and engineering. The revised edition was greatly influenced by the statements that Lord Rayleigh and Richard Feynman made as follows:

"In the mathematical investigation I have usually employed such methods as present themselves naturally to a physicist. The pure mathematician will complain, and (it must be confessed) sometimes with justice, of deficient rigor. But to this question there are two sides. For, however important it may be to maintain a uniformly high standard in pure mathematics, the physicist may occasionally do well to rest content with arguments, which are fairly satisfactory and conclusive from his point of view. To his mind, exercised in a different order of ideas, the more severe procedure of the pure mathematician may appear not more but less demonstrative. And further, in many cases of difficulty to insist upon highest standard would mean the exclusion of the subject altogether in view of the space that would be required."

Lord Rayleigh

"... However, the emphasis should be somewhat more on how to do the mathematics quickly and easily, and what formulas are true, rather than the mathematicians' interest in methods of rigorous proof."

Richard P. Feynman

We have made many additions and changes in order to modernize the contents and to improve the clarity of the previous edition. We have also taken advantage of this new edition to correct typographical errors, and to update the bibliography, to include additional topics, examples of applications, exercises, comments and observations, and in some cases, to entirely rewrite and reorganize many sections. This is plenty of material in the book for a year-long course. Some of the material need not be covered in a course work and can be left for the readers to study on their own in order to prepare them for further study and research. This edition contains a collection of over 900 worked examples and exercises with answers and hints to selected exercises. Some of the major changes and additions include the following:

1. Chapter 1 on Introduction has been completely revised and a new section on historical comments was added to provide information about the historical developments of the subject. These changes have been made to provide the reader to see the direction in which the subject has developed and find those contributed to its developments.
2. A new Chapter 2 on first-order, quasi-linear, and linear partial differential equations, and method of characteristics has been added with many new examples and exercises.

3. Two sections on conservation laws, Burgers' equation, the Schrödinger and the Korteweg-de Vries equations have been included in Chapter 3.

4. Chapter 6 on Fourier series and integrals with applications has been completely revised and new material added, including a proof of the pointwise convergence theorem.

5. A new section on fractional partial differential equations has been added to Chapter 12 with many new examples of applications.

6. A new section on the Lax pair and the Zakharov and Shabat Scheme has been added to Chapter 13 to modernize its contents.

7. Some sections of Chapter 14 have been revised and a new short section on the finite element method has been added to this chapter.

8. A new Chapter 15 on tables of integral transforms has been added in order to make the book self-contained.

9. The whole section on Answers and Hints to Selected Exercises has been expanded to provide additional help to students. All figures have been redrawn and many new figures have been added for a clear understanding of physical explanations.

10. An Appendix on special functions and their properties has been expanded.

Some of the highlights in this edition include the following:

• The book offers a detailed and clear explanation of every concept and method that is introduced, accompanied by carefully selected worked examples, with special emphasis given to those topics in which students experience difficulty.

• A wide variety of modern examples of applications has been selected from areas of integral and ordinary differential equations, generalized functions and partial differential equations, quantum mechanics, fluid dynamics and solid mechanics, calculus of variations, linear and nonlinear stability analysis.

• The book is organized with sufficient flexibility to enable instructors to select chapters appropriate for courses of differing lengths, emphases, and levels of difficulty.

• A wide spectrum of exercises has been carefully chosen and included at the end of each chapter so the reader may further develop both rigorous skills in the theory and applications of partial differential equations and a deeper insight into the subject.

• Many new research papers and standard books have been added to the bibliography to stimulate new interest in future study and research. Index of the book has also been completely revised in order to include a wide variety of topics.

• The book provides information that puts the reader at the forefront of current research.

With the improvements and many challenging worked-out problems and exercises, we hope this edition will continue to be a useful textbook for students as well as a research reference for professionals in mathematics, applied mathematics, physics and engineering.

It is our pleasure to express our grateful thanks to many friends, colleagues, and students around the world who offered their suggestions and help at various stages of the preparation of the book. We offer special thanks to Dr. Andras Balogh, Mr. Kanadpriya Basu, and Dr. Dambaru Bhatta for drawing all figures, and to Mrs. Veronica Martinez for typing the manuscript with constant changes and revisions. In spite of the best efforts of everyone involved, some typographical errors doubtless remain. Finally, we wish to express our special thanks to Tom Grasso and the staff of Birkhäuser Boston for their help and cooperation.

Tyn Myint-U

Lokenath Debnath

Preface to the Third Edition

The theory of partial differential equations has long been one of the most important fields in mathematics. This is essentially due to the frequent occurrence and the wide range of applications of partial differential equations in many branches of physics, engineering, and other sciences. With much interest and great demand for theory and applications in diverse areas of science and engineering, several excellent books on PDEs have been published. This book is written to present an approach based mainly on the mathematics, physics, and engineering problems and their solutions, and also to construct a course appropriate for all students of mathematical, physical, and engineering sciences. Our primary objective, therefore, is not concerned with an elegant exposition of general theory, but rather to provide students with the fundamental concepts, the underlying principles, a wide range of applications, and various methods of solution of partial differential equations.

This book, a revised and expanded version of the second edition published in 1980, was written for a one-semester course in the theory and applications of partial differential equations. It has been used by advanced undergraduate or beginning graduate students in applied mathematics, physics, engineering, and other applied sciences. The prerequisite for its study is a standard calculus sequence with elementary ordinary differential equations. This revised edition is in part based on lectures given by Tyn Myint-U at Manhattan College and by Lokenath Debnath at the University of Central Florida. This revision preserves the basic content and style of the earlier editions, which were written by Tyn Myint-U alone. However, the authors have made some major additions and changes in this third edition in order to modernize the contents and to improve clarity. Two new chapters added are on nonlinear PDEs, and on numerical and approximation methods. New material emphasizing applications has been inserted. New examples and exercises have been provided. Many physical interpretations of mathematical solutions have been added. Also, the authors have improved the exposition by reorganizing some material and by making examples, exercises, and applications more prominent in the text. These additions and changes have been made with the student uppermost in mind.

The first chapter gives an introduction to partial differential equations. The second chapter deals with the mathematical models representing physical and engineering problems that yield the three basic types of PDEs. Included are only important equations of most common interest in physics and engineering. The third chapter constitutes an account of the classification of linear PDEs of second order in two independent variables into hyperbolic, parabolic, and elliptic types and, in addition, illustrates the determination of the general solution for a class of relatively simple equations.

Cauchy's problem, the Goursat problem, and the initial boundary-value problems involving hyperbolic equations of the second order are presented in Chapter 4. Special attention is given to the physical significance of solutions and the methods of solution of the wave equation in Cartesian, spherical polar, and cylindrical polar coordinates. The fifth chapter contains a fuller treatment of Fourier series and integrals essential for the study of PDEs. Also included are proofs of several important theorems concerning Fourier series and integrals.

Separation of variables is one of the simplest methods, and the most widely used method, for solving PDEs. The basic concept and separability conditions necessary for its application are discussed in the sixth chapter. This is followed by some well-known problems of applied mathematics, mathematical physics, and engineering sciences along with a detailed analysis of each problem. Special emphasis is also given to the existence and uniqueness of the solutions and to the fundamental similarities and differences in the properties of the solutions to the various PDEs. In Chapter 7, self-adjoint eigenvalue problems are treated in depth, building on their introduction in the preceding chapter. In addition, Green's function and its applications to eigenvalue problems and boundary-value problems for ordinary differential equations are presented. Following the general theory of eigenvalues and eigenfunctions, the most common special functions, including the Bessel, Legendre, and Hermite functions, are discussed as examples of the major role of special functions in the physical and engineering sciences. Applications to heat conduction problems and the Schrödinger equation for the linear harmonic oscillator are also included.

Boundary-value problems and the maximum principle are described in Chapter 8, and emphasis is placed on the existence, uniqueness, and well-posedness of solutions. Higher-dimensional boundary-value problems and the method of eigenfunction expansion are treated in the ninth chapter, which also includes several applications to the vibrating membrane, waves in three dimensions, heat conduction in a rectangular volume, the three-dimensional Schrödinger equation in a central field of force, and the hydrogen atom. Chapter 10 deals with the basic concepts and construction of Green's function and its application to boundary-value problems.

Chapter 11 provides an introduction to the use of integral transform methods and their applications to numerous problems in applied mathematics, mathematical physics, and engineering sciences. The fundamental properties and the techniques of Fourier, Laplace, Hankel, and Mellin transforms are discussed in some detail. Applications to problems concerning heat flows, fluid flows, elastic waves, current and potential electric transmission lines are included in this chapter.

Chapters 12 and 13 are entirely new. First-order and second-order nonlinear PDEs are covered in Chapter 12. Most of the contents of this chapter have been developed during the last twenty-five years. Several new nonlinear PDEs including the one-dimensional nonlinear wave equation, Whitham's equation, Burgers' equation, the Korteweg–de Vries equation, and the nonlinear Schrödinger equation are solved. The solutions of these equations are then discussed with physical significance. Special emphasis is given to the fundamental similarities and differences in the properties of the solutions to the corresponding linear and nonlinear equations under consideration.

The final chapter is devoted to the major numerical and approximation methods for finding solutions of PDEs. A fairly detailed treatment of explicit and implicit finite difference methods is given with applications The variational method and the Euler–Lagrange equations are described with many applications. Also included are the Rayleigh–Ritz, the Galerkin, and the Kantorovich methods of approximation with many illustrations and applications.

This new edition contains almost four hundred examples and exercises, which are either directly associated with applications or phrased in terms of the physical and engineering contexts in which they arise. The exercises truly complement the text, and answers to most exercises are provided at the end of the book. The Appendix has been expanded to include some basic properties of the Gamma function and the tables of Fourier, Laplace, and Hankel transforms. For students wishing to know more about the subject or to have further insight into the subject matter, important references are listed in the Bibliography.

The chapters on mathematical models, Fourier series and integrals, and eigenvalue problems are self-contained, so these chapters can be omitted for those students who have prior knowledge of the subject.

An attempt has been made to present a clear and concise exposition of the mathematics used in analyzing a variety of problems. With this in mind, the chapters are carefully organized to enable students to view the material in an orderly perspective. For example, the results and theorems in the chapters on Fourier series and integrals and on eigenvalue problems are explicitly mentioned, whenever necessary, to avoid confusion with their use in the development of PDEs. A wide range of problems subject to various boundary conditions has been included to improve the student's understanding.

In this third edition, specific changes and additions include the following:

1. Chapter 2 on mathematical models has been revised by adding a list of the most common linear PDEs in applied mathematics, mathematical physics, and engineering science.

2. The chapter on the Cauchy problem has been expanded by including the wave equations in spherical and cylindrical polar coordinates. Examples and exercises on these wave equations and the energy equation have been added.

3. Eigenvalue problems have been revised with an emphasis on Green's functions and applications. A section on the Schrödinger equation for the linear harmonic oscillator has been added. Higher-dimensional boundary-value problems with an emphasis on applications, and a section on the hydrogen atom and on the three-dimensional Schrödinger equation in a central field of force have been added to Chapter 9.

4. Chapter 11 has been extensively reorganized and revised in order to include Hankel and Mellin transforms and their applications, and has new sections on the asymptotic approximation method and the finite Hankel transform with applications. Many new examples and exercises, some new material with applications, and physical interpretations of mathematical solutions have also been included.

5. A new chapter on nonlinear PDEs of current interest and their applications has been added with considerable emphasis on the fundamental similarities and the distinguishing differences in the properties of the solutions to the nonlinear and corresponding linear equations.

6. Chapter 13 is also new. It contains a fairly detailed treatment of explicit and implicit finite difference methods with their stability analysis. A large section on the variational methods and the Euler–Lagrange equations has been included with many applications. Also included are the Rayleigh–Ritz, the Galerkin, and the Kantorovich methods of approximation with illustrations and applications.

7. Many new applications, examples, and exercises have been added to deepen the reader's understanding. Expanded versions of the tables of Fourier, Laplace, and Hankel transforms are included. The bibliography has been updated with more recent and important references.

As a text on partial differential equations for students in applied mathematics, physics, engineering, and applied sciences, this edition provides the student with the art of combining mathematics with intuitive and physical thinking to develop the most effective approach to solving problems.

In preparing this edition, the authors wish to express their sincere thanks to those who have read the manuscript and offered many valuable suggestions and comments. The authors also wish to express their thanks to the editor and the staff of Elsevier–North Holland, Inc. for their kind help and cooperation.

Tyn Myint-U
Lokenath Debnath

1

Introduction

"If you wish to foresee the future of mathematics, our proper course is to study the history and present condition of the science."

Henri Poincaré

"However varied may be the imagination of man, nature is a thousand times richer, ... Each of the theories of physics ... presents (partial differential) equations under a new aspect ... without the theories, we should not know partial differential equations."

Henri Poincaré

1.1 Brief Historical Comments

Historically, partial differential equations originated from the study of surfaces in geometry and a wide variety of problems in mechanics. During the second half of the nineteenth century, a large number of famous mathematicians became actively involved in the investigation of numerous problems presented by partial differential equations. The primary reason for this research was that partial differential equations both express many fundamental laws of nature and frequently arise in the mathematical analysis of diverse problems in science and engineering.

The next phase of the development of linear partial differential equations was characterized by efforts to develop the general theory and various methods of solution of linear equations. In fact, partial differential equations have been found to be essential to the theory of surfaces on the one hand and to the solution of physical problems on the other. These two areas of mathematics can be seen as linked by the bridge of the calculus of variations. With the discovery of the basic concepts and properties of distributions, the modern theory of linear partial differential equations is now well established. The subject plays a central role in modern mathematics, especially in physics, geometry, and analysis.

Almost all physical phenomena obey mathematical laws that can be formulated by differential equations. This striking fact was first discovered by Isaac Newton (1642–1727) when he formulated the laws of mechanics and applied them to describe the motion of the planets. During the three centuries since Newton's fundamental discoveries, many partial differential equations that govern physical, chemical, and biological phenomena have been found and successfully solved by numerous methods. These equations include Euler's equations for the dynamics of rigid bodies and for the motion of an ideal fluid, Lagrange's equations of motion, Hamilton's equations of motion in analytical mechanics, Fourier's equation for the diffusion of heat, Cauchy's equation of motion and Navier's equation of motion in elasticity, the Navier–Stokes equations for the motion of viscous fluids, the Cauchy–Riemann equations in complex function theory, the Cauchy–Green equations for the static and dynamic behavior of elastic solids, Kirchhoff's equations for electrical circuits, Maxwell's equations for electromagnetic fields, and the Schrödinger equation and the Dirac equation in quantum mechanics. This is only a sampling, and the recent mathematical and scientific literature reveals an almost unlimited number of differential equations that have been discovered to model physical, chemical and biological systems and processes.

From the very beginning of the study, considerable attention has been given to the geometric approach to the solution of differential equations. The fact that families of curves and surfaces can be defined by a differential equation means that the equation can be studied geometrically in terms of these curves and surfaces. The curves involved, known as *characteristic curves*, are very useful in determining whether it is or is not possible to find a surface containing a given curve and satisfying a given differential equation. This geometric approach to differential equations was begun by Joseph-Louis Lagrange (1736–1813) and Gaspard Monge (1746–1818). Indeed, Monge first introduced the ideas of characteristic surfaces and characteristic cones (or Monge cones). He also did some work on second-order linear, homogeneous partial differential equations.

The study of first-order partial differential equations began to receive some serious attention as early as 1739, when Alex-Claude Clairaut (1713–1765) encountered these equations in his work on the shape of the earth. On the other hand, in the 1770s Lagrange first initiated a systematic study of the first-order nonlinear partial differential equations in the form

$$f\left(x, y, u, u_x, u_y\right) = 0, \tag{1.1.1}$$

where $u = u\left(x, y\right)$ is a function of two independent variables.

Motivated by research on gravitational effects on bodies of different shapes and mass distributions, another major impetus for work in partial differential equations originated from potential theory. Perhaps the most important partial differential equation in applied mathematics is the *potential equation*, also known as the *Laplace equation* $u_{xx} + u_{yy} = 0$, where subscripts denote partial derivatives. This equation arose in steady state heat conduction problems involving homogeneous solids. James Clerk Maxwell (1831–1879) also gave a new initiative to potential theory through his famous equations, known as *Maxwell's equations* for electromagnetic fields.

Lagrange developed analytical mechanics as the application of partial differential equations to the motion of rigid bodies. He also described the geometrical content of a first-order partial differential equation and developed the method of characteristics for finding the general solution of quasi-linear equations. At the same time, the specific solution of physical interest was obtained by formulating an *initial-value problem* (or a *Cauchy Problem*) that satisfies certain supplementary conditions. The solution of an initial-value problem still plays an important role in applied mathematics, science and engineering. The fundamental role of characteristics was soon recognized in the study of quasi-linear and nonlinear partial differential equations. Physically, the first-order, quasi-linear equations often represent conservation laws which describe the conservation of some physical quantities of a system.

In its early stages of development, the theory of second-order linear partial differential equations was concentrated on applications to mechanics and physics. All such equations can be classified into three basic categories: the wave equation, the heat equation, and the Laplace equation (or potential equation). Thus, a study of these three different kinds of equations yields much information about more general second-order linear partial differential equations. Jean d'Alembert (1717–1783) first derived the one-dimensional wave equation for vibration of an elastic string and solved this equation in 1746. His solution is now known as the *d'Alembert solution*. The wave equation is one of the oldest equations in mathematical physics. Some form of this equation, or its various generalizations, almost inevitably arises in any mathematical analysis of phenomena involving the propagation of waves in a continuous medium. In fact, the studies of water waves, acoustic waves, elastic waves in solids, and electromagnetic waves are all based on this equation. A technique known as the *method of separation of variables* is perhaps one of the oldest systematic methods for solving partial differential equations including the wave equation. The wave equation and its methods of solution attracted the attention of many famous mathematicians including Leonhard Euler (1707–1783), James Bernoulli (1667–1748), Daniel Bernoulli (1700–1782), J.L. Lagrange (1736–1813), and Jacques Hadamard (1865–1963). They discovered solutions in several different forms, and the merit of their solutions and relations among these solutions were argued in a series of papers extending over more than twenty-five years; most concerned the nature of the kinds of functions that can be represented by trigonometric (or Fourier) series. These controversial problems were finally resolved during the nineteenth century.

It was Joseph Fourier (1768–1830) who made the first major step toward developing a general method of solutions of the equation describing the conduction of heat in a solid body in the early 1800s. Although Fourier is most celebrated for his work on the conduction of heat, the mathematical methods involved, particularly trigonometric series, are important and very useful in many other situations. He created a coherent mathematical method by which the different components of an equation and its solution in series were neatly identified with the different aspects of the physical solution being analyzed. In spite of the striking success of Fourier analysis as one of the most useful mathematical methods, J.L. Lagrange and S.D. Poisson (1781–1840) hardly recognized Fourier's work because of its lack of rigor. Nonetheless, Fourier was eventually recognized for his pioneering work after publication of his monumental treatise entitled *La Théorie Auatytique de la Chaleur* in 1822.

It is generally believed that the concept of an integral transform originated from the Integral Theorem as stated by Fourier in his 1822 treatise. It was the work of Augustin Cauchy (1789–1857) that contained the exponential form of the Fourier Integral Theorem as

$$f(x) = \frac{1}{2\pi} \int_{-\infty}^{\infty} e^{ikx} \left[\int_{-\infty}^{\infty} e^{-ik\xi} f(\xi) \, d\xi \right] dk. \qquad (1.1.2)$$

This theorem has been expressed in several slightly different forms to better adapt it for particular applications. It has been recognized, almost from the start, however, that the form which best combines mathematical simplicity and complete generality makes use of the exponential oscillating function $\exp(ikx)$. Indeed, the Fourier integral formula (1.1.2) is regarded as one of the most fundamental results of modern mathematical analysis, and it has widespread physical and engineering applications. The generality and importance of the theorem is well expressed by Kelvin and Tait who said: " ... Fourier's Theorem, which is not only one of the most beautiful results of modern analysis, but may be said to furnish an indispensable instrument in the treatment of nearly every recondite question in modern physics. To mention only sonorous vibrations, the propagation of electric signals along a telegraph wire, and the conduction of heat by the earth's crust, as subjects in their generality intractable without it, is to give but a feeble idea of its importance." This integral formula (1.1.2) is usually used to define the classical Fourier transform of a function and the inverse Fourier transform. No doubt, the scientific achievements of Joseph Fourier have not only provided the fundamental basis for the study of heat equation, Fourier series, and Fourier integrals, but for the modern developments of the theory and applications of the partial differential equations.

One of the most important of all the partial differential equations involved in applied mathematics and mathematical physics is that associated with the name of Pierre-Simon Laplace (1749–1827). This equation was first discovered by Laplace while he was involved in an extensive study of gravitational attraction of arbitrary bodies in space. Although the main field of Laplace's research was celestial mechanics, he also made important contributions to the theory of probability and its applications. This work introduced the method known later as the Laplace transform, a simple and elegant method of solving differential and integral equations. Laplace first introduced the concept of *potential*, which is invaluable in a wide range of subjects, such as gravitation, electromagnetism, hydrodynamics, and acoustics. Consequently, the Laplace equation is often referred to as the *potential equation*. This equation is also an important special case of both the wave equation and the heat equation in two or three dimensions. It arises in the study of many physical phenomena including electrostatic or gravitational potential, the velocity potential for an imcompossible fluid flows, the steady state heat equation, and the equilibrium (time independent) displacement field of a two- or three-dimensional elastic membrane. The Laplace equation also occurs in other branches of applied mathematics and mathematical physics.

Since there is no time dependence in any of the mathematical problems stated above, there are no initial data to be satisfied by the solutions of the Laplace equation. They must, however, satisfy certain boundary conditions on the boundary curve or surface of a region in which the Laplace equation is to be solved. The problem of finding a solution of Laplace's equation that takes on the given boundary values is known as the *Dirichlet boundary-value problem*, after Peter Gustav Lejeune Dirichlet (1805–1859). On the other hand, if the values of the normal derivative are prescribed on the boundary, the problem is known as *Neumann boundary-value problem*, in honor of Karl Gottfried Neumann (1832–1925). Despite great efforts by many mathematicians including Gaspard Monge (1746–1818), Adrien-Marie Legendre (1752–1833), Carl Friedrich Gauss (1777–1855), Simeon-Denis Poisson (1781–1840), and Jean Victor Poncelet (1788–1867), very little was known about the general properties of the solutions of Laplace's equation until 1828, when George Green (1793–1841) and Mikhail Ostrogradsky (1801–1861) independently investigated properties of a class of solutions known as *harmonic functions*. On the other hand, Augustin Cauchy (1789–1857) and Bernhard Riemann (1826–1866) derived a set of first-order partial differential equations, known as the *Cauchy–Riemann equations*, in their independent work on functions of complex variables. These equations led to the Laplace equation, and functions satisfying this equation in a domain are called *harmonic functions* in that domain. Both Cauchy and Riemann occupy a special place in the history of mathematics. Riemann made enormous contributions to almost all areas of pure and applied mathematics. His extraordinary achievements stimulated further developments, not only in mathematics, but also in mechanics, physics, and the natural sciences as a whole.

Augustin Cauchy is universally recognized for his fundamental contributions to complex analysis. He also provided the first systematic and rigorous investigation of differential equations and gave a rigorous proof for the existence of power series solutions of a differential equation in the 1820s. In 1841 Cauchy developed what is known as the *method of majorants* for proving that a solution of a partial differential equation exists in the form of a power series in the independent variables. The method of majorants was also introduced independently by Karl Weierstrass (1815–1896) in that same year in application to a system of differential equations. Subsequently, Weierstrass's student Sophie Kowalewskaya (1850–1891) used the method of majorants and a normalization theorem of Carl Gustav Jacobi (1804–1851) to prove an exceedingly elegant theorem, known as the *Cauchy–Kowalewskaya theorem*. This theorem quite generally asserts the local existence of solutions of a system of partial differential equations with initial conditions on a noncharacteristic surface. This theorem seems to have little practical importance because it does not distinguish between well-posed and ill-posed problems; it covers situations where a small change in the initial data leads to a large change in the solution. Historically, however, it is the first existence theorem for a general class of partial differential equations.

The general theory of partial differential equations was initiated by A.R. Forsyth (1858–1942) in the fifth and sixth volumes of his *Theory of Differential Equations* and by E.J.B. Goursat (1858–1936) in his book entitled *Cours d' analyse mathematiques* (1918) and his *Lecons sur l' integration des equations aux dérivées*, volume 1 (1891) and volume 2 (1896). Another notable contribution to this subject was made by E. Cartan's book, *Lecons sur les invariants intégraux*, published in 1922. Joseph Liouville (1809–1882) formulated a more tractable partial differential equation in the form

$$u_{xx} + u_{yy} = k \exp(au), \tag{1.1.3}$$

and obtained a general solution of it. This equation has a large number of applications. It is a special case of the equation derived by J.L. Lagrange for the stream function ψ in the case of two-dimensional steady vortex motion in an incompossible fluid, that is,

$$\psi_{xx} + \psi_{yy} = F(\psi), \tag{1.1.4}$$

where $F(\psi)$ is an arbitrary function of ψ. When $\psi = u$ and $F(u) = ke^{au}$, equation (1.1.4) reduces to the Liouville equation (1.1.3). In view of the special mathematical interest in the nonhomogeneous nonlinear equation of the type (1.1.4), a number of famous mathematicians including Henri Poincaré, E. Picard (1856–1941), Cauchy (1789–1857), Sophus Lie (1842–1899), L.M.H. Navier (1785–1836), and G.G. Stokes (1819–1903) made many major contributions to partial differential equations.

Historically, Euler first solved the eigenvalue problem when he developed a simple mathematical model for describing the the 'buckling' modes of a vertical elastic beam. The general theory of eigenvalue problems for second-order differential equations, now known as the *Sturm–Liouville Theory*, originated from the study of a class of boundary-value problems due to

Charles Sturm (1803–1855) and Joseph Liouville (1809–1882). They showed that, in general, there is an infinite set of eigenvalues satisfying the given equation and the associated boundary conditions, and that these eigenvalues increase to infinity. Corresponding to these eigenvalues, there is an infinite set of orthogonal eigenfunctions so that the linear superposition principle can be applied to find the convergent infinite series solution of the given problem. Indeed, the Sturm–Liouville theory is a natural generalization of the theory of Fourier series that greatly extends the scope of the method of separation of variables. In 1926, the WKB approximation method was developed by Gregor Wentzel, Hendrik Kramers, and Marcel-Louis Brillouin for finding the approximate eigenvalues and eigenfunctions of the one-dimensional Schrödinger equation in quantum mechanics. This method is now known as the *short-wave approximation* or the *geometrical optics approximation* in wave propagation theory.

At the end of the seventeenth century, many important questions and problems in geometry and mechanics involved minimizing or maximizing of certain integrals for two reasons. The first of these were several existence problems, such as, Newton's problem of missile of least resistance, Bernoulli's isoperimetric problem, Bernoulli's problem of the brachistochrone (*brachistos* means shortest, *chronos* means time), the problem of minimal surfaces due to Joseph Plateau (1801–1883), and Fermat's principle of least time. Indeed, the variational principle as applied to the propagation and reflection of light in a medium was first enunciated in 1662 by one of the greatest mathematicians of the seventeenth century, Pierre Fermat (1601–1665). According to his principle, a ray of light travels in a homogeneous medium from one point to another along a path in a minimum time. The second reason is somewhat philosophical, that is, how to discover a minimizing principle in nature. The following 1744 statement of Euler is characteristic of the philosophical origin of what is known as the *principle of least action*: "As the construction of the universe is the most perfect possible, being the handiwork of all-wise Maker, nothing can be met with in the world in which some maximal or minimal property is not displayed. There is, consequently, no doubt but all the effects of the world can be derived by the method of maxima and minima from their final causes as well as from their efficient ones." In the middle of the eighteenth century, Pierre de Maupertius (1698–1759) stated a fundamental principle, known as the *principle of least action*, as a guide to the nature of the universe. A still more precise and general formulation of Maupertius' principle of least action was given by Lagrange in his *Analytical Mechanics* published in 1788. He formulated it as

$$\delta S = \delta \int_{t_1}^{t_2} (2T) \, dt = 0, \tag{1.1.5}$$

where T is the kinematic energy of a dynamical system with the constraint that the total energy, $(T + V)$, is constant along the trajectories, and V is the potential energy of the system. He also derived the celebrated equation of motion for a holonomic dynamical system

$$\frac{d}{dt}\left(\frac{\partial T}{\partial \dot{q}_i}\right) - \frac{\partial T}{\partial q_i} = Q_i, \tag{1.1.6}$$

where q_i are the generalized coordinates, $\dot{q}_i$ is the velocity, and Q_i is the force. For a conservative dynamical system, $Q_i = -\frac{\partial V}{\partial q_i}$, $V = V(q_i)$, $\frac{\partial V}{\partial \dot{q}_i} = 0$, then (1.1.6) can be expressed in terms of the Lagrangian, $L = T - V$, as

$$\frac{d}{dt}\left(\frac{\partial L}{\partial \dot{q}_i}\right) - \frac{\partial L}{\partial q_i} = 0. \tag{1.1.7}$$

This principle was then reformulated by Euler in a way that made it useful in mathematics and physics.

The work of Lagrange remained unchanged for about half a century until William R. Hamilton (1805–1865) published his research on the general method in analytical dynamics which gave a new and very appealing form to the Lagrange equations. Hamilton's work also included his own variational principle. In his work on optics during 1834–1835, Hamilton elaborated a new principle of mechanics, known as *Hamilton's principle*, describing the stationary action for a conservative dynamical system in the form

$$\delta A = \delta \int_{t_0}^{t_1} (T - V)\, dt = \delta \int_{t_0}^{t_1} L\, dt = 0. \tag{1.1.8}$$

Hamilton's principle (1.1.8) readily led to the Lagrange equation (1.1.6). In terms of time t, the generalized coordinates q_i, and the generalized momenta $p_i = (\partial L / \dot{q}_i)$ which characterize the state of a dynamical system, Hamilton introduced the function

$$H(q_i, p_i, t) = p_i \dot{q}_i - L(q_i, p_i, t), \tag{1.1.9}$$

and then used it to represent the equation of motion (1.1.6) as a system of first order partial differential equations

$$\dot{q}_i = \frac{\partial H}{\partial p_i}, \qquad \dot{p}_i = -\frac{\partial H}{\partial \dot{q}_i}. \tag{1.1.10}$$

These equations are known as the celebrated *Hamilton canonical equations of motion*, and the function $H(q_i, p_i, t)$ is referred to as the *Hamiltonian* which is equal to the total energy of the system. Following the work of Hamilton, Karl Jacobi, Mikhail Ostrogradsky (1801–1862), and Henri Poincaré (1854–1912) put forth new modifications of the variational principle. Indeed, the action integral S can be regarded as a function of generalized coordinates and time provided the terminal point is not fixed. In 1842, Jacobi showed that S satisfies the first-order partial differential equation

$$\frac{\partial S}{\partial t} + H\left(q_i, \frac{\partial S}{\partial q_i}, t\right) = 0, \tag{1.1.11}$$

which is known as the *Hamilton–Jacobi equation*. In 1892, Poincaré defined the action integral on the trajectories in phase space of the variable q_i and p_i as

$$S = \int_{t_0}^{t_1} [p_i\dot{q}_i - H(p_i, q_i)]\, dt, \tag{1.1.12}$$

and then formulated another modification of the Hamilton variational principle which also yields the Hamilton canonical equations (1.1.10). From (1.1.12) also follows the celebrated Poincaré–Cartan invariant

$$I = \oint_C (p_i\delta q_i - H\delta t), \tag{1.1.13}$$

where C is an arbitrary closed contour in the phase space.

Indeed, the discovery of the calculus of variations in a modern sense began with the independent work of Euler and Lagrange. The first necessary condition for the existence of an extremum of a functional in a domain leads to the celebrated *Euler–Lagrange equation*. This equation in its various forms now assumes primary importance, and more emphasis is given to the first variation, mainly due to its power to produce significant equations, than to the second variation, which is of fundamental importance in answering the question of whether or not an extremal actually provides a minimum (or a maximum). Thus, the fundamental concepts of the calculus of variations were developed in the eighteenth century in order to obtain the differential equations of applied mathematics and mathematical physics. During its early development, the problems of the calculus of variations were reduced to questions of the existence of differential equations problems until David Hilbert developed a new method in which the existence of a minimizing function was established directly as the limit of a sequence of approximations.

Considerable attention has been given to the problem of finding a necessary and sufficient condition for the existence of a function which extremized the given functional. Although the problem of finding a sufficient condition is a difficult one, Legendre and C.G.J. Jacobi (1804–1851) discovered a second necessary condition and a third necessary condition respectively. Finally, it was Weierstrass who first provided a satisfactory foundation to the theory of calculus of variations in his lectures at Berlin between 1856 and 1870. His lectures were essentially concerned with a complete review of the work of Legendre and Jacobi. At the same time, he reexamined the concepts of the first and second variations and looked for a sufficient condition associated with the problem. In contrast to the work of his predecessors, Weierstrass introduced the ideas of 'strong variations' and 'the excess function' which led him to discover a fourth necessary condition and a satisfactory sufficient condition. Some of his outstanding discoveries announced in his lectures were published in his collected work. At the conclusion of his famous lecture on 'Mathematical Problems' at the Paris International Congress of Mathematicians in 1900, David Hilbert (1862–1943), perhaps the most brilliant mathematician of the late nineteenth century, gave a new method for the discussion of the minimum value of a functional. He obtained another derivation of Weierstrass's excess function and a new approach to Jacobi's problem of determining necessary and sufficient conditions for the existence of a minimum of a functional; all this without the use of the second variation. Finally, the calculus of variations entered the new and wider field of 'global' problems with the original work of George D. Birkhoff (1884–1944) and his associates. They succeeded in liberating the theory of calculus of variations from the limitations imposed by the restriction to 'small variations', and gave a general treatment of the global theory of the subject with large variations.

In 1880, George Fitzgerald (1851–1901) probably first employed the variational principle in electromagnetic theory to derive Maxwell's equations for an electromagnetic field in a vacuum. Moreover, the variational principle received considerable attention in electromagnetic theory after the work of Karl Schwarzchild in 1903 as well as the work of Max Born (1882–1970) who formulated the principle of stationary action in electrodynamics in a symmetric four-dimensional form. On the other hand, Poincaré showed in 1905 that the action integral is invariant under the Lorentz transformations. With the development of the special theory of relativity and the relativistic theory of gravitation in the beginning of the twentieth century, the variational principles received tremendous attention from many great mathematicians and physicists including Albert Einstein (1879–1955), Hendrix Lorentz (1853–1928), Hermann Weyl (1885–1955), Felix Klein (1849–1925), Amalie Noether (1882–1935), and David Hilbert. Even before the use of variational principles in electrodynamics, Lord Rayleigh (1842–1919) employed variational methods in his famous book, *The Theory of Sound*, for the derivation of equations for oscillations in plates and rods in order to calculate frequencies of natural oscillations of elastic systems. In his pioneering work in the 1960's, Gerald Whitham first developed a general approach to linear and nonlinear dispersive waves using a Lagrangian. He successfully formulated the averaged variational principle, which is now known as *the Whitham averaged variational principle*, which was employed to derive the basic equations for linear and nonlinear dispersive wave propagation problems. In 1967, Luke first explicitly formulated a variational principle for nonlinear water waves. In 1968, Bretherton and Garret generalized the Whitham averaged variational principle to describe the conservation law for the wave action in a moving medium. Subsequently, Ostrovsky and Pelinovsky (1972) also generalized the Whitham averaged variational principle to nonconservative systems.

With the rapid development of the theory and applications of differential equations, the closed form analytical solutions of many different types of equations were hardly possible. However, it is extremely important and absolutely necessary to provide some insight into the qualitative and quantitative nature of solutions subject to initial and boundary conditions. This insight usually takes the form of numerical and graphical representatives of the solutions. It was E. Picard (1856–1941) who first developed the method of successive approximations for the solutions of differential equations in most general form and later made it an essential part of his treatment of differential equations in the second volume of his *Traité d'Analyse* published in 1896. During the last two centuries, the calculus of finite differences in various forms played a significant role in finding the numerical solutions of differential equations. Historically, many well known integration formulas and numerical methods including the Euler–Maclaurin formula, Gregory integration formula, the Gregory–Newton formula, Simpson's rule, Adam–Bashforth's method, the Jacobi iteration, the Gauss–Seidel method, and the Runge–Kutta method have been developed and then generalized in various forms.

With the development of modern calculators and high-speed electronic computers, there has been an increasing trend in research toward the numerical solution of ordinary and partial differential equations during the twentieth century. Special attention has also given to in depth studies of convergence, stability, error analysis, and accuracy of numerical solutions. Many well-known numerical methods including the Crank–Nicolson methods, the Lax–Wendroff method, Richtmyer's method, and Stone's implicit iterative technique have been developed in the second half of the twentieth century. All finite difference methods reduce differential equations to discrete forms. In recent years, more modern and powerful computational methods such as the finite element method and the boundary element method have been developed in order to handle curved or irregularly shaped domains. These methods are distinguished by their more general character, which makes them more capable of dealing with complex geometries, allows them to use non-structured grid systems, and allows more natural imposition of the boundary conditions.

During the second half of the nineteenth century, considerable attention was given to problems concerning the existence, uniqueness, and stability of solutions of partial differential equations. These studies involved not only the Laplace equation, but the wave and diffusion equations as well, and were eventually extended to partial differential equations with variable coefficients. Through the years, tremendous progress has been made on the general theory of ordinary and partial differential equations. With the advent of new ideas and methods, new results and applications, both analytical and numerical studies are continually being added to this subject. Partial differential equations have been the subject of vigorous mathematical research for over three centuries and remain so today. This is an active area of research for mathematicians and scientists. In part, this is motivated by the large number of problems in partial differential equations that mathematicians, scientists, and engineers are faced with that are seemingly intractable. Many of these equations are nonlinear and come from such areas of applications as fluid mechanics, plasma physics, nonlinear optics, solid mechanics, biomathematics, and quantum field theory. Owing to the ever increasing need in mathematics, science, and engineering to solve more and more complicated real world problems, it seems quite likely that partial differential equations will remain a major area of research for many years to come.

1.2 Basic Concepts and Definitions

A differential equation that contains, in addition to the dependent variable and the independent variables, one or more partial derivatives of the dependent variable is called a *partial differential equation*. In general, it may be written in the form

$$f(x, y, \ldots, u, u_x, u_y, \ldots, u_{xx}, u_{xy}, \ldots) = 0, \qquad (1.2.1)$$

involving several independent variables x, y, ..., an unknown function u of these variables, and the partial derivatives u_x, u_y, ..., u_{xx}, u_{xy}, ..., of the function. Subscripts on dependent variables denote differentiations, e.g.,

$$u_x = \partial u / \partial x, \qquad u_{xy} = \partial^2 / \partial y \, \partial x.$$

Here equation (1.2.1) is considered in a suitable domain D of the n-dimensional space R^n in the independent variables x, y, ... We seek functions $u = u(x, y, \ldots)$ which satisfy equation (1.2.1) identically in D. Such functions, if they exist, are called *solutions* of equation (1.2.1). From these many possible solutions we attempt to select a particular one by introducing suitable additional conditions.

For instance,

$$uu_{xy} + u_x = y,$$
$$u_{xx} + 2yu_{xy} + 3xu_{yy} = 4\sin x, \qquad (1.2.2)$$
$$(u_x)^2 + (u_y)^2 = 1,$$
$$u_{xx} - u_{yy} = 0,$$

are partial differential equations. The functions

$$u(x, y) = (x + y)^3,$$
$$u(x, y) = \sin(x - y),$$

are solutions of the last equation of (1.2.2), as can easily be verified.

The *order* of a partial differential equation is the order of the highest-ordered partial derivative appearing in the equation. For example

$$u_{xx} + 2xu_{xy} + u_{yy} = e^y$$

is a second-order partial differential equation, and

$$u_{xxy} + xu_{yy} + 8u = 7y$$

is a third-order partial differential equation.

A partial differential equation is said to be *linear* if it is linear in the unknown function and all its derivatives with coefficients depending only on the independent variables; it is said to be *quasi-linear* if it is linear in the highest-ordered derivative of the unknown function. For example, the equation

$$yu_{xx} + 2xyu_{yy} + u = 1$$

is a second-order linear partial differential equation, whereas

$$u_x u_{xx} + xuu_y = \sin y$$

is a second-order quasi-linear partial differential equation. The equation which is not linear is called a *nonlinear* equation.

We shall be primarily concerned with linear second-order partial differential equations, which frequently arise in problems of mathematical physics. The most general second-order linear partial differential equation in n independent variables has the form

$$\sum_{i,j=1}^{n} A_{ij} u_{x_i x_j} + \sum_{i=1}^{n} B_i u_{x_i} + Fu = G, \qquad (1.2.3)$$

where we assume without loss of generality that $A_{ij} = A_{ji}$. We also assume that B_i, F, and G are functions of the n independent variables x_i.

If G is identically zero, the equation is said to be *homogeneous*; otherwise it is *nonhomogeneous*.

The general solution of a linear ordinary differential equation of nth order is a family of functions depending on n independent arbitrary constants. In the case of partial differential equations, the general solution depends on arbitrary functions rather than on arbitrary constants. To illustrate this, consider the equation

$$u_{xy} = 0.$$

If we integrate this equation with respect to y, we obtain

$$u_x(x, y) = f(x).$$

A second integration with respect to x yields

$$u(x, y) = g(x) + h(y),$$

where $g(x)$ and $h(y)$ are arbitrary functions.

Suppose u is a function of three variables, x, y, and z. Then, for the equation

$$u_{yy} = 2,$$

one finds the general solution

$$u(x, y, z) = y^2 + yf(x, z) + g(x, z),$$

where f and g are arbitrary functions of two variables x and z.

We recall that in the case of ordinary differential equations, the first task is to find the general solution, and then a particular solution is determined by finding the values of arbitrary constants from the prescribed conditions. But, for partial differential equations, selecting a particular solution satisfying the supplementary conditions from the general solution of a partial differential equation may be as difficult as, or even more difficult than, the problem of finding the general solution itself. This is so because the general solution of a partial differential equation involves arbitrary functions; the specialization of such a solution to the particular form which satisfies supplementary conditions requires the determination of these arbitrary functions, rather than merely the determination of constants.

For linear homogeneous ordinary differential equations of order n, a linear combination of n linearly independent solutions is a solution. Unfortunately, this is not true, in general, in the case of partial differential equations. This is due to the fact that the solution space of every homogeneous linear partial differential equation is infinite dimensional. For example, the partial differential equation

$$u_x - u_y = 0 \qquad (1.2.4)$$

can be transformed into the equation

$$2u_\eta = 0$$

by the transformation of variables

$$\xi = x + y, \qquad \eta = x - y.$$

The general solution is

$$u(x, y) = f(x + y),$$

where $f(x + y)$ is an arbitrary function. Thus, we see that each of the functions

$$(x+y)^n ,$$
$$\sin n\,(x+y),$$
$$\cos n\,(x+y),$$
$$\exp n\,(x+y), \quad n = 1,2,3,\dots$$

is a solution of equation (1.2.4). The fact that a simple equation such as (1.2.4) yields infinitely many solutions is an indication of an added difficulty which must be overcome in the study of partial differential equations. Thus, we generally prefer to directly determine the particular solution of a partial differential equation satisfying prescribed supplementary conditions.

1.3 Mathematical Problems

A problem consists of finding an unknown function of a partial differential equation satisfying appropriate supplementary conditions. These conditions may be *initial conditions (I.C.)* and/or *boundary conditions (B.C.)*. For example, the partial differential equation (PDE)

$$
\begin{array}{llll}
& u_t - u_{xx} = 0, & 0 < x < l, & t > 0, \\
\text{with} \quad I.C. & u(x,0) = \sin x, & 0 \le x \le l, & t > 0, \\
B.C. & u(0,t) = 0, & & t \ge 0, \\
B.C. & u(l,t) = 0, & & t \ge 0,
\end{array}
$$

constitutes a problem which consists of a partial differential equation and three supplementary conditions. The equation describes the heat conduction in a rod of length l. The last two conditions are called the *boundary conditions* which describe the function at two prescribed boundary points. The first condition is known as the *initial condition* which prescribes the unknown function $u(x,t)$ throughout the given region at some initial time t, in this case $t = 0$. This problem is known as the *initial boundary-value problem*. Mathematically speaking, the time and the space coordinates are regarded as independent variables. In this respect, the initial condition is merely a point prescribed on the t-axis and the boundary conditions are prescribed, in this case, as two points on the x-axis. Initial conditions are usually prescribed at a certain time $t = t_0$ or $t = 0$, but it is not customary to consider the other end point of a given time interval.

In many cases, in addition to prescribing the unknown function, other conditions such as their derivatives are specified on the boundary and/or at time t_0.

In considering the problem of unbounded domain, the solution can be determined uniquely by prescribing initial conditions only. The corresponding problem is called the *initial-value problem* or the *Cauchy problem*. The mathematical definition is given in Chapter 5. The solution of such a problem may be interpreted physically as the solution unaffected by the boundary conditions at infinity. For problems affected by the boundary at infinity, boundedness conditions on the behavior of solutions at infinity must be prescribed.

A mathematical problem is said to be *well-posed* if it satisfies the following requirements:

1. Existence: There is at least one solution.
2. Uniqueness: There is at most one solution.
3. Continuity: The solution depends continuously on the data.

The first requirement is an obvious logical condition, but we must keep in mind that we cannot simply state that the mathematical problem has a solution just because the physical problem has a solution. We may well be erroneously developing a mathematical model, say, consisting of a partial differential equation whose solution may not exist at all. The same can be said about the uniqueness requirement. In order to really reflect the physical problem that has a unique solution, the mathematical problem must have a unique solution.

For physical problems, it is not sufficient to know that the problem has a unique solution. Hence the last requirement is not only useful but also essential. If the solution is to have physical significance, a small change in the initial data must produce a small change in the solution. The data in a physical problem are normally obtained from experiment, and are approximated in order to solve the problem by numerical or approximate methods. It is essential to know that the process of making an approximation to the data produces only a small change in the solution.

1.4 Linear Operators

An operator is a mathematical rule which, when applied to a function, produces another function. For example, in the expressions

$$L\left[u\right] = \frac{\partial^2 u}{\partial x^2} + \frac{\partial^2 u}{\partial y^2},$$

$$M\left[u\right] = \frac{\partial^2 u}{\partial x^2} - \frac{\partial u}{\partial x} + x\frac{\partial u}{\partial y},$$

$L = \left(\partial^2/\partial x^2 + \partial^2/\partial y^2\right)$ and $M = \left(\partial^2/\partial x^2 - \partial/\partial x\right) + x\left(\partial/\partial y\right)$ are called the *differential operators*.

An operator is said to be linear if it satisfies the following:

1. A constant c may be taken outside the operator:

$$L\left[cu\right] = cL\left[u\right]. \tag{1.4.1}$$

2. The operator operating on the sum of two functions gives the sum of the operator operating on the individual functions:

$$L[u_1 + u_2] = L[u_1] + L[u_2].\qquad(1.4.2)$$

We may combine (1.4.1) and (1.4.2) as

$$L[c_1 u_1 + c_2 u_2] = c_1 L[u_1] + c_2 L[u_2],\qquad(1.4.3)$$

where c_1 and c_2 are any constants. This can be extended to a finite number of functions. If $u_1, u_2, \ldots, u_k$ are k functions and $c_1, c_2, \ldots, c_k$ are k constants, then by repeated application of equation (1.4.3)

$$L\left[\sum_{j=1}^{k} c_j u_j\right] = \sum_{j=1}^{k} c_j L[u_j].\qquad(1.4.4)$$

We may now define the sum of two linear differential operators formally. If L and M are two linear operators, then the sum of L and M is defined as

$$(L + M)[u] = L[u] + M[u],\qquad(1.4.5)$$

where u is a sufficiently differentiable function. It can be readily shown that $L + M$ is also a linear operator.

The product of two linear differential operators L and M is the operator which produces the same result as is obtained by the successive operations of the operators L and M on u, that is,

$$LM[u] = L(M[u]),\qquad(1.4.6)$$

in which we assume that $M[u]$ and $L(M[u])$ are defined. It can be readily shown that LM is also a linear operator.

In general, linear differential operators satisfy the following:

1. $L + M = M + L$ (commutative) (1.4.7)
2. $(L + M) + N = L + (M + N)$ (associative) (1.4.8)
3. $(LM)N = L(MN)$ (associative) (1.4.9)
4. $L(c_1 M + c_2 N) = c_1 LM + c_2 LN$ (distributive). (1.4.10)

For linear differential operators with constant coefficients,

5. $LM = ML$ (commutative). (1.4.11)

Example 1.4.1. Let $L = \frac{\partial^2}{\partial x^2} + x\frac{\partial}{\partial y}$ and $M = \frac{\partial^2}{\partial y^2} - y\frac{\partial}{\partial y}$.

$$LM\left[u\right] = \left(\frac{\partial^2}{\partial x^2} + x\frac{\partial}{\partial y}\right)\left(\frac{\partial^2 u}{\partial y^2} - y\frac{\partial u}{\partial y}\right)$$

$$= \frac{\partial^4 u}{\partial x^2 \partial y^2} - y\frac{\partial^3 u}{\partial x^2 \partial y} + x\frac{\partial^3 u}{\partial y^3} - xy\frac{\partial^2 u}{\partial y^2},$$

$$ML\left[u\right] = \left(\frac{\partial^2}{\partial y^2} - y\frac{\partial}{\partial y}\right)\left(\frac{\partial^2 u}{\partial x^2} + x\frac{\partial u}{\partial y}\right)$$

$$= \frac{\partial^4 u}{\partial y^2 \partial x^2} + x\frac{\partial^3 u}{\partial y^3} - y\frac{\partial^3 u}{\partial y \partial x^2} - xy\frac{\partial^2 u}{\partial y^2},$$

which shows that $LM \neq ML$.

Now let us consider a linear second-order partial differential equation. In the case of two independent variables, such an equation takes the form

$$A\left(x,y\right)u_{xx} + B\left(x,y\right)u_{xy} + C\left(x,y\right)u_{yy}$$
$$+ D\left(x,y\right)u_x + E\left(x,y\right)u_y + F\left(x,y\right)u = G\left(x,y\right), \qquad (1.4.12)$$

where A, B, C, D, E, and F are the coefficients, and G is the nonhomogeneous term.

If we denote

$$L = A\frac{\partial^2}{\partial x^2} + B\frac{\partial^2}{\partial x \partial y} + C\frac{\partial^2}{\partial y^2} + D\frac{\partial}{\partial x} + E\frac{\partial}{\partial y} + F,$$

then equation (1.4.12) may be written in the form

$$L\left[u\right] = G. \qquad (1.4.13)$$

Very often the square bracket is omitted and we simply write

$$Lu = G.$$

Let v_1, v_2, ..., v_n be n functions which satisfy

$$L\left[v_j\right] = G_j, \qquad j = 1, 2, \ldots, n$$

and let w_1, w_2, ..., w_n be n functions which satisfy

$$L\left[w_j\right] = 0, \qquad j = 1, 2, \ldots, n.$$

If we let

$$u_j = v_j + w_j$$

then, the function

$$u = \sum_{j=1}^{n} u_j$$

satisfies the equation

$$L\left[u\right] = \sum_{j=1}^{n} G_j.$$

This is called the *principle of linear superposition.*

In particular, if v is a particular solution of equation (1.4.13), that is, $L\left[v\right] = G$, and w is a solution of the associated homogeneous equation, that is, $L\left[w\right] = 0$, then $u = v + w$ is a solution of $L\left[u\right] = G$.

The principle of linear superposition is of fundamental importance in the study of partial differential equations. This principle is used extensively in solving linear partial differential equations by the method of separation of variables.

Suppose that there are infinitely many solutions $u_1\left(x,y\right)$, $u_2\left(x,y\right)$, ... $u_n\left(x,y\right)$, ... of a linear homogeneous partial differential equation $Lu = 0$. Can we say that every infinite linear combination $c_1 u_1 + c_2 u_2 + \cdots + c_n u_n + \cdots$ of these solutions, where c_1, c_2, ..., c_n, ... are any constants, is again a solution of the equation? Of course, by an infinite linear combination, we mean an infinite series and we must require that the infinite series

$$\sum_{k=0}^{\infty} c_k\, u_k = \lim_{n \to \infty} \sum_{k=0}^{n} c_k\, u_k \qquad (1.4.14)$$

must be convergent to u. In general, we state that the infinite series is a solution of the homogeneous equation.

There is another kind of infinite linear combination which is also used to find the solution of a given linear equation. This is concerned with a family of solutions $u\left(x,y;\, k\right)$ of the linear equation, where k is any real number, not just the values $1, 2, 3, \ldots$. If $c_k = c\left(k\right)$ is any function of the real parameter k such that

$$\int_a^b c\left(k\right) u\left(x,y;\, k\right)\, dk \quad \text{or} \quad \int_{-\infty}^{\infty} c\left(k\right) u\left(x,y;\, k\right)\, dk \qquad (1.4.15)$$

is convergent, then, under suitable conditions, the integral (1.4.15), again, is a solution. This may be called the *linear integral superposition principle.*

To illustrate these ideas, we consider the equation

$$Lu = u_x + 2u_y = 0. \qquad (1.4.16)$$

It is easy to verify that, for every real k, the function

$$u\left(x,y;\, k\right) = e^{k(2x-y)} \qquad (1.4.17)$$

is a solution of (1.4.16).

Multiplying (1.4.17) by e^{-k} and integrating with respect to k over $-1 \leq k \leq 1$ gives

$$u\left(x,y\right) = \int_{-1}^{1} e^{-k}\, e^{k(2x-y)}\, dk = \frac{e^{2x-y-1}}{2x-y-1} \qquad (1.4.18)$$

It is easy to verify that $u\left(x,y\right)$ given by (1.4.18) is also a solution of (1.4.16).

It is also easy to verify that $u\left(x,y;\,k\right) = e^{-ky}\cos\left(k\,x\right)$, $k \in \mathbb{R}$ is a one-parameter family of solutions of the Laplace equation

$$\nabla^2 u \equiv u_{xx} + u_{yy} = 0 \qquad (1.4.19)$$

It is also easy to check that

$$v\left(x,y;\,k\right) = \frac{\partial}{\partial k}\, u\left(x,y;\,k\right) \qquad (1.4.20)$$

is also a one-parameter family of solutions of (1.4.19), $k \in \mathbb{R}$. Further, for any (x,y) in the upper half-plane $y > 0$, the integral

$$v\left(x,y\right) \equiv \int_0^\infty u\left(x,y,k\right)\, dk = \int_0^\infty e^{-ky}\cos\left(k\,x\right)\, dk, \qquad (1.4.21)$$

is convergent, and $v\left(x,y\right)$ is a solution of (1.4.19) for $x \in \mathbb{R}$ and $y > 0$. This follows from direct computation of v_{xx} and v_{yy}. The solution (1.4.21) is another example of the linear integral superposition principle.

1.5 Superposition Principle

We may express supplementary conditions using the operator notation. For instance, the initial boundary-value problem

$$\begin{aligned}
u_{tt} - c^2 u_{xx} &= G\left(x,t\right) & 0 < x < l, & \qquad t > 0, \\
u\left(x,0\right) &= g_1\left(x\right) & 0 \le x \le l, & \\
u_t\left(x,0\right) &= g_2\left(x\right) & 0 \le x \le l, & \qquad (1.5.1) \\
u\left(0,t\right) &= g_3\left(t\right) & & \qquad t \ge 0, \\
u\left(l,t\right) &= g_4\left(t\right) & & \qquad t \ge 0,
\end{aligned}$$

may be written in the form

$$\begin{aligned}
L\left[u\right] &= G, \\
M_1\left[u\right] &= g_1, \\
M_2\left[u\right] &= g_2, \qquad (1.5.2) \\
M_3\left[u\right] &= g_3, \\
M_4\left[u\right] &= g_4,
\end{aligned}$$

where g_i are the prescribed functions and the subscripts on operators are assigned arbitrarily.

Now let us consider the problem

$$L[u] = G,$$
$$M_1[u] = g_1,$$
$$M_2[u] = g_2,$$ (1.5.3)
$$\vdots$$
$$M_n[u] = g_n.$$

By virtue of the linearity of the equation and the supplementary conditions, we may divide problem (1.5.3) into a series of problems as follows:

$$L[u_1] = G,$$
$$M_1[u_1] = 0,$$
$$M_2[u_1] = 0,$$ (1.5.4)
$$\vdots$$
$$M_n[u_1] = 0,$$

$$L[u_2] = 0,$$
$$M_1[u_2] = g_1,$$
$$M_2[u_2] = 0,$$ (1.5.5)
$$\vdots$$
$$M_n[u_2] = 0,$$

$$L[u_n] = 0,$$
$$M_1[u_n] = 0,$$
$$M_2[u_n] = 0,$$ (1.5.6)
$$\vdots$$
$$M_n[u_n] = g_n.$$

Then the solution of problem (1.5.3) is given by

$$u = \sum_{i=1}^{n} u_i.$$ (1.5.7)

Let us consider one of the subproblems, say, (1.5.5). Suppose we find a sequence of functions $\phi_1, \phi_2, \ldots$, which may be finite or infinite, satisfying the homogeneous system

$$L[\phi_i] = 0,$$
$$M_2[\phi_i] = 0,$$

$$\vdots$$

$$M_n[\phi_i] = 0, \qquad i = 1, 2, 3, \ldots$$

(1.5.8)

and suppose we can express g_1 in terms of the series

$$g_1 = c_1 M_1[\phi_1] + c_2 M_1[\phi_2] + \ldots .$$

(1.5.9)

Then the linear combination

$$u_2 = c_1 \phi_1 + c_2 \phi_2 + \ldots ,$$

(1.5.10)

is the solution of problem (1.5.5). In the case of an infinite number of terms in the linear combination (1.5.10), we require that the infinite series be uniformly convergent and sufficiently differentiable, and that all the series $N_k(u_i)$ where $N_0 = L$, $N_j = M_j$ for $j = 1, 2, \ldots, n$ convergence uniformly.

1.6 Exercises

1. For each of the following, state whether the partial differential equation is linear, quasi-linear or nonlinear. If it is linear, state whether it is homogeneous or nonhomogeneous, and gives its order.

   (a) $u_{xx} + xu_y = y,$

   (b) $uu_x - 2xyu_y = 0,$

   (c) $u_x^2 + uu_y = 1,$

   (d) $u_{xxxx} + 2u_{xxyy} + u_{yyyy} = 0,$

   (e) $u_{xx} + 2u_{xy} + u_{yy} = \sin x,$

   (f) $u_{xxx} + u_{xyy} + \log u = 0,$

   (g) $u_{xx}^2 + u_x^2 + \sin u = e^y,$

   (h) $u_t + uu_x + u_{xxx} = 0.$

2. Verify that the functions

$$u(x, y) = x^2 - y^2$$
$$u(x, y) = e^x \sin y$$
$$u(x, y) = 2xy$$

   are the solutions of the equation

$$u_{xx} + u_{yy} = 0.$$

3. Show that $u = f(xy)$, where f is an arbitrary differentiable function satisfies

$$xu_x - yu_y = 0$$

and verify that the functions $\sin(xy)$, $\cos(xy)$, $\log(xy)$, e^{xy}, and $(xy)^3$ are solutions.

4. Show that $u = f(x)g(y)$ where f and g are arbitrary twice differentiable functions satisfies

$$uu_{xy} - u_x u_y = 0.$$

5. Determine the general solution of the differential equation

$$u_{yy} + u = 0.$$

6. Find the general solution of

$$u_{xx} + u_x = 0,$$

by setting $u_x = v$.

7. Find the general solution of

$$u_{xx} - 4u_{xy} + 3u_{yy} = 0,$$

by assuming the solution to be in the form $u(x, y) = f(\lambda x + y)$, where λ is an unknown parameter.

8. Find the general solution of

$$u_{xx} - u_{yy} = 0.$$

9. Show that the general solution of

$$\frac{\partial^2 u}{\partial t^2} - c^2 \frac{\partial^2 u}{\partial x^2} = 0,$$

is $u(x, t) = f(x - ct) + g(x + ct)$, where f and g are arbitrary twice differentiable functions.

10. Verify that the function

$$u = \phi(xy) + x\psi\left(\frac{y}{x}\right),$$

is the general solution of the equation

$$x^2 u_{xx} - y^2 u_{yy} = 0.$$

11. If $u_x = v_y$ and $v_x = -u_y$, show that both u and v satisfy the Laplace equations

$$\nabla^2 u = 0 \quad \text{and} \quad \nabla^2 v = 0.$$

12. If $u(x, y)$ is a homogeneous function of degree n, show that u satisfies the first-order equation

$$xu_x + yu_y = nu.$$

13. Verify that

$$u(x, y, t) = A \cos(kx) \cos(ly) \cos(nct) + B \sin(kx) \sin(ly) \sin(nct),$$

where $k^2 + l^2 = n^2$, is a solution of the equation

$$u_{tt} = c^2 (u_{xx} + u_{yy}).$$

14. Show that

$$u(x, y; k) = e^{-ky} \sin(kx), \quad x \in \mathbb{R}, \quad y > 0,$$

is a solution of the equation

$$\nabla^2 u \equiv u_{xx} + u_{yy} = 0$$

for any real parameter k. Verify that

$$u(x, y) = \int_0^\infty c(k) e^{-ky} \sin(kx) \, dk$$

is also a solution of the above equation.

15. Show, by differentiation that,

$$u(x, t) = \frac{1}{\sqrt{4\pi kt}} \exp\left(-\frac{x^2}{4kt}\right), \quad x \in \mathbb{R}, \quad t > 0,$$

is a solution of the diffusion equation

$$u_t = k\, u_{xx},$$

where k is a constant.

16. (a) Verify that

$$u(x, y) = \log\left(\sqrt{x^2 + y^2}\right),$$

satisfies the equation

$$u_{xx} + u_{yy} = 0$$

for all $(x, y) \neq (0, 0)$.

(b) Show that

$$u\left(x, y, z\right) = \left(x^2 + y^2 + z^2\right)^{-\frac{1}{2}}$$

is a solution of the Laplace equation

$$u_{xx} + u_{yy} + u_{zz} = 0$$

except at the origin.

(c) Show that

$$u\left(r\right) = a\, r^n$$

satisfies the equation

$$r^2 u'' + 2ru' - n\left(n + 1\right)u = 0.$$

17. Show that

$$u_n\left(r, \theta\right) = r^n \cos\left(n\theta\right) \quad \text{and} \quad u_n\left(r, \theta\right) = r^n \sin\left(n\theta\right), \quad n = 0, 1, 2, 3, \cdots$$

are solutions of the Laplace equation

$$\nabla^2 u \equiv u_{rr} + \frac{1}{r}u_r + \frac{1}{r^2}u_{\theta\theta} = 0.$$

18. Verify by differentiation that $u\left(x, y\right) = \cos x \cosh y$ satisfies the Laplace equation

$$u_{xx} + u_{yy} = 0.$$

19. Show that $u\left(x, y\right) = f\left(2y + x^2\right) + g\left(2y - x^2\right)$ is a general solution of the equation

$$u_{xx} - \frac{1}{x}u_x - x^2 u_{yy} = 0.$$

20. If u satisfies the Laplace equation $\nabla^2 u \equiv u_{xx} + u_{yy} = 0$, show that both xu and yu satisfy the biharmonic equation

$$\nabla^4 \begin{pmatrix} xu \\ \\ yu \end{pmatrix} = 0,$$

but xu and yu will not satisfy the Laplace equation.

21. Show that

$$u\left(x, y, t\right) = f\left(x + iky - i\omega t\right) + g\left(x - iky - i\omega t\right)$$

is a general solution of the wave equation

$$u_{tt} = c^2\left(u_{xx} + u_{yy}\right),$$

where f and g are arbitrary twice differentiable functions, and $\omega^2 = c^2\left(k^2 - 1\right)$, k, ω, c are constants.

22. Verify that

$$u(x, y) = x^3 + y^2 + e^x (\cos x \sin y \cosh y - \sin x \cos y \sinh y)$$

is a classical solution of the Poisson equation

$$u_{xx} + u_{yy} = (6x + 2).$$

23. Show that

$$u(x, y) = \exp\left(-\frac{x}{b}\right) f(ax - by)$$

satisfies the equation

$$b\, u_x + a\, u_y + u = 0.$$

24. Show that

$$u_{tt} - c^2 u_{xx} + 2b\, u_t = 0$$

has solutions of the form

$$u(x, t) = (A \cos kx + B \sin kx) V(t),$$

where c, b, A and B are constants.

25. Show that

$$c^2 \left(u_{rr} + \frac{1}{r} u_r \right) - u_{tt} = 0$$

has solutions of the form

$$u(r, t) = \frac{V(r)}{r} \cos(nct), \qquad n = 0, 1, 2, \ldots.$$

Find a differential equation for $V(r)$.

2

First-Order, Quasi-Linear Equations and Method of Characteristics

"As long as a branch of knowledge offers an abundance of problems, it is full of vitality."

David Hilbert

"Since a general solution must be judged impossible from want of analysis, we must be content with the knowledge of some special cases, and that all the more, since the development of various cases seems to be the only way to bringing us at last to a more perfect knowledge."

Leonhard Euler

2.1 Introduction

Many problems in mathematical, physical, and engineering sciences deal with the formulation and the solution of first-order partial differential equations. From a mathematical point of view, first-order equations have the advantage of providing a conceptual basis that can be utilized for second-, third-, and higher-order equations.

This chapter is concerned with first-order, quasi-linear and linear partial differential equations and their solution by using the Lagrange method of characteristics and its generalizations.

2.2 Classification of First-Order Equations

The most general, first-order, partial differential equation in two independent variables x and y is of the form

$$F\left(x, y, u, u_x, u_y\right) = 0, \qquad (x, y) \in D \subset R^2, \qquad (2.2.1)$$

where F is a given function of its arguments, and $u = u\left(x, y\right)$ is an unknown function of the independent variables x and y which lie in some given domain D in R^2, $u_x = \frac{\partial u}{\partial x}$ and $u_y = \frac{\partial u}{\partial y}$. Equation (2.2.1) is often written in terms of standard notation $p = u_x$ and $q = u_y$ so that (2.2.1) takes the form

$$F\left(x, y, u, p, q\right) = 0. \qquad (2.2.2)$$

Similarly, the most general, first-order, partial differential equation in three independent variables x, y, z can be written as

$$F\left(x, y, z, u, u_x, u_y, u_z\right) = 0. \qquad (2.2.3)$$

Equation (2.2.1) or (2.2.2) is called a *quasi-linear partial differential equation* if it is linear in first-partial derivatives of the unknown function $u\left(x, y\right)$. So, the most general quasi-linear equation must be of the form

$$a\left(x, y, u\right) u_x + b\left(x, y, u\right) u_y = c\left(x, y, u\right), \qquad (2.2.4)$$

where its coefficients a, b, and c are functions of x, y, and u.

The following are examples of quasi-linear equations:

$$x\left(y^2 + u\right) u_x - y\left(x^2 + u\right) u_y = \left(x^2 - y^2\right) u, \qquad (2.2.5)$$

$$u u_x + u_t + n u^2 = 0, \qquad (2.2.6)$$

$$\left(y^2 - u^2\right) u_x - x y\, u_y = x u. \qquad (2.2.7)$$

Equation (2.2.4) is called a *semilinear partial differential equation* if its coefficients a and b are independent of u, and hence, the semilinear equation can be expressed in the form

$$a\left(x, y\right) u_x + b\left(x, y\right) u_y = c\left(x, y, u\right). \qquad (2.2.8)$$

Examples of semilinear equations are

$$x u_x + y u_y = u^2 + x^2, \qquad (2.2.9)$$

$$\left(x + 1\right)^2 u_x + \left(y - 1\right)^2 u_y = \left(x + y\right) u^2, \qquad (2.2.10)$$

$$u_t + a u_x + u^2 = 0, \qquad (2.2.11)$$

where a is a constant.

Equation (2.2.1) is said to be *linear* if F is linear in each of the variables u, u_x, and u_y, and the coefficients of these variables are functions only of the independent variables x and y. The most general, first-order, *linear* partial differential equation has the form

$$a\left(x, y\right) u_x + b\left(x, y\right) u_y + c\left(x, y\right) u = d\left(x, y\right), \qquad (2.2.12)$$

where the coefficients a, b, and c, in general, are functions of x and y and $d(x, y)$ is a given function. Unless stated otherwise, these functions are assumed to be continuously differentiable. Equations of the form (2.2.12) are called *homogeneous* if $d(x, y) \equiv 0$ or *nonhomogeneous* if $d(x, y) \neq 0$.

Obviously, linear equations are a special kind of the quasi-linear equation (2.2.4) if a, b are independent of u and c is a linear function in u. Similarly, semilinear equation (2.2.8) reduces to a linear equation if c is linear in u.

Examples of linear equations are

$$xu_x + yu_y - nu = 0, \tag{2.2.13}$$

$$nu_x + (x + y)\, u_y - u = e^x, \tag{2.2.14}$$

$$yu_x + xu_y = xy, \tag{2.2.15}$$

$$(y - z)\, u_x + (z - x)\, u_y + (x - y)\, u_z = 0. \tag{2.2.16}$$

An equation which is *not* linear is often called a *nonlinear equation*. So, first-order equations are often classified as linear and nonlinear.

2.3 Construction of a First-Order Equation

We consider a system of geometrical surfaces described by the equation

$$f(x, y, z, a, b) = 0, \tag{2.3.1}$$

where a and b are arbitrary parameters. We differentiate (2.3.1) with respect to x and y to obtain

$$f_x + p\, f_z = 0, \qquad f_y + q\, f_z = 0, \tag{2.3.2}$$

where $p = \frac{\partial z}{\partial x}$ and $q = \frac{\partial z}{\partial y}$.

The set of three equations (2.3.1) and (2.3.2) involves two arbitrary parameters a and b. In general, these two parameters can be eliminated from this set to obtain a first-order equation of the form

$$F(x, y, z, p, q) = 0. \tag{2.3.3}$$

Thus the system of surfaces (2.3.1) gives rise to a first-order partial differential equation (2.3.3). In other words, an equation of the form (2.3.1) containing two arbitrary parameters is called a *complete solution* or a *complete integral* of equation (2.3.3). Its role is somewhat similar to that of a general solution for the case of an ordinary differential equation.

On the other hand, any relationship of the form

$$f(\phi, \psi) = 0, \tag{2.3.4}$$

which involves an arbitrary function f of two known functions $\phi = \phi(x, y, z)$ and $\psi = \psi(x, y, z)$ and provides a solution of a first-order partial differential equation is called a *general solution* or *general integral* of this equation. Clearly, the general solution of a first-order partial differential equation depends on an arbitrary function. This is in striking contrast to the situation for ordinary differential equations where the general solution of a first-order ordinary differential equation depends on one arbitrary constant. The general solution of a partial differential equation can be obtained from its complete integral. We obtain the general solution of (2.3.3) from its complete integral (2.3.1) as follows.

First, we prescribe the second parameter b as an arbitrary function of the first parameter a in the complete solution (2.3.1) of (2.3.3), that is, $b = b(a)$. We then consider the envelope of the one-parameter family of solutions so defined. This envelope is represented by the two simultaneous equations

$$f(x, y, z, a, b(a)) = 0, \qquad (2.3.5)$$

$$f_a(x, y, z, a, b(a)) + f_b(x, y, z, b(a)) b'(a) = 0, \qquad (2.3.6)$$

where the second equation (2.3.6) is obtained from the first equation (2.3.5) by partial differentiation with respect to a. In principle, equation (2.3.5) can be solved for $a = a(x, y, z)$ as a function of x, y, and z. We substitute this result back in (2.3.5) to obtain

$$f\{x, y, z, a(x, y, z),\ b(a(x, y, z))\} = 0, \qquad (2.3.7)$$

where b is an arbitrary function. Indeed, the two equations (2.3.5) and (2.3.6) together define the general solution of (2.3.3). When a definite $b(a)$ is prescribed, we obtain a *particular solution* from the general solution. Since the general solution depends on an arbitrary function, there are infinitely many solutions. In practice, only one solution satisfying prescribed conditions is required for a physical problem. Such a solution may be called a *particular solution.*

In addition to the general and particular solutions of (2.3.3), if the envelope of the two-parameter system (2.3.1) of surfaces exists, it also represents a solution of the given equation (2.3.3); the envelope is called the *singular solution* of equation (2.3.3). The singular solution can easily be constructed from the complete solution (2.3.1) representing a two-parameter family of surfaces. The envelope of this family is given by the system of three equations

$$f(x, y, z, a, b) = 0, \quad f_a(x, y, z, a, b) = 0, \quad f_b(x, y, z, a, b) = 0. \ (2.3.8)$$

In general, it is possible to eliminate a and b from (2.3.8) to obtain the equation of the envelope which gives the singular solution. It may be pointed out that the singular solution *cannot* be obtained from the general solution. Its nature is similar to that of the singular solution of a first-order ordinary differential equation.

Finally, it is important to note that solutions of a partial differential equation are expected to be represented by smooth functions. A function is called *smooth* if all of its derivatives exist and are continuous. However, in general, solutions are not always smooth. A solution which is *not* everywhere differentiable is called a *weak solution*. The most common weak solution is the one that has discontinuities in its first partial derivatives across a curve, so that the solution can be represented by shock waves as surfaces of discontinuity. In the case of a first-order partial differential equation, there are discontinuous solutions where z itself and *not* merely $p = \frac{\partial z}{\partial x}$ and $q = \frac{\partial z}{\partial y}$ are discontinuous. In fact, this kind of discontinuity is usually known as a *shock wave*. An important feature of quasi-linear and nonlinear partial differential equations is that their solutions may develop discontinuities as they move away from the initial state. We close this section by considering some examples.

Example 2.3.1. Show that a family of spheres

$$x^2 + y^2 + (z - c)^2 = r^2, \tag{2.3.9}$$

satisfies the first-order linear partial differential equation

$$yp - xq = 0. \tag{2.3.10}$$

Differentiating the equation (2.3.9) with respect to x and y gives

$$x + p(z - c) = 0 \quad \text{and} \quad y + q(z - c) = 0.$$

Eliminating the arbitrary constant c from these equations, we obtain the first-order, partial differential equation

$$yp - xq = 0.$$

Example 2.3.2. Show that the family of spheres

$$(x - a)^2 + (y - b)^2 + z^2 = r^2 \tag{2.3.11}$$

satisfies the first-order, nonlinear, partial differential equation

$$z^2 \left(p^2 + q^2 + 1\right) = r^2. \tag{2.3.12}$$

We differentiate the equation of the family of spheres with respect to x and y to obtain

$$(x - a) + z\,p = 0, \quad (y - b) + z\,q = 0.$$

Eliminating the two arbitrary constants a and b, we find the nonlinear partial differential equation

$$z^2 \left(p^2 + q^2 + 1 \right) = r^2.$$

All surfaces of revolution with the z-axis as the axis of symmetry satisfy the equation

$$z = f \left(x^2 + y^2 \right), \tag{2.3.13}$$

where f is an arbitrary function. Writing $u = x^2 + y^2$ and differentiating (2.3.13) with respect to x and y, respectively, we obtain

$$p = 2x \, f'(u), \quad q = 2y \, f'(u).$$

Eliminating the arbitrary function $f(u)$ from these results, we find the equation

$$yp - xq = 0.$$

Theorem 2.3.1. If $\phi = \phi(x, y, z)$ and $\psi = \psi(x, y, z)$ are two given functions of x, y, and z and if $f(\phi, \psi) = 0$, where f is an arbitrary function of ϕ and ψ, then $z = z(x, y)$ satisfies a first-order, partial differential equation

$$p \, \frac{\partial(\phi, \psi)}{\partial(y, z)} + q \, \frac{\partial(\phi, \psi)}{\partial(z, x)} = \frac{\partial(\phi, \psi)}{\partial(x, y)}, \tag{2.3.14}$$

where

$$\frac{\partial(\phi, \psi)}{\partial(x, y)} = \begin{vmatrix} \phi_x & \phi_y \\ \psi_x & \psi_y \end{vmatrix}. \tag{2.3.15}$$

Proof. We differentiate $f(\phi, \psi) = 0$ with respect to x and y respectively to obtain the following equations:

$$\frac{\partial f}{\partial \phi} \left(\frac{\partial \phi}{\partial x} + p \frac{\partial \phi}{\partial z} \right) + \frac{\partial f}{\partial \psi} \left(\frac{\partial \psi}{\partial x} + p \frac{\partial \psi}{\partial z} \right) = 0, \tag{2.3.16}$$

$$\frac{\partial f}{\partial \phi} \left(\frac{\partial \phi}{\partial y} + q \frac{\partial \phi}{\partial z} \right) + \frac{\partial f}{\partial \psi} \left(\frac{\partial \psi}{\partial y} + q \frac{\partial \psi}{\partial z} \right) = 0. \tag{2.3.17}$$

Nontrivial solutions for $\frac{\partial f}{\partial \phi}$ and $\frac{\partial f}{\partial \psi}$ can be found if the determinant of the coefficients of these equations vanishes, that is,

$$\begin{vmatrix} \phi_x + p\phi_z & \psi_x + p\psi_z \\ \phi_y + q\phi_z & \psi_y + q\psi_z \end{vmatrix} = 0. \tag{2.3.18}$$

Expanding this determinant gives the first-order, quasi-linear equation (2.3.14).

2.4 Geometrical Interpretation of a First-Order Equation

To investigate the geometrical content of a first-order, partial differential equation, we begin with a general, quasi-linear equation

$$a\left(x,y,u\right)u_x + b\left(x,y,u\right)u_y - c\left(x,y,u\right) = 0. \tag{2.4.1}$$

We assume that the possible solution of (2.4.1) in the form $u = u\left(x,y\right)$ or in an implicit form

$$f\left(x,y,u\right) \equiv u\left(x,y\right) - u = 0 \tag{2.4.2}$$

represents a possible *solution surface* in (x,y,u) space. This is often called an *integral surface* of the equation (2.4.1). At any point (x,y,u) on the solution surface, the gradient vector $\nabla f = (f_x, f_y, f_u) = (u_x, u_y, -1)$ is normal to the solution surface. Clearly, equation (2.4.1) can be written as the dot product of two vectors

$$a\,u_x + b\,u_y - c = (a,b,c)\cdot(u_x, u_y - 1) = 0. \tag{2.4.3}$$

This clearly shows that the vector (a,b,c) must be a tangent vector of the integral surface (2.4.2) at the point (x,y,u), and hence, it determines a direction field called the the *characteristic direction* or *Monge axis*. This direction is of fundamental importance in determining a solution of equation (2.4.1). To summarize, we have shown that $f\left(x,y,u\right) = u\left(x,y\right) - u = 0$, as a surface in the (x,y,u)-space, is a solution of (2.4.1) if and only if the direction vector field (a,b,c) lies in the tangent plane of the integral surface $f\left(x,y,u\right) = 0$ at each point (x,y,u), where $\nabla f \neq 0$, as shown in Figure 2.4.1.

A curve in (x,y,u)-space, whose tangent at every point coincides with the characteristic direction field (a,b,c), is called a *characteristic curve*. If the parametric equations of this characteristic curve are

$$x = x\left(t\right), \quad y = y\left(t\right), \quad u = u\left(t\right), \tag{2.4.4}$$

then the tangent vector to this curve is $\left(\frac{dx}{dt}, \frac{dy}{dt}, \frac{du}{dt}\right)$ which must be equal to (a,b,c). Therefore, the system of ordinary differential equations of the characteristic curve is given by

$$\frac{dx}{dt} = a\left(x,y,u\right), \quad \frac{dy}{dt} = b\left(x,y,u\right), \quad \frac{du}{dt} = c\left(x,y,u\right). \tag{2.4.5}$$

These are called the *characteristic equations* of the quasi-linear equation (2.4.1).

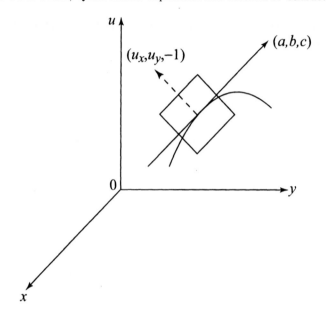

Figure 2.4.1 Tangent and normal vector fields of solution surface at a point (x, y, u).

In fact, there are only two independent ordinary differential equations in the system (2.4.5); therefore, its solutions consist of a two-parameter family of curves in (x, y, u)-space.

The projection on $u = 0$ of a characteristic curve on the (x, t)-plane is called a *characteristic base curve* or simply *characteristic*.

Equivalently, the characteristic equations (2.4.5) in the nonparametric form are

$$\frac{dx}{a} = \frac{dy}{b} = \frac{du}{c}. \tag{2.4.6}$$

The typical problem of solving equation (2.4.1) with a prescribed u on a given plane curve C is equivalent to finding an integral surface in (x, y, u) space, satisfying the equation (2.4.1) and containing the three-dimensional space curve Γ defined by the values of u on C, which is the projection on $u = 0$ of Γ.

Remark 1. The above geometrical interpretation can be generalized for higher-order partial differential equations. However, it is not easy to visualize geometrical arguments that have been described for the case of three space dimensions.

Remark 2. The geometrical interpretation is more complicated for the case of nonlinear partial differential equations, because the normals to possible solution surfaces through a point do not lie in a plane. The tangent planes no longer intersect along one straight line, but instead, they envelope along a curved surface known as the *Monge cone*. Any further discussion is beyond the scope of this book.

We conclude this section by adding an important observation regarding the nature of the characteristics in the (x, t)-plane. For a quasi-linear equation, characteristics are determined by the first two equations in (2.4.5) with their slopes

$$\frac{dy}{dx} = \frac{b(x, y, u)}{a(x, y, u)}. \tag{2.4.7}$$

If (2.4.1) is a linear equation, then a and b are independent of u, and the characteristics of (2.4.1) are *plane curves* with slopes

$$\frac{dy}{dx} = \frac{b(x, y)}{a(x, y)}. \tag{2.4.8}$$

By integrating this equation, we can determine the characteristics which represent a one-parameter family of *curves* in the (x, t)-plane. However, if a and b are constant, the characteristics of equation (2.4.1) are straight lines.

2.5 Method of Characteristics and General Solutions

We can use the geometrical interpretation of first-order, partial differential equations and the properties of characteristic curves to develop a method for finding the general solution of quasi-linear equations. This is usually referred to as *the method of characteristics* due to Lagrange. This method of solution of quasi-linear equations can be described by the following result.

Theorem 2.5.1. The general solution of a first-order, quasi-linear partial differential equation

$$a(x, y, u) u_x + b(x, y, u) u_y = c(x, y, u) \tag{2.5.1}$$

is

$$f(\phi, \psi) = 0, \tag{2.5.2}$$

where f is an arbitrary function of $\phi(x, y, u)$ and $\psi(x, y, u)$, and $\phi = $ constant $= c_1$ and $\psi = $ constant $= c_2$ are solution curves of the characteristic equations

$$\frac{dx}{a} = \frac{dy}{b} = \frac{du}{c}. \tag{2.5.3}$$

The solution curves defined by $\phi(x, y, u) = c_1$ and $\psi(x, y, u) = c_2$ are called the families of *characteristic curves* of equation (2.5.1).

Proof. Since $\phi\left(x, y, u\right) = c_1$ and $\psi\left(x, y, u\right) = c_2$ satisfy equations (2.5.3), these equations must be compatible with the equation

$$d\phi = \phi_x dx + \phi_y dy + \phi_u du = 0. \qquad (2.5.4)$$

This is equivalent to the equation

$$a\,\phi_x + b\,\phi_y + c\,\phi_u = 0. \qquad (2.5.5)$$

Similarly, equation (2.5.3) is also compatible with

$$a\,\psi_x + b\,\psi_y + c\,\psi_u = 0. \qquad (2.5.6)$$

We now solve (2.5.5), (2.5.6) for a, b, and c to obtain

$$\frac{a}{\frac{\partial(\phi,\psi)}{\partial(y,u)}} = \frac{b}{\frac{\partial(\phi,\psi)}{\partial(u,x)}} = \frac{c}{\frac{\partial(\phi,\psi)}{\partial(x,y)}}. \qquad (2.5.7)$$

It has been shown earlier that $f\left(\phi, \psi\right) = 0$ satisfies an equation similar to (2.3.14), that is,

$$p\,\frac{\partial\left(\phi, \psi\right)}{\partial\left(y, u\right)} + q\,\frac{\partial\left(\phi, \psi\right)}{\partial\left(u, x\right)} = \frac{\partial\left(\phi, \psi\right)}{\partial\left(x, y\right)}. \qquad (2.5.8)$$

Substituting, (2.5.7) in (2.5.8), we find that $f\left(\phi, \psi\right) = 0$ is a solution of (2.5.1). This completes the proof.

Note that an analytical method has been used to prove Theorem 2.5.1. Alternatively, a geometrical argument can be used to prove this theorem. The geometrical method of proof is left to the reader as an exercise.

Many problems in applied mathematics, science, and engineering involve partial differential equations. We seldom try to find or discuss the properties of a solution to these equations in its most general form. In most cases of interest, we deal with those solutions of partial differential equations which satisfy certain supplementary conditions. In the case of a first-order partial differential equation, we determine the specific solution by formulating an *initial-value problem* or a *Cauchy problem*.

Theorem 2.5.2. (*The Cauchy Problem for a First-Order Partial Differential Equation*). Suppose that C is a given curve in the (x, y)-plane with its parametric equations

$$x = x_0\left(t\right), \quad y = y_0\left(t\right), \qquad (2.5.9)$$

where t belongs to an interval $I \subset R$, and the derivatives $x_0'\left(t\right)$ and $y_0'\left(t\right)$ are piecewise continuous functions, such that $\left(x_0'\right)^2 + \left(y_0'\right)^2 \neq 0$. Also, suppose that $u = u_0\left(t\right)$ is a given function on the curve C. Then, there exists a solution $u = u\left(x, y\right)$ of the equation

$$F(x, y, u, u_x, u_y) = 0 \tag{2.5.10}$$

in a domain D of R^2 containing the curve C for all $t \in I$, and the solution $u(x, y)$ satisfies the given initial data, that is,

$$u(x_0(t), y_0(t)) = u_0(t) \tag{2.5.11}$$

for all values of $t \in I$.

In short, the Cauchy problem is to determine a solution of equation (2.5.10) in a neighborhood of C, such that the solution $u = u(x, y)$ takes a prescribed value $u_0(t)$ on C. The curve C is called the *initial curve* of the problem, and $u_0(t)$ is called the *initial data*. Equation (2.5.11) is called the *initial condition* of the problem.

The solution of the Cauchy problem also deals with such questions as the conditions on the functions F, $x_0(t)$, $y_0(t)$, and $u_0(t)$ under which a solution exists and is unique.

We next discuss a method for solving a Cauchy problem for the first-order, quasi-linear equation (2.5.1). We first observe that geometrically $x = x_0(t)$, $y = y_0(t)$, and $u = u_0(t)$ represent an initial curve Γ in (x, y, u)-space. The curve C, on which the Cauchy data is prescribed, is the projection of Γ on the (x, y)-plane. We now present a precise formulation of the Cauchy problem for the first-order, quasi-linear equation (2.5.1).

Theorem 2.5.3. (*The Cauchy Problem for a Quasi-linear Equation*). Suppose that $x_0(t)$, $y_0(t)$, and $u_0(t)$ are continuously differentiable functions of t in a closed interval, $0 \le t \le 1$, and that a, b, and c are functions of x, y, and u with continuous first-order partial derivatives with respect to their arguments in some domain D of (x, y, u)-space containing the initial curve

$$\Gamma: x = x_0(t), \quad y = y_0(t), \quad u = u_0(t), \tag{2.5.12}$$

where $0 \le t \le 1$, and satisfying the condition

$$y_0'(t) a(x_0(t), y_0(t), u_0(t)) - x_0'(t) b(x_0(t), y_0(t), u_0(t)) \ne 0. \tag{2.5.13}$$

Then there exists a unique solution $u = u(x, y)$ of the quasi-linear equation (2.5.1) in the neighborhood of $C: x = x_0(t)$, $y = y_0(t)$, and the solution satisfies the initial condition

$$u_0(t) = u(x_0(t), y_0(t)), \quad \text{for} \quad 0 \le t \le 1. \tag{2.5.14}$$

Note: The condition (2.5.13) excludes the possibility that C could be a characteristic.

Example 2.5.1. Find the general solution of the first-order linear partial differential equation.

$$x\, u_x + y\, u_y = u. \tag{2.5.15}$$

The characteristic curves of this equation are the solutions of the characteristic equations

$$\frac{dx}{x} = \frac{dy}{y} = \frac{du}{u}. \tag{2.5.16}$$

This system of equations gives the integral surfaces

$$\phi = \frac{y}{x} = C_1 \quad \text{and} \quad \psi = \frac{u}{x} = C_2,$$

where C_1 and C_2 are arbitrary constants. Thus, the general solution of (2.5.15) is

$$f\left(\frac{y}{x}, \frac{u}{x}\right) = 0, \tag{2.5.17}$$

where f is an arbitrary function. This general solution can also be written as

$$u(x, y) = x\, g\left(\frac{y}{x}\right), \tag{2.5.18}$$

where g is an arbitrary function.

Example 2.5.2. Obtain the general solution of the linear Euler equation

$$x\, u_x + y\, u_y = nu. \tag{2.5.19}$$

The integral surfaces are the solutions of the characteristic equations

$$\frac{dx}{x} = \frac{dy}{y} = \frac{du}{nu}. \tag{2.5.20}$$

From these equations, we get

$$\frac{y}{x} = C_1, \quad \frac{u}{x^n} = C_2,$$

where C_1 and C_2 are arbitrary constants. Hence, the general solution of (2.5.19) is

$$f\left(\frac{y}{x}, \frac{u}{x^n}\right) = 0. \tag{2.5.21}$$

This can also be written as

$$\frac{u}{x^n} = g\left(\frac{y}{x}\right)$$

or

$$u(x, y) = x^n g\left(\frac{y}{x}\right). \tag{2.5.22}$$

This shows that the solution $u(x, y)$ is a homogeneous function of x and y of degree n.

Example 2.5.3. Find the general solution of the linear equation

$$x^2 \, u_x + y^2 \, u_y = (x + y) \, u. \tag{2.5.23}$$

The characteristic equations associated with (2.5.23) are

$$\frac{dx}{x^2} = \frac{dy}{y^2} = \frac{du}{(x + y) \, u}. \tag{2.5.24}$$

From the first two of these equations, we find

$$x^{-1} - y^{-1} = C_1, \tag{2.5.25}$$

where C_1 is an arbitrary constant.
It follows from (2.5.24) that

$$\frac{dx - dy}{x^2 - y^2} = \frac{du}{(x + y) \, u}$$

or

$$\frac{d \, (x - y)}{x - y} = \frac{du}{u}.$$

This gives

$$\frac{x - y}{u} = C_2, \tag{2.5.26}$$

where C_2 is a constant. Furthermore, (2.5.25) and (2.5.26) also give

$$\frac{xy}{u} = C_3, \tag{2.5.27}$$

where C_3 is a constant.
Thus, the general solution (2.5.23) is given by

$$f \left(\frac{xy}{u}, \frac{x - y}{u} \right) = 0, \tag{2.5.28}$$

where f is an arbitrary function. This general solution representing the integral surface can also be written as

$$u \, (x, y) = xy \, g \left(\frac{x - y}{u} \right), \tag{2.5.29}$$

where g is an arbitrary function, or, equivalently,

$$u \, (x, y) = xy \, h \left(\frac{x - y}{xy} \right), \tag{2.5.30}$$

where h is an arbitrary function.

Example 2.5.4. Show that the general solution of the linear equation

$$(y - z)\,u_x + (z - x)\,u_y + (x - y)\,u_z = 0 \qquad (2.5.31)$$

is

$$u\,(x, y, z) = f\left(x + y + z, x^2 + y^2 + z^2\right), \qquad (2.5.32)$$

where f is an arbitrary function.

The characteristic curves satisfy the characteristic equations

$$\frac{dx}{y - z} = \frac{dy}{z - x} = \frac{dz}{x - y} = \frac{du}{0} \qquad (2.5.33)$$

or

$$du = 0, \quad dx + dy + dz = 0, \quad x\,dx + y\,dy + z\,dz = 0.$$

Integration of these equations gives

$$u = C_1, \quad x + y + z = C_2, \quad \text{and} \quad x^2 + y^2 + z^2 = C_3,$$

where C_1, C_2 and C_3 are arbitrary constants.

Thus, the general solution can be written in terms of an arbitrary function f in the form

$$u\,(x, y, z) = f\left(x + y + z, x^2 + y^2 + z^2\right).$$

We next verify that this is a general solution by introducing three independent variables ξ, η, ζ defined in terms of x, y, and z as

$$\xi = x + y + z, \quad \eta = x^2 + y^2 + z^2, \quad \text{and} \quad \zeta = y + z, \qquad (2.5.34)$$

where ζ is an arbitrary combination of y and z. Clearly the general solution becomes

$$u = f\,(\xi, \eta),$$

and hence,

$$u_\zeta = u_x \frac{\partial x}{\partial \zeta} + u_y \frac{\partial y}{\partial \zeta} + u_z \frac{\partial z}{\partial \zeta}. \qquad (2.5.35)$$

It follows from (2.5.34) that

$$0 = \frac{\partial x}{\partial \zeta} + \frac{\partial y}{\partial \zeta} + \frac{\partial z}{\partial \zeta}, \qquad 0 = 2\left(x \frac{\partial x}{\partial \zeta} + y \frac{\partial y}{\partial \zeta} + z \frac{\partial z}{\partial \zeta}\right), \qquad \frac{\partial y}{\partial \zeta} + \frac{\partial z}{\partial \zeta} = 1.$$

It follows from the first and the third results that $\frac{\partial x}{\partial \zeta} = -1$ and, therefore,

$$x = y \frac{\partial y}{\partial \zeta} + z \frac{\partial z}{\partial \zeta}, \qquad y = y \frac{\partial y}{\partial \zeta} + y \frac{\partial z}{\partial \zeta}, \qquad z = z \frac{\partial y}{\partial \zeta} + z \frac{\partial z}{\partial \zeta}.$$

Clearly, it follows by subtracting that

$$x - y = (z - y)\frac{\partial z}{\partial \zeta}, \quad x - z = (y - z)\frac{\partial y}{\partial \zeta}.$$

Using the values for $\frac{\partial x}{\partial \zeta}$, $\frac{\partial z}{\partial \zeta}$, and $\frac{\partial y}{\partial \zeta}$ in (2.5.35), we obtain

$$(z - y)\frac{\partial u}{\partial \zeta} = (y - z)\frac{\partial u}{\partial x} + (z - x)\frac{\partial u}{\partial y} + (x - y)\frac{\partial u}{\partial z}. \quad (2.5.36)$$

If $u = u(\xi, \eta)$ satisfies (2.5.31), then $\frac{\partial u}{\partial \zeta} = 0$ and, hence, (2.5.36) reduces to (2.5.31). This shows that the general solution (2.5.32) satisfies equation (2.5.31).

Example 2.5.5. Find the solution of the equation

$$u(x + y)u_x + u(x - y)u_y = x^2 + y^2, \quad (2.5.37)$$

with the Cauchy data $u = 0$ on $y = 2x$.

The characteristic equations are

$$\frac{dx}{u(x + y)} = \frac{dy}{u(x - y)} = \frac{du}{x^2 + y^2} = \frac{ydx + xdy - udu}{0} = \frac{xdx - ydy - udu}{0}.$$

Consequently,

$$d\left[\left(xy - \frac{1}{2}u^2\right)\right] = 0 \quad \text{and} \quad d\left[\frac{1}{2}\left(x^2 - y^2 - u^2\right)\right] = 0. \quad (2.5.38)$$

These give two integrals

$$u^2 - x^2 + y^2 = C_1 \quad \text{and} \quad 2xy - u^2 = C_2, \quad (2.5.39)$$

where C_1 and C_2 are constants. Hence, the general solution is

$$f\left(x^2 - y^2 - u^2, \quad 2xy - u^2\right) = 0,$$

where f is an arbitrary function.

Using the Cauchy data in (2.5.39), we obtain $4C_1 = 3C_2$. Therefore

$$4\left(u^2 - x^2 + y^2\right) = 3\left(2xy - u^2\right).$$

Thus, the solution of equation (2.5.37) is given by

$$7u^2 = 6xy + 4\left(x^2 - y^2\right). \quad (2.5.40)$$

Example 2.5.6. Obtain the solution of the linear equation

$$u_x - u_y = 1, \quad (2.5.41)$$

with the Cauchy data

$$u(x,0) = x^2.$$

The characteristic equations are

$$\frac{dx}{1} = \frac{dy}{-1} = \frac{du}{1}. \tag{2.5.42}$$

Obviously,

$$\frac{dy}{dx} = -1 \quad \text{and} \quad \frac{du}{dx} = 1.$$

Clearly,

$$x + y = \text{constant} = C_1 \quad \text{and} \quad u - x = \text{constant} = C_2.$$

Thus, the general solution is given by

$$u - x = f(x + y), \tag{2.5.43}$$

where f is an arbitrary function.

We now use the Cauchy data to find $f(x) = x^2 - x$, and hence, the solution is

$$u(x,y) = (x+y)^2 - y. \tag{2.5.44}$$

The characteristics $x + y = C_1$ are drawn in Figure 2.5.1. The value of u must be given at one point on each characteristic which intersects the line $y = 0$ only at one point, as shown in Figure 2.5.1.

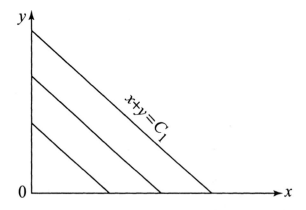

Figure 2.5.1 Characteristics of equation (2.5.41).

Example 2.5.7. Obtain the solution of the equation

$$(y - u)\, u_x + (u - x)\, u_y = x - y, \tag{2.5.45}$$

with the condition $u = 0$ on $xy = 1$.

The characteristic equations for equation (2.5.45) are

$$\frac{dx}{y - u} = \frac{dy}{u - x} = \frac{du}{x - y}. \tag{2.5.46}$$

The parametric forms of these equations are

$$\frac{dx}{dt} = y - u, \qquad \frac{dy}{dt} = u - x, \qquad \frac{du}{dt} = x - y.$$

These lead to the following equations:

$$\dot{x} + \dot{y} + \dot{u} = 0 \quad \text{and} \quad x\dot{x} + y\dot{y} + u\dot{u} = 0, \tag{2.5.47}$$

where the dot denotes the derivative with respect to t.
   Integrating (2.5.47), we obtain

$$x + y + u = \text{const.} = C_1 \quad \text{and} \quad x^2 + y^2 + u^2 = \text{const.} = C_2. \tag{2.5.48}$$

These equations represent circles.
   Using the Cauchy data, we find that

$$C_1^2 = (x + y)^2 = x^2 + y^2 + 2xy = C_2 + 2.$$

Thus, the integral surface is described by

$$(x + y + u)^2 = x^2 + y^2 + u^2 + 2.$$

Hence, the solution is given by

$$u(x, y) = \frac{1 - xy}{x + y}. \tag{2.5.49}$$

Example 2.5.8. Solve the linear equation

$$y\, u_x + x\, u_y = u, \tag{2.5.50}$$

with the Cauchy data

$$u(x, 0) = x^3 \quad \text{and} \quad u(0, y) = y^3. \tag{2.5.51}$$

The characteristic equations are

$$\frac{dx}{y} = \frac{dy}{x} = \frac{du}{u}$$

or

$$\frac{du}{u} = \frac{dx - dy}{y - x} = \frac{dx + dy}{y + x}.$$

Solving these equations, we obtain

$$u = \frac{C_1}{x - y} = C_2 (x + y)$$

or

$$u = C_2 (x + y), \quad x^2 - y^2 = \frac{C_1}{C_2} = \text{constant} = C.$$

So the characteristics are rectangular hyperbolas for $C > 0$ or $C < 0$.

Thus, the general solution is given by

$$f\left(\frac{u}{x + y}, x^2 - y^2\right) = 0$$

or, equivalently,

$$u(x, y) = (x + y) g (x^2 - y^2). \tag{2.5.52}$$

Using the Cauchy data, we find that $g(x^2) = x^2$, that is, $g(x) = x$.

Consequently, the solution becomes

$$u(x, y) = (x + y)(x^2 - y^2) \quad \text{on} \quad x^2 - y^2 = C > 0.$$

Similarly,

$$u(x, y) = (x + y)(y^2 - x^2) \quad \text{on} \quad y^2 - x^2 = C > 0.$$

It follows from these results that $u \to 0$ in all regions, as $x \to \pm y$ (or $y \to \pm x$), and hence, u is continuous across $y = \pm x$ which represent asymptotes of the rectangular hyperbolas $x^2 - y^2 = C$. However, u_x and u_y are *not* continuous, as $y \to \pm x$. For $x^2 - y^2 = C > 0$,

$$u_x = 3x^2 + 2xy - y^2 = (x + y)(3x - y) \to 0, \quad \text{as} \quad y \to -x.$$
$$u_y = -3y^2 - 2xy + x^2 = (x + y)(x - 3y) \to 0, \quad \text{as} \quad y \to -x.$$

Hence, both u_x and u_y are continuous as $y \to -x$. On the other hand,

$$u_x \to 4x^2, \ u_y \to -4x^2 \quad \text{as} \quad y \to x.$$

This implies that u_x and u_y are discontinuous across $y = x$.

Combining all these results, we conclude that $u(x, y)$ is continuous everywhere in the (x, t)-plane, and u_x, u_y are continuous everywhere in the (x, t)-plane except on the line $y = x$. Hence, the partial derivatives u_x, u_y are discontinuous on $y = x$. Thus, the development of *discontinuities* across characteristics is a significant feature of the solutions of partial differential equations.

Example 2.5.9. Determine the integral surfaces of the equation

$$x \left(y^2 + u \right) u_x - y \left(x^2 + u \right) u_y = \left(x^2 - y^2 \right) u, \qquad (2.5.53)$$

with the data

$$x + y = 0, \qquad u = 1.$$

The characteristic equations are

$$\frac{dx}{x \left(y^2 + u \right)} = \frac{dy}{-y \left(x^2 + u \right)} = \frac{du}{\left(x^2 - y^2 \right) u} \qquad (2.5.54)$$

or

$$\frac{\frac{dx}{x}}{\left(y^2 + u \right)} = \frac{\frac{dy}{y}}{-\left(x^2 + u \right)} = \frac{\frac{du}{u}}{\left(x^2 - y^2 \right)} = \frac{\frac{dx}{x} + \frac{dy}{y} + \frac{du}{u}}{0}.$$

Consequently,

$$\log \left(xyu \right) = \log C_1$$

or

$$xyu = C_1.$$

From (2.5.54), we obtain

$$\frac{x dx}{x^2 \left(y^2 + u \right)} = \frac{y dy}{-y^2 \left(x^2 + u \right)} = \frac{du}{\left(x^2 - y^2 \right) u} = \frac{x dx + y dy - du}{0},$$

whence we find that

$$x^2 + y^2 - 2u = C_2.$$

Using the given data, we obtain

$$C_1 = -x^2 \quad \text{and} \quad C_2 = 2x^2 - 2,$$

so that

$$C_2 = -2 \left(C_1 + 1 \right).$$

Thus the integral surface is given by

$$x^2 + y^2 - 2u = -2 - 2xyu$$

or

$$2xyu + x^2 + y^2 - 2u + 2 = 0. \qquad (2.5.55)$$

Example 2.5.10. Obtain the solution of the equation

$$x\, u_x + y\, u_y = x \exp(-u) \tag{2.5.56}$$

with the data

$$u = 0 \quad \text{on} \quad y = x^2.$$

The characteristic equations are

$$\frac{dx}{x} = \frac{dy}{y} = \frac{du}{x \exp(-u)} \tag{2.5.57}$$

or

$$\frac{y}{x} = C_1.$$

We also obtain from (2.5.57) that $dx = e^u du$ which can be integrated to find

$$e^u = x + C_2.$$

Thus, the general solution is given by

$$f\left(e^u - x, \frac{y}{x}\right) = 0$$

or, equivalently,

$$e^u = x + g\left(\frac{y}{x}\right). \tag{2.5.58}$$

Applying the Cauchy data, we obtain $g(x) = 1 - x$. Thus, the solution of (2.5.56) is given by

$$e^u = x + 1 - \frac{y}{x}$$

or

$$u = \log\left(x + 1 - \frac{y}{x}\right). \tag{2.5.59}$$

Example 2.5.11. Solve the initial-value problem

$$u_t + u\, u_x = x, \qquad u(x,0) = f(x), \tag{2.5.60}$$

where (a) $f(x) = 1$ and (b) $f(x) = x$.
   The characteristic equations are

$$\frac{dt}{1} = \frac{dx}{u} = \frac{du}{x} = \frac{d(x+u)}{x+u}. \tag{2.5.61}$$

Integration gives

$$t = \log\left(x + u\right) - \log C_1$$

or

$$\left(u + x\right) e^{-t} = C_1.$$

Similarly, we get

$$u^2 - x^2 = C_2.$$

For case (a), we obtain

$$1 + x = C_1 \quad \text{and} \quad 1 - x^2 = C_2, \quad \text{and hence} \quad C_2 = 2C_1 - C_1^2.$$

Thus,

$$\left(u^2 - x^2\right) = 2\left(u + x\right) e^{-t} - \left(u + x\right)^2 e^{-2t}$$

or

$$u - x = 2e^{-t} - \left(u + x\right) e^{-2t}.$$

A simple manipulation gives the solution

$$u\left(x, t\right) = x \tanh t + \operatorname{sech} t. \tag{2.5.62}$$

Case (b) is left to the reader as an exercise.

Example 2.5.12. Find the integral surface of the equation

$$u\, u_x + u_y = 1, \tag{2.5.63}$$

so that the surface passes through an initial curve represented parametrically by

$$x = x_0\left(s\right), \quad y = y_0\left(s\right), \quad u = u_0\left(s\right), \tag{2.5.64}$$

where s is a parameter.

The characteristic equations for the given equations are

$$\frac{dx}{u} = \frac{dy}{1} = \frac{du}{1},$$

which are, in the parametric form,

$$\frac{dx}{d\tau} = u, \quad \frac{dy}{d\tau} = 1, \quad \frac{du}{d\tau} = 1, \tag{2.5.65}$$

where τ is a parameter. Thus the solutions of this parametric system in general depend on two parameters s and τ. We solve this system (2.5.65) with the initial data

$$x(s,0) = x_0(s), \quad y(s,0) = y_0(s), \quad u(s,0) = u_0(s).$$

The solutions of (2.5.65) with the given initial data are

$$\left.\begin{array}{l} x(s,\tau) = \frac{\tau^2}{2} + \tau u_0(s) + x_0(s) \\ y(s,\tau) = \tau + y_0(s) \\ u(s,\tau) = \tau + u_0(s) \end{array}\right\} . \tag{2.5.66}$$

We choose a particular set of values for the initial data as

$$x(s,0) = 2s^2, \quad y(s,0) = 2s, \quad u(s,0) = 0, \quad s > 0.$$

Therefore, the solutions are given by

$$x = \frac{1}{2}\tau^2 + 2s^2, \quad y = \tau + 2s, \quad u = \tau. \tag{2.5.67}$$

Eliminating τ and s from (2.5.67) gives the integral surface

$$(u - y)^2 + u^2 = 2x$$

or

$$2u = y \pm \left(4x - y^2\right)^{\frac{1}{2}}. \tag{2.5.68}$$

The solution surface satisfying the data $u = 0$ on $y^2 = 2x$ is given by

$$2u = y - \left(4x - y^2\right)^{\frac{1}{2}}. \tag{2.5.69}$$

This represents the solution surface only when $y^2 < 4x$. Thus, the solution does not exist for $y^2 > 4x$ and is *not* differentiable when $y^2 = 4x$. We verify that $y^2 = 4x$ represents the *envelope* of the family of characteristics in the (x,t)-plane given by the τ-eliminant of the first two equations in (2.5.67), that is,

$$F(x,y,s) = 2x - (y - 2s)^2 - 4s^2 = 0. \tag{2.5.70}$$

This represents a family of parabolas for different values of the parameter s. Thus, the envelope is obtained by eliminating s from equations $\frac{\partial F}{\partial s} = 0$ and $F = 0$. This gives $y^2 = 4x$, which is the envelope of the characteristics for different s, as shown in Figure 2.5.2.

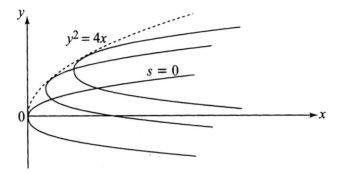

Figure 2.5.2 Dotted curve is the envelope of the characteristics.

2.6 Canonical Forms of First-Order Linear Equations

It is often convenient to transform the more general first-order linear partial differential equation (2.2.12)

$$a\left(x,y\right)u_x + b\left(x,y\right)u_y + c\left(x,y\right)u = d\left(x,y\right), \qquad (2.6.1)$$

into a *canonical* (or *standard*) form which can be easily integrated to find the general solution of (2.6.1). We use the characteristics of this equation (2.6.1) to introduce the new transformation by equations

$$\xi = \xi\left(x,y\right), \qquad \eta = \eta\left(x,y\right), \qquad (2.6.2)$$

where ξ and η are once continuously differentiable and their Jacobian $J\left(x,y\right) \equiv \xi_x\eta_y - \xi_y\eta_x$ is nonzero in a domain of interest so that x and y can be determined uniquely from the system of equations (2.6.2). Thus, by chain rule,

$$u_x = u_\xi\xi_x + u_\eta\eta_x, \qquad u_y = u_\xi\xi_y + u_\eta\eta_y, \qquad (2.6.3)$$

we substitute these partial derivatives (2.6.3) into (2.6.1) to obtain the equation

$$A\,u_\xi + B\,u_\eta + cu = d, \qquad (2.6.4)$$

where

$$A = u\xi_x + b\xi_y, \qquad B = a\eta_x + b\eta_y. \qquad (2.6.5)$$

From (2.6.5) we see that $B = 0$ if η is a solution of the first-order equation

$$a\eta_x + b\eta_y = 0. \qquad (2.6.6)$$

This equation has infinitely many solutions. We can obtain one of them by assigning initial condition on a non-characteristic initial curve and solving the resulting initial-value problem according to the method described earlier. Since $\eta(x, y)$ satisfies equation (2.6.6), the level curves $\eta(x, y) = $ constant are always characteristic curves of equation (2.6.1). Thus, one set of the new transformations are the characteristic curves of (2.6.1). The second set, $\xi(x, y) = $ constant, can be chosen to be any one parameter family of smooth curves which are nowhere tangent to the family of the characteristic curves. We next assert that $A \neq 0$ in a neighborhood of some point in the domain D in which $\eta(x, y)$ is defined and $J \neq 0$. For, if $A = 0$ at some point of D, then $B = 0$ at the same point. Consequently, equations (2.6.5) would form a system of linear homogeneous equations in a and b, where the Jacobian J is the determinant of its coefficient matrix. Since $J \neq 0$, both a and b must be zero at that point which contradicts the original assumption that a and b do not vanish simultaneously. Finally, since $B = 0$ and $A \neq 0$ in D, we can divide (2.6.4) by A to obtain the canonical form

$$u_\xi + \alpha(\xi, \eta)\, u = \beta(\xi, \eta),\qquad(2.6.7)$$

where $\alpha(\xi, \eta) = \frac{c}{A}$ and $\beta(\xi, \eta) = \frac{d}{A}$.

Equation (2.6.7) represents an ordinary differential equation with ξ as the independent variable and η as a parameter which may be treated as constant. This equation (2.6.7) is called the *canonical form* of equation (2.6.1) in terms of the coordinates (ξ, η). Generally, the canonical equation (2.6.7) can easily be integrated and the general solution of (2.6.1) can be obtained after replacing ξ and η by the original variables x and y.

We close this section by considering some examples that illustrate this procedure. In practice, it is convenient to choose $\xi = \xi(x, y)$ and $\eta(x, y) = y$ or $\xi = x$ and $\eta = \eta(x, y)$ so that $J \neq 0$.

Example 2.6.1. Reduce each of the following equations

$$u_x - u_y = u,\qquad(2.6.8)$$
$$yu_x + u_y = x,\qquad(2.6.9)$$

to canonical form, and obtain the general solution.

In (2.6.8), $a = 1$, $b = -1$, $c = -1$ and $d = 0$. The characteristic equations are

$$\frac{dx}{1} = \frac{dy}{-1} = \frac{du}{u}.$$

The characteristic curves are $\xi = x + y = c_1$, and we choose $\eta = y = c_2$ where c_1 and c_2 are constants. Consequently, $u_x = u_\xi$ and $u_y = u_\xi + u_\eta$, and hence, equation (2.6.8) becomes

$$u_\eta = u.$$

Integrating this equation gives

$$\ln u\left(\xi,\eta\right) = -\eta + \ln f\left(\xi\right),$$

where $f\left(\xi\right)$ is an arbitrary function of ξ only.
Equivalently,

$$u\left(\xi,\eta\right) = f\left(\xi\right)e^{-\eta}.$$

In terms of the original variables x and y, the general solution of equation (2.6.8) is

$$u\left(x,y\right) = f\left(x+y\right)e^{-y}, \tag{2.6.10}$$

where f is an arbitrary function.

The characteristic equations of (2.6.9) are

$$\frac{dx}{y} = \frac{dy}{1} = \frac{du}{x}.$$

It follows from the first two equations that $\xi\left(x,y\right) = x - \frac{y^2}{2} = c_1$; we choose $\eta\left(x,y\right) = y = c_2$. Consequently, $u_x = u_\xi$ and $u_y = -y\,u_\xi + u_\eta$ and hence, equation (2.6.9) reduces to

$$u_\eta = \xi + \frac{1}{2}\eta^2.$$

Integrating this equation gives the general solution

$$u\left(\xi,\eta\right) = \xi\eta + \frac{1}{6}\eta^3 + f\left(\xi\right),$$

where f is an arbitrary function.

Thus, the general solution of (2.6.9) in terms of x and y is

$$u\left(x,y\right) = xy - \frac{1}{3}y^3 + f\left(x - \frac{y^2}{2}\right).$$

2.7 Method of Separation of Variables

During the last two centuries several methods have been developed for solving partial differential equations. Among these, a technique known as the *method of separation of variables* is perhaps the oldest systematic method for solving partial differential equations. Its essential feature is to transform the partial differential equations by a set of ordinary differential equations. The required solution of the partial differential equations is then exposed as a product $u\left(x,y\right) = X\left(x\right)Y\left(y\right) \neq 0$, or as a sum $u\left(x,y\right) = X\left(x\right) + Y\left(y\right)$, where $X\left(x\right)$ and $Y\left(y\right)$ are functions of x and y, respectively. Many significant problems in partial differential equations can be solved by the method of separation of variables. This method has been considerably refined and generalized over the last two centuries and is one of the classical techniques of applied mathematics, mathematical physics and engineering science.

Usually, the first-order partial differential equation can be solved by separation of variables without the need for Fourier series. The main purpose of this section is to illustrate the method by examples.

Example 2.7.1. Solve the initial-value problem

$$u_x + 2u_y = 0, \qquad u(0, y) = 4\,e^{-2y}. \qquad (2.7.1ab)$$

We seek a separable solution $u(x, y) = X(x)\,Y(y) \neq 0$ and substitute into the equation to obtain

$$X'(x)\,Y(y) + 2X(x)\,Y'(y) = 0.$$

This can also be expressed in the form

$$\frac{X'(x)}{2X(x)} = -\frac{Y'(y)}{Y(y)}. \qquad (2.7.2)$$

Since the left-hand side of this equation is a function of x only and the right-hand is a function of y only, it follows that (2.7.2) can be true if both sides are equal to the same constant value λ which is called an arbitrary separation constant. Consequently, (2.7.2) gives two ordinary differential equations

$$X'(x) - 2\lambda X(x) = 0, \qquad Y'(y) + \lambda Y(y) = 0. \qquad (2.7.3)$$

These equations have solutions given, respectively, by

$$X(x) = A\,e^{2\lambda x} \quad \text{and} \quad Y(y) = B\,e^{-\lambda y}, \qquad (2.7.4)$$

where A and B are arbitrary integrating constants.

Consequently, the general solution is given by

$$u(x, y) = AB \exp(2\lambda x - \lambda y) = C \exp(2\lambda x - \lambda y), \qquad (2.7.5)$$

where $C = AB$ is an arbitrary constant.

Using the condition (2.7.1b), we find

$$4\,e^{-2y} = u(0, y) = Ce^{-\lambda y},$$

and hence, we deduce that $C = 4$ and $\lambda = 2$. Therefore, the final solution is

$$u(x, y) = 4 \exp(4x - 2y). \qquad (2.7.6)$$

Example 2.7.2. Solve the equation

$$y^2 u_x^2 + x^2 u_y^2 = (xyu)^2 . \tag{2.7.7}$$

We assume $u(x, y) = f(x) g(y) \neq 0$ is a separable solution of (2.7.7), and substitute into the equation. Consequently, we obtain

$$y^2 \{ f'(x) g(y) \}^2 + x^2 \{ f(x) g'(y) \}^2 = x^2 y^2 \{ f(x) g(y) \}^2 ,$$

or, equivalently,

$$\frac{1}{x^2} \left\{ \frac{f'(x)}{f(x)} \right\}^2 + \frac{1}{y^2} \left\{ \frac{g'(y)}{g(y)} \right\}^2 = 1,$$

or

$$\frac{1}{x^2} \left\{ \frac{f'(x)}{f(x)} \right\}^2 = 1 - \frac{1}{y^2} \left\{ \frac{g'(y)}{g(y)} \right\}^2 = \lambda^2 ,$$

where λ^2 is a separation constant. Thus,

$$\frac{1}{x} \frac{f'(x)}{f(x)} = \lambda \quad \text{and} \quad \frac{g'(y)}{y\,g(y)} = \sqrt{1 - \lambda^2} . \tag{2.7.8}$$

Solving these ordinary differential equations, we find

$$f(x) = A \exp \left(\frac{\lambda}{2} x^2 \right) \quad \text{and} \quad g(y) = B \exp \left(\frac{1}{2} y \sqrt{1 - \lambda^2} \right),$$

where A and B are arbitrary constant. Thus, the general solution is

$$u(x, y) = C \exp \left(\frac{\lambda}{2} x^2 + \frac{1}{2} y^2 \sqrt{1 - \lambda^2} \right), \tag{2.7.9}$$

where $C = AB$ is an arbitrary constant.

Using the condition $u(x, 0) = 3 \exp (x^2/4)$, we can determine both C and λ in (2.7.9). It turns out that $C = 3$ and $\lambda = (1/2)$, and the solution becomes

$$u(x, y) = 3 \exp \left[\frac{1}{4} \left(x^2 + y^2 \sqrt{3} \right) \right]. \tag{2.7.10}$$

Example 2.7.3. Use the separation of variables $u(x, y) = f(x) + g(y)$ to solve the equation

$$u_x^2 + u_y^2 = 1. \tag{2.7.11}$$

Obviously,

$$\{f'(x)\}^2 = 1 - \{g'(y)\}^2 = \lambda^2,$$

where λ^2 is a separation constant. Thus, we obtain

$$f'(x) = \lambda \quad \text{and} \quad g'(y) = \sqrt{1 - \lambda^2}.$$

Solving these ordinary differential equations, we find

$$f(x) = \lambda x + A \quad \text{and} \quad g(y) = y\sqrt{1 - \lambda^2} + B,$$

where A and B are constants of integration. Finally, the solution of (2.7.11) is given by

$$u(x, y) = \lambda x + y\sqrt{1 - \lambda^2} + C, \tag{2.7.12}$$

where $C = A + B$ is an arbitrary constant.

Example 2.7.4. Use $u(x, y) = f(x) + g(y)$ to solve the equation

$$u_x^2 + u_y + x^2 = 0. \tag{2.7.13}$$

Obviously, equation (2.7.13) has the separable form

$$\{f'(x)\}^2 + x^2 = -g'(y) = \lambda^2,$$

where λ^2 is a separation constant.
Consequently,

$$f'(x) = \sqrt{\lambda^2 - x^2} \quad \text{and} \quad g'(y) = -\lambda^2.$$

These can be integrated to obtain

$$
\begin{aligned}
f(x) &= \int \sqrt{\lambda^2 - x^2}\, dx + A \\
&= \lambda^2 \int \cos^2 \theta\, d\theta + A, \quad (x = \lambda \sin \theta) \\
&= \frac{1}{2}\lambda^2 \left[\sin^{-1}\left(\frac{x}{\lambda}\right) + \frac{x}{\lambda}\sqrt{1 - \frac{x^2}{\lambda^2}} \right] + A
\end{aligned}
$$

and

$$g(y) = -\lambda^2 y + B.$$

Finally, the general solution is given by

$$u(x, y) = \frac{1}{2}\lambda^2 \sin^{-1}\left(\frac{x}{\lambda}\right) + \frac{x}{2}\sqrt{\lambda^2 - x^2} - \lambda^2 y + C, \tag{2.7.14}$$

where $C = A + B$ is an arbitrary constant.

Example 2.7.5. Use $v = \ln u$ and $v = f(x) + g(y)$ to solve the equation

$$x^2 u_x^2 + y^2 u_y^2 = u^2. \tag{2.7.15}$$

In view of $v = \ln u$, $v_x = \frac{1}{u} u_x$ and $v_y = \frac{1}{u} u_y$, and hence, equation (2.7.15) becomes

$$x^2 v_x^2 + y^2 v_y^2 = 1. \tag{2.7.16}$$

Substitute $v(x, y) = f(x) + g(y)$ into (2.7.16) to obtain

$$x^2 \left\{ f'(x) \right\}^2 + y^2 \left\{ g'(y) \right\}^2 = 1$$

or

$$x^2 \left\{ f'(x) \right\}^2 = 1 - y^2 \left\{ g'(y) \right\}^2 = \lambda^2,$$

where λ^2 is a separation constant. Thus, we obtain

$$f'(x) = \frac{\lambda}{x} \quad \text{and} \quad g'(y) = \frac{1}{y}\sqrt{1 - \lambda^2}\, .$$

Integrating these equations gives

$$f(x) = \lambda \ln x + A \quad \text{and} \quad g(y) = \sqrt{1 - \lambda^2}\, \ln y + B,$$

where A and B are integrating constants. Therefore, the general solution of (2.7.16) is given by

$$\begin{aligned} v(x, y) &= \lambda \ln x + \sqrt{1 - \lambda^2}\, \ln y + \ln C \\ &= \ln \left(x^\lambda \cdot y^{\sqrt{1-\lambda^2}} \cdot C \right), \end{aligned} \tag{2.7.17}$$

where $\ln C = A + B$. The final solution is

$$u(x, y) = e^v = C\, x^\lambda \cdot y^{\sqrt{1-\lambda^2}}, \tag{2.7.18}$$

where C is an integrating constant.

2.8 Exercises

1. (a) Show that the family of right circular cones whose axes coincide with the z-axis

$$x^2 + y^2 = (z - c)^2 \tan^2 \alpha$$

satisfies the first-order, partial differential equation

$$yp - xq = 0.$$

(b) Show that all the surfaces of revolution, $z = f\left(x^2 + y^2\right)$ with the z-axis as the axis of symmetry, where f is an arbitrary function, satisfy the partial differential equation

$$yp - xq = 0.$$

(c) Show that the two-parameter family of curves $u - ax - by - ab = 0$ satisfies the nonlinear equation

$$xp + yq + pq = u.$$

2. Find the partial differential equation arising from each of the following surfaces:

(a) $z = x + y + f\left(xy\right)$,   (b) $z = f\left(x - y\right)$,

(c) $z = xy + f\left(x^2 + y^2\right)$,   (d) $2z = \left(\alpha x + y\right)^2 + \beta$.

3. Find the general solution of each of the following equations:

(a) $u_x = 0$,   (b) $a\,u_x + b\,u_y = 0$; a, b, are constant,

(c) $u_x + y\,u_y = 0$,   (d) $\left(1 + x^2\right) u_x + u_y = 0$,

(e) $2xy\,u_x + \left(x^2 + y^2\right) u_y = 0$, (f) $\left(y + u\right) u_x + y\,u_y = x - y$,

(g) $y^2 u_x - xy\,u_y = x\left(u - 2y\right)$,   (h) $yu_y - xu_x = 1$,

(i) $y^2 up + u^2 xq = -xy^2$,   (j) $\left(y - xu\right) p + \left(x + yu\right) q = x^2 + y^2$.

4. Find the general solution of the equation

$$u_x + 2xy^2 u_y = 0.$$

5. Find the solution of the following Cauchy problems:

(a) $3u_x + 2u_y = 0$, with $u\left(x, 0\right) = \sin x$,

(b) $y\,u_x + x\,u_y = 0$, with $u\left(0, y\right) = \exp\left(-y^2\right)$,

(c) $x\,u_x + y\,u_y = 2xy$, with $u = 2$ on $y = x^2$,

(d) $u_x + x\,u_y = 0$, with $u\left(0, y\right) = \sin y$,

(e) $y\,u_x + x\,u_y = xy$, $x \geq 0$, $y \geq 0$, with $u\left(0, y\right) = \exp\left(-y^2\right)$

for $y > 0$, and $u\left(x, 0\right) = \exp\left(-x^2\right)$ for $x > 0$,

(f) $u_x + x\,u_y = \left(y - \frac{1}{2}x^2\right)^2$, with $u\,(0, y) = \exp\,(y)$,

(g) $x\,u_x + y\,u_y = u + 1$, with $u\,(x, y) = x^2$ on $y = x^2$,

(h) $u\,u_x - u\,u_y = u^2 + (x + y)^2$, with $u = 1$ on $y = 0$,

(i) $x\,u_x + (x + y)\,u_y = u + 1$, with $u\,(x, y) = x^2$ on $y = 0$.

(j) $\sqrt{x}\,u_x + u\,u_y + u^2 = 0$, $u\,(x, 0) = 1,\, 0 < x < \infty$.

(k) $u\,x^2 u_x + e^{-y} u_y + u^2 = 0$, $u\,(x, 0) = 1,\, 0 < x < \infty$.

6. Solve the initial-value problem

$$u_t + u\,u_x = 0$$

with the initial curve

$$x = \frac{1}{2}\tau^2,\ t = \tau, u = \tau.$$

7. Find the solution of the Cauchy problem

$$2xy\,u_x + \left(x^2 + y^2\right) u_y = 0,\ \text{with } u = \exp\left(\frac{x}{x - y}\right) \text{on } x + y = 1.$$

8. Solve the following equations:

(a) $x\,u_x + y\,u_y + z\,u_z = 0$,

(b) $x^2\,u_x + y^2\,u_y + z\,(x + y)\,u_z = 0$,

(c) $x\,(y - z)\,u_x + y\,(z - x)\,u_y + z\,(x - y)\,u_z = 0$,

(d) $yz\,u_x - xz\,u_y + xy\left(x^2 + y^2\right) u_z = 0$,

(e) $x\left(y^2 - z^2\right) u_x + y\left(z^2 - y^2\right) u_y + z\left(x^2 - y^2\right) u_z = 0$.

9. Solve the equation

$$u_x + x\,u_y = y$$

with the Cauchy data

$$(a)\ u\,(0, y) = y^2,\quad (b)\ u\,(1, y) = 2y.$$

10. Show that $u_1 = e^x$ and $u_2 = e^{-y}$ are solutions of the nonlinear equation

$$(u_x + u_y)^2 - u^2 = 0$$

but that their sum $(e^x + e^{-y})$ is not a solution of the equation.

11. Solve the Cauchy problem

$$(y + u)\, u_x + y\, u_y = (x - y), \quad \text{with} \quad u = 1 + x \quad \text{on} \quad y = 1.$$

12. Find the integral surfaces of the equation $u\, u_x + u_y = 1$ for each of the following initial data:

(a) $x\,(s,0) = s$, $y\,(s,0) = 2s$, $u\,(s,0) = s$,

(b) $x\,(s,0) = s^2$, $y\,(s,0) = 2s$, $u\,(s,0) = s$,

(c) $x\,(s,0) = \frac{s^2}{2}$, $y\,(s,0) = s$, $u\,(s,0) = s$.

Draw characteristics in each case.

13. Show that the solution of the equation

$$y\, u_x - x\, u_y = 0$$

containing the curve $x^2 + y^2 = a^2$, $u = y$, does not exist.

14. Solve the following Cauchy problems:

(a) $x^2 u_x - y^2 u_y = 0$, $u \to e^x$ as $y \to \infty$,

(b) $y\, u_x + x\, u_y = 0$, $u = \sin x$ on $x^2 + y^2 = 1$,

(c) $-x\, u_x + y\, u_y = 1$ for $0 < x < y$, $u = 2x$ on $y = 3x$,

(d) $2x\, u_x + (x + 1)\, u_y = y$ for $x > 0$, $u = 2y$ on $x = 1$,

(e) $x\, u_x - 2y\, u_y = x^2 + y^2$ for $x > 0$, $y > 0$, $u = x^2$ on $y = 1$,

(f) $u_x + 2\, u_y = 1 + u$, $u\,(x,y) = \sin x$ on $y = 3x + 1$,

(g) $u_x + 3u_y = u$, $u\,(x,y) = \cos x$ on $y = x$,

(h) $u_x + 2x\, u_y = 2xu$, $u\,(x,0) = x^2$,

(i) $u\, u_x + u_y = u$, $u\,(x,0) = 2x$, $1 \le x \le 2$,

(j) $u_x + u_y = u^2$, $u\,(x,0) = \tanh x$.

Show that the solution of (j) is unbounded on the critical curve $y \tanh (x - y) = 1$.

15. Find the solution surface of the equation

$$\left(u^2 - y^2\right) u_x + xy\, u_y + xu = 0, \quad \text{with} \quad u = y = x, \ x > 0.$$

16. (a) Solve the Cauchy problem

$$u_x + uu_y = 1, \quad u\left(0, y\right) = ay,$$

where a is a constant.

(b) Find the solution of the equation in (a) with the data

$$x\left(s, 0\right) = 2s, \quad y\left(s, 0\right) = s^2, \quad u\left(0, s^2\right) = s.$$

17. Solve the following equations:

(a) $\left(y + u\right) u_x + \left(x + u\right) u_y = x + y,$

(b) $x\, u \left(u^2 + xy\right) u_x - y\, u \left(u^2 + xy\right) u_y = x^4,$

(c) $\left(x + y\right) u_x + \left(x - y\right) u_y = 0,$

(d) $y\, u_x + x\, u_y = xy \left(x^2 - y^2\right),$

(e) $\left(cy - bz\right) z_x + \left(az - cx\right) z_y = bx - ay.$

18. Solve the equation

$$x\, z_x + y\, z_y = z,$$

and find the curves which satisfy the associated characteristic equations and intersect the helix $x + y^2 = a^2$, $z = b \, \tan^{-1}\left(\frac{y}{x}\right)$.

19. Obtain the family of curves which represent the general solution of the partial differential equation

$$\left(2x - 4y + 3u\right) u_x + \left(x - 2y - 3u\right) u_y = -3\left(x - 2y\right).$$

Determine the particular member of the family which contains the line $u = x$ and $y = 0$.

20. Find the solution of the equation

$$y\, u_x - 2xy\, u_y = 2xu$$

with the condition $u\left(0, y\right) = y^3$.

21. Obtain the general solution of the equation

$$(x + y + 5z)\, p + 4zq + (x + y + z) = 0 \quad (p = z_x, \ q = z_y),$$

and find the particular solution which passes through the circle

$$z = 0, \quad x^2 + y^2 = a^2.$$

22. Obtain the general solution of the equation

$$(z^2 - 2yz - y^2)\, p + x\, (y + z)\, q = x\, (y - z) \quad (p = z_x, \ q = z_y).$$

Find the integral surfaces of this equation passing through

(a) the x-axis, (b) the y-axis, and (c) the z-axis.

23. Solve the Cauchy problem

$$(x + y)\, u_x + (x - y)\, u_y = 1, \quad u\, (1, y) = \frac{1}{\sqrt{2}}.$$

24. Solve the following Cauchy problems:

(a) $3\, u_x + 2\, u_y = 0$, $u\, (x, 0) = f\, (x)$,

(b) $a\, u_x + b\, u_y = c\, u$, $u\, (x, 0) = f\, (x)$, where a, b, c are constants,

(c) $x\, u_x + y\, u_y = c\, u$, $u\, (x, 0) = f\, (x)$,

(d) $u\, u_x + u_y = 1$, $u\, (s, 0) = \alpha s$, $x\, (s, 0) = s$, $y\, (s, 0) = 0$.

25. Apply the method of separation of variables $u\, (x, y) = f\, (x)\, g\, (y)$ to solve the following equations:

(a) $u_x + u = u_y$, $\quad u\, (x, 0) = 4e^{-3x}$, $\qquad$ (b) $u_x u_y = u^2$,

(c) $u_x + 2u_y = 0$, $\quad u\, (0, y) = 3e^{-2y}$, $\qquad$ (d) $y^2 u_x^2 + x^2 u_y^2 = (xyu)^2$,

(e) $x^2 u_{xy} + 9y^2 u = 0$, $\quad u\, (x, 0) = \exp\left(\frac{1}{x}\right)$, (f) $y\, u_x - x\, u_y = 0$,

(g) $u_t = c^2\, (u_{xx} + u_{yy})$, $\qquad\qquad\qquad\quad$ (h) $u_{xx} + u_{yy} = 0$.

26. Use a separable solution $u\, (x, y) = f\, (x) + g\, (y)$ to solve the following equations:

(a) $u_x^2 + u_y^2 = 1$, $\qquad\qquad\qquad\qquad$ (b) $u_x^2 + u_y^2 = u$,

(c) $u_x^2 + u_y + x^2 = 0$, $\qquad\qquad\qquad$ (d) $x^2 u_x^2 + y^2 u_y^2 = 1$,

(e) $y\, u_x + x\, u_y = 0$, $\quad u\, (0, y) = y^2$.

27. Apply $v = \ln u$ and then $v(x, y) = f(x) + g(y)$ to solve the following equations:

(a) $x^2 u_x^2 + y^2 u_y^2 = u^2$,

(b) $x^2 u_x^2 + y^2 u_y^2 = (xyu)^2$.

28. Apply $\sqrt{u} = v$ and $v(x, y) = f(x) + g(y)$ to solve the equation

$$x^4 u_x^2 + y^2 u_y^2 = 4u.$$

29. Using $v = \ln u$ and $v = f(x) + g(y)$, show that the solution of the Cauchy problem

$$y^2 u_x^2 + x^2 u_y^2 = (xyu)^2, \quad u(x, 0) = e^{x^2}$$

is

$$u(x, y) = \exp\left(x^2 + i\frac{\sqrt{3}}{2} y^2 \right).$$

30. Reduce each of the following equations into canonical form and find the general solution:

(a) $u_x + u_y = u$,   (b) $u_x + x\, u_y = y$,

(c) $u_x + 2xy\, u_y = x$,   (d) $u_x - y\, u_y - u = 1$.

31. Find the solution of each of the following equations by the method of separation of variables:

(a) $u_x - u_y = 0$, $u(0, y) = 2e^{3y}$,

(b) $u_x - u_y = u$, $u(x, 0) = 4e^{-3x}$,

(c) $a\, u_x + b\, u_y = 0$, $u(x, 0) = \alpha e^{\beta x}$,

   where a, b, α and β are constants.

32. Find the solution of the following initial-value systems

(a) $u_t + 3uu_x = v - x$, $v_t - cv_x = 0$ with $u(x, 0) = x$ and $v(x, 0) = x$.

(b) $u_t + 2uu_x = v - x$, $v_t - cv_x = 0$ with $u(x, 0) = x$ and $v(x, 0) = x$.

33. Solve the following initial-value systems

    (a) $u_t + uu_x = e^{-x}v$, $v_t - av_x = 0$ with $u(x,0) = x$ and $v(x,0) = e^x$.

    (b) $u_t - 2uu_x = v - x$, $v_t + cv_x = 0$ with $u(x,0) = x$ and $v(x,0) = x$.

34. Consider the *Fokker–Planck equation* (See Reif (1965)) in statistical mechanics to describe the evolution of the probability distribution function in the form

$$u_t = u_{xx} + (x\,u)_x\,,$$
$$u(x,0) = f(x)\,.$$

    Neglecting the term u_{xx}, solve the first-order linear equation

$$u_t - x\,u_x = u \quad \text{with} \quad u(x,0) = f(x)\,.$$

3

Mathematical Models

"Physics can't exist without mathematics which provides it with the only language in which it can speak. Thus, services are continuously exchanged between pure mathematical analysis and physics. It is really remarkable that among works of analysis most useful for physics were those cultivated for their own beauty. In exchange, physics, exposing new problems, is as useful for mathematics as it is a model for an artist."

Henri Poincaré

"It is no paradox to say in our most theoretical models we may be nearest to our most practical applications."

A. N. Whitehead

"... builds models based on data from all levels: gene expression, protein location in the cell, models of cell function, and computer representations of organs and organisms."

E. Pennisi

3.1 Classical Equations

Partial differential equations arise frequently in formulating fundamental laws of nature and in the study of a wide variety of physical, chemical, and biological models. We start with a special type of second-order linear partial differential equation for the following reasons. First, second-order linear equations arise more frequently in a wide variety of applications. Second, their mathematical treatment is simpler and easier to understand than that of first-order equations in general. Usually, in almost all physical phenomena (or physical processes), the dependent variable $u = u\,(x, y, z, t)$ is a function of three space variables, x, y, z and time variable t.

The three basic types of second-order partial differential equations are:

(a) The wave equation

$$u_{tt} - c^2\,(u_{xx} + u_{yy} + u_{zz}) = 0. \tag{3.1.1}$$

(b) The heat equation

$$u_t - k\,(u_{xx} + u_{yy} + u_{zz}) = 0. \tag{3.1.2}$$

(c) The Laplace equation

$$u_{xx} + u_{yy} + u_{zz} = 0. \tag{3.1.3}$$

In this section, we list a few more common linear partial differential equations of importance in applied mathematics, mathematical physics, and engineering science. Such a list naturally cannot ever be complete. Included are only equations of most common interest:

(d) The Poisson equation

$$\nabla^2 u = f\,(x, y, z)\,. \tag{3.1.4}$$

(e) The Helmholtz equation

$$\nabla^2 u + \lambda u = 0. \tag{3.1.5}$$

(f) The biharmonic equation

$$\nabla^4 u = \nabla^2\,(\nabla^2 u) = 0. \tag{3.1.6}$$

(g) The biharmonic wave equation

$$u_{tt} + c^2 \nabla^4 u = 0. \tag{3.1.7}$$

(h) The telegraph equation

$$u_{tt} + a u_t + b u = c^2 u_{xx}. \tag{3.1.8}$$

(i) The Schrödinger equations in quantum physics

$$i\hbar\psi_t = \left[\left(-\frac{\hbar^2}{2m}\right)\nabla^2 + V\,(x, y, z)\right]\psi, \tag{3.1.9}$$

$$\nabla^2 \Psi + \frac{2m}{\hbar^2}\,[E - V\,(x, y, z)]\,\Psi = 0. \tag{3.1.10}$$

(j) The Klein–Gordon equation

$$\Box u + \lambda^2 u = 0, \tag{3.1.11}$$

where

$$\nabla^2 \equiv \frac{\partial^2}{\partial x^2} + \frac{\partial^2}{\partial y^2} + \frac{\partial^2}{\partial z^2}, \qquad (3.1.12)$$

is the Laplace operator in rectangular Cartesian coordinates (x, y, z),

$$\Box \equiv \nabla^2 - \frac{1}{c^2}\frac{\partial^2}{\partial t^2}, \qquad (3.1.13)$$

is the d'Alembertian, and in all equations λ, a, b, c, m, E are constants and $h = 2\pi\hbar$ is the Planck constant.

(k) For a compressible fluid flow, Euler's equations

$$\mathbf{u}_t + (\mathbf{u} \cdot \nabla)\,\mathbf{u} = -\frac{1}{\rho}\nabla p, \qquad \rho_t + \mathrm{div}\,(\rho\mathbf{u}) = 0, \qquad (3.1.14)$$

where $\mathbf{u} = (u, v, w)$ is the fluid velocity vector, ρ is the fluid density, and $p = p\,(\rho)$ is the pressure that relates p and ρ (the *constitutive equation* or *equation of state*).

Many problems in mathematical physics reduce to the solving of partial differential equations, in particular, the partial differential equations listed above. We will begin our study of these equations by first examining in detail the mathematical models representing physical problems.

3.2 The Vibrating String

One of the most important problems in mathematical physics is the vibration of a stretched string. Simplicity and frequent occurrence in many branches of mathematical physics make it a classic example in the theory of partial differential equations.

Let us consider a stretched string of length l fixed at the end points. The problem here is to determine the equation of motion which characterizes the position $u\,(x, t)$ of the string at time t after an initial disturbance is given.

In order to obtain a simple equation, we make the following assumptions:

1. The string is flexible and elastic, that is the string cannot resist bending moment and thus the tension in the string is always in the direction of the tangent to the existing profile of the string.
2. There is no elongation of a single segment of the string and hence, by Hooke's law, the tension is constant.
3. The weight of the string is small compared with the tension in the string.
4. The deflection is small compared with the length of the string.
5. The slope of the displaced string at any point is small compared with unity.

6. There is only pure transverse vibration.

We consider a differential element of the string. Let T be the tension at the end points as shown in Figure 3.2.1. The forces acting on the element of the string in the vertical direction are

$$T \sin \beta - T \sin \alpha.$$

By Newton's second law of motion, the resultant force is equal to the mass times the acceleration. Hence,

$$T \sin \beta - T \sin \alpha = \rho \, \delta s \, u_{tt} \tag{3.2.1}$$

where ρ is the line density and δs is the smaller arc length of the string. Since the slope of the displaced string is small, we have

$$\delta s \simeq \delta x.$$

Since the angles α and β are small

$$\sin \alpha \simeq \tan \alpha, \qquad \sin \beta \simeq \tan \beta.$$

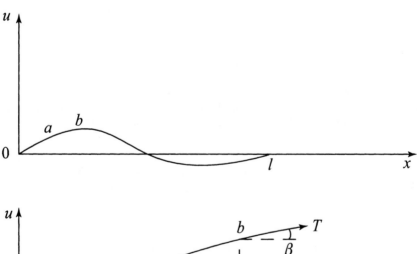

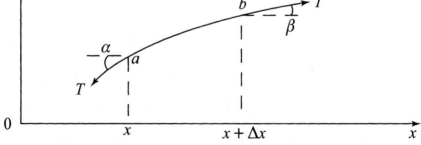

Figure 3.2.1 An Element of a vertically displaced string.

Thus, equation (3.2.1) becomes

$$\tan \beta - \tan \alpha = \frac{\rho \, \delta x}{T} u_{tt}. \tag{3.2.2}$$

But, from calculus we know that $\tan \alpha$ and $\tan \beta$ are the slopes of the string at x and $x + \delta x$:

$$\tan \alpha = u_x \, (x, t)$$

and

$$\tan \beta = u_x \, (x + \delta x, t)$$

at time t. Equation (3.2.2) may thus be written as

$$\frac{1}{\delta x} \left[(u_x)_{x+\delta x} - (u_x)_x \right] = \frac{\rho}{T} u_{tt}, \quad \frac{1}{\delta x} \left[u_x \, (x + \delta x, t) - u_x \, (x, t) \right] = \frac{\rho}{T} u_{tt}.$$

In the limit as δx approaches zero, we find

$$u_{tt} = c^2 u_{xx} \tag{3.2.3}$$

where $c^2 = T/\rho$. This is called the *one-dimensional wave equation*.

If there is an external force f per unit length acting on the string. Equation (3.2.3) assumes the form

$$u_{tt} = c^2 u_{xx} + F, \quad F = f/\rho, \tag{3.2.4}$$

where f may be pressure, gravitation, resistance, and so on.

3.3 The Vibrating Membrane

The equation of the vibrating membrane occurs in a large number of problems in applied mathematics and mathematical physics. Before we derive the equation for the vibrating membrane we make certain simplifying assumptions as in the case of the vibrating string:

1. The membrane is flexible and elastic, that is, the membrane cannot resist bending moment and the tension in the membrane is always in the direction of the tangent to the existing profile of the membrane.
2. There is no elongation of a single segment of the membrane and hence, by Hooke's law, the tension is constant.
3. The weight of the membrane is small compared with the tension in the membrane.
4. The deflection is small compared with the minimal diameter of the membrane.

5. The slope of the displayed membrane at any point is small compared with unity.
6. There is only pure transverse vibration.

We consider a small element of the membrane. Since the deflection and slope are small, the area of the element is approximately equal to $\delta x \delta y$. If T is the tensile force per unit length, then the forces acting on the sides of the element are $T\,\delta x$ and $T\,\delta y$, as shown in Figure 3.3.1.

The forces acting on the element of the membrane in the vertical direction are

$$T\,\delta x \sin\beta - T\,\delta x \sin\alpha + T\,\delta y \sin\delta - T\,\delta y \sin\gamma.$$

Since the slopes are small, sines of the angles are approximately equal to their tangents. Thus, the resultant force becomes

$$T\,\delta x \left(\tan\beta - \tan\alpha\right) + T\,\delta y \left(\tan\delta - \tan\gamma\right).$$

By Newton's second law of motion, the resultant force is equal to the mass times the acceleration. Hence,

$$T\,\delta x \left(\tan\beta - \tan\alpha\right) + T\,\delta y \left(\tan\delta - \tan\gamma\right) = \rho\,\delta A\, u_{tt} \qquad (3.3.1)$$

where ρ is the mass per unit area, $\delta A \simeq \delta x \delta y$ is the area of this element, and u_{tt} is computed at some point in the region under consideration. But from calculus, we have

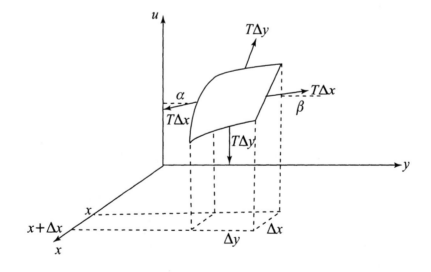

Figure 3.3.1 An element of vertically displaced membrane.

$$\tan \alpha = u_y\,(x_1, y)$$
$$\tan \beta = u_y\,(x_2, y + \delta y)$$
$$\tan \gamma = u_x\,(x, y_1)$$
$$\tan \delta = u_x\,(x + \delta x, y_2)$$

where x_1 and x_2 are the values of x between x and $x + \delta x$, and y_1 and y_2 are the values of y between y and $y + \delta y$. Substituting these values in (3.3.1), we obtain

$$T\,\delta x\left[u_y\,(x_2, y + \delta y) - u_y\,(x_1, y)\right] + T\,\delta y\left[u_x\,(x + \delta x, y_2) - u_x\,(x, y_1)\right]$$
$$= \rho\,\delta x \delta y\, u_{tt}.$$

Division by $\rho\,\delta x \delta y$ yields

$$\frac{T}{\rho}\left[\frac{u_y\,(x_2, y + \delta y) - u_y\,(x_1, y)}{\delta y} + \frac{u_x\,(x + \delta x, y_2) - u_x\,(x, y_1)}{\delta x}\right] = u_{tt}.$$
$$(3.3.2)$$

In the limit as δx approaches zero and δy approaches zero, we obtain

$$u_{tt} = c^2\left(u_{xx} + u_{yy}\right), \tag{3.3.3}$$

where $c^2 = T/\rho$. This equation is called the *two-dimensional wave equation*.

If there is an external force f per unit area acting on the membrane. Equation (3.3.3) takes the form

$$u_{tt} = c^2\left(u_{xx} + u_{yy}\right) + F, \tag{3.3.4}$$

where $F = f/\rho$.

3.4 Waves in an Elastic Medium

If a small disturbance is originated at a point in an elastic medium, neighboring particles are set into motion, and the medium is put under a state of strain. We consider such states of motion to extend in all directions. We assume that the displacements of the medium are small and that we are not concerned with translation or rotation of the medium as a whole.

Let the body under investigation be homogeneous and isotropic. Let δV be a differential volume of the body, and let the stresses acting on the faces of the volume be τ_{xx}, τ_{yy}, τ_{zz}, τ_{xy}, τ_{xz}, τ_{yx}, τ_{yz}, τ_{zx}, τ_{zy}. The first three stresses are called the *normal stresses* and the rest are called the *shear stresses*. (See Figure 3.4.1).

We shall assume that the stress tensor τ_{ij} is symmetric describing the condition of the rotational equilibrium of the volume element, that is,

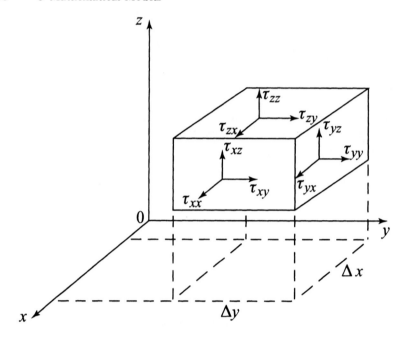

Figure 3.4.1 Volume element of an elastic body.

$$\tau_{ij} = \tau_{ji}, \quad i \neq j, \quad i,j = x,y,z. \tag{3.4.1}$$

Neglecting the body forces, the sum of all the forces acting on the volume element in the x-direction is

$$\left[(\tau_{xx})_{x+\delta x} - (\tau_{xx})_x\right] \delta y \delta z + \left[(\tau_{xy})_{y+\delta y} - (\tau_{xy})_y\right] \delta z \delta x$$
$$+ \left[(\tau_{xz})_{z+\delta z} - (\tau_{xz})_z\right] \delta x \delta y.$$

By Newton's law of motion this resultant force is equal to the mass times the acceleration. Thus, we obtain

$$\left[(\tau_{xx})_{x+\delta x} - (\tau_{xx})_x\right] \delta y \delta z + \left[(\tau_{xy})_{y+\delta y} - (\tau_{xy})_y\right] \delta z \delta x$$
$$+ \left[(\tau_{xz})_{z+\delta z} - (\tau_{xz})_z\right] \delta x \delta y = \rho \, \delta x \delta y \delta z \, u_{tt} \tag{3.4.2}$$

where ρ is the density of the body and u is the displacement component in the x-direction. Hence, in the limit as δV approaches zero, we obtain

$$\frac{\partial \tau_{xx}}{\partial x} + \frac{\partial \tau_{xy}}{\partial y} + \frac{\partial \tau_{xz}}{\partial z} = \rho \frac{\partial^2 u}{\partial t^2}. \tag{3.4.3}$$

Similarly, the following two equations corresponding to y and z directions are obtained:

$$\frac{\partial \tau_{yx}}{\partial x} + \frac{\partial \tau_{yy}}{\partial y} + \frac{\partial \tau_{yz}}{\partial z} = \rho \frac{\partial^2 v}{\partial t^2}, \tag{3.4.4}$$

$$\frac{\partial \tau_{zx}}{\partial x} + \frac{\partial \tau_{zy}}{\partial y} + \frac{\partial \tau_{zz}}{\partial z} = \rho \frac{\partial^2 w}{\partial t^2}, \tag{3.4.5}$$

where v and w are the displacement components in the y and z directions respectively.

We may now define linear strains [see Sokolnikoff (1956)] as

$$\varepsilon_{xx} = \frac{\partial u}{\partial x}, \quad \varepsilon_{yz} = \frac{1}{2}\left(\frac{\partial w}{\partial y} + \frac{\partial v}{\partial z}\right),$$

$$\varepsilon_{yy} = \frac{\partial v}{\partial y}, \quad \varepsilon_{zx} = \frac{1}{2}\left(\frac{\partial u}{\partial z} + \frac{\partial w}{\partial x}\right), \tag{3.4.6}$$

$$\varepsilon_{zz} = \frac{\partial w}{\partial z}, \quad \varepsilon_{xy} = \frac{1}{2}\left(\frac{\partial v}{\partial x} + \frac{\partial u}{\partial y}\right),$$

in which ε_{xx}, ε_{yy}, ε_{zz} represent unit elongations and ε_{yz}, ε_{zx}, ε_{xy} represent unit shearing strains.

In the case of an isotropic body, generalized Hooke's law takes the form

$$\begin{aligned}
\tau_{xx} &= \lambda\theta + 2\mu\varepsilon_{xx}, & \tau_{yz} &= 2\mu\varepsilon_{yz}, \\
\tau_{yy} &= \lambda\theta + 2\mu\varepsilon_{yy}, & \tau_{zx} &= 2\mu\varepsilon_{zx}, \\
\tau_{zz} &= \lambda\theta + 2\mu\varepsilon_{zz}, & \tau_{xy} &= 2\mu\varepsilon_{xy},
\end{aligned} \tag{3.4.7}$$

where $\theta = \varepsilon_{xx} + \varepsilon_{yy} + \varepsilon_{zz}$ is called the *dilatation*, and λ and μ are *Lame's constants*.

Expressing stresses in terms of displacements, we obtain

$$\tau_{xx} = \lambda\theta + 2\mu\frac{\partial u}{\partial x},$$

$$\tau_{xy} = \mu\left(\frac{\partial v}{\partial x} + \frac{\partial u}{\partial y}\right), \tag{3.4.8}$$

$$\tau_{xz} = \mu\left(\frac{\partial w}{\partial x} + \frac{\partial u}{\partial z}\right).$$

By differentiating equations (3.4.8), we obtain

$$\frac{\partial \tau_{xx}}{\partial x} = \lambda\frac{\partial \theta}{\partial x} + 2\mu\frac{\partial^2 u}{\partial x^2},$$

$$\frac{\partial \tau_{xy}}{\partial y} = \mu\frac{\partial^2 v}{\partial x \partial y} + \mu\frac{\partial^2 u}{\partial y^2}, \tag{3.4.9}$$

$$\frac{\partial \tau_{xz}}{\partial z} = \mu\frac{\partial^2 w}{\partial x \partial z} + \mu\frac{\partial^2 u}{\partial z^2}.$$

Substituting equation (3.4.9) into equation (3.4.3) yields

$$\lambda\frac{\partial\theta}{\partial x} + \mu\left(\frac{\partial^2 u}{\partial x^2} + \frac{\partial^2 v}{\partial x\partial y} + \frac{\partial^2 w}{\partial x\partial z}\right) + \mu\left(\frac{\partial^2 u}{\partial x^2} + \frac{\partial^2 u}{\partial y^2} + \frac{\partial^2 u}{\partial z^2}\right) = \rho\frac{\partial^2 u}{\partial t^2}.$$

(3.4.10)

We note that

$$\frac{\partial^2 u}{\partial x^2} + \frac{\partial^2 v}{\partial x\partial y} + \frac{\partial^2 w}{\partial x\partial z} = \frac{\partial}{\partial x}\left(\frac{\partial u}{\partial x} + \frac{\partial v}{\partial y} + \frac{\partial w}{\partial z}\right) = \frac{\partial\theta}{\partial x},$$

and introduce the notation

$$\triangle = \nabla^2 = \frac{\partial^2}{\partial x^2} + \frac{\partial^2}{\partial y^2} + \frac{\partial^2}{\partial z^2}.$$

The symbol $\triangle$ or ∇^2 is called the *Laplace operator*. Hence, equation (3.4.10) becomes

$$(\lambda + \mu)\frac{\partial\theta}{\partial x} + \mu\nabla^2 u = \rho\frac{\partial^2 u}{\partial t^2}.$$

(3.4.11)

In a similar manner, we obtain the other two equations which are

$$(\lambda + \mu)\frac{\partial\theta}{\partial y} + \mu\nabla^2 v = \rho\frac{\partial^2 v}{\partial t^2}.$$

(3.4.12)

$$(\lambda + \mu)\frac{\partial\theta}{\partial z} + \mu\nabla^2 w = \rho\frac{\partial^2 w}{\partial t^2}.$$

(3.4.13)

The set of equations (3.4.11)–(3.4.13) is called the *Navier equations* of motion. In vector form, the Navier equations of motion assume the form

$$(\lambda + \mu)\,\mathrm{grad}\,\mathrm{div}\,\mathbf{u} + \mu\nabla^2\mathbf{u} = \rho\,\mathbf{u}_{tt},$$

(3.4.14)

where $\mathbf{u} = u\mathbf{i} + v\mathbf{j} + w\mathbf{k}$ and $\theta = \mathrm{div}\,\mathbf{u}$.

(i) If $\mathrm{div}\,\mathbf{u} = 0$, the general equation becomes

$$\mu\nabla^2\mathbf{u} = \rho\,\mathbf{u}_{tt},$$

or

$$\mathbf{u}_{tt} = c_T^2\nabla^2\mathbf{u},$$

(3.4.15)

where c_T is called the *transverse wave velocity* given by

$$c_T = \sqrt{\mu/\rho}.$$

This is the case of an equivoluminal wave propagation, since the volume expansion θ is zero for waves moving with this velocity. Sometimes these waves are called *waves of distortion* because the velocity of propagation depends on μ and ρ; the shear modulus μ characterizes the distortion and rotation of the volume element.

(ii) When curl $\mathbf{u} = 0$, the vector identity

$$\text{curl curl } \mathbf{u} = \text{grad div } \mathbf{u} - \nabla^2 \mathbf{u},$$

gives

$$\text{grad div } \mathbf{u} = \nabla^2 \mathbf{u},$$

Then the general equation becomes

$$(\lambda + 2\mu)\, \nabla^2 \mathbf{u} = \rho\, \mathbf{u}_{tt},$$

or

$$\mathbf{u}_{tt} = c_L^2 \nabla^2 \mathbf{u}, \qquad (3.4.16)$$

where c_L is called the *longitudinal wave velocity* given by

$$c_L = \sqrt{\frac{\lambda + 2\mu}{\rho}}.$$

This is the case of *irrotational* or *dilatational* wave propagation, since curl $\mathbf{u} = 0$ describes irrotational motion. Equations (3.4.15) and (3.4.16) are called the *three-dimensional wave equations*.

In general, the wave equation may be written as

$$u_{tt} = c^2 \nabla^2 u, \qquad (3.4.17)$$

where the Laplace operator may be one, two, or three dimensional. The importance of the wave equation stems from the facts that this type of equation arises in many physical problems; for example, sound waves in space, electrical vibration in a conductor, torsional oscillation of a rod, shallow water waves, linearized supersonic flow in a gas, waves in an electric transmission line, waves in magnetohydrodynamics, and longitudinal vibrations of a bar.

To give a more general method of decomposing elastic waves into transverse and longitudinal wave forms, we write the Navier equations of motion in the form

$$c_T^2 \nabla^2 \mathbf{u} + \left(c_L^2 - c_T^2\right) \text{grad}\,(\text{div }\mathbf{u}) = \mathbf{u}_{tt}. \qquad (3.4.18)$$

We now decompose this equation into two vector equations by defining $\mathbf{u} = \mathbf{u}_T + \mathbf{u}_L$, where $\mathbf{u}_T$ and $\mathbf{u}_L$ satisfy the equations

$$\text{div } \mathbf{u}_T = 0 \quad \text{and} \quad \text{curl } \mathbf{u}_L = \mathbf{0}. \qquad (3.4.19\text{ab})$$

Since $\mathbf{u}_T$ is defined by (3.4.19a) that is divergenceless, it follows from vector analysis that there exists a rotation vector $\boldsymbol{\psi}$ such that

$$\mathbf{u}_T = \text{curl } \boldsymbol{\psi}, \tag{3.4.20}$$

where $\boldsymbol{\psi}$ is called the *vector potential*.

On the other hand, $\mathbf{u}_L$ is irrotational as given by (3.4.19b), so there exists a scalar function $\phi\,(\mathbf{x}, t)$, called the *scalar potential* such that

$$\mathbf{u}_L = \text{grad } \phi. \tag{3.4.21}$$

Using (3.4.20) and (3.4.21), we can write

$$\mathbf{u} = \text{curl } \boldsymbol{\psi} + \text{grad } \phi. \tag{3.4.22}$$

This means that the displacement vector field is decomposed into a divergenceless vector and irrotational vector.

Inserting $\mathbf{u} = \mathbf{u}_T + \mathbf{u}_L$ into (3.4.18), taking the divergence of each term of the resulting equation, and then using (3.4.19a) gives

$$\text{div}\left[c_L^2 \nabla^2 \mathbf{u}_L - (\mathbf{u}_L)_{tt}\right] = 0. \tag{3.4.23}$$

It is noted that the curl of the square bracket in (3.4.23) is also zero. Clearly, any vector whose divergence and curl both vanish is identically a zero vector. Consequently,

$$c_L^2 \nabla^2 \mathbf{u}_L = (\mathbf{u}_L)_{tt}. \tag{3.4.24}$$

This shows that $\mathbf{u}_L$ satisfies the vector wave equation with the wave velocity c_L. Since $\mathbf{u}_L = \text{grad } \phi$, it is clear that the scalar potential ϕ also satisfies the wave equation with the same wave speed. All solutions of (3.4.24) represent longitudinal waves that are irrotational (since $\boldsymbol{\psi} = \mathbf{0}$).

Similarly, we substitute $\mathbf{u} = \mathbf{u}_L + \mathbf{u}_T$ into (3.4.18), take the curl of the resulting equation, and use the fact that curl $\mathbf{u}_L = \mathbf{0}$ to obtain

$$\text{curl}\left[c_T^2 \nabla^2 \mathbf{u}_T - (\mathbf{u}_T)_{tt}\right] = \mathbf{0}. \tag{3.4.25}$$

Since the divergence of the expression inside the square bracket is also zero, it follows that

$$c_T^2 \nabla^2 \mathbf{u}_T = (\mathbf{u}_T)_{tt}. \tag{3.4.26}$$

This is a vector wave equation for $\mathbf{u}_T$ whose solutions represent transverse waves that are irrotational but are accompanied by no change in volume (*equivoluminal, transverse, rotational waves*). These waves propagate with a wave velocity c_T.

We close this section by seeking time-harmonic solutions of (3.4.18) in the form

$$\mathbf{u} = \text{Re}\left[\mathbf{U}\,(x, y, z)\,e^{i\omega t}\right]. \tag{3.4.27}$$

Invoking (3.4.27) into equation (3.4.18) gives the following equation for the function $\mathbf{U}$

$$c_T \nabla^2 \mathbf{U} + (c_L - c_T)\operatorname{grad}(\operatorname{div}\mathbf{U}) + \omega^2 \mathbf{U} = 0. \qquad (3.4.28)$$

Inserting, $\mathbf{u} = \mathbf{u}_T + \mathbf{u}_L$, and using the above method of taking the divergence and curl of (3.4.28) respectively leads to equation for $\mathbf{U}_L$ and $\mathbf{U}_T$ as follows

$$\nabla^2 \mathbf{U}_L + k_L^2 \nabla^2 \mathbf{U}_L = 0, \quad \nabla^2 \mathbf{U}_T + k_T^2 \mathbf{U}_T = 0, \qquad (3.4.29)$$

where

$$k_L^2 = \frac{\omega^2}{c_L^2} \quad \text{and} \quad k_T^2 = \frac{\omega^2}{c_T^2}. \qquad (3.4.30)$$

Equations (3.4.29) are called the *reduced wave equations* (or the *Helmholtz equations*) for $\mathbf{U}_L$ and $\mathbf{U}_T$. Obviously, equations (3.4.29) can also be derived by assuming time-harmonic solutions for $\mathbf{u}_L$ and $\mathbf{u}_T$ in the form

$$\begin{pmatrix} \mathbf{u}_L \\ \mathbf{u}_T \end{pmatrix} = e^{i\omega t} \begin{pmatrix} \mathbf{U}_L \\ \mathbf{U}_T \end{pmatrix}, \qquad (3.4.31)$$

and substituting these results into (3.4.24) and (3.4.26) respectively.

3.5 Conduction of Heat in Solids

We consider a domain D^* bounded by a closed surface B^*. Let $u(x, y, z, t)$ be the temperature at a point (x, y, z) at time t. If the temperature is not constant, heat flows from places of higher temperature to places of lower temperature. Fourier's law states that the rate of flow is proportional to the gradient of the temperature. Thus the velocity of the heat flow in an isotropic body is

$$\mathbf{v} = -K \operatorname{grad} u, \qquad (3.5.1)$$

where K is a constant, called the *thermal conductivity* of the body.

Let D be an arbitrary domain bounded by a closed surface B in D^*. Then the amount of heat leaving D per unit time is

$$\iint_B v_n ds,$$

where $v_n = \mathbf{v} \cdot \mathbf{n}$ is the component of $\mathbf{v}$ in the direction of the outer unit normal $\mathbf{n}$ of B. Thus, by Gauss' theorem (Divergence theorem)

$$\iint_B v_n\, ds = \iiint_D \operatorname{div}(-K \operatorname{grad} u)\, dx\, dy\, dz$$

$$= -K \iiint_D \nabla^2 u\, dx\, dy\, dz. \tag{3.5.2}$$

But the amount of heat in D is given by

$$\iiint_D \sigma \rho u\, dx\, dy\, dz, \tag{3.5.3}$$

where ρ is the density of the material of the body and σ is its specific heat. Assuming that integration and differentiation are interchangeable, the rate of decrease of heat in D is

$$-\iiint_D \sigma \rho \frac{\partial u}{\partial t}\, dx\, dy\, dz. \tag{3.5.4}$$

Since the rate of decrease of heat in D must be equal to the amount of heat leaving D per unit time, we have

$$-\iiint_D \sigma \rho u_t\, dx\, dy\, dz = -K \iiint_D \nabla^2 u\, dx\, dy\, dz,$$

or

$$-\iiint_D \left[\sigma \rho u_t - K \nabla^2 u\right] dx\, dy\, dz = 0, \tag{3.5.5}$$

for an arbitrary D in D^*. We assume that the integrand is continuous. If we suppose that the integrand is not zero at a point (x_0, y_0, z_0) in D, then, by continuity, the integrand is not zero in a small region surrounding the point (x_0, y_0, z_0). Continuing in this fashion we extend the region encompassing D. Hence the integral must be nonzero. This contradicts (3.5.5). Thus, the integrand is zero everywhere, that is,

$$u_t = \kappa \nabla^2 u, \tag{3.5.6}$$

where $\kappa = K/\sigma\rho$. This is known as the *heat equation*.

   This type of equation appears in a great variety of problems in mathematical physics, for example the concentration of diffusing material, the motion of a tidal wave in a long channel, transmission in electrical cables, and unsteady boundary layers in viscous fluid flows.

3.6 The Gravitational Potential

In this section, we shall derive one of the most well-known equations in the theory of partial differential equations, the Laplace equation.

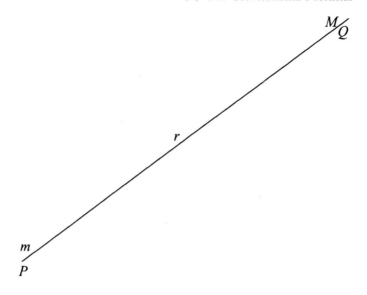

Figure 3.6.1 Two particles at P and Q.

We consider two particles of masses m and M, at P and Q as shown in Figure 3.6.1. Let r be the distance between them. Then, according to Newton's law of gravitation, a force proportional to the product of their masses, and inversely proportional to the square of the distance between them, is given in the form

$$F = G\frac{mM}{r^2},\tag{3.6.1}$$

where G is the gravitational constant.

It is customary in potential theory to choose the unit of force so that $G = 1$. Thus, F becomes

$$F = \frac{mM}{r^2}.\tag{3.6.2}$$

If $\mathbf{r}$ represents the vector PQ, the force per unit mass at Q due to the mass at P may be written as

$$\mathbf{F} = \frac{-m\mathbf{r}}{r^3} = \nabla\left(\frac{m}{r}\right),\tag{3.6.3}$$

which is called the *intensity* of the gravitational field of force.

We suppose that a particle of unit mass moves under the attraction of the particle of mass m at P from infinity up to Q. The work done by the force $\mathbf{F}$ is

$$\int_{\infty}^{r} \mathbf{F} \, dr = \int_{\infty}^{r} \nabla \left(\frac{m}{r} \right) dr = \frac{m}{r}. \tag{3.6.4}$$

This is called the *potential* at Q due to the particle at P. We denote this by

$$V = -\frac{m}{r}, \tag{3.6.5}$$

so that the intensity of force at P is

$$\mathbf{F} = \nabla \left(\frac{m}{r} \right) = -\nabla V. \tag{3.6.6}$$

We shall now consider a number of masses m_1, m_2, ..., m_n, whose distances from Q are $r_1, r_2, \ldots, r_n$, respectively. Then the force of attraction per unit mass at Q due to the system is

$$\mathbf{F} = \sum_{k=1}^{n} \nabla \frac{m_k}{r_k} = \nabla \sum_{k=1}^{n} \frac{m_k}{r_k}. \tag{3.6.7}$$

The work done by the forces acting on a particle of unit mass is

$$\int_{\infty}^{r} \mathbf{F} \cdot dr = \sum_{k=1}^{n} \frac{m_k}{r_k} = -V. \tag{3.6.8}$$

Then the potential satisfies the equation

$$\nabla^2 V = -\nabla^2 \sum_{k=1}^{n} \frac{m_k}{r_k} = -\sum_{k=1}^{n} \nabla^2 \left(\frac{m_k}{r_k} \right) = 0, \ r_k \neq 0. \tag{3.6.9}$$

In the case of a continuous distribution of mass in some volume R, we have, as in Figure 3.6.2.

$$V(x, y, z) = \iiint_{R} \frac{\rho(\xi, \eta, \zeta)}{r} dR, \tag{3.6.10}$$

where $r = \sqrt{(x - \xi)^2 + (y - \eta)^2 + (z - \zeta)^2}$ and Q is outside the body. It immediately follows that

$$\nabla^2 V = 0. \tag{3.6.11}$$

This equation is called the *Laplace equation*, also known as the *potential equation*. It appears in many physical problems, such as those of electrostatic potentials, potentials in hydrodynamics, and harmonic potentials in the theory of elasticity. We observe that the Laplace equation can be viewed as the special case of the heat and the wave equations when the dependent variables involved are independent of time.

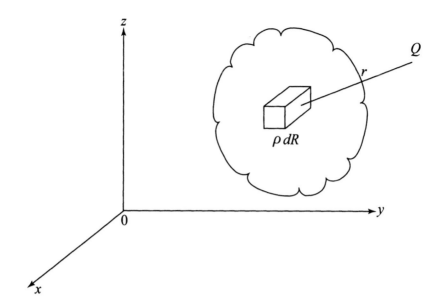

Figure 3.6.2 Continuous Mass Distribution.

3.7 Conservation Laws and The Burgers Equation

A conservation law states that the rate of change of the total amount of material contained in a fixed domain of volume V is equal to the flux of that material across the closed bounding surface S of the domain. If we denote the density of the material by $\rho(\mathbf{x}, t)$ and the flux vector by $\mathbf{q}(\mathbf{x}, t)$, then the conservation law is given by

$$\frac{d}{dt} \int_V \rho \, dV = - \int_S (\mathbf{q} \cdot \mathbf{n}) \, dS, \qquad (3.7.1)$$

where dV is the volume element and dS is the surface element of the boundary surface S, $\mathbf{n}$ denotes the outward unit normal vector to S as shown in Figure 3.7.1, and the right-hand side measures the *outward* flux — hence, the minus sign is used.

Applying the Gauss divergence theorem and taking $\frac{d}{dt}$ inside the integral sign, we obtain

$$\int_V \left(\frac{\partial \rho}{\partial t} + \operatorname{div} \mathbf{q} \right) dV = 0. \qquad (3.7.2)$$

This result is true for any arbitrary volume V, and, if the integrand is continuous, it must vanish everywhere in the domain. Thus, we obtain the differential form of the conservation law

$$\rho_t + \operatorname{div} \mathbf{q} = 0. \qquad (3.7.3)$$

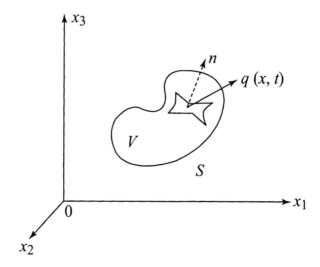

Figure 3.7.1 Volume V of a closed domain bounded by a surface S with surface element dS and outward normal vector **n**.

The one-dimensional version of the conservation law (3.7.3) is

$$\frac{\partial \rho}{\partial t} + \frac{\partial q}{\partial x} = 0. \tag{3.7.4}$$

To investigate the nature of the discontinuous solution or shock waves, we assume a functional relation $q = Q(\rho)$ and allow a jump discontinuity for ρ and q. In many physical problems of interest, it would be a better approximation to assume that q is a function of the density gradient ρ_x as well as ρ. A simple model is to take

$$q = Q(\rho) - \nu\rho_x, \tag{3.7.5}$$

where ν is a positive constant. Substituting (3.7.5) into (3.7.4), we obtain the *nonlinear diffusion equation*

$$\rho_t + c(\rho)\rho_x = \nu\rho_{xx}, \tag{3.7.6}$$

where $c(\rho) = Q'(\rho)$.

We multiply (3.7.6) by $c'(\rho)$ to obtain

$$c_t + c\,c_x = \nu\,c'(\rho)\,\rho_{xx},$$
$$= \nu\left\{c_{xx} - c''(\rho)\,\rho_x^2\right\}. \tag{3.7.7}$$

If $Q(\rho)$ is a quadratic function in ρ, then $c(\rho)$ is linear in ρ, and $c''(\rho) = 0$. Consequently, (3.7.7) becomes

$$c_t + c\,c_x = \nu\,c_{xx}. \tag{3.7.8}$$

As a simple model of turbulence, c is replaced by the fluid velocity field $u(x, t)$ to obtain the well-known *Burgers equation*

$$u_t + u\, u_x = \nu\, u_{xx}, \qquad (3.7.9)$$

where ν is the kinematic viscosity.

Thus the Burgers equation is a balance between time evolution, non-linearity, and diffusion. This is the simplest nonlinear model equation for diffusive waves in fluid dynamics. Burgers (1948) first developed this equation primarily to shed light on the study of turbulence described by the interaction of the two opposite effects of convection and diffusion. However, turbulence is more complex in the sense that it is both three dimensional and statistically random in nature. Equation (3.7.9) arises in many physical problems including one-dimensional turbulence (where this equation had its origin), sound waves in a viscous medium, shock waves in a viscous medium, waves in fluid-filled viscous elastic tubes, and magnetohydrodynamic waves in a medium with finite electrical conductivity. We note that (3.7.9) is *parabolic* provided the coefficient of u_x is constant, whereas the resulting (3.7.9) with $\nu = 0$ is hyperbolic. More importantly, the properties of the solution of the parabolic equation are significantly different from those of the hyperbolic equation.

3.8 The Schrödinger and the Korteweg–de Vries Equations

We consider the following Fourier integral representation of a quasi-monochromatic plane wave solution

$$u(x, t) = \int_{-\infty}^{\infty} F(k) \exp\left[i\left\{kx - \omega(k) t\right\}\right] dk, \qquad (3.8.1)$$

where the spectrum function $F(k)$ is determined from the given initial or boundary conditions and has the property $F(-k) = F^*(k)$, and $\omega = \omega(k)$ is the dispersion relation. We assume that the initial wave is slowly modulated as it propagates in a dispersive medium. For such a quasi-monochromatic wave, most of the energy is confined in a neighborhood of a specified wave number $k = k_0$, so that spectrum function $F(k)$ has a sharp peak around the point $k = k_0$ with a narrow wave number width $k - k_0 = \delta k = O(\varepsilon)$, and the dispersion relation $\omega(k)$ can be expanded about k_0 in the form

$$\omega = \omega_0 + \omega_0'(\delta k) + \frac{1}{2!}\omega_0''(\delta k)^2 + \frac{1}{3!}\omega_0'''(\delta k)^3 + \cdots, \qquad (3.8.2)$$

where $\omega_0 = \omega(k_0)$, $\omega_0' = \omega'(k_0)$, $\omega_0'' = \omega''(k_0)$, and $\omega_0''' = \omega'''(k_0)$.

Substituting (3.8.2) into (3.8.1) gives a new form

$$u(x,t) = A(x,t) \exp\left[i\left(k_0 x - \omega_0 t\right)\right] + c.c., \qquad (3.8.3)$$

where c.c. stands for the complex conjugate and $A(x,t)$ is the complex wave amplitude given by

$$A(x,t) = \int_0^\infty F(k_0 + \delta k) \exp\left[i\left\{(x - \omega_0' t)(\delta k) - \frac{1}{2}\omega_0''(\delta k)^2 t\right.\right.$$
$$\left.\left. - \frac{1}{3}\omega_0'''(\delta k)^3 t\right\}\right] d(\delta k), \quad (3.8.4)$$

where it has been assumed that $\omega(-k) = -\omega(k)$. Since (3.8.4) depends on $(x - \omega_0' t)\,\delta k$, $(\delta k)^2 t$, $(\delta k)^3 t$ where $\delta k = O(\varepsilon)$ is small, the wave amplitude $A(x,t)$ is a slowly varying function of $x^* = (x - \omega_0' t)$ and time t.

We keep only the term with (δk) in (3.8.4) and neglect all terms with $(\delta k)^n$, $n = 2, 3, \cdots$, so that (3.8.4) becomes

$$A(x,t) = \int_0^\infty F(k_0 + \delta k) \exp\left[i\left\{(x - \omega_0' t)\right\}(\delta k)\right] d(\delta k). \quad (3.8.5)$$

A simple calculation reveals that $A(x,t)$ satisfies the evolution equation

$$\frac{\partial A}{\partial t} + c_g \frac{\partial A}{\partial x} = 0, \qquad (3.8.6)$$

where $c_g = \omega_0'$ is the group velocity.

In the next step, we retain only terms with (δk) and $(\delta k)^2$ in (3.8.4) to obtain

$$A(x,t) = \int_0^\infty F(k_0 + \delta k) \exp\left[i\left\{(x - \omega_0' t)(\delta k) - \frac{1}{2}\omega_0''(\delta k)^2\right\}\right] d(\delta k). \quad (3.8.7)$$

A simple calculation shows that $A(x,t)$ satisfies the *linear Schrödinger equation*

$$i\left(\frac{\partial A}{\partial t} + \omega_0' \frac{\partial A}{\partial x}\right) + \frac{1}{2}\omega_0'' \frac{\partial^2 A}{\partial x^2} = 0. \qquad (3.8.8)$$

Using the slow variables

$$\xi = \varepsilon(x - \omega_0' t), \qquad \tau = \varepsilon^2 t, \qquad (3.8.9)$$

the modulated wave amplitude $A(\xi, \tau)$ satisfies the linear Schrödinger equation

$$i A_\tau + \frac{1}{2}\omega_0'' A_{\xi\xi} = 0. \qquad (3.8.10)$$

On the other hand, for the frequencies at which the group velocity ω_0' reaches an extremum, $\omega_0'' = 0$. In this case, the cubic term in the dispersion relation (3.7.2) plays an important role. Consequently, equation (3.8.4) reduces to a form similar to (3.8.7) with $\omega_0'' = 0$ in the exponential factor. Once again, a simple calculation from the resulting integral (3.8.4) reveals that $A(x, t)$ satisfies the *linearized Korteweg–de Vries (KdV) equation*

$$\frac{\partial A}{\partial t} + \omega_0' \frac{\partial A}{\partial x} + \frac{1}{6} \omega_0''' \frac{\partial^3 A}{\partial x^3} = 0. \tag{3.8.11}$$

By transferring to the new variables $\xi = x - \omega_0' t$ and $\tau = t$ which correspond to a reference system moving with the group velocity ω_0', we obtain the linearized KdV equation

$$\frac{\partial A}{\partial \tau} + \frac{1}{6} \omega_0''' \frac{\partial^3 A}{\partial \xi^3} = 0. \tag{3.8.12}$$

This describes waves in a dispersive medium with a weak high frequency dispersion.

One of the remarkable *nonlinear* model equations is the *Korteweg–de Vries (KdV) equation* in the form

$$u_t + \alpha u \, u_x + \beta u_{xxx} = 0, \qquad -\infty < x < \infty, \quad t > 0. \tag{3.8.13}$$

This equation arises in many physical problems including water waves, ion acoustic waves in a plasma, and longitudinal dispersive waves in elastic rods. The exact solution of this equation is called the *soliton* which is remarkably stable. We shall discuss the soliton solution in Chapter 13.

Another remarkable nonlinear model equation describing solitary waves is known as the *nonlinear Schrödinger (NLS) equation* written in the standard form

$$i \, u_t + \frac{1}{2} \omega_0'' u_{xx} + \gamma |u|^2 u = 0, \qquad -\infty < x < \infty, \quad t > 0. \tag{3.8.14}$$

This equation admits a solution called the *solitary waves* and describes the evolution of the water waves; it arises in many other physical systems that include nonlinear optics, hydromagnetic and plasma waves, propagation of heat pulse in a solid, and nonlinear instability problems. The solution of this equation will be discussed in Chapter 13.

3.9 Exercises

1. Show that the equation of motion of a long string is

$$u_{tt} = c^2 u_{xx} - g,$$

where g is the gravitational acceleration.

2. Derive the *damped wave equation* of a string

$$u_{tt} + a\,u_t = c^2 u_{xx},$$

where the damping force is proportional to the velocity and a is a constant. Considering a restoring force proportional to the displacement of a string, show that the resulting equation is

$$u_{tt} + au_t + bu = c^2 u_{xx},$$

where b is a constant. This equation is called the *telegraph equation*.

3. Consider the transverse vibration of a uniform beam. Adopting Euler's beam theory, the moment M at a point can be written as

$$M = -EI\,u_{xx},$$

where EI is called the flexural rigidity, E is the elastic modulus, and I is the moment of inertia of the cross section of the beam. Show that the transverse motion of the beam may be described by

$$u_{tt} + c^2 u_{xxxx} = 0,$$

where $c^2 = EI/\rho A$, ρ is the density, and A is the cross-sectional area of the beam.

4. Derive the deflection equation of a thin elastic plate

$$\nabla^4 u = q/D,$$

where q is the uniform load per unit area, D is the flexural rigidity of the plate, and

$$\nabla^4 u = u_{xxxx} + 2u_{xxyy} + u_{yyyy}.$$

5. Derive the one-dimensional heat equation

$$u_t = \kappa u_{xx}, \quad \text{where } \kappa \text{ is a constant.}$$

Assuming that heat is also lost by radioactive exponential decay of the material in the bar, show that the above equation becomes

$$u_t = \kappa u_{xx} + he^{-\alpha x},$$

where h and α are constants.

6. Starting from *Maxwell's equations* in electrodynamics, show that in a conducting medium electric intensity **E**, magnetic intensity **H**, and current density **J** satisfy

$$\nabla^2 \mathbf{X} = \mu\varepsilon \mathbf{X}_{tt} + \mu\sigma \mathbf{X}_t,$$

where **X** represents **E**, **H**, and **J**, μ is the magnetic inductive capacity, ε is the electric inductive capacity, and σ is the electrical conductivity.

7. Derive the *continuity equation*

$$\rho_t + \operatorname{div}(\rho \mathbf{u}) = 0,$$

and *Euler's equation of motion*

$$\rho[\mathbf{u}_t + (\mathbf{u} \cdot \operatorname{grad})\mathbf{u}] + \operatorname{grad} p = 0,$$

in fluid dynamics.

8. In the derivation of the Laplace equation (3.6.11), the potential at Q which is outside the body is ascertained. Now determine the potential at Q when it is inside the body, and show that it satisfies the *Poisson equation*

$$\nabla^2 u = -4\pi\rho,$$

where ρ is the density of the body.

9. Setting $U = e^{ikt}u$ in the wave equation $U_{tt} = \nabla^2 U$ and setting $U = e^{-k^2 t}u$ in the heat equation $U_t = \nabla^2 U$, show that $u(x, y, z)$ satisfies the *Helmholtz equation*

$$\nabla^2 u + k^2 u = 0.$$

10. The Maxwell equations in vacuum are

$$\nabla \times \mathbf{E} = -\frac{\partial \mathbf{B}}{\partial t}, \qquad \nabla \times \mathbf{B} = \mu\varepsilon\frac{\partial \mathbf{E}}{\partial t},$$
$$\nabla \cdot \mathbf{E} = 0, \qquad \nabla \cdot \mathbf{B} = 0,$$

where μ and ε are universal constants. Show that the magnetic field $\mathbf{B} = (0, B_y(x, t), 0)$ and the electric field $\mathbf{E} = (0, 0, E_z(x, t))$ satisfy the wave equation

$$\frac{\partial^2 u}{\partial t^2} = c^2 \frac{\partial^2 u}{\partial x^2},$$

where $u = B_y$ or E_z and $c = (\mu\varepsilon)^{-\frac{1}{2}}$ is the speed of light.

11. The equations of gas dynamics are linearized for small perturbations about a constant state $u = 0$, $\rho = \rho_0$, and $p_0 = p(\rho_0)$ with $c_0^2 = p'(\rho_0)$. In terms of velocity potential ϕ defined by $\mathbf{u} = \nabla\phi$, the perturbation equations are

$$\rho_t + \rho_0 \operatorname{div} \mathbf{u} = 0,$$
$$p - p_0 = -\rho_0\phi_t = c_0^2(\rho - \rho_0),$$
$$\rho - \rho_0 = -\frac{\rho_0}{c_0^2}\phi_t.$$

Show that f and $\mathbf{u}$ satisfy the three dimensional wave equations

$$f_{tt} = c_0^2 \nabla^2 f, \quad \text{and} \quad \mathbf{u}_{tt} = c_0^2 \nabla^2 \mathbf{u},$$

where $f = p$, ρ, or ϕ and

$$\nabla^2 \equiv \frac{\partial^2}{\partial x^2} + \frac{\partial^2}{\partial y^2} + \frac{\partial^2}{\partial z^2}.$$

12. Consider a slender body moving in a gas with arbitrary constant velocity U, and suppose (x_1, x_2, x_3) represents the frame of reference in which the motion of the gas is small and described by the equations of problem 11. The body moves in the negative x_1 direction, and (x, y, z) denotes the coordinates fixed with respect to the body so that the coordinate transformation is $(x, y, z) = (x_1 + Ut, x_2, x_3)$. Show that the wave equation $\phi_{tt} = c_0^2 \nabla^2 \phi$ reduces to the form

$$\left(M^2 - 1\right) \Phi_{xx} = \Phi_{yy} + \Phi_{zz},$$

where $M \equiv U/c_0$ is the Mach number and Φ is the potential in the new frame of reference (x, y, z).

13. Consider the motion of a gas in a taper tube of cross section $A(x)$. Show that the equation of continuity and the equation of motion are

$$\rho = \rho_0 \left(1 - \frac{\partial \xi}{\partial x} - \frac{\xi}{A} \frac{\partial A}{\partial x}\right) = \rho_0 \left[1 - \frac{1}{A} \frac{\partial}{\partial x}(A\xi)\right],$$

and

$$\rho_0 \frac{\partial^2 \xi}{\partial t^2} = -\frac{\partial p}{\partial x},$$

where x is the distance along the length of the tube, $\xi(x)$ is the displacement function, $p = p(\rho)$ is the pressure-density relation, ρ_0 is the average density, and ρ is the local density of the gas.
Hence derive the equation of motion

$$\xi_{tt} = c^2 \frac{\partial}{\partial x}\left[\frac{1}{A} \frac{\partial}{\partial x}(A\xi)\right], \qquad c^2 = \frac{\partial p}{\partial \rho}.$$

Find the equation of motion when A is constant. If $A(x) = a_0 \exp(2\alpha x)$ where a_0 and α are constants, show that the above equation takes the form

$$\xi_{tt} = c^2 \left(\xi_{xx} + 2\alpha \xi_x\right).$$

14. Consider the current $I(x, t)$ and the potential $V(x, t)$ at a point x and time t of a uniform electric transmission line with resistance R,

inductance L, capacity C, and leakage conductance G per unit length.
(a) Show that both I and V satisfy the system of equations

$$LI_t + RI = -V_x,$$
$$CV_t + GV = -I_x.$$

Derive the telegraph equation

$$u_{xx} - c^{-2} u_{tt} - a\,u_t - bu = 0, \quad \text{for} \quad u = I \quad \text{or} \quad V,$$

where $c^2 = (LC)^{-1}$, $a = RC + LG$ and $b = RG$.

(b) Show that the telegraph equation can be written in the form

$$u_{tt} - c^2 u_{xx} + (p + q)\,u_t + pq\,u = 0,$$

where $p = \frac{G}{C}$ and $q = \frac{R}{L}$.

(c) Apply the transformation

$$u = v \exp\left[-\frac{1}{2}\,(p + q)\,t \right]$$

to transform the above equation into the form

$$v_{tt} - c^2 v_{xx} = \frac{1}{4}\,(p - q)^2\, v.$$

(d) When $p = q$, show that there exists an undisturbed wave solution in the form

$$u(x, t) = e^{-pt} f(x \pm ct),$$

which propagates in either direction, where f is an arbitrary twice differentiable function of its argument.
If $u(x, t) = A \exp\left[i\,(kx - \omega t) \right]$ is a solution of the telegraph equation

$$u_{tt} - c^2 u_{xx} - \alpha u_t - \beta u = 0, \quad \alpha = p + q, \quad \beta = pq,$$

show that the dispersion relation holds

$$\omega^2 + i\alpha\omega - \left(c^2 k^2 + \beta^2 \right) = 0.$$

Solve the dispersion relation to show that

$$u(x, t) = \exp\left(-\frac{1}{2} p t \right) \exp\left[i\left(kx - \frac{t}{2}\sqrt{4c^2 k^2 + (4q - p^2)} \right) \right].$$

When $p^2 = 4q$, show that the solution represents attenuated nondispersive waves.

(e) Find the equations for I and V in the following cases:

(i) Lossless transmission line $(R = G = 0)$,

(ii) Ideal submarine cable $(L = G = 0)$,

(iii) Heaviside's distortionless line $(R/L = G/C = \text{constant} = k)$.

15. The Fermi–Pasta–Ulam model is used to describe waves in an anharmonic lattice of length l consisting of a row of n identical masses m, each connected to the next by nonlinear springs of constant κ. The masses are at a distance $h = l/n$ apart, and the springs when extended or compressed by an amount d exert a force $F = \kappa \left(d + \alpha d^2\right)$ where α measures the strength of nonlinearity. The equation of motion of the ith mass is

$$m\ddot{y}_i = \kappa \left[(y_{i+1} - y_i) - (y_i - y_{i-1}) + \alpha \left\{(y_{i+1} - y_i)^2 - (y_i - y_{i-1})^2\right\}\right],$$

where $i = 1, 2, 3 ... n$, y_i is the displacement of the ith mass from its equilibrium position, and κ, α are constants with $y_0 = y_n = 0$.
Assume a continuum approximation of this discrete system so that the Taylor expansions

$$y_{i+1} - y_i = hy_x + \frac{h^2}{2!}y_{xx} + \frac{h^3}{3!}y_{xxx} + \frac{h^4}{4!}y_{xxxx} + o\left(h^5\right),$$

$$y_i - y_{i-1} = hy_x - \frac{h^2}{2!}y_{xx} + \frac{h^3}{3!}y_{xxx} - \frac{h^4}{4!}y_{xxxx} + o\left(h^5\right),$$

can be used to derive the nonlinear differential equation

$$y_{tt} = c^2 \left[1 + 2\alpha h y_x\right] y_{xx} + o\left(h^4\right),$$

$$y_{tt} = c^2 \left[1 + 2\alpha h y_x\right] y_{xx} + \frac{c^2 h^2}{12} y_{xxxx} + o\left(h^5\right),$$

where

$$c^2 = \frac{\kappa h^2}{m}.$$

Using a change of variables $\xi = x - ct$, $\tau = c\alpha ht$, show that $u = y_\xi$ satisfies the Korteweg–de Vries (KdV) equation

$$u_\tau + uu_\xi + \beta u_{\xi\xi\xi} = o\left(\varepsilon^2\right), \qquad \varepsilon = \alpha h, \qquad \beta = \frac{h}{24\alpha}.$$

16. The one-dimensional isentropic fluid flow is obtained from Euler's equations (3.1.14) in the form

$$u_t + u\,u_x = -\frac{1}{\rho}p_x, \quad \rho_t + (\rho u)_x = 0, \quad p = p\left(\rho\right).$$

(a) Show that u and ρ satisfy the one-dimensional wave equation

$$\begin{pmatrix} u \\ \rho \end{pmatrix}_{tt} - c^2 \begin{pmatrix} u \\ \rho \end{pmatrix}_{xx} = 0,$$

where $c^2 = \frac{dp}{d\rho}$ is the velocity of sound.

(b) For a compressible adiabatic gas, the equation of state is $p = A\rho^\gamma$, where A and γ are constants; show that

$$c^2 = \frac{\gamma p}{\rho}.$$

17. (a) Obtain the two-dimensional unsteady fluid flow equations from (3.1.14).

(b) Find the two-dimensional steady fluid flow equations from (3.1.14). Hence or otherwise, show that

$$\left(c^2 - u^2\right) u_x - u v \left(u_y + v_x\right) + \left(c^2 - v^2\right) v_y = 0,$$

where

$$c^2 = p'\left(\rho\right).$$

(c) Show that, for an irrotational fluid flow ($\mathbf{u} = \nabla \phi$), the above equation reduces to the quasi-linear partial differential equations

$$\left(c^2 - \phi_x^2\right) \phi_{xx} - 2\phi_x \phi_y \phi_{xy} + \left(c^2 - \phi_y^2\right) \phi_{yy} = 0.$$

(d) Show that the slope of the characteristic C satisfies the quadratic equation

$$\left(c^2 - u^2\right) \left(\frac{dy}{dx}\right)^2 + 2uv \left(\frac{dy}{dx}\right) + \left(c^2 - v^2\right) = 0.$$

Hence or otherwise derive

$$\left(c^2 - v^2\right) \left(\frac{dv}{du}\right)^2 - 2uv \left(\frac{dv}{du}\right) + \left(c^2 - u^2\right) = 0.$$

18. For an inviscid incompressible fluid flow under the body force, $\mathbf{F} = -\nabla\Phi$, the Euler equations are

$$\frac{\partial \mathbf{u}}{\partial t} + \mathbf{u} \cdot \nabla \mathbf{u} = -\nabla\Phi - \frac{1}{\rho}\nabla p, \quad \mathrm{div}\,\mathbf{u} = 0.$$

(a) Show that the vorticity $\boldsymbol{\omega} = \nabla \times \mathbf{u}$ satisfies the vorticity equation

$$\frac{D\boldsymbol{\omega}}{Dt} = \frac{\partial \boldsymbol{\omega}}{\partial t} + \mathbf{u} \cdot \nabla \boldsymbol{\omega} = \boldsymbol{\omega} \cdot \nabla \mathbf{u}.$$

(b) Give the interpretation of this vorticity equation.

(c) In two dimensions, show that $\frac{D\boldsymbol{\omega}}{Dt} = 0$ (conservation of vorticity).

19. The evolution of the probability distribution function $u(x, t)$ in nonequilibrium statistical mechanics is described by the *Fokker–Planck equation* (See Reif (1965))

$$\frac{\partial u}{\partial t} = \frac{\partial}{\partial x}\left(\frac{\partial u}{\partial t} + x\right) u.$$

(a) Use the change of variables

$$\xi = x e^t \quad \text{and} \quad v = u e^{-t}$$

to show that the Fokker–Planck equation assumes the form with $u(x, t) = e^t v(\xi, \tau)$

$$v_t = e^{2t} v_{\xi\xi}.$$

(b) Make a suitable change of variable t to $\tau(t)$, and transform the above equation into the standard diffusion equation

$$v_t = v_{\xi\xi}.$$

20. The electric field $\mathbf{E}(\mathbf{x})$ and the electromagnetic field $\mathbf{H}(\mathbf{x})$ in free space (a vacuum) satisfy the Maxwell equations $\mathbf{E}_t = c\operatorname{curl}\mathbf{H}$, $\mathbf{H}_t = -c\operatorname{curl}\mathbf{H}$, $\operatorname{div}\mathbf{E} = 0 = \operatorname{div}\mathbf{H}$, where c is the constant speed of light in a vacuum. Show that both $\mathbf{E}$ and $\mathbf{H}$ the *three-dimensional wave equations*

$$\mathbf{E}_{tt} = c^2 \nabla^2 \mathbf{E} \quad \text{and} \quad \mathbf{H}_{tt} = c^2 \nabla^2 \mathbf{H},$$

where $\mathbf{x} = (x, y, z)$ and ∇^2 is the three-dimensional Laplacian.

21. Consider longitudinal vibrations of a free elastic rod with a variable cross section $A(x)$ with x measured along the axis of the rod from the origin. Assuming that the material of the rod satisfies Hooke's law, show that the displacement function $u(x, t)$ satisfies the generalized wave equation

$$u_{tt} = c^2 u_{xx} + \frac{c^2}{A(x)}\left(\frac{dA}{dx}\right) u_x,$$

where $c^2 = (\lambda/\rho)$, λ is a constant that describes the elastic nature of the material, and ρ is the line density of the rod. When $A(x)$ is constant, the above equation reduces to one-dimensional wave equation.

4

Classification of Second-Order Linear Equations

"When we have a good understanding of the problem, we are able to clear it of all auxiliary notions and to reduce it to simplest element."

René Descartes

"The first process ... in the effectual study of sciences must be one of simplification and reduction of the results of previous investigations to a form in which the mind can grasp them."

James Clerk Maxwell

4.1 Second-Order Equations in Two Independent Variables

The general linear second-order partial differential equation in one dependent variable u may be written as

$$\sum_{i,j=1}^{n} A_{ij} u_{x_i x_j} + \sum_{i=1}^{n} B_i u_{x_i} + Fu = G, \qquad (4.1.1)$$

in which we assume $A_{ij} = A_{ji}$ and A_{ij}, B_i, F, and G are real-valued functions defined in some region of the space $(x_1, x_2, \ldots, x_n)$.

Here we shall be concerned with second-order equations in the dependent variable u and the independent variables x, y. Hence equation (4.1.1) can be put in the form

$$A u_{xx} + B u_{xy} + C u_{yy} + D u_x + E u_y + Fu = G, \qquad (4.1.2)$$

where the coefficients are functions of x and y and do not vanish simultaneously. We shall assume that the function u and the coefficients are twice continuously differentiable in some domain in $\mathbb{R}^2$.

The classification of partial differential equations is suggested by the classification of the quadratic equation of conic sections in analytic geometry. The equation

$$Ax^2 + Bxy + Cy^2 + Dx + Ey + F = 0,$$

represents hyperbola, parabola, or ellipse accordingly as $B^2 - 4AC$ is positive, zero, or negative.

The classification of second-order equations is based upon the possibility of reducing equation (4.1.2) by coordinate transformation to *canonical* or *standard* form at a point. An equation is said to be *hyperbolic, parabolic,* or *elliptic* at a point (x_0, y_0) accordingly as

$$B^2(x_0, y_0) - 4A(x_0, y_0) C(x_0, y_0) \tag{4.1.3}$$

is positive, zero, or negative. If this is true at all points, then the equation is said to be hyperbolic, parabolic, or elliptic in a domain. In the case of two independent variables, a transformation can always be found to reduce the given equation to canonical form in a given domain. However, in the case of several independent variables, it is not, in general, possible to find such a transformation.

To transform equation (4.1.2) to a canonical form we make a change of independent variables. Let the new variables be

$$\xi = \xi(x, y), \qquad \eta = \eta(x, y). \tag{4.1.4}$$

Assuming that ξ and η are twice continuously differentiable and that the Jacobian

$$J = \begin{vmatrix} \xi_x & \xi_y \\ \eta_x & \eta_y \end{vmatrix}, \tag{4.1.5}$$

is nonzero in the region under consideration, then x and y can be determined uniquely from the system (4.1.4). Let x and y be twice continuously differentiable functions of ξ and η. Then we have

$$
\begin{aligned}
u_x &= u_\xi \xi_x + u_\eta \eta_x, & u_y &= u_\xi \xi_y + u_\eta \eta_y, \\
u_{xx} &= u_{\xi\xi} \xi_x^2 + 2u_{\xi\eta} \xi_x \eta_x + u_{\eta\eta} \eta_x^2 + u_\xi \xi_{xx} + u_\eta \eta_{xx}, \\
u_{xy} &= u_{\xi\xi} \xi_x \xi_y + u_{\xi\eta}(\xi_x \eta_y + \xi_y \eta_x) + u_{\eta\eta} \eta_x \eta_y + u_\xi \xi_{xy} + u_\eta \eta_{xy}, \\
u_{yy} &= u_{\xi\xi} \xi_y^2 + 2u_{\xi\eta} \xi_y \eta_y + u_{\eta\eta} \eta_y^2 + u_\xi \xi_{yy} + u_\eta \eta_{yy}.
\end{aligned}
\tag{4.1.6}
$$

Substituting these values in equation (4.1.2) we obtain

$$A^* u_{\xi\xi} + B^* u_{\xi\eta} + C^* u_{\eta\eta} + D^* u_\xi + E^* u_\eta + F^* u = G^*, \tag{4.1.7}$$

where

$$A^* = A\xi_x^2 + B\xi_x\xi_y + C\xi_y^2,$$
$$B^* = 2A\xi_x\eta_x + B\left(\xi_x\eta_y + \xi_y\eta_x\right) + 2C\xi_y\eta_y,$$
$$C^* = A\eta_x^2 + B\eta_x\eta_y + C\eta_y^2,$$
$$D^* = A\xi_{xx} + B\xi_{xy} + C\xi_{yy} + D\xi_x + E\xi_y, \qquad (4.1.8)$$
$$E^* = A\eta_{xx} + B\eta_{xy} + C\eta_{yy} + D\eta_x + E\eta_y,$$
$$F^* = F, \qquad G^* = G.$$

The resulting equation (4.1.7) is in the same form as the original equation (4.1.2) under the general transformation (4.1.4). The nature of the equation remains invariant under such a transformation if the Jacobian does not vanish. This can be seen from the fact that the sign of the discriminant does not alter under the transformation, that is,

$$B^{*2} - 4A^*C^* = J^2\left(B^2 - 4AC\right), \qquad (4.1.9)$$

which can be easily verified. It should be noted here that the equation can be of a different type at different points of the domain, but for our purpose we shall assume that the equation under consideration is of the single type in a given domain.

The classification of equation (4.1.2) depends on the coefficients $A\left(x,y\right)$, $B\left(x,y\right)$, and $C\left(x,y\right)$ at a given point (x,y). We shall, therefore, rewrite equation (4.1.2) as

$$Au_{xx} + Bu_{xy} + Cu_{yy} = H\left(x,y,u,u_x,u_y\right), \qquad (4.1.10)$$

and equation (4.1.7) as

$$A^*u_{\xi\xi} + B^*u_{\xi\eta} + C^*u_{\eta\eta} = H^*\left(\xi,\eta,u,u_\xi,u_\eta\right). \qquad (4.1.11)$$

4.2 Canonical Forms

In this section we shall consider the problem of reducing equation (4.1.10) to canonical form.

We suppose first that none of A, B, C, is zero. Let ξ and η be new variables such that the coefficients A^* and C^* in equation (4.1.11) vanish. Thus, from (4.1.8), we have

$$A^* = A\xi_x^2 + B\xi_x\xi_y + C\xi_y^2 = 0,$$
$$C^* = A\eta_x^2 + B\eta_x\eta_y + C\eta_y^2 = 0.$$

These two equations are of the same type and hence we may write them in the form

$$A\zeta_x^2 + B\zeta_x\zeta_y + C\zeta_y^2 = 0, \qquad (4.2.1)$$

in which ζ stand for either of the functions ξ or η. Dividing through by ζ_y^2, equation (4.2.1) becomes

$$A \left(\frac{\zeta_x}{\zeta_y} \right)^2 + B \left(\frac{\zeta_x}{\zeta_y} \right) + C = 0. \qquad (4.2.2)$$

Along the curve $\zeta = $ constant, we have

$$d\zeta = \zeta_x dx + \zeta_y dy = 0.$$

Thus,

$$\frac{dy}{dx} = -\frac{\zeta_x}{\zeta_y}, \qquad (4.2.3)$$

and therefore, equation (4.2.2) may be written in the form

$$A \left(\frac{dy}{dx} \right)^2 - B \left(\frac{dy}{dx} \right) + C = 0, \qquad (4.2.4)$$

the roots of which are

$$\frac{dy}{dx} = \left(B + \sqrt{B^2 - 4AC} \right) / 2A, \qquad (4.2.5)$$

$$\frac{dy}{dx} = \left(B - \sqrt{B^2 - 4AC} \right) / 2A. \qquad (4.2.6)$$

These equations, which are known as the *characteristic equations*, are ordinary differential equations for families of curves in the xy-plane along which $\xi = $ constant and $\eta = $ constant. The integrals of equations (4.2.5) and (4.2.6) are called the *characteristic curves*. Since the equations are first-order ordinary differential equations, the solutions may be written as

$$\phi_1 (x, y) = c_1, \qquad c_1 = \text{constant},$$
$$\phi_2 (x, y) = c_2, \qquad c_2 = \text{constant}.$$

Hence the transformations

$$\xi = \phi_1 (x, y), \qquad \eta = \phi_2 (x, y),$$

will transform equation (4.1.10) to a canonical form.

(A) Hyperbolic Type

If $B^2 - 4AC > 0$, then integration of equations (4.2.5) and (4.2.6) yield two real and distinct families of characteristics. Equation (4.1.11) reduces to

$$u_{\xi\eta} = H_1, \qquad (4.2.7)$$

where $H_1 = H^*/B^*$. It can be easily shown that $B^* \neq 0$. This form is called the *first canonical form of the hyperbolic equation.*

Now if new independent variables

$$\alpha = \xi + \eta, \qquad \beta = \xi - \eta, \qquad (4.2.8)$$

are introduced, then equation (4.2.7) is transformed into

$$u_{\alpha\alpha} - u_{\beta\beta} = H_2\left(\alpha, \beta, u, u_\alpha, u_\beta\right). \qquad (4.2.9)$$

This form is called the *second canonical form of the hyperbolic equation.*

(B) Parabolic Type

In this case, we have $B^2 - 4AC = 0$, and equations (4.2.5) and (4.2.6) coincide. Thus, there exists one real family of characteristics, and we obtain only a single integral $\xi = $ constant (or $\eta = $ constant).

Since $B^2 = 4AC$ and $A^* = 0$, we find that

$$A^* = A\xi_x^2 + B\xi_x\xi_y + C\xi_y^2 = \left(\sqrt{A}\,\xi_x + \sqrt{C}\,\xi_y\right)^2 = 0.$$

From this it follows that

$$A^* = 2A\xi_x\eta_x + B\left(\xi_x\eta_y + \xi_y\eta_x\right) + 2C\xi_y\eta_y$$
$$= 2\left(\sqrt{A}\,\xi_x + \sqrt{C}\,\xi_y\right)\left(\sqrt{A}\,\eta_x + \sqrt{C}\,\eta_y\right) = 0,$$

for arbitrary values of $\eta\left(x, y\right)$ which is functionally independent of $\xi\left(x, y\right)$; for instance, if $\eta = y$, the Jacobian does not vanish in the domain of parabolicity.

Division of equation (4.1.11) by C^* yields

$$u_{\eta\eta} = H_3\left(\xi, \eta, u, u_\xi, u_\eta\right), \qquad C^* \neq 0. \qquad (4.2.10)$$

This is called the *canonical form of the parabolic equation.*

Equation (4.1.11) may also assume the form

$$u_{\xi\xi} = H_3^*\left(\xi, \eta, u, u_\xi, u_\eta\right), \qquad (4.2.11)$$

if we choose $\eta = $ constant as the integral of equation (4.2.5).

(C) Elliptic Type

For an equation of elliptic type, we have $B^2 - 4AC < 0$. Consequently, the quadratic equation (4.2.4) has no real solutions, but it has two complex conjugate solutions which are continuous complex-valued functions of the real variables x and y. Thus, in this case, there are no real characteristic curves. However, if the coefficients A, B, and C are analytic functions of x and y, then one can consider equation (4.2.4) for complex x and y. A function of two real variables x and y is said to be analytic in a certain domain if in some neighborhood of every point (x_0, y_0) of this domain, the function can be represented as a Taylor series in the variables $(x - x_0)$ and $(y - y_0)$.

Since ξ and η are complex, we introduce new real variables

$$\alpha = \frac{1}{2}(\xi + \eta), \qquad \beta = \frac{1}{2i}(\xi - \eta), \qquad (4.2.12)$$

so that

$$\xi = \alpha + i\beta, \qquad \eta = \alpha - i\beta. \qquad (4.2.13)$$

First, we transform equations (4.1.10). We then have

$$A^{**}(\alpha, \beta)\, u_{\alpha\alpha} + B^{**}(\alpha, \beta)\, u_{\alpha\beta} + C^{**}(\alpha, \beta)\, u_{\beta\beta} = H_4(\alpha, \beta, u, u_\alpha, u_\beta),$$
$$(4.2.14)$$

in which the coefficients assume the same form as the coefficients in equation (4.1.11). With the use of (4.2.13), the equations $A^* = C^* = 0$ become

$$\left(A\alpha_x^2 + B\alpha_x\alpha_y + C\alpha_y^2\right) - \left(A\beta_x^2 + B\beta_x\beta_y + C\beta_y^2\right)$$
$$+ i\left[2A\alpha_x\beta_x + B\left(\alpha_x\beta_y + \alpha_y\beta_x\right) + 2C\alpha_y\beta_y\right] = 0,$$

$$\left(A\alpha_x^2 + B\alpha_x\alpha_y + C\alpha_y^2\right) - \left(A\beta_x^2 + B\beta_x\beta_y + C\beta_y^2\right)$$
$$- i\left[2A\alpha_x\beta_x + B\left(\alpha_x\beta_y + \alpha_y\beta_x\right) + 2C\alpha_y\beta_y\right] = 0,$$

or,

$$(A^{**} - C^{**}) + iB^{**} = 0, \qquad (A^{**} - C^{**}) - iB^{**} = 0.$$

These equations are satisfied if and only if

$$A^{**} = C^{**} \quad \text{and} \quad B^{**} = 0.$$

Hence, equation (4.2.14) transforms into the form

$$A^{**}u_{\alpha\alpha} + A^{**}u_{\beta\beta} = H_4(\alpha, \beta, u, u_\alpha, u_\beta).$$

Dividing through by A^{**}, we obtain

$$u_{\alpha\alpha} + u_{\beta\beta} = H_5(\alpha, \beta, u, u_\alpha, u_\beta), \qquad (4.2.15)$$

where $H_5 = (H_4/A^{**})$. This is called the *canonical form of the elliptic equation*.

We close this discussion of canonical forms by adding an important comment. From mathematical and physical points of view, characteristics or characteristic coordinates play a very important physical role in hyperbolic equations. However, they do not play a particularly physical role in parabolic and elliptic equations, but their role is somewhat mathematical in solving these equations. In general, first-order partial differential equations such as advection-reaction equations are regarded as hyperbolic because they describe propagation of waves like the wave equation. On the other hand, second-order linear partial differential equations with constant coefficients are sometimes classified by the associated dispersion relation $\omega = \omega(\boldsymbol{\kappa})$ as defined in Section 13.3. In one-dimensional case, $\omega = \omega(k)$. If $\omega(k)$ is real and $\omega''(k) \neq 0$, the equation is called *dispersive*. The word dispersive simply means that the phase velocity $c_p = (\omega/k)$ of a plane wave solution, $u(x,t) = A \exp[i(kx - \omega t)]$ depends on the wavenumber k. This means that waves of different wavelength propagate with different phase velocities and hence, disperse in the medium. If $\omega = \omega(k) = \sigma(k) + i\mu(k)$ is complex, the associated partial differential equation is called *diffusive*. From a physical point of view, such a classification of equations is particularly useful. Both dispersive and diffusive equations are physically important, and such equations will be discussed in Chapter 13.

Example 4.2.1. Consider the equation

$$y^2 u_{xx} - x^2 u_{yy} = 0.$$

Here

$$A = y^2, \quad B = 0, \quad C = -x^2.$$

Thus,

$$B^2 - 4AC = 4x^2 y^2 > 0.$$

The equation is hyperbolic everywhere except on the coordinate axes $x = 0$ and $y = 0$. From the characteristic equations (4.2.5) and (4.2.6), we have

$$\frac{dy}{dx} = \frac{x}{y}, \qquad \frac{dy}{dx} = -\frac{x}{y}.$$

After integration of these equations, we obtain

$$\frac{1}{2}y^2 - \frac{1}{2}x^2 = c_1, \qquad \frac{1}{2}y^2 + \frac{1}{2}x^2 = c_2.$$

The first of these curves is a family of hyperbolas

$$\frac{1}{2}y^2 - \frac{1}{2}x^2 = c_1,$$

and the second is a family of circles

$$\frac{1}{2}y^2 + \frac{1}{2}x^2 = c_2.$$

To transform the given equation to canonical form, we consider

$$\xi = \frac{1}{2}y^2 - \frac{1}{2}x^2, \qquad \eta = \frac{1}{2}y^2 + \frac{1}{2}x^2.$$

From the relations (4.1.6), we have

$$u_x = u_\xi \xi_x + u_\eta \eta_x = -xu_\xi + xu_\eta,$$
$$u_y = u_\xi \xi_y + u_\eta \eta_y = yu_\xi + yu_\eta,$$
$$u_{xx} = u_{\xi\xi}\xi_x^2 + 2u_{\xi\eta}\xi_x\eta_x + u_{\eta\eta}\eta_x^2 + u_\xi\xi_{xx} + u_\eta\eta_{xx}$$
$$= x^2 u_{\xi\xi} - 2x^2 u_{\xi\eta} + x^2 u_{\eta\eta} - u_\xi + u_\eta.$$
$$u_{yy} = u_{\xi\xi}\xi_y^2 + 2u_{\xi\eta}\xi_y\eta_y + u_{\eta\eta}\eta_y^2 + u_\xi\xi_{yy} + u_\eta\eta_{yy}$$
$$= y^2 u_{\xi\xi} + 2y^2 u_{\xi\eta} + y^2 u_{\eta\eta} + u_\xi + u_\eta.$$

Thus, the given equation assumes the canonical form

$$u_{\xi\eta} = \frac{\eta}{2\left(\xi^2 - \eta^2\right)} u_\xi - \frac{\xi}{2\left(\xi^2 - \eta^2\right)} u_\eta.$$

Example 4.2.2. Consider the partial differential equation

$$x^2 u_{xx} + 2xy\, u_{xy} + y^2 u_{yy} = 0.$$

In this case, the discriminant is

$$B^2 - 4AC = 4x^2y^2 - 4x^2y^2 = 0.$$

The equation is therefore parabolic everywhere. The characteristic equation is

$$\frac{dy}{dx} = \frac{y}{x},$$

and hence, the characteristics are

$$\frac{y}{x} = c,$$

which is the equation of a family of straight lines.

Consider the transformation

$$\xi = \frac{y}{x}, \qquad \eta = y,$$

where η is chosen arbitrarily. The given equation is then reduced to the canonical form

$$y^2 u_{\eta\eta} = 0.$$

Thus,

$$u_{\eta\eta} = 0 \quad \text{for} \quad y \neq 0.$$

Example 4.2.3. The equation

$$u_{xx} + x^2 u_{yy} = 0,$$

is elliptic everywhere except on the coordinate axis $x = 0$ because

$$B^2 - 4AC = -4x^2 < 0, \quad x \neq 0.$$

The characteristic equations are

$$\frac{dy}{dx} = ix, \qquad \frac{dy}{dx} = -ix.$$

Integration yields

$$2y - ix^2 = c_1, \qquad 2y + ix^2 = c_2.$$

Thus, if we write

$$\xi = 2y - ix^2, \qquad \eta = 2y + ix^2,$$

and hence,

$$\alpha = \frac{1}{2}(\xi + \eta) = 2y, \qquad \beta = \frac{1}{2i}(\xi - \eta) = -x^2,$$

we obtain the canonical form

$$u_{\alpha\alpha} + u_{\beta\beta} = -\frac{1}{2\beta} u_\beta.$$

It should be remarked here that a given partial differential equation may be of a different type in a different domain. Thus, for example, *Tricomi's equation*

$$u_{xx} + x u_{yy} = 0, \tag{4.2.16}$$

is elliptic for $x > 0$ and hyperbolic for $x < 0$, since $B^2 - 4AC = -4x$. For a detailed treatment, see Hellwig (1964).

4.3 Equations with Constant Coefficients

In this case of an equation with real constant coefficients, the equation is of a single type at all points in the domain. This is because the discriminant $B^2 - 4AC$ is a constant.

From the characteristic equations

$$\frac{dy}{dx} = \left(B \pm \sqrt{B^2 - 4AC}\right)/2A, \tag{4.3.1}$$

we can see that the characteristics

$$y = \left(\frac{B + \sqrt{B^2 - 4AC}}{2A}\right) x + c_1, \quad y = \left(\frac{B - \sqrt{B^2 - 4AC}}{2A}\right) x + c_2, \quad (4.3.2)$$

are two families of straight lines. Consequently, the characteristic coordinates take the form

$$\xi = y - \lambda_1 x, \qquad \eta = y - \lambda_2 x, \qquad (4.3.3)$$

where

$$\lambda_{1,2} = \frac{B \pm \sqrt{B^2 - 4AC}}{2A}. \qquad (4.3.4)$$

The linear second-order partial differential equation with constant coefficients may be written in the general form as

$$A u_{xx} + B u_{xy} + C u_{yy} + D u_x + E u_y + F u = G\left(x, y\right). \qquad (4.3.5)$$

In particular, the equation

$$A u_{xx} + B u_{yy} + C u_{yy} = 0, \qquad (4.3.6)$$

is called the *Euler equation*.

(A) Hyperbolic Type

If $B^2 - 4AC > 0$, the equation is of hyperbolic type, in which case the characteristics form two distinct families.

Using (4.3.3), equation (4.3.5) becomes

$$u_{\xi\eta} = D_1 u_\xi + E_1 u_\eta + F_1 u + G_1\left(\xi, \eta\right), \qquad (4.3.7)$$

where D_1, E_1, and F_1 are constants. Here, since the coefficients are constants, the lower order terms are expressed explicitly.

When $A = 0$, equation (4.3.1) does not hold. In this case, the characteristic equation may be put in the form

$$-B\left(dx/dy\right) + C\left(dx/dy\right)^2 = 0,$$

which may again be rewritten as

$$dx/dy = 0, \quad \text{and} \quad -B + C\left(dx/dy\right) = 0.$$

Integration gives

$$x = c_1, \qquad x = \left(B/C\right) y + c_2,$$

where c_1 and c_2 are integration constants. Thus, the characteristic coordinates are

$$\xi = x, \qquad \eta = x - (B/C)\, y. \tag{4.3.8}$$

Under this transformation, equation (4.3.5) reduces to the canonical form

$$u_{\xi\eta} = D_1^* u_\xi + E_1^* u_\eta + F_1^* u + G_1^* \left(\xi, \eta \right), \tag{4.3.9}$$

where D_1^*, E_1^*, and F_1^* are constants.

The canonical form of the Euler equation (4.3.6) is

$$u_{\xi\eta} = 0. \tag{4.3.10}$$

Integrating this equation gives the general solution

$$u = \phi\left(\xi \right) + \psi\left(\eta \right) = \phi\left(y - \lambda_1, x \right) + \psi\left(y - \lambda_2, x \right), \tag{4.3.11}$$

where ϕ and ψ are arbitrary functions, and λ_1 and λ_2 are given by (4.3.3).

(B) Parabolic Type

When $B^2 - 4AC = 0$, the equation is of parabolic type, in which case only one real family of characteristics exists. From equation (4.3.4), we find that

$$\lambda_1 = \lambda_2 = (B/2A),$$

so that the single family of characteristics is given by

$$y = (B/2A)\, x + c_1,$$

where c_1 is an integration constant. Thus, we have

$$\xi = y - (B/2A)\, x, \qquad \eta = hy + kx, \tag{4.3.12}$$

where η is chosen arbitrarily such that the Jacobian of the transformation is not zero, and h and k are constants.

With the proper choice of the constants h and k in the transformation (4.3.12), equation (4.3.5) reduces to

$$u_{\eta\eta} = D_2 u_\xi + E_2 u_\eta + F_2 u + G_2 \left(\xi, \eta \right), \tag{4.3.13}$$

where D_2, E_2, and F_2 are constants.

If $B = 0$, we can see at once from the relation

$$B^2 - 4AC = 0,$$

that C or A vanishes. The given equation is then already in the canonical form. Similarly, in the other cases when A or C vanishes, B vanishes. The given equation is is then also in canonical form.

The canonical form of the Euler equation (4.3.6) is

$$u_{\eta\eta} = 0. \tag{4.3.14}$$

Integrating twice gives the general solution

$$u = \phi(\xi) + \eta\psi(\xi), \tag{4.3.15}$$

where ξ and η are given by (4.3.12). Choosing $h = 1$, $k = 0$ and $\lambda = \left(\frac{B}{2A}\right)$ for simplicity, the general solution of the Euler equation in the parabolic case is

$$u = \phi(y - \lambda x) + y\psi(y - \lambda x). \tag{4.3.16}$$

(C) Elliptic Type

When $B^2 - 4AC < 0$, the equation is of elliptic type. In this case, the characteristics are complex conjugates.

The characteristic equations yield

$$y = \lambda_1 x + c_1, \qquad y = \lambda_2 x + c_2, \tag{4.3.17}$$

where λ_1 and λ_2 are complex numbers. Accordingly, c_1 and c_2 are allowed to take on complex values. Thus,

$$\xi = y - (a + ib)x, \qquad \eta = y - (a - ib)x, \tag{4.3.18}$$

where $\lambda_{1,2} = a \pm ib$ in which a and b are real constants, and

$$a = \frac{B}{2A}, \quad \text{and} \quad b = \frac{1}{2A}\sqrt{4AC - B^2}.$$

Introduce the new variables

$$\alpha = \frac{1}{2}(\xi + \eta) = y - ax, \qquad \beta = \frac{1}{2i}(\xi - \eta) = -bx. \tag{4.3.19}$$

Application of this transformation readily reduces equation (4.3.5) to the canonical form

$$u_{\alpha\alpha} + u_{\beta\beta} = D_3 u_\alpha + E_3 u_\beta + F_3 u + G_3(\alpha, \beta), \tag{4.3.20}$$

where D_3, E_3, F_3 are constants.

We note that $B^2 - AC < 0$, so neither A nor C is zero.

In this elliptic case, the Euler equation (4.3.6) gives the complex characteristics (4.3.18) which are

$$\xi = (y - ax) - ibx, \qquad \eta = (y - ax) + ibx = \bar{\xi}. \tag{4.3.21}$$

Consequently, the Euler equation becomes

$$u_{\xi\xi} = 0, \tag{4.3.22}$$

with the general solution

$$u = \phi(\xi) + \psi(\bar{\xi}). \tag{4.3.23}$$

The appearance of complex arguments in the general solution (4.3.23) is a general feature of elliptic equations.

Example 4.3.1. Consider the equation

$$4\,u_{xx} + 5\,u_{xy} + u_{yy} + u_x + u_y = 2.$$

Since $A = 4$, $B = 5$, $C = 1$, and $B^2 - 4AC = 9 > 0$, the equation is hyperbolic. Thus, the characteristic equations take the form

$$\frac{dy}{dx} = 1, \qquad \frac{dy}{dx} = \frac{1}{4},$$

and hence, the characteristics are

$$y = x + c_1, \qquad y = (x/4) + c_2.$$

The linear transformation

$$\xi = y - x, \qquad \eta = y - (x/4),$$

therefore reduces the given equation to the canonical form

$$u_{\xi\eta} = \frac{1}{3}\,u_\eta - \frac{8}{9}.$$

This is the first canonical form.

The second canonical form may be obtained by the transformation

$$\alpha = \xi + \eta, \qquad \beta = \xi - \eta,$$

in the form

$$u_{\alpha\alpha} - u_{\beta\beta} = \frac{1}{3}\,u_\alpha - \frac{1}{3}\,u_\beta - \frac{8}{9}.$$

Example 4.3.2. The equation

$$u_{xx} - 4\,u_{xy} + 4\,u_{yy} = e^y,$$

is parabolic since $A = 1$, $B = -4$, $C = 4$, and $B^2 - 4AC = 0$. Thus, we have from equation (4.3.12)

$$\xi = y + 2x, \qquad \eta = y,$$

in which η is chosen arbitrarily. By means of this mapping, the equation transforms into

$$u_{\eta\eta} = \frac{1}{4}\,e^\eta.$$

Example 4.3.3. Consider the equation

$$u_{xx} + u_{xy} + u_{yy} + u_x = 0.$$

Since $A = 1$, $B = 1$, $C = 1$, and $B^2 - 4AC = -3 < 0$, the equation is elliptic.

We have

$$\lambda_{1,2} = \frac{B \pm \sqrt{B^2 - 4AC}}{2A} = \frac{1}{2} \pm i\frac{\sqrt{3}}{2},$$

and hence,

$$\xi = y - \left(\frac{1}{2} + i\frac{\sqrt{3}}{2}\right)x, \qquad \eta = y - \left(\frac{1}{2} - i\frac{\sqrt{3}}{2}\right)x.$$

Introducing the new variables

$$\alpha = \frac{1}{2}(\xi + \eta) = y - \frac{1}{2}x, \qquad \beta = \frac{1}{2i}(\xi - \eta) = -\frac{\sqrt{3}}{2}x,$$

the given equation is then transformed into canonical form

$$u_{\alpha\alpha} + u_{\beta\beta} = \frac{2}{3}u_\alpha + \frac{2}{\sqrt{3}}u_\beta.$$

Example 4.3.4. Consider the wave equation

$$u_{tt} - c^2 u_{xx} = 0, \quad c \text{ is constant.}$$

Since $A = -c^2$, $B = 0$, $C = 1$, and $B^2 - 4AC = 4c^2 > 0$, the wave equation is hyperbolic everywhere. According to (4.2.4), the equation of characteristics is

$$-c^2 \left(\frac{dt}{dx}\right)^2 + 1 = 0,$$

or

$$dx^2 - c^2 dt^2 = 0.$$

Therefore,

$$x + ct = \xi = \text{ constant}, \qquad x - ct = \eta = \text{ constant}.$$

Thus, the characteristics are straight lines, which are shown in Figure 4.3.1. The characteristics form a natural set of coordinates for the hyperbolic equation.

In terms of new coordinates ξ and η defined above, we obtain

$$u_{xx} = u_{\xi\xi} + 2u_{\xi\eta} + u_{\eta\eta},$$
$$u_{tt} = c^2 \left(u_{\xi\xi} - 2u_{\xi\eta} + u_{\eta\eta}\right),$$

so that the wave equation becomes

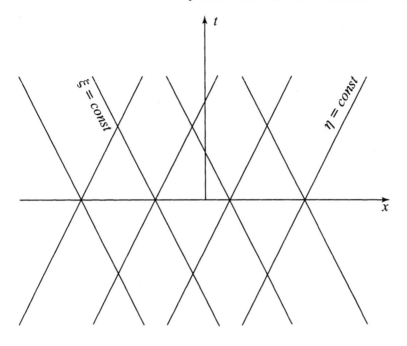

Figure 4.3.1 Characteristics for the wave equation.

$$-4c^2 u_{\xi\eta} = 0.$$

Since $c \neq 0$, we have

$$u_{\xi\eta} = 0.$$

Integrating with respect to ξ, we obtain

$$u_\eta = \psi_1(\eta).$$

where ψ_1 is the arbitrary function of η. Integrating with respect to η, we obtain

$$u(\xi, \eta) = \int \psi_1(\eta)\, d\eta + \phi(\xi).$$

If we set $\psi(\eta) = \int \psi_1(\eta)\, d\eta$, the general solution becomes

$$u(\xi, \eta) = \phi(\xi) + \psi(\eta),$$

which is, in terms of the original variables x and t,

$$u(x, t) = \phi(x + ct) + \psi(x - ct),$$

provided ϕ and ψ are arbitrary but twice differentiable functions.

Note that ϕ is constant on "wavefronts" $x = -ct + \xi$ that travel toward decreasing x as t increases, whereas ψ is constant on wavefronts $x = ct + \eta$ that travel toward increasing x as t increases. Thus, any general solution can be expressed as the sum of two waves, one traveling to the right with constant velocity c and the other traveling to the left with the same velocity c.

Example 4.3.5. Find the characteristic equations and characteristics, and then reduce the equations

$$u_{xx} \mp \left(\text{sech}^4 x \right) u_{yy} = 0, \qquad (4.3.24ab)$$

to the canonical forms.

In equation (4.3.24a), $A = 1$, $B = 0$ and $C = -\text{sech}^4 x$. Hence,

$$B^2 - 4AC = 4 \text{sech}^4 x > 0.$$

Hence, the equation is hyperbolic. The characteristic equations are

$$\frac{dy}{dx} = \frac{B \pm \sqrt{B^2 - 4AC}}{2A} = \pm \text{sech}^2 x.$$

Integration gives

$$y \mp \tanh x = \text{constant}.$$

Hence,

$$\xi = y + \tanh x, \qquad \eta = y - \tanh x.$$

Using these characteristic coordinates, the given equation can be transformed into the canonical form

$$u_{\xi\eta} = \frac{(\eta - \xi)}{\left[4 - (\xi - \eta)^2 \right]} \left(u_\xi - u_\eta \right). \qquad (4.3.25)$$

In equation (4.3.24b), $A = 1$, $B = 0$ and $C = \text{sech}^4 x$. Hence,

$$B^2 - 4AC = \pm i \, \text{sech}^2 x.$$

Integrating gives

$$y \mp i \tanh x = \text{constant}.$$

Thus,

$$\xi = y + i \tanh x, \qquad \eta = y - i \tanh x.$$

The new real variables α and β are

$$\alpha = \frac{1}{2}\left(\xi + \eta\right) = y, \qquad \beta = \frac{1}{2i}\left(\xi - \eta\right) = \tanh x.$$

In terms of these new variables, equation (4.3.24b) can be transformed into the canonical form

$$u_{\alpha\alpha} + u_{\beta\beta} = \frac{2\beta}{1 - \beta^2}\, u_\beta, \quad |\beta| < 1. \tag{4.3.26}$$

Example 4.3.6. Consider the equation

$$u_{xx} + \left(2\cosec y\right) u_{xy} + \left(\cosec^2 y\right) u_{yy} = 0. \tag{4.3.27}$$

In this case, $A = 1$, $B = 2\cosec y$ and $C = \cosec^2 y$. Hence, $B^2 - 4AC = 0$, and

$$\frac{dy}{dx} = \frac{B}{2A} = \cosec y.$$

The characteristic curves are therefore given by

$$\xi = x + \cos y \quad \text{and} \quad \eta = y.$$

Using these variables, the canonical form of (4.3.27) is

$$u_{\eta\eta} = \left(\sin^2 \eta \, \cos \eta\right) u_\xi. \tag{4.3.28}$$

4.4 General Solutions

In general, it is not so simple to determine the general solution of a given equation. Sometimes further simplification of the canonical form of an equation may yield the general solution. If the canonical form of the equation is simple, then the general solution can be immediately ascertained.

Example 4.4.1. Find the general solution of

$$x^2 u_{xx} + 2xy\, u_{xy} + y^2 u_{yy} = 0.$$

In Example 4.2.2, using the transformation $\xi = y/x$, $\eta = y$, this equation was reduced to the canonical form

$$u_{\eta\eta} = 0, \quad \text{for} \quad y \neq 0.$$

Integrating twice with respect to η, we obtain

$$u\left(\xi, \eta\right) = \eta f\left(\xi\right) + g\left(\xi\right),$$

where $f\left(\xi\right)$ and $g\left(\xi\right)$ are arbitrary functions. In terms of the independent variables x and y, we have

$$u\left(x, y\right) = y f\left(\frac{y}{x}\right) + g\left(\frac{y}{x}\right).$$

Example 4.4.2. Determine the general solution of

$$4\,u_{xx} + 5\,u_{xy} + u_{yy} + u_x + u_y = 2.$$

Using the transformation $\xi = y - x$, $\eta = y - (x/4)$, the canonical form of this equation is (see Example 4.3.1)

$$u_{\xi\eta} = \frac{1}{3}\,u_\eta - \frac{8}{9}.$$

By means of the substitution $v = u_\eta$, the preceding equation reduces to

$$v_\xi = \frac{1}{3}\,v - \frac{8}{9}.$$

This can be easily integrated by separating the variables. Integrating with respect to ξ, we have

$$v = \frac{8}{3} + \frac{1}{3}\,e^{(\xi/3)} F\,(\eta)\,.$$

Integrating with respect to η, we obtain

$$u\,(\xi, \eta) = \frac{8}{3}\,\eta + \frac{1}{3} g\,(\eta)\,e^{\xi/3} + f\,(\xi)\,,$$

where $f\,(\xi)$ and $g\,(\eta)$ are arbitrary functions. The general solution of the given equation becomes

$$u\,(x, y) = \frac{8}{3}\left(y - \frac{x}{4}\right) + \frac{1}{3} g\left(y - \frac{x}{4}\right) e^{\frac{1}{3}(y-x)} + f\,(y - x)\,.$$

Example 4.4.3. Obtain the general solution of

$$3\,u_{xx} + 10\,u_{xy} + 3\,u_{yy} = 0.$$

Since $B^2 - 4AC = 64 > 0$, the equation is hyperbolic. Thus, from equation (4.3.2), the characteristics are

$$y = 3x + c_1, \qquad y = \frac{1}{3}x + c_2.$$

Using the transformations

$$\xi = y - 3x, \qquad \eta = y - \frac{1}{3}x,$$

the given equation can be reduced to the form

$$\left(\frac{64}{3}\right) u_{\xi\eta} = 0.$$

Hence, we obtain

$$u_{\xi\eta} = 0.$$

Integration yields

$$u(\xi, \eta) = f(\xi) + g(\eta).$$

In terms of the original variables, the general solution is

$$u(x, y) = f(y - 3x) + g\left(y - \frac{x}{3}\right).$$

Example 4.4.4. Find the general solution of the following equations

$$y\,u_{xx} + 3\,y\,u_{xy} + 3\,u_x = 0, \qquad y \neq 0, \qquad (4.4.1)$$
$$u_{xx} + 2\,u_{xy} + u_{yy} = 0, \qquad (4.4.2)$$
$$u_{xx} + 2\,u_{xy} + 5\,u_{yy} + u_x = 0. \qquad (4.4.3)$$

In equation (4.4.1), $A = y$, $B = 3y$, $C = 0$, $D = 3$, $E = F = G = 0$. Hence $B^2 - 4AC = 9y^2 > 0$ and the equation is hyperbolic for all points (x, y) with $y \neq 0$. Consequently, the characteristic equations are

$$\frac{dy}{dx} = \frac{B \pm \sqrt{B^2 - 4AC}}{2A} = \frac{3y \pm 3y}{2y} = 3, \quad 0.$$

Integrating gives

$$y = c_1 \quad \text{and} \quad y = 3x + c_2.$$

The characteristic curves are

$$\xi = y \quad \text{and} \quad \eta = y - 3x.$$

In terms of these variables, the canonical form of (4.4.1) is

$$\xi\,u_{\xi\eta} + u_\eta = 0.$$

Writing $v = u_\eta$ and using the integrating factor gives

$$v = u_\eta = \frac{1}{\xi}C(\eta),$$

where $C(\eta)$ is an arbitrary function.

Integrating again with respect to η gives

$$u(\xi, \eta) = \frac{1}{\xi}\int C(\eta)\,d\eta + g(\xi) = \frac{1}{\xi}f(\eta) + g(\xi),$$

where f and g are arbitrary functions. Finally, in terms of the original variables, the general solution is

$$u(x,y) = \frac{1}{y} f(y - 3x) + g(y). \tag{4.4.4}$$

Equation (4.4.2) has coefficients $A = 1$, $B = 2$, $C = 1$, $D = E = F = G = 0$. Hence, $B^2 - 4AC = 0$, the equation is parabolic. The characteristic equation is

$$\frac{dy}{dx} = 1,$$

and the characteristics are

$$\xi = y - x = c_1 \quad \text{and} \quad \eta = y.$$

Using these variables, equation (4.4.2) takes the canonical form

$$u_{\eta\eta} = 0.$$

Integrating twice gives the general solution

$$u(\xi, \eta) = \eta f(\xi) + g(\xi),$$

where f and g are arbitrary functions.

In terms of x and y, this solution becomes

$$u(x,y) = y f(y - x) + g(y - x). \tag{4.4.5}$$

The coefficients of equation (4.4.3) are $A = 1$, $B = 2$, $C = 5$, $E = 1$, $F = G = 0$ and hence $B^2 - 4AC = -16 < 0$, equation (4.4.3) is elliptic. The characteristic equations are

$$\frac{dy}{dx} = (1 \pm 2i).$$

The characteristics are

$$y = (1 - 2i) x + c_1, \qquad y = (1 + 2i) x + c_2,$$

and hence,

$$\xi = y - (1 - 2i) x, \qquad \eta = y - (1 + 2i) x,$$

and new real variables α and β are

$$\alpha = \frac{1}{2}(\xi + \eta) = y - x, \qquad \eta = \frac{1}{2i}(\xi - \eta) = 2x.$$

The canonical form is given by

$$(u_{\alpha\alpha} + u_{\beta\beta}) = \frac{1}{4}(u_\alpha - 2u_\beta). \tag{4.4.6}$$

It is not easy to find a general solution of (4.4.6).

Example 4.4.5. Use $u = f(\xi)$, $\xi = \frac{x}{\sqrt{4\kappa t}}$ to solve the parabolic system

$$u_t = \kappa\, u_{xx}, \quad -\infty < x < \infty, \quad t > 0, \tag{4.4.7}$$
$$u(x,0) = 0, \quad x < 0; \quad u(x,0) = u_0, \quad x > 0, \tag{4.4.8}$$

where κ and u_0 are constant.

We use the given transformations to obtain

$$u_t = f'(\xi)\,\xi_t = -\frac{1}{2}\frac{x}{\sqrt{4\kappa t^3}}\,f'(\xi),$$

$$u_{xx} = \frac{\partial}{\partial x}(u_x) = \frac{\partial}{\partial x}(f'(\xi)\cdot \xi_x) = \frac{1}{4\kappa t}\,f''(\xi).$$

Consequently, equation (4.4.7) becomes

$$f''(\xi) + 2\,\xi f'(\xi) = 0.$$

The solution of this equation is

$$f'(\xi) = A\exp\left(-\xi^2\right),$$

where A is a constant of integration. Integrating again gives

$$f(\xi) = A\int_0^\xi e^{-\alpha^2}\,d\alpha + B,$$

where B is an integrating constant.

Using the given conditions yields

$$0 = A\int_0^{-\infty} e^{-\alpha^2}\,d\alpha + B, \qquad u_0 = A\int_0^\infty e^{-\alpha^2}\,d\alpha + B,$$

which give

$$A = \frac{u_0}{\sqrt{\pi}} \quad \text{and} \quad B = \frac{1}{2}\,u_0.$$

Thus, the final solution is

$$u(x,t) = u_0\left[\frac{1}{\sqrt{\pi}}\int_0^{\frac{x}{\sqrt{4\kappa t}}} e^{-\alpha^2}\,d\alpha + \frac{1}{2}\right].$$

4.5 Summary and Further Simplification

We summarize the classification of linear second-order partial differential equations with constant coefficients in two independent variables.

hyperbolic: $$u_{rs} = a_1 u_r + a_2 u_s + a_3 u + f_1, \quad (4.5.1)$$
$$u_{rr} - u_{ss} = a_1^* u_r + a_2^* u_s + a_3^* u + f_1^*, \quad (4.5.2)$$
parabolic: $$u_{rs} = b_1 u_r + b_2 u_s + b_3 u + f_2, \quad (4.5.3)$$
elliptic: $$u_{rr} + u_{ss} = c_1 u_r + c_2 u_s + c_3 u + f_3, \quad (4.5.4)$$

where r and s represent the new independent variables in the linear transformations

$$r = r(x, y), \qquad s = s(x, y), \qquad (4.5.5)$$

and the Jacobian $J \neq 0$.

To simplify equation (4.5.1) further, we introduce the new dependent variable

$$v = u e^{-(ar+bs)}, \qquad (4.5.6)$$

where a and b are undetermined coefficients. Finding the derivatives, we obtain

$$u_r = (v_r + av) e^{ar+bs},$$
$$u_s = (v_s + bv) e^{ar+bs},$$
$$u_{rr} = (v_{rr} + 2av_r + a^2 v) e^{ar+bs},$$
$$u_{rs} = (v_{rs} + av_s + bv_r + abv) e^{ar+bs},$$
$$u_{ss} = (v_{ss} + 2bv_s + b^2 v) e^{ar+bs}.$$

Substitution of these equation (4.5.1) yields

$$v_{rs} + (b - a_1) v_r + (a - a_2) v_s + (ab - a_1 a - a_2 b - a_3) v = f_1 e^{-(ar+bs)}.$$

In order that the first derivatives vanish, we set

$$b = a_1 \quad \text{and} \quad a = a_2.$$

Thus, the above equation becomes

$$v_{rs} = (a_1 a_2 + a_3) v + g_1,$$

where $g_1 = f_1 e^{-(a_2 r + a_1 s)}$. In a similar manner, we can transform equations (4.5.2)–(4.5.4). Thus, we have the following transformed equations corresponding to equations (4.5.1)–(4.5.4).

hyperbolic: $$v_{rs} = h_1 v + g_1,$$
$$v_{rr} - v_{ss} = h_1^* v + g_1^*, \qquad (4.5.7)$$
parabolic: $$v_{ss} = h_2 v + g_2,$$
elliptic: $$v_{rr} + v_{ss} = h_3 v + g_3.$$

In the case of partial differential equations in several independent variables or in higher order, the classification is considerably more complex. For further reading, see Courant and Hilbert (1953, 1962).

4.6 Exercises

1. Determine the region in which the given equation is hyperbolic, parabolic, or elliptic, and transform the equation in the respective region to canonical form.

(a) $xu_{xx} + u_{yy} = x^2$,

(b) $u_{xx} + y^2 u_{yy} = y$,

(c) $u_{xx} + xy u_{yy} = 0$,

(d) $x^2 u_{xx} - 2xy u_{xy} + y^2 u_{yy} = e^x$,

(e) $u_{xx} + u_{xy} - xu_{yy} = 0$,

(f) $e^x u_{xx} + e^y u_{yy} = u$,

(g) $u_{xx} - \sqrt{y}\, u_{xy} + \left(\frac{x}{4}\right) u_{yy} + 2x\, u_x - 3y\, u_y + 2u = \exp\left(x^2 - 2y\right)$,
$$y \geq 0,$$

(h) $u_{xx} - \sqrt{y}\, u_{xy} + x u_{yy} = \cos\left(x^2 - 2y\right)$, $\quad y \geq 0$,

(i) $u_{xx} - yu_{xy} + xu_x + yu_y + u = 0$,

(j) $\sin^2 x\, u_{xx} + \sin 2x\, u_{xy} + \cos^2 x\, u_{yy} = x$,

2. Obtain the general solution of the following equations:

(i) $\quad x^2 u_{xx} + 2xy u_{xy} + y^2 u_{yy} + xy u_x + y^2 u_y = 0$,

(ii) $\quad ru_{tt} - c^2 ru_{rr} - 2c^2 u_r = 0$, $c = $ constant,

(iii) $\quad 4u_x + 12u_{xy} + 9u_{yy} - 9u = 9$,

(iv) $\quad u_{xx} + u_{xy} - 2u_{yy} - 3u_x - 6u_y = 9\left(2x - y\right)$,

(v) $\quad yu_x + 3y\, u_{xy} + 3u_x = 0$, $\quad y \neq 0$.

(vi) $\quad u_{xx} + u_{yy} = 0$,

(vii) $4\, u_{xx} + u_{yy} = 0$,

(viii) $u_{xx} - 2\, u_{xy} + u_{yy} = 0$,

(ix) $\quad 2\, u_{xx} + u_{yy} = 0$,

(x) $\quad u_{xx} + 4\, u_{xy} + 4\, u_{yy} = 0$,

(xi) $\quad 3\, u_{xx} + 4\, u_{xy} - \frac{3}{4}\, u_{yy} = 0$.

3. Find the characteristics and characteristic coordinates, and reduce the following equations to canonical form:

(a) $u_{xx} + 2u_{xy} + 3u_{yy} + 4u_x + 5u_y + u = e^x$,

(b) $2u_{xx} - 4u_{xy} + 2u_{yy} + 3u = 0$,

(c) $u_{xx} + 5u_{xy} + 4u_{yy} + 7u_y = \sin x$, (d) $u_{xx} + u_{yy} + 2u_x + 8u_y + u = 0$,

(e) $u_{xy} + 2u_{yy} + 9u_x + u_y = 2$,   (f) $6u_{xx} - u_{xy} + u = y^2$,

(g) $u_{xy} + u_x + u_y = 3x$,   (h) $u_{yy} - 9u_x + 7u_y = \cos y$,

(i) $x^2 u_{xx} - y^2 u_{yy} - u_x = 1 + 2y^2$, (j) $u_{xx} + yu_{yy} + \frac{1}{2}u_y + 4yu_x = 0$,

(k) $x^2 y^2 u_{xx} + 2xyu_{xy} + u_{yy} = 0$,   (l) $u_{xx} + yu_{yy} = 0$.

4. Determine the general solutions of the following equations:

(i) $u_{xx} - \frac{1}{c^2} u_{yy} = 0$, $c = $ constant, (ii) $u_{xx} + u_{yy} = 0$,

(iii) $u_{xxxx} + 2u_{xxyy} + u_{yyyy} = 0$,   (iv) $u_{xx} - 3u_{xy} + 2u_{yy} = 0$,

(v) $u_{xx} + u_{xy} = 0$,   (vi) $u_{xx} + 10u_{xy} + 9u_{yy} = y$.

5. Transform the following equations to the form $v_{\xi\eta} = cv$, $c = $ constant,

(i) $u_{xx} - u_{yy} + 3u_x - 2u_y + u = 0$,

(ii) $3u_{xx} + 7u_{xy} + 2u_{yy} + u_y + u = 0$,

by introducing the new variables $v = u e^{-(a\xi + b\eta)}$, where a and b are undetermined coefficients.

6. Given the parabolic equation

$$u_{xx} = au_t + bu_x + cu + f,$$

where the coefficients are constants, by the substitution $u = v e^{\frac{1}{2}bx}$, for the case $c = -\left(b^2/4\right)$, show that the given equation is reduced to the heat equation

$$v_{xx} = av_t + g, \qquad g = fe^{-bx/2}.$$

7. Reduce the Tricomi equation

$$u_{xx} + xu_{yy} = 0,$$

to the canonical form

(i) $u_{\xi\eta} - [6(\xi - \eta)]^{-1} (u_\xi - u_\eta) = 0,$ for $x < 0,$

(ii) $u_{\alpha\alpha} + u_{\beta\beta} + \frac{1}{3\beta} = 0,$ $x > 0.$

Show that the characteristic curves for $x < 0$ are cubic parabolas.

8. Use the polar coordinates r and θ ($x = r\cos\theta$, $y = r\sin\theta$) to transform the Laplace equation $u_{xx} + u_{yy} = 0$ into the polar form

$$\nabla^2 u = u_{rr} + \frac{1}{r} u_r + \frac{1}{r^2} u_{\theta\theta} = 0.$$

9. (a) Using the cylindrical polar coordinates $x = r\cos\theta$, $y = r\sin\theta$, $z = z$, transform the three-dimensional Laplace equation $u_{xx} + u_{yy} + u_{zz} = 0$ into the form

$$u_{rr} + \frac{1}{r} u_r + \frac{1}{r^2} u_{\theta\theta} + u_{zz} = 0.$$

(b) Use the spherical polar coordinates (r, θ, ϕ) so that $x = r\sin\phi\cos\theta$, $y = r\sin\phi\sin\theta$, $z = r\cos\phi$ to transform the three-dimensional Laplace equation $u_{xx} + u_{yy} + u_{zz} = 0$ into the form

$$u_{rr} + \frac{2}{r} u_r + \frac{1}{r^2 \sin\phi} (\sin\phi\, u_\phi)_\phi + \frac{1}{r^2 \sin^2\phi} u_{\theta\theta} = 0.$$

(c) Transform the diffusion equation

$$u_t = \kappa (u_{xx} + u_{yy}),$$

into the axisymmetric form

$$u_t = \kappa \left(u_{rr} + \frac{1}{r} u_r \right).$$

10. (a) Apply a linear transformation $\xi = ax + by$ and $\eta = cx + dy$, to transform the Euler equation

$$A\, u_{xx} + 2B\, u_{xy} + C\, u_{yy} = 0$$

into canonical form, where a, b, c, d, A, B and C are constants .
(b) Show that the same transformation as in (a) can be used to transform the nonhomogeneous Euler equation

$$A\, u_{xx} + 2B\, u_{xy} + C\, u_{yy} = F(x, y, u, u_x, u_y)$$

into canonical form.

11. Obtain the solution of the Cauchy problem

$$u_{xx} + u_{yy} = 0,$$
$$u(x,0) = f(x) \quad \text{and} \quad u_y(x,0) = g(x).$$

12. Classify each of the following equations and reduce it to canonical form:

(a) $y\,u_{xx} - x\,u_{yy} = 0, \quad x > 0, \quad y > 0;$ (b) $u_{xx} + \left(\text{sech}^4 x\right) u_{yy} = 0,$

(c) $y^2 u_{xx} + x^2 u_{yy} = 0,$ (d) $u_{xx} - \left(\text{sech}^4 x\right) u_{yy} = 0,$

(e) $u_{xx} + 6u_{xy} + 9u_{yy} + 3y\,u_y = 0,$

(f) $y^2 u_{xx} + 2xy\,u_{xy} + 2x^2 u_{yy} + xu_x = 0,$

(g) $u_{xx} - (2\cos x)\,u_{xy} + \left(1 + \cos^2 x\right) u_{yy} + u = 0,$

(h) $u_{xx} + (2\,\text{cosec}\,y)\,u_{xy} + \left(\text{cosec}^2 y\right) u_{yy} = 0.$

(i) $u_{xx} - 2\,u_{xy} + u_{yy} + 3\,u_x - u + 1 = 0,$

(j) $u_{xx} - y^2 u_{yy} + u_x - u + x^2 = 0,$

(k) $u_{xx} + y\,u_{yy} - x\,u_y + y = 0.$

13. Transform the equation

$$u_{xy} + y\,u_{yy} + \sin(x+y) = 0$$

into the canonical form. Use the canonical form to find the general solution.

14. Classify each of the following equations for $u(x,t)$:

(a) $u_t = (p\,u_x)_x,$ (b) $u_{tt} - c^2 u_{xx} + \alpha u = 0,$

(c) $(a\,u_x)_x + (a\,u_t)_t = 0,$ (d) $u_{xt} - a\,u_t = 0,$

where $p(x)$, $c(x,t)$, $a(x,t)$, and $\alpha(x)$ are given functions that take only positive values in the (x,t) plane. Find the general solution of the equation in (d).

5

The Cauchy Problem and Wave Equations

"Since a general solution must be judged impossible from want of analysis, we must be content with the knowledge of some special cases, and that all the more, since the development of various cases seems to be the only way to bringing us at last to a more perfect knowledge."

Leonhard Euler

"What would geometry be without Gauss, mathematical logic without Boole, algebra without Hamilton, analysis without Cauchy?"

George Temple

5.1 The Cauchy Problem

In the theory of ordinary differential equations, by the initial-value problem we mean the problem of finding the solutions of a given differential equation with the appropriate number of initial conditions prescribed at an initial point. For example, the second-order ordinary differential equation

$$\frac{d^2u}{dt^2} = f\left(t, u, \frac{du}{dt}\right)$$

and the initial conditions

$$u(t_0) = \alpha, \qquad \left(\frac{du}{dt}\right)(t_0) = \beta,$$

constitute an initial-value problem.

An analogous problem can be defined in the case of partial differential equations. Here we shall state the problem involving second-order partial differential equations in two independent variables.

We consider a second-order partial differential equation for the function u in the independent variables x and y, and suppose that this equation can be solved explicitly for u_{yy}, and hence, can be represented in the from

$$u_{yy} = F\left(x, y, u, u_x, u_y, u_{xx}, u_{xy}\right). \tag{5.1.1}$$

For some value $y = y_0$, we prescribe the initial values of the unknown function and of the derivative with respect to y

$$u\left(x, y_0\right) = f\left(x\right), \qquad u_y\left(x, y_0\right) = g\left(x\right). \tag{5.1.2}$$

The problem of determining the solution of equation (5.1.1) satisfying the initial conditions (5.1.2) is known as the *initial-value problem*. For instance, the initial-value problem of a vibrating string is the problem of finding the solution of the wave equation

$$u_{tt} = c^2 u_{xx},$$

satisfying the initial conditions

$$u\left(x, t_0\right) = u_0\left(x\right), \qquad u_t\left(x, t_0\right) = v_0\left(x\right),$$

where $u_0\left(x\right)$ is the initial displacement and $v_0\left(x\right)$ is the initial velocity.

In initial-value problems, the initial values usually refer to the data assigned at $y = y_0$. It is not essential that these values be given along the line $y = y_0$; they may very well be prescribed along some curve L_0 in the xy plane. In such a context, the problem is called the *Cauchy problem* instead of the initial-value problem, although the two names are actually synonymous.

We consider the Euler equation

$$A u_{xx} + B u_{xy} + C u_{yy} = F\left(x, y, u, u_x, u_y\right), \tag{5.1.3}$$

where A, B, C are functions of x and y. Let (x_0, y_0) denote points on a smooth curve L_0 in the xy plane. Also let the parametric equations of this curve L_0 be

$$x_0 = x_0\left(\lambda\right), \qquad y_0 = y_0\left(\lambda\right), \tag{5.1.4}$$

where λ is a parameter.

We suppose that two functions $f\left(\lambda\right)$ and $g\left(\lambda\right)$ are prescribed along the curve L_0. The Cauchy problem is now one of determining the solution $u\left(x, y\right)$ of equation (5.1.3) in the neighborhood of the curve L_0 satisfying the Cauchy conditions

$$u = f\left(\lambda\right), \tag{5.1.5a}$$

$$\frac{\partial u}{\partial n} = g\left(\lambda\right), \tag{5.1.5b}$$

on the curve L_0 where n is the direction of the normal to L_0 which lies to the left of L_0 in the counterclockwise direction of increasing arc length. The function $f(\lambda)$ and $g(\lambda)$ are called the *Cauchy data*.

For every point on L_0, the value of u is specified by equation (5.1.5a). Thus, the curve L_0 represented by equation (5.1.4) with the condition (5.1.5a) yields a twisted curve L in (x, y, u) space whose projection on the xy plane is the curve L_0. Thus, the solution of the Cauchy problem is a surface, called an *integral surface*, in the (x, y, u) space passing through L and satisfying the condition (5.1.5b), which represents a tangent plane to the integral surface along L.

If the function $f(\lambda)$ is differentiable, then along the curve L_0, we have

$$\frac{du}{d\lambda} = \frac{\partial u}{\partial x}\frac{dx}{d\lambda} + \frac{\partial u}{\partial y}\frac{dy}{d\lambda} = \frac{df}{d\lambda}, \qquad (5.1.6)$$

and

$$\frac{\partial u}{\partial n} = \frac{\partial u}{\partial x}\frac{dx}{dn} + \frac{\partial u}{\partial y}\frac{dy}{dn} = g, \qquad (5.1.7)$$

but

$$\frac{dx}{dn} = -\frac{dy}{ds} \quad \text{and} \quad \frac{dy}{dn} = \frac{dx}{ds}. \qquad (5.1.8)$$

Equation (5.1.7) may be written as

$$\frac{\partial u}{\partial n} = -\frac{\partial u}{\partial x}\frac{dy}{ds} + \frac{\partial u}{\partial y}\frac{dx}{ds} = g. \qquad (5.1.9)$$

Since

$$\begin{vmatrix} \frac{dx}{d\lambda} & \frac{dy}{d\lambda} \\ -\frac{dy}{ds} & \frac{dx}{ds} \end{vmatrix} = \frac{(dx)^2 + (dy)^2}{ds\,d\lambda} \neq 0, \qquad (5.1.10)$$

it is possible to find u_x and u_y on L_0 from the system of equations (5.1.6) and (5.1.9). Since u_x and u_y are known on L_0, we find the higher derivatives by first differentiating u_x and u_y with respect to λ. Thus, we have

$$\frac{\partial^2 u}{\partial x^2}\frac{dx}{d\lambda} + \frac{\partial^2 u}{\partial x\,\partial y}\frac{dy}{d\lambda} = \frac{d}{d\lambda}\left(\frac{\partial u}{\partial x}\right), \qquad (5.1.11)$$

$$\frac{\partial^2 u}{\partial x\,\partial y}\frac{dx}{d\lambda} + \frac{\partial^2 u}{\partial y^2}\frac{dy}{d\lambda} = \frac{d}{d\lambda}\left(\frac{\partial u}{\partial y}\right). \qquad (5.1.12)$$

From equation (5.1.3), we have

$$A\frac{\partial^2 u}{\partial x^2} + B\frac{\partial^2 u}{\partial x\,\partial y} + C\frac{\partial^2 u}{\partial y^2} = F, \qquad (5.1.13)$$

where F is known since u_x and u_y have been found. The system of equations can be solved for u_{xx}, u_{xy}, and u_{yy}, if

$$\begin{vmatrix} \frac{dx}{d\lambda} & \frac{dy}{d\lambda} & 0 \\ 0 & \frac{dx}{d\lambda} & \frac{dy}{d\lambda} \\ A & B & C \end{vmatrix} = C\left(\frac{dx}{d\lambda}\right)^2 - B\left(\frac{dx}{d\lambda}\right)\left(\frac{dy}{d\lambda}\right) + A\left(\frac{dy}{d\lambda}\right)^2 \neq 0. \quad (5.1.14)$$

The equation

$$A\left(\frac{dy}{dx}\right)^2 - B\left(\frac{dy}{dx}\right) + C = 0, \quad (5.1.15)$$

is called the *characteristic equation*. It is then evident that the necessary condition for obtaining the second derivatives is that the curve L_0 must not be a characteristic curve.

If the coefficients of equation (5.1.3) and the function (5.1.5) are analytic, then all the derivatives of higher orders can be computed by the above process. The solution can then be represented in the form of a Taylor series:

$$u(x,y) = \sum_{n=0}^{\infty} \sum_{k=0}^{\infty} \frac{1}{k!\,(n-k)!} \frac{\partial^n u_0}{\partial x_0^k \partial y_0^{n-k}} (x-x_0)^k (y-y_0)^{n-k}, \quad (5.1.16)$$

which can be shown to converge in the neighborhood of the curve L_0. Thus, we may state the famous Cauchy–Kowalewskaya theorem.

5.2 The Cauchy–Kowalewskaya Theorem

Let the partial differential equation be given in the form

$$u_{yy} = F(y, x_1, x_2, \ldots, x_n, u, u_y, u_{x_1}, u_{x_2} \ldots, u_{x_n},$$
$$u_{x_1 y}, u_{x_2 y}, \ldots, u_{x_n y}, u_{x_1 x_1}, u_{x_2 x_2}, \ldots, u_{x_n x_n}), \quad (5.2.1)$$

and let the initial conditions

$$u = f(x_1, x_2, \ldots, x_n), \quad (5.2.2)$$
$$u_y = g(x_1, x_2, \ldots, x_n), \quad (5.2.3)$$

be given on the noncharacteristic manifold $y = y_0$.

If the function F is analytic in some neighborhood of the point $(y^0, x_1^0, x_2^0, \ldots, x_n^0, u^0, u_y^0, \ldots)$ and if the functions f and g are analytic in some neighborhood of the point $(x_1^0, x_2^0, \ldots, x_n^0)$, then the Cauchy problem has a unique analytic solution in some neighborhood of the point $(y^0, x_1^0, x_2^0, \ldots, x_n^0)$.

For the proof, see Petrovsky (1954).

The preceding statement seems equally applicable to hyperbolic, parabolic, or elliptic equations. However, we shall see that difficulties arise in formulating the Cauchy problem for nonhyperbolic equations. Consider, for instance, the famous Hadamard (1952) example.

The problem consists of the elliptic (or Laplace) equation

$$u_{xx} + u_{yy} = 0,$$

and the initial conditions on $y = 0$

$$u(x, 0) = 0, \qquad u_y(x, 0) = n^{-1} \sin nx.$$

The solution of this problem is

$$u(x, y) = n^{-2} \sinh ny \, \sin nx,$$

which can be easily verified.

It can be seen that, when n tends to infinity, the function $n^{-1} \sin nx$ tends uniformly to zero. But the solution $n^{-2} \sinh ny \, \sin nx$ does not become small, as n increases for any nonzero y. Physically, the solution represents an oscillation with unbounded amplitude $(n^{-2} \sinh ny)$ as $y \to \infty$ for any fixed x. Even if n is a fixed number, this solution is unstable in the sense that $u \to \infty$ as $y \to \infty$ for any fixed x for which $\sin nx \neq 0$. It is obvious then that the solution does not depend continuously on the data. Thus, it is not a properly posed problem.

In addition to existence and uniqueness, the question of continuous dependence of the solution on the initial data arises in connection with the Cauchy–Kowalewskaya theorem. It is well known that any continuous function can accurately be approximated by polynomials. We can apply the Cauchy–Kowalewskaya theorem with continuous data by using polynomial approximations only if a small variation in the initial data leads to a small change in the solution.

5.3 Homogeneous Wave Equations

To study Cauchy problems for hyperbolic partial differential equations, it is quite natural to begin investigating the simplest and yet most important equation, the one-dimensional wave equation, by the method of characteristics. The essential characteristic of the solution of the general wave equation is preserved in this simplified case.

We shall consider the following Cauchy problem of an infinite string with the initial condition

$$u_{tt} - c^2 u_{xx} = 0, \qquad x \in \mathbb{R}, \quad t > 0, \tag{5.3.1}$$
$$u(x, 0) = f(x), \qquad x \in \mathbb{R}, \tag{5.3.2}$$
$$u_t(x, 0) = g(x), \qquad x \in \mathbb{R}. \tag{5.3.3}$$

By the method of characteristics described in Chapter 4, the characteristic equation according to equation (4.2.4) is

$$dx^2 - c^2 dt^2 = 0,$$

which reduces to

$$dx + c\, dt = 0, \qquad dx - c\, dt = 0.$$

The integrals are the straight lines

$$x + ct = c_1, \qquad x - ct = c_2.$$

Introducing the characteristic coordinates

$$\xi = x + ct, \qquad \eta = x - ct,$$

we obtain

$$u_{xx} = u_{\xi\xi} + 2\, u_{\xi\eta} + u_{\eta\eta}, \qquad u_{tt} = c^2 \left(u_{\xi\xi} - 2\, u_{\xi\eta} + u_{\eta\eta} \right).$$

Substitution of these in equation (5.3.1) yields

$$-4c^2 u_{\xi\eta} = 0.$$

Since $c \neq 0$, we have

$$u_{\xi\eta} = 0.$$

Integrating with respect to ξ, we obtain

$$u_\eta = \psi^*(\eta),$$

where $\psi^*(\eta)$ is an arbitrary function of η. Integrating again with respect to η, we obtain

$$u(\xi, \eta) = \int \psi^*(\eta)\, d\eta + \phi(\xi).$$

If we set $\psi(\eta) = \int \psi^*(\eta)\, d\eta$, we have

$$u(\xi, \eta) = \phi(\xi) + \psi(\eta),$$

where ϕ and ψ are arbitrary functions. Transforming to the original variables x and t, we find the general solution of the wave equation

$$u\left(x,t\right) = \phi\left(x+ct\right) + \psi\left(x-ct\right), \qquad (5.3.4)$$

provided ϕ and ψ are twice differentiable functions.

Now applying the initial conditions (5.3.2) and (5.3.3), we obtain

$$u\left(x,0\right) = f\left(x\right) = \phi\left(x\right) + \psi\left(x\right), \qquad (5.3.5)$$
$$u_t\left(x,0\right) = g\left(x\right) = c\,\phi'\left(x\right) - c\,\psi'\left(x\right). \qquad (5.3.6)$$

Integration of equation (5.3.6) gives

$$\phi\left(x\right) - \psi\left(x\right) = \frac{1}{c}\int_{x_0}^{x} g\left(\tau\right) d\tau + K, \qquad (5.3.7)$$

where x_0 and K are arbitrary constants. Solving for ϕ and ψ from equations (5.3.5) and (5.3.7), we obtain

$$\phi\left(x\right) = \frac{1}{2}f\left(x\right) + \frac{1}{2c}\int_{x_0}^{x} g\left(\tau\right) d\tau + \frac{K}{2},$$
$$\psi\left(x\right) = \frac{1}{2}f\left(x\right) - \frac{1}{2c}\int_{x_0}^{x} g\left(\tau\right) d\tau - \frac{K}{2}.$$

The solution is thus given by

$$
\begin{aligned}
u\left(x,t\right) &= \frac{1}{2}\left[f\left(x+ct\right) + f\left(x-ct\right)\right] + \frac{1}{2c}\left[\int_{x_0}^{x+ct} g\left(\tau\right) d\tau - \int_{x_0}^{x-ct} g\left(\tau\right) d\tau\right] \\
&= \frac{1}{2}\left[f\left(x+ct\right) + f\left(x-ct\right)\right] + \frac{1}{2c}\int_{x-ct}^{x+ct} g\left(\tau\right) d\tau. \qquad (5.3.8)
\end{aligned}
$$

This is called the celebrated *d'Alembert solution* of the Cauchy problem for the one-dimensional wave equation.

It is easy to verify by direct substitution that $u\left(x,t\right)$, represented by (5.3.8), is the unique solution of the wave equation (5.3.1) provided $f\left(x\right)$ is twice continuously differentiable and $g\left(x\right)$ is continuously differentiable. This essentially proves the existence of the d'Alembert solution. By direct substitution, it can also be shown that the solution (5.3.8) is uniquely determined by the initial conditions (5.3.2) and (5.3.3). It is important to note that the solution $u\left(x,t\right)$ depends only on the initial values of f at points $x-ct$ and $x+ct$ and values of g between these two points. In other words, the solution does not depend at all on initial values outside this interval, $x-ct \leq x \leq x+ct$. This interval is called the *domain of dependence* of the variables $\left(x,t\right)$.

Moreover, the solution depends continuously on the initial data, that is, the problem is well posed. In other words, a small change in either f or g results in a correspondingly small change in the solution $u\left(x,t\right)$. Mathematically, this can be stated as follows:

For every $\varepsilon > 0$ and for each time interval $0 \leq t \leq t_0$, there exists a number $\delta\left(\varepsilon, t_0\right)$ such that

$$|u\left(x,t\right) - u^*\left(x,t\right)| < \varepsilon,$$

whenever

$$|f\left(x\right) - f^*\left(x\right)| < \delta, \qquad |g\left(x\right) - g^*\left(x\right)| < \delta.$$

The proof follows immediately from equation (5.3.8). We have

$$
\begin{aligned}
|u\left(x,t\right) - u^*\left(x,t\right)| \leq &\frac{1}{2} |f\left(x + ct\right) - f^*\left(x + ct\right)| \\
&+ \frac{1}{2} |f\left(x - ct\right) - f^*\left(x - ct\right)| \\
&+ \frac{1}{2c} \int_{x-ct}^{x+ct} |g\left(\tau\right) - g^*\left(\tau\right)| \, d\tau < \varepsilon,
\end{aligned}
$$

where $\varepsilon = \delta\left(1 + t_0\right)$.

For any finite time interval $0 < t < t_0$, a small change in the initial data only produces a small change in the solution. This shows that the problem is well posed.

Example 5.3.1. Find the solution of the initial-value problem

$$
\begin{aligned}
u_{tt} &= c^2 u_{xx}, \quad x \in \mathbb{R}, \quad t > 0, \\
u\left(x,0\right) &= \sin x, \quad u_t\left(x,0\right) = \cos x.
\end{aligned}
$$

From (5.3.8), we have

$$
\begin{aligned}
u\left(x,t\right) &= \frac{1}{2} \left[\sin\left(x + ct\right) + \sin\left(x - ct\right)\right] + \frac{1}{2c} \int_{x-ct}^{x+ct} \cos \tau \, d\tau \\
&= \sin x \cos ct + \frac{1}{2c} \left[\sin\left(x + ct\right) - \sin\left(x - ct\right)\right] \\
&= \sin x \cos ct + \frac{1}{c} \cos x \sin ct.
\end{aligned}
$$

It follows from the d'Alembert solution that, if an initial displacement or an initial velocity is located in a small neighborhood of some point (x_0, t_0), it can influence only the area $t > t_0$ bounded by two characteristics $x - ct = $ constant and $x + ct = $ constant with slope $\pm(1/c)$ passing through the point (x_0, t_0), as shown in Figure 5.3.1. This means that the initial displacement propagates with the speed $\frac{dx}{dt} = c$, whereas the effect of the initial velocity propagates at all speeds up to c. This infinite sector R in this figure is called the *range of influence* of the point (x_0, t_0).

According to (5.3.8), the value of $u\left(x_0, t_0\right)$ depends on the initial data f and g in the interval $[x_0 - ct_0, x_0 + ct_0]$ which is cut out of the initial line by the two characteristics $x - ct = $ constant and $x + ct = $ constant with slope $\pm(1/c)$ passing through the point (x_0, t_0). The interval $[x_0 - ct_0, x_0 + ct_0]$

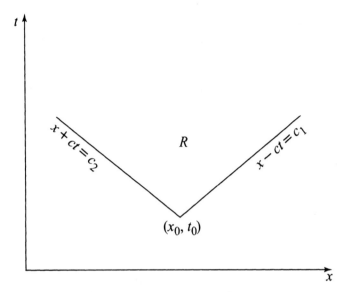

Figure 5.3.1 Range of influence

on the line $t = 0$ is called the *domain of dependence* of the solution at the point (x_0, t_0), as shown in Figure 5.3.2.

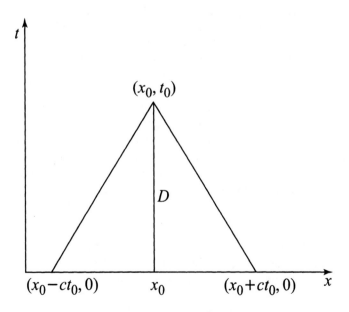

Figure 5.3.2 Domain of dependence

Since the solution $u(x, t)$ at every point (x, t) inside the triangular region D in this figure is completely determined by the Cauchy data on the interval $[x_0 - ct_0, x_0 + ct_0]$, the region D is called the *region of determinancy* of the solution.

We will now investigate the physical significance of the d'Alembert solution (5.3.8) in greater detail. We rewrite the solution in the form

$$u(x, t) = \frac{1}{2}f(x + ct) + \frac{1}{2c}\int_0^{x+ct} g(\tau)\, d\tau + \frac{1}{2}f(x - ct) - \frac{1}{2c}\int_0^{x-ct} g(\tau)\, d\tau.$$

$$(5.3.9)$$

Or, equivalently,

$$u(x, t) = \phi(x + ct) + \psi(x - ct), \tag{5.3.10}$$

where

$$\phi(\xi) = \frac{1}{2}f(\xi) + \frac{1}{2c}\int_0^{\xi} g(\tau)\, d\tau, \tag{5.3.11}$$

$$\psi(\eta) = \frac{1}{2}f(\eta) - \frac{1}{2c}\int_0^{\eta} g(\tau)\, d\tau. \tag{5.3.12}$$

Evidently, $\phi(x + ct)$ represents a progressive wave traveling in the negative x-direction with speed c without change of shape. Similarly, $\psi(x - ct)$ is also a progressive wave propagating in the positive x-direction with the same speed c without change of shape. We shall examine this point in greater detail. Treat $\psi(x - ct)$ as a function of x for a sequence of times t. At $t = 0$, the shape of this function of $u = \psi(x)$. At a subsequent time, its shape is given by $u = \psi(x - ct)$ or $u = \psi(\xi)$, where $\xi = x - ct$ is the new coordinate obtained by translating the origin a distance ct to the right. Thus, the shape of the curve remains the same as time progresses, but moves to the right with velocity c as shown in Figure 5.3.3. This shows that $\psi(x - ct)$ represents a progressive wave traveling in the positive x-direction with velocity c without change of shape. Similarly, $\phi(x + ct)$ is also a progressive wave propagating in the negative x-direction with the same speed c without change of shape. For instance,

$$u(x, t) = \sin(x \mp ct) \tag{5.3.13}$$

represent sinusoidal waves traveling with speed c in the positive and negative directions respectively without change of shape. The propagation of waves without change of shape is common to all linear wave equations.

To interpret the d'Alembert formula we consider two cases:

Case 1. We first consider the case when the initial velocity is zero, that is,

$$g(x) = 0.$$

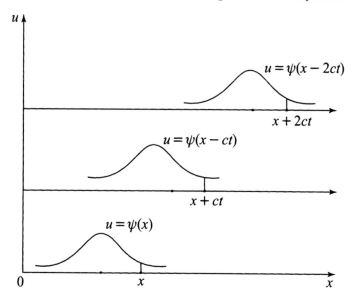

Figure 5.3.3 Progressive Waves.

Then, the d'Alembert solution has the form

$$u\left(x,t\right) = \frac{1}{2}\left[f\left(x+ct\right) + f\left(x-ct\right)\right].$$

Now suppose that the initial displacement $f\left(x\right)$ is different from zero in an interval $(-b, b)$. Then, in this case the forward and the backward waves are represented by

$$u = \frac{1}{2}f\left(x\right).$$

The waves are initially superimposed, and then they separate and travel in opposite directions.

We consider $f\left(x\right)$ which has the form of a triangle. We draw a triangle with the ordinate $x = 0$ one-half that of the given function at that point, as shown in Figure 5.3.4. If we displace these graphs and then take the sum of the ordinates of the displaced graphs, we obtain the shape of the string at any time t.

As can be seen from the figure, the waves travel in opposite directions away from each other. After both waves have passed the region of initial disturbance, the string returns to its rest position.

Case 2. We consider the case when the initial displacement is zero, that is,

$$f\left(x\right) = 0,$$

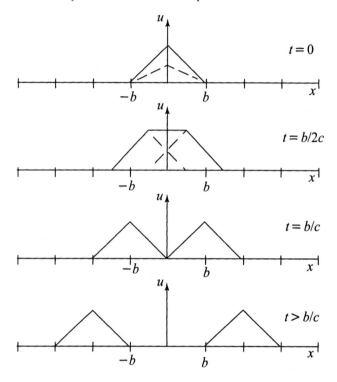

Figure 5.3.4 Triangular Waves.

and the d'Alembert solution assumes the form

$$u\left(x,t\right) = \frac{1}{2} \int_{x-ct}^{x+ct} g\left(\tau\right) d\tau = \frac{1}{2} \left[G\left(x+ct\right) - G\left(x-ct\right)\right],$$

where

$$G\left(x\right) = \frac{1}{c} \int_{x_0}^{x} g\left(\tau\right) d\tau.$$

If we take for the initial velocity

$$g\left(x\right) = \begin{cases} 0 & |x| > b \\ g_0 & |x| \leq b, \end{cases}$$

then, the function $G\left(x\right)$ is equal to zero for values of x in the interval $x \leq -b$, and

$$G\left(x\right) = \begin{cases} \dfrac{1}{c} \displaystyle\int_{-b}^{x} g_0 \, d\tau = \dfrac{g_0}{c}\left(x+b\right) & \text{for } -b \leq x \leq b, \\[4mm] \dfrac{1}{c} \displaystyle\int_{-b}^{x} g_0 \, d\tau = \dfrac{2bg_0}{c} & \text{for } \quad x > b. \end{cases}$$

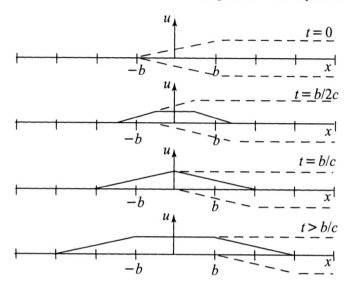

Figure 5.3.5 Graph of $u(x, t)$ at time t.

As in the previous case, the two waves which differ in sign travel in opposite directions on the x-axis. After some time t the two functions $(1/2)\,G(x)$ and $-(1/2)\,G(x)$ move a distance ct. Thus, the graph of u at time t is obtained by summing the ordinates of the displaced graphs as shown in Figure 5.3.5. As t approaches infinity, the string will reach a state of rest, but it will not, in general, assume its original position. This displacement is known as the *residual displacement*.

In the preceding examples, we note that $f(x)$ is continuous, but not continuously differentiable and $g(x)$ is discontinuous. To these initial data, there corresponds a generalized solution. By a generalized solution we mean the following:

Let us suppose that the function $u(x, t)$ satisfies the initial conditions (5.3.2) and (5.3.3). Let $u(x, t)$ be the limit of a uniformly convergent sequence of solutions $u_n(x, t)$ which satisfy the wave equation (5.3.1) and the initial conditions

$$u_n(x, 0) = f_n(x), \qquad \left(\frac{\partial u_n}{\partial t}\right)(x, 0) = g_n(x).$$

Let $f_n(x)$ be a continuously differentiable function, and let the sequence converge uniformly to $f(x)$; let $g_n(x)$ be a continuously differentiable function, and $\int_{x_0}^{x} g_n(\tau)\,d\tau$ approach uniformly to $\int_{x_0}^{x} g(\tau)\,d\tau$. Then, the function $u(x, t)$ is called the *generalized solution* of the problem (5.3.1)–(5.3.3).

In general, it is interesting to discuss the effect of discontinuity of the function $f(x)$ at a point $x = x_0$, assuming that $g(x)$ is a smooth function. Clearly, it follows from (5.3.8) that $u(x, t)$ will be discontinuous at each point (x, t) such that $x + ct = x_0$ or $x - ct = x_0$, that is, at each point of the two characteristic lines intersecting at the point $(x_0, 0)$. This means that discontinuities are propagated along the characteristic lines. At each point of the characteristic lines, the partial derivatives of the function $u(x, t)$ fail to exist, and hence, u can no longer be a solution of the Cauchy problem in the usual sense. However, such a function may be called a *generalized solution* of the Cauchy problem. Similarly, if $f(x)$ is continuous, but either $f'(x)$ or $f''(x)$ has a discontinuity at some point $x = x_0$, the first- or second-order partial derivatives of the solution $u(x, t)$ will be discontinuous along the characteristic lines through $(x_0, 0)$. Finally, a discontinuity in $g(x)$ at $x = x_0$ would lead to a discontinuity in the first- or second-order partial derivatives of u along the characteristic lines through $(x_0, 0)$, and a discontinuity in $g'(x)$ at x_0 will imply a discontinuity in the second-order partial derivatives of u along the characteristic lines through $(x_0, 0)$. The solution given by (5.3.8) with f, f', f'', g, and g' piecewise continuous on $-\infty < x < \infty$ is usually called the *generalized solution* of the Cauchy problem.

5.4 Initial Boundary-Value Problems

We have just determined the solution of the initial-value problem for the infinite vibrating string. We will now study the effect of a boundary on the solution.

(A) Semi-infinite String with a Fixed End

Let us first consider a semi-infinite vibrating string with a fixed end, that is,

$$
\begin{aligned}
u_{tt} &= c^2 u_{xx}, & 0 < x < \infty, && t > 0, \\
u(x, 0) &= f(x), & 0 \le x < \infty, \\
u_t(x, 0) &= g(x), & 0 \le x < \infty, \\
u(0, t) &= 0, & 0 \le t < \infty.
\end{aligned}
\tag{5.4.1}
$$

It is evident here that the boundary condition at $x = 0$ produces a wave moving to the right with the velocity c. Thus, for $x > ct$, the solution is the same as that of the infinite string, and the displacement is influenced only by the initial data on the interval $[x - ct, \, x + ct]$, as shown in Figure 5.4.1.

When $x < ct$, the interval $[x - ct, \, x + ct]$ extends onto the negative x-axis where f and g are not prescribed.

But from the d'Alembert formula

$$
u(x, t) = \phi(x + ct) + \psi(x - ct),
\tag{5.4.2}
$$

where

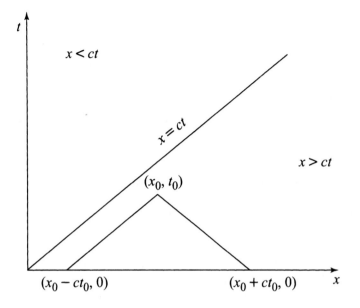

Figure 5.4.1 Displacement influenced by the initial data on $[x - ct, \; x + ct]$.

$$\phi\left(\xi\right) = \frac{1}{2}f\left(\xi\right) + \frac{1}{2c}\int_0^\xi g\left(\tau\right)d\tau + \frac{K}{2}, \tag{5.4.3}$$

$$\psi\left(\eta\right) = \frac{1}{2}f\left(\eta\right) - \frac{1}{2c}\int_0^\eta g\left(\tau\right)d\tau - \frac{K}{2}, \tag{5.4.4}$$

we see that

$$u\left(0,t\right) = \phi\left(ct\right) + \psi\left(-ct\right) = 0.$$

Hence,

$$\psi\left(-ct\right) = -\phi\left(ct\right).$$

If we let $\alpha = -ct$, then

$$\psi\left(\alpha\right) = -\phi\left(-\alpha\right).$$

Replacing α by $x - ct$, we obtain for $x < ct$,

$$\psi\left(x - ct\right) = -\phi\left(ct - x\right),$$

and hence,

$$\psi\left(x - ct\right) = -\frac{1}{2}f\left(ct - x\right) - \frac{1}{2c}\int_0^{ct-x} g\left(\tau\right)d\tau - \frac{K}{2}.$$

The solution of the initial boundary-value problem, therefore, is given by

$$u\left(x,t\right) = \frac{1}{2}\left[f\left(x + ct\right) + f\left(x - ct\right)\right] + \frac{1}{2c}\int_{x-ct}^{x+ct} g\left(\tau\right) d\tau \quad \text{for } x > ct, \text{ (5.4.5)}$$

$$u\left(x,t\right) = \frac{1}{2}\left[f\left(x + ct\right) - f\left(ct - x\right)\right] + \frac{1}{2c}\int_{ct-x}^{x+ct} g\left(\tau\right) d\tau \quad \text{for } x < ct. \text{ (5.4.6)}$$

In order for this solution to exist, f must be twice continuously differentiable and g must be continuously differentiable, and in addition

$$f\left(0\right) = f''\left(0\right) = g\left(0\right) = 0.$$

Solution (5.4.6) has an interesting physical interpretation. If we draw the characteristics through the point (x_0, t_0) in the region $x > ct$, we see, as pointed out earlier, that the displacement at (x_0, t_0) is determined by the initial values on $[x_0 - ct_0, x_0 + ct_0]$.

If the point (x_0, t_0) lies in the region $x > ct$ as shown in Figure 5.4.1, we see that the characteristic $x + ct = x_0 + ct_0$ intersects the x-axis at $(x_0 + ct_0, 0)$. However, the characteristic $x - ct = x_0 - ct_0$ intersects the t-axis at $(0, t_0 - x_0/c)$, and the characteristic $x + ct = ct_0 - x_0$ intersects the x-axis at $(ct_0 - x_0, 0)$. Thus, the disturbance at $(ct_0 - x_0, 0)$ travels along the backward characteristic $x + ct = ct_0 - x_0$, and is reflected at $(0, t_0 - x_0/c)$ as a forward moving wave represented by $-\phi\left(ct_0 - x_0\right)$.

Example 5.4.1. Determine the solution of the initial boundary-value problem

$$\begin{aligned} u_{tt} &= 4\,u_{xx}, & x &> 0, & t &> 0, \\ u\left(x,0\right) &= |\sin x|, & x &> 0, \\ u_t\left(x,0\right) &= 0, & x &\geq 0, \\ u\left(x,0\right) &= 0, & t &\geq 0. \end{aligned}$$

For $x > 2t$,

$$\begin{aligned} u\left(x,t\right) &= \frac{1}{2}\left[f\left(x + 2t\right) + f\left(x - 2t\right)\right] \\ &= \frac{1}{2}\left[|\sin\left(x + 2t\right)| - |\sin\left(x - 2t\right)|\right], \end{aligned}$$

and for $x < 2t$,

$$\begin{aligned} u\left(x,t\right) &= \frac{1}{2}\left[f\left(x + 2t\right) - f\left(2t - x\right)\right] \\ &= \frac{1}{2}\left[|\sin\left(x + 2t\right)| - |\sin\left(2t - x\right)|\right]. \end{aligned}$$

Notice that $u\left(0,t\right) = 0$ is satisfied by $u\left(x,t\right)$ for $x < 2t$ (that is, $t > 0$).

(B) Semi-infinite String with a Free End

We consider a semi-infinite string with a free end at $x = 0$. We will determine the solution of

$$u_{tt} = c^2 u_{xx}, \qquad 0 < x < \infty, \qquad t > 0,$$
$$u(x,0) = f(x), \qquad 0 \le x < \infty, \tag{5.4.7}$$
$$u_t(x,0) = g(x), \qquad 0 \le x < \infty,$$
$$u_x(0,t) = 0, \qquad 0 \le t < \infty.$$

As in the case of the fixed end, for $x > ct$ the solution is the same as that of the infinite string. For $x < ct$, from the d'Alembert solution (5.4.2)

$$u(x,t) = \phi(x+ct) + \psi(x-ct),$$

we have

$$u_x(x,t) = \phi'(x+ct) + \psi'(x-ct).$$

Thus,

$$u_x(0,t) = \phi'(ct) + \psi'(-ct) = 0.$$

Integration yields

$$\phi(ct) - \psi(-ct) = K,$$

where K is a constant. Now, if we let $\alpha = -ct$, we obtain

$$\psi(\alpha) = \phi(-\alpha) - K.$$

Replacing α by $x - ct$, we have

$$\psi(x-ct) = \phi(ct-x) - K,$$

and hence,

$$\psi(x-ct) = \frac{1}{2} f(ct-x) + \frac{1}{2c} \int_0^{ct-x} g(\tau)\, d\tau - \frac{K}{2}.$$

The solution of the initial boundary-value problem, therefore, is given by

$$u(x,t) = \frac{1}{2}\left[f(x+ct) + f(x-ct)\right] + \frac{1}{2c} \int_{x-ct}^{x+ct} g(\tau)\, d\tau \quad \text{for } x > ct. \tag{5.4.8}$$

$$u(x,t) = \frac{1}{2}\left[f(x+ct) + f(ct-x)\right] + \frac{1}{2c}\left[\int_0^{x+ct} g(\tau)\, d\tau + \int_0^{ct-x} g(\tau)\, d\tau\right]$$
$$\text{for } x < ct. \tag{5.4.9}$$

We note that for this solution to exist, f must be twice continuously differentiable and g must be continuously differentiable, and in addition,

$$f'(0) = g'(0) = 0.$$

Example 5.4.2. Find the solution of the initial boundary-value problem

$$u_{tt} = u_{xx}, \qquad 0 < x < \infty, \qquad t > 0,$$
$$u(x,0) = \cos\left(\frac{\pi x}{2}\right), \qquad 0 \le x < \infty,$$
$$u_t(x,0) = 0, \qquad 0 \le x < \infty,$$
$$u_x(x,0) = 0, \qquad t \ge 0.$$

For $x > t$

$$u(x,t) = \frac{1}{2}\left[\cos\frac{\pi}{2}(x+t) + \cos\frac{\pi}{2}(x-t)\right]$$
$$= \cos\left(\frac{\pi}{2}x\right)\cos\left(\frac{\pi}{2}t\right),$$

and for $x < t$

$$u(x,t) = \frac{1}{2}\left[\cos\frac{\pi}{2}(x+t) + \cos\frac{\pi}{2}(t-x)\right]$$
$$= \cos\left(\frac{\pi}{2}x\right)\cos\left(\frac{\pi}{2}t\right).$$

5.5 Equations with Nonhomogeneous Boundary Conditions

In the case of the initial boundary-value problems with nonhomogeneous boundary conditions, such as

$$u_{tt} = c^2 u_{xx}, \qquad x > 0, \qquad t > 0,$$
$$u(x,0) = f(x), \qquad x \ge 0, \qquad\qquad (5.5.1)$$
$$u_t(x,0) = g(x), \qquad x \ge 0,$$
$$u(0,t) = p(t), \qquad t \ge 0,$$

we proceed in a manner similar to the case of homogeneous boundary conditions. Using equation (5.4.2), we apply the boundary condition to obtain

$$u(0,t) = \phi(ct) + \psi(-ct) = p(t).$$

If we let $\alpha = -ct$, we have

$$\psi(\alpha) = p\left(-\frac{\alpha}{c}\right) - \phi(-\alpha).$$

Replacing α by $x - ct$, the preceding relation becomes

$$\psi(x - ct) = p\left(t - \frac{x}{c}\right) - \phi(ct - x).$$

Thus, for $0 \leq x < ct$,

$$u(x,t) = p\left(t - \frac{x}{c}\right) + \frac{1}{2}\left[f(x+ct) - f(ct-x)\right] + \frac{1}{2c}\int_{ct-x}^{x+ct} g(\tau)\,d\tau$$

$$= p\left(t - \frac{x}{c}\right) + \phi(x+ct) - \psi(ct-x), \qquad (5.5.2)$$

where $\phi(x+ct = \xi)$ is given by (5.3.11), and $\psi(\eta)$ is given by

$$\psi(\eta) = \frac{1}{2}f(\eta) + \frac{1}{2c}\int_0^\eta g(\tau)\,d\tau. \qquad (5.5.3)$$

The solution for $x > ct$ is given by the solution (5.4.5) of the infinite string.

In this case, in addition to the differentiability conditions satisfied by f and g, as in the case of the problem with the homogeneous boundary conditions, p must be twice continuously differentiable in t and

$$p(0) = f(0), \qquad p'(0) = g(0), \qquad p''(0) = c^2 f''(0).$$

We next consider the initial boundary-value problem

$$\begin{aligned}
u_{tt} &= c^2 u_{xx}, & x &> 0, & t &> 0, \\
u(x,0) &= f(x), & x &\geq 0, \\
u_t(x,0) &= g(x), & x &\geq 0, \\
u_x(0,t) &= q(t), & t &\geq 0.
\end{aligned}$$

Using (5.4.2), we apply the boundary condition to obtain

$$u_x(0,t) = \phi'(ct) + \psi'(-ct) = q(t).$$

Then, integrating yields

$$\phi(ct) - \psi(-ct) = c\int_0^t q(\tau)\,d\tau + K.$$

If we let $\alpha = -ct$, then

$$\psi(\alpha) = \phi(-\alpha) - c\int_0^{-\alpha/c} q(\tau)\,d\tau - K.$$

Replacing α by $x - ct$, we obtain

$$\psi(x-ct) = \phi(ct-x) - c\int_0^{t-x/c} q(\tau)\,d\tau - K.$$

The solution of the initial boundary-value problem for $x < ct$, therefore, is given by

$$u\left(x,t\right) = \frac{1}{2}\left[f\left(x+ct\right) + f\left(ct-x\right)\right] + \frac{1}{2c}\left[\int_{0}^{x+ct} g\left(\tau\right)d\tau + \int_{0}^{ct-x} g\left(\tau\right)d\tau\right]$$

$$-c\int_{0}^{t-x/c} q\left(\tau\right)d\tau. \quad (5.5.4)$$

Here f and g must satisfy the differentiability conditions, as in the case of the problem with the homogeneous boundary conditions. In addition

$$f'\left(0\right) = q\left(0\right), \qquad g'\left(0\right) = q'\left(0\right).$$

The solution for the initial boundary-value problem involving the boundary condition

$$u_x\left(0,t\right) + h\,u\left(0,t\right) = 0, \qquad h = \text{constant}$$

can also be constructed in a similar manner from the d'Alembert solution.

5.6 Vibration of Finite String with Fixed Ends

The problem of the finite string is more complicated than that of the infinite string due to the repeated reflection of waves from the boundaries

We first consider the vibration of the string of length l fixed at both ends. The problem is that of finding the solution of

$$\begin{aligned}
u_{tt} &= c^2 u_{xx}, & 0 < x < l, & \qquad t > 0, \\
u\left(x,0\right) &= f\left(x\right), & 0 \le x \le l, & \\
u_t\left(x,0\right) &= g\left(x\right), & 0 \le x \le l, & \qquad (5.6.1) \\
u\left(0,t\right) &= 0, & u\left(l,t\right) = 0, & \qquad t \ge 0,
\end{aligned}$$

From the previous results, we know that the solution of the wave equation is

$$u\left(x,t\right) = \phi\left(x+ct\right) + \psi\left(x-ct\right).$$

Applying the initial conditions, we have

$$\begin{aligned}
u\left(x,0\right) &= \phi\left(x\right) + \psi\left(x\right) = f\left(x\right), & 0 \le x \le l, \\
u_t\left(x,0\right) &= c\,\phi'\left(x\right) - c\,\psi'\left(x\right) = g\left(x\right), & 0 \le x \le l.
\end{aligned}$$

Solving for ϕ and ψ, we find

$$\phi\left(\xi\right) = \frac{1}{2}f\left(\xi\right) + \frac{1}{2c}\int_{0}^{\xi} g\left(\tau\right)d\tau + \frac{K}{2}, \qquad 0 \le \xi \le l, \qquad (5.6.2)$$

$$\psi\left(\eta\right) = \frac{1}{2}f\left(\eta\right) - \frac{1}{2c}\int_{0}^{\eta} g\left(\tau\right)d\tau - \frac{K}{2}, \qquad 0 \le \eta \le l. \qquad (5.6.3)$$

Hence,

$$u\left(x,t\right) = \frac{1}{2}\left[f\left(x+ct\right) + f\left(x-ct\right)\right] + \frac{1}{2c}\int_{x-ct}^{x+ct} g\left(\tau\right)d\tau, \quad (5.6.4)$$

for $0 \le x + ct \le l$ and $0 \le x - ct \le l$. The solution is thus uniquely determined by the initial data in the region

$$t \le \frac{x}{c}, \qquad t \le \frac{l-x}{c}, \qquad t \ge 0.$$

For larger times, the solution depends on the boundary conditions. Applying the boundary conditions, we obtain

$$u\left(0,t\right) = \phi\left(ct\right) + \psi\left(-ct\right) = 0, \qquad t \ge 0, \qquad (5.6.5)$$
$$u\left(l,t\right) = \phi\left(l+ct\right) + \psi\left(l-ct\right) = 0, \qquad t \ge 0. \qquad (5.6.6)$$

If we set $\alpha = -ct$, equation (5.6.5) becomes

$$\psi\left(\alpha\right) = -\phi\left(-\alpha\right), \qquad \alpha \le 0, \qquad (5.6.7)$$

and if we set $\alpha = l + ct$, equation (5.6.6) takes the form

$$\phi\left(\alpha\right) = -\psi\left(2l-\alpha\right), \qquad \alpha \ge l. \qquad (5.6.8)$$

With $\xi = -\eta$, we may write equation (5.6.2) as

$$\phi\left(-\eta\right) = \frac{1}{2}f\left(-\eta\right) + \frac{1}{2c}\int_0^{-\eta} g\left(\tau\right)d\tau + \frac{K}{2}, \qquad 0 \le -\eta \le l. \quad (5.6.9)$$

Thus, from (5.6.7) and (5.6.9), we have

$$\psi\left(\eta\right) = -\frac{1}{2}f\left(-\eta\right) - \frac{1}{2c}\int_0^{-\eta} g\left(\tau\right)d\tau - \frac{K}{2}, \qquad -l \le \eta \le 0. \quad (5.6.10)$$

We see that the range of $\psi\left(\eta\right)$ is extended to $-l \le \eta \le l$.

If we put $\alpha = \xi$ in equation (5.6.8), we obtain

$$\phi\left(\xi\right) = -\psi\left(2l-\xi\right), \qquad \xi \ge l. \qquad (5.6.11)$$

Then, by putting $\eta = 2l - \xi$ in equation (5.6.3), we obtain

$$\psi\left(2l-\xi\right) = \frac{1}{2}f\left(2l-\xi\right) - \frac{1}{2c}\int_0^{2l-\xi} g\left(\tau\right)d\tau - \frac{K}{2}, \qquad 0 \le 2l - \xi \le l.$$
$$(5.6.12)$$

Substitution of this in equation (5.6.11) yields

$$\phi\left(\xi\right) = -\frac{1}{2}f\left(2l - \xi\right) + \frac{1}{2c}\int_0^{2l-\xi} g\left(\tau\right) d\tau + \frac{K}{2}, \quad l \le \xi \le 2l. \quad (5.6.13)$$

The range of $\phi\left(\xi\right)$ is thus extended to $0 \le \xi \le 2l$. Continuing in this manner, we obtain $\phi\left(\xi\right)$ for all $\xi \ge 0$ and $\psi\left(\eta\right)$ for all $\eta \le l$. Hence, the solution is determined for all $0 \le x \le l$ and $t \ge 0$.

In order to observe the effect of the boundaries on the propagation of waves, the characteristics are drawn through the end point until they meet the boundaries and then continue inward as shown in Figure 5.6.1. It can be seen from the figure that only direct waves propagate in region 1. In regions 2 and 3, both direct and reflected waves propagate. In regions, 4,5,6, ... , several waves propagate along the characteristics reflected from both of the boundaries $x = 0$ and $x = l$.

Example 5.6.1. Determine the solution of the following problem

$$
\begin{aligned}
u_{tt} &= c^2 u_{xx}, & 0 < x < l, \quad t > 0, \\
u\left(x,0\right) &= \sin\left(\pi x/l\right), & 0 \le x \le l, \\
u_t\left(x,0\right) &= 0, & 0 \le x \le l, \\
u\left(0,t\right) &= 0, \quad u\left(l,t\right) = 0, & t \ge 0.
\end{aligned}
$$

From equations (5.6.2) and (5.6.3), we have

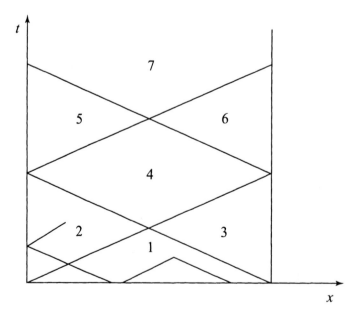

Figure 5.6.1 Regions of wave propagation.

$$\phi(\xi) = \frac{1}{2}\sin\left(\frac{\pi\xi}{l}\right) + \frac{K}{2}, \qquad 0 \le \xi \le l.$$

$$\psi(\eta) = \frac{1}{2}\sin\left(\frac{\pi\eta}{l}\right) - \frac{K}{2}, \qquad 0 \le \eta \le l.$$

Using equation (5.6.10), we obtain

$$\psi(\eta) = -\frac{1}{2}\sin\left(-\frac{\pi\eta}{l}\right) - \frac{K}{2}, \qquad -l \le \eta \le 0$$

$$= \frac{1}{2}\sin\left(\frac{\pi\eta}{l}\right) - \frac{K}{2}.$$

From equation (5.6.13), we find

$$\phi(\xi) = -\frac{1}{2}\sin\left\{\frac{\pi}{l}(2l - \xi)\right\} + \frac{K}{2}, \qquad l \le \xi \le 2l.$$

Again by equation (5.6.7) and from the preceding $\phi(\xi)$, we have

$$\phi(\eta) = \frac{1}{2}\sin\left(\frac{\pi\eta}{l}\right) - \frac{K}{2}, \qquad -2l \le \eta \le -l.$$

Proceeding in this manner, we determine the solution

$$u(x,t) = \phi(\xi) + \psi(\eta)$$
$$= \frac{1}{2}\left[\sin\frac{\pi}{l}(x + ct) + \sin\frac{\pi}{l}(x - ct)\right]$$

for all x in $(0, l)$ and for all $t > 0$.

Similarly, the solution of the finite initial boundary-value problem

$$\begin{aligned}
u_{tt} &= c^2 u_{xx}, & 0 < x < l, & \qquad t > 0, \\
u(x,0) &= f(x), & 0 \le x \le l, & \\
u_t(x,0) &= g(x), & 0 \le x \le l, & \\
u(0,t) &= p(t), & u(l,t) = q(t), & \qquad t \ge 0,
\end{aligned}$$

can be determined by the same method.

5.7 Nonhomogeneous Wave Equations

We shall consider next the Cauchy problem for the nonhomogeneous wave equation

$$u_{tt} = c^2 u_{xx} + h^*(x, t), \tag{5.7.1}$$

with the initial conditions

$$u(x,0) = f(x), \qquad u_t(x,0) = g^*(x). \tag{5.7.2}$$

By the coordinate transformation

$$y = ct, \tag{5.7.3}$$

the problem is reduced to

$$u_{xx} - u_{yy} = h\,(x,y)\,, \tag{5.7.4}$$
$$u\,(x,0) = f\,(x)\,, \tag{5.7.5}$$
$$u_y\,(x,0) = g\,(x)\,, \tag{5.7.6}$$

where $h\,(x,y) = -h^*/c^2$ and $g\,(x) = g^*/c$.

Let $P_0\,(x_0,y_0)$ be a point of the plane, and let Q_0 be the point $(x_0,0)$ on the initial line $y = 0$. Then the characteristics, $x + y = $ constant, of equation (5.7.4) are two straight lines drawn through the point P_0 with slopes ± 1. Obviously, they intersect the x-axis at the points $P_1\,(x_0 - y_0,0)$ and $P_2\,(x_0 + y_0,0)$, as shown in Figure 5.7.1. Let the sides of the triangle $P_0 P_1 P_2$ be designated by B_0, B_1, and B_2, and let D be the region representing the interior of the triangle and its boundaries B. Integrating both sides of equation (5.7.4), we obtain

$$\iint_R (u_{xx} - u_{yy})\,dR = \iint_R h\,(x,y)\,dR. \tag{5.7.7}$$

Now we apply Green's theorem to obtain

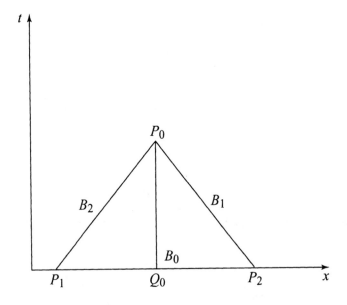

Figure 5.7.1 Triangular Region.

$$\iint_R (u_{xx} - u_{yy})\, dR = \oint_B (u_x dy + u_y dx).\tag{5.7.8}$$

Since B is composed of B_0, B_1, and B_2, we note that

$$\int_{B_0} (u_x\, dy + u_y\, dx) = \int_{x_0-y_0}^{x_0+y_0} u_y\, dx,$$

$$\int_{B_1} (u_x\, dy + u_y\, dx) = \int_{B_1} (-u_x\, dx - u_y\, dy),$$

$$= u(x_0 + y_0, 0) - u(x_0, y_0),$$

$$\int_{B_2} (u_x\, dy + u_y\, dx) = \int_{B_2} (u_x\, dx + u_y\, dy),$$

$$= u(x_0 - y_0, 0) - u(x_0, y_0).$$

Hence,

$$\oint_B (u_x\, dy + u_y\, dx) = -2\,u(x_0, y_0) + u(x_0 - y_0, 0)$$

$$+ u(x_0 + y_0, 0) + \int_{x_0-y_0}^{x_0+y_0} u_y\, dx.\tag{5.7.9}$$

Combining equations (5.7.7), (5.7.8) and (5.7.9), we obtain

$$u(x_0, y_0) = \frac{1}{2}[u(x_0 + y_0, 0) + u(x_0 - y_0, 0)]$$

$$+ \frac{1}{2}\int_{x_0-y_0}^{x_0+y_0} u_y\, dx - \frac{1}{2}\iint_R h(x, y)\, dR.\tag{5.7.10}$$

We have chosen x_0, y_0 arbitrarily, and as a consequence, we replace x_0 by x and y_0 by y. Equation (5.7.10) thus becomes

$$u(x, y) = \frac{1}{2}[f(x + y) + f(x - y)] + \frac{1}{2}\int_{x-y}^{x+y} g(\tau)\, d\tau - \frac{1}{2}\iint_R h(x, y)\, dR.$$

In terms of the original variables

$$u(x, t) = \frac{1}{2}[f(x + ct) + f(x - ct)] + \frac{1}{2c}\int_{x-ct}^{x+ct} g^*(\tau)\, d\tau - \frac{1}{2}\iint_R h(x, t)\, dR.$$

$$\tag{5.7.11}$$

Example 5.7.1. Determine the solution of

$$u_{xx} - u_{yy} = 1,$$
$$u(x, 0) = \sin x,$$
$$u_y(x, 0) = x.$$

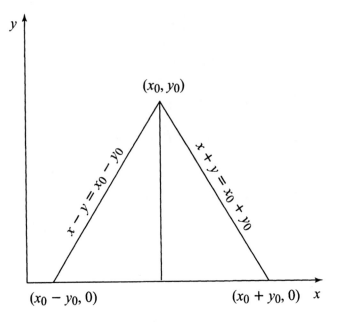

Figure 5.7.2 Triangular Region.

It is easy to see that the characteristics are $x + y = \text{constant} = x_0 + y_0$ and $x - y = \text{constant} = x_0 - y_0$, as shown in Figure 5.7.2. Thus,

$$u\left(x_0, y_0\right) = \frac{1}{2}\left[\sin\left(x_0 + y_0\right) + \sin\left(x_0 - y_0\right)\right]$$

$$+ \frac{1}{2}\int_{x_0-y_0}^{x_0+y_0} \tau \, d\tau - \frac{1}{2}\int_0^{y_0}\int_{y+x_0-y_0}^{-y+x_0+y_0} dx \, dy$$

$$= \frac{1}{2}\left[\sin\left(x_0 + y_0\right) + \sin\left(x_0 - y_0\right)\right] + x_0 y_0 - \frac{1}{2}y_0^2.$$

Now dropping the subscript zero, we obtain the solution

$$u\left(x, y\right) = \frac{1}{2}\left[\sin\left(x + y\right) + \sin\left(x - y\right)\right] + xy - \frac{1}{2}y^2.$$

5.8 The Riemann Method

We shall discuss Riemann's method of integrating the linear hyperbolic equation

$$L\left[u\right] \equiv u_{xy} + au_x + bu_y + cu = f\left(x, y\right), \qquad (5.8.1)$$

where L denotes the linear operator, and $a(x, y)$, $b(x, y)$, $c(x, y)$, and $f(x, y)$ are differentiable functions in some domain D^*. The method consists essentially of the derivation of an integral formula which represents the solution of the Cauchy problem.

Let $v(x, y)$ be a function having continuous second-order partial derivatives. Then, we may write

$$vu_{xy} - uv_{xy} = (vu_x)_y - (vu_y)_x \,,$$
$$vau_x = (avu)_x - u(av)_x \,, \qquad (5.8.2)$$
$$vbu_y = (bvu)_y - u(bv)_y \,,$$

so that

$$vL[u] - uM[v] = U_x + V_y, \qquad (5.8.3)$$

where M is the operator represented by

$$M[v] = v_{xy} - (av)_x - (bv)_y + cv, \qquad (5.8.4)$$

and

$$U = auv - uv_y, \qquad V = buv + vu_x. \qquad (5.8.5)$$

The operator M is called the *adjoint operator* of L. If $M = L$, then the operator L is said to be *self-adjoint*. Now applying Green's theorem, we have

$$\iint_D (U_x + V_y)\, dx\, dy = \oint_C (U\, dy - V\, dx), \qquad (5.8.6)$$

where C is the closed curve bounding the region of integration D which is in D^*.

Let Λ be a smooth initial curve which is continuous, as shown in Figure 5.8.1. Since equation (5.8.1) is in first canonical form, x and y are the characteristic coordinates. We assume that the tangent to Λ is nowhere parallel to the x or y axis. Let $P(\alpha, \beta)$ be a point at which the solution to the Cauchy problem is sought. Line PQ parallel to the x axis intersects the initial curve Λ at Q, and line PR parallel to the y axis intersects the curve Λ at R. We suppose that u and u_x or u_y are prescribed along Λ.

Let C be the closed contour $PQRP$ bounding D. Since $dy = 0$ on PQ and $dx = 0$ on PR, it follows immediately from equations (5.8.3) and (5.8.6) that

$$\iint_D (vL[u] - uM[v])\, dx\, dy = \int_Q^R (U\, dy - V\, dx) + \int_R^P U\, dy - \int_P^Q V\, dx.$$
$$(5.8.7)$$

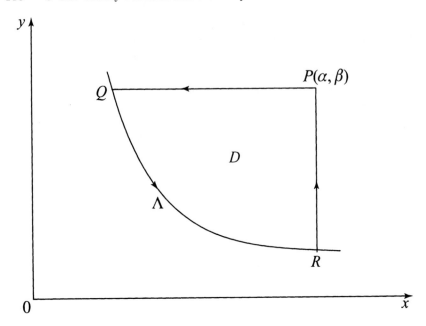

Figure 5.8.1 Smooth initial curve.

From equation (5.8.5), we find

$$\int_P^Q V\,dx = \int_P^Q bvu\,dx + \int_P^Q vu_x\,dx.$$

Integrating by parts, we obtain

$$\int_P^Q vu_x dx = [uv]_P^Q - \int_P^Q uv_x dx.$$

Hence, we may write

$$\int_P^Q V\,dx = [uv]_P^Q + \int_P^Q u\,(bv - v_x)\,dx.$$

Substitution of this integral in equation (5.8.7) yields

$$[uv]_P = [uv]_Q + \int_P^Q u\,(bv - v_x)\,dx - \int_R^P u\,(av - v_y)\,dy - \int_Q^R (U\,dy - V\,dx)$$

$$+ \iint_D (vL\,[u] - uM\,[v])\,dx\,dy. \qquad (5.8.8)$$

Suppose we can choose the function $v\,(x, y; \alpha, \beta)$ to be the solution of the adjoint equation

$$M[v] = 0, \tag{5.8.9}$$

satisfying the conditions

$$\begin{aligned}
v_x &= bv &\text{when} \quad y = \beta, \\
v_y &= av &\text{when} \quad x = \alpha, \\
v &= 1 &\text{when} \quad x = \alpha \quad \text{and} \quad y = \beta.
\end{aligned} \tag{5.8.10}$$

The function $v(x, y; \alpha, \beta)$ is called the *Riemann function*. Since $L[u] = f$, equation (5.8.8) reduces to,

$$[u]_P = [uv]_Q - \int_Q^R uv\,(a\,dy - b\,dx) + \int_Q^R (uv_y dy + vu_x dx) + \iint_D vf\,dx\,dy. \tag{5.8.11}$$

This gives us the value of u at the point P when u and u_x are prescribed along the curve Λ. When u and u_y are prescribed, the identity

$$[uv]_R - [uv]_Q = \int_Q^R \left\{ (uv)_x \, dx + (uv)_y \, dy \right\},$$

may be used to put equation (5.8.8) in the form

$$[u]_P = [uv]_R - \int_Q^R uv\,(a\,dy - b\,dx) - \int_Q^R (uv_x dx + vu_y dy)$$
$$+ \iint_D vf\,dx\,dy. \tag{5.8.12}$$

By adding equations (5.8.11) and (5.8.12), the value of u at P is given by

$$[u]_P = \frac{1}{2} \left([uv]_Q + [uv]_R \right) - \int_Q^R uv\,(a\,dy - b\,dx) - \frac{1}{2} \int_Q^R u\,(v_x dx - v_y dy)$$
$$+ \frac{1}{2} \int_Q^R v\,(u_x dx - u_y dy) + \iint_D vf\,dx\,dy \tag{5.8.13}$$

which is the solution of the Cauchy problem in terms of the Cauchy data given along the curve Λ. It is easy to see that the solution at the point (α, β) depends only on the Cauchy data along the arc QR on Λ. If the initial data were to change outside this arc QR, the solution would change only outside the triangle PQR. Thus, from Figure 5.8.2, we can see that each characteristic separates the region in which the solution remains unchanged from the region in which it varies. Because of this fact, the unique continuation of the solution across any characteristic is not possible. This is evident from Figure 5.8.2. The solution on the right of the characteristic $P_1 R_1$ is determined by the initial data given in $Q_1 R_2$, whereas the solution

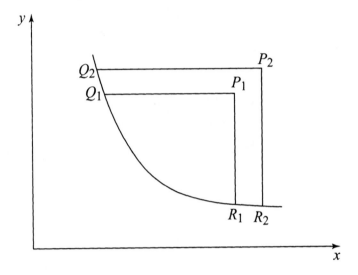

Figure 5.8.2 Solution on the right and left of the characteristic.

on the left is determined by the initial data given on Q_1R_1. If the initial data on R_1R_2 were changed, the solution on the right of P_1R_1 only will be affected.

It should be remarked here that the initial curve can intersect each characteristic at only one point. Suppose, for example, the initial curve Λ intersects the characteristic at two points, as shown in Figure 5.8.3. Then, the solution at P obtained from the initial data on QR will be different from the solution obtained from the initial data on RS. Hence, the Cauchy problem, in this case, is not solvable.

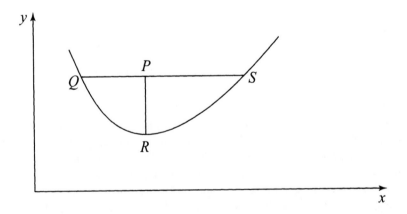

Figure 5.8.3 Initial curve intersects the characteristic at two points.

Example 5.8.1. The telegraph equation

$$w_{tt} + a^* w_t + b^* w = c^2 w_{xx},$$

may be transformed into canonical form

$$L[u] = u_{\xi\eta} + ku = 0,$$

by the successive transformations

$$w = u\, e^{-a^* t/2},$$

and

$$\xi = x + ct, \qquad \eta = x - ct,$$

where $k = \left(a^{*2} - 4b^*\right)/16c^2$.

We apply Riemann's method to determine the solution satisfying the initial conditions

$$u(x, 0) = f(x), \qquad u_t(x, 0) = g(x).$$

Since

$$t = \frac{1}{2c}(\xi - \eta),$$

the line $t = 0$ corresponds to the straight line $\xi = \eta$ in the $\xi - \eta$ plane. The initial conditions may thus be transformed into

$$[u]_{\xi=\eta} = f(\xi), \tag{5.8.14}$$

$$[u_\xi - u_\eta]_{\xi=\eta} = c^{-1} g(\xi). \tag{5.8.15}$$

We next determine the Riemann function $v(\xi, \eta; \alpha, \beta)$ which satisfies

$$v_{\xi\eta} + kv = 0, \tag{5.8.16}$$
$$v_\xi(\xi, \beta; \alpha, \beta) = 0, \tag{5.8.17}$$
$$v_\eta(\alpha, \eta; \alpha, \beta) = 0, \tag{5.8.18}$$
$$v(\alpha, \beta; \alpha, \beta) = 1. \tag{5.8.19}$$

The differential equation (5.8.16) is self-adjoint, that is,

$$L[v] = M[v] = v_{\xi\eta} + kv.$$

We assume that the Riemann function is of the form

$$v(\xi, \eta; \alpha, \beta) = F(s),$$

with the argument $s = (\xi - \alpha)(\eta - \beta)$. Substituting this value in equation (5.8.16), we obtain

$$sF_{ss} + F_s + kF = 0.$$

If we let $\lambda = \sqrt{4ks}$, the above equation becomes

$$F''(\lambda) + \frac{1}{\lambda} F'(\lambda) + F(\lambda) = 0.$$

This is the Bessel equation of order zero, and the solution is

$$F(\lambda) = J_0(\lambda),$$

disregarding $Y_0(\lambda)$ which is unbounded at $\lambda = 0$. Thus, the Riemann function is

$$v(\xi, \eta; \alpha, \beta) = J_0\left(\sqrt{4k(\xi - \alpha)(\eta - \beta)}\right)$$

which satisfies equation (5.8.16) and is equal to one on the characteristics $\xi = \alpha$ and $\eta = \beta$. Since $J_0'(0) = 0$, equations (5.8.17) and (5.8.18) are satisfied. From this, it immediately follows that

$$[v_\xi]_{\xi=\eta} = \frac{\sqrt{k}(\xi - \beta)}{\sqrt{(\xi - \alpha)(\eta - \beta)}} [J_0'(\lambda)]_{\xi=\eta},$$

$$[v_\eta]_{\xi=\eta} = \frac{\sqrt{k}(\xi - \alpha)}{\sqrt{(\xi - \alpha)(\eta - \beta)}} [J_0'(\lambda)]_{\xi=\eta}.$$

Thus, we have

$$[v_\xi - u_\eta]_{\xi=\eta} = \frac{\sqrt{k}(\alpha - \beta)}{\sqrt{(\xi - \alpha)(\xi - \beta)}} [J_0'(\lambda)]_{\xi=\eta}. \qquad (5.8.20)$$

From the initial condition

$$u(Q) = f(\beta) \quad \text{and} \quad u(R) = f(\alpha), \qquad (5.8.21)$$

and substituting equations (5.8.15), (5.8.19), and (5.8.20) into equation (5.8.13), we obtain

$$u(\alpha, \beta) = \frac{1}{2}[f(\alpha) + f(\beta)]$$

$$-\frac{1}{2}\int_\beta^\alpha \frac{\sqrt{k}(\alpha - \beta)}{\sqrt{(\tau - \alpha)(\tau - \beta)}} J_0'\left(\sqrt{4k(\tau - \alpha)(\tau - \beta)}\right) f(\tau)\, d\tau$$

$$+\frac{1}{2c}\int_\beta^\alpha J_0\left(\sqrt{4k(\tau - \alpha)(\tau - \beta)}\right) g(\tau)\, d\tau. \qquad (5.8.22)$$

Replacing α and β by ξ and η, and substituting the original variables x and t, we obtain

$$u\left(x,t\right) = \frac{1}{2}\left[f\left(x+ct\right)+f\left(x-ct\right)\right] + \frac{1}{2}\int_{x-ct}^{x+ct} G\left(x,t,\tau\right) d\tau, \quad (5.8.23)$$

where

$$G\left(x,t,\tau\right)$$
$$= \left\{-2\sqrt{k}\,ctf\left(\tau\right)J_0\left(\sqrt{4k\left[\left(\tau-x\right)^2-c^2t^2\right]}\right)\right\}\Big/\sqrt{\left(\tau-x\right)^2-c^2t^2}$$
$$+ c^{-1}g\left(\tau\right)J_0\left(\sqrt{4k\left[\left(\tau-x\right)^2-c^2t^2\right]}\right).$$

If we set $k = 0$, we arrive at the d'Alembert solution for the wave equation

$$u\left(x,t\right) = \frac{1}{2}\left[f\left(x+ct\right)+f\left(x-ct\right)\right] + \frac{1}{2c}\int_{x-ct}^{x+ct} g\left(\tau\right) d\tau.$$

5.9 Solution of the Goursat Problem

The Goursat problem is that of finding the solution of a linear hyperbolic equation

$$u_{xy} = a_1\left(x,y\right)u_x + a_2\left(x,y\right)u_y + a_3\left(x,y\right)u + h\left(x,y\right), \quad (5.9.1)$$

satisfying the prescribed conditions

$$u\left(x,y\right) = f\left(x\right), \quad (5.9.2)$$

on a characteristic, say, $y = 0$, and

$$u\left(x,y\right) = g\left(x\right) \quad (5.9.3)$$

on a monotonic increasing curve $y = y\left(x\right)$ which, for simplicity, is assumed to intersect the characteristic at the origin.

The solution in the region between the x-axis and the monotonic curve in the first quadrant can be determined by the method of successive approximations. The proof is given in Garabedian (1964).

Example 5.9.1. Determine the solution of the Goursat problem

$$u_{tt} = c^2 u_{xx}, \quad (5.9.4)$$
$$u\left(x,t\right) = f\left(x\right), \quad \text{on} \quad x - ct = 0, \quad (5.9.5)$$
$$u\left(x,t\right) = g\left(x\right), \quad \text{on} \quad t = t\left(x\right), \quad (5.9.6)$$

where $f\left(0\right) = g\left(0\right)$.

The general solution of the wave equation is

$$u\left(x,t\right) = \phi\left(x + ct\right) + \psi\left(x - ct\right).$$

Applying the prescribed conditions, we obtain

$$f\left(x\right) = \phi\left(2x\right) + \psi\left(0\right), \tag{5.9.7}$$
$$g\left(x\right) = \phi\left(x + ct\left(x\right)\right) + \psi\left(x - ct\left(x\right)\right). \tag{5.9.8}$$

It is evident that

$$f\left(0\right) = \phi\left(0\right) + \psi\left(0\right) = g\left(0\right).$$

Now, if $s = x - ct\left(x\right)$, the inverse of it is $x = \alpha\left(s\right)$. Thus, equation (5.9.8) may be written as

$$g\left(\alpha\left(s\right)\right) = \phi\left(x + ct\left(x\right)\right) + \psi\left(s\right). \tag{5.9.9}$$

Replacing x by $\left(x + ct\left(x\right)\right)/2$ in equation (5.9.7), we obtain

$$f\left(\frac{x + ct\left(x\right)}{2}\right) = \phi\left(x + ct\left(x\right)\right) + \psi\left(0\right). \tag{5.9.10}$$

Thus, using (5.9.10), equation (5.9.9) becomes

$$\psi\left(s\right) = g\left(\alpha\left(s\right)\right) - f\left(\frac{\alpha\left(s\right) + ct\left(\alpha\left(s\right)\right)}{2}\right) + \psi\left(0\right).$$

Replacing s by $x - ct$, we have

$$\psi\left(x - ct\right) = g\left(\alpha\left(x - ct\right)\right) - f\left(\frac{\alpha\left(x - ct\right) + ct\left(\alpha\left(x - ct\right)\right)}{2}\right) + \psi\left(0\right).$$

Hence, the solution is given by

$$u\left(x,t\right) = f\left(\frac{x + ct}{2}\right) - f\left(\frac{\alpha\left(x - ct\right) + ct\left(\alpha\left(x - ct\right)\right)}{2}\right) + g\left(\alpha\left(x - ct\right)\right).$$
$$\tag{5.9.11}$$

Let us consider a special case when the curve $t = t\left(x\right)$ is a straight line represented by $t - kx = 0$ with a constant $k > 0$. Then $s = x - ckx$ and hence $x = s/\left(1 - ck\right)$. Using these values in (5.9.11), we obtain

$$u\left(x,t\right) = f\left(\frac{x + ct}{2}\right) - f\left(\frac{\left(1 + ck\right)\left(x - ct\right)}{2\left(1 - ck\right)}\right) + g\left(\frac{x - ct}{1 - ck}\right). \tag{5.9.12}$$

When the values of u are prescribed on both characteristics, the problem of finding u of a linear hyperbolic equation is called a *characteristic initial-value problem*. This is a degenerate case of the Goursat problem.

Consider the characteristic initial-value problem

$$u_{xy} = h(x, y),\qquad\qquad (5.9.13)$$
$$u(x, 0) = f(x),\qquad\qquad (5.9.14)$$
$$u(0, y) = g(y),\qquad\qquad (5.9.15)$$

where f and g are continuously differentiable, and $f(0) = g(0)$.
Integrating equation (5.9.13), we obtain

$$u(x, y) = \int_0^x \int_0^y h(\xi, \eta)\, d\eta\, d\xi + \phi(x) + \psi(y),\qquad (5.9.16)$$

where ϕ and ψ are arbitrary functions. Applying the prescribed conditions (5.9.14) and (5.9.15), we have

$$u(x, 0) = \phi(x) + \psi(0) = f(x),\qquad\qquad (5.9.17)$$
$$u(0, y) = \phi(0) + \psi(y) = g(y).\qquad\qquad (5.9.18)$$

Thus,

$$\phi(x) + \psi(y) = f(x) + g(y) - \phi(0) - \psi(0).\qquad (5.9.19)$$

But from (5.9.17), we have

$$\phi(0) + \psi(0) = f(0).\qquad\qquad (5.9.20)$$

Hence, from (5.9.16), (5.9.19) and (5.9.20), we obtain

$$u(x, y) = f(x) + g(y) - f(0) + \int_0^x \int_0^y h(\xi, \eta)\, d\eta\, d\xi.\quad (5.9.21)$$

Example 5.9.2. Determine the solution of the characteristic initial-value problem

$$u_{tt} = c^2 u_{xx},$$
$$u(x, t) = f(x)\quad \text{on}\quad x + ct = 0,$$
$$u(x, t) = g(x)\quad \text{on}\quad x - ct = 0,$$

where $f(0) = g(0)$.

Here it is not necessary to reduce the given equation to canonical form. The general solution of the wave equation is

$$u(x, t) = \phi(x + ct) + \psi(x - ct).$$

The characteristics are

$$x + ct = 0,\qquad x - ct = 0.$$

Applying the prescribed conditions, we have

$$u\left(x,t\right) = \phi\left(2x\right) + \psi\left(0\right) = f\left(x\right) \quad \text{on} \quad x + ct = 0, \qquad (5.9.22)$$
$$u\left(x,t\right) = \phi\left(0\right) + \psi\left(2x\right) = g\left(x\right) \quad \text{on} \quad x - ct = 0. \qquad (5.9.23)$$

We observe that these equations are compatible, since $f\left(0\right) = g\left(0\right)$.

Now, replacing x by $\left(x + ct\right)/2$ in equation (5.9.22) and replacing x by $\left(x - ct\right)/2$ in equation (5.9.23), we have

$$\phi\left(x + ct\right) = f\left(\frac{x + ct}{2}\right) - \psi\left(0\right),$$

$$\phi\left(x - ct\right) = g\left(\frac{x - ct}{2}\right) - \phi\left(0\right).$$

Hence, the solution is given by

$$u\left(x,t\right) = f\left(\frac{x + ct}{2}\right) + g\left(\frac{x - ct}{2}\right) - f\left(0\right). \qquad (5.9.24)$$

We note that this solution can be obtained by substituting $k = -1/c$ into (5.9.12).

Example 5.9.3. Find the solution of the characteristic initial-value problem

$$y^3 u_{xx} - y u_{yy} + u_y = 0, \qquad (5.9.25)$$

$$u\left(x,y\right) = f\left(x\right) \quad \text{on} \quad x + \frac{y^2}{2} = 4 \quad \text{for} \quad 2 \le x \le 4,$$

$$u\left(x,y\right) = g\left(x\right) \quad \text{on} \quad x - \frac{y^2}{2} = 0 \quad \text{for} \quad 0 \le x \le 2,$$

with $f\left(2\right) = g\left(2\right)$.

Since the equation is hyperbolic except for $y = 0$, we reduce it to the canonical form

$$u_{\xi\eta} = 0,$$

where $\xi = x + \left(y^2/2\right)$ and $\eta = x - \left(y^2/2\right)$. Thus, the general solution is

$$u\left(x,y\right) = \phi\left(x + \frac{y^2}{2}\right) + \psi\left(x - \frac{y^2}{2}\right). \qquad (5.9.26)$$

Applying the prescribed conditions, we have

$$f\left(x\right) = \phi\left(4\right) + \psi\left(2x - 4\right), \qquad (5.9.27)$$
$$g\left(x\right) = \phi\left(2x\right) + \psi\left(0\right). \qquad (5.9.28)$$

Now, if we replace $\left(2x - 4\right)$ by $\left(x - y^2/2\right)$ in (5.9.27) and $\left(2x\right)$ by $\left(x + y^2/2\right)$ in (5.9.28), we obtain

$$\psi\left(x - \frac{y^2}{2}\right) = f\left(\frac{x}{2} - \frac{y^2}{4} + 2\right) - \phi\left(4\right),$$

$$\phi\left(x + \frac{y^2}{2}\right) = g\left(\frac{x}{2} + \frac{y^2}{4}\right) - \psi\left(0\right).$$

Thus,

$$u\left(x, y\right) = f\left(\frac{x}{2} - \frac{y^2}{4} + 2\right) + g\left(\frac{x}{2} + \frac{y^2}{4}\right) - \phi\left(4\right) - \psi\left(0\right).$$

But from (5.9.27) and (5.9.28), we see that

$$f\left(2\right) = \phi\left(4\right) + \psi\left(0\right) = g\left(2\right).$$

Hence,

$$u\left(x, y\right) = f\left(\frac{x}{2} - \frac{y^2}{4} + 2\right) + g\left(\frac{x}{2} + \frac{y^2}{4}\right) - f\left(2\right).$$

5.10 Spherical Wave Equation

In spherical polar coordinates (r, θ, ϕ), the wave equation (3.1.1) takes the form

$$\frac{1}{r^2}\frac{\partial}{\partial r}\left(r^2 \frac{\partial u}{\partial r}\right) + \frac{1}{r^2 \sin\theta}\frac{\partial}{\partial \theta}\left(\sin\theta \frac{\partial u}{\partial \theta}\right) + \frac{1}{r^2 \sin^2\theta}\frac{\partial^2 u}{\partial \phi^2} = \frac{1}{c^2}\frac{\partial^2 u}{\partial t^2} \quad (5.10.1)$$

Solutions of this equation are called *spherical symmetric waves* if u depends on r and t only. Thus, the solution $u = u\left(r, t\right)$ which satisfies the wave equation with spherical symmetry in three-dimensional space is

$$\frac{1}{r^2}\frac{\partial}{\partial r}\left(r^2 \frac{\partial u}{\partial r}\right) = \frac{1}{c^2}\frac{\partial^2 u}{\partial t^2}. \quad (5.10.2)$$

Introducing a new dependent variable $U = ru\left(r, t\right)$, this equation reduces to a simple form

$$U_{tt} = c^2 U_{rr}. \quad (5.10.3)$$

This is identical with the one-dimensional wave equation (5.3.1) and has the general solution in the form

$$U\left(r, t\right) = \phi\left(r + ct\right) + \psi\left(r - ct\right), \quad (5.10.4)$$

or, equivalently,

$$u\left(r, t\right) = \frac{1}{r}\left[\phi\left(r + ct\right) + \psi\left(r - ct\right)\right]. \quad (5.10.5)$$

This solution consists of two progressive spherical waves traveling with constant velocity c. The terms involving ϕ and ψ represent the incoming waves to the origin and the outgoing waves from the origin respectively.

Physically, the solution for only outgoing waves generated by a source is of most interest, and has the form

$$u(r,t) = \frac{1}{r}\psi(r - ct),\qquad\qquad (5.10.6)$$

where the explicit form of ψ is to be determined from the properties of the source. In the context of fluid flows, u represents the velocity potential so that the limiting total flux through a sphere of center at the origin and radius r is

$$Q(t) = \lim_{r \to 0} 4\pi r^2 u_r(r,t) = -4\pi\psi(-ct).\qquad\qquad (5.10.7)$$

In physical terms, we say that there is a simple (or monopole) point source of strength $Q(t)$ located at the origin. Thus, the solution (5.10.6) can be expressed in terms of Q as

$$u(r,t) = -\frac{1}{4\pi r}Q\left(t - \frac{r}{c}\right).\qquad\qquad (5.10.8)$$

This represents the velocity potential of the point source, and u_r is called the *radial velocity*. In fluid flows, the difference between the pressure at any time t and the equilibrium value is given by

$$p - p_0 = \rho u_t = -\frac{\rho}{4\pi r}\dot{Q}\left(t - \frac{r}{c}\right),\qquad\qquad (5.10.9)$$

where ρ is the density of the fluid.

Following an analysis similar to Section 5.3, the solution of the initial-value problem with the initial data

$$u(r,0) = f(r),\quad u_t(r,0) = g(r),\quad r \geq 0,\qquad\qquad (5.10.10)$$

where f and g are continuously differentiable, is given by

$$u(r,t) = \frac{1}{2r}\left[(r + ct)f(r + ct) + (r - ct)f(r - ct) + \frac{1}{c}\int_{r-ct}^{r+ct}\tau g(\tau)\,d\tau\right],$$

$$(5.10.11)$$

provided $r \geq ct$. However, when $r < ct$, this solution fails because f and g are not defined for $r < 0$. This initial data at $t = 0$, $r \geq 0$ determine the solution $u(r,t)$ only up to the characteristic $r = ct$ in the r-t plane. To find u for $r < ct$, we require u to be finite at $r = 0$ for all $t \geq 0$, that is, $U = 0$ at $r = 0$. Thus, the solution for $U(r,t)$ is

$$U(r,t) = \frac{1}{2}\left[(r+ct)f(r+ct) + (r-ct)f(r-ct) + \frac{1}{c}\int_{r-ct}^{r+ct} \tau g(\tau)\, d\tau\right],$$

$$(5.10.12)$$

provided $r \geq ct \geq 0$, and

$$U(r,t) = \frac{1}{2}\left[\phi(ct+r) + \psi(ct-r)\right], \qquad ct \geq r \geq 0, \quad (5.10.13)$$

where

$$\phi(ct) + \psi(ct) = 0, \quad \text{for} \quad ct \geq 0. \tag{5.10.14}$$

In view of the fact that $U_r + \frac{1}{c}U_t$ is constant on each characteristic $r + ct = \text{constant}$, it turns out that

$$\phi'(ct+r) = (r+ct)f'(r+ct) + f(r+ct) + \frac{1}{c}(r+ct)g(r+ct),$$

or

$$\phi'(ct) = ctf'(ct) + f(ct) + t\,g(ct).$$

Integration gives

$$\phi(t) = tf(t) + \frac{1}{c}\int_0^t \tau g(\tau)\, d\tau + \phi(0),$$

so that

$$\psi(t) = -tf(t) - \frac{1}{c}\int_0^t \tau g(\tau)\, d\tau - \phi(0).$$

Substituting these values into (5.10.13) and using $U(r,t) = ru(r,t)$, we obtain, for $ct > r$,

$$u(r,t) = \frac{1}{2r}\left[(ct+r)f(ct+r) - (ct-r)f(ct-r) + \frac{1}{c}\int_{ct-r}^{ct+r} \tau g(\tau)\, d\tau\right].$$

$$(5.10.15)$$

5.11 Cylindrical Wave Equation

In cylindrical polar coordinates (R, θ, z), the wave equation (3.1.1) assumes the form

$$u_{RR} + \frac{1}{R}u_R + \frac{1}{R^2}u_{\theta\theta} + u_{zz} = \frac{1}{c^2}u_{tt}. \tag{5.11.1}$$

If u depends only on R and t, this equation becomes

$$u_{RR} + \frac{1}{R} u_R = \frac{1}{c^2} u_{tt}. \tag{5.11.2}$$

Solutions of (5.11.2) are called *cylindrical waves*.

In general, it is not easy to find the solution of (5.11.1). However, we shall solve this equation by using the method of separation of variables in Chapter 7. Here we derive the solution for outgoing cylindrical waves from the spherical wave solution (5.10.8). We assume that sources of constant strength $Q(t)$ per unit length are distributed uniformly on the z-axis. The solution for the cylindrical waves produced by the line source is given by the total disturbance

$$u(R,t) = -\frac{1}{4\pi} \int_{-\infty}^{\infty} \frac{1}{r} Q\left(t - \frac{r}{c}\right) dz = -\frac{1}{2\pi} \int_{0}^{\infty} \frac{1}{r} Q\left(t - \frac{r}{c}\right) dz, \tag{5.11.3}$$

where R is the distance from the z-axis so that $R^2 = (r^2 - z^2)$.

Substitution of $z = R \sinh \xi$ and $r = R \cosh \xi$ in (5.11.3) gives

$$u(R,t) = -\frac{1}{2\pi} \int_{0}^{\infty} Q\left(t - \frac{R}{c} \cosh \xi\right) d\xi. \tag{5.11.4}$$

This is usually considered as the cylindrical wave function due to a source of strength $Q(t)$ at $R = 0$. It follows from (5.11.4) that

$$u_{tt} = -\frac{1}{2\pi} \int_{0}^{\infty} Q''\left(t - \frac{R}{c} \cosh \xi\right) d\xi, \tag{5.11.5}$$

$$u_R = \frac{1}{2\pi c} \int_{0}^{\infty} \cosh \xi \, Q'\left(t - \frac{R}{c} \cosh \xi\right) d\xi, \tag{5.11.6}$$

$$u_{RR} = -\frac{1}{2\pi c^2} \int_{0}^{\infty} \cosh^2 \xi \, Q''\left(t - \frac{R}{c} \cosh \xi\right) d\xi, \tag{5.11.7}$$

which give

$$c^2 \left(u_{RR} + \frac{1}{R} u_R\right) - u_{tt} = \frac{1}{2\pi} \int_{0}^{\infty} \frac{d}{d\xi} \left[\frac{c}{R} Q'\left(t - \frac{R}{c} \cosh \xi\right) \sinh \xi\right] d\xi$$

$$= \lim_{\xi \to \infty} \left[\frac{c}{2\pi R} Q'\left(t - \frac{R}{c} \cosh \xi\right) \sinh \xi\right] = 0,$$

provided the differentiation under the sign of integration is justified and the above limit is zero. This means that $u(R,t)$ satisfies the cylindrical wave equation (5.11.2).

In order to find the asymptotic behavior of the solution as $R \to 0$, we substitute $\cosh \xi = \frac{c(t-\zeta)}{R}$ into (5.11.4) and (5.11.6) to obtain

$$u = -\frac{1}{2\pi} \int_{-\infty}^{t-R/c} \frac{Q(\zeta) \, d\zeta}{\left[(t-\zeta)^2 - \frac{R^2}{c^2}\right]^{\frac{1}{2}}}, \tag{5.11.8}$$

$$u_R = \frac{1}{2\pi} \int_{-\infty}^{t-R/c} \left(\frac{t-\zeta}{R}\right) \frac{Q'(\zeta) \, d\zeta}{\left[(t-\zeta)^2 - \frac{R^2}{c^2}\right]^{\frac{1}{2}}}, \tag{5.11.9}$$

which, in the limit $R \to 0$, give

$$u_R \sim \frac{1}{2\pi R} \int_{-\infty}^{t} Q'(\zeta)\, d\zeta = \frac{1}{2\pi R} Q(t). \qquad (5.11.10)$$

This leads to the result

$$\lim_{R \to 0} 2\pi R\, u_R = Q(t), \qquad (5.11.11)$$

or

$$u(R,t) \sim \frac{1}{2\pi} Q(t) \log R \quad \text{as} \quad R \to 0. \qquad (5.11.12)$$

We next investigate the nature of the cylindrical wave solution near the waterfront ($R = ct$) and in the far field ($R \to \infty$). We assume $Q(t) = 0$ for $t < 0$ so that the lower limit of integration in (5.11.8) may be taken to be zero, and the solution is non-zero for $\tau = t - \frac{R}{c} > 0$, where τ is the time passed after the arrival of the wavefront. Consequently, (5.11.8) becomes

$$u(R,t) = -\frac{1}{2\pi} \int_{0}^{\tau} \frac{Q(\zeta)\, d\zeta}{\left[(t-\zeta)\left(t-\zeta+\frac{2R}{c}\right)\right]^{\frac{1}{2}}}. \qquad (5.11.13)$$

Since $0 < \zeta < \tau$, $\frac{2R}{c} > \frac{R}{c} > \tau > \tau - \zeta > 0$, so that the second factor under the radical is approximately equal to $\frac{2R}{c}$ when $R \gg c\tau$, and hence,

$$
\begin{aligned}
u(R,t) &\sim -\frac{1}{2\pi} \left(\frac{c}{2R}\right)^{\frac{1}{2}} \int_{0}^{\tau} \frac{Q(\zeta)\, d\zeta}{(t-\zeta)^{\frac{1}{2}}} = -\left(\frac{c}{2R}\right)^{\frac{1}{2}} q(\tau) \\
&= -\left(\frac{c}{2R}\right)^{\frac{1}{2}} q\left(t - \frac{R}{c}\right), \qquad R \gg \frac{ct}{2}, \qquad (5.11.14)
\end{aligned}
$$

where

$$q(\tau) = \frac{1}{2\pi} \int_{0}^{\tau} \frac{Q(\zeta)\, d\zeta}{\sqrt{\tau-\zeta}}. \qquad (5.11.15)$$

Evidently, the amplitude involved in the solution (5.11.14) decays like $R^{-\frac{1}{2}}$ for large R ($R \to \infty$).

Example 5.11.1. Determine the asymptotic form of the solution (5.11.4) for a harmonically oscillating source of frequency ω.

We take the source in the form $Q(t) = q_0 \exp\left[-i(\omega+i\varepsilon)t\right]$, where ε is positive and small so that $Q(t) \to 0$ as $t \to -\infty$. The small imaginary part ε of ω will make insignificant contributions to the solution at finite time as $\varepsilon \to 0$. Thus, the solution (5.11.4) becomes

$$
\begin{aligned}
u(R,t) &= -\left(\frac{q_0}{2\pi}\right) e^{-i\omega t} \int_{0}^{\infty} \exp\left(\frac{i\omega R}{c} \cosh \xi\right) d\xi \\
&= -\left(\frac{i q_0}{4}\right) e^{-i\omega t} H_0^{(1)}\left(\frac{\omega R}{c}\right), \qquad (5.11.16)
\end{aligned}
$$

where $H_0^{(1)}(z)$ is the Hankel function given by

$$H_0^{(1)}(z) = \frac{2}{\pi i} \int_0^\infty \exp(iz \cosh \xi)\, d\xi. \qquad (5.11.17)$$

In view of the asymptotic expansion of $H_0^{(1)}(z)$ in the form

$$H_0^{(1)}(z) \sim \left(\frac{2}{\pi z}\right)^{\frac{1}{2}} \exp\left[i\left(z - \frac{\pi}{4}\right)\right], \qquad z \to \infty, \qquad (5.11.18)$$

the asymptotic solution for $u(R,t)$ in the limit $\left(\frac{\omega R}{c}\right) \to \infty$ is

$$u(R,t) \sim -\left(\frac{iq_0}{4}\right)\left(\frac{2c}{\pi \omega R}\right)^{\frac{1}{2}} \exp\left[-i\left(\omega t - \frac{\omega R}{c} - \frac{\pi}{4}\right)\right].$$

This represents the cylindrical wave propagating with constant velocity c. The amplitude of the wave decays like $R^{-\frac{1}{2}}$ as $R \to \infty$.

Example 5.11.2. For a supersonic flow $(M > 1)$ past a solid body of revolution, the perturbation potential Φ satisfies the cylindrical wave equation

$$\Phi_{RR} + \frac{1}{R}\Phi_R = N^2 \Phi_{xx}, \qquad N^2 = M^2 - 1,$$

where R is the distance from the path of the moving body and x is the distance from the nose of the body.

It follows from problem 12 in 3.9 Exercises that Φ satisfies the equation

$$\Phi_{yy} + \Phi_{zz} = N^2 \Phi_{xx}.$$

This represents a two-dimensional wave equation with $x \leftrightarrow t$ and $N^2 \leftrightarrow \frac{1}{c^2}$. For a body of revolution with $(y, z) \leftrightarrow (R, \theta)$, $\frac{\partial}{\partial \theta} \equiv 0$, the above equation reduces to the cylindrical wave equation

$$\Phi_{RR} + \frac{1}{R}\Phi_R = \frac{1}{c^2}\Phi_{tt}.$$

5.12 Exercises

1. Determine the solution of each of the following initial-value problems:

(a) $u_{tt} - c^2 u_{xx} = 0$, $u(x,0) = 0$, $u_t(x,0) = 1$.

(b) $u_{tt} - c^2 u_{xx} = 0$, $u(x,0) = \sin x$, $u_t(x,0) = x^2$.

(c) $u_{tt} - c^2 u_{xx} = 0$, $u(x,0) = x^3$, $u_t(x,0) = x$.

(d) $u_{tt} - c^2 u_{xx} = 0$, $u(x,0) = \cos x$, $u_t(x,0) = e^{-1}$.

(e) $u_{tt} - c^2 u_{xx} = 0$, $u(x,0) = \log(1 + x^2)$, $u_t(x,0) = 2$.

(f) $u_{tt} - c^2 u_{xx} = 0$, $u(x,0) = x$, $u_t(x,0) = \sin x$.

2. Determine the solution of each of the following initial-value problems:

(a) $u_{tt} - c^2 u_{xx} = x$, $u(x,0) = 0$, $u_t(x,0) = 3$.

(b) $u_{tt} - c^2 u_{xx} = x + ct$, $u(x,0) = x$, $u_t(x,0) = \sin x$.

(c) $u_{tt} - c^2 u_{xx} = e^x$, $u(x,0) = 5$, $u_t(x,0) = x^2$.

(d) $u_{tt} - c^2 u_{xx} = \sin x$, $u(x,0) = \cos x$, $u_t(x,0) = 1 + x$.

(e) $u_{tt} - c^2 u_{xx} = xe^t$, $u(x,0) = \sin x$, $u_t(x,0) = 0$.

(f) $u_{tt} - c^2 u_{xx} = 2$, $u(x,0) = x^2$, $u_t(x,0) = \cos x$.

3. A gas which is contained in a sphere of radius R is at rest initially, and the initial condensation is given by s_0 inside the sphere and zero outside the sphere. The condensation is related to the velocity potential by

$$s(t) = \left(1/c^2\right) u_t,$$

at all times, and the velocity potential satisfies the wave equation

$$u_{tt} = \nabla^2 u.$$

Determine the condensation $s(t)$ for all $t > 0$.

4. Solve the initial-value problem

$$u_{xx} + 2u_{xy} - 3u_{yy} = 0,$$
$$u(x,0) = \sin x, \qquad u_y(x,0) = x.$$

5. Find the longitudinal oscillation of a rod subject to the initial conditions

$$u(x,0) = \sin x,$$
$$u_t(x,0) = x.$$

6. By using the Riemann method, solve the following problems:

(a) $\sin^2 \mu\, \phi_{xx} - \cos^2 \mu\, \phi_{yy} - \left(\lambda^2 \sin^2 \mu\, \cos^2 \mu\right) \phi = 0$,

$$\phi(0,y) = f_1(y), \qquad \phi(x,0) = g_1(x),$$
$$\phi_x(0,y) = f_2(y), \qquad \phi_y(x,0) = g_2(x).$$

(b) $x^2 u_{xx} - t^2 u_{tt} = 0,$

$$u(x, t_1) = f(x), \qquad u_t(x, t_2) = g(x).$$

7. Determine the solution of the initial boundary-value problem

$$\begin{aligned}
u_{tt} &= 4\,u_{xx}, & 0 < x < \infty, \quad t > 0, \\
u(x,0) &= x^4, & 0 \le x < \infty, \\
u_t(x,0) &= 0, & 0 \le x < \infty, \\
u(0,t) &= 0, & t \ge 0.
\end{aligned}$$

8. Determine the solution of the initial boundary-value problem

$$\begin{aligned}
u_{tt} &= 9\,u_{xx}, & 0 < x < \infty, \quad t > 0, \\
u(x,0) &= 0, & 0 \le x < \infty, \\
u_t(x,0) &= x^3, & 0 \le x < \infty, \\
u_x(0,t) &= 0, & t \ge 0.
\end{aligned}$$

9. Determine the solution of the initial boundary-value problem

$$\begin{aligned}
u_{tt} &= 16\,u_{xx}, & 0 < x < \infty, \quad t > 0, \\
u(x,0) &= \sin x, & 0 \le x < \infty, \\
u_t(x,0) &= x^2, & 0 \le x < \infty, \\
u(0,t) &= 0, & t \ge 0.
\end{aligned}$$

10. In the initial boundary-value problem

$$\begin{aligned}
u_{tt} &= c^2 u_{xx}, & 0 < x < l, \quad t > 0, \\
u(x,0) &= f(x), & 0 \le x \le l, \\
u_t(x,0) &= g(x), & 0 \le x \le l, \\
u(0,t) &= 0, & t \ge 0,
\end{aligned}$$

if f and g are extended as odd functions, show that $u(x,t)$ is given by the solution (5.4.5) for $x > ct$ and solution (5.4.6) for $x < ct$.

11. In the initial boundary-value problem

$$\begin{aligned}
u_{tt} &= c^2 u_{xx}, & 0 < x < l, \quad t > 0, \\
u(x,0) &= f(x), & 0 \le x \le l, \\
u_t(x,0) &= g(x), & 0 \le x \le l, \\
u_x(0,t) &= 0, & t \ge 0,
\end{aligned}$$

if f and g are extended as even functions, show that $u(x,t)$ is given by solution (5.4.8) for $x > ct$, and solution (5.4.9) for $x < ct$.

12. Determine the solution of the initial boundary-value problem

$$u_{tt} = c^2 u_{xx}, \qquad 0 < x < \infty, \quad t > 0,$$
$$u(x,0) = f(x), \qquad 0 \le x < \infty,$$
$$u_t(x,0) = 0, \qquad 0 \le x < \infty,$$
$$u_x(0,t) + h\,u(0,t) = 0, \qquad\qquad t \ge 0, \quad h = \text{constant.}$$

State the compatibility condition of f.

13. Find the solution of the problem

$$u_{tt} = c^2 u_{xx}, \qquad at < x < \infty, \quad t > 0,$$
$$u(x,0) = f(x), \qquad 0 < x < \infty,$$
$$u_t(x,0) = 0, \qquad 0 < x < \infty,$$
$$u(at,t) = 0, \qquad\qquad t > 0,$$

where $f(0) = 0$ and a is constant.

14. Find the solution of the initial boundary-value problem

$$u_{tt} = u_{xx}, \qquad 0 < x < 2, \quad t > 0,$$
$$u(x,0) = \sin(\pi x/2), \qquad 0 \le x \le 2,$$
$$u_t(x,0) = 0, \qquad 0 \le x \le 2,$$
$$u(0,t) = 0, \ u(2,t) = 0, \quad t \ge 0.$$

15. Find the solution of the initial boundary-value problem

$$u_{tt} = 4\,u_{xx}, \qquad 0 < x < 1, \quad t > 0,$$
$$u(x,0) = 0, \qquad 0 \le x \le 1,$$
$$u_t(x,0) = x(1-x), \qquad 0 \le x \le 1,$$
$$u(0,t) = 0, \ u(1,t) = 0, \quad t \ge 0.$$

16. Determine the solution of the initial boundary-value problem

$$u_{tt} = c^2 u_{xx}, \qquad 0 < x < l, \quad t > 0,$$
$$u(x,0) = f(x), \qquad 0 \le x \le l,$$
$$u_t(x,0) = g(x), \qquad 0 \le x \le l,$$
$$u_x(0,t) = 0, \ u_x(l,t) = 0, \ t \ge 0,$$

by extending f and g as even functions about $x = 0$ and $x = l$.

17. Determine the solution of the initial boundary-value problem

$$u_{tt} = c^2 u_{xx}, \qquad 0 < x < l, \quad t > 0,$$
$$u(x,0) = f(x), \qquad 0 \le x \le l,$$
$$u_t(x,0) = g(x), \qquad 0 \le x \le l,$$
$$u(0,t) = p(t), \ u(l,t) = q(t), \ t \ge 0.$$

18. Determine the solution of the initial boundary-value problem

$$u_{tt} = c^2 u_{xx}, \qquad 0 < x < l, \qquad t > 0,$$
$$u(x,0) = f(x), \qquad 0 \le x \le l,$$
$$u_t(x,0) = g(x), \qquad 0 \le x \le l,$$
$$u_x(0,t) = p(t), \quad u_x(l,t) = q(t), \quad t \ge 0.$$

19. Solve the characteristic initial-value problem

$$xy^3 u_{xx} - x^3 y\, u_{yy} - y^3 u_x + x^3 u_y = 0,$$
$$u(x,y) = f(x) \quad \text{on} \quad y^2 - x^2 = 8 \quad \text{for} \quad 0 \le x \le 2,$$
$$u(x,y) = g(x) \quad \text{on} \quad y^2 + x^2 = 16 \quad \text{for} \quad 2 \le x \le 4,$$

with $f(2) = g(2)$.

20. Solve the Goursat problem

$$xy^3 u_{xx} - x^3 y\, u_{yy} - y^3 u_x + x^3 u_y = 0,$$
$$u(x,y) = f(x) \quad \text{on} \quad y^2 + x^2 = 16 \quad \text{for} \quad 0 \le x \le 4,$$
$$u(x,y) = g(y) \quad \text{on} \quad x = 0 \quad \text{for} \quad 0 \le y \le 4,$$

where $f(0) = g(4)$.

21. Solve

$$u_{tt} = c^2 u_{xx},$$
$$u(x,t) = f(x) \quad \text{on} \quad t = t(x),$$
$$u(x,t) = g(x) \quad \text{on} \quad x + ct = 0,$$

where $f(0) = g(0)$.

22. Solve the characteristic initial-value problem

$$xu_{xx} - x^3 u_{yy} - u_x = 0, \qquad x \ne 0,$$
$$u(x,y) = f(y) \quad \text{on} \quad y - \frac{x^2}{2} = 0 \quad \text{for} \quad 0 \le y \le 2,$$
$$u(x,y) = g(y) \quad \text{on} \quad y + \frac{x^2}{2} = 4 \quad \text{for} \quad 2 \le y \le 4,$$

where $f(2) = g(2)$.

23. Solve

$$u_{xx} + 10\, u_{xy} + 9\, u_{yy} = 0,$$
$$u(x,0) = f(x),$$
$$u_y(x,0) = g(x).$$

24. Solve

$$4\,u_{xx} + 5\,u_{xy} + u_{yy} + u_x + u_y = 2,$$
$$u\,(x,0) = f\,(x)\,,$$
$$u_y\,(x,0) = g\,(x)\,.$$

25. Solve

$$3\,u_{xx} + 10\,u_{xy} + 3\,u_{yy} = 0,$$
$$u\,(x,0) = f\,(x)\,, \quad u_y\,(x,0) = g\,(x)\,.$$

26. Solve

$$u_{xx} - 3\,u_{xy} + 2\,u_{yy} = 0,$$
$$u\,(x,0) = f\,(x)\,, \quad u_y\,(x,0) = g\,(x)\,.$$

27. Solve

$$x^2 u_{xx} - t^2 u_{tt} = 0 \quad x > 0, \quad t > 0,$$
$$u\,(x,1) = f\,(x)\,,$$
$$u_t\,(x,1) = g\,(x)\,.$$

28. Consider the initial boundary-value problem for a string of length l under the action of an external force $q\,(x,t)$ per unit length. The displacement $u\,(x,t)$ satisfies the wave equation

$$\rho\,u_{tt} = T u_{xx} + \rho\,q\,(x,t)\,,$$

where ρ is the line density of the string and T is the constant tension of the string. The initial and boundary conditions of the problem are

$$u\,(x,0) = f\,(x)\,, \qquad u_t\,(x,0) = g\,(x)\,, \qquad 0 \le x \le l,$$
$$u\,(0,t) = u\,(l,t) = 0, \qquad t > 0.$$

Show that the energy equation is

$$\frac{dE}{dt} = [T u_x u_t]_0^l + \int_0^l \rho\,q\,u_t\,dx,$$

where E represents the energy integral

$$E\,(t) = \frac{1}{2} \int_0^l \left(\rho\,u_t^2 + T u_x^2 \right) dx.$$

Explain the physical significance of the energy equation.

Hence or otherwise, derive the principle of conservation of energy, that is, that the total energy is constant for all $t \ge 0$ provided that the string has free or fixed ends and there are no external forces.

29. Show that the solution of the signaling problem governed by the wave equation

$$u_{tt} = c^2 u_{xx}, \qquad x > 0, \qquad t > 0,$$
$$u(x,0) = u_t(x,0) = 0, \qquad x > 0,$$
$$u(0,t) = U(t), \qquad t > 0,$$

is

$$u(x,t) = U\left(t - \frac{x}{c}\right) H\left(t - \frac{x}{c}\right),$$

where H is the Heaviside unit step function.

30. Obtain the solution of the initial-value problem of the homogeneous wave equation

$$u_{tt} - c^2 u_{xx} = \sin(kx - \omega t), \qquad -\infty < x < \infty, \qquad t > 0,$$
$$u(x,0) = 0 = u_t(x,0), \qquad \text{for all} \quad x \in \mathbb{R},$$

where c, k and ω are constants.

Discuss the non-resonance case, $\omega \neq ck$ and the resonance case, $\omega = ck$.

31. In each of the following Cauchy problems, obtain the solution of the system

$$u_{tt} - c^2 u_{xx} = 0, \qquad x \in \mathbb{R}, \qquad t > 0,$$
$$u(x,0) = f(x) \quad \text{and} \quad u_t(x,0) = g(x) \quad \text{for} \quad x \in \mathbb{R},$$

for the given c, $f(x)$ and $g(x)$:

(a) $c = 3$, $\quad f(x) = \cos x$, $\qquad g(x) = \sin 2x$.

(b) $c = 1$, $\quad f(x) = \sin 3x$, $\qquad g(x) = \cos 3x$.

(c) $c = 7$, $\quad f(x) = \cos 3x$, $\qquad g(x) = x$.

(d) $c = 2$, $\quad f(x) = \cosh x$, $\qquad g(x) = 2x$.

(e) $c = 3$, $\quad f(x) = x^3$, $\qquad g(x) = x \cos x$.

(f) $c = 4$, $\quad f(x) = \cos x$, $\qquad g(x) = xe^{-x}$.

32. If $u(x,t)$ is the solution of the nonhomogeneous Cauchy problem

$$u_{tt} - c^2 u_{xx} = p(x,t), \qquad \text{for} \quad x \in \mathbb{R}, \qquad t > 0,$$
$$u(x,0) = 0 = u_t(x,0), \qquad \text{for} \quad x \in \mathbb{R},$$

and if $v(x,t,\tau)$ is the solution of the nonhomogeneous Cauchy problem

$$v_{tt} - c^2 v_{xx} = 0, \quad \text{for} \quad x \in \mathbb{R}, \quad t > 0,$$
$$v(x,0;\tau) = 0, \quad v_t(x,0;\tau) = p(x,\tau), \quad x \in \mathbb{R},$$

show that

$$u(x,t) = \int_0^t v(x,t;\tau)\, d\tau.$$

This is known as the *Duhamel principle* for the wave equation.

33. Show that the solution of the nonhomogeneous diffusion equation with homogeneous boundary and initial data

$$u_t = \kappa u_{xx} + p(x,t), \quad 0 < x < l, \quad t > 0,$$
$$u(0,t) = 0 = u(l,t), \quad t > 0,$$
$$u(x,0) = 0, \quad 0 < x < l,$$

is

$$u(x,t) = \int_0^t v(x,t;\tau)\, d\tau,$$

where $v = v(x,t;\tau)$ satisfies the homogeneous diffusion equation with nonhomogeneous boundary and initial data

$$v_{tt} = \kappa v_{xx} + p(x,t), \quad 0 < x < l, \quad t > 0,$$
$$v(0,t;\tau) = 0 = v(l,t;\tau), \quad t > 0,$$
$$v(x,\tau;\tau) = p(x,\tau).$$

This is known as the *Duhamel principle* for the diffusion equation.

34. Use the Duhamel principle to solve the nonhomogeneous diffusion equation

$$u_t = \kappa u_{xx} + e^{-t}\sin \pi x, \quad 0 < x < l, \quad t > 0,$$

with the homogeneous boundary and initial data

$$u(0,t) = 0, \quad u(1,t) = 0, \quad t > 0,$$
$$u(x,0) = 0, \quad 0 \le x \le 1.$$

35. (a) Verify that

$$u_n(x,y) = \exp\left(ny - \sqrt{n}\right)\sin nx,$$

is the solution of the Laplace equation

$$u_{xx} + u_{yy} = 0, \quad x \in \mathbb{R}, \quad y > 0,$$
$$u(x, 0) = 0, \quad u_y(x, 0) = n \exp\left(-\sqrt{n}\right) \sin nx,$$

where n is a positive integer.
(b) Show that this Cauchy problem is not well posed.

36. Show that the following Cauchy problems are not well posed:

(a) $\qquad u_t = u_{xx}, \quad x \in \mathbb{R}, \quad t > 0,$

$$u(0, t) = \left(\tfrac{2}{n}\right) \sin\left(2n^2 t\right), \quad u_x(0, t) = 0, \quad t > 0.$$

(b) $u_{xx} + u_{yy} = 0, \quad x \in \mathbb{R}, \quad t > 0,$

$$u_n(x, 0) \to 0, \quad (u_n)_y(x, 0) \to 0, \quad \text{as} \quad n \to \infty.$$

6

Fourier Series and Integrals with Applications

"The thorough study of nature is the most ground for mathematical discoveries."

Joseph Fourier

"Nearly fifty years had passed without any progress on the question of analytic representation of an arbitrary function, when an assertion of Fourier threw new light on the subject. Thus a new era began for the development of this part of Mathematics and this was heralded in a stunning way by major developments in mathematical Physics."

Bernhard Riemann

"Fourier created a coherent method by which the different components of an equation and its solution in series were neatly identified with different aspects of physical solution being analyzed. He also had a uniquely sure instinct for interpreting the asymptotic properties of the solutions of his equations for their physical meaning. So powerful was his approach that a full century passed before non-linear equations regained prominence in mathematical physics."

Ioan James

6.1 Introduction

This chapter is devoted to the theory of Fourier series and integrals. Although the treatment can be extensive, the exposition of the theory here will be concise, but sufficient for its application to many problems of applied mathematics and mathematical physics.

The Fourier theory of trigonometric series is of great practical importance because certain types of discontinuous functions which cannot be expanded in power series can be expanded in Fourier series. More importantly, a wide class of problems in physics and engineering possesses periodic phenomena and, as a consequence, Fourier's trigonometric series become an indispensable tool in the analysis of these problems.

We shall begin our study with the basic concepts and definitions of some properties of real-valued functions.

6.2 Piecewise Continuous Functions and Periodic Functions

A single-valued function f is said to be *piecewise continuous* in an interval $[a, b]$ if there exist finitely many points $a = x_1 < x_2 < \ldots < x_n = b$, such that f is continuous in the intervals $x_j < x < x_{j+1}$ and the one-sided limits $f(x_j+)$ and $f(x_{j+1}-)$ exist for all $j = 1, 2, 3, \ldots, n - 1$.

A piecewise continuous function is shown in Figure 6.2.1. Functions such as $1/x$ and $\sin(1/x)$ fail to be piecewise continuous in the closed interval $[0, 1]$ because the one-sided limit $f(0+)$ does not exist in either case.

If f is piecewise continuous in an interval $[a, b]$, then it is necessarily bounded and integrable over that interval. Also, it follows immediately that the product of two piecewise continuous functions is piecewise continuous on a common interval.

If f is piecewise continuous in an interval $[a, b]$ and if, in addition, the first derivative f' is continuous in each of the intervals $x_j < x < x_{j+1}$, and the limits $f'(x_j+)$ and $f'(x_j-)$ exist, then f is said to be *piecewise smooth*; if, in addition, the second derivative f'' is continuous in each of

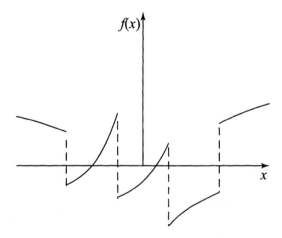

Figure 6.2.1 Graph of a Piecewise Continuous Function.

the intervals $x_j < x < x_{j+1}$, and the limits $f''(x_j+)$ and $f''(x_j-)$ exist, then f is said to be *piecewise very smooth*.

A piecewise continuous function $f(x)$ in an interval $[a, b]$ is said to be *periodic* if there exists a real positive number p such that

$$f(x + p) = f(x), \tag{6.2.1}$$

for all x, p is called the *period* of f, and the smallest value of p is termed the *fundamental period*. A sample graph of a periodic function is given in Figure 6.2.2.

If f is periodic with period p, then

$$f(x + p) = f(x),$$
$$f(x + 2p) = f(x + p + p) = f(x + p),$$
$$f(x + 3p) = f(x + 2p + p) = f(x + 2p),$$
$$f(x + np) = f(x + (n - 1)p + p) = f(x + (n - 1)p) = f(x),$$

for any integer n. Hence, for all integral values of n

$$f(x + np) = f(x). \tag{6.2.2}$$

It can be readily shown that if $f_1, f_2, \ldots, f_k$ have the period p and c_k are the constants, then

$$f = c_1 f_1 + c_2 f_2 + \ldots + c_k f_k, \tag{6.2.3}$$

has the period p.

Well known examples of periodic functions are the sine and cosine functions. As a special case, a constant function is also a periodic function with arbitrary period p. Thus, by the relation (6.2.3), the series

$$a_0 + a_1 \cos x + a_2 \cos 2x + \ldots + b_1 \sin x + b_2 \sin 2x + \ldots$$

if it converges, obviously has the period 2π. Such types of series, which occur frequently in problems of applied mathematics and mathematical physics, will be treated later.

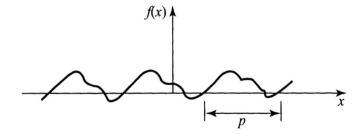

Figure 6.2.2 Periodic Function.

6.3 Systems of Orthogonal Functions

A sequence of functions $\{\phi_n(x)\}$ is said to be *orthogonal* with respect to the weight function $q(x)$ on the interval $[a, b]$ if

$$\int_a^b \phi_m(x)\,\phi_n(x)\,q(x)\,dx = 0, \qquad m \neq n. \qquad (6.3.1)$$

If $m = n$, then we have

$$\|\phi_n\| = \left[\int_a^b \phi_n^2(x)\,q(x)\,dx\right]^{\frac{1}{2}} \qquad (6.3.2)$$

which is called the *norm* of the orthogonal system $\{\phi_n(x)\}$.

Example 6.3.1. The sequence of functions $\{\sin mx\}$, $m = 1, 2, \ldots$, form an orthogonal system on the interval $[-\pi, \pi]$, because

$$\int_{-\pi}^{\pi} \sin mx \, \sin nx \, dx = \begin{cases} 0, & m \neq n, \\ \pi, & m = n. \end{cases}$$

In this example we notice that the weight function is equal to unity, and the value of the norm is $\sqrt{\pi}$.

An orthogonal system $\phi_1, \phi_2, \ldots, \phi_n$, where n may be finite or infinite, which satisfies the relations

$$\int_a^b \phi_m(x)\,\phi_n(x)\,q(x)\,dx = \begin{cases} 0, & m \neq n, \\ 1, & m = n, \end{cases} \qquad (6.3.3)$$

is called an *orthonormal* system of functions on $[a, b]$. It is evident that an orthonormal system can be obtained from an orthogonal system by dividing each function by its norm on $[a, b]$.

Example 6.3.2. The sequence of functions

$$1, \cos x, \sin x, \ldots, \cos nx, \sin nx$$

forms an orthogonal system on $[-\pi, \pi]$ since

$$\int_{-\pi}^{\pi} \sin mx \, \sin nx \, dx = \begin{cases} 0, & m \neq n, \\ \pi, & m = n, \end{cases}$$

$$\int_{-\pi}^{\pi} \sin mx \, \cos nx \, dx = 0, \text{ for all } m, n, \qquad (6.3.4)$$

$$\int_{-\pi}^{\pi} \cos mx \, \cos nx \, dx = \begin{cases} 0, & m \neq n, \\ \pi, & m = n, \end{cases}$$

for positive integers m and n. To normalize this system, we divide the elements of the original orthogonal system by their norms. Hence,

$$\frac{1}{\sqrt{2\pi}}, \frac{\cos x}{\sqrt{\pi}}, \frac{\sin x}{\sqrt{\pi}}, \dots, \frac{\cos nx}{\sqrt{\pi}}, \frac{\sin nx}{\sqrt{\pi}}$$

forms an orthogonal system.

6.4 Fourier Series

The functions
$$1, \cos x, \sin x, \dots, \cos 2x, \sin 2x, \dots$$

are mutually orthogonal to each other in the interval $[-\pi, \pi]$ and are linearly independent. Thus, we *formally* associate a trigonometric series with any piecewise continuous periodic function $f(x)$ of period 2π and write

$$f(x) \sim \frac{a_0}{2} + \sum_{k=1}^{\infty} (a_k \cos kx + b_k \sin kx), \tag{6.4.1}$$

where the symbol $\sim$ indicates an association of a_0, a_k, and b_k to f in some unique manner. The coefficients a_0, a_k and b_k will be determined soon. The coefficient $(a_0/2)$ instead of a_0 is used for convenience in the representation. However, it is not easy to say that the series on the right hand side of (6.4.1) itself converges and also represents the function $f(x)$. Indeed, the series may converge or diverge.

Let $f(x)$ be a Riemann integrable function defined on the interval $[-\pi, \pi]$. Suppose that we define the nth partial sum

$$s_n(x) = \frac{a_0}{2} + \sum_{k=1}^{n} (a_k \cos kx + b_k \sin kx), \tag{6.4.2}$$

to represent $f(x)$ on $[-\pi, \pi]$. We shall seek the coefficients a_0, a_k, and b_k such that $s_n(x)$ represents the best approximation to $f(x)$ in the sense of least squares, that is, we seek to minimize the integral

$$I(a_0, a_k, b_k) = \int_{-\pi}^{\pi} [f(x) - s_n(x)]^2 \, dx. \tag{6.4.3}$$

This is an extremal problem. A necessary condition for a_0, a_k, b_k, so that I be minimum, that is the first partial derivatives of I with respect to these coefficients vanish. Thus, substituting equation (6.4.2) into (6.4.3) and differentiating with respect to a_0, a_k, and b_k, we obtain

$$\frac{\partial I}{\partial a_0} = -\int_{-\pi}^{\pi} \left[f(x) - \frac{a_0}{2} - \sum_{j=1}^{n} (a_j \cos jx + b_j \sin jx) \right] dx. \qquad (6.4.4)$$

$$\frac{\partial I}{\partial a_k} = -2\int_{-\pi}^{\pi} \left[f(x) - \frac{a_0}{2} - \sum_{j=1}^{n} (a_j \cos jx + b_j \sin jx) \right] \cos kx\, dx. (6.4.5)$$

$$\frac{\partial I}{\partial b_k} = -2\int_{-\pi}^{\pi} \left[f(x) - \frac{a_0}{2} - \sum_{j=1}^{n} (a_j \cos jx + b_j \sin jx) \right] \sin kx\, dx. (6.4.6)$$

Using the orthogonality relations of the trigonometric functions (6.3.4) and noting that

$$\int_{-\pi}^{\pi} \cos mx\, dx = \int_{-\pi}^{\pi} \sin mx\, dx = 0, \qquad (6.4.7)$$

where m and n are positive integers, equations (6.4.4), (6.4.5), and (6.4.6) become

$$\frac{\partial I}{\partial a_0} = \pi a_0 - \int_{-\pi}^{\pi} f(x)\, dx, \qquad (6.4.8)$$

$$\frac{\partial I}{\partial a_k} = 2\pi a_k - 2\int_{-\pi}^{\pi} f(x) \cos kx\, dx, \qquad (6.4.9)$$

$$\frac{\partial I}{\partial b_k} = 2\pi b_k - 2\int_{-\pi}^{\pi} f(x) \sin kx\, dx, \qquad (6.4.10)$$

which must vanish for I to have an extremal value. Thus, we have

$$a_0 = \frac{1}{\pi} \int_{-\pi}^{\pi} f(x)\, dx, \qquad (6.4.11)$$

$$a_k = \frac{1}{\pi} \int_{-\pi}^{\pi} f(x) \cos kx\, dx, \qquad (6.4.12)$$

$$b_k = \frac{1}{\pi} \int_{-\pi}^{\pi} f(x) \sin kx\, dx. \qquad (6.4.13)$$

Note that a_0 is the special case of a_k which is the reason for writing $(a_0/2)$ rather than a_0 in equation (6.4.1). It immediately follows from equations (6.4.8), (6.4.9), and (6.4.10) that

$$\frac{\partial^2 I}{\partial a_0^2} = \pi, \qquad (6.4.14)$$

$$\frac{\partial^2 I}{\partial a_k^2} = \frac{\partial^2 I}{\partial b_k^2} = 2\pi, \qquad (6.4.15)$$

and all mixed second order and all remaining higher order derivatives vanish. Now if we expand I in a Taylor series about $(a_0, a_1, \ldots, a_n, b_1, \ldots, b_n)$, we have

$$I\left(a_0 + \Delta a_0, \ldots, b_n + \Delta b_n\right) = I\left(a_0, \ldots, b_n\right) + \Delta I, \qquad (6.4.16)$$

where ΔI stands for the remaining terms. Since the first derivatives, all mixed second derivatives, and all remaining higher derivatives vanish, we obtain

$$\Delta I = \frac{1}{2!}\left[\frac{\partial^2 I}{\partial a_0^2}\,\Delta a_0^2 + \sum_{k=1}^{n}\left(\frac{\partial^2 I}{\partial a_k^2}\,\Delta a_k^2 + \frac{\partial^2 I}{\partial b_k^2}\,\Delta b_k^2\right)\right]. \qquad (6.4.17)$$

By virtue of equations (6.4.14) and (6.4.15), ΔI is positive. Hence, for I to have a minimum value, the coefficients a_0, a_k, b_k must be given by equations (6.4.11), (6.4.12), and (6.4.13) respectively. These coefficients are called the *Fourier coefficients* of $f\left(x\right)$ and the series in (6.4.1) is said to be the *Fourier series* corresponding to $f\left(x\right)$, where its coefficients a_0, a_k and b_k are given by (6.4.11), (6.4.12), and (6.4.13) respectively. Thus, the correspondence (6.4.1) asserts nothing about the convergence or divergence of the formally constructed Fourier series. The question arises whether it is possible to represent *all continuous functions* by Fourier series. The investigation of the sufficient conditions for such a representation to be possible turns out to be a difficult problem.

We remark that the possibility of representing the given function $f\left(x\right)$ by a Fourier series does not imply that the Fourier series converges to the function $f\left(x\right)$. If the Fourier series of a continuous function converges uniformly, then it represents the function. As a matter of fact, there exist Fourier series which diverge. A convergent trigonometric series need not be a Fourier series. For instance, the trigonometric series

$$\sum_{n=2}^{\infty} \frac{\sin nx}{\log n},$$

which is convergent for all values of x, is not a Fourier series, for there is no integrable function corresponding to this series.

6.5 Convergence of Fourier Series

We introduce three kinds of convergence of a Fouriers Series: (i) Pointwise Convergence, (ii) Uniform Convergence, and (iii) Mean-Square Convergence.

Definition 6.5.1. *(Pointwise Convergence). An infinite series $\sum_{n=1}^{\infty} f_n\left(x\right)$ is called* pointwise convergent *in $a < x < b$ to $f\left(x\right)$ if it converges to $f\left(x\right)$ for each x in $a < x < b$. In other words, for each x in $a < x < b$, we have*

$$\left|f\left(x\right) - s_n\left(x\right)\right| \to 0 \qquad as \quad n \to \infty,$$

where $s_n\left(x\right)$ is the nth partial sum defined by $s_n\left(x\right) = \sum_{k=1}^{n} f_k\left(x\right)$.

Definition 6.5.2. *(Uniform Convergence). The series $\sum_{n=1}^{\infty} f_n(x)$ is said to* converge uniformly *to $f(x)$ in $a \leq x \leq b$ if*

$$\max_{a \leq x \leq b} |f(x) - s_n(x)| \to 0 \qquad as \quad n \to \infty.$$

Evidently, uniform convergence implies pointwise convergence, but the converse is not necessarily true.

Definition 6.5.3. *(Mean-Square Convergence). The series $\sum_{n=1}^{\infty} f_n(x)$ converges in the* mean-square *(or L^2) sense to $f(x)$ in $a \leq x \leq b$ if*

$$\int_a^b |f(x) - s_n(x)|^2 \, dx \to 0 \qquad as \quad n \to \infty.$$

It is noted that uniform convergence is stronger than both pointwise convergence and mean-square convergence.

The study of convergence of Fourier series has a long and complex history. The fundamental question is whether the Fourier series of a periodic function f converge to f. The answer is certainly *not* obvious. If $f(x)$ is 2π-periodic continuous function, then the Fourier series (6.4.1) may converge to f for a given x in $-\pi \leq x \leq \pi$, but *not* for all x in $-\pi \leq x \leq \pi$. This leads to the questions of *local convergence* or the behavior of f near a given point x, and of *global convergence* or the overall behavior of a function f over the entire interval $[-\pi, \pi]$.

There is another question that deals with the mean-square convergence of the Fourier series to $f(x)$ in $(-\pi, \pi)$, that is, if $f(x)$ is integrable on $(-\pi, \pi)$, then

$$\frac{1}{2\pi} \int_{-\pi}^{\pi} |f(x) - s_n(x)|^2 \, dx \to 0 \qquad as \quad n \to \infty.$$

This is known as the *mean-square convergence* theorem which does not provide any insight into the problem of pointwise convergence. Indeed, the mean-square convergence theorem does not guarantee the convergence of the Fourier series for any x. On the other hand, if $f(x)$ is 2π-periodic and piecewise smooth on $\mathbb{R}$, then the Fourier series (6.4.1) of the function f converges for every x in $-\pi \leq x \leq \pi$. It has been known since 1876 that there are periodic continuous functions whose Fourier series *diverge* at certain points. It was an open question for a period of a century whether a Fourier series of a continuous function converges at any point. In 1966, Lennart Carleson (1966) provided an affirmative answer with a deep theorem which states that the Fourier series of any square integrable function f converges to f at almost every point.

Let $f(x)$ be piecewise continuous and periodic with period 2π. It is obvious that

$$\int_{-\pi}^{\pi} [f(x) - s_n(x)]^2 \, dx \geq 0, \tag{6.5.1}$$

Expanding (6.5.1) gives

$$\int_{-\pi}^{\pi} [f(x) - s_n(x)]^2 \, dx = \int_{-\pi}^{\pi} [f(x)]^2 \, dx - 2 \int_{-\pi}^{\pi} f(x) s_n(x) \, dx$$
$$+ \int_{-\pi}^{\pi} [s_n(x)]^2 \, dx.$$

But, by the definitions of the Fourier coefficients (6.4.11), (6.4.12), and (6.4.13) and by the orthogonal relations for the trigonometric series (6.3.4), we have

$$\int_{-\pi}^{\pi} f(x) s_n(x) \, dx = \int_{-\pi}^{\pi} f(x) \left[\frac{a_0}{2} + \sum_{k=1}^{n} (a_k \cos kx + b_k \sin kx) \right] dx$$
$$= \frac{\pi a_0^2}{2} + \pi \sum_{k=1}^{n} (a_k^2 + b_k^2), \tag{6.5.2}$$

and

$$\int_{-\pi}^{\pi} s_n^2(x) \, dx = \int_{-\pi}^{\pi} \left[\frac{a_0}{2} + \sum_{k=1}^{n} (a_k \cos kx + b_k \sin kx) \right]^2 dx$$
$$= \frac{\pi a_0^2}{2} + \pi \sum_{k=1}^{n} (a_k^2 + b_k^2). \tag{6.5.3}$$

Consequently,

$$\int_{-\pi}^{\pi} [f(x) - s_n(x)]^2 \, dx = \int_{-\pi}^{\pi} f^2(x) \, dx - \left[\frac{\pi a_0^2}{2} + \pi \sum_{k=1}^{n} (a_k^2 + b_k^2) \right] \geq 0. \tag{6.5.4}$$

It follows from (6.5.4) that

$$\frac{a_0^2}{2} + \sum_{k=1}^{n} (a_k^2 + b_k^2) \leq \frac{1}{\pi} \int_{-\pi}^{\pi} f^2(x) \, dx \tag{6.5.5}$$

for all values of n. Since the right hand of equation (6.5.5) is independent of n, we obtain

$$\frac{a_0^2}{2} + \sum_{k=1}^{\infty} (a_k^2 + b_k^2) \leq \frac{1}{\pi} \int_{-\pi}^{\pi} f^2(x) \, dx. \tag{6.5.6}$$

This is known as *Bessel's inequality*.

We see that the left side is nondecreasing and is bounded above, and therefore, the series

$$\frac{a_0^2}{2} + \sum_{k=1}^{\infty} \left(a_k^2 + b_k^2 \right), \tag{6.5.7}$$

converges. Thus, the necessary condition for the convergence of series (6.5.7) is that

$$\lim_{k \to \infty} a_k = 0, \qquad \lim_{k \to \infty} b_k = 0. \tag{6.5.8}$$

The Fourier series is said to *converge in the mean* to $f(x)$ when

$$\lim_{n \to \infty} \int_{-\pi}^{\pi} \left[f(x) - \left(\frac{a_0}{2} + \sum_{k=1}^{n} a_k \cos kx + b_k \sin kx \right) \right]^2 dx = 0. \tag{6.5.9}$$

If the Fourier series converges in the mean to $f(x)$, then

$$\frac{a_0^2}{2} + \sum_{k=1}^{\infty} \left(a_k^2 + b_k^2 \right) = \frac{1}{\pi} \int_{-\pi}^{\pi} f^2(x)\, dx. \tag{6.5.10}$$

This is called *Parseval's relation* and is one of the central results in the theory of Fourier series. This relation is frequently used to derive the sum of many important numerical infinite series. Furthermore, if the relation (6.5.9) holds true, the set of trigonometric functions 1, $\cos x$, $\sin x$, $\cos 2x$, $\sin 2x$, ... is said to be *complete*.

The Parseval relation (6.5.10) can formally be derived from the convergence of Fourier series to $f(x)$ in $[-\pi, \pi]$. In other words, if

$$f(x) = \frac{1}{2} a_0 + \sum_{k=1}^{\infty} \left(a_k \cos kx + b_k \sin kx \right), \tag{6.5.11}$$

where a_0, a_k and b_k are given by (6.4.11), (6.4.12) and (6.4.13) respectively, we multiply (6.4.11) by $\frac{1}{\pi} f(x)$ and integrate the resulting expression from $-\pi$ to π to obtain

$$\frac{1}{\pi} \int_{-\pi}^{\pi} f^2(x)\, dx = \frac{a_0}{2\pi} \int_{-\pi}^{\pi} f(x)\, dx + \sum_{k=1}^{\infty} \left[\frac{a_k}{\pi} \int_{-\pi}^{\pi} f(x) \cos kx\, dx \right.$$
$$\left. + \frac{b_k}{\pi} \int_{-\pi}^{\pi} f(x) \sin kx\, dx \right]. \tag{6.5.12}$$

Replacing all integrals on the right hand side of (6.5.12) by the Fourier coefficients gives the Parseval relation (6.5.10).

6.6 Examples and Applications of Fourier Series

The Fourier coefficients (6.4.11), (6.4.12), and (6.4.13) of Section 6.4 may be obtained in a different way. Suppose the function $f(x)$ of period 2π has the Fourier series expansion

$$f(x) = \frac{a_0}{2} + \sum_{k=1}^{\infty} (a_k \cos kx + b_k \sin kx). \tag{6.6.1}$$

If we assume that the infinite series is term-by-term integrable (we will see later that uniform convergence of the series is a sufficient condition for this), then

$$\int_{-\pi}^{\pi} f(x)\, dx = \int_{-\pi}^{\pi} \left[\frac{a_0}{2} + \sum_{k=1}^{\infty} (a_k \cos kx + b_k \sin kx) \right] dx = \pi a_0.$$

Hence,

$$a_0 = \frac{1}{\pi} \int_{-\pi}^{\pi} f(x)\, dx. \tag{6.6.2}$$

Again, we multiply both sides of equations (6.6.1) by $\cos nx$ and integrate the resulting expression from $-\pi$ to π. We obtain

$$\int_{-\pi}^{\pi} f(x) \cos nx\, dx = \int_{-\pi}^{\pi} \left[\frac{a_0}{2} + \sum_{k=1}^{\infty} (a_k \cos kx + b_k \sin kx) \right] \cos nx\, dx = \pi a_k.$$

Thus,

$$a_k = \frac{1}{\pi} \int_{-\pi}^{\pi} f(x) \cos kx\, dx. \tag{6.6.3}$$

In a similar manner, we find that

$$b_k = \frac{1}{\pi} \int_{-\pi}^{\pi} f(x) \sin kx\, dx. \tag{6.6.4}$$

The coefficients a_0, a_k, b_k just found are exactly the same as those obtained in Section 6.4.

Example 6.6.1. Find the Fourier series expansion for the function shown in Figure 6.6.1.

$$f(x) = x + x^2, \qquad -\pi < x < \pi.$$

Here

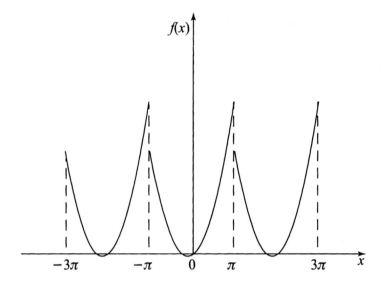

Figure 6.6.1 Graph of $f(x) = x + x^2$.

$$a_0 = \frac{1}{\pi} \int_{-\pi}^{\pi} f(x) \, dx = \frac{1}{\pi} \int_{-\pi}^{\pi} \left(x + x^2\right) dx = \frac{2\pi^2}{3},$$

and

$$a_k = \frac{1}{\pi} \int_{-\pi}^{\pi} f(x) \cos kx \, dx$$

$$= \frac{1}{\pi} \int_{-\pi}^{\pi} \left(x + x^2\right) \cos kx \, dx$$

$$= \frac{1}{\pi} \left[\frac{x \sin kx}{k} \Big|_{-\pi}^{\pi} - \int_{-\pi}^{\pi} \frac{\sin kx}{k} \, dx \right]$$

$$+ \frac{1}{\pi} \left[\frac{x^2 \sin kx}{k} \Big|_{-\pi}^{\pi} - \int_{-\pi}^{\pi} \frac{2x \sin kx}{k} \, dx \right]$$

$$= -\frac{2}{k\pi} \left[-\frac{x \cos kx}{k} \Big|_{-\pi}^{\pi} + \int_{-\pi}^{\pi} \frac{\cos kx}{k} \, dx \right]$$

$$= \frac{4}{k^2} \cos k\pi = \frac{4}{k^2} (-1)^k \quad \text{for } k = 1, 2, 3, \dots.$$

Similarly,

$$b_k = \frac{1}{\pi} \int_{-\pi}^{\pi} f(x) \sin kx \, dx$$

$$= \frac{1}{\pi} \int_{-\pi}^{\pi} (x + x^2) \sin kx \, dx$$

$$= -\frac{2}{k} \cos k\pi = -\frac{2}{k} (-1)^k, \quad \text{for } k = 1, 2, 3, \dots.$$

Therefore, the Fourier series expansion for f is

$$f(x) = \frac{\pi^2}{3} + \sum_{k=1}^{\infty} \left[\frac{4}{k^2} (-1)^k \cos kx - \frac{2}{k} (-1)^k \sin kx \right]$$

$$= \frac{\pi^2}{3} - 4\cos x + 2\sin x + \cos 2x - \sin 2x - \dots. \qquad (6.6.5)$$

Example 6.6.2. Consider the periodic function shown in Figure 6.6.2.

$$f(x) = \begin{cases} -\pi, & -\pi < x < 0, \\ x, & 0 < x < \pi. \end{cases}$$

In this case,

$$a_0 = \frac{1}{\pi} \int_{-\pi}^{\pi} f(x) \, dx$$

$$= \frac{1}{\pi} \left[\int_{-\pi}^{0} -\pi dx + \int_{0}^{\pi} x \, dx \right] = -\frac{\pi}{2},$$

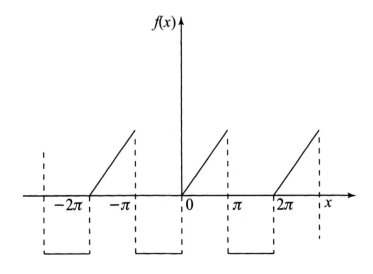

Figure 6.6.2 Graph of $f(x)$.

and

$$a_k = \frac{1}{\pi} \int_{-\pi}^{\pi} f(x) \cos kx\, dx$$

$$= \frac{1}{\pi} \left[\int_{-\pi}^{0} -\pi \cos kx\, dx + \int_{0}^{\pi} x \cos kx\, dx \right]$$

$$= \frac{1}{k^2 \pi} (\cos k\pi - 1) = \frac{1}{k^2 \pi} \left[(-1)^k - 1 \right].$$

Also

$$b_k = \frac{1}{\pi} \int_{-\pi}^{\pi} f(x) \sin kx\, dx$$

$$= \frac{1}{\pi} \left[\int_{-\pi}^{0} -\pi \sin kx\, dx + \int_{0}^{\pi} x \sin kx\, dx \right]$$

$$= \frac{1}{k} (1 - 2 \cos k\pi) = \frac{1}{k} \left[1 - 2(-1)^k \right].$$

Hence, the Fourier series is

$$f(x) = -\frac{\pi}{4} + \sum_{k=1}^{\infty} \left\{ \frac{1}{k^2 \pi} \left[(-1)^k - 1 \right] \cos kx + \frac{1}{k} \left[1 - 2(-1)^k \right] \sin kx \right\}.$$

$$(6.6.6)$$

Example 6.6.3. Consider the sawtooth wave function $f(x) = x$ in the interval $-\pi < x < \pi$, $f(x) = f(x \pm 2k\pi)$ for $k = 1, 2, \ldots$.

This is a periodic function with period 2π and represents a sawtooth wave function as shown in Figure 6.6.3 and it is piecewise continuous.

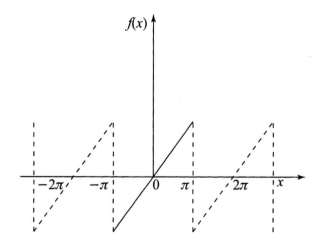

Figure 6.6.3 The sawtooth wave function.

We can readily find that

$$a_k = 0, \qquad k = 0, 1, 2, \dots.$$

The coefficients b_k are given by

$$b_k = \frac{1}{\pi} \int_{-\pi}^{\pi} f(x) \sin kx \, dx$$

$$= \frac{1}{\pi} \int_{-\pi}^{\pi} x \sin kx \, dx = \frac{2}{k} (-1)^{k+1}.$$

Hence, the Fourier series is

$$f(x) = 2 \sum_{k=1}^{\infty} (-1)^{k+1} \frac{\sin kx}{k} = 2 \left(\frac{\sin x}{1} - \frac{\sin 2x}{2} + \frac{\sin 3x}{3} - \frac{\sin 4x}{4} + \dots \right).$$

$$(6.6.7)$$

This Fourier series does not agree with the function at every point in $[-\pi, \pi]$. At the endpoints $x = \pm \pi$, the series is equal to zero, but the function does not vanish at either endpoint. However, the Fourier series (6.6.7) converges to the value x at each point x in $-\pi < x < \pi$, but it converges to $0 = \frac{1}{2} [f(\pi - 0) + f(\pi + 0)] = \frac{1}{2} (\pi - \pi)$ which is the mean value of the two limits as $x \to \pm \pi$. Thus, the convergence is *not* uniform. The partial sum of the series is

$$s_n(x) = 2 \sum_{k=1}^{n} \frac{1}{k} (-1)^{k+1} \sin kx. \qquad (6.6.8)$$

In the neighborhood of $x = \pm \pi$, the difference between $f(x)$ and $s_n(x)$ seems not to become smaller as n increases, but the size of the region where this occurs decreases indicating nonuniform convergence. This nonuniform oscillatory nature close to discontinuities is known as the *Gibbs phenomenon*.

It is important to point out that the sawtooth wave function f is not continuous in each point in $[-\pi, \pi]$. What happens in the neighborhood of $x = \pm \pi$?

We consider the partial sum $s_n(x)$ given by (6.6.8), and put $x_n = \pi - \frac{\pi}{n}$ to approximate the value $s_n(x)$. We have

$$s_n(x_n) = \sum_{k=1}^{n} \frac{2(-1)^{k+1}}{k} \sin k \left(\pi - \frac{\pi}{n} \right) = \sum_{k=1}^{n} \frac{2}{k} \sin \left(\frac{k\pi}{n} \right)$$

which can be rewritten in the form

$$= 2 \sum_{k=1}^{n} \frac{\sin \left(\frac{\pi k}{n} \right)}{\left(\frac{\pi k}{n} \right)} \cdot \left(\frac{\pi}{n} \right). \qquad (6.6.9)$$

This sum can be identified with a Riemann sum of the definite integral

$$2 \int_0^\pi \left(\frac{\sin x}{x} \right) dx$$

which is obtained by dividing the interval $[0, \pi]$ into the n subintervals $\left[\frac{(k-1)\pi}{n}, \frac{k\pi}{n} \right]$, $1 \leq k \leq n$. Obviously, each subinterval is of length $\left(\frac{\pi}{n} \right)$, and we evaluate the function in each such subinterval at the right-hand endpoint $\frac{k\pi}{n}$, $1 \leq k \leq n$. Consequently,

$$\lim_{n \to \infty} s_n(x) = 2 \int_0^\pi \frac{\sin x}{x} dx \approx 1.18\pi.$$

The point x_n, as $n \to \infty$, approaches $x = \pi$ from the left. Hence, $f(x_n)$ tends to the value π. The jump at the point $x = \pi$ is $f(\pi-) - f(\pi+) = 2\pi$, and, therefore, for sufficiently large n,

$$\frac{s_n(x_n) - f(x_n)}{f(\pi-) - f(\pi+)} \approx \frac{1.18\pi - \pi}{2\pi} = 0.09.$$

We next draw the graphs of $s_7(x)$ and $s_{10}(x)$ which exhibit oscillations over the graph of f as shown in Figures 6.6.4 (a) and (b). These graphs show the so-called overshooting in a neighborhood of $x = \pm\pi$ and $x = \pm 3\pi$, and hence, the Gibbs phenomenon.

We next derive the sum of several numerical series from Example 6.6.3. Substituting $x = \frac{\pi}{2}$ in (6.6.7) gives

$$\frac{\pi}{2} = 2 \sum_{k=1}^\infty \frac{(-1)^{k+1}}{k} \sin\left(\frac{\pi k}{2} \right) = 2 \left[1 - \frac{1}{3} + \frac{1}{5} - \frac{1}{7} + \dots \right].$$

This gives the well-known slowly convergent numerical series

$$1 - \frac{1}{3} + \frac{1}{5} - \frac{1}{7} + \dots = \frac{\pi}{4}. \tag{6.6.10}$$

On the other hand, putting $x = \frac{\pi}{4}$ in (6.6.7) gives another numerical series

$$\frac{\pi}{8} = \frac{1}{\sqrt{2}} - \frac{1}{2} + \frac{1}{3\sqrt{2}} - \frac{1}{5\sqrt{2}} - \frac{1}{6} - \frac{1}{7\sqrt{2}} + \frac{1}{9\sqrt{2}} - \frac{1}{10} + \frac{1}{11\sqrt{2}} - \frac{1}{13\sqrt{2}}$$
$$+ \frac{1}{14} - \dots$$

$$= \frac{1}{\sqrt{2}} \left(1 + \frac{1}{3} - \frac{1}{5} - \frac{1}{7} + \frac{1}{9} + \frac{1}{11} - \dots \right)$$
$$- \frac{1}{2} \left(1 - \frac{1}{3} + \frac{1}{5} - \frac{1}{7} + \dots \right)$$

In view of (6.6.10), we obtain another numerical series

$$1 + \frac{1}{3} - \frac{1}{5} - \frac{1}{7} + \frac{1}{9} + \frac{1}{11} - \dots = \frac{\pi}{2\sqrt{2}}. \tag{6.6.11}$$

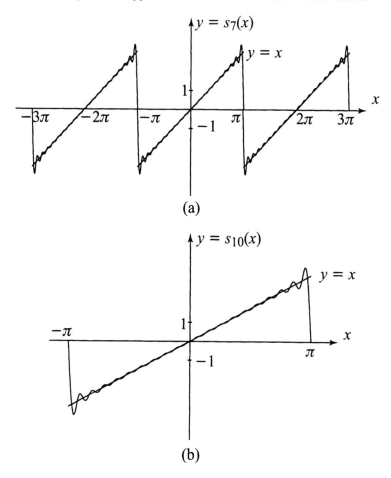

Figure 6.6.4 Graphs of $s_7(x)$ in (a) and $s_{10}(x)$ in (b).

6.7 Examples and Applications of Cosine and Sine Fourier Series

Let $f(x)$ be an even function defined on the interval $[-\pi, \pi]$. Since $\cos kx$ is an even function, and $\sin kx$ an odd function, the function $f(x) \cos kx$ is an even function and the function $f(x) \sin kx$ an odd function. Thus, we find that the Fourier coefficients of $f(x)$ are

$$a_k = \frac{1}{\pi} \int_{-\pi}^{\pi} f(x) \cos kx \, dx = \frac{2}{\pi} \int_{0}^{\pi} f(x) \cos kx \, dx, \qquad k = 0, 1, 2, \ldots,$$

(6.7.1)

$$b_k = \frac{1}{\pi} \int_{-\pi}^{\pi} f(x) \sin kx \, dx = 0, \qquad k = 1, 2, 3, \ldots.$$

Hence, the Fourier series of an even function can be written as

$$f(x) \sim \frac{a_0}{2} + \sum_{k=1}^{\infty} a_k \cos kx, \qquad (6.7.2)$$

where the coefficients a_k are given by formula (6.7.1).

In a similar manner, if $f(x)$ is an odd function, the function $f(x)\cos kx$ is an odd function and the function $f(x)\sin kx$ is an even function. As a consequence, the Fourier coefficients of $f(x)$, in this case, are

$$a_k = \frac{1}{\pi}\int_{-\pi}^{\pi} f(x)\cos kx\, dx = 0, \qquad k = 0, 1, 2, \ldots, \qquad (6.7.3)$$

$$b_k = \frac{1}{\pi}\int_{-\pi}^{\pi} f(x)\sin kx\, dx = \frac{2}{\pi}\int_{0}^{\pi} f(x)\sin kx\, dx, \qquad k = 1, 2, \ldots. \quad (6.7.4)$$

Therefore, the Fourier series of an odd function can be written as

$$f(x) = \sum_{k=1}^{\infty} b_k \sin kx, \qquad (6.7.5)$$

where the coefficients b_k are given by formula (6.7.4).

Example 6.7.1. Obtain the Fourier series of the function

$$f(x) = \operatorname{sgn} x = \begin{cases} -1, & -\pi < x < 0, \\ 0, & x = 0, \\ +1, & 0 < x < \pi, \end{cases} \quad \text{and} \quad f(x \pm 2k\pi) = f(x).$$

In this case, f is an odd function with period 2π and represents square wave function as shown in Figure 6.7.1. Clearly, $a_k = 0$ for $k = 0, 1, 2, 3, \ldots$ and

$$b_k = \frac{2}{\pi}\int_{0}^{\pi} f(x)\sin kx\, dx$$

$$= \frac{2}{\pi}\int_{0}^{\pi} \sin kx\, dx = \frac{2}{k\pi}\left[1 - (-1)^k\right].$$

Thus, $b_{2k} = 0$ and $b_{2k-1} = [(4/\pi)(2k-1)]$. Therefore, the Fourier series of the function $f(x)$ is

$$f(x) = \frac{4}{\pi}\sum_{k=1}^{\infty} \frac{\sin(2k-1)x}{(2k-1)}. \qquad (6.7.6)$$

This Fourier series consists of only odd harmonics. The loss of even harmonics is due to the fact that $\operatorname{sgn}\left(\frac{1}{2} + x\right) = \operatorname{sgn}\left(\frac{1}{2} - x\right)$.

Putting $x = \frac{\pi}{4}$ in (6.7.6) gives (6.6.10).

The nth partial sum of the series (6.7.6) is

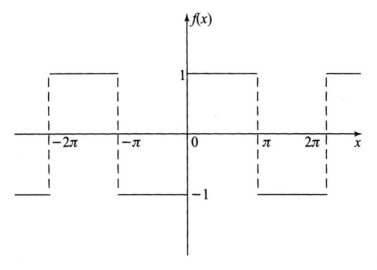

Figure 6.7.1 The square wave function.

$$s_n\left(x\right) = \frac{4}{\pi} \sum_{k=1}^{n} \frac{\sin\left(2k-1\right)x}{\left(2k-1\right)}. \tag{6.7.7}$$

We now examine the manner in which the first n terms of the series (6.7.7) tend to $f\left(x\right)$ as $n \to \infty$. The graphs of $s_n\left(x\right)$ for $n = 1,3,5$ is shown in Figure 6.7.2. To investigate the oscillations as $n \to \infty$, we locate the first peak to the right of the origin and calculate its height as $n \to \infty$.

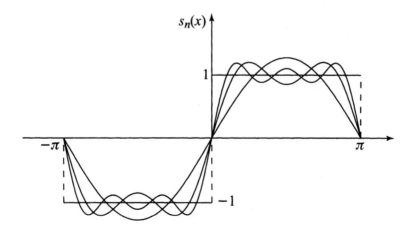

Figure 6.7.2 The Gibbs phenomenon.

The peak overshoot occurs at a local maximum of the partial sum $s_n(x)$ so that

$$
0 = s'_n(x) = \frac{4}{\pi} \sum_{k=1}^{n} \cos(2k-1)x
$$

$$
= \frac{2}{\pi \sin x} \sum_{k=1}^{n} 2 \sin x \cos(2k-1)x
$$

$$
= \frac{2}{\pi \sin x} \sum_{k=1}^{n} [\sin 2kx - \sin(2k-2)x]
$$

$$
= \frac{2 \sin nx}{\pi \sin x}. \tag{6.7.8}
$$

This leads to the points of maximum at $x = \frac{\pi}{2n}, \frac{2\pi}{2n}, \ldots, (2n-1)\frac{\pi}{2n}$ so that the first peak to the right of the origin occurs at $x = \frac{\pi}{2n}$. So, the value of $s_n(x)$ at this point is

$$
s_n\left(\frac{\pi}{2n}\right) = \frac{2}{\pi} \cdot \frac{\pi}{n} \sum_{k=1}^{n} \frac{\sin(2k-1)\frac{\pi}{2n}}{(2k-1)\frac{\pi}{2n}}.
$$

In the limit as $n \to \infty$, this becomes

$$
\lim_{n \to \infty} s_n\left(\frac{\pi}{2n}\right) = \left(\frac{2}{\pi}\right) \lim_{n \to \infty} \frac{\pi}{n} \sum_{k=1}^{n} \frac{\sin(2k-1)\frac{\pi}{2n}}{(2k-1)\frac{\pi}{2n}}
$$

$$
= \frac{2}{\pi} \int_0^{\pi} \frac{\sin x}{x} dx = \frac{2}{\pi}(1.852) = 1.179.
$$

Instead of a maximum value 1, it turns out to be 1.179 approximately. If the overshoot is compared to the jump in the function it is about $\frac{1}{2}(1.179 - 1) \times 100\% \sim 9\%$. The onset of the Gibbs phenomenon is shown in the Figure 6.7.3.

Historically, this phenomenon was first observed by a physicist A. Michelson (1852–1931) at the end of the nineteenth century. In order to calculate some Fourier coefficients of a function graphically, he developed equipment that is called a *harmonic analyser and synthesizer*. He calculated some partial sums, s_n, of a given function f graphically. Michelson also observed from this graphical representation that, for some functions, the graphs of s_n were very close to the function. However, for the *sgn* function, the graphs of the partial sums exhibit a large error in the neighborhood of the origin, $x = 0$ and $x = \pm \pi$ (the jump discontinuities of the function) independent of the number of terms in the partial sums. It was J.W. Gibbs (1839–1903) who first provided the explanation for this strikingly new phenomenon and showed that these large errors were not associated with numerical computations. Indeed, he further showed that the

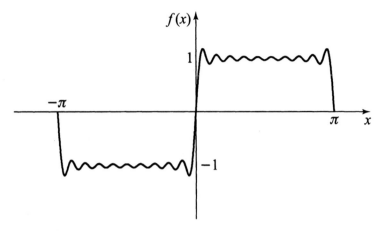

Figure 6.7.3 The Gibbs phenomenon.

large errors occur at every jump discontinuity for every piecewise continuous function f in $[-\pi, \pi]$ with certain derivative properties near the jump discontinuity.

Example 6.7.2. Expand $|\sin x|$ in Fourier series. Since $|\sin x|$ is an even function, as shown in Figure 6.7.4, $b_k = 0$ for $k = 1, 2, \ldots$ and

$$
\begin{aligned}
a_k &= \frac{2}{\pi} \int_0^\pi f(x) \cos kx \, dx \\
&= \frac{2}{\pi} \int_0^\pi \sin x \cos kx \, dx \\
&= \frac{1}{\pi} \int_0^\pi \left[\sin(1+k)x + \sin(1-k)x \right] dx \\
&= \frac{2\left[1 + (-1)^k\right]}{\pi(1 - k^2)} \qquad \text{for} \quad k = 0, 2, 3, \ldots.
\end{aligned}
$$

For $k = 1$,

$$
a_1 = \frac{2}{\pi} \int_0^\pi \sin x \cos x \, dx = 0.
$$

Hence, the Fourier series of $f(x)$ is

$$
f(x) = \frac{2}{\pi} + \frac{4}{\pi} \sum_{k=1}^\infty \frac{\cos(2kx)}{(1 - 4k^2)}.
$$

Example 6.7.3. Find the Fourier series of the triangular wave function which is defined by

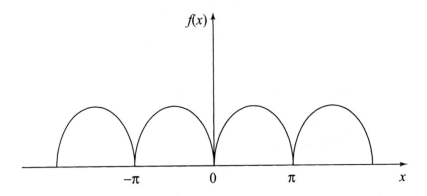

Figure 6.7.4 The rectified sine function.

$$f\left(x\right) = \left|x\right| = \begin{cases} -x, & -\pi \leq x \leq 0 \\ \\ x, & 0 \leq x \leq \pi \end{cases} \qquad (6.7.9)$$

and $f\left(x\right) = f\left(x \pm 2n\pi\right)$, $n = 1, 2, 3, \ldots$.

This is an even periodic function with period 2π as shown in Figure 6.7.5. This function gives a Fourier cosine series with $b_n = 0$ for all n and

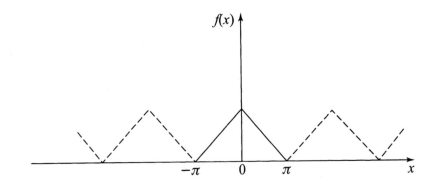

Figure 6.7.5 Triangular wave function.

$$a_0 = \frac{1}{\pi} \int_{-\pi}^{\pi} |x| \, dx = \frac{2}{\pi} \int_0^{\pi} x \, dx = \pi$$

$$a_n = \frac{1}{\pi} \int_{-\pi}^{\pi} |x| \cos nx \, dx = \frac{2}{\pi} \int_0^{\pi} |x| \cos nx \, dx$$

$$= \frac{2}{\pi} \int_0^{\pi} x \cos nx \, dx = \frac{2}{\pi} \left(\left[\frac{x \sin nx}{n} \right]_0^{\pi} - \int_0^{\pi} \frac{\sin nx}{n} dx \right),$$

integrating by parts

$$= \frac{2}{\pi n^2} [\cos nx]_0^{\pi} = \frac{2}{\pi n^2} [(-1)^n - 1] = \begin{cases} 0, & \text{if } n \text{ is even,} \\ -\frac{4}{\pi n^2} & \text{if } n \text{ is odd.} \end{cases}$$

Thus, the Fourier cosine series is given by

$$f(x) = \frac{1}{2} a_0 + \sum_{n=1}^{\infty} a_n \cos nx$$

$$= \frac{\pi}{2} - \frac{4}{\pi} \left(\frac{\cos x}{1^2} + \frac{\cos 3x}{3^2} + \frac{\cos 5x}{5^2} + \dots \right). \qquad (6.7.10)$$

Substituting $x = 0$ in (6.7.10) yields the following numerical series

$$\frac{\pi^2}{8} = \frac{1}{1^2} + \frac{1}{3^2} + \frac{1}{7^2} + \dots = \sum_{n=1}^{\infty} \frac{1}{(2n-1)^2}. \qquad (6.7.11)$$

This series can be used to derive the sum of reciprocals of squares of all positive integers,

$$S = \sum_{n=1}^{\infty} \frac{1}{n^2} = \frac{\pi^2}{6} \qquad (6.7.12)$$

and vise versa.

We have

$$\frac{\pi^2}{8} = \sum_{n=1}^{\infty} \frac{1}{(2n-1)^2} = \sum_{n=1}^{\infty} \frac{1}{n^2} - \sum_{n=1}^{\infty} \frac{1}{(2n)^2}$$

$$= S - \frac{1}{4} \sum_{n=1}^{\infty} \frac{1}{n^2} = S - \frac{1}{4} S.$$

This gives (6.7.12).

The series (6.7.12) is just the value of $s = 2$ of the *Riemann zeta function* defined by

$$\zeta(s) = \sum_{n=1}^{\infty} \frac{1}{n^s}, \qquad (6.7.13)$$

where s may be complex. This definition of $\zeta(s)$ can be extended in a natural way and this extension is called the *analytic continuation* of $\zeta(s)$ to include all complex numbers s except for $s = 1$. This function was introduced by Bernhard Riemann in 1841. He proved many properties of this function and made several conjectures, some of which are still open problems in mathematics. It is well known that $\zeta(s)$ has zeros on the real axis only at the even negative integers. Riemann conjectured that all complex zeros of $\zeta(s)$ lie on the line $\mathrm{Re}(s) = \frac{1}{2}$. This is called the *Riemann Hypothesis*; it is still an unsolved problem in mathematics. However, it is proved that there are an infinite number of zeros on the line. Indeed, by the fall of 2002, fifty billion complex zeros have been found — all of them lie on the stated line.

Example 6.7.4. Obtain the Fourier series of $f(x) = x^2$, $-\pi \le x \le \pi$, with $f(x) = f(x \pm 2n\pi)$, for $n = 1, 2, \ldots$.

Obviously, this is an even function, and its periodic extensions are shown in Figure 6.7.6.

Since $f(x)$ is an even function, $b_n \equiv 0$ for all $n \ge 1$. It turns out that

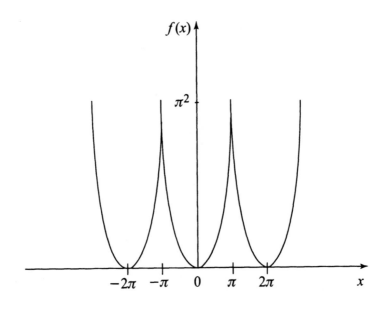

Figure 6.7.6   Periodic extension of x^2.

$$a_0 = \frac{1}{\pi} \int_{-\pi}^{\pi} x^2 \, dx = \frac{2}{\pi} \int_0^{\pi} x \, dx = \frac{2\pi^2}{3},$$

$$a_n = \frac{1}{\pi} \int_{-\pi}^{\pi} x^2 \cos nx \, dx = \frac{2}{\pi} \int_0^{\pi} x^2 \cos nx \, dx$$

$$= \frac{2}{\pi} \left(\left[\frac{x^2}{n} \sin nx \right]_0^{\pi} - \frac{2}{\pi} \int_0^{\pi} x \sin nx \, dx \right),$$

$$= -\frac{4}{n\pi} \int_0^x x \sin nx \, dx$$

$$= -\frac{4}{n\pi} \left(\left[-\frac{x}{n} \cos nx \right]_0^{\pi} + \frac{1}{\pi} \int_0^{\pi} \cos nx \, dx \right),$$

$$= -\frac{4}{n\pi} \left(-\frac{\pi}{n} (-1)^n - \frac{1}{n^2} [\sin nx]_0^{\pi} \right)$$

$$= \frac{4}{n^2} (-1)^n, \quad n \geq 1.$$

Consequently, the Fourier cosine series for x^2 is given by

$$x^2 = \frac{\pi^2}{3} + 4 \sum_{n=1}^{\infty} \frac{1}{n^2} (-1)^n \cos nx$$

$$= \frac{\pi^2}{3} - 4 \left(\frac{\cos x}{1^2} - \frac{\cos 2x}{2^2} + \frac{\cos 3x}{3^2} - \cdots \right). \qquad (6.7.14)$$

Putting $x = 0$ in (6.7.14) gives the following numerical series

$$\frac{1}{1^2} - \frac{1}{2^2} + \frac{1}{3^2} - \frac{1}{4^2} + \cdots = \frac{\pi^2}{12}. \qquad (6.7.15)$$

Substituting $x = \pi$ in (6.7.14) yields (6.7.12) which can also be obtained from (6.7.15) and vice versa. Thus, we have

$$S = \sum_{n=1}^{\infty} \frac{1}{n^2} = \frac{1}{1^2} + \frac{1}{2^2} + \frac{1}{3^2} + \frac{1}{4^2} + \frac{1}{5^2} + \frac{1}{6^2} + \cdots$$

$$= \left(\frac{1}{1^1} - \frac{1}{2^2} + \frac{1}{3^2} - \frac{1}{4^2} + \cdots \right) + 2 \left(\frac{1}{2^2} + \frac{1}{4^2} + \frac{1}{6^2} + \cdots \right)$$

$$= \frac{\pi^2}{12} + \frac{2}{4} \left(\frac{1}{1^2} + \frac{1}{2^2} + \frac{1}{3^2} + \cdots \right) = \frac{\pi^2}{12} + \frac{1}{2} S,$$

which gives the value of $S = \frac{\pi^2}{6}$. Adding (6.7.12) and (6.7.15) gives

$$\frac{1}{1^2} + \frac{1}{3^2} + \frac{1}{5^2} + \cdots = \frac{\pi^2}{8}, \qquad (6.7.16)$$

and then, subtracting (6.7.15) from (6.7.12) yields

$$\frac{1}{2^2} + \frac{1}{4^2} + \frac{1}{6^2} + \frac{1}{8^2} + \cdots = \frac{\pi^2}{24}. \qquad (6.7.17)$$

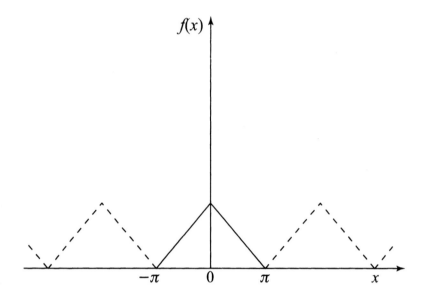

Figure 6.7.7 Periodic extension of $f(x)$.

In the preceding sections, we have prescribed the function $f(x)$ in the interval $(-\pi, \pi)$ and assumed $f(x)$ to be periodic with period 2π in the entire interval $(-\infty, \infty)$. In practice, we frequently encounter problems in which a function is defined only in the interval $(-\pi, \pi)$. In such a case, we simply extend the function periodically with period 2π, as in Figure 6.7.7. In this way, we are able to represent the function $f(x)$ by the Fourier series expansion, although we are interested only in the expansion on $(-\pi, \pi)$.

If the function f is defined only in the interval $(0, \pi)$, we may extend f in two ways. The first is the *even extension* of f, denoted and defined by (see Figure 6.7.8)

$$F_e(x) = \begin{cases} f(x), & 0 < x < \pi, \\ f(-x), & -\pi < x < 0, \end{cases}$$

while the second is the *odd extension* of f, denoted and defined by (see Figure 6.7.9)

$$F_0(x) = \begin{cases} f(x), & 0 < x < \pi, \\ -f(-x), & -\pi < x < 0. \end{cases}$$

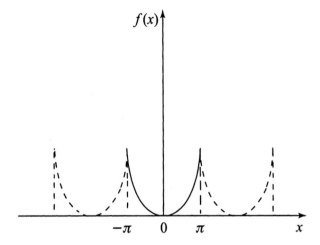

Figure 6.7.8 Even extension of $f(x)$.

Since $F_e(x)$ and $F_0(x)$ are even and odd functions with period 2π respectively, the Fourier series expansions of $F_e(x)$ and $F_0(x)$ are

$$F_e(x) = \frac{a_0}{2} + \sum_{k=1}^{\infty} a_k \cos kx,$$

where

$$a_k = \frac{2}{\pi} \int_0^{\pi} f(x) \cos kx \, dx,$$

and

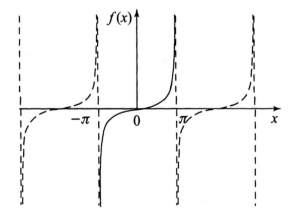

Figure 6.7.9 Odd extension of $f(x)$.

$$F_0(x) = \sum_{k=1}^{\infty} b_k \sin kx,$$

where

$$b_k = \frac{2}{\pi} \int_0^{\pi} f(x) \sin kx \, dx.$$

6.8 Complex Fourier Series

It is sometimes convenient to represent a function by an expansion in complex form. This expansion can easily be derived from the Fourier series

$$f(x) = \frac{a_0}{2} + \sum_{k=1}^{\infty} (a_k \cos kx + b_k \sin kx).$$

Using Euler's formulas

$$\cos x = \frac{e^{ix} + e^{-ix}}{2}, \qquad \sin x = \frac{e^{ix} - e^{-ix}}{2i},$$

we write

$$f(x) = \frac{a_0}{2} + \sum_{k=1}^{\infty} \left[a_k \left(\frac{e^{ikx} + e^{-ikx}}{2} \right) + b_k \left(\frac{e^{ikx} - e^{-ikx}}{2i} \right) \right]$$

$$= \frac{a_0}{2} + \sum_{k=1}^{\infty} \left[\left(\frac{a_k - ib_k}{2} \right) e^{ikx} + \left(\frac{a_k + ib_k}{2} \right) e^{-ikx} \right]$$

$$= c_0 + \sum_{k=1}^{\infty} \left(c_k \, e^{ikx} + c_{-k} \, e^{-ikx} \right),$$

where

$$c_0 = \frac{a_0}{2} = \frac{1}{2\pi} \int_{-\pi}^{\pi} f(x) \, dx$$

$$c_k = \frac{a_k - ib_k}{2} = \frac{1}{2\pi} \int_{-\pi}^{\pi} f(x) \left(\cos kx - i \sin kx \right) dx$$

$$= \frac{1}{2\pi} \int_{-\pi}^{\pi} f(x) \, e^{-ikx} dx$$

$$c_{-k} = \frac{a_k + ib_k}{2} = \frac{1}{2\pi} \int_{-\pi}^{\pi} f(x) \left(\cos kx + i \sin kx \right) dx$$

$$= \frac{1}{2\pi} \int_{-\pi}^{\pi} f(x) \, e^{ikx} dx.$$

Thus, we obtain the Fourier series expansion for $f(x)$ in complex form

$$f(x) = \sum_{k=-\infty}^{\infty} c_k\, e^{ikx}, \qquad -\pi < x < \pi, \tag{6.8.1}$$

where

$$c_k = \frac{1}{2\pi} \int_{-\pi}^{\pi} f(x)\, e^{-ikx} dx. \tag{6.8.2}$$

We derive the following *Parseval formula*

$$\frac{1}{2\pi} \int_{-\pi}^{\pi} f^2(x)\, dx = \sum_{k=-\infty}^{\infty} |c_k|^2. \tag{6.8.3}$$

from formulas (6.8.1)–(6.8.2).

Multiplying (6.8.1) by $\frac{1}{2\pi} f(x)$ and integrating from $-\pi < x < \pi$ yields

$$\frac{1}{2\pi} \int_{-\pi}^{\pi} f^2(x)\, dx = \sum_{k=-\infty}^{\infty} c_k \cdot \frac{1}{2\pi} \int_{-\pi}^{\pi} f(x)\, e^{ikx} dx$$

$$= \sum_{k=-\infty}^{\infty} c_k \cdot c_{-k} = \sum_{k=-\infty}^{\infty} c_k\, \bar{c}_k = \sum_{k=-\infty}^{\infty} |c_k|^2.$$

Example 6.8.1. Obtain the complex Fourier series expansion for the function

$$f(x) = e^x, \qquad -\pi < x < \pi.$$

We find

$$c_k = \frac{1}{2\pi} \int_{-\pi}^{\pi} f(x)\, e^{-ikx} dx$$

$$= \frac{1}{2\pi} \int_{-\pi}^{\pi} e^x e^{-ikx} dx$$

$$= \frac{(1+ik)(-1)^k}{\pi(1+k^2)} \sinh \pi,$$

and hence, the Fourier series is

$$f(x) = \sum_{k=-\infty}^{\infty} \frac{(1+ik)(-1)^k}{\pi(1+k^2)} \sinh \pi\, e^{ikx}. \tag{6.8.4}$$

We apply the Parseval formula (6.8.3) to this example and obtain

$$\sum_{k=-\infty}^{\infty} \frac{|1+ik|^2 \sinh^2 \pi}{\pi^2 (1+k^2)^2} = \frac{1}{4\pi} \left(e^{2\pi} - e^{-2\pi} \right) = \frac{1}{2\pi} \sinh 2\pi.$$

Simplifying this result gives

$$\sum_{k=-\infty}^{\infty} \frac{1}{(1+k^2)} = \pi \coth \pi. \tag{6.8.5}$$

Example 6.8.2. Show that, for $-1 < a < 1$,

$$\sum_{n=0}^{\infty} a^n \cos nx = \frac{1 - a \cos x}{1 - 2a \cos x + a^2}, \tag{6.8.6}$$

$$\sum_{n=0}^{\infty} a^n \sin nx = \frac{a \sin x}{1 - 2a \cos x + a^2}. \tag{6.8.7}$$

We denote the cosine series by C and sine series by S so that

$$C + iS = \sum_{n=0}^{\infty} a^n \left(\cos nx + i \sin nx \right) = \sum_{n=0}^{\infty} \left(a\, e^{ix} \right)^n$$

$$= \frac{1}{1 - a\, e^{ix}}, \quad \text{since } \left| a\, e^{ix} \right| < 1.$$

$$= \frac{1}{1 - a \cos x - ia \sin x} = \frac{(1 - a \cos x) + ia \sin x}{(1 - a \cos x)^2 + a^2 \sin^2 x}$$

$$= \frac{(1 - a \cos x) + ia \sin x}{1 - 2a \cos x + a^2}. \tag{6.8.8}$$

Equating the real and imaginary part gives the desired results.

6.9 Fourier Series on an Arbitrary Interval

So far we have been concerned with functions defined on the interval $[-\pi, \pi]$. In many applications, however, this interval is restrictive, and the interval of interest may be arbitrary, say $[a, b]$.

If we introduce the new variable t by the transformation

$$x = \frac{1}{2} (b + a) + \frac{(b - a)}{2\pi} t, \tag{6.9.1}$$

then, the interval $a \leq x \leq b$ becomes $-\pi \leq t \leq \pi$. Thus, the function $f\left[(b + a)/2 + ((b - a)/2\pi) t \right] = F(t)$ obviously has period 2π. Expanding this function in a Fourier series, we obtain

$$F(t) = \frac{a_0}{2} + \sum_{k=1}^{\infty} (a_k \cos kt + b_k \sin kt), \tag{6.9.2}$$

where

$$a_k = \frac{1}{\pi} \int_{-\pi}^{\pi} F(t) \cos kt \, dt, \qquad k = 0, 1, 2, \ldots,$$

and

$$b_k = \frac{1}{\pi} \int_{-\pi}^{\pi} F(t) \sin kt \, dt, \qquad k = 1, 2, 3, \ldots.$$

On changing t into x, we find the expansion for $f(x)$ in $[a, b]$

$$f(x) = \frac{a_0}{2} + \sum_{k=1}^{\infty} \left[a_k \cos \frac{k\pi(2x - b - a)}{(b - a)} + b_k \sin \frac{k\pi(2x - b - a)}{(b - a)} \right], \quad (6.9.3)$$

where

$$a_k = \frac{2}{b - a} \int_a^b f(x) \cos \left[\frac{kx(2x - b - a)}{(b - a)} \right] dx, \qquad k = 0, 1, 2, \ldots, \quad (6.9.4)$$

$$b_k = \frac{2}{b - a} \int_a^b f(x) \sin \left[\frac{kx(2x - b - a)}{(b - a)} \right] dx, \qquad k = 1, 2, 3, \ldots. \quad (6.9.5)$$

It is sometimes convenient to take the interval in which the function f is defined as $[-l, l]$. It follows at once from the result just obtained that by letting $a = -l$ and $b = l$, the Fourier expansion for f in $[-l, l]$ takes the form

$$f(x) = \frac{a_0}{2} + \sum_{k=1}^{\infty} \left[a_k \cos \left(\frac{k\pi x}{l} \right) + b_k \sin \left(\frac{k\pi x}{l} \right) \right], \quad (6.9.6)$$

where

$$a_k = \frac{1}{l} \int_{-l}^{l} f(x) \cos \left(\frac{k\pi x}{l} \right) dx, \qquad k = 0, 1, 2, \ldots, \quad (6.9.7)$$

$$b_k = \frac{1}{l} \int_{-l}^{l} f(x) \sin \frac{k\pi x}{l} dx, \qquad k = 1, 2, 3, \ldots. \quad (6.9.8)$$

If f is an even function of period $2l$, then, from equation (6.9.6), we can readily determine the Fourier cosine expansion in the form

$$f(x) = \frac{a_0}{2} + \sum_{k=1}^{\infty} a_k \cos \left(\frac{k\pi x}{l} \right), \quad (6.9.9)$$

where

$$a_k = \frac{2}{l} \int_0^l f(x) \cos \left(\frac{k\pi x}{l} \right) dx, \qquad k = 0, 1, 2, \ldots. \quad (6.9.10)$$

If f is an odd function of period $2l$, then, from equation (6.9.6), the Fourier sine expansion for f is

$$f(x) = \sum_{k=1}^{\infty} b_k \sin \left(\frac{k\pi x}{l} \right), \quad (6.9.11)$$

where

$$b_k = \frac{2}{l} \int_0^l f(x) \cos \left(\frac{k \pi x}{l} \right) dx. \qquad (6.9.12)$$

Finally, we make a change of variable to obtain the complex Fourier series of $f(x)$ on the interval $-l < x < l$. Suppose $f(x)$ is a periodic function with period $2l$. We put $a = -l$ and $b = l$ so that the change of variable formula becomes $x = \frac{lt}{\pi}$ and

$$f(x) = f\left(\frac{lt}{\pi} \right) = F(t). \qquad (6.9.13)$$

Clearly, $F(t)$ is periodic with period 2π. If it is piecewise smooth, it can be expanded in a complex Fourier series in the form

$$F(t) = \sum_{k=-\infty}^{\infty} c_k e^{ikt}, \qquad c_k = \frac{1}{2\pi} \int_{-\pi}^{\pi} F(t) e^{-ikt} dt. \qquad (6.9.14)$$

Substituting $t = \frac{\pi x}{l}$ into (6.9.14) gives the complex Fourier series expansion of the original function f in the form

$$f(x) = \sum_{k=-\infty}^{\infty} c_k \exp \left(\frac{i x \pi k}{l} \right), \quad c_k = \frac{1}{2l} \int_{-l}^{l} f(x) \exp \left(-\frac{i x \pi k}{l} \right) dx. (6.9.15)$$

In particular, if $f(t)$ is a periodic function of time variable t with period $T = \frac{2\pi}{\omega}$ and $\omega \left(= \frac{2\pi}{T} = \frac{\pi}{l} \right)$ is the frequency, then

$$f(t) = \frac{a_0}{2} + \sum_{n=1}^{\infty} [a_n \cos (n \omega t) + b_n \sin (n \omega t)], \qquad (6.9.16)$$

where a_n and b_n are given by (6.9.7) and (6.9.8) with $\frac{\pi}{l}$ replaced by ω and $k = n$. The terms $(a_1 \cos \omega t + b_1 \sin \omega t)$, $(a_2 \cos 2\omega t + b_2 \sin 2\omega t)$, ..., $(a_n \cos n\omega t + b_n \sin n\omega t)$, ... are called the *first*, the *second*, and the *nth harmonic* respectively.

Example 6.9.1. Consider the odd periodic function

$$f(x) = x, \qquad -2 < x < 2,$$

as shown in Figure 6.9.1. Here $l = 2$. Since f is odd, $a_k = 0$, and

$$b_k = \frac{2}{l} \int_0^l f(x) \sin \left(\frac{k \pi x}{l} \right) dx,$$

$$= \frac{2}{2} \int_0^2 x \sin \left(\frac{k \pi x}{2} \right) dx = -\frac{4}{k \pi} (-1)^k \quad \text{for} \quad k = 1, 2, 3, \ldots.$$

Therefore, the Fourier sine series of f is

$$f(x) = \sum_{k=1}^{\infty} \frac{4}{k \pi} (-1)^{k+1} \sin \left(\frac{k \pi x}{2} \right).$$

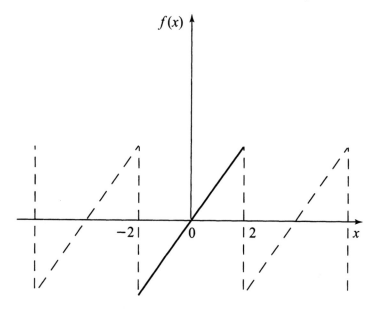

Figure 6.9.1 Odd periodic function of $f(x) = x$ in $-2 < x < 2$.

Example 6.9.2. Consider the function

$$f(x) = \begin{cases} 1, & 0 < x < \frac{1}{2} \\ 0, & \frac{1}{2} < x < 1. \end{cases}$$

In this case, the period is $2l = 2$ or $l = 1$. Extend f as shown in Figure 6.9.2. Since the extension is even, we have $b_k = 0$ and

$$a_0 = \frac{2}{l} \int_0^l f(x)\, dx = \frac{2}{1} \int_0^{\frac{1}{2}} dx = 1$$

$$a_k = \frac{2}{l} \int_0^l f(x) \cos\left(\frac{k\pi x}{l}\right) dx$$

$$= \frac{2}{1} \int_0^1 \cos(k\pi x)\, dx = \left(\frac{2}{k\pi}\right) \sin\left(\frac{k\pi}{2}\right).$$

Hence,

$$f(x) = \frac{1}{2} + \sum_{k=1}^{\infty} \frac{2}{(2k-1)\pi} (-1)^{k-1} \cos(2k-1)\pi x.$$

Example 6.9.3. Find the Fourier series of $f(x) = x^2$ in $(-l, l)$ from the corresponding Fourier series in $(-\pi, \pi)$.

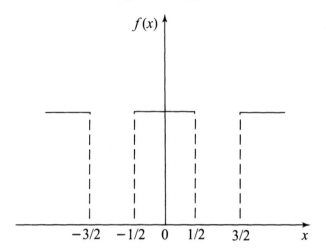

Figure 6.9.2 Even extension of $f(x)$ in Example 6.9.2.

It directly follows from (6.7.14) and (6.9.9) as

$$f(x) = x^2 = \frac{\pi^2}{3} + 4 \sum_{k=1}^{\infty} \frac{1}{k^2} (-1)^k \cos\left(\frac{\pi kx}{l}\right), \quad -l < x < l.$$

Example 6.9.4. Find the sine and cosine Fourier series of $f(x) = x$ in $(0, l)$ from the corresponding Fourier series in $(0, \pi)$.

We know the Fourier sine series for x in $(0, \pi)$

$$f(x) = x = 2 \sum_{k=1}^{\infty} \frac{(-1)^{k+1}}{k} \sin kx, \quad 0 < x < \pi.$$

Using the transformation $x = \left(\frac{\pi t}{l}\right)$, we obtain

$$t = \frac{2l}{\pi} \sum_{k=1}^{\infty} \frac{(-1)^{k+1}}{k} \sin\left(\frac{\pi kt}{l}\right), \quad 0 < x < l.$$

We have the Fourier cosine series for x in $(0, \pi)$

$$f(x) = x = \frac{\pi}{2} - \frac{4}{\pi} \sum_{k=1}^{\infty} \frac{1}{(2k-1)^2} \cos(2k-1)x, \quad 0 < x < \pi.$$

Similarly, using the transformation $x = \frac{\pi t}{l}$ gives

$$t = \frac{l}{2} - \frac{4l}{\pi^2} \sum_{k=1}^{\infty} \frac{1}{(2k-1)^2} \cos\left[(2k-1)\frac{\pi t}{l}\right], \quad 0 < x < l.$$

Example 6.9.5. From the solution to problem 25 in 6.14 Exercises, we obtain the complex Fourier series of $f(t)$ in $-l < t < l$.

We use the transformation $x = \frac{lt}{\pi}$ in Exercise 25 so that

$$\left.\begin{array}{c} -\frac{1}{2}\left(\pi + \frac{lt}{\pi}\right) \\[2mm] \frac{1}{2}\left(\pi - \frac{lt}{\pi}\right) \end{array}\right\} = \frac{1}{2i}\sum_{n\neq 0}\frac{1}{n}\exp\left(\frac{inlt}{\pi}\right).$$

6.10 The Riemann–Lebesgue Lemma and Pointwise Convergence Theorem

We have stated earlier that if $f(x)$ is piecewise continuous on the interval $[-\pi, \pi]$, then there exists a Fourier series expansion which converges in the mean to $f(x)$.

In this section, we shall discuss the Pointwise Convergence Theorem with a proof using the Riemann–Lebesgue Lemma.

Lemma 6.10.1. (*Riemann–Lebesgue Lemma*) *If $g(x)$ is piecewise continuous on the interval $[a, b]$, then*

$$\lim_{\lambda \to \infty}\int_a^b g(x)\sin\lambda x\,dx = 0. \tag{6.10.1}$$

Proof. Consider the integral

$$I(\lambda) = \int_a^b g(x)\sin\lambda x\,dx. \tag{6.10.2}$$

With the change of variable

$$x = t + \pi/\lambda,$$

we have

$$\sin\lambda x = \sin\lambda(t + \pi/\lambda) = -\sin\lambda t,$$

and

$$I(\lambda) = -\int_{a-\pi/\lambda}^{b-\pi/\lambda} g(t + \pi/\lambda)\sin\lambda t\,dt. \tag{6.10.3}$$

Since t is a dummy variable, we write the above integral as

$$I(\lambda) = -\int_{a-\pi/\lambda}^{b-\pi/\lambda} g(x + \pi/\lambda)\sin\lambda x\,dx. \tag{6.10.4}$$

Addition of equations (6.10.2) and (6.10.4) yields

$$2I(\lambda) = \int_a^b g(x) \sin \lambda x \, dx - \int_{a-\pi/\lambda}^{b-\pi/\lambda} g(x+\pi/\lambda) \sin \lambda x \, dx$$

$$= -\int_{a-\pi/\lambda}^a g(x+\pi/\lambda) \sin \lambda x \, dx + \int_{b-\pi/\lambda}^b g(x) \sin \lambda x \, dx$$

$$+ \int_a^{b-\pi/\lambda} [g(x) - g(x+\pi/\lambda)] \sin \lambda x \, dx. \qquad (6.10.5)$$

First, let $g(x)$ be a continuous function in $[a, b]$. Then $g(x)$ is necessarily bounded, that is, there exists an M such that $|g(x)| \le M$. Hence,

$$\left| \int_{a-\pi/\lambda}^a g(x+\pi/\lambda) \sin \lambda x \, dx \right| = \left| \int_a^{a+\pi/\lambda} g(x) \sin \lambda x \, dx \right| \le \frac{\pi M}{\lambda},$$

and

$$\left| \int_{b-\pi/\lambda}^b g(x) \sin \lambda x \, dx \right| \le \frac{\pi M}{\lambda}.$$

Consequently,

$$|I(\lambda)| \le \frac{\pi M}{\lambda} + \int_a^{b-\pi/\lambda} |g(x) - g(x+\pi/\lambda)| \, dx. \qquad (6.10.6)$$

Since $g(x)$ is a continuous function on a closed interval $[a, b]$, it is uniformly continuous on $[a, b]$ so that

$$|g(x) - g(x+\pi/\lambda)| < \varepsilon/(b-a), \qquad (6.10.7)$$

for all $\lambda > \Lambda$ and all x in $[a, b]$. We now choose λ such that $\pi M/\lambda < \varepsilon/2$, whenever $\lambda > \Lambda$. Then

$$|I(\lambda)| < \frac{\varepsilon}{2} + \frac{\varepsilon}{2} = \varepsilon.$$

If $g(x)$ is piecewise continuous in $[a, b]$, then the proof consists of a repeated application of the preceding argument to every subinterval of $[a, b]$ in which $g(x)$ is continuous.

Theorem 6.10.1. (Pointwise Convergence Theorem). *If $f(x)$ is piecewise smooth and periodic function with period 2π in $[-\pi, \pi]$, then for any* x

$$\frac{a_0}{2} + \sum_{k=1}^{\infty} (a_k \cos kx + b_k \sin kx) = \frac{1}{2}[f(x+) + f(x-)], \qquad (6.10.8)$$

where

$$a_k = \frac{1}{\pi} \int_{-\pi}^{\pi} f(t) \cos kt \, dt, \qquad k = 0, 1, 2, \ldots, \qquad (6.10.9)$$

$$b_k = \frac{1}{\pi} \int_{-\pi}^{\pi} f(t) \sin kt \, dt, \qquad k = 1, 2, 3, \ldots. \qquad (6.10.10)$$

Proof. The nth partial sum $s_n(x)$ of the series (6.10.8) is

$$s_n(x) = \frac{1}{2}a_0 + \sum_{k=1}^{n} (a_k \cos kx + b_k \sin kx). \qquad (6.10.11)$$

We use integrals in (6.10.9)–(6.10.10) to replace a_k and b_k in (6.10.11) so that

$$s_n(x) = \frac{1}{2\pi} \int_{-\pi}^{\pi} \left[1 + 2\sum_{k=1}^{n} (\cos kt \cos kx + \sin kt \sin kx) \right] f(t)\, dt$$

$$= \frac{1}{2\pi} \int_{-\pi}^{\pi} \left[1 + 2\sum_{k=1}^{n} \cos k(x - t) \right] f(t)\, dt$$

$$= \frac{1}{2\pi} \int_{-\pi}^{\pi} D_n(x - t) f(t)\, dt, \qquad (6.10.12)$$

where $D_n(\theta)$ is called the *Dirichlet kernel* defined by

$$D_n(\theta) = 1 + 2\sum_{k=1}^{n} \cos k\theta. \qquad (6.10.13)$$

The next step is to study the properties of this kernel $D_n(\theta)$ which is an even function with period 2π and satisfies the condition

$$\frac{1}{2\pi} \int_{-\pi}^{\pi} D_n(\theta)\, d\theta = 1 + 0 + 0 + \ldots + 0 = 1. \qquad (6.10.14)$$

We find the value of the sum in (6.10.13) by Euler's formula so that

$$D_n(\theta) = 1 + \sum_{k=1}^{n} \left(e^{ik\theta} + e^{-ik\theta} \right) = \sum_{k=-n}^{n} e^{ik\theta}$$

$$= e^{-in\theta} + \ldots + 1 + \ldots + e^{in\theta}.$$

This is a finite geometric series with the first term $e^{-in\theta}$, the ratio $e^{i\theta}$, and the last term $e^{in\theta}$, and hence, its sum is given by

$$D_n(\theta) = \frac{e^{-in\theta} - e^{i(n+1)\theta}}{1 - e^{i\theta}}$$

$$= \frac{\exp\left[-\left(n + \tfrac{1}{2}\right)i\theta \right] - \exp\left[\left(n + \tfrac{1}{2}\right)i\theta \right]}{\exp\left(-\tfrac{1}{2}i\theta\right) - \exp\left(+\tfrac{1}{2}i\theta\right)}$$

$$= \frac{\sin\left(n + \tfrac{1}{2}\right)\theta}{\sin\tfrac{1}{2}\theta}. \qquad (6.10.15)$$

The graph of $D_n(\theta)$ is shown in Figure 6.10.1. It looks similar to that of the diffusion kernel as drawn in Figure 12.4.1 in Chapter 12 except for its symmetric oscillatory trail.

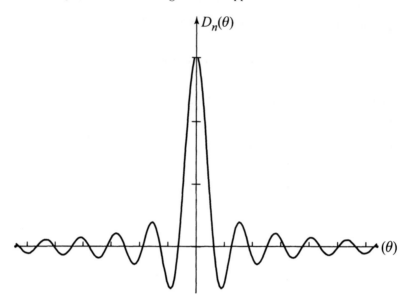

Figure 6.10.1 Graph of $D_n(\theta)$ against θ.

We next put $t - x = \theta$ in (6.10.12) to obtain

$$s_n(x) = \frac{1}{2\pi} \int_{x-\pi}^{x+\pi} D_n(\theta) f(x + \theta) \, d\theta. \tag{6.10.16}$$

Since both D_n and f have period 2π, the limits of the integral can be taken from $-\pi$ to π, and hence, (6.10.16) assumes the form

$$s_n(x) = \frac{1}{2\pi} \int_{-\pi}^{\pi} D_n(\theta) f(x + \theta) \, d\theta. \tag{6.10.17}$$

We next use (6.10.14) to express the difference of $s_n(x)$ and $\frac{1}{2}[f(x+) + f(x-)]$ in the form

$$s_n(x) - \frac{1}{2}[f(x+) + f(x-)]$$

$$= \frac{1}{2\pi} \int_{-\pi}^{0} D_n(\theta) [f(x + \theta) - f(x-)] \, d\theta$$

$$+ \frac{1}{2\pi} \int_{0}^{\pi} D_n(\theta) [f(x + \theta) - f(x+)] \, d\theta$$

which is, by (6.10.15),

$$= \frac{1}{2\pi} \int_{-\pi}^{0} g_-(\theta) \sin\left(n + \frac{1}{2}\right) \theta \, d\theta + \frac{1}{2\pi} \int_{0}^{\pi} g_+(\theta) \sin\left(n + \frac{1}{2}\right) \theta \, d\theta,$$

$$\tag{6.10.18}$$

where

$$g_{\pm}(\theta) = \left(\sin\frac{\theta}{2}\right)^{-1} [f(x+\theta) - f(x\pm)].\qquad(6.10.19)$$

Since the denominators of the functions $g_{\pm}(\theta)$ vanish at $\theta = 0$, integrals in (6.10.18) may diverge at this point. However, by assumption, f is piecewise smooth, and hence,

$$\lim_{\theta\to 0\pm} g_{\pm}(\theta) = \lim_{\theta\to 0\pm} \frac{f(x+\theta) - f(x)}{\frac{\theta}{2}} \cdot \left(\frac{\frac{\theta}{2}}{\sin\frac{\theta}{2}}\right) = 2f'(x\pm).(6.10.20)$$

Evidently, the above limits exist, and $g_{\pm}(\theta)$ are piecewise continuous elsewhere in the interval $(-\pi,\pi)$. Therefore, by the Riemann–Lebesgue Lemma 6.10.1, both integrals in (6.10.18) vanish as $n \to \infty$. Thus,

$$\lim_{n\to\infty} s_n(x) = \frac{1}{2}[f(x+) + f(x-)].$$

This proves that the Fourier series converges for each x in $(-\pi,\pi)$.

Remark 3. At a point of continuity the series converges to the function $f(x)$.

Remark 4. At a point of discontinuity, the series is equal to the arithmetic mean of the limits of the function on both sides of the discontinuity.

Remark 5. The condition of piecewise smoothness under which the Fourier series converges pointwise is a *sufficient* condition. A large number of examples of applications is covered by this case. However, the pointwise convergence Theorem 6.10.1 can be proved under weaker conditions.

Example 6.10.1. In Example 6.6.1, we obtained that the Fourier series expansion for $(x + x^2)$ in $[-\pi,\pi]$, as shown in Figure 6.10.2, is

$$f(x) \sim \frac{\pi^2}{3} + \sum_{k=1}^{\infty}\left[\frac{4}{k^2}(-1)^k\cos kx - \frac{2}{k}(-1)^k\sin kx\right].$$

Since $f(x) = x + x^2$ is piecewise smooth, the series converges, and hence, we write

$$x + x^2 = \frac{\pi^2}{3} + \sum_{k=1}^{\infty}\left[\frac{4}{k^2}(-1)^k\cos kx - \frac{2}{k}(-1)^k\sin kx\right],$$

at points of continuity. At points of discontinuity, such as $x = \pi$, by virtue of the Pointwise Convergence Theorem,

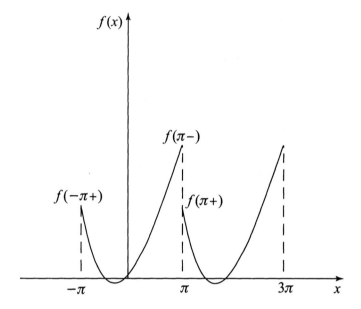

Figure 6.10.2 Graph of $f(x)$.

$$\frac{1}{2}\left[\left(\pi + \pi^2\right) + \left(-\pi + \pi^2\right)\right] = \frac{\pi^2}{3} + \sum_{k=1}^{\infty} \frac{4}{k^2} (-1)^k \cos k\pi, \quad (6.10.21)$$

since

$$f(\pi-) = \pi + \pi^2 \quad \text{and} \quad f(\pi+) = f(-\pi+) = -\pi + \pi^2.$$

Simplification of equation (6.10.21) gives

$$\pi^2 = \frac{\pi^2}{3} + \sum_{k=1}^{\infty} \frac{4}{k^2} (-1)^{2k},$$

or

$$\frac{\pi^2}{6} = \sum_{k=1}^{\infty} \frac{1}{k^2}.$$

The series can be used to obtain the sum of reciprocals of squares of odd positive integers, that is,

$$\sum_{n=1}^{\infty} \frac{1}{(2n-1)^2}.$$

We have

$$\frac{\pi^2}{6} = \sum_{n=1}^{\infty} \frac{1}{n^2} = \sum_{n=1}^{\infty} \frac{1}{(2n)^2} + \sum_{n=1}^{\infty} \frac{1}{(2n-1)^2} = \frac{1}{4} \cdot \frac{\pi^2}{6} + \sum_{n=1}^{\infty} \frac{1}{(2n-1)^2},$$

or,

$$\sum_{n=1}^{\infty} \frac{1}{(2n-1)^2} = \frac{\pi^2}{6}\left(1 - \frac{1}{4}\right) = \frac{\pi^2}{8}.$$

Conversely, this series can be used to find the sum of reciprocals of squares of all positive integers.

Example 6.10.2. Find the Fourier series of the following function

$$f(x) = \begin{cases} 0, & -2 \leq x < 0 \\[2mm] 2-x, & 0 < x \leq 2. \end{cases}$$

This function is defined over the interval $-2 \leq x \leq 2$, where it is piecewise smooth with a finite discontinuity at $x = 0$. We use (6.9.7) and (6.9.8) to calculate the Fourier coefficients

$$a_0 = \frac{1}{2}\int_0^2 (2-x)\,dx = 1$$

$$a_k = \frac{1}{2}\int_0^2 (2-x)\cos\left(\frac{\pi k x}{2}\right)dx = \frac{2}{\pi^2 k^2}\left[1 - (-1)^k\right], \quad k = 1,2,3,\ldots.$$

$$b_k = \frac{1}{2}\int_0^2 (2-x)\sin\left(\frac{\pi k x}{2}\right)dx = \frac{2}{\pi k}, \quad k = 1,2,3,\ldots.$$

Consequently, the Fourier series (6.9.6) becomes

$$f(x) = \frac{1}{2} + \frac{2}{\pi}\sum_{k=1}^{\infty}\left[\frac{\left\{1-(-1)^k\right\}}{\pi k^2}\cos\left(\frac{\pi k x}{2}\right) + \frac{1}{k}\sin\left(\frac{\pi k x}{2}\right)\right]. \quad (6.10.22)$$

The function $f(x)$ is continuous at $x = 1$ where $f(1) = 1$, so that the Fourier series (6.10.22) gives

$$1 = \frac{1}{2} + \frac{2}{\pi}\sum_{n=1}^{\infty}\left[\frac{\left\{1-(-1)^n\right\}}{\pi n^2}\cos\left(\frac{n\pi}{2}\right) + \frac{1}{n}\sin\left(\frac{n\pi}{2}\right)\right].$$

Since the factor $1 - (-1)^n = 0$ for even n, and $\cos\frac{n\pi}{2} = 0$ when n is odd, every term of the cosine series vanishes for all n. Consequently,

$$\frac{\pi}{4} = \sum_{n=1}^{\infty}\frac{1}{n}\sin\left(\frac{n\pi}{2}\right) = \sum_{n=1}^{\infty}\frac{(-1)^n}{(2n+1)}. \quad (6.10.23)$$

On the other hand, $f(x)$ is discontinuous at $x = 0$ and the Fourier series must converge to $\frac{1}{2}(0+2) = 1$. Thus,

$$\frac{\pi^2}{4} = \sum_{n=1}^{\infty} \frac{[1 - (-1)^n]}{n^2} = 2 \sum_{n=1}^{\infty} \frac{1}{(2n-1)^2},$$

or

$$\sum_{n=1}^{\infty} \frac{1}{(2n-1)^2} = \frac{\pi^2}{8}. \qquad (6.10.24)$$

6.11 Uniform Convergence, Differentiation, and Integration

In the preceding section, we have proved the pointwise convergence of the Fourier series for a piecewise smooth function. Here, we shall consider several theorems without proof concerning uniform convergence, term-by-term differentiation, and integration of Fourier series.

Theorem 6.11.1. (Uniform and Absolute Convergence Theorem)
Let $f(x)$ be a continuous function with period 2π, and let $f'(x)$ be piecewise continuous in the interval $[-\pi, \pi]$. If, in addition, $f(-\pi) = f(\pi)$, then the Fourier series expansion for $f(x)$ is uniformly and absolutely convergent.

In the preceding theorem, we have assumed that $f(x)$ is continuous and $f'(x)$ is piecewise continuous. With less stringent conditions on f, the following theorem can be proved.

Theorem 6.11.2. *Let $f(x)$ be piecewise smooth in the interval $[-\pi, \pi]$. If $f(x)$ is periodic with period 2π, then the Fourier series for f converges uniformly to f in every closed interval containing no discontinuity.*

We note that the partial sums $s_n(x)$ of a Fourier series cannot approach the function $f(x)$ uniformly over any interval containing a point of discontinuity of f. The behavior of the deviation of $s_n(x)$ from $f(x)$ in such an interval is known as the *Gibbs phenomenon*. For instance, in the Example 6.7.1, the Fourier series of the function is given by

$$f(x) = \frac{4}{\pi} \sum_{k=1}^{\infty} \frac{\sin(2k-1)x}{(2k-1)}. \qquad (6.11.1)$$

From graphs of the partial sums $s_n(x)$ against the x-axis, as shown in Figures 6.7.2 and 6.7.3, we find that $s_n(x)$ oscillate above and below the value of f. It can be observed that, near the discontinuous points $x = 0$ and $x = \pi$, s_n deviate from the function rather significantly. Although the magnitude of oscillation decreases at all points in the interval for large n, very near the points of discontinuity the amplitude remains practically independent of n as n increases. This illustrates the fact that the Fourier series of a function f does not converge uniformly on any interval which contains a discontinuity.

Termwise differentiation of Fourier series is, in general, not permissible. From Example 6.6.3, the Fourier series for $f(x) = x$ is given by

$$x = 2\left[\sin x - \frac{\sin 2x}{2} + \frac{\sin 3x}{3} - \ldots\right], \qquad (6.11.2)$$

which converges for all x, whereas the series after formal term-by-term differentiation,

$$1 \sim 2\left[\cos x - \cos 2x + \cos 3x - \ldots\right].$$

This series is *not* the Fourier series of $f'(x) = 1$, since the Fourier series of $f'(x) = 1$ is the function 1. In fact, this series is *not* a Fourier series of any piecewise continuous function defined in $[-\pi, \pi]$ as the coefficients do not tend to zero which contradicts the Riemann–Lebesque lemma.

In fact, the series of $f'(x) = 1$ diverges for all x since the nth term, $\cos nx$ does not tend to zero as $n \to \infty$. The difficulty arises from the fact that the given function $f(x) = x$ in $[-\pi, \pi]$ when extended periodically is discontinuous at the points $\pm \pi$, $\pm 3\pi$, ... We shall see below that the continuity of the periodic function is one of the conditions that must be met for the termwise differentiation of a Fourier series.

Theorem 6.11.3. (Differentiation Theorem) *Let $f(x)$ be a continuous function in the interval $[-\pi, \pi]$ with $f(-\pi) = f(\pi)$, and let $f'(x)$ be piecewise smooth in that interval. Then Fourier series for f' can be obtained by termwise differentiation of the series for f, and the differentiated series converges pointwise to f' at points of continuity and to $[f'(x) + f'(-x)]/2$ at discontinuous points.*

The termwise integration of Fourier series is possible under more general conditions than termwise differentiation. We recall that in calculus, the series of functions to be integrated must converge uniformly in order to assure the convergence of a termwise integrated series. However, in the case of Fourier series, this condition is not necessary.

Theorem 6.11.4. (Integration Theorem) *Let $f(x)$ be piecewise continuous in $[-\pi, \pi]$, and periodic with period 2π. Then the Fourier series of $f(x)$*

$$\frac{a_0}{2} + \sum_{k=1}^{\infty}\left(a_k \cos kx + b_k \sin kx\right),$$

whether convergent or not, can be integrated term by term between any limits.

Example 6.11.1. In Example 6.7.2, we have found that $f(x) = |\sin x|$ is represented by the Fourier series

$$\sin x = \frac{2}{\pi} + \frac{4}{\pi} \sum_{k=1}^{\infty} \frac{\cos(2kx)}{(1-4k^2)}, \qquad -\pi < x < \pi. \qquad (6.11.3)$$

Since $f(x) = |\sin x|$ is continuous in the interval $[-\pi, \pi]$ and $f(-\pi) = f(\pi)$, we differentiate the series term by term, obtaining

$$\cos x = -\frac{8}{\pi} \sum_{k=1}^{\infty} \frac{k \sin(2kx)}{(1-4k^2)}, \qquad (6.11.4)$$

by use of Theorem 6.11.3, since $f'(x)$ is piecewise smooth in $[-\pi, \pi]$. In this way, we obtain the Fourier sine series expansion of the cosine function in $(-\pi, \pi)$. Note that the reverse process is not permissible.

Example 6.11.2. Consider the function $f(x) = x$ in the interval $-\pi < x \leq \pi$. As shown in Example 6.6.3, the Fourier series of $f(x) = x$ is

$$x = 2\left[\sin x - \frac{\sin 2x}{2} + \frac{\sin 3x}{3} - \cdots\right].$$

By Theorem 6.11.4, we can integrate the series term by term from a to x to obtain

$$\frac{1}{2}\left(x^2 - a^2\right) = 2\left[-\left(\cos x - \frac{\cos 2x}{2^2} + \frac{\cos 3x}{3^3} - \cdots\right) \right.$$
$$\left. + \left(\cos a - \frac{\cos 2a}{2^2} + \frac{\cos 3a}{3^3} - \cdots\right)\right].$$

To determine the sum of the series of constants, we write

$$\frac{x^2}{4} = C - \sum_{k=1}^{\infty} (-1)^{k+1} \frac{\cos kx}{k^2},$$

where C is a constant. Since the series on the right is the Fourier series which converges uniformly, we can integrate the series term by term from $-\pi$ to π to obtain

$$\int_{-\pi}^{\pi} \frac{x^2}{2}\,dx = 2\left[\int_{-\pi}^{\pi} C\,dx - \sum_{k=1}^{\infty} \frac{(-1)^{k+1}}{k^2} \int_{-\pi}^{\pi} \cos kx\,dx\right],$$

or,

$$\frac{\pi^3}{3} = 2(2\pi C).$$

Hence,

$$C = \frac{\pi^2}{12}.$$

Therefore, by integrating the Fourier series of $f(x) = x$ in $(-\pi, \pi)$, we obtain the Fourier series expansion for the function $f(x) = x^2$ as

$$x^2 = 4 \left[\frac{\pi^2}{12} - \sum_{k=1}^{\infty} (-1)^{k+1} \frac{\cos kx}{k^2} \right] = \frac{\pi^2}{3} + 4 \sum_{k=1}^{\infty} \frac{(-1)^k}{k^2} \cos kx. \quad (6.11.5)$$

Obviously,

$$\frac{a_0}{2} = \frac{\pi^2}{3} \quad \text{and} \quad a_k = \frac{4(-1)^k}{k^2}.$$

Substituting these results into the Parseval relation (6.5.10) gives

$$\frac{2\pi^4}{9} + 16 \sum_{k=1}^{\infty} \frac{1}{k^4} = \frac{1}{\pi} \int_{-\pi}^{\pi} x^4 \, dx = \frac{2}{5} \pi^4.$$

This leads to another well-known numerical series

$$\sum_{k=1}^{\infty} \frac{1}{k^4} = \frac{\pi^4}{90}. \quad (6.11.6)$$

Example 6.11.3. **(Parseval's relation for the cosine Fourier series)** If $f(x)$ is continuous with continuous derivatives in $0 \le x \le \pi$ with $f'(0) = 0 = f'(\pi)$, and its cosine Fourier series

$$f(x) \sim \frac{1}{2} a_0 + \sum_{k=1}^{\infty} a_k \cos kx \quad (6.11.7)$$

converges uniformly in $0 \le x \le \pi$, and $g(x)$ is piecewise continuous on $[0, \pi]$ with its Fourier cosine series given by

$$g(x) \sim \frac{1}{2} b_0 + \sum_{k=1}^{\infty} b_k \cos kx, \quad (6.11.8)$$

then the following Parseval relation holds

$$\frac{1}{\pi} \int_0^{\pi} f(x) g(x) \, dx = \frac{1}{4} a_0 b_0 + \frac{1}{2} \sum_{k=1}^{\infty} a_k b_k. \quad (6.11.9)$$

It follows from the assumptions that term-by-term integration holds. Thus,

$$\int_0^{\pi} f(x) g(x) \, dx = \int_0^{\pi} \left[\frac{1}{2} a_0 + \sum_{k=1}^{\infty} a_k \cos kx \right] g(x) \, dx$$

$$= \frac{1}{2} a_0 \int_0^{\pi} g(x) \, dx + \sum_{k=1}^{\infty} a_k \int_0^{\pi} g(x) \cos kx \, dx$$

$$= \frac{\pi}{4} a_0 b_0 + \frac{\pi}{2} \sum_{k=1}^{\infty} a_k b_k.$$

This gives the desired relation (6.11.9). An argument similar to this example yields the *Parseval relation* for the Fourier sine series in $0 \le x \le \pi$

$$\frac{1}{\pi} \int_0^\pi f(x) g(x) \, dx = \frac{1}{2} \sum_{k=1}^\infty b_k \beta_k, \qquad (6.11.10)$$

where b_k and β_k are the Fourier coefficients involved in the Fourier sine series of $f(x)$ and $g(x)$ respectively with $f'(0) = 0 = f'(\pi)$.

6.12 Double Fourier Series

The theory of Fourier series expansions for functions of two variables is analogous to that of Fourier series expansions for functions of one variable. Here we shall present a short description of double Fourier series.

We have seen earlier that, if $f(x)$ is piecewise continuous and periodic with period 2π, then the Fourier series

$$f(x) \sim \frac{a_0}{2} + \sum_{m=1}^\infty (a_m \cos mx + b_m \sin mx),$$

converges in the mean to $f(x)$. If f is continuously differentiable, then its Fourier series converges uniformly.

For the sake of simplicity and convenience, let us consider the function $f(x, y)$ which is continuously differentiable (a stronger condition than necessary). Let $f(x, y)$ be periodic with period 2π, that is,

$$f(x + 2\pi, y) = f(x, y + 2\pi) = f(x, y).$$

Then, if we hold y fixed, we can expand $f(x, y)$ in a uniformly convergent Fourier series

$$f(x, y) = \frac{a_0(y)}{2} + \sum_{m=1}^\infty [a_m(y) \cos mx + b_m(y) \sin mx] \quad (6.12.1)$$

in which the coefficients are functions of y, namely,

$$a_m(y) = \frac{1}{\pi} \int_{-\pi}^\pi f(x, y) \cos mx \, dx,$$

$$b_m(y) = \frac{1}{\pi} \int_{-\pi}^\pi f(x, y) \sin mx \, dx.$$

These coefficients are continuously differentiable in y, and therefore, we can expand them in uniformly convergent series

$$a_m\left(y\right) = \frac{a_{m0}}{2} + \sum_{n=1}^{\infty}\left(a_{mn}\cos ny + b_{mn}\sin ny\right),$$

$$(6.12.2)$$

$$b_m\left(y\right) = \frac{c_{m0}}{2} + \sum_{n=1}^{\infty}\left(c_{mn}\cos ny + d_{mn}\sin ny\right),$$

where

$$a_{mn} = \frac{1}{\pi^2}\int_{-\pi}^{\pi}\int_{-\pi}^{\pi}f\left(x,y\right)\cos mx\cos ny\,dx\,dy,$$

$$b_{mn} = \frac{1}{\pi^2}\int_{-\pi}^{\pi}\int_{-\pi}^{\pi}f\left(x,y\right)\cos mx\sin ny\,dx\,dy,$$

$$(6.12.3)$$

$$c_{mn} = \frac{1}{\pi^2}\int_{-\pi}^{\pi}\int_{-\pi}^{\pi}f\left(x,y\right)\sin mx\cos ny\,dx\,dy,$$

$$d_{mn} = \frac{1}{\pi^2}\int_{-\pi}^{\pi}\int_{-\pi}^{\pi}f\left(x,y\right)\sin mx\sin ny\,dx\,dy.$$

Substitution of a_m and b_m into equation (6.12.1) yields

$$f\left(x,y\right) = \frac{a_{00}}{4} + \frac{1}{2}\sum_{n=1}^{\infty}\left(a_{0n}\cos ny + b_{0n}\sin ny\right)$$

$$+\frac{1}{2}\sum_{m=1}^{\infty}\left(a_{m0}\cos mx + c_{m0}\sin mx\right)$$

$$+\sum_{m=1}^{\infty}\sum_{n=1}^{\infty}\left(a_{mn}\cos mx\cos ny + b_{mn}\cos mx\sin ny\right.$$

$$\left. + c_{mn}\sin mx\cos ny + d_{mn}\sin mx\sin ny\right). \quad (6.12.4)$$

This is called the *double Fourier series* for $f\left(x,y\right)$.

(a) When $f\left(-x,y\right) = f\left(x,y\right)$ and $f\left(x,-y\right) = f\left(x,y\right)$, all the coefficients vanish except a_{mn}, and the double Fourier series reduces to

$$f\left(x,y\right) = \sum_{m=1}^{\infty}\sum_{n=1}^{\infty}a_{mn}\cos mx\cos ny, \qquad (6.12.5)$$

where

$$a_{mn} = \frac{4}{\pi^2}\int_0^{\pi}\int_0^{\pi}f\left(x,y\right)\cos mx\cos ny\,dx\,dy.$$

(b) When $f\left(-x,y\right) = f\left(x,y\right)$ and $f\left(x,-y\right) = -f\left(x,y\right)$, we have

$$f(x,y) = \frac{1}{\pi^2} \sum_{n=1}^{\infty} b_{0n} \sin ny + \sum_{m=1}^{\infty} \sum_{n=1}^{\infty} b_{mn} \cos mx \sin ny, \quad (6.12.6)$$

where

$$b_{mn} = \frac{4}{\pi^2} \int_0^\pi \int_0^\pi f(x,y) \cos mx \sin ny \, dx \, dy.$$

(c) When $f(-x,y) = -f(x,y)$ and $f(x,-y) = f(x,y)$, we have

$$f(x,y) = \frac{1}{2} \sum_{m=1}^{\infty} c_{m0} \sin mx + \sum_{m=1}^{\infty} \sum_{n=1}^{\infty} c_{mn} \sin mx \cos ny, \quad (6.12.7)$$

where

$$c_{mn} = \frac{4}{\pi^2} \int_0^\pi \int_0^\pi f(x,y) \sin mx \cos ny \, dx \, dy.$$

(d) When $f(-x,y) = -f(x,y)$ and $f(x,-y) = -f(x,y)$, we have

$$f(x,y) = \sum_{m=1}^{\infty} \sum_{n=1}^{\infty} d_{mn} \sin mx \sin ny, \quad (6.12.8)$$

where

$$d_{mn} = \frac{4}{\pi^2} \int_0^\pi \int_0^\pi f(x,y) \sin mx \sin ny \, dx \, dy.$$

Example 6.12.1. Expand the function $f(x,y) = xy$ into double Fourier series in the interval $-\pi < x < \pi, \pi < y < \pi$.

Since $f(-x,y) = -xy = -f(x,y)$ and $f(x,-y) = -xy = -f(x,y)$, we find

$$d_{mn} = \frac{4}{\pi^2} \int_0^\pi \int_0^\pi xy \sin mx \sin ny \, dx \, dy = (-1)^{(m+n)} \left(\frac{4}{mn}\right).$$

Thus, the double Fourier series for f in $-\pi < x < \pi, \pi < y < \pi$ is

$$f(x,y) = 4 \sum_{m=1}^{\infty} \sum_{n=1}^{\infty} (-1)^{m+n} \frac{\sin mx \sin ny}{mn}.$$

6.13 Fourier Integrals

In earlier sections of this chapter, we have described Fourier series for functions which are periodic with period 2π in the interval $(-\infty, \infty)$. However, functions which are not periodic cannot be represented by Fourier series.

In many problems of physical interest, it is desirable to develop an integral representation for such a function that is analogous to a Fourier series.

We have seen in Section 6.9 that the Fourier series for $f(x)$ in the interval $[-l, l]$ is represented by

$$f(x) = \frac{a_0}{2} + \sum_{k=1}^{\infty} \left[a_k \cos \left(\frac{k\pi x}{l} \right) + b_k \sin \left(\frac{k\pi x}{l} \right) \right], \qquad (6.13.1)$$

where

$$a_k = \frac{1}{l} \int_{-l}^{l} f(t) \cos \left(\frac{k\pi t}{l} \right) dt, \qquad k = 0, 1, 2, \ldots, \qquad (6.13.2)$$

$$b_k = \frac{1}{l} \int_{-l}^{l} f(t) \sin \left(\frac{k\pi t}{l} \right) dt, \qquad k = 1, 2, 3, \ldots. \qquad (6.13.3)$$

Substituting (6.13.2) and (6.13.3) into (6.13.1), we have

$$f(x) = \frac{1}{2l} \int_{-l}^{l} f(t)\, dt + \frac{1}{l} \sum_{k=1}^{\infty} \left[\int_{-l}^{l} f(t) \cos \left(\frac{k\pi t}{l} \right) \cdot \cos \left(\frac{k\pi x}{l} \right) dt \right.$$

$$\left. + \left[\int_{-l}^{l} f(t) \sin \left(\frac{k\pi t}{l} \right) \cdot \sin \left(\frac{k\pi x}{l} \right) dt \right] \right]$$

$$= \frac{1}{2l} \int_{-l}^{l} f(t)\, dt + \frac{1}{l} \sum_{k=1}^{\infty} \int_{-l}^{l} f(t) \cos \left[\frac{k\pi}{l} (t - x) \right] dt. \qquad (6.13.4)$$

Suppose that $f(x)$ is absolutely integrable, that is,

$$\int_{-\infty}^{\infty} |f(x)|\, dx$$

converges. Then,

$$\frac{|a_0|}{2} = \frac{1}{2l} \left| \int_{-l}^{l} f(t)\, dt \right| \leq \frac{1}{2l} \int_{-\infty}^{\infty} |f(t)|\, dt$$

which approaches zero as $l \to \infty$. Thus, holding x fixed, as l approaches infinity, equation (6.13.4) becomes

$$f(x) = \lim_{l \to \infty} \frac{1}{l} \sum_{k=1}^{\infty} \int_{-l}^{l} f(t) \cos \left[\frac{k\pi}{l} (t - x) \right] dt.$$

Now let

$$\alpha_k = \frac{k\pi}{l}, \qquad \delta\alpha = (\alpha_{k+1} - \alpha_k) = \frac{\pi}{l}.$$

Then, the function $f(x)$ can be written as

$$f(x) = \lim_{l \to \infty} \sum_{k=1}^{\infty} F(\alpha_k)\, \delta\alpha,$$

where

$$F(\alpha) = \frac{1}{\pi} \int_{-l}^{l} f(t) \cos\left[\alpha\,(t - x)\right] dt.$$

If we plot $F(\alpha)$ against α, we can clearly see that the sum

$$\sum_{k=1}^{\infty} F(\alpha_k)\, \delta\alpha$$

is an approximation to the area under the curve $y = F(\alpha)$ (see Figure 6.13.1). As $l \to \infty$ and $\delta\alpha \to 0$ the infinite sum formally approaches the definite integral. We therefore have

$$f(x) = \int_{0}^{\infty} \left[\frac{1}{\pi} \int_{-\infty}^{\infty} f(t) \cos\left[\alpha\,(t - x)\right] dt\right] d\alpha \qquad (6.13.5)$$

which is the *Fourier integral* representation for the function $f(x)$. Its convergence to $f(x)$ is suggested, but by no means established by the preceding arguments. We shall now prove that this representation is indeed valid if $f(x)$ satisfies certain conditions.

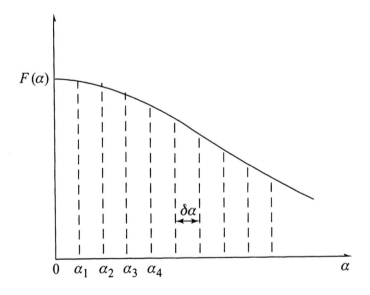

Figure 6.13.1 Area under the curve $F(\alpha)$.

Lemma 6.13.1. *If f is piecewise smooth in the interval $[0, b]$, then for $b > 0$,*

$$\lim_{\lambda \to \infty} \int_0^b f(x) \frac{\sin \lambda x}{x} dx = \frac{\pi}{2} f(0+).$$

Proof.

$$\int_0^b f(x) \frac{\sin \lambda x}{x} dx = \int_0^b f(0+) \frac{\sin \lambda x}{x} dx + \int_0^b \frac{f(x) - f(0+)}{x} \sin \lambda x \, dx$$

$$= f(0+) \int_0^{\lambda b} \frac{\sin t}{t} dt + \int_0^b \frac{f(x) - f(0+)}{x} \sin \lambda x \, dx.$$

Since f is piecewise smooth, the integrand of the last integral is bounded as $\lambda \to \infty$, and thus, by the Riemann–Lebesgue lemma 6.10.1, the last integral tends to zero as $\lambda \to \infty$. Hence,

$$\lim_{\lambda \to \infty} \int_0^b f(x) \frac{\sin \lambda x}{x} dx = \frac{\pi}{2} f(0+), \qquad (6.13.6)$$

since

$$\int_0^\infty \frac{\sin t}{t} dt = \frac{\pi}{2}.$$

Theorem 6.13.1. *(Fourier Integral Theorem) If f is piecewise smooth in every finite interval, and absolutely integrable on $(-\infty, \infty)$, then*

$$\frac{1}{\pi} \int_0^\infty \left[\int_{-\infty}^\infty f(t) \cos k(t - x) \, dt \right] dk = \frac{1}{2} [f(x+) + f(x-)].$$

Proof. Noting that $|\cos k(t - x)| \le 1$ and that by hypothesis

$$\int_{-\infty}^\infty f(t) \, dt < \infty,$$

we see that the integral

$$\int_{-\infty}^\infty f(t) \cos k(t - x) \, dt$$

converges independently of k and x. It therefore follows that in the double integral

$$I = \int_0^\lambda \left[\int_{-\infty}^\infty f(t) \cos k(t - x) \, dt \right] dk,$$

the order of integration may be interchanged. We then have

$$I = \int_{-\infty}^{\infty} f(t) \left[\int_0^{\lambda} \cos k\,(t - x)\,dk \right] dt$$

$$= \int_{-\infty}^{\infty} f(t) \left[\frac{\sin \lambda\,(t - x)}{(t - x)} \right] dt$$

$$= \left[\int_{-\infty}^{-M} + \int_{-M}^{x} + \int_{x}^{M} + \int_{M}^{\infty} \right] f(t)\, \frac{\sin \lambda\,(t - x)}{(t - x)}\,dt.$$

If we substitute $u = t - x$, we have

$$\int_{x}^{M} f(t)\, \frac{\sin \lambda\,(t - x)}{(t - x)}\,dt = \int_0^{M - x} f(u + x) \left(\frac{\sin \lambda u}{u} \right) du$$

which is equal to $\pi f(x+)/2$ in the limit $\lambda \to \infty$, by Lemma 6.13.1. Similarly, the second integral tends to $\pi f(x-)/2$ when $\lambda \to \infty$. If we make M sufficiently large, the absolute values of the first and the last integrals are each less than $\varepsilon/2$. Consequently, as $\lambda \to \infty$

$$\int_0^{\infty} \left[\int_{-\infty}^{\infty} f(t) \cos k\,(t - x)\,dt \right] dk = \frac{\pi}{2}\,[f(x+) + f(x-)]. \quad (6.13.7)$$

If f is continuous at the point x, then

$$f(x+) = f(x-) = f(x)$$

so that integral (6.13.7) reduces to the *Fourier integral representation* for f as

$$f(x) = \frac{1}{\pi} \int_0^{\infty} \left[\int_{-\infty}^{\infty} f(t) \cos k\,(t - x)\,dt \right] dk. \quad (6.13.8)$$

We may express the Fourier integral representation (6.13.8) in complex form. In this case, we substitute

$$\cos k\,(t - x) = \cos k\,(x - t) = \frac{1}{2} \left[e^{ik(x-t)} + e^{-ik(x-t)} \right]$$

into equation (6.13.8) and write it as the sum of two integrals

$$f(x) = \frac{1}{2\pi} \int_0^{\infty} \int_{-\infty}^{\infty} f(t)\, e^{ik(x-t)}\,dt\,dk + \frac{1}{2\pi} \int_0^{\infty} \int_{-\infty}^{\infty} f(t)\, e^{-ik(x-t)}\,dt\,dk.$$

Changing the integration variable from k to $-k$ in the second integral, we obtain

$$f(x) = \frac{1}{2\pi} \left[\int_0^{\infty} \int_{-\infty}^{\infty} f(t)\, e^{ik(x-t)}\,dt\,dk - \int_0^{-\infty} \int_{-\infty}^{\infty} f(t)\, e^{ik(x-t)}\,dt\,dk \right]$$

$$= \frac{1}{2\pi} \left[\int_0^{\infty} \int_{-\infty}^{\infty} f(t)\, e^{ik(x-t)}\,dt\,dk + \int_{-\infty}^{0} \int_{-\infty}^{\infty} f(t)\, e^{ik(x-t)}\,dt\,dk \right]$$

$$= \frac{1}{2\pi} \int_{-\infty}^{\infty} \int_{-\infty}^{\infty} f(t)\, e^{ik(x-t)}\,dt\,dk. \quad (6.13.9)$$

Or, equivalently,

$$f(x) = \frac{1}{\sqrt{2\pi}} \int_{-\infty}^{\infty} e^{ikx} dk \left[\frac{1}{\sqrt{2\pi}} \int_{-\infty}^{\infty} e^{-ikt} f(t) dt \right]$$
$$= \frac{1}{\sqrt{2\pi}} \int_{-\infty}^{\infty} F(k) e^{ikx} dk, \qquad (6.13.10)$$

where

$$F(k) = \frac{1}{\sqrt{2\pi}} \int_{-\infty}^{\infty} e^{-ikt} f(t) dt. \qquad (6.13.11)$$

Either (6.13.9) or (6.13.10) with coefficient $F(k)$ is called the *complex form of the Fourier integral representation* for $f(x)$.

Now we assume that $f(x)$ is either an even or an odd function. Any function that is not even or odd can be expressed as a sum of two such functions. Expanding the cosine function in (6.13.8), we obtain the *Fourier cosine formula*

$$f(x) = f(-x) = \frac{2}{\pi} \int_0^{\infty} \cos kx\, dk \int_0^{\infty} \cos kt\, f(t)\, dt. \qquad (6.13.12)$$

Similarly, for an odd function, we obtain the *Fourier sine formula*

$$f(x) = -f(-x) = \frac{2}{\pi} \int_0^{\infty} \sin kx\, dk \int_0^{\infty} \sin kt\, f(t)\, dt. \qquad (6.13.13)$$

Example 6.13.1. The rectangular pulse can be expressed as a sum of Heaviside functions

$$f(x) = H(x+1) - H(x-1).$$

Find its Fourier integral representation.

From (6.13.5) we find

$$f(x) = \frac{1}{\pi} \int_0^{\infty} \left[\int_{-1}^1 \cos[k(t-x)] dt \right] dk$$
$$= \frac{1}{\pi} \int_0^{\infty} \left[\cos kx \int_{-1}^1 \cos kt\, dt + \sin kx \int_{-1}^1 \sin kt\, dt \right] dk$$
$$= \frac{2}{\pi} \int_0^{\infty} \left(\frac{\sin k}{k} \right) \cos kx\, dk.$$

Example 6.13.2. Find the Fourier cosine integral representation of the function

$$f(x) = \begin{cases} 1, & 0 < x < 1, \\ \\ 0, & x \geq 1. \end{cases}$$

We have, from (6.13.12),

$$f(x) = \frac{2}{\pi} \int_0^\infty \cos kx \, dk \int_0^1 \cos kt \, dt = \frac{2}{\pi} \int_0^\infty \left(\frac{\sin k}{k} \right) \cos kx \, dk,$$

or,

$$1 = \frac{2}{\pi} \int_0^\infty \frac{\sin k}{k} \cos kx \, dk.$$

6.14 Exercises

1. Find the Fourier series of the following functions:

(a) $f(x) = \begin{cases} x & -\pi < x < 0 \\ h & 0 < x < \pi, \end{cases}$ h is a constant

(b) $f(x) = \begin{cases} 1 & -\pi < x < 0 \\ x^2 & 0 < x < \pi, \end{cases}$

(c) $f(x) = x + \sin x$ $-\pi < x < \pi,$

(d) $f(x) = 1 + x$ $-\pi < x < \pi,$

(e) $f(x) = e^x$ $-\pi < x < \pi,$

(f) $f(x) = 1 + x + x^2$ $-\pi < x < \pi.$

2. Determine the Fourier sine series of the following functions:

(a) $f(x) = \pi - x$ $0 < x < \pi,$

(b) $f(x) = \begin{cases} 1 & 0 < x < \pi/2 \\ 2 & \pi/2 < x < \pi, \end{cases}$

(c) $f(x) = x^2$ $0 < x < \pi,$

(d) $f(x) = \cos x$ $0 < x < \pi,$

(e) $f(x) = x^3$ $0 < x < \pi,$

(f) $f(x) = e^x$ $0 < x < \pi.$

3. Obtain the Fourier cosine series representation for the following functions:

(a) $f(x) = \pi + x$ $\qquad\qquad$ $0 < x < \pi$,

(b) $f(x) = x$ $\qquad\qquad$ $0 < x < \pi$,

(c) $f(x) = x^2$ $\qquad\qquad$ $0 < x < \pi$,

(d) $f(x) = \sin 3x$ $\qquad\qquad$ $0 < x < \pi$,

(e) $f(x) = e^x$ $\qquad\qquad$ $0 < x < \pi$,

(f) $f(x) = \cosh x$ $\qquad\qquad$ $0 < x < \pi$.

4. Expand the following functions in a Fourier series:

(a) $f(x) = x^2 + x$ $\qquad\qquad$ $-1 < x < 1$,

(b) $f(x) = \begin{cases} 1 & 0 < x < 3 \\ 0 & 3 < x < 6, \end{cases}$

(c) $f(x) = \sin(\pi x/l)$ $\qquad\qquad$ $0 < x < l$,

(d) $f(x) = x^3$ $\qquad\qquad$ $-2 < x < 2$,

(e) $f(x) = e^{-x}$ $\qquad\qquad$ $0 < x < 1$,

(f) $f(x) = \sinh x$ $\qquad\qquad$ $-1 < x < 1$.

5. Expand the following functions in a complex Fourier series:

(a) $f(x) = e^{2x}$ $\qquad\qquad$ $-\pi < x < \pi$,

(b) $f(x) = \cosh x$ $\qquad\qquad$ $-\pi < x < \pi$,

(c) $f(x) = \begin{cases} 1 & -\pi < x < 0 \\ \cos x & 0 < x < \pi, \end{cases}$

(d) $f(x) = x$ $\qquad\qquad$ $-1 < x < 1$,

(e) $f(x) = x^2$ $\qquad\qquad$ $-\pi < x < \pi$,

(f) $f(x) = \sinh(\pi x/2)$ $\qquad\qquad$ $-2 < x < 2$.

6. (a) Find the Fourier series expansion of the function

$$f(x) = \begin{cases} 0, & -\pi < x < 0 \\ x/2, & 0 < x < \pi. \end{cases}$$

(b) With the use of the Fourier series of $f(x)$ in 6(a), show that

$$\frac{\pi^2}{8} = 1 + \frac{1}{3^2} + \frac{1}{5^2} + \frac{1}{7^2} + \ldots.$$

7. (a) Determine the Fourier series of the function

$$f(x) = x^2, \qquad -l < x < l.$$

(b) With the use of the Fourier series of $f(x)$ in 7(a), show that

$$\frac{\pi^2}{12} = 1 - \frac{1}{2^2} + \frac{1}{3^2} - \frac{1}{4^2} + \ldots.$$

8. Determine the Fourier series expansion of each of the following functions by performing the differentiation of the appropriate Fourier series:

(a) $\sin^2 x$ $0 < x < \pi$,

(b) $\cos^2 x$ $0 < x < \pi$,

(c) $\sin x \cos x$ $0 < x < \pi$,

(d) $\cos x + \cos 2x$ $0 < x < \pi$,

(e) $\cos x + \cos 2x$ $0 < x < \pi$.

9. Find the functions represented by the new series which are obtained by termwise integration of the following series from 0 to x:

(a) $\displaystyle\sum_{k=1}^{\infty} \frac{(-1)^{k+1}}{k} \sin kx = x/2$ $-\pi < x < \pi$,

(b) $\displaystyle\frac{3}{2} + \frac{1}{\pi}\sum_{k=1}^{\infty} \frac{1-(-1)^k}{k} \sin kx = \begin{cases} 1 & -\pi < x < 0 \\ 2 & 0 < x < \pi, \end{cases}$

(c) $\displaystyle\sum_{k=1}^{\infty} (-1)^{k+1} \frac{\cos kx}{k} = \ln\left(2\cos\frac{x}{2}\right)$ $-\pi < x < \pi$,

(d) $\sum\limits_{k=1}^{\infty} \frac{\sin(2k+1)x}{(2k+1)^3} = \frac{\pi^2 x - \pi x^2}{8}$ $0 < x < 2\pi,$

(e) $\left(\frac{4}{\pi}\right) \sum\limits_{k=1}^{\infty} \frac{\sin(2k-1)x}{(2k-1)} = \begin{cases} -1 & -\pi < x < 0 \\ 1 & 0 < x < \pi. \end{cases}$

10. Determine the double Fourier series of the following functions:

(a) $f(x,y) = 1$ $0 < x < \pi$ $0 < y < \pi,$

(b) $f(x,y) = xy^2$ $0 < x < \pi$ $0 < y < \pi,$

(c) $f(x,y) = x^2 y^2$ $0 < x < \pi$ $0 < y < \pi,$

(d) $f(x,y) = x^2 + y$ $-\pi < x < \pi$ $-\pi < y < \pi,$

(e) $f(x,y) = x \sin y$ $-\pi < x < \pi$ $-\pi < y < \pi,$

(f) $f(x,y) = e^{x+y}$ $-\pi < x < \pi$ $-\pi < y < \pi,$

(g) $f(x,y) = xy$ $0 < x < 1$ $0 < y < 2,$

(h) $f(x,y) = 1$ $0 < x < a$ $0 < y < b,$

(i) $f(x,y) = x \cos y$ $-1 < x < 1$ $-2 < y < 2,$

(j) $f(x,y) = xy^2$ $-\pi < x < \pi$ $-\pi < y < \pi,$

(k) $f(x,y) = x^2 y^2$ $-\pi < x < \pi$ $-\pi < y < \pi.$

11. Deduce the general double Fourier series expansion formula for the function $f(x,y)$ in the rectangle $-a < x < a,\ -b < y < b$.

12. Prove the *Weierstrass Approximation Theorem*: If f is a continuous function on the interval $-\pi \le x \le \pi$ and if $f(-\pi) = f(\pi)$, then, for any $\varepsilon > 0$, there exists a trigonometric polynomial

$$T(x) = \frac{a_0}{2} + \sum_{k=1}^{n} (a_k \cos kx + b_k \sin kx)$$

such that

$$|f(x) - T(x)| < \varepsilon$$

for all x in $[-\pi, \pi]$.

13. Use the Fourier cosine or sine integral formula to show that

(a) $e^{-ax} = \frac{2}{\pi} \int_0^\infty \frac{\alpha}{\alpha^2+\beta^2} \cos \beta x \, d\beta$, $\qquad x \geq 0, \qquad \alpha > 0,$

(b) $e^{-ax} = \frac{2}{\pi} \int_0^\infty \frac{\beta}{\alpha^2+\beta^2} \sin \beta x \, d\beta$, $\qquad x > 0, \qquad \alpha > 0.$

14. Show that the Fourier integral representation of the function

$$f(x) = \begin{cases} x^2, & 0 < x < a \\ 0, & x > a \end{cases}$$

is

$$f(x) = \frac{2}{\pi} \int_0^\infty \left[\left(a^2 - \frac{2}{k^2} \right) \sin ak + \frac{2a}{k} \cos ak \right] \frac{\cos kx}{k} \, dk.$$

15. Apply the Parseval relation (6.5.10) to Example 6.7.3 or Example 6.7.4 to show that

(a) $\displaystyle\sum_{n=1}^\infty \frac{1}{(2n-1)^4} = \frac{\pi^4}{96}$ and (b) $\displaystyle\sum_{n=1}^\infty \frac{1}{n^4} = \frac{\pi^4}{90}.$

16. (a) Obtain the Fourier series for the 2π-periodic odd function $f(x) = x(\pi - x)$ on $[0, \pi]$.

(b) Use the Parseval relation (6.5.10) to show that

$$\sum_{n=1}^\infty \frac{1}{(2n-1)^6} = \frac{\pi^6}{960} \quad \text{and} \quad \sum_{n=1}^\infty \frac{1}{n^6} = \frac{\pi^6}{945}.$$

17. If the 2π-periodic even function is given by $f(x) = |x|$ for $-\pi \leq x \leq \pi$, show that

$$f(x) = \frac{\pi}{2} - \frac{4}{\pi} \sum_{n=1}^\infty \frac{\cos(2n-1)x}{(2n-1)^2}.$$

18. Consider the sawtooth function defined by $f(x) = \pi - x$, $0 < x < 2\pi$, and $f(x + 2n\pi) = f(x)$ with $f(0) = 0$.

(a) Show that the Fourier series for this function is

$$f(x) = \sum_{k=0}^\infty \frac{2}{k} \sin kx,$$

and f has a jump discontinuity at the origin with

$$f\left(0+\right) = \frac{\pi}{2}, \qquad f\left(0-\right) = -\frac{\pi}{2}, \qquad \text{and} \quad f\left(0+\right) - f\left(0-\right) = \pi.$$

(b) Show that
$$\max_{0 \leq x \leq \frac{\pi}{n}} s_n\left(x\right) - \frac{\pi}{2} = \int_0^\pi \frac{\sin\theta}{\theta}\,d\theta - \frac{\pi}{2}.$$

(c) The result (b) is a manifestation of the Gibbs phenomenon, that is, near a jump discontinuity, the Fourier series of f overshoots (or undershoots) it by approximately 9% of the jump.

(d) If $d_n\left(x\right) = s_n\left(x\right) - f\left(x\right)$ in $0 \leq x \leq 2\pi$, show that
$$d_n'\left(x\right) = 2\pi\, D_n\left(x\right),$$
where $D_n\left(x\right)$ is given by (6.10.15).

(e) Using $D_n\left(x\right) = \displaystyle\sum_{k=-n}^{n} e^{ikx}$, show that the first critical point of $d_n\left(x\right)$ to the right of the origin occurs at $x_n = \pi/\left(n+\frac{1}{2}\right)$, and that
$$\lim_{n\to\infty} d_n\left(x_n\right) = 2\int_0^\pi \frac{\sin\theta}{\theta}\,d\theta - \pi.$$

(f) Draw the graph of the fortieth partial sum
$$s_{40}\left(x\right) = \sum_{k=1}^{40} \frac{2\sin kx}{k}, \qquad -2\pi < x < 2\pi,$$
and then examine the Gibbs phenomenon for the function $f\left(x\right)$.

19. Consider the characteristic function of the interval $[a,b] \subset [-\pi,\pi]$ defined by
$$f\left(x\right) = \chi_{[a,b]}\left(x\right) = \begin{cases} 1, & a \leq x \leq b \\[2mm] 0, & \text{otherwise.} \end{cases}$$
Show that the Fourier series in $a \leq x \leq b$ is given by
$$f\left(x\right) \sim \frac{1}{2\pi}\left(b-a\right) + \sum_{k\neq 0} \frac{\exp\left(-ika\right) - \exp\left(-ikb\right)}{2\pi i k} \cdot e^{ikx}.$$

20. (a) Obtain the Fourier series of $f\left(x\right) = x\left(\pi - x\right)$, $0 \leq x \leq \pi$.

(b) Derive the following the numerical series
$$1 - \frac{1}{3^3} + \frac{1}{5^3} - \frac{1}{7^3} + \ldots = \frac{\pi^3}{12},$$
$$1 + \frac{1}{3^3} - \frac{1}{5^3} - \frac{1}{7^3} + \frac{1}{9^3} + \frac{1}{11^3} - \ldots = \frac{3\sqrt{2}\,\pi^3}{128}.$$

21. (a) Show that the Fourier series of the triangular function with vertices at $(0,0)$, $(\pi/2, 1)$ and $(\pi, 0)$ defined by

$$f(x) = \begin{cases} \frac{2x}{\pi}, & 0 \le x \le \frac{\pi}{2} \\ \frac{2}{\pi}(\pi - x), & \pi/2 \le x \le \pi \end{cases}$$

is

$$f(x) = \frac{8}{\pi^2}\left\{ \frac{\sin x}{1^2} - \frac{\sin 3x}{3^2} + \frac{\sin 5x}{5^2} - \frac{\sin 7x}{7^2} + \cdots \right\}.$$

(b) Show that

$$\sum_{n=1}^{\infty} \frac{1}{(2n-1)^2} = \frac{\pi^2}{8}.$$

22. Obtain the Fourier sine series and the Fourier cosine series of the following functions:

(a) $f(x) = 1$, $0 < x < a$,

(b) $f(x) = x$, $0 < x < a$,

(c) $f(x) = x^2$, $0 < x < a$.

23. Find the full Fourier series of the following functions

(a) $f(x) = x$, $0 < x < a$,

(b) $f(x) = \begin{cases} -1 - x, & -1 < x < 0, \\ +1 - x, & 0 < x < 1. \end{cases}$

(c) $f(x) = \begin{cases} 0, & 0 \le x \le \pi, \\ 1, & \pi \le x \le 2\pi. \end{cases}$

24. Obtain the Fourier cosine series for the function

$$f(x) = \begin{cases} \cos x, & 0 < x \le \frac{\pi}{2}, \\ 0, & \frac{\pi}{2} \le x < \pi. \end{cases}$$

25. Show that $\frac{1}{2i}\sum_{n\neq 0} \frac{1}{n}e^{inx}$ is the complex Fourier series of the 2π-periodic sawtooth function defined by $f(0) = 0$, and

$$f(x) = \begin{cases} -\frac{1}{2}(\pi + x), & -\pi < x < 0, \\ \frac{1}{2}(\pi - x), & 0 < x < \pi. \end{cases}$$

26. Suppose $f(x)$ and $g(x)$ have the following Fourier series expansion in $-\pi \leq x \leq \pi$:

$$f(x) \sim \frac{1}{2}a_0 + \sum_{k=1}^{\infty} (a_k \cos kx + b_k \sin kx),$$

$$g(x) \sim \frac{1}{2}\alpha_0 + \sum_{k=1}^{\infty} (\alpha_k \cos kx + \beta_k \sin kx),$$

where $f(x)$ and $g(x)$ together with their first two derivatives are continuous on $-\pi \leq x \leq \pi$, and $f(-\pi) = f(\pi)$, $f'(-\pi) = f'(\pi)$, $g(-\pi) = g(\pi)$, $g'(-\pi) = g'(\pi)$ hold.

Prove that the following general Parseval relation holds:

$$\frac{1}{\pi}\int_{-\pi}^{\pi} f(x)g(x)\,dx = \frac{1}{2}a_0\alpha_0 + \sum_{k=1}^{\infty}(a_k\alpha_k + b_k\beta_k).$$

When $f(x) = g(x)$, then the Parseval relation (6.5.10) is a special case of the above result.

27. Obtain the Fourier integral representation for the following functions:

(a) $f(x) = H(a - |x|) = \begin{cases} 1, & |x| < a \\ 0, & |x| > a, \end{cases}$

(b) $f(x) = \begin{cases} \sin x, & |x| < \pi \\ 0, & |x| > \pi. \end{cases}$

28. If $f(x)$, $x \in \mathbb{R}$, is defined by

$$f(x) = \begin{cases} -1, & -a < x < 0, \\ +1, & 0 \leq x < a, \\ 0, & \text{otherwise}, \end{cases}$$

show that $f(x)$ has the Fourier sine integral representation

$$f(x) = \frac{2}{\pi}\int_0^{\infty} \frac{1}{k}(1 - \cos ka)\sin kx\,dk.$$

29. If $f(x) = e^{-x}$, $0 < x < \infty$, show that

(a) the Fourier sine integral representation is

$$f(x) = \frac{2}{\pi} \int_0^\infty \frac{k \sin kx}{1 + k^2} dk,$$

(b) the Fourier cosine integral representation is

$$f(x) = \frac{2}{\pi} \int_0^\infty \frac{\cos kx}{1 + k^2} dk.$$

30. If $f(x)$ is defined by

$$f(x) = \begin{cases} 0, & x < 0, \\ e^{-x}, & x > 0, \end{cases}$$

show that

(a) the Fourier integral representation of $f(x)$ is

$$f(x) = \frac{1}{\pi} \int_0^\infty \frac{(\cos kx + k \sin kx)}{(1 + k^2)} dk,$$

(b) the Fourier cosine integral representation of $f(x)$ is

$$f(x) = \frac{1}{2\pi} \int_{-\infty}^\infty \left(\frac{1 - ik}{1 + k^2}\right) e^{ikx} dk.$$

31. (a) Obtain both the complex Fourier series and the usual Fourier series of $f(x) = \exp[x(1 + 2\pi i)]$ on the interval $[-1, 1]$.

(b) Find the sum of each of the series

$$\sum_{k=1}^\infty \frac{1}{(1 + \pi^2 k^2)} \quad \text{and} \quad \sum_{k=1}^\infty \frac{(-1)^k}{(1 + \pi^2 k^2)}.$$

32. Use Example 6.7.2 to calculate the value of the following series:

(a) $\displaystyle\sum_{k=1}^\infty \frac{1}{(4k^2 - 1)},$ (b) $\displaystyle\sum_{k=1}^\infty \frac{(-1)^k}{(4k^2 - 1)},$

(c) $\displaystyle\sum_{k=1}^\infty \frac{1}{(4k^2 - 1)^2},$ and   (d) $\displaystyle\sum_{k=1}^\infty \frac{k^2}{(4k^2 - 1)^2}.$

33. Show that the complex Fourier series of $f(x) = x$ is given by

$$x \sim \sum_{k=1}^\infty \frac{(-1)^k}{k} \left(i e^{ikx}\right) + \sum_{k=-1}^{-\infty} \frac{(-1)^k}{k} \left(i e^{ikx}\right).$$

34. (a) Show that the Fourier series for $f(x)$ is defined by

$$f(x) = \begin{cases} \sin 2x, & 0 \le x \le \frac{\pi}{2} \\ \\ 0, & \frac{\pi}{2} \le x \le \pi, \end{cases}$$

is

$$f(x) = \frac{1}{\pi} + \frac{1}{2}\sin 2x - \left(\frac{2}{\pi}\right)\sum_{k=1}^{\infty} \frac{\cos 4kx}{(4k^2 - 1)}.$$

(b) Show that

$$\sum_{k=1}^{\infty} \frac{1}{(4k^2 - 1)^2} = \frac{1}{16}\left(\pi^2 - 8\right).$$

(c) Find the sum of the infinite series

$$\frac{\sin(4x)}{1.2.3} + \frac{\sin(2.4x)}{3.4.5} + \frac{\sin(3.4x)}{5.6.7} + \cdots, \quad 0 \le x \le \pi.$$

35. (a) Obtain the complex Fourier series of

$$f(x) = \cos(ax), \qquad -\pi \le x \le \pi,$$

where a is real but *not* an integer.

(b) Hence, show that

$$\pi \cot \pi x = \frac{1}{x} - \sum_{k=1}^{\infty} \frac{2x}{(k^2 - x^2)}.$$

(c) Derive the product formula

$$\sin \pi x = \pi x \prod_{n=1}^{\infty}\left(1 - \frac{x^2}{n^2}\right).$$

(d) Show that

$$\frac{\pi}{2} = \prod_{n=1}^{\infty} \frac{2n}{(2n-1)} \cdot \frac{2n}{(2n+1)} = \left(\frac{2}{1}\cdot\frac{2}{3}\right)\cdot\left(\frac{4}{3}\cdot\frac{4}{5}\right)\cdot\left(\frac{6}{5}\cdot\frac{6}{7}\right)\cdot\left(\frac{8}{7}\cdot\frac{8}{9}\right)\cdots.$$

36. Obtain the Fourier series of the following functions:

(a) $f(x) = e^x$, $\qquad 0 \le x \le 2\pi$, $\quad f(x + 2\pi) = f(x)$.

(b) $f(x) = \begin{cases} +1, & -\pi < x < -\frac{\pi}{2}, \ 0 < x < \frac{\pi}{2}, \\ \\ -1, & -\frac{\pi}{2} < x < 0, \ \frac{\pi}{2} < x < \pi, \\ \\ 0, & x = \frac{n\pi}{2}, \quad n = 0, \pm 1, \pm 2, \ldots \end{cases}$

and draw the graph of this function.

(c) $f(x) = x - [x]$,

where $[x]$ is the greatest integer not exceeding x.

37. Find the Fourier series for each of the functions $f(x)$ in $-l < x < l$ and $f(x)$ is defined outside this interval so that $f(x + 2l) = f(x)$ for all x:

(a) $f(x) = \begin{cases} 0, & -l < x < 0, \\ l, & 0 < x < l. \end{cases}$
$\qquad$
(b) $f(x) = \begin{cases} -x, & -l \le x < 0, \\ x, & 0 \le x < l. \end{cases}$

(c) $f(x) = \begin{cases} l + x, & -l \le x < 0, \\ l - x, & 0 \le x < l. \end{cases}$
$\qquad$
(d) $f(x) = \begin{cases} 0, & -l \le x < 0, \\ x^2, & 0 \le x < l. \end{cases}$

Examine the Gibbs phenomenon at the points of discontinuity at $x = 0$ and $x = l$ for the function in (a).

38. Prove the following identities:

(a) $\dfrac{1}{2} + \displaystyle\sum_{k=1}^{n} \cos kx = \dfrac{\sin\left(n + \frac{1}{2}\right) x}{2 \sin \frac{x}{2}}$.

(b) $s_n(x) = \dfrac{1}{\pi} \displaystyle\int_{-\pi}^{\pi} f(\xi + x) \dfrac{\sin\left(n + \frac{1}{2}\right) \xi}{2 \sin \frac{1}{2}\xi} \, d\xi$,

where $s_n(x)$ is the nth partial sum of a Fourier series of $f(x)$ in $(-\pi, \pi)$.

7

Method of Separation of Variables

"However, the emphasis should be somewhat more on how to do the mathematics quickly and easily, and what formulas are true, rather than the mathematicians' interest in methods of rigorous proof."

Richard Feynman

"As a *science*, mathematics has been adapted to the description of natural phenomena, and the great practitioners in this field, such as von Kármán, Taylor and Lighthill, have never concerned themselves with the logical foundations of mathematics, but have boldly taken a pragmatic view of mathematics as an intellectual machine which works successfully. Description has been verified by further observation, still more strikingly by prediction, ... "

George Temple

7.1 Introduction

The method of separation of variables combined with the principle of superposition is widely used to solve initial boundary-value problems involving linear partial differential equations. Usually, the dependent variable $u(x, y)$ is expressed in the separable form $u(x, y) = X(x) Y(y)$, where X and Y are functions of x and y respectively. In many cases, the partial differential equation reduces to two ordinary differential equations for X and Y. A similar treatment can be applied to equations in three or more independent variables. However, the question of separability of a partial differential equation into two or more ordinary differential equations is by no means a trivial one. In spite of this question, the method is widely used in finding solutions of a large class of initial boundary-value problems. This method of solution is also known as the *Fourier method* (or *the method of eigenfunction expansion*). Thus, the procedure outlined above leads to the important ideas of eigenvalues, eigenfunctions, and orthogonality, all of which are very general and powerful for dealing with linear problems. The following examples illustrate the general nature of this method of solution.

7.2 Separation of Variables

In this section, we shall introduce one of the most common and elementary methods, called the *method of separation of variables*, for solving initial boundary-value problems. The class of problems for which this method is applicable contains a wide range of problems of mathematical physics, applied mathematics, and engineering science.

We now describe the method of separation of variables and examine the conditions of applicability of the method to problems which involve second-order partial differential equations in two independent variables.

We consider the second-order homogeneous partial differential equation

$$a^* u_{x^* x^*} + b^* u_{x^* y^*} + c^* u_{y^* y^*} + d^* u_{x^*} + e^* u_{y^*} + f^* u = 0 \quad (7.2.1)$$

where a^*, b^*, c^*, d^*, e^* and f^* are functions of x^* and y^*.

We have stated in Chapter 4 that by the transformation

$$x = x\left(x^*, y^*\right), \qquad y = y\left(x^*, y^*\right), \qquad (7.2.2)$$

where

$$\frac{\partial\left(x, y\right)}{\partial\left(x^*, y^*\right)} \neq 0,$$

we can always transform equation (7.2.1) into canonical form

$$a\left(x, y\right) u_{xx} + c\left(x, y\right) u_{yy} + d\left(x, y\right) u_x + e\left(x, y\right) u_y + f\left(x, y\right) u = 0, (7.2.3)$$

which when

(i) $a = -c$ is hyperbolic,
(ii) $a = 0$ or $c = 0$ is parabolic,
(iii) $a = c$ is elliptic.

We assume a separable solution of (7.2.3) in the form

$$u\left(x, y\right) = X\left(x\right) Y\left(y\right) \neq 0, \qquad (7.2.4)$$

where X and Y are, respectively, functions of x and of y alone, and are twice continuously differentiable. Substituting equations (7.2.4) into equation (7.2.3), we obtain

$$a\,X''Y + c\,XY'' + d\,X'Y + e\,XY' + f\,XY = 0, \qquad (7.2.5)$$

where the primes denote differentiation with respect to the appropriate variables. Let there exist a function $p(x, y)$, such that, if we divide equation (7.2.5) by $p(x, y)$, we obtain

$$a_1(x)\,X''Y + b_1(y)\,XY'' + a_2(x)\,X'Y + b_2(y)\,XY'$$
$$+ \left[a_3(x) + b_3(y)\right] XY = 0. \quad (7.2.6)$$

Dividing equation (7.2.6) again by XY, we obtain

$$\left[a_1 \frac{X''}{X} + a_2 \frac{X'}{X} + a_3\right] = -\left[b_1 \frac{Y''}{Y} + b_2 \frac{Y'}{Y} + b_3\right]. \qquad (7.2.7)$$

The left side of equation (7.2.7) is a function of x only. The right side of equation (7.2.7) depends only upon y. Thus, we differentiate equation (7.2.7) with respect to x to obtain

$$\frac{d}{dx}\left[a_1 \frac{X''}{X} + a_2 \frac{X'}{X} + a_3\right] = 0. \qquad (7.2.8)$$

Integration of equation (7.2.8) yields

$$a_1 \frac{X''}{X} + a_2 \frac{X'}{X} + a_3 = \lambda, \qquad (7.2.9)$$

where λ is a separation constant. From equations (7.2.7) and (7.2.9), we have

$$b_1 \frac{Y''}{Y} + b_2 \frac{Y'}{Y} + b_3 = -\lambda. \qquad (7.2.10)$$

We may rewrite equations (7.2.9) and (7.2.10) in the form

$$a_1 X'' + a_2 X' + (a_3 - \lambda)\,X = 0, \qquad (7.2.11)$$

and

$$b_1 Y'' + b_2 Y' + (b_3 + \lambda)\,Y = 0. \qquad (7.2.12)$$

Thus, $u(x, y)$ is the solution of equation (7.2.3) if $X(x)$ and $Y(y)$ are the solutions of the ordinary differential equations (7.2.11) and (7.2.12) respectively.

If the coefficients in equation (7.2.1) are constant, then the reduction of equation (7.2.1) to canonical form is no longer necessary. To illustrate this, we consider the second-order equation

$$Au_{xx} + Bu_{xy} + Cu_{yy} + Du_x + Eu_y + Fu = 0, \qquad (7.2.13)$$

where A, B, C, D, E, and F are constants which are not all zero.

As before, we assume a separable solution in the form

$$u(x,y) = X(x)Y(y) \neq 0.$$

Substituting this in equation (7.2.13), we obtain

$$AX''Y + BX'Y' + CXY'' + DX'Y + EXY' + FXY = 0. \quad (7.2.14)$$

Division of this equation by AXY yields

$$\frac{X''}{X} + \frac{B}{A}\frac{X'}{X}\frac{Y'}{Y} + \frac{C}{A}\frac{Y''}{Y} + \frac{D}{A}\frac{X'}{X} + \frac{E}{A}\frac{Y'}{Y} + \frac{F}{A} = 0, \quad A \neq 0. \quad (7.2.15)$$

We differentiate this equation with respect to x to obtain

$$\left(\frac{X''}{X}\right)' + \frac{B}{A}\left(\frac{X'}{X}\right)'\frac{Y'}{Y} + \frac{D}{A}\left(\frac{X'}{X}\right)' = 0. \quad (7.2.16)$$

Thus, we have

$$\frac{\left(\frac{X''}{X}\right)'}{\frac{B}{A}\left(\frac{X'}{X}\right)'} + \frac{D}{B} = -\frac{Y'}{Y}. \quad (7.2.17)$$

This equation is obviously separable, so that both sides must be equal to a constant λ. Therefore, we obtain

$$Y' + \lambda Y = 0, \quad (7.2.18)$$

$$\left(\frac{X''}{X}\right)' + \left(\frac{D}{B} - \lambda\right)\frac{B}{A}\left(\frac{X'}{X}\right)' = 0. \quad (7.2.19)$$

Integrating equation (7.2.19) with respect to x, we obtain

$$\frac{X''}{X} + \left(\frac{D}{B} - \lambda\right)\frac{B}{A}\left(\frac{X'}{X}\right) = -\beta, \quad (7.2.20)$$

where β is a constant to be determined. Substituting equation (7.2.18) into the original equation (7.2.15), we obtain

$$X'' + \left(\frac{D}{B} - \lambda\right)\frac{B}{A}X' + \left(\lambda^2 - \frac{E}{C}\lambda + \frac{F}{C}\right)\frac{C}{A}X = 0. \quad (7.2.21)$$

Comparing equations (7.2.20) and (7.2.21), we clearly find

$$\beta = \left(\lambda^2 - \frac{E}{C}\lambda + \frac{F}{C}\right)\frac{C}{A}.$$

Therefore, $u(x,y)$ is a solution of equations (7.2.13) if $X(x)$ and $Y(y)$ satisfy the ordinary differential equations (7.2.21) and (7.2.18) respectively.

We have just described the conditions on the separability of a given partial differential equation. Now, we shall take a look at the boundary conditions involved. There are several types of boundary conditions. The ones that appear most frequently in problems of applied mathematics and mathematical physics include

(i) Dirichlet condition: u is prescribed on a boundary
(ii) Neumann condition: $(\partial u/\partial n)$ is prescribed on a boundary
(iii) Mixed condition: $(\partial u/\partial n) + hu$ is prescribed on a boundary, where $(\partial u/\partial n)$ is the directional derivative of u along the outward normal to the boundary, and h is a given continuous function on the boundary. For details, see Chapter 9 on boundary-value problems.

Besides these three boundary conditions, also known as, the *first*, *second*, and *third boundary conditions*, there are other conditions, such as the Robin condition; one condition is prescribed on one portion of a boundary and another is given on the remainder of the boundary. We shall consider a variety of boundary conditions as we treat problems later.

To separate boundary conditions, such as the ones listed above, it is best to choose a coordinate system suitable to a boundary. For instance, we choose the Cartesian coordinate system (x, y) for a rectangular region such that the boundary is described by the coordinate lines $x = $ constant and $y = $ constant, and the polar coordinate system (r, θ) for a circular region so that the boundary is described by the lines $r = $ constant and $\theta = $ constant.

Another condition that must be imposed on the separability of boundary conditions is that boundary conditions, say at $x = x_0$, must contain the derivatives of u with respect to x only, and their coefficients must depend only on x. For example, the boundary condition

$$[u + u_y]_{x=x_0} = 0$$

cannot be separated. Needless to say, a mixed condition, such as $u_x + u_y$, cannot be prescribed on an axis.

7.3 The Vibrating String Problem

As a first example, we shall consider the problem of a vibrating string of constant tension T^* and density ρ with $c^2 = T^*/\rho$ stretched along the x-axis from 0 to l, fixed at its end points. We have seen in Chapter 5 that the problem is given by

$$u_{tt} - c^2 u_{xx} = 0, \qquad 0 < x < l, \qquad t > 0, \qquad (7.3.1)$$
$$u(x, 0) = f(x), \qquad 0 \le x \le l, \qquad\qquad (7.3.2)$$
$$u_t(x, 0) = g(x), \qquad 0 \le x \le l, \qquad\qquad (7.3.3)$$
$$u(0, t) = 0, \qquad\qquad\qquad t \ge 0, \qquad (7.3.4)$$
$$u(l, t) = 0, \qquad\qquad\qquad t \ge 0, \qquad (7.3.5)$$

where f and g are the initial displacement and initial velocity respectively.

By the method of separation of variables, we assume a solution in the form

$$u(x,t) = X(x) T(t) \neq 0. \qquad (7.3.6)$$

If we substitute equation (7.3.6) into equation (7.3.1), we obtain

$$XT'' = c^2 X''T,$$

and hence,

$$\frac{X''}{X} = \frac{1}{c^2} \frac{T''}{T}, \qquad (7.3.7)$$

whenever $XT \neq 0$. Since the left side of equation (7.3.7) is independent of t and the right side is independent of x, we must have

$$\frac{X''}{X} = \frac{1}{c^2} \frac{T''}{T} = \lambda,$$

where λ is a separation constant. Thus,

$$X'' - \lambda X = 0, \qquad (7.3.8)$$
$$T'' - \lambda c^2 T = 0. \qquad (7.3.9)$$

We now separate the boundary conditions. From equations (7.3.4) and (7.3.6), we obtain

$$u(0,t) = X(0) T(t) = 0.$$

We know that $T(t) \neq 0$ for all values of t, therefore,

$$X(0) = 0. \qquad (7.3.10)$$

In a similar manner, boundary condition (7.3.5) implies

$$X(l) = 0. \qquad (7.3.11)$$

To determine $X(x)$ we first solve the *eigenvalue problem* (eigenvalue problems are also treated in Chapter 8)

$$X'' - \lambda X = 0, \qquad X(0) = 0, \qquad X(l) = 0. \qquad (7.3.12)$$

We look for values of λ which gives us nontrivial solutions. We consider three possible cases

$$\lambda > 0, \qquad \lambda = 0, \qquad \lambda < 0.$$

Case 1. $\lambda > 0$. The general solution in this case is of the form

$$X\left(x\right) = Ae^{-\sqrt{\lambda}\,x} + Be^{\sqrt{\lambda}\,x}$$

where A and B are arbitrary constants. To satisfy the boundary conditions, we must have

$$A + B = 0, \qquad Ae^{-\sqrt{\lambda}\,l} + Be^{\sqrt{\lambda}\,l} = 0. \tag{7.3.13}$$

We see that the determinant of the system (7.3.13) is different from zero. Consequently, A and B must both be zero, and hence, the general solution $X\left(x\right)$ is identically zero. The solution is trivial and hence, is no interest.

Case 2. $\lambda = 0$. Here, the general solution is

$$X\left(x\right) = A + Bx.$$

Applying the boundary conditions, we have

$$A = 0, \qquad A + Bl = 0.$$

Hence $A = B = 0$. The solution is thus identically zero.

Case 3. $\lambda < 0$. In this case, the general solution assumes the form

$$X\left(x\right) = A \cos \sqrt{-\lambda}\,x + B \sin \sqrt{-\lambda}\,x.$$

From the condition $X\left(0\right) = 0$, we obtain $A = 0$. The condition $X\left(l\right) = 0$ gives

$$B \sin \sqrt{-\lambda}\,l = 0.$$

If $B = 0$, the solution is trivial. For nontrivial solutions, $B \ne 0$, hence,

$$\sin \sqrt{-\lambda}\,l = 0.$$

This equation is satisfied when

$$\sqrt{-\lambda}\,l = n\pi \quad \text{for} \quad n = 1, 2, 3, \ldots,$$

or

$$-\lambda_n = \left(n\pi/l\right)^2. \tag{7.3.14}$$

For this infinite set of discrete values of λ, the problem has a nontrivial solution. These values of λ_n are called the *eigenvalues* of the problem, and the functions

$$\sin\left(n\pi/l\right)x, \qquad n = 1, 2, 3, \ldots$$

are the corresponding *eigenfunctions*.

We note that it is not necessary to consider negative values of n since

$$\sin\left(-n\right)\pi x/l = -\sin n\pi x/l.$$

No new solution is obtained in this way.

The solutions of problems (7.3.12) are, therefore,

$$X_n\left(x\right) = B_n \sin\left(n\pi x/l\right). \tag{7.3.15}$$

For $\lambda = \lambda_n$, the general solution of equation (7.3.9) may be written in the form

$$T_n\left(t\right) = C_n \cos\left(\frac{n\pi c}{l}\right)t + D_n \sin\left(\frac{n\pi c}{l}\right)t, \tag{7.3.16}$$

where C_n and D_n are arbitrary constants.

Thus, the functions

$$u_n\left(x,t\right) = X_n\left(x\right)T_n\left(t\right) = \left(a_n \cos\frac{n\pi c}{l}t + b_n \sin\frac{n\pi c}{l}t\right)\sin\left(\frac{n\pi x}{l}\right) \tag{7.3.17}$$

satisfy equation (7.3.1) and the boundary conditions (7.3.4) and (7.3.5), where $a_n = B_n C_n$ and $b_n = B_n D_n$.

Since equation (7.3.1) is linear and homogeneous, by the superposition principle, the infinite series

$$u\left(x,t\right) = \sum_{n=1}^{\infty}\left(a_n \cos\frac{n\pi c}{l}t + b_n \sin\frac{n\pi c}{l}t\right)\sin\left(\frac{n\pi x}{l}\right) \tag{7.3.18}$$

is also a solution, provided it converges and is twice continuously differentiable with respect to x and t. Since each term of the series satisfies the boundary conditions (7.3.4) and (7.3.5), the series satisfies these conditions. There remain two more initial conditions to be satisfied. From these conditions, we shall determine the constants a_n and b_n.

First we differentiate the series (7.3.18) with respect to t. We have

$$u_t = \sum_{n=1}^{\infty}\frac{n\pi c}{l}\left(-a_n \sin\frac{n\pi c}{l}t + b_n \cos\frac{n\pi c}{l}t\right)\sin\left(\frac{n\pi x}{l}\right). \tag{7.3.19}$$

Then applying the initial conditions (7.3.2) and (7.3.3), we obtain

$$u\left(x,0\right) = f\left(x\right) = \sum_{n=1}^{\infty}a_n \sin\left(\frac{n\pi x}{l}\right), \tag{7.3.20}$$

$$u_t\left(x,0\right) = g\left(x\right) = \sum_{n=1}^{\infty}b_n\left(\frac{n\pi c}{l}\right)\sin\left(\frac{n\pi x}{l}\right). \tag{7.3.21}$$

These equations will be satisfied if $f\left(x\right)$ and $g\left(x\right)$ can be represented by Fourier sine series. The coefficients are given by

$$a_n = \frac{2}{l}\int_0^l f\left(x\right)\sin\left(\frac{n\pi x}{l}\right)dx, \qquad b_n = \frac{2}{n\pi c}\int_0^l g\left(x\right)\sin\left(\frac{n\pi x}{l}\right)dx,$$

$$\tag{7.3.22ab}$$

The solution of the vibrating string problem is therefore given by the series
(7.3.18) where the coefficients a_n and b_n are determined by the formulae
(7.3.22ab).

We examine the physical significance of the solution (7.3.17) in the
context of the free vibration of a string of length l. The eigenfunctions

$$u_n\,(x,t) = (a_n \cos \omega_n t + b_n \sin \omega_n t) \sin \left(\frac{n\pi x}{l}\right), \quad \omega_n = \frac{n\pi c}{l}, \quad (7.3.23)$$

are called the nth *normal modes* of vibration or the nth harmonic, and
ω_n represent the discrete spectrum of *circular* (or *radian*) *frequencies* or
$\nu_n = \frac{\omega_n}{2\pi} = \frac{nc}{2l}$, which are called the *angular frequencies*. The first harmonic
($n = 1$) is called the *fundamental harmonic* and all other harmonics ($n > 1$)
are called *overtones*. The frequency of the fundamental mode is given by

$$\omega_1 = \frac{\pi c}{l}, \quad \nu_1 = \frac{1}{2l}\sqrt{\frac{T^*}{\rho}}. \quad (7.3.24)$$

Result (7.3.24) is considered the fundamental law (or *Mersenne law*) of
a stringed musical instrument. The angular frequency of the fundamental
mode of transverse vibration of a string varies as the square root of the
tension, inversely as length, and inversely as the square root of the density.
The period of the fundamental mode is $T_1 = \frac{2c}{\omega_1} = \frac{2l}{c}$, which is called the
fundamental period. Finally, the solution (7.3.18) describes the motion of a
plucked string as a superposition of all normal modes of vibration with frequencies which are all integral multiples ($\omega_n = n\omega_1$ or $\nu_n = n\nu_1$) of the
fundamental frequency. This is the main reason that stringed instruments
produce sweeter musical sounds (or tones) than drum instruments.

In order to describe waves produced in the plucked string with zero
initial velocity ($u_t\,(x,0) = 0$), we write the solution (7.3.23) in the form

$$u_n\,(x,t) = a_n \sin \left(\frac{n\pi x}{l}\right) \cos \left(\frac{n\pi ct}{l}\right), \quad n = 1,2,3,\dots. \quad (7.3.25)$$

These solutions are called *standing waves* with amplitude $a_n \sin \left(\frac{n\pi x}{l}\right)$,
which vanishes at

$$x = 0, \frac{l}{n}, \frac{2l}{n}, \dots, l.$$

These are called the *nodes* of the nth harmonic. The string displays n loops
separated by the nodes as shown in Figure 7.3.1.

It follows from elementary trigonometry that (7.3.25) takes the form

$$u_n\,(x,t) = \frac{1}{2}a_n \left[\sin \frac{n\pi}{l}\,(x-ct) + \sin \frac{n\pi}{l}\,(x+ct)\right]. \quad (7.3.26)$$

This shows that a standing wave is expressed as a sum of two progressive
waves of equal amplitude traveling in opposite directions. This result is in
agreement with the d'Alembert solution.

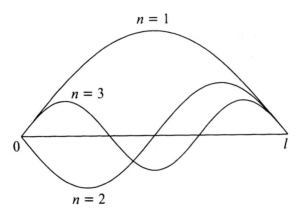

Figure 7.3.1 Several modes of vibration in a string.

Finally, we can rewrite the solution (7.3.23) of the nth normal modes in the form

$$u_n(x,t) = c_n \sin\left(\frac{n\pi x}{l}\right) \cos\left(\frac{n\pi ct}{l} - \varepsilon_n\right), \qquad (7.3.27)$$

where $c_n = \left(a_n^2 + b_n^2\right)^{\frac{1}{2}}$ and $\tan \varepsilon_n = \left(\frac{b_n}{a_n}\right)$.

This solution represents transverse vibrations of the string at any point x and at any time t with amplitude $c_n \sin\left(\frac{n\pi x}{l}\right)$ and circular frequency $\omega_n = \frac{n\pi c}{l}$. This form of the solution enables us to calculate the kinetic and potential energies of the transverse vibrations. The total kinetic energy (K.E.) is obtained by integrating with respect to x from 0 to l, that is,

$$K_n = K.E. = \int_0^l \frac{1}{2}\rho \left(\frac{\partial u_n}{\partial t}\right)^2 dx, \qquad (7.3.28)$$

where ρ is the line density of the string. Similarly, the total potential energy (P.E.) is given by

$$V_n = P.E. = \frac{1}{2}T^* \int_0^l \left(\frac{\partial u_n}{\partial x}\right)^2 dx. \qquad (7.3.29)$$

Substituting (7.3.27) in (7.3.28) and (7.3.29) gives

$$K_n = \frac{1}{2}\rho \left(\frac{n\pi c}{l}c_n\right)^2 \sin^2\left(\frac{n\pi ct}{l} - \varepsilon_n\right) \int_0^l \sin^2\left(\frac{n\pi x}{l}\right) dx$$

$$= \frac{\rho c^2 \pi^2}{4l}\left(n\, c_n\right)^2 \sin^2\left(\frac{n\pi ct}{l} - \varepsilon_n\right) = \frac{1}{4}\rho l \omega_n^2 c_n^2 \sin^2\left(\omega_n t - \varepsilon_n\right), (7.3.30)$$

where $\omega_n = \frac{n\pi c}{l}$.

Similarly,

$$V_n = \frac{1}{2}T^* \left(\frac{n\pi c_n}{l}\right)^2 \cos^2\left(\frac{n\pi ct}{l} - \varepsilon_n\right) \int_0^l \cos^2\left(\frac{n\pi x}{l}\right) dx$$

$$= \frac{\pi^2 T^*}{4l}(n\,c_n)^2 \cos^2\left(\frac{n\pi ct}{l} - \varepsilon_n\right) = \frac{1}{4}\rho l\omega_n^2 c_n^2 \cos^2\left(\omega_n t - \varepsilon_n\right). \quad (7.3.31)$$

Thus, the total energy of the nth normal mode of vibrations is given by

$$E_n = K_n + V_n = \frac{1}{4}\rho l\left(\omega_n c_n\right)^2 = \text{constant.} \qquad (7.3.32)$$

For a given string oscillating in a normal mode, the total energy is proportional to the square of the circular frequency and to the square of the amplitude.

Finally, the total energy of the system is given by

$$E = \sum_{n=1}^{\infty} E_n = \frac{1}{4}\rho l \sum_{n=1}^{\infty} \omega_n^2 c_n^2, \qquad (7.3.33)$$

which is constant because $E_n = \text{constant}$.

Example 7.3.1. The Plucked String of length l

As a special case of the problem just treated, consider a stretched string fixed at both ends. Suppose the string is raised to a height h at $x = a$ and then released. The string will oscillate freely. The initial conditions, as shown in Figure 7.3.2, may be written

$$u(x,0) = f(x) = \begin{cases} hx/a, & 0 \le x \le a \\ h(l-x)/(l-a), & a \le x \le l. \end{cases}$$

Since $g(x) = 0$, the coefficients b_n are identically equal to zero. The coefficients a_n, according to equation (7.3.22a), are given by

$$a_n = \frac{2}{l}\int_0^l f(x)\sin\left(\frac{n\pi x}{l}\right) dx$$

$$= \frac{2}{l}\int_0^a \frac{hx}{a}\sin\left(\frac{n\pi x}{l}\right) dx + \frac{2}{l}\int_a^l \frac{h(l-x)}{(l-a)}\sin\left(\frac{n\pi x}{l}\right) dx.$$

Integration and simplification yields

$$a_n = \frac{2hl^2}{\pi^2 a(l-a)}\frac{1}{n^2}\sin\left(\frac{n\pi a}{l}\right).$$

Thus, the displacement of the plucked string is

$$u(x,t) = \frac{2hl^2}{\pi^2 a(l-a)}\sum_{n=1}^{\infty}\frac{1}{n^2}\sin\left(\frac{n\pi a}{l}\right)\sin\left(\frac{n\pi x}{l}\right)\cos\left(\frac{n\pi c}{l}\right)t.$$

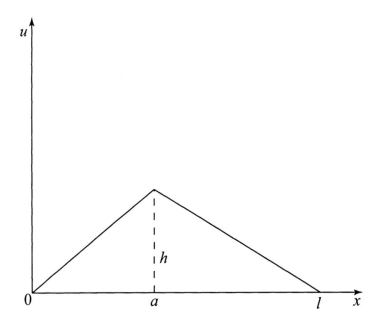

Figure 7.3.2 Plucked String

Example 7.3.2. The struck string of length l

Here, we consider the string with no initial displacement. Let the string be struck at $x = a$ so that the initial velocity is given by

$$u_t(x,0) = \begin{cases} \frac{v_0}{a} x, & 0 \le x \le a \\[2mm] v_0 (l-x)/(l-a), & a \le x \le l \end{cases}.$$

Since $u(x,0) = 0$, we have $a_n = 0$. By applying equation (7.3.22b), we find that

$$b_n = \frac{2}{n\pi c} \int_0^a \frac{v_0}{a} x \sin\left(\frac{n\pi x}{l}\right) dx + \frac{2}{n\pi c} \int_a^l v_0 \frac{(l-x)}{(l-a)} \sin\left(\frac{n\pi x}{l}\right) dx$$

$$= \frac{2 v_0 l^3}{\pi^3 ca (l-a)} \frac{1}{n^3} \sin\left(\frac{n\pi a}{l}\right).$$

Hence, the displacement of the struck string is

$$u(x,t) = \frac{2 v_0 l^3}{\pi^3 ca (l-a)} \sum_{n=1}^{\infty} \frac{1}{n^3} \sin\left(\frac{n\pi a}{l}\right) \sin\left(\frac{n\pi x}{l}\right) \cos\left(\frac{n\pi c}{l}\right) t.$$

7.4 Existence and Uniqueness of Solution of the Vibrating String Problem

In the preceding section we found that the initial boundary-value problem (7.3.1)–(7.3.5) has a formal solution given by (7.3.18). We shall now show that the expression (7.3.18) is the solution of the problem under certain conditions.

First we see that

$$u_1(x,t) = \sum_{n=1}^{\infty} a_n \cos\left(\frac{n\pi c}{l}t\right) \sin\left(\frac{n\pi x}{l}\right) \tag{7.4.1}$$

is the formal solution of the problem (7.3.1)–(7.3.5) with $g(x) \equiv 0$, and

$$u_2(x,t) = \sum_{n=1}^{\infty} b_n \sin\left(\frac{n\pi c}{l}t\right) \sin\left(\frac{n\pi x}{l}\right) \tag{7.4.2}$$

is the formal solution of the above problem with $f(x) \equiv 0$. By linearity of the problem, the solution (7.3.18) may be considered as the sum of the two formal solutions (7.4.1) and (7.4.2).

We first assume that $f(x)$ and $f'(x)$ are continuous on $[0, l]$, and $f(0) = f(l) = 0$. Then by Theorem 6.10.1, the series for the function $f(x)$ given by (7.3.20) converges absolutely and uniformly on the interval $[0, l]$.

Using the trigonometric identity

$$\sin\left(\frac{n\pi x}{l}\right) \cos\left(\frac{n\pi c}{l}t\right) = \frac{1}{2}\sin\frac{n\pi}{l}(x - ct) + \frac{1}{2}\sin\frac{n\pi}{l}(x + ct), \tag{7.4.3}$$

$u_1(x,t)$ may be written as

$$u_1(x,t) = \frac{1}{2}\sum_{n=1}^{\infty} a_n \sin\frac{n\pi}{l}(x - ct) + \frac{1}{2}\sum_{n=1}^{\infty} a_n \sin\frac{n\pi}{l}(x + ct).$$

Define

$$F(x) = \sum_{n=1}^{\infty} a_n \sin\left(\frac{n\pi x}{l}\right) \tag{7.4.4}$$

and assume that $F(x)$ is the odd periodic extension of $f(x)$, that is,

$$F(x) = f(x) \quad 0 \le x \le l$$
$$F(-x) = -F(x) \quad \text{for all } x$$
$$F(x \pm 2l) = F(x).$$

We can now rewrite $u_1(x,t)$ in the form

$$u_1(x,t) = \frac{1}{2}[F(x - ct) + F(x + ct)]. \tag{7.4.5}$$

To show that the boundary conditions are satisfied, we note that

$$u_1(0,t) = \frac{1}{2}[F(-ct) + F(ct)]$$

$$= \frac{1}{2}[-F(ct) + F(ct)] = 0$$

$$u_1(l,t) = \frac{1}{2}[F(l-ct) + F(l+ct)]$$

$$= \frac{1}{2}[F(-l-ct) + F(l+ct)]$$

$$= \frac{1}{2}[-F(l+ct) + F(l+ct)] = 0.$$

Since

$$u_1(x,0) = \frac{1}{2}[F(x) + F(x)]$$

$$= F(x) = f(x), \qquad 0 \leq x \leq l,$$

we see that the initial condition $u_1(x,0) = f(x)$ is satisfied. Thus, equation (7.3.1) and conditions (7.3.2)–(7.3.3) with $g(x) \equiv 0$ are satisfied. Since f' is continuous in $[0, l]$, F' exists and is continuous for all x. Thus, if we differentiate $u_1(x,t)$ with respect to t, we obtain

$$\frac{\partial u_1}{\partial t} = \frac{1}{2}[-cF'(x-ct) + cF'(x+ct)],$$

and

$$\frac{\partial u_1}{\partial t}(x,0) = \frac{1}{2}[-cF'(x) + cF'(x)] = 0.$$

We therefore see that initial condition (7.3.3) is also satisfied.

In order to show that $u_1(x,t)$ satisfies the differential equation (7.3.1), we impose additional restrictions on f. Let f'' be continuous on $[0, l]$ and $f''(0) = f''(l) = 0$. Then, F'' exists and is continuous everywhere, and therefore,

$$\frac{\partial^2 u_1}{\partial t^2} = \frac{1}{2}c^2[F''(x-ct) + F''(x+ct)],$$

$$\frac{\partial^2 u_1}{\partial x^2} = \frac{1}{2}[F''(x-ct) + F''(x+ct)].$$

We find therefore that

$$\frac{\partial^2 u_1}{\partial t^2} = c^2 \frac{\partial^2 u_1}{\partial x^2}.$$

Next, we shall state the assumptions which must be imposed on g to make $u_2(x,t)$ the solution of problem (7.3.1)–(7.3.5) with $f(x) \equiv 0$. Let g

and g' be continuous on $[0, l]$ and let $g(0) = g(l) = 0$. Then the series for the function $g(x)$ given by (7.3.21) converges absolutely and uniformly in the interval $[0, l]$. Introducing the new coefficients $c_n = (n\pi c/l) b_n$, we have

$$u_2(x, t) = \left(\frac{l}{\pi c}\right) \sum_{n=1}^{\infty} \frac{c_n}{n} \sin\left(\frac{n\pi c}{l} t\right) \sin\left(\frac{n\pi x}{l}\right). \tag{7.4.6}$$

We shall see that term-by-term differentiation with respect to t is permitted, and hence,

$$\frac{\partial u_2}{\partial t} = \sum_{n=1}^{\infty} c_n \cos\left(\frac{n\pi c}{l} t\right) \sin\left(\frac{n\pi x}{l}\right). \tag{7.4.7}$$

Using the trigonometric identity (7.4.3), we obtain

$$\frac{\partial u_2}{\partial t} = \frac{1}{2} \sum_{n=1}^{\infty} c_n \sin \frac{n\pi}{l} (x - ct) + \frac{1}{2} \sum_{n=1}^{\infty} c_n \sin \frac{n\pi}{l} (x + ct). \tag{7.4.8}$$

These series are absolutely and uniformly convergent because of the assumptions on g, and hence, the series (7.4.6) and (7.4.7) converge absolutely and uniformly on $[0, l]$. Thus, the term-by-term differentiation is justified.

Let

$$G(x) = \sum_{n=1}^{\infty} c_n \sin\left(\frac{n\pi x}{l}\right)$$

be the odd periodic extension of the function $g(x)$. Then, equation (7.4.8) can be written in the form

$$\frac{\partial u_2}{\partial t} = \frac{1}{2} \left[G(x - ct) + G(x + ct) \right].$$

Integration yields

$$u_2(x, t) = \frac{1}{2} \int_0^t G(x - ct')\, dt' + \frac{1}{2} \int_0^t G(x + ct')\, dt'$$

$$= \frac{1}{2c} \int_{x-ct}^{x+ct} G(\tau)\, d\tau. \tag{7.4.9}$$

It immediately follows that $u_2(x, 0) = 0$, and

$$\frac{\partial u_2}{\partial t}(x, 0) = G(x) = g(x), \quad 0 \le x \le l.$$

Moreover,

$$u_2(0, t) = \frac{1}{2} \int_0^t G(-ct')\, dt' + \frac{1}{2} \int_0^t G(ct')\, dt'$$

$$= -\frac{1}{2} \int_0^t G(ct')\, dt' + \frac{1}{2} \int_0^t G(ct')\, dt' = 0$$

and

$$u_2\left(l,t\right) = \frac{1}{2}\int_0^t G\left(l - ct'\right)dt' + \frac{1}{2}\int_0^t G\left(l + ct'\right)dt'$$

$$= \frac{1}{2}\int_0^t G\left(-l - ct'\right)dt' + \frac{1}{2}\int_0^t G\left(l + ct'\right)dt'$$

$$= -\frac{1}{2}\int_0^t G\left(l + ct'\right)dt' + \frac{1}{2}\int_0^t G\left(l + ct'\right)dt' = 0.$$

Finally, $u_2\left(x,t\right)$ must satisfy the differential equation. Since g' is continuous on $[0,l]$, G' exists so that

$$\frac{\partial^2 u_2}{\partial t^2} = \frac{c}{2}\left[-G'\left(x - ct\right) + G'\left(x + ct\right)\right].$$

Differentiating $u_2\left(x,t\right)$ represented by equation (7.4.6) with respect to x, we obtain

$$\frac{\partial u_2}{\partial x} = \frac{1}{c}\sum_{n=1}^{\infty} c_n \sin\left(\frac{n\pi c}{l}t\right)\cos\left(\frac{n\pi x}{l}\right)$$

$$= \frac{1}{2c}\sum_{n=1}^{\infty} c_n\left[-\sin\frac{n\pi}{l}\left(x - ct\right) + \sin\frac{n\pi}{l}\left(x + ct\right)\right]$$

$$= \frac{1}{2c}\left[-G\left(x - ct\right) + G\left(x + ct\right)\right].$$

Differentiating again with respect to x, we obtain

$$\frac{\partial^2 u_2}{\partial x^2} = \frac{1}{2c}\left[-G'\left(x - ct\right) + G'\left(x + ct\right)\right].$$

It is quite evident that

$$\frac{\partial^2 u_2}{\partial t^2} = c^2\frac{\partial^2 u_2}{\partial x^2}.$$

Thus, the solution of the initial boundary-value problem (7.3.1)–(7.3.5) is established.

Theorem 7.4.2. (Uniqueness Theorem) *There exists at most one solution of the wave equation*

$$u_{tt} = c^2 u_{xx}, \quad 0 < x < l, \quad t > 0,$$

satisfying the initial conditions

$$u\left(x,0\right) = f\left(x\right), \quad u_t\left(x,0\right) = g\left(x\right), \quad 0 \le x \le l,$$

and the boundary conditions

$$u\left(0,t\right) = 0, \quad u\left(l,t\right) = 0, \quad t \ge 0,$$

where $u\left(x,t\right)$ is a twice continuously differentiable function with respect to both x and t.

Proof. Suppose that there are two solutions u_1 and u_2 and let $v = u_1 - u_2$. It can readily be seen that $v(x,t)$ is the solution of the problem

$$
\begin{aligned}
v_{tt} &= c^2 v_{xx}, && 0 < x < l, \quad t > 0, \\
v(0,t) &= 0, && t \geq 0, \\
v(l,t) &= 0, && t \geq 0, \\
v(x,0) &= 0, && 0 \leq x \leq l, \\
v_t(x,0) &= 0, && 0 \leq x \leq l.
\end{aligned}
$$

We shall prove that the function $v(x,t)$ is identically zero. To do so, consider the energy integral

$$
E(t) = \frac{1}{2} \int_0^l \left(c^2 v_x^2 + v_t^2 \right) dx \tag{7.4.10}
$$

which physically represents the total energy of the vibrating string at time t.

Since the function $v(x,t)$ is twice continuously differentiable, we differentiate $E(t)$ with respect to t. Thus,

$$
\frac{dE}{dt} = \int_0^l \left(c^2 v_x v_{xt} + v_t v_{tt} \right) dx. \tag{7.4.11}
$$

Integrating the first integral in (7.4.11) by parts, we have

$$
\int_0^l c^2 v_x v_{xt} dx = \left[c^2 v_x v_t \right]_0^l - \int_0^l c^2 v_t v_{xx} dx.
$$

But from the condition $v(0,t) = 0$ we have $v_t(0,t) = 0$, and similarly, $v_t(l,t) = 0$ for $x = l$. Hence, the expression in the square brackets vanishes, and equation (7.4.11) becomes

$$
\frac{dE}{dt} = \int_0^l v_t \left(v_{tt} - c^2 v_{xx} \right) dx. \tag{7.4.12}
$$

Since $v_{tt} - c^2 v_{xx} = 0$, equation (7.4.12) reduces to

$$
\frac{dE}{dt} = 0
$$

which means

$$
E(t) = \text{constant} = C.
$$

Since $v(x,0) = 0$ we have $v_x(x,0) = 0$. Taking into account the condition $v_t(x,0) = 0$, we evaluate C to obtain

$$E\left(0\right) = C = \frac{1}{2}\int_0^l \left[c^2 v_x^2 + v_t^2\right]_{t=0} dx = 0.$$

This implies that $E\left(t\right) = 0$ which can happen only when $v_x = 0$ and $v_t = 0$ for $t > 0$. To satisfy both of these conditions, we must have $v\left(x, t\right) =$ constant. Employing the condition $v\left(x, 0\right) = 0$, we then find $v\left(x, t\right) = 0$. Therefore, $u_1\left(x, t\right) = u_2\left(x, t\right)$ and the solution $u\left(x, t\right)$ is unique.

7.5 The Heat Conduction Problem

We consider a homogeneous rod of length l. The rod is sufficiently thin so that the heat is distributed equally over the cross section at time t. The surface of the rod is insulated, and therefore, there is no heat loss through the boundary. The temperature distribution of the rod is given by the solution of the initial boundary-value problem

$$
\begin{aligned}
u_t &= ku_{xx}, & 0 < x < l, \quad t > 0, \\
u\left(0, t\right) &= 0, & t \geq 0, \\
u\left(l, t\right) &= 0, & t \geq 0, \\
u\left(x, 0\right) &= f\left(x\right), & 0 \leq x \leq l.
\end{aligned}
\tag{7.5.1}
$$

If we assume a solution in the form

$$u\left(x, t\right) = X\left(x\right) T\left(t\right) \neq 0.$$

Equation (7.5.1) yields

$$XT' = kX''T.$$

Thus, we have

$$\frac{X''}{X} = \frac{T'}{kT} = -\alpha^2,$$

where α is a positive constant. Hence, X and T must satisfy

$$X'' + \alpha^2 X = 0, \tag{7.5.2}$$
$$T' + \alpha^2 kT = 0. \tag{7.5.3}$$

From the boundary conditions, we have

$$u\left(0, t\right) = X\left(0\right) T\left(t\right) = 0, \qquad u\left(l, t\right) = X\left(l\right) T\left(t\right) = 0.$$

Thus,

$$X\left(0\right) = 0, \qquad X\left(l\right) = 0,$$

for an arbitrary function $T(t)$. Hence, we must solve the eigenvalue problem

$$X'' + \alpha^2 X = 0,$$
$$X(0) = 0, \qquad X(l) = 0.$$

The solution of equation (7.5.2) is

$$X(x) = A \cos \alpha x + B \sin \alpha x.$$

Since $X(0) = 0$, $A = 0$. To satisfy the second condition, we have

$$X(l) = B \sin \alpha l = 0.$$

Since $B = 0$ yields a trivial solution, we must have $B \neq 0$ and hence,

$$\sin \alpha l = 0.$$

Thus,

$$\alpha = \frac{n\pi}{l} \quad \text{for} \quad n = 1, 2, 3 \ldots.$$

Substituting these eigenvalues, we have

$$X_n(x) = B_n \sin\left(\frac{n\pi x}{l}\right).$$

Next, we consider equation (7.5.3), namely,

$$T' + \alpha^2 kT = 0,$$

the solution of which is

$$T(t) = Ce^{-\alpha^2 kt}.$$

Substituting $\alpha = (n\pi/l)$, we have

$$T_n(t) = C_n e^{-(n\pi/l)^2 kt}.$$

Hence, the nontrivial solution of the heat equation which satisfies the two boundary conditions is

$$u_n(x,t) = X_n(x) T_n(t) = a_n e^{-(n\pi/l)^2 kt} \sin\left(\frac{n\pi x}{l}\right), \qquad n = 1, 2, 3 \ldots,$$

where $a_n = B_n C_n$ is an arbitrary constant.

By the principle of superposition, we obtain a formal series solution as

$$u(x,t) = \sum_{n=1}^{\infty} u_n(x,t),$$

$$= \sum_{n=1}^{\infty} a_n e^{-(n\pi/l)^2 kt} \sin\left(\frac{n\pi x}{l}\right), \qquad (7.5.4)$$

which satisfies the initial condition if

$$u\left(x,0\right) = f\left(x\right) = \sum_{n=1}^{\infty} a_n \sin\left(\frac{n\pi x}{l}\right).$$

This holds true if $f\left(x\right)$ can be represented by a Fourier sine series with Fourier coefficients

$$a_n = \frac{2}{l} \int_0^l f\left(x\right) \sin\left(\frac{n\pi x}{l}\right) dx. \qquad (7.5.5)$$

Hence,

$$u\left(x,t\right) = \sum_{n=1}^{\infty} \left[\frac{2}{l} \int_0^l f\left(\tau\right) \sin\left(\frac{n\pi\tau}{l}\right) d\tau\right] e^{-\left(n\pi/l\right)^2 kt} \sin\left(\frac{n\pi x}{l}\right) \quad (7.5.6)$$

is the formal series solution of the heat conduction problem.

Example 7.5.1. (a) Suppose the initial temperature distribution is $f\left(x\right) = x\left(l - x\right)$. Then, from equation (7.5.5), we have

$$a_n = \frac{8l^2}{n^3\pi^3}, \qquad n = 1, 3, 5, \ldots.$$

Thus, the solution is

$$u\left(x,t\right) = \left(\frac{8l^2}{\pi^3}\right) \sum_{n=1,3,5,\ldots}^{\infty} \frac{1}{n^3} e^{-\left(n\pi/l\right)^2 kt} \sin\left(\frac{n\pi x}{l}\right).$$

(b) Suppose the temperature at one end of the rod is held constant, that is,

$$u\left(l,t\right) = u_0, \qquad t \geq 0.$$

The problem here is

$$\begin{aligned} u_t &= k\,u_{xx}, & 0 < x < l, \quad t > 0, \\ u\left(0,t\right) &= 0, & u\left(l,t\right) = u_0, \\ u\left(x,0\right) &= f\left(x\right), & 0 < x < l. \end{aligned} \qquad (7.5.7)$$

Let

$$u\left(x,t\right) = v\left(x,t\right) + \frac{u_0 x}{l}.$$

Substitution of $u\left(x,t\right)$ in equations (7.5.7) yields

$$\begin{aligned} v_t &= k\,v_{xx}, & 0 < x < l, \quad t > 0, \\ v\left(0,t\right) &= 0, & v\left(l,t\right) = 0, \\ v\left(x,0\right) &= f\left(x\right) - \frac{u_0 x}{l}, & 0 < x < l. \end{aligned}$$

Hence, with the knowledge of solution (7.5.6), we obtain the solution

$$u\left(x,t\right) = \sum_{n=1}^{\infty} \left[\frac{2}{l} \int_{0}^{l} \left(f\left(\tau\right) - \frac{u_0\tau}{l}\right) \sin\left(\frac{n\pi\tau}{l}\right) d\tau\right] e^{-(n\pi/l)^2 kt} \sin\left(\frac{n\pi x}{l}\right)$$

$$+ \left(\frac{u_0 x}{l}\right). \quad (7.5.8)$$

7.6 Existence and Uniqueness of Solution of the Heat Conduction Problem

In the preceding section, we found that (7.5.4) is the formal solution of the heat conduction problem (7.5.1), where a_n is given by (7.5.5).

We shall prove the existence of this formal solution if $f\left(x\right)$ is continuous in $[0,l]$ and $f\left(0\right) = f\left(l\right) = 0$, and $f'\left(x\right)$ is piecewise continuous in $(0,l)$. Since $f\left(x\right)$ is bounded, we have

$$|a_n| = \frac{2}{l} \left|\int_{0}^{l} f\left(x\right) \sin\left(\frac{n\pi x}{l}\right) dx\right| \leq \frac{2}{l} \int_{0}^{l} |f\left(x\right)| \, dx \leq C,$$

where C is a positive constant. Thus, for any finite $t_0 > 0$,

$$\left|a_n e^{-(n\pi/l)^2 kt} \sin\left(\frac{n\pi x}{l}\right)\right| \leq C e^{-(n\pi/l)^2 kt_0} \quad \text{when} \quad t \geq t_0.$$

According to the ratio test, the series of terms $\exp\left[-\left(n\pi/l\right)^2 kt_0\right]$ converges. Hence, by the Weierstrass M-test, the series (7.5.4) converges uniformly with respect to x and t whenever $t \geq t_0$ and $0 \leq x \leq l$.

Differentiating equation (7.5.4) termwise with respect to t, we obtain

$$u_t = -\sum_{n=1}^{\infty} a_n \left(\frac{n\pi}{l}\right)^2 k e^{-(n\pi/l)^2 kt} \sin\left(\frac{n\pi x}{l}\right). \quad (7.6.1)$$

We note that

$$\left|-a_n \left(\frac{n\pi}{l}\right)^2 k e^{-(n\pi/l)^2 kt} \sin\left(\frac{n\pi x}{l}\right)\right| \leq C \left(\frac{n\pi}{l}\right)^2 k e^{-(n\pi/l)^2 kt_0}$$

when $t \geq t_0$, and the series of terms $C \left(n\pi/l\right)^2 k \exp\left[-\left(n\pi/l\right)^2 kt_0\right]$ converges by the ratio test. Hence, equation (7.6.1) is uniformly convergent in the region $0 \leq x \leq l$, $t \geq t_0$. In a similar manner, the series (7.5.4) can be differentiated twice with respect to x, and as a result

$$u_{xx} = -\sum_{n=1}^{\infty} a_n \left(\frac{n\pi}{l}\right)^2 e^{-(n\pi/l)^2 kt} \sin\left(\frac{n\pi x}{l}\right). \quad (7.6.2)$$

Evidently, from equations (7.6.1) and (7.6.2),

$$u_t = k\, u_{xx}.$$

Hence, equation (7.5.4) is a solution of the one-dimensional heat equation in the region $0 \le x \le l$, $t \ge 0$.

Next, we show that the boundary conditions are satisfied. Here, we note that the series (7.5.4) representing the function $u(x,t)$ converges uniformly in the region $0 \le x \le l$, $t \ge 0$. Since the function represented by a uniformly convergent series of continuous functions is continuous, $u(x,t)$ is continuous at $x = 0$ and $x = l$. As a consequence, when $x = 0$ and $x = l$, solution (7.5.4) satisfies

$$u(0,t) = 0, \qquad u(l,t) = 0,$$

for all $t > 0$.

It remains to show that $u(x,t)$ satisfies the initial condition

$$u(x,0) = f(x), \qquad 0 \le x \le l.$$

Under the assumptions stated earlier, the series for $f(x)$ given by

$$f(x) = \sum_{n=1}^{\infty} a_n \sin\left(\frac{n\pi x}{l}\right)$$

is uniformly and absolutely convergent. By Abel's test of convergence the series formed by the product of the terms of a uniformly convergent series

$$\sum_{n=1}^{\infty} a_n \sin\left(\frac{n\pi x}{l}\right)$$

and a uniformly bounded and monotone sequence $\exp\left[-(n\pi/l)^2 \, kt\right]$ converges uniformly with respect to t. Hence,

$$u(x,t) = \sum_{n=1}^{\infty} a_n \, e^{-(n\pi/l)^2 kt} \, \sin\left(\frac{n\pi x}{l}\right)$$

converges uniformly for $0 \le x \le l$, $t \ge 0$, and by the same reasoning as before, $u(x,t)$ is continuous for $0 \le x \le l$, $t \ge 0$. Thus, the initial condition

$$u(x,0) = f(x), \qquad 0 \le x \le l$$

is satisfied. The existence of solution is therefore established.

In the above discussion the condition imposed on $f(x)$ is stronger than necessary. The solution can be obtained with a less stringent condition on $f(x)$ (see Weinberger (1965)).

Theorem 7.6.1. (*Uniqueness Theorem*) *Let $u(x,t)$ be a continuously differentiable function. If $u(x,t)$ satisfies the differential equation*

$$u_t = k\,u_{xx}, \quad 0 < x < l, \quad t > 0,$$

the initial conditions

$$u(x,0) = f(x), \quad 0 \le x \le l,$$

and the boundary conditions

$$u(0,t) = 0, \quad u(l,t) = 0, \quad t \ge 0,$$

then, the solution is unique.

Proof. Suppose that there are two distinct solutions $u_1(x,t)$ and $u_2(x,t)$. Let

$$v(x,t) = u_1(x,t) - u_2(x,t).$$

Then,

$$
\begin{aligned}
v_t &= k\,v_{xx}, & 0 < x < l, \quad & t > 0, \\
v(0,t) &= 0, & v(l,t) = 0, \quad & t \ge 0, \\
v(x,0) &= 0, & 0 \le x \le l, &
\end{aligned}
\tag{7.6.3}
$$

Consider the function defined by the integral

$$J(t) = \frac{1}{2k} \int_0^l v^2\,dx.$$

Differentiating with respect to t, we have

$$J'(t) = \frac{1}{k} \int_0^l v v_t\,dx = \int_0^l v v_{xx}\,dx,$$

by virtue of equation (7.6.3). Integrating by parts, we have

$$\int_0^l v v_{xx}\,dx = [v v_x]_0^l - \int_0^l v_x^2\,dx.$$

Since $v(0,t) = v(l,t) = 0$,

$$J'(t) = - \int_0^l v_x^2\,dx \le 0.$$

From the condition $v(x,0) = 0$, we have $J(0) = 0$. This condition and $J'(t) \le 0$ implies that $J(t)$ is a nonincreasing function of t. Thus,

$$J(t) \leq 0.$$

But by definition of $J(t)$,

$$J(t) \geq 0.$$

Hence,

$$J(t) = 0, \quad \text{for } t \geq 0.$$

Since $v(x,t)$ is continuous, $J(t) = 0$ implies

$$v(x,t) = 0$$

in $0 \leq x \leq l$, $t \geq 0$. Therefore, $u_1 = u_2$ and the solution is unique.

7.7 The Laplace and Beam Equations

Example 7.7.1. Consider the steady state temperature distribution in a thin rectangular slab. Two sides are insulated, one side is maintained at zero temperature, and the temperature of the remaining side is prescribed to be $f(x)$. Thus, we are required to solve

$$\nabla^2 u = 0, \quad 0 < x < a, \quad 0 < y < b,$$
$$u(x,0) = f(x), \quad 0 \leq x \leq a,$$
$$u(x,b) = 0, \quad 0 \leq x \leq a,$$
$$u_x(0,y) = 0, \quad u_x(a,y) = 0.$$

Let $u(x,y) = X(x)Y(y)$. Substitution of this into the Laplace equation yields

$$X'' - \lambda X = 0, \qquad Y'' + \lambda X = 0.$$

Since the boundary conditions are homogeneous on $x = 0$ and $x = a$, we have $\lambda = -\alpha^2$ with $\alpha \geq 0$ for nontrivial solutions of the eigenvalue problem

$$X'' + \alpha^2 X = 0,$$
$$X'(0) = X'(a) = 0.$$

The solution is

$$X(x) = A \cos \alpha x + B \sin \alpha x.$$

Application of the boundary conditions then yields $B = 0$ and $\alpha = (n\pi/a)$ with $n = 0, 1, 2, \ldots$. Hence,

$$X_n(x) = A \cos \left(\frac{n\pi x}{a} \right).$$

The solution of the Y equation is clearly

$$Y(y) = C \cosh \alpha y + D \sinh \alpha y$$

which can be written in the form

$$Y(y) = E \sinh \alpha (y + F),$$

where $E = \left(D^2 - C^2\right)^{\frac{1}{2}}$ and $F = \left[\tanh^{-1}(C/D)\right]/\alpha$.

Applying the homogeneous boundary condition $Y(b) = 0$, we obtain

$$Y(b) = E \sinh \alpha (b + F) = 0$$

which implies

$$F = -b, \qquad E \neq 0$$

for nontrivial solutions. Hence, we have

$$u(x,y) = \frac{(b-y)}{b} \frac{a_0}{2} + \sum_{n=1}^{\infty} a_n \cos\left(\frac{n\pi x}{a}\right) \sinh\left\{\frac{n\pi}{a}(y-b)\right\}.$$

Now we apply the remaining nonhomogeneous condition to obtain

$$u(x,0) = f(x) = \frac{a_0}{2} + \sum_{n=1}^{\infty} a_n \cos\left(\frac{n\pi x}{a}\right) \sinh\left(-\frac{n\pi b}{a}\right).$$

Since this is a Fourier cosine series, the coefficients are given by

$$a_0 = \frac{2}{a} \int_0^a f(x)\, dx,$$

$$a_n = \frac{-2}{a \sinh\left(\frac{n\pi b}{a}\right)} \int_0^a f(x) \cos\left(\frac{n\pi x}{a}\right) dx, \qquad n = 1, 2, \ldots.$$

Thus, the solution is

$$u(x,y) = \left(\frac{b-y}{b}\right) \frac{a_0}{2} + \sum_{n=1}^{\infty} a_n^* \frac{\sinh\frac{n\pi}{a}(b-y)}{\sinh\frac{n\pi b}{a}} \cos\left(\frac{n\pi x}{a}\right),$$

where

$$a_n^* = \frac{2}{a} \int_0^a f(x) \cos\left(\frac{n\pi x}{a}\right) dx.$$

If, for example $f(x) = x$ in $0 < x < \pi$, $0 < y < \pi$, then we find (note that $a = \pi$)

$$a_0 = \pi, \qquad a_n^* = \frac{2}{\pi n^2} \left[(-1)^n - 1\right], \qquad n = 1, 2, \ldots$$

and hence, the solution has the final form

$$u(x,y) = \frac{1}{2}(\pi - y) + \sum_{n=1}^{\infty} \frac{2}{\pi n^2} \left[(-1)^n - 1\right] \frac{\sinh n(\pi - y)}{\sinh n\pi} \cos nx.$$

Example 7.7.2. As another example, we consider the transverse vibration of a beam. The equation of motion is governed by

$$u_{tt} + a^2 u_{xxxx} = 0, \qquad 0 < x < l, \quad t > 0,$$

where $u(x,t)$ is the displacement and a is the physical constant. Note that the equation is of the fourth order in x. Let the initial and boundary conditions be

$$\begin{aligned}
u(x,0) &= f(x), & 0 \le x \le l, \\
u_t(x,0) &= g(x), & 0 \le x \le l, \\
u(0,t) &= u(l,t) = 0, & t > 0, \\
u_{xx}(0,t) &= u_{xx}(l,t) = 0, & t > 0.
\end{aligned} \qquad (7.7.1)$$

The boundary conditions represent the beam being simple supported, that is, the displacements and the bending moments at the ends are zero.

Assume a nontrivial solution in the form

$$u(x,t) = X(x) T(t),$$

which transforms the equation of motion into the forms

$$T'' + a^2 \alpha^4 T = 0, \qquad X^{(iv)} - \alpha^4 X = 0, \qquad \alpha > 0.$$

The equation for $X(x)$ has the general solution

$$X(x) = A \cosh \alpha x + B \sinh \alpha x + C \cos \alpha x + D \sin \alpha x.$$

The boundary conditions require that

$$X(0) = X(l) = 0, \qquad X''(0) = X''(l) = 0.$$

Differentiating X twice with respect to x, we obtain

$$X''(x) = A\alpha^2 \cosh \alpha x + B\alpha^2 \sinh \alpha x - C\alpha^2 \cos \alpha x - D\alpha^2 \sin \alpha x.$$

Now applying the conditions $X(0) = X''(0) = 0$, we obtain

$$A + C = 0, \qquad \alpha^2 (A - C) = 0,$$

and hence,

$$A = C = 0.$$

The conditions $X(l) = X''(l) = 0$ yield

$$\begin{aligned}
B \sinh \alpha l + D \sin \alpha l &= 0, \\
B \sinh \alpha l - D \sin \alpha l &= 0.
\end{aligned}$$

These equations are satisfied if

$$B \sinh \alpha l = 0, \qquad D \sin \alpha l = 0.$$

Since $\sinh \alpha l \neq 0$, B must vanish. For nontrivial solutions, $D \neq 0$,

$$\sin \alpha l = 0,$$

and hence,

$$\alpha = \left(\frac{n\pi}{l} \right), \qquad n = 1, 2, 3, \ldots.$$

We then obtain

$$X_n (x) = D_n \sin \left(\frac{n\pi x}{l} \right).$$

The general solution for $T(t)$ is

$$T(t) = E \cos \left(a\alpha^2 t \right) + F \sin \left(a\alpha^2 t \right).$$

Inserting the values of α^2, we obtain

$$T_n (t) = E_n \cos \left\{ a \left(\frac{n\pi}{l} \right)^2 t \right\} + F_n \sin \left\{ a \left(\frac{n\pi}{l} \right)^2 t \right\}.$$

Thus, the general solution for the transverse vibrations of a beam is

$$u(x, t) = \sum_{n=1}^{\infty} \left[a_n \cos \left\{ a \left(\frac{n\pi}{l} \right)^2 t \right\} + b_n \sin \left\{ a \left(\frac{n\pi}{l} \right)^2 t \right\} \right] \sin \left(\frac{n\pi x}{l} \right).$$

$$(7.7.2)$$

To satisfy the initial condition $u(x, 0) = f(x)$, we must have

$$u(x, 0) = f(x) = \sum_{n=1}^{\infty} a_n \sin \left(\frac{n\pi x}{l} \right)$$

from which we find

$$a_n = \frac{2}{l} \int_0^l f(x) \sin \left(\frac{n\pi x}{l} \right) dx. \qquad (7.7.3)$$

Now the application of the second initial condition gives

$$u_t (x, 0) = g(x) = \sum_{n=1}^{\infty} b_n a \left(\frac{n\pi}{l} \right)^2 \sin \left(\frac{n\pi x}{l} \right)$$

and hence,

$$b_n = \frac{2}{al} \left(\frac{l}{n\pi} \right)^2 \int_0^l g(x) \sin \left(\frac{n\pi x}{l} \right) dx. \qquad (7.7.4)$$

Thus, the solution of the initial boundary-value problem is given by equations (7.7.2)–(7.7.4).

7.8 Nonhomogeneous Problems

The partial differential equations considered so far in this chapter are homogeneous. In practice, there is a very important class of problems involving nonhomogeneous equations. First, we shall illustrate a problem involving a time-independent nonhomogeneous equations.

Example 7.8.1. Consider the initial boundary-value problem

$$\begin{aligned}
u_{tt} &= c^2 u_{xx} + F\left(x\right), \quad 0 < x < l, \quad t > 0, \\
u\left(x,0\right) &= f\left(x\right), \qquad\qquad\quad 0 \le x \le l, \\
u_t\left(x,0\right) &= g\left(x\right), \qquad\qquad\quad 0 \le x \le l, \\
u\left(0,t\right) &= A, \quad u\left(l,t\right) = B, \quad t > 0.
\end{aligned} \tag{7.8.1}$$

We assume a solution in the form

$$u\left(x,t\right) = v\left(x,t\right) + U\left(x\right).$$

Substitution of $u\left(x,t\right)$ in equation (7.8.1) yields

$$v_{tt} = c^2 \left(v_{xx} + U_{xx}\right) + F\left(x\right),$$

and if $U\left(x\right)$ satisfies the equation

$$c^2 U_{xx} + F\left(x\right) = 0,$$

then $v\left(x,t\right)$ satisfies the wave equation

$$v_{tt} = c^2 v_{xx}.$$

In a similar manner, if $u\left(x,t\right)$ is inserted in the initial and boundary conditions, we obtain

$$\begin{aligned}
u\left(x,0\right) &= v\left(x,0\right) + U\left(x\right) = f\left(x\right), \\
u_t\left(x,0\right) &= v_t\left(x,0\right) \qquad\quad = g\left(x\right), \\
u\left(0,t\right) &= v\left(0,t\right) + U\left(0\right) \; = A, \\
u\left(l,t\right) &= v\left(l,t\right) + U\left(l\right) \; = B.
\end{aligned}$$

Thus, if $U\left(x\right)$ is the solution of the problem

$$\begin{aligned}
c^2 U_{xx} + F &= 0, \\
U\left(0\right) = A, \quad U\left(l\right) &= B,
\end{aligned}$$

then $v\left(x,t\right)$ must satisfy

$$\begin{aligned}
v_{tt} &= c^2 v_{xx}, \\
v\left(x,0\right) &= f\left(x\right) - U\left(x\right), \\
v_t\left(x,0\right) &= g\left(x\right), \\
v\left(0,t\right) &= 0, \quad v\left(l,t\right) = 0.
\end{aligned} \tag{7.8.2}$$

Now $v(x, t)$ can be solved easily since $U(x)$ is known. It can be seen that

$$U(x) = A + (B - A)\frac{x}{l} + \frac{x}{l} \int_0^l \left[\frac{1}{c^2} \int_0^\eta F(\xi)\, d\xi \right] d\eta$$
$$- \int_0^x \left[\frac{1}{c^2} \int_0^\eta F(\xi)\, d\xi \right] d\eta.$$

As a specific example, consider the problem

$$u_{tt} = c^2 u_{xx} + h, \quad h \text{ is a constant}$$
$$u(x, 0) = 0, \quad u_t(x, 0) = 0, \quad (7.8.3)$$
$$u(0, t) = 0, \quad u(l, t) = 0.$$

Then, the solution of the system

$$c^2 U_{xx} + h = 0,$$
$$U(0) = 0, \quad U(l) = 0,$$

is

$$U(x) = \frac{h}{2c^2}\left(lx - x^2\right).$$

The function $v(x, t)$ must satisfy

$$v_{tt} = c^2 v_{xx},$$
$$v(x, 0) = -\frac{h}{2c^2}\left(lx - x^2\right), \quad v_t(x, 0) = 0,$$
$$v(0, t) = 0, \quad v(l, t) = 0.$$

The solution is given (see Section 7.3 with $g(x) = 0$) by

$$v(x, t) = \sum_{n=1}^\infty a_n \cos\left(\frac{n\pi c}{l} t\right) \sin\left(\frac{n\pi x}{l}\right),$$

and the coefficient is

$$a_n = \frac{2}{l} \int_0^l \left[-\frac{h}{2c^2}\left(lx - x^2\right)\right] \sin\left(\frac{n\pi x}{l}\right) dx$$
$$a_n = -\frac{4l^2 h}{n^3 \pi^3 c^2} \quad \text{for } n \text{ odd}$$
$$a_n = 0 \quad \text{for } n \text{ even.}$$

The solution of the given initial boundary-value problem is, therefore, given by

$$u(x,t) = v(x,t) + U(x)$$

$$= \frac{hx}{2c^2}(l-x) + \sum_{n=1}^{\infty}\left(-\frac{4l^2h}{c^2\pi^3}\right)\frac{\cos(2n-1)(\pi ct/l)}{(2n-1)^3}$$

$$\times \sin(2n-1)(\pi x/l).\,(7.8.4)$$

Let us now consider the problem of a finite string with an external force acting on it. If the ends are fixed, we have

$$
\begin{aligned}
u_{tt} - c^2 u_{xx} &= h(x,t), & 0 < x < l, \quad t > 0, \\
u(x,0) &= f(x), & 0 \le x \le l, \\
u_t(x,0) &= g(x), & 0 \le x \le l, \\
u(0,t) &= 0, & u(l,t) = 0, \quad t \ge 0.
\end{aligned}
\tag{7.8.5}
$$

We assume a solution involving the eigenfunctions, $\sin(n\pi x/l)$, of the associated eigenvalue problem in the form

$$u(x,t) = \sum_{n=1}^{\infty} u_n(t)\sin\left(\frac{n\pi x}{l}\right), \tag{7.8.6}$$

where the functions $u_n(t)$ are to be determined. It is evident that the boundary conditions are satisfied. Let us also assume that

$$h(x,t) = \sum_{n=1}^{\infty} h_n(t)\sin\left(\frac{n\pi x}{l}\right). \tag{7.8.7}$$

Thus,

$$h_n(t) = \frac{2}{l}\int_0^l h(x,t)\sin\left(\frac{n\pi x}{l}\right)dx. \tag{7.8.8}$$

We assume that the series (7.8.6) is convergent. We then find u_{tt} and u_{xx} from (7.8.6) and substitution of these values into (7.8.5) yields

$$\sum_{n=1}^{\infty}\left[u_n''(t) + \lambda_n^2\, u_n(t)\right]\sin\left(\frac{n\pi x}{l}\right) = \sum_{n=1}^{\infty} h_n(t)\sin\left(\frac{n\pi x}{l}\right),$$

where $\lambda_n = (n\pi c/l)$. Multiplying both sides of this equation by $\sin(m\pi x/l)$, where $m = 1, 2, 3, \ldots$, and integrating from $x = 0$ to $x = l$, we obtain

$$u_n''(t) + \lambda_n^2\, u_n(t) = h_n(t)$$

the solution of which is given by

$$u_n(t) = a_n\cos\lambda_n t + b_n\sin\lambda_n t + \frac{1}{\lambda_n}\int_0^t h_n(\tau)\sin[\lambda_n(t-\tau)]\,d\tau. \tag{7.8.9}$$

Hence, the formal solution (7.8.6) takes the final form

$$u\left(x,t\right) = \sum_{n=1}^{\infty} \left\{ a_n \cos \lambda_n t + b_n \sin \lambda_n t \right.$$

$$\left. + \frac{1}{\lambda_n} \int_0^t h_n\left(\tau\right) \sin\left[\lambda_n\left(t-\tau\right)\right] d\tau \right\} \cdot \sin\left(\frac{n\pi x}{l}\right). \quad (7.8.10)$$

Applying the initial conditions, we have

$$u\left(x,0\right) = f\left(x\right) = \sum_{n=1}^{\infty} a_n \sin\left(\frac{n\pi x}{l}\right).$$

Thus,

$$a_n = \frac{2}{l} \int_0^l f\left(x\right) \sin\left(\frac{n\pi x}{l}\right) dx. \quad (7.8.11)$$

Similarly,

$$u_t\left(x,0\right) = g\left(x\right) = \sum_{n=1}^{\infty} b_n \lambda_n \sin\left(\frac{n\pi x}{l}\right).$$

Thus,

$$b_n = \left(\frac{2}{l\lambda_n}\right) \int_0^l g\left(x\right) \sin\left(\frac{n\pi x}{l}\right) dx. \quad (7.8.12)$$

Hence, the formal solution of the initial boundary-value problem (7.8.5) is given by (7.8.10) with a_n given by (7.8.11) and b_n given by (7.8.12).

Example 7.8.2. Determine the solution of the initial boundary-value problem

$$\begin{aligned}
u_{tt} - u_{xx} &= h, & 0 < x < 1, \quad t > 0, \quad h = \text{constant}, \\
u\left(x,0\right) &= x\left(1-x\right), & 0 \le x \le 1, \\
u_t\left(x,0\right) &= 0, & 0 \le x \le 1, \\
u\left(0,t\right) &= 0, & u\left(1,t\right) = 0, \quad t \ge 0.
\end{aligned} \quad (7.8.13)$$

In this case, $c = 1$, $\lambda_n = n\pi$, $b_n = 0$ and a_n is given by

$$a_n = 2 \int_0^1 x\left(1-x\right) \sin n\pi x \, dx = \frac{4}{\left(n\pi\right)^3} \left[1 - \left(-1\right)^n\right].$$

We also have

$$h_n = 2 \int_0^1 h \sin\left(\frac{n\pi x}{l}\right) dx = \frac{2h}{n\pi} \left[1 - \left(-1\right)^n\right].$$

Hence, the integral term in (7.8.9) represents $\phi_n(t)$ given by

$$\phi_n(t) = \frac{1}{\lambda_n} \int_0^t h_n(\tau) \sin\left[\lambda_n(t-\tau)\right] d\tau = \frac{2h}{n\pi\lambda_n^2}\left[1 - (-1)^n\right](1 - \cos\lambda_n t).$$

The solution (7.8.10) is thus given by

$$u(x,t) = \sum_{n=1}^{\infty} \left\{ \frac{4}{n^3\pi^3}\left[1 - (-1)^n\right]\cos n\pi t \right.$$

$$\left. + \frac{2h}{n^3\pi^3}\left[1 - (-1)^n\right](1 - \cos n\pi t)\right\} \cdot \sin n\pi x. \quad (7.8.14)$$

We have treated the initial boundary-value problem with the fixed end conditions. Problems with other boundary conditions can also be solved in a similar manner.

We will now consider the initial boundary-value problem with time-dependent boundary conditions, namely,

$$\begin{aligned}
u_{tt} - u_{xx} &= h(x,t), & 0 < x < l, \quad t > 0, \\
u(x,0) &= f(x), & 0 \le x \le l, \\
u_t(x,0) &= g(x), & 0 \le x \le l, \quad\quad (7.8.15) \\
u(0,t) &= p(t), & u(l,t) = q(t), \quad t \ge 0.
\end{aligned}$$

We assume a solution in the form

$$u(x,t) = v(x,t) + U(x,t). \quad (7.8.16)$$

Substituting this into equation (7.8.15), we obtain

$$v_{tt} - c^2 v_{xx} = h - U_{tt} + c^2 U_{xx}.$$

For the initial and boundary conditions, we have

$$\begin{aligned}
v(x,0) &= f(x) - U(x,0), \\
v_t(x,0) &= g(x) - U_t(x,0), \\
v(0,t) &= p(t) - U(0,t), \\
v(l,t) &= q(t) - U(l,t).
\end{aligned}$$

In order to make the boundary conditions homogeneous, we set

$$U(0,t) = p(t), \qquad U(l,t) = q(t).$$

Thus, $U(x,t)$ must take the form

$$U(x,t) = p(t) + \frac{x}{l}\left[q(t) - p(t)\right]. \quad (7.8.17)$$

The problem now is to find the function $v(x,t)$ which satisfies

$$
\begin{aligned}
v_{tt} - c^2 v_{xx} &= h - U_{tt} = H(x,t), \\
v(x,0) &= f(x) - U(x,0) = F(x), \\
v_t(x,0) &= g(x) - U_t(x,0) = G(x), \\
v(0,t) &= 0, \qquad v(l,t) = 0.
\end{aligned}
\qquad (7.8.18)
$$

This is the same type of problem as the one with homogeneous boundary condition that has previously been treated.

Example 7.8.3. Find the solution of the problem

$$
\begin{aligned}
u_{tt} - u_{xx} &= h, & 0 < x < 1, \quad t > 0, \quad h = \text{constant}, \\
u(x,0) &= x(1-x), & 0 \le x \le 1, \\
u_t(x,0) &= 0, & 0 \le x \le 1, \\
u(0,t) &= t, & u(1,t) = \sin t, \quad t \ge 0.
\end{aligned}
\qquad (7.8.19)
$$

In this case, we use (7.8.16) and (7.8.17) with $c = 1$ and $\lambda_n = n\pi$ so that

$$
u(x,t) = v(x,t) + U(x,t), \qquad U(x,t) = t + x(\sin t - t). \ (7.8.20)
$$

Then, v must satisfy

$$
\begin{aligned}
v_{tt} - v_{xx} &= h + x\sin t, \\
v(x,0) &= x(1-x), \\
v_t(x,0) &= -1, \\
v(0,t) &= 0, \quad v(1,t) = 0.
\end{aligned}
\qquad (7.8.21)
$$

It follows from (7.8.8) that

$$
\begin{aligned}
h_n(t) &= 2\int_0^1 (h + x\sin t)\sin n\pi x \, dx \\
&= \frac{2h}{n\pi}[1 - (-1)^n] + \frac{2(-1)^{n+1}}{n\pi}\sin t = a + b\sin t \ (\text{say}).
\end{aligned}
\qquad (7.8.22)
$$

We also find

$$
a_n = 2\int_0^1 x(1-x)\sin n\pi x \, dx = \frac{4}{(n\pi)^3}[1 - (-1)^n],
$$

and

$$
b_n = \frac{2}{n\pi}\int_0^1 \sin n\pi x \, dx = \frac{2}{(n\pi)^2}[1 - (-1)^n].
$$

Then, we determine the integral term in (7.8.9) so that

$$\phi_n(t) = \frac{1}{n\pi} \int_0^t (a + b \sin \tau) \sin [n\pi (t - \tau)] \, d\tau$$

$$= \frac{1}{n\pi} \left\{ \frac{a}{n\pi} (1 - \cos n\pi t) + \frac{b}{4} [(\sin 2t - 2t) \cos n\pi t \right.$$

$$\left. - (\cos 2t - 1) \sin n\pi t] \right\}. \quad (7.8.23)$$

Hence, the solution of the problem (7.8.21) is

$$v(x,t) = \sum_{n=1}^{\infty} [a_n \cos n\pi t + b_n \sin n\pi t + \phi_n(t)] \sin n\pi x. \quad (7.8.24)$$

Thus, the solution of problem (7.8.19) is given by

$$u(x,t) = v(x,t) + U(x,t),$$

where $v(x,t)$ is given by (7.8.24) and $U(x,t)$ is given by (7.8.20)

Example 7.8.4. Use the method of separation of variables to derive the Hermite equation from the *Fokker–Planck equation* of nonequilibrium statistical mechanics

$$u_t - u_{xx} = (x\, u)_x. \quad (7.8.25)$$

We seek a nontrivial separable solution $u(x,t) = X(x) T(t)$ so that equation (7.8.25) reduces to a pair of ordinary differential equations

$$X'' + xX' + (1 + n) X = 0 \quad \text{and} \quad T' + nT = 0, \quad (7.8.26ab)$$

where $(-n)$ is a separation constant.

We next use

$$X(x) = \exp \left(-\frac{1}{2} x^2 \right) f(x) \quad (7.8.27)$$

and rescale the independent variable to obtain the Hermite equation for f in the form

$$\frac{d^2 f}{d\xi^2} - 2\xi \frac{df}{d\xi} + 2nf = 0.$$

The solution of (7.8.26b) gives

$$T(t) = c_n \exp(-nt), \quad (7.8.28)$$

where the coefficients c_n are constants.

Thus, the solution of the Fokker–Planck equation is given by

$$u\left(x,t\right) = \sum_{n=1}^{\infty} a_n \exp\left(-nt - \frac{1}{2}x^2\right) H_n\left(\frac{x}{\sqrt{2}}\right), \qquad (7.8.29)$$

where H_n is the Hermite function and a_n are arbitrary constants to be determined from the given initial condition

$$u\left(x,0\right) = f\left(x\right). \qquad (7.8.30)$$

We make the change of variables

$$\xi = x\,e^t \quad \text{and} \quad u = e^t v, \qquad (7.8.31)$$

in equation (7.8.25). Consequently, equation (7.8.25) becomes

$$\frac{\partial v}{\partial t} = e^{2t}\frac{\partial^2 v}{\partial \xi^2}. \qquad (7.8.32)$$

Making another change of variable t to $\tau\left(t\right)$, we transform (7.8.32) into the linear diffusion equation

$$\frac{\partial v}{\partial \tau} = \frac{\partial^2 v}{\partial \xi^2}. \qquad (7.8.33)$$

Finally, we note that the asymptotic behavior of the solution $u\left(x,t\right)$ as $t \to \infty$ is of special interest. The reader is referred to Reif (1965) for such behavior.

7.9 Exercises

1. Solve the following initial boundary-value problems:

(a) $\quad u_{tt} = c^2 u_{xx}, \qquad\quad 0 < x < 1, \qquad t > 0,$

$\quad u\left(x,0\right) = x\left(1-x\right), \qquad u_t\left(x,0\right) = 0, \quad 0 \le x \le 1,$

$\quad u\left(0,t\right) = u\left(1,t\right) = 0, \quad t > 0.$

(b) $\quad u_{tt} = c^2 u_{xx}, \qquad\quad 0 < x < \pi, \qquad t > 0,$

$\quad u\left(x,0\right) = 3\sin x, \qquad\quad u_t\left(x,0\right) = 0, \quad 0 \le x \le \pi,$

$\quad u\left(0,t\right) = u\left(1,t\right) = 0, \quad t > 0.$

2. Determine the solutions of the following initial boundary-value problems:

(a) $\quad u_{tt} = c^2 u_{xx}, \qquad 0 < x < \pi, \qquad\qquad t > 0,$

$\quad u(x,0) = 0, \qquad\qquad u_t(x,0) = 8\sin^2 x, \qquad 0 \le x \le \pi,$

$\quad u(0,t) = u(\pi,t) = 0, \qquad t > 0.$

(b) $\quad u_{tt} = c^2 u_{xx} = 0, \qquad 0 < x < 1, \qquad\qquad t > 0,$

$\quad u(x,0) = 0, \qquad\qquad u_t(x,0) = x\sin\pi x, \qquad 0 \le x \le 1,$

$\quad u(0,t) = u(1,t) = 0, \qquad t > 0.$

3. Find the solution of each of the following problems:

(a) $\quad u_{tt} = c^2 u_{xx} = 0, \qquad 0 < x < 1, \qquad\qquad t > 0,$

$\quad u(x,0) = x(1-x), \qquad u_t(x,0) = x - \tan\frac{\pi x}{4}, \qquad 0 \le x \le 1,$

$\quad u(0,t) = u(\pi,t) = 0, \qquad t > 0.$

(b) $\quad u_{tt} = c^2 u_{xx} = 0, \qquad 0 < x < \pi, \qquad\qquad t > 0,$

$\quad u(x,0) = \sin x, \qquad\qquad u_t(x,0) = x^2 - \pi x, \qquad 0 \le x \le \pi,$

$\quad u(0,t) = u(\pi,t) = 0, \qquad t > 0.$

4. Solve the following problems:

(a) $\quad u_{tt} = c^2 u_{xx} = 0, \qquad 0 < x < \pi, \qquad t > 0,$

$\quad u(x,0) = x + \sin x, \qquad u_t(x,0) = 0, \qquad 0 \le x \le \pi,$

$\quad u(0,t) = u_x(\pi,t) = 0, \qquad t > 0.$

(b) $\quad u_{tt} = c^2 u_{xx} = 0, \qquad 0 < x < \pi, \qquad t > 0,$

$\quad u(x,0) = \cos x, \qquad\qquad u_t(x,0) = 0, \qquad 0 \le x \le \pi,$

$\quad u_x(0,t) = 0, \qquad\qquad u_x(\pi,t) = 0, \qquad t > 0.$

5. By the method of separation of variables, solve the telegraph equation:

$$u_{tt} + au_t + bu = c^2 u_{xx}, \qquad 0 < x < l, \qquad t > 0,$$
$$u(x,0) = f(x), \qquad u_t(x,0) = 0,$$
$$u(0,t) = u(l,t) = 0, \qquad\qquad t > 0.$$

6. Obtain the solution of the damped wave motion problem:

$$u_{tt} + au_t = c^2 u_{xx}, \qquad 0 < x < l, \qquad t > 0,$$
$$u(x,0) = 0, \qquad u_t(x,0) = g(x),$$
$$u(0,t) = u(l,t) = 0.$$

7. The torsional oscillation of a shaft of circular cross section is governed by the partial differential equation

$$\theta_{tt} = a^2 \theta_{xx},$$

where $\theta(x,t)$ is the angular displacement of the cross section and a is a physical constant. The ends of the shaft are fixed elastically, that is,

$$\theta_x(0,t) - h\theta(0,t) = 0, \qquad \theta_x(l,t) + h\theta(l,t) = 0.$$

Determine the angular displacement if the initial angular displacement is $f(x)$.

8. Solve the initial boundary-value problem of the longitudinal vibration of a truncated cone of length l and base of radius a. The equation of motion is given by

$$\left(1 - \frac{x}{h}\right)^2 \frac{\partial^2 u}{\partial t^2} = c^2 \frac{\partial}{\partial x}\left[\left(1 - \frac{x}{h}\right)^2 \frac{\partial u}{\partial x}\right], \qquad 0 < x < l, \quad t > 0,$$

where $c^2 = (E/\rho)$, E is the elastic modulus, ρ is the density of the material and $h = la/(a - l)$. The two ends are rigidly fixed. If the initial displacement is $f(x)$, that is, $u(x,0) = f(x)$, find $u(x,t)$.

9. Establish the validity of the formal solution of the initial boundary-value problems:

$$u_{tt} = c^2 u_{xx}, \qquad 0 < x < \pi, \qquad\qquad t > 0,$$
$$u(x,0) = f(x), \qquad u_t(x,0) = g(x), \qquad 0 \le x \le \pi,$$
$$u_x(0,t) = 0, \qquad u_x(\pi,t) = 0, \qquad\qquad t > 0.$$

10. Prove the uniqueness of the solution of the initial boundary-value problem:

$$u_{tt} = c^2 u_{xx}, \qquad 0 < x < \pi, \qquad\qquad t > 0,$$
$$u(x,0) = f(x), \qquad u_t(x,0) = g(x), \qquad 0 \le x \le \pi,$$
$$u_x(0,t) = 0, \qquad u_x(\pi,t) = 0, \qquad\qquad t > 0.$$

11. Determine the solution of

$$u_{tt} = c^2 u_{xx} + A \sinh x, \quad 0 < x < l, \qquad t > 0,$$
$$u(x,0) = 0, \quad u_t(x,0) = 0, \quad 0 \le x \le l,$$
$$u(0,t) = h, \qquad u(l,t) = k, \qquad\qquad\qquad t > 0,$$

where h, k, and A are constants.

12. Solve the problem:

$$u_{tt} = c^2 u_{xx} + Ax, \qquad 0 < x < 1, \quad t > 0, \quad A = \text{constant},$$
$$u(x,0) = 0, \quad u_t(x,0) = 0, \quad 0 \le x \le 1,$$
$$u(0,t) = 0, \qquad u(1,t) = 0, \qquad\qquad t > 0.$$

13. Solve the problem:

$$u_{tt} = c^2 u_{xx} + x^2, \qquad 0 < x < 1, \qquad t > 0,$$
$$u(x,0) = x, \quad u_t(x,0) = 0, \quad 0 \le x \le 1,$$
$$u(0,t) = 0, \qquad u(1,t) = 1, \qquad\qquad t \ge 0.$$

14. Find the solution of the following problems:

(a) $u_t = k u_{xx} + h, \qquad\qquad 0 < x < 1, \qquad t > 0, \quad h = \text{constant},$
$u(x,0) = u_0(1 - \cos \pi x), \quad 0 \le x \le 1, \qquad\qquad\qquad u_0 = \text{constant},$
$u(0,t) = 0, \qquad\qquad\qquad u(l,t) = 2u_0, \qquad t \ge 0.$

(b) $u_t = k u_{xx} - hu, \qquad\qquad 0 < x < l, \qquad t > 0, \quad h = \text{constant},$
$u(x,0) = f(x), \qquad\qquad\qquad 0 \le x \le l,$
$u_x(0,t) = u_x(l,t) = 0, \qquad\quad t > 0.$

15. Obtain the solution of each of the following initial boundary-value problems:

(a) $u_t = 4 u_{xx}, \qquad\qquad 0 < x < 1, \qquad t > 0,$

$u(x,0) = x^2(1 - x), \qquad 0 \le x \le 1,$

$u(0,t) = 0, \qquad\qquad\qquad u(l,t) = 0, \qquad t \ge 0.$

(b) $u_t = k u_{xx}, \qquad\qquad 0 < x < \pi, \qquad t > 0,$

$u(x,0) = \sin^2 x, \qquad\qquad 0 \le x \le \pi,$

$u(0,t) = 0, \qquad\qquad\qquad u(\pi,t) = 0, \qquad t \ge 0.$

(c) $u_t = u_{xx},$ $0 < x < 2,$ $t > 0,$

   $u(x,0) = x,$ $0 \le x \le 2,$

   $u(0,t) = 0,$ $u_x(2,t) = 1,$ $t \ge 0.$

(d) $u_t = k\,u_{xx},$ $0 < x < l,$ $t > 0,$

   $u(x,0) = \sin(\pi x/2l),$ $0 \le x \le l,$

   $u(0,t) = 0,$ $u(l,t) = 1,$ $t \ge 0.$

16. Find the temperature distribution in a rod of length l. The faces are insulated, and the initial temperature distribution is given by $x\,(l - x)$.

17. Find the temperature distribution in a rod of length π, one end of which is kept at zero temperature and the other end of which loses heat at a rate proportional to the temperature at that end $x = \pi$. The initial temperature distribution is given by $f(x) = x$.

18. The voltage distribution in an electric transmission line is given by

$$v_t = k\,v_{xx}, \quad 0 < x < l, \quad t > 0.$$

A voltage equal to zero is maintained at $x = l$, while at the end $x = 0$, the voltage varies according to the law

$$v(0,t) = Ct, \qquad t > 0,$$

where C is a constant. Find $v(x,t)$ if the initial voltage distribution is zero.

19. Establish the validity of the formal solution of the initial boundary-value problem:

$$u_t = k\,u_{xx}, \qquad 0 < x < l, \qquad t > 0,$$
$$u(x,0) = f(x), \qquad 0 \le x \le l,$$
$$u(0,t) = 0, \qquad u_x(l,t) = 0, \quad t \ge 0.$$

20. Prove the uniqueness of the solution of the problem:

$$u_t = k\,u_{xx}, \qquad 0 < x < l, \qquad t > 0,$$
$$u(x,0) = f(x), \qquad 0 \le x \le l,$$
$$u_x(0,t) = 0, \qquad u_x(l,t) = 0, \quad t \ge 0.$$

21. Solve the radioactive decay problem:

$$u_t - k\,u_{xx} = Ae^{-ax}, \qquad 0 < x < \pi, \qquad t > 0,$$
$$u(x,0) = \sin x, \qquad 0 \le x \le \pi,$$
$$u(0,t) = 0, \qquad u(\pi,t) = 0, \qquad t \ge 0.$$

22. Determine the solution of the initial boundary-value problem:

$$u_t - k\,u_{xx} = h(x,t), \qquad 0 < x < l, \qquad t > 0, \quad k = \text{constant},$$
$$u(x,0) = f(x), \qquad 0 \le x \le l,$$
$$u(0,t) = p(t), \qquad u(l,t) = q(t), \qquad t \ge 0.$$

23. Determine the solution of the initial boundary-value problem:

$$u_t - k\,u_{xx} = h(x,t), \qquad 0 < x < l, \qquad t > 0,$$
$$u(x,0) = f(x), \qquad 0 \le x \le l,$$
$$u(0,t) = p(t), \qquad u_x(l,t) = q(t), \qquad t \ge 0.$$

24. Solve the problem:

$$\dot{u}_t - k\,u_{xx} = 0, \qquad 0 < x < 1, \qquad t > 0,$$
$$u(x,0) = x(1-x), \qquad 0 \le x \le 1,$$
$$u(0,t) = t, \qquad u(1,t) = \sin t, \qquad t \ge 0.$$

25. Solve the problem:

$$u_t - 4u_{xx} = xt, \qquad 0 < x < 1, \qquad t \ge 0,$$
$$u(x,0) = \sin \pi x, \qquad 0 \le x \le 1,$$
$$u(0,t) = t, \qquad u(1,t) = t^2, \qquad t \ge 0.$$

26. Solve the problem:

$$u_t - k\,u_{xx} = x \cos t, \qquad 0 < x < \pi, \qquad t > 0,$$
$$u(x,0) = \sin x, \qquad 0 \le x \le \pi,$$
$$u(0,t) = t^2, \qquad u(\pi,t) = 2t, \qquad t \ge 0.$$

27. Solve the problem:

$$u_t - u_{xx} = 2x^2 t, \qquad 0 < x < 1, \qquad t > 0,$$
$$u(x,0) = \cos(3\pi x/2), \qquad 0 \le x \le 1,$$
$$u(0,t) = 1, \qquad u_x(1,t) = \frac{3\pi}{2}, \qquad t \ge 0.$$

28. Solve the problem:

$$u_t - 2u_{xx} = h, \qquad 0 < x < 1, \qquad t > 0, \qquad h = \text{constant},$$
$$u(x,0) = x, \qquad 0 \le x \le 1,$$
$$u(0,t) = \sin t, \qquad u_x(1,t) + u(1,t) = 2, \qquad t \ge 0.$$

29. Determine the solution of the initial boundary-value problem:

$$u_{tt} - c^2 u_{xx} = h(x,t), \quad 0 < x < l, \qquad t > 0,$$
$$u(x,0) = f(x), \qquad 0 \le x \le l,$$
$$u_t(x,0) = g(x), \qquad 0 \le x \le l,$$
$$u(0,t) = p(t), \qquad u_x(l,t) = q(t), \quad t \ge 0.$$

30. Determine the solution of the initial boundary-value problem:

$$u_{tt} - c^2 u_{xx} = h(x,t), \quad 0 < x < l, \qquad t > 0,$$
$$u(x,0) = f(x), \qquad 0 \le x \le l,$$
$$u_t(x,0) = g(x), \qquad 0 \le x \le l,$$
$$u_x(0,t) = p(t), \qquad u_x(l,t) = q(t), \quad t \ge 0.$$

31. Solve the problem:

$$u_{tt} - u_{xx} = 0, \qquad 0 < x < 1, \qquad t > 0,$$
$$u(x,0) = x, \qquad u_t(x,0) = 0, \qquad 0 \le x \le 1,$$
$$u(0,t) = t^2, \qquad u(1,t) = \cos t, \quad t \ge 0.$$

32. Solve the problem:

$$u_{tt} - 4 u_{xx} = xt, \qquad 0 < x < 1, \qquad t > 0,$$
$$u(x,0) = x, \qquad u_t(x,0) = 0, \qquad 0 \le x \le 1,$$
$$u(0,t) = 0, \qquad u_x(1,t) = 1+t, \quad t \ge 0.$$

33. Solve the problem:

$$u_{tt} - 9 u_{xx} = 0, \qquad 0 < x < 1, \qquad t > 0,$$
$$u(x,0) = \sin\left(\frac{\pi x}{2}\right), \quad u_t(x,0) = 1 + x, \quad 0 \le x \le 1,$$
$$u_x(0,t) = \pi/2, \qquad u_x(1,t) = 0, \qquad t \ge 0.$$

34. Find the solution of the problem:

$$u_{tt} + 2k u_t - c^2 u_{xx} = 0, \qquad 0 < x < l, \qquad t > 0,$$
$$u(x,0) = 0, \qquad u_t(x,0) = 0, \qquad 0 \le x \le l,$$
$$u_x(0,t) = 0, \qquad u(l,t) = h, \qquad t \ge 0, \quad h = \text{constant}.$$

35. Solve the problem:

$$u_t - c^2 u_{xx} + hu = hu_0, \qquad -\pi < x < \pi, \qquad t > 0,$$
$$u(x,0) = f(x), \qquad -\pi \le x \le \pi,$$
$$u(-\pi,t) = u(\pi,t), \qquad u_x(-\pi,t) = u_x(\pi,t), \qquad t \ge 0,$$

where h and u_0 are constants.

36. Prove the uniqueness theorem for the boundary-value problem involving the Laplace equation:

$$u_{xx} + u_{yy} = 0, \qquad 0 < x < a, \qquad\qquad 0 < y < b,$$
$$u(x,0) = f(x), \quad u(x,b) = 0, \qquad\qquad 0 \le x \le a,$$
$$u_x(0,y) = 0 = u_x(a,y), \qquad\qquad\qquad 0 \le y \le b.$$

37. Consider the telegraph equation problem:

$$u_{tt} - c^2 u_{xx} + a u_t + b u = 0, \qquad 0 < x < l, \qquad\qquad t > 0,$$
$$u(x,0) = f(x), \qquad u_t(x,0) = g(x) \quad \text{for} \quad 0 \le x \le l,$$
$$u(0,t) = 0 = u(l,t) \quad \text{for} \quad t \ge 0,$$

where a and b are positive constants.

(a) Show that, for any $T > 0$,

$$\int_0^l \left(u_t^2 + c^2 u_x^2 + b u^2 \right)_{t=T} dx \le \int_0^l \left(u_t^2 + c^2 u_x^2 + b u^2 \right)_{t=0} dx.$$

(b) Use the above integral inequality from (a) to show that the initial boundary-value problem for the telegraph equation can have only one solution.

8

Eigenvalue Problems and Special Functions

"The tool which serves as intermediary between theory and practice, between thought and observation, is mathematics; it is mathematics which builds the linking bridges and gives the ever more reliable forms. From this it has come about that our entire contemporary culture, in as much as it is based on the intellectual penetration and the exploitation of nature, has its foundations in mathematics."

David Hilbert

"In 1836/7, he published some important joint work with his friend Sturm on what became known as Sturm–Liouville theory, which became important in physics."

Ioan James

8.1 Sturm–Liouville Systems

In the preceding chapter, we determined the solutions of partial differential equations by the method of separation of variables. In this chapter, we generalize the method of separation of variables and the associated eigenvalue problems. This generalization, usually known as the *Sturm–Liouville theory,* greatly extends the scope of the method of separation of variables.

Under separable conditions we transformed the second-order homogeneous partial differential equation into two ordinary differential equations (7.2.11) and (7.2.12) which are of the form

$$a_1\left(x\right)\frac{d^2y}{dx^2} + a_2\left(x\right)\frac{dy}{dx} + \left[a_3\left(x\right) + \lambda\right]y = 0. \qquad (8.1.1)$$

If we introduce

$$p\left(x\right) = \exp\left[\int^{x} \frac{a_2\left(t\right)}{a_1\left(t\right)} dt\right], \; q\left(x\right) = \frac{a_3\left(x\right)}{a_1\left(x\right)} p\left(x\right), \; s\left(x\right) = \frac{p\left(x\right)}{a_1\left(x\right)}, \quad (8.1.2)$$

into equation (8.1.1), we obtain

$$\frac{d}{dx}\left(p\frac{dy}{dx}\right) + \left(q + \lambda s\right) y = 0, \qquad (8.1.3)$$

which is known as the *Sturm–Liouville equation*. In terms of the Sturm–Liouville operator

$$L \equiv \frac{d}{dx}\left(p\frac{d}{dx}\right) + q,$$

equation (8.1.3) can be written as

$$L\left[y\right] + \lambda s\left(x\right) y = 0, \qquad (8.1.4)$$

where λ is a parameter independent of x, and p, q, and s are real-valued functions of x. To ensure the existence of solutions, we let q and s be continuous and p be continuously differentiable in a closed finite interval $[a, b]$.

The Sturm–Liouville equation is called *regular* in the interval $[a, b]$ if the functions $p\left(x\right)$ and $s\left(x\right)$ are positive in the interval $[a, b]$. Thus, for a given λ, there exist two linearly independent solutions of a regular Sturm–Liouville equation in the interval $[a, b]$.

The Sturm–Liouville equation

$$L\left[y\right] + \lambda s\left(x\right) y = 0, \qquad a \le x \le b,$$

together with the separated end conditions

$$a_1 y\left(a\right) + a_2 y'\left(a\right) = 0, \quad b_1 y\left(b\right) + b_2 y'\left(b\right) = 0, \qquad (8.1.5)$$

where a_1 and a_2, and likewise b_1 and b_2, are not both zero and are given real numbers, is called a *regular Sturm–Liouville (RSL) system*.

The values of λ for which the Sturm–Liouville system has a nontrivial solution are called the *eigenvalues*, and the corresponding solutions are called the *eigenfunctions*. For a regular Sturm–Liouville problem, we denote the domain of L by $D\left(L\right)$, that is, $D\left(L\right)$ is the space of all complex-valued functions y defined on $[a, b]$ for which $y'' \in L^2\left(\left[a, b\right]\right)$ and which satisfy boundary conditions (8.1.5).

Example 8.1.1. Consider the regular Sturm–Liouville problem

$$y'' + \lambda y = 0, \qquad 0 \le x \le \pi,$$
$$y\left(0\right) = 0, \quad y'\left(\pi\right) = 0.$$

When $\lambda \leq 0$, it can be readily shown that λ is not an eigenvalue. However, when $\lambda > 0$, the solution of the Sturm–Liouville equation is

$$y(x) = A \cos \sqrt{\lambda}\, x + B \sin \sqrt{\lambda}\, x.$$

Applying the condition $y(0) = 0$, we obtain $A = 0$. The condition $y'(\pi) = 0$ yields

$$B\sqrt{\lambda} \cos \sqrt{\lambda}\, \pi = 0.$$

Since $\lambda \neq 0$ and $B = 0$ yields a trivial solution, we must have

$$\cos \sqrt{\lambda}\, \pi = 0, \quad B \neq 0.$$

This equation is satisfied if

$$\sqrt{\lambda} = \frac{2n-1}{2}, \qquad n = 1, 2, 3, \ldots,$$

and hence, the eigenvalues are $\lambda_n = (2n-1)^2/4$, and the corresponding eigenfunctions are

$$\sin\left(\frac{2n-1}{2}\right) x, \qquad n = 1, 2, 3, \ldots.$$

Example 8.1.2. Consider the Euler equation

$$x^2 y'' + xy' + \lambda u = 0, \qquad 1 \leq x \leq e$$

with the end conditions

$$y(1) = 0, \quad y(e) = 0.$$

By using the transformation (8.1.2), the Euler equation can be put into the Sturm–Liouville form:

$$\frac{d}{dx}\left(x \frac{dy}{dx}\right) + \frac{1}{x} \lambda y = 0.$$

The solution of the Euler equation is

$$y(x) = c_1 x^{i\sqrt{\lambda}} + c_2 x^{-i\sqrt{\lambda}}.$$

Noting that $x^{ia} = e^{ia \ln x} = \cos(a \ln x) + i \sin(a \ln x)$, the solution $y(x)$ becomes

$$y(x) = A \cos\left(\sqrt{\lambda} \ln x\right) + B \sin\left(\sqrt{\lambda} \ln x\right),$$

where A and B are constants related to c_1 and c_2. The end condition $y(1) = 0$ gives $A = 0$, and the end condition $y(e) = 0$ gives

$$\sin \sqrt{\lambda} = 0, \qquad B \neq 0,$$

which in turn yields the eigenvalues

$$\lambda_n = n^2 \pi^2, \qquad n = 1, 2, 3, \ldots,$$

and the corresponding eigenfunctions

$$\sin(n\pi \ln x), \quad n = 1, 2, 3, \ldots.$$

Another type of problem that often occurs in practice is the periodic Sturm–Liouville system.

The Sturm–Liouville equation

$$\frac{d}{dx}\left(p(x)\frac{dy}{dx}\right) + [q(x) + \lambda s(x)]y = 0, \quad a \leq x \leq b,$$

in which $p(a) = p(b)$, together with the periodic end conditions

$$y(a) = y(b), \qquad y'(a) = y'(b)$$

is called a *periodic Sturm–Liouville system.*

Example 8.1.3. Consider the periodic Sturm–Liouville system

$$y'' + \lambda y = 0, \qquad -\pi \leq x \leq \pi,$$
$$y(-\pi) = y(\pi), \qquad y'(-\pi) = y'(\pi).$$

Here we note that $p(x) = 1$, hence $p(-\pi) = p(\pi)$. When $\lambda > 0$, we see that the solution of the Sturm–Liouville equation is

$$y(x) = A\cos\sqrt{\lambda}\,x + B\sin\sqrt{\lambda}\,x.$$

Application of the periodic end conditions yields

$$\left(2\sin\sqrt{\lambda}\,\pi\right)B = 0,$$
$$\left(2\sqrt{\lambda}\sin\sqrt{\lambda}\,\pi\right)A = 0.$$

Thus, to obtain a nontrivial solution, we must have

$$\sin\left(\sqrt{\lambda}\right)\pi = 0, \qquad A \neq 0, \qquad B \neq 0.$$

Consequently,

$$\lambda_n = n^2, \qquad n = 1, 2, 3, \ldots.$$

Since $\sin\sqrt{\lambda}\,\pi = 0$ is satisfied for arbitrary A and B, we obtain two linearly independent eigenfunctions $\cos nx$, and $\sin nx$ corresponding to the same eigenvalue n^2.

It can be readily shown that if $\lambda < 0$, the solution of the Sturm–Liouville equation does not satisfy the periodic end conditions. However, when $\lambda = 0$ the corresponding eigenfunction is 1. Thus, the eigenvalues of the periodic Sturm–Liouville system are 0, $\{n^2\}$, and the corresponding eigenfunctions are 1, $\{\cos nx\}$, $\{\sin nx\}$, where n is a positive integer.

8.2 Eigenvalues and Eigenfunctions

In Examples 8.1.1 and 8.1.2 of the regular Sturm–Liouville systems in the preceding section, we see that there exists only one linearly independent eigenfunction corresponding to the eigenvalue λ, which is called an *eigenvalue of multiplicity one* (or a *simple eigenvalue*). An eigenvalue is said to be of multiplicity k if there exist k linearly independent eigenfunctions corresponding to the same eigenvalue. In Example 8.1.3 of the periodic Sturm–Liouville system, the eigenfunctions $\cos nx$, $\sin nx$ correspond to the same eigenvalue n^2. Thus, this eigenvalue is of multiplicity two.

In the preceding examples, we see that the eigenfunctions are $\cos nx$ and $\sin nx$ for $n = 1, 2, 3, \ldots$. It can be easily shown by using trigonometric identities that

$$\int_{-\pi}^{\pi} \cos mx \cos nx \, dx = 0, \qquad m \neq n,$$

$$\int_{-\pi}^{\pi} \cos mx \sin nx \, dx = 0, \qquad \text{for all integers } m, n,$$

$$\int_{-\pi}^{\pi} \sin mx \sin nx \, dx = 0, \qquad m \neq n.$$

We say that these functions are orthogonal to each other in the interval $[-\pi, \pi]$. The orthogonality relation holds in general for the eigenfunctions of Sturm–Liouville systems

Let $\phi(x)$ and $\psi(x)$ be any real-valued integrable functions on an interval I. Then ϕ and ψ are said to be *orthogonal* on I with respect to a weight function $\rho(x) > 0$, if and only if,

$$\langle \phi, \psi \rangle = \int_{I} \phi(x) \psi(x) \rho(x) \, dx = 0. \tag{8.2.1}$$

The interval I may be of infinite extent, or it may be either open or closed at one or both ends of the finite interval.

When $\phi = \psi$ in (8.2.1) we define the *norm* of ϕ by

$$\|\phi\| = \left[\int_{I} \phi^2(x) \rho(x) \, dx \right]^{\frac{1}{2}}. \tag{8.2.2}$$

Theorem 8.2.1. *Let the coefficients p, q, and s in the Sturm–Liouville system be continuous in $[a, b]$. Let the eigenfunctions ϕ_j and ϕ_k, corresponding to λ_j and λ_k, be continuously differentiable. Then ϕ_j and ϕ_k are orthogonal with respect to the weight function $s(x)$ in $[a, b]$.*

Proof. Since ϕ_j corresponding to λ_j satisfies the Sturm–Liouville equation, we have

$$\frac{d}{dx}\left(p \phi_j'\right) + (q + \lambda_j s) \phi_j = 0 \tag{8.2.3}$$

and for the same reason

$$\frac{d}{dx}\left(p\,\phi'_k\right) + \left(q + \lambda_k s\right)\phi_k = 0. \qquad (8.2.4)$$

Multiplying equation (8.2.3) by ϕ_k and equation (8.2.4) by ϕ_j, and subtracting, we obtain

$$(\lambda_j - \lambda_k)\,s\,\phi_j\phi_k = \phi_j\,\frac{d}{dx}\left(p\,\phi'_k\right) - \phi_k\,\frac{d}{dx}\left(p\,\phi'_j\right)$$

$$= \frac{d}{dx}\left[(p\,\phi'_k)\,\phi_j - (p\,\phi'_j)\,\phi_k\right]$$

and integrating yields

$$(\lambda_j - \lambda_k)\int_a^b s\,\phi_j\phi_k dx = \left[p\left(\phi_j\phi'_k - \phi'_j\phi_k\right)\right]_a^b$$

$$= p\,(b)\left[\phi_j\,(b)\,\phi'_k\,(b) - \phi'_j\,(b)\,\phi_k\,(b)\right]$$

$$-p\,(a)\left[\phi_j\,(a)\,\phi'_k\,(a) - \phi'_j\,(a)\,\phi_k\,(a)\right]\ (8.2.5)$$

the right side of which is called the *boundary term* of the Sturm–Liouville system. The end conditions for the eigenfunctions ϕ_j and ϕ_k are

$$b_1\phi_j\,(b) + b_2\phi'_j\,(b) = 0,$$
$$b_1\phi_k\,(b) + b_2\phi'_k\,(b) = 0.$$

If $b_2 \neq 0$, we multiply the first condition by $\phi_k\,(b)$ and the second condition by $\phi_j\,(b)$, and subtract to obtain

$$\left[\phi_j\,(b)\,\phi'_k\,(b) - \phi'_j\,(b)\,\phi_k\,(b)\right] = 0. \qquad (8.2.6)$$

In a similar manner, if $a_2 \neq 0$, we obtain

$$\left[\phi_j\,(a)\,\phi'_k\,(a) - \phi'_j\,(a)\,\phi_k\,(a)\right] = 0. \qquad (8.2.7)$$

We see by virtue of (8.2.6) and (8.2.7) that

$$(\lambda_j - \lambda_k)\int_a^b s\,\phi_j\,\phi_k\,dx = 0. \qquad (8.2.8)$$

If λ_j and λ_k are distinct eigenvalues, then

$$\int_a^b s\,\phi_j\,\phi_k\,dx = 0. \qquad (8.2.9)$$

Theorem 8.2.2. *The eigenfunctions of a periodic Sturm–Liouville system in $[a, b]$ are orthogonal with respect to the weight function $s\,(x)$ in $[a, b]$.*

Proof. The periodic conditions for the eigenfunctions ϕ_j and ϕ_k are

$$\phi_j(a) = \phi_j(b), \qquad \phi'_j(a) = \phi'_j(b),$$
$$\phi_k(a) = \phi_k(b), \qquad \phi'_k(a) = \phi'_k(b).$$

Substitution of these into equation (8.2.5) yields

$$(\lambda_j - \lambda_k) \int_a^b s\,\phi_j\,\phi_k\,dx = [p(b) - p(a)]\,[\phi_j(a)\,\phi'_k(a) - \phi'_j(a)\,\phi_k(a)].$$

Since $p(a) = p(b)$, we have

$$(\lambda_j - \lambda_k) \int_a^b s\,\phi_j\,\phi_k\,dx = 0. \tag{8.2.10}$$

For distinct eigenvalues $\lambda_j \neq \lambda_k$, $(\lambda_j - \lambda_k) \neq 0$ and thus,

$$\int_a^b s\,\phi_j\,\phi_k\,dx = 0. \tag{8.2.11}$$

Theorem 8.2.3. *For any y, $z \in D(L)$, we have the* Lagrange identity

$$yL[z] - zL[y] = \frac{d}{dx}[p(yz' - zy')]. \tag{8.2.12}$$

Proof. Using the definition of the Sturm–Liouville operator, we have

$$yL[z] - zL[y] = y\frac{d}{dx}\left(p\frac{dz}{dx}\right) + qyz - z\frac{d}{dx}\left(p\frac{dy}{dx}\right) - qyz$$
$$= \frac{d}{dx}[p(yz' - zy')].$$

Theorem 8.2.4. *The Sturm–Liouville operator L is self-adjoint. In other words, for any y, $z \in D(L)$, we have*

$$\langle L[y], z \rangle = \langle y, L[z] \rangle, \tag{8.2.13}$$

where $<,>$ is the inner product in $L^2([a,b])$ defined by

$$\langle f, g \rangle = \int_a^b f(x)\,\overline{g(x)}\,dx. \tag{8.2.14}$$

Proof. Since all constants involved in the boundary conditions of Sturm–Liouville system are real, if $z \in D(L)$, then $\overline{z} \in D(L)$.

Also since p, q and s are real-valued, $\overline{L[z]} = L[\overline{z}]$. Consequently, we have

$$\langle L[y], z \rangle - \langle y, L[z] \rangle = \int_a^b (\overline{z}\,L[y] - y\,L[\overline{z}])\,dx$$
$$= [p(\overline{z}\,y' - y\,\overline{z}')]_a^b, \quad \text{by } (8.2.12). \tag{8.2.15}$$

We next show that the right hand side of this equality vanishes for a regular RSL system. If $p(a) = 0$, the result follows immediately. If $p(a) > 0$, then y and z satisfy the boundary conditions of the form (8.1.5) at $x = a$. That is,

$$\begin{bmatrix} y(a) & y'(a) \\ \bar{z}(a) & \bar{z}'(a) \end{bmatrix} \begin{bmatrix} a_1 \\ a_2 \end{bmatrix} = 0.$$

Since a_1 and a_2 are not both zero, we have

$$\bar{z}(a)\, y'(a) - y(a)\, \bar{z}'(a) = 0.$$

A similar argument can be used to the other end point $x = b$, so that the right-hand side of (8.2.15) vanishes. This proves the theorem.

Theorem 8.2.5. *All the eigenvalues of a regular Sturm–Liouville system with $s(x) > 0$ are real.*

Proof. Let λ be an eigenvalue of a RSL system and let $y(x)$ be the corresponding eigenfunction. This means that $y \neq 0$ and $L[y] = -\lambda s y$. Then

$$0 = \langle L[y], y \rangle - \langle y, L[y] \rangle = (\bar{\lambda} - \lambda) \int_a^b s(x)\, |y(x)|^2\, dx.$$

Since $s(x) > 0$ in $[a, b]$ and $y \neq 0$, the integrand is a positive number. Thus, $\lambda = \bar{\lambda}$, and hence, the eigenvalues are real. This completes the proof.

Theorem 8.2.6. *If $\phi_1(x)$ and $\phi_2(x)$ are any two solutions of the equation $L[y] + \lambda s y = 0$ on $[a, b]$, then $p(x) W(x; \phi_1, \phi_2) = constant$, where W is the Wronskian.*

Proof. Since ϕ_1 and ϕ_2 are solutions of $L[y] + \lambda s y = 0$, we have

$$\frac{d}{dx}\left(p\frac{d\phi_1}{dx} \right) + (q + \lambda s)\,\phi_1 = 0,$$

$$\frac{d}{dx}\left(p\frac{d\phi_2}{dx} \right) + (q + \lambda s)\,\phi_2 = 0.$$

Multiplying the first equation by ϕ_2 and the second equation by ϕ_1, and subtracting, we obtain

$$\phi_1 \frac{d}{dx}\left(p\frac{d\phi_2}{dx} \right) - \phi_2 \frac{d}{dx}\left(p\frac{d\phi_1}{dx} \right) = 0.$$

Integrating this equation from a to x, we obtain

$$p(x)\left[\phi_1(x)\,\phi_2'(x) - \phi_1'(x)\,\phi_2(x)\right] = p(a)\left[\phi_1(a)\,\phi_2'(a) - \phi_1'(a)\,\phi_2(a)\right]$$

$$p(x)\, W(x; \phi, \phi_2) = \text{constant}. \tag{8.2.16}$$

This is called *Abel's formula* where W is the Wronskian.

Theorem 8.2.7. *An eigenfunction of a regular Sturm–Liouville system is unique except for a constant factor.*

Proof. Let $\phi_1(x)$ and $\phi_2(x)$ be eigenfunctions corresponding to an eigenvalue λ. Then, according to Abel's formula (8.2.16), we have

$$p(x)\,W(x;\phi_1,\phi_2) = \text{constant}, \quad p(x) > 0,$$

where W is the Wronskian. Thus, if W vanishes at a point in $[a, b]$, it must vanish for all $x \in [a, b]$.

Since ϕ_1 and ϕ_2 satisfy the end condition at $x = a$, we have

$$a_1\phi_1(a) + a_2\phi_1'(a) = 0,$$
$$a_1\phi_2(a) + a_2\phi_2'(a) = 0.$$

Since a_1 and a_2 are not both zero, we have

$$\begin{vmatrix} \phi_1(a) & \phi_1'(a) \\ \phi_1(a) & \phi_2'(a) \end{vmatrix} = W(a;\phi_1,\phi_2) = 0.$$

Therefore, $W(x;\phi_1\phi_2) = 0$ for all $x \in [a, b]$, which is a sufficient condition for the linear dependence of two functions ϕ_1 and ϕ_2. Hence, $\phi_1(x)$ differs from $\phi_2(x)$ only by a constant factor.

Theorem 8.2.5 states that all eigenvalues of a regular Sturm–Liouville system are real, but it does not guarantee that any eigenvalue exists. However, it can be proved that a self-adjoint regular Sturm–Liouville system has a denumerably infinite number of eigenvalues. To illustrate this, we consider the following example.

Example 8.2.1. Consider the Sturm–Liouville system

$$y'' + \lambda y = 0, \qquad 0 \le x \le 1,$$
$$y(0) = 0, \quad y(1) + hy'(1) = 0, \quad h > 0 \text{ a constant.}$$

Here $p = 1$, $q = 0$, $s = 1$. The solution of the Sturm–Liouville equation is

$$y(x) = A\cos\sqrt{\lambda}\,x + B\sin\sqrt{\lambda}\,x.$$

Since $y(0) = 0$, gives $A = 0$, we have

$$y(x) = B\sin\sqrt{\lambda}\,x.$$

Applying the second end condition, we have

$$\sin\sqrt{\lambda} + h\sqrt{\lambda}\cos\sqrt{\lambda} = 0, \qquad B \ne 0$$

which can be rewritten as

$$\tan \sqrt{\lambda} = -h\sqrt{\lambda}.$$

If $\alpha = \sqrt{\lambda}$ is introduced in this equation, we have

$$\tan \alpha = -h\,\alpha.$$

This equation does not possess an explicit solution. Thus, we determine the solution graphically by plotting the functions $\xi = \tan \alpha$ and $\xi = -h\alpha$ against α, as shown in Figure 8.2.1. The roots are given by the intersection of two curves, and as is evident from the graph, there are infinitely many roots α_n for $n = 1, 2, 3, \ldots$. To each root α_n, there corresponds an eigenvalue

$$\lambda_n = \alpha_n^2, \qquad n = 1, 2, 3, \ldots.$$

Thus, there exists an ordered sequence of eigenvalues

$$\lambda_0 < \lambda_1 < \lambda_2 < \lambda_3 < \ldots$$

with

$$\lim_{n \to \infty} \lambda_n = \infty.$$

The corresponding eigenfunctions are $\sin \left(\sqrt{\lambda_n}\, x \right)$.

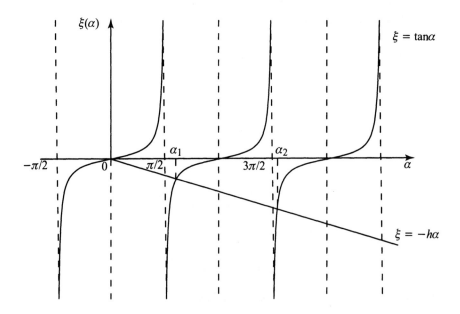

Figure 8.2.1 Intersection of $\xi = \tan \alpha$ and $\xi = -h\,\alpha$.

Theorem 8.2.8. *A self-adjoint regular Sturm–Liouville system has an infinite sequence of real eigenvalues*

$$\lambda_1 < \lambda_2 < \lambda_3 < \ldots$$

with

$$\lim_{n \to \infty} \lambda_n = \infty.$$

For each n the corresponding eigenfunction $\phi_n(x)$, uniquely determined up to a constant factor, has exactly n zeros in the interval (a, b).

Proof of this theorem can be found in the book by Myint-U (1978).

8.3 Eigenfunction Expansions

A real-valued function $\phi(x)$ is said to be *square-integrable* with respect to a weight function $\rho(x) > 0$, if, on an interval I,

$$\int_I \phi^2(x) \rho(x) \, dx < +\infty. \tag{8.3.1}$$

An immediate consequence of this definition is the *Schwarz inequality*

$$\left| \int_I \phi(x) \psi(x) \rho(x) \, dx \right|^2 \le \int_I \phi^2(x) \rho(x) \, dx \int_I \psi^2(x) \rho(x) \, dx \tag{8.3.2}$$

for square-integrable functions $\phi(x)$ and $\psi(x)$.

Let $\{\phi_n(x)\}$, for positive integers n, be an orthogonal set of square-integrable functions with a positive weight function $\rho(x)$ on an interval I. Let $f(x)$ be a given function that can be represented by a uniformly convergent series of the form

$$f(x) = \sum_{n=1}^{\infty} c_n \phi_n(x), \tag{8.3.3}$$

where the coefficients c_n are constants. Now multiplying both sides of (8.3.3) by $\phi_m(x) \rho(x)$ and integrating term by term over the interval I (uniform convergence of the series is a sufficient condition for this), we obtain

$$\int_I f(x) \phi_m(x) \rho(x) \, dx = \sum_{n=1}^{\infty} \int_I c_n \phi_n(x) \phi_m(x) \rho(x) \, dx,$$

and hence, for $n = m$,

$$\int_I f(x) \phi_n(x) \rho(x) \, dx = c_n \int_I \phi_n^2(x) \rho(x) \, dx.$$

Thus,

$$c_n = \frac{\int_I f \, \phi_n \, \rho \, dx}{\int_I \phi_n^2 \, \rho \, dx}. \tag{8.3.4}$$

Hence, we have the following theorem:

Theorem 8.3.1. *If f is represented by a uniformly convergent series*

$$f(x) = \sum_{n=1}^{\infty} c_n \phi_n(x)$$

on an interval I, where ϕ_n are square-integrable functions orthogonal with respect to a positive weight function $\rho(x)$, then c_n are determined by

$$c_n = \frac{\int_I f \, \phi_n \, \rho \, dx}{\int_I \phi_n^2 \, \rho \, dx}.$$

Example 8.3.1. The Legendre polynomials $P_n(x)$ are orthogonal with respect to the weight function $\rho(x) = 1$ on $(-1, 1)$. If we assume that $f(x)$ can be represented by the *Fourier–Legendre series*

$$f(x) = \sum_{n=1}^{\infty} c_n P_n(x)$$

then, c_n are given by

$$c_n = \frac{\int_{-1}^{1} f(x) \, P_n(x) \, dx}{\int_{-1}^{1} P_n^2(x) \, dx}$$

$$= \left(\frac{2n+1}{2}\right) \int_{-1}^{1} f(x) \, P_n(x) \, dx.$$

In the above discussion, we assumed that the given function $f(x)$ is represented by a uniformly convergent series. This is rather restrictive, and we will show in the following section that $f(x)$ can be represented by a mean-square convergent series.

8.4 Convergence in the Mean

Let $\{\phi_n\}$ be the set of square-integrable functions orthogonal with respect to a weight function $\rho(x)$ on $[a, b]$. Let

$$s_n(x) = \sum_{k=1}^{n} c_k \phi_k(x)$$

be the nth partial sum of the series $\sum\limits_{k=1}^{\infty} c_k \phi_k(x)$.

Let f be a square-integrable function. The sequence $\{s_n\}$ is said to *converge in the mean* to $f(x)$ on the interval I with respect to the weight function $\rho(x)$ if

$$\lim_{n \to +\infty} \int_I [f(x) - s_n(x)]^2 \rho(x)\, dx = 0. \tag{8.4.1}$$

We shall now seek the coefficients c_k such that $s_n(x)$ represents the best approximation to $f(x)$ in the sense of least squares, that is, we seek to minimize the integral

$$E(c_k) = \int_I [f(x) - s_n(x)]^2 \rho(x)\, dx$$
$$= \int_I f^2 \rho\, dx - 2 \sum_{k=1}^{n} c_k \int_I f \phi_k \rho\, dx + \sum_{k=1}^{n} c_k^2 \int_I \phi_k^2 \rho\, dx. \tag{8.4.2}$$

This is an extremal problem. A necessary condition on the c_k for E to be minimum is that the first partial derivatives of E with respect to these coefficients vanish. Thus, differentiating (8.4.2) with respect to c_k, we obtain

$$\frac{\partial E}{\partial c_k} = -2 \int_I f \phi_k \rho\, dx + 2c_k \int_I \phi_k^2 \rho\, dx = 0 \tag{8.4.3}$$

and hence,

$$c_k = \frac{\int_I f \phi_k \rho\, dx}{\int_I \phi_k^2 \rho\, dx}. \tag{8.4.4}$$

Now if we complete the square, the right side of (8.4.2) becomes

$$E = \int_I f^2 \rho\, dx + \sum_{k=1}^{n} \int_I \phi_k^2 \rho\, dx \left[c_k - \frac{\int_I f \phi_k \rho\, dx}{\int_I \phi_k^2 \rho\, dx} \right]^2 - \sum_{k=1}^{n} \frac{\left(\int_I f \phi_k \rho\, dx \right)^2}{\int_I \phi_k^2 \rho\, dx}.$$

The right side shows that E is a minimum if and only if c_k is given by (8.4.4). Therefore, this choice of c_k yields the best approximation to $f(x)$ in the sense of least squares.

For series convergent in the mean to $f(x)$, we conventionally write

$$f(x) \sim \sum_{k=1}^{\infty} c_k \phi_k(x),$$

where the coefficients c_k are the *generalized Fourier coefficients* and the series is the *generalized Fourier series*. This series may or may not be pointwise or uniformly convergent.

8.5 Completeness and Parseval's Equality

Substituting the Fourier coefficients (8.4.4) into (8.4.2), we obtain

$$\int_I \left(f(x) - \sum_{k=1}^n c_k \phi_k(x) \right)^2 \rho(x) \, dx = \int_I f^2 \rho \, dx - \sum_{k=1}^n c_k^2 \int_I \phi_k^2 \, \rho \, dx.$$

Since the left side is nonnegative, we have

$$\sum_{k=1}^n c_k^2 \int_I \phi_k^2 \, \rho \, dx \leq \int_I f^2 \rho \, dx. \tag{8.5.1}$$

The integral on the right side is finite, and hence, the series on the left side is bounded above for any n. Thus, as $n \to \infty$, the inequality (8.5.1) may be written as

$$\sum_{k=1}^\infty c_k^2 \int_I \phi_k^2 \, \rho \, dx \leq \int_I f^2 \rho \, dx. \tag{8.5.2}$$

This is called *Bessel's inequality*.

If the series converges in the mean to $f(x)$, that is,

$$\lim_{n \to \infty} \int_I \left(f(x) - \sum_{k=1}^n c_k \phi_k(x) \right)^2 \rho(x) \, dx = 0,$$

then, it follows from the above derivation that

$$\sum_{k=1}^\infty c_k^2 \int_I \phi_k^2 \rho \, dx = \int_I f^2 \rho \, dx$$

which is called *Parseval's equality*. Sometimes it is known as the *completeness relation*. Thus, when every continuous square-integrable function $f(x)$ can be expanded into an infinite series

$$f(x) = \sum_{k=1}^\infty c_k \phi_k(x),$$

the sequence of continuous square-integrable functions $\{\phi_k\}$ orthogonal with respect to the weight function ρ is said to be *complete*.

Next we state the following theorem:

Theorem 8.5.1. *The eigenfunctions of any regular Sturm–Liouville system are complete in the space of functions that are piecewise continuous on the interval $[a, b]$ with respect to the weight function $s(x)$. Moreover, any piecewise smooth function on $[a, b]$ that satisfies the end conditions of* *the regular Sturm–Liouville system can be expanded in an absolutely and uniformly convergent series*

$$f(x) = \sum_{k=1}^{\infty} c_k \phi_k(x),$$

where c_k are given by

$$c_k = \int_a^b f \, \phi_k \, s(x) \, dx \bigg/ \int_a^b \phi_k^2 \, s(x) \, dx.$$

Proof of a more general theorem can be found in Coddington and Levinson (1955).

Example 8.5.1. Consider a cylindrical wire of length l whose surface is perfectly insulated against the flow of heat. The end $l = 0$ is maintained at the zero degree temperature, while the other end radiates freely into the surrounding medium of zero degree temperature. Let the initial temperature distribution in the wire be $f(x)$. Find the temperature distribution $u(x, t)$.

The initial boundary-value problem is

$$u_t = k \, u_{xx}, \qquad 0 < x < l, \quad t > 0, \qquad\qquad (8.5.3)$$
$$u(x, 0) = f(x), \qquad 0 < x \le l, \qquad\qquad\qquad\quad (8.5.4)$$
$$u(0, t) = 0, \qquad\qquad\qquad\qquad t > 0, \qquad\qquad\qquad (8.5.5)$$
$$h u(l, t) + u'(l, t) = 0, \qquad\qquad\quad t > 0, \qquad h > 0. \; (8.5.6)$$

By the method of separation of variables, we assume a nontrivial solution in the form

$$u(x, t) = X(x) T(t)$$

and substituting it in the heat equation, we obtain

$$X'' + \lambda X = 0, \qquad T' + k\lambda T = 0,$$

where $\lambda > 0$ is a separation constant. The solution of the latter equation is

$$T(t) = C e^{-k\lambda t} \qquad\qquad\qquad\qquad\qquad\qquad (8.5.7)$$

where C is an arbitrary constant. The former equation has to be solved subject to the boundary conditions

$$X(0) = 0, \qquad h X(l) + X'(l) = 0.$$

This is a Sturm–Liouville system which gives the solution with $X(0) = 0$

$$X(x) = B \sin \sqrt{\lambda} \, x, \qquad\qquad\qquad\qquad\qquad (8.5.8)$$

where B is a constant to be determined.

Application of the second end condition (8.5.6) yields

$$h \sin \sqrt{\lambda} l + \sqrt{\lambda} \cos \sqrt{\lambda} l = 0 \quad \text{for} \quad B \neq 0$$

which can be rewritten as

$$\tan \sqrt{\lambda} l = -\sqrt{\lambda}/h.$$

If $\alpha = \sqrt{\lambda} l$ is introduced in the preceding equation, we have

$$\tan \alpha = -a\,\alpha,$$

where $a = (1/hl)$. As in Example 8.2.1, there exists a sequence of eigenvalues

$$\lambda_1 < \lambda_2 < \lambda_3 < \cdots$$

with $\lim_{n \to \infty} \lambda_n = \infty$. The corresponding eigenfunctions are $\sin \sqrt{\lambda_n}\, x$, and hence,

$$X_n(x) = B_n \sin \sqrt{\lambda_n}\, x. \tag{8.5.9}$$

Therefore, combining (8.5.7) with $C = C_n$ and (8.5.9), the solution takes the form

$$u_n(x,t) = a_n\, e^{-k\lambda_n t} \sin \sqrt{\lambda_n}\, x, \qquad a_n = B_n C_n$$

which satisfies the heat equation and the boundary conditions. Since the heat equation is linear and homogeneous, we form the series solution

$$u(x,t) = \sum_{n=1}^{\infty} a_n\, e^{-k\lambda_n t} \sin \sqrt{\lambda_n}\, x, \tag{8.5.10}$$

which is also a solution, provided it converges and is twice differentiable with respect to x and once differentiable with respect to t. According to Theorem 8.2.1, the eigenfunctions $\sin \sqrt{\lambda_n}\, x$ form an orthogonal system over the interval $(0, l)$. Application of the initial condition yields

$$u(x,0) = f(x) \sim \sum_{n=1}^{\infty} a_n \sin \sqrt{\lambda_n}\, x.$$

If we assume that f is a piecewise smooth function on $[a, b]$, then, by Theorem 8.5.1, we can expand $f(x)$ in terms of the eigenfunctions, and formally write

$$f(x) = \sum_{n=1}^{\infty} a_n \sin \sqrt{\lambda_n}\, x,$$

where the coefficient a_n is given by

$$a_n = \int_0^l f(x) \sin \sqrt{\lambda_n}\, x \, dx \bigg/ \int_0^l \sin^2 \sqrt{\lambda_n}\, x \, dx.$$

With this value of a_n, the temperature distribution is given by (8.5.10).

8.6 Bessel's Equation and Bessel's Function

Bessel's equation frequently occurs in problems of applied mathematics and mathematical physics involving cylindrical symmetry.

The standard form of Bessel's equation is given by

$$x^2 y'' + x y' + \left(x^2 - \nu^2\right) y = 0, \tag{8.6.1}$$

where ν is a nonnegative real number. We shall first restrict our attention to $x > 0$. Since $x = 0$ is the regular singular point, a solution is taken in accordance with the Frobenius method to be

$$y(x) = \sum_{n=0}^{\infty} a_n x^{s+n}, \tag{8.6.2}$$

where the index s is to be determined. Substitution of this series into equation (8.6.1) then yields

$$\left(s^2 - \nu^2\right) a_0 x^s + \left[(s+1)^2 - \nu^2\right] a_1 x^{s+1}$$

$$+ \sum_{n=2}^{\infty} \left\{ \left[(s+n)^2 - \nu^2\right] a_n + a_{n-2} \right\} x^{s+n} = 0. \tag{8.6.3}$$

The requirement that the coefficient of x^s vanish leads to the initial equation

$$\left(s^2 - \nu^2\right) a_0 = 0, \tag{8.6.4}$$

from which it follows that $s = \pm\nu$ for arbitrary $a_0 \neq 0$. Since the leading term in the series (8.6.2) is $a_0 x^s$, it is clear that for $\nu > 0$ the solution of Bessel's equation corresponding to the choice $s = \nu$ vanishes at the origin, whereas the solution corresponding to $s = -\nu$ is infinite at that point.

We consider first the regular solution of Bessel's equation, that is, the solution corresponding to the choice $s = \nu$. The vanishing of the coefficient of x^{s+1} in equation (8.6.3) requires that

$$(2\nu + 1) a_1 = 0, \tag{8.6.5}$$

which in turn implies that $a_1 = 0$ (since $\nu \geq 0$). From the requirement that the coefficient of x^{s+n} in equation (8.6.3) be zero, we obtain the two-term recurrence relation

$$a_n = -\frac{a_{n-2}}{n(2\nu + n)}. \tag{8.6.6}$$

Since $a_1 = 0$, it is obvious that $a_n = 0$ for $n = 3, 5, 7, \ldots$. The remaining coefficients are given by

$$a_{2k} = \frac{(-1)^k a_0}{2^{2k} k! (\nu + k)(\nu + k - 1)\ldots(\nu + 1)} \tag{8.6.7}$$

for $k = 1, 2, 3, \ldots$. This relation may also be written as

$$a_{2k} = \frac{(-1)^k 2^\nu \Gamma(\nu + 1) a_0}{2^{2k+\nu} k! \Gamma(\nu + k + 1)}, \qquad k = 1, 2, \ldots, \tag{8.6.8}$$

where $\Gamma(\alpha)$ is the gamma function, whose properties are described in the Appendix.

Hence, the regular solution of Bessel's equation takes the form

$$y(x) = a_0 \sum_{k=0}^{\infty} \frac{(-1)^k 2^\nu \Gamma(\nu + 1)}{2^{2k+\nu} k! \Gamma(\nu + k + 1)} x^{2k+\nu}. \tag{8.6.9}$$

It is customary to choose

$$a_0 = \frac{1}{2^\nu \Gamma(\nu + 1)} \tag{8.6.10}$$

and to denote the corresponding solution by $J_\nu(x)$. This solution, called the *Bessel function of the first kind of order* ν, is therefore given by

$$J_\nu(x) = \sum_{k=0}^{\infty} \frac{(-1)^k x^{2k+\nu}}{2^{2k+\nu} k! \, \Gamma(\nu + k + 1)}. \tag{8.6.11}$$

To determine the irregular solution of the Bessel equation for $s = -\nu$, we proceed as above. In this way, we obtain as the analogue of equation (8.6.5) the relation

$$(-2\nu + 1) a_1 = 0$$

from which it follows, without loss of generality, that $a_1 = 0$. Using the recurrence relation

$$a_n = -\frac{a_{n-2}}{n(n - 2\nu)}, \qquad n \geq 2 \tag{8.6.12}$$

we obtain the irregular solution of the Bessel function of the first kind of order $-\nu$ as

$$J_{-\nu}(x) = \sum_{k=0}^{\infty} \frac{(-1)^k x^{2k-\nu}}{2^{2k-\nu} k! \, \Gamma(-\nu + k + 1)}. \tag{8.6.13}$$

It can be easily proved that, if ν is not an integer, J_ν and $J_{-\nu}$ converge for all values of x, and are linearly independent. Thus, the general solution of the Bessel equation for nonintegral ν is

$$y(x) = c_1 J_\nu(x) + c_2 J_{-\nu}(x). \tag{8.6.14}$$

If ν is an integer, say $\nu = n$, then from equation (8.6.13), noting that, when gamma functions in the coefficients of the first n terms become infinite, the coefficients become zero, hence we have

$$J_{-n}(x) = \sum_{k=n}^{\infty} \frac{(-1)^k x^{2k-n}}{2^{2k-n} k! \, \Gamma(-n+k+1)}.$$

$$= (-1)^n \sum_{k=0}^{\infty} \frac{(-1)^k x^{2k+n}}{2^{2k+n} k! \, \Gamma(n+k+1)}.$$

$$= (-1)^n J_n(x). \tag{8.6.15}$$

This shows that J_{-n} is not independent of J_n, and therefore, a second linearly independent solution is required.

A number of distinct irregular solutions are discussed in the literature, but the one most commonly used, as defined by Watson (1966), is

$$Y_\nu(x) = \frac{(\cos \nu\pi) J_\nu(x) - J_{-\nu}(x)}{\sin \nu\pi}. \tag{8.6.16}$$

For nonintegral ν, it is obvious that $Y_\nu(x)$, being a linear combination of $J_\nu(x)$ and $J_{-\nu}(x)$, is linearly independent of $J_\nu(x)$. When ν is a nonnegative integer n, $Y_\nu(x)$ is indeterminate. But

$$Y_n(x) = \lim_{\nu \to n} Y_\nu(x)$$

exists and is a solution of the Bessel equation. Moreover, it is linearly independent of $J_n(x)$. (For an extended treatment, see Watson (1966)). The function $Y_\nu(x)$ is called *the Bessel function of the second kind of order ν*. Thus, the general solution of the Bessel equation is

$$y(x) = c_1 J_\nu(x) + c_2 Y_\nu(x), \quad \text{for} \quad \nu \geq 0. \tag{8.6.17}$$

Like elementary functions, the Bessel functions are tabulated (see Jahnke et al. (1960)). For illustration, the functions J_0, J_1, Y_0 and Y_1 are plotted for small values of x in Figure 8.6.1.

It should be noted that $J_\nu(x)$ for $\nu \geq 0$ and $J_{-\nu}(x)$ for a positive integer ν are finite at the origin, but $J_{-\nu}(x)$ for nonintegral ν and $Y_\nu(x)$ for $\nu \geq 0$ approach infinity as x tends to zero.

Some of the useful recurrence relations are

$$J_{\nu-1}(x) + J_{\nu+1}(x) = \frac{2\nu}{x} J_\nu(x), \tag{8.6.18}$$

$$\nu J_\nu(x) + x J_\nu'(x) = x J_{\nu-1}(x), \tag{8.6.19}$$

$$J_{\nu-1}(x) - J_{\nu+1}(x) = 2 J_\nu'(x), \tag{8.6.20}$$

$$\nu J_\nu(x) - x J_\nu'(x) = x J_{\nu+1}(x), \tag{8.6.21}$$

$$\frac{d}{dx}\left[x^\nu J_\nu(x)\right] = x^\nu J_{\nu-1}(x), \tag{8.6.22}$$

$$\frac{d}{dx}\left[x^{-\nu} J_\nu(x)\right] = -x^{-\nu} J_{\nu+1}(x). \tag{8.6.23}$$

All of these relations also hold true for $Y_\nu(x)$.

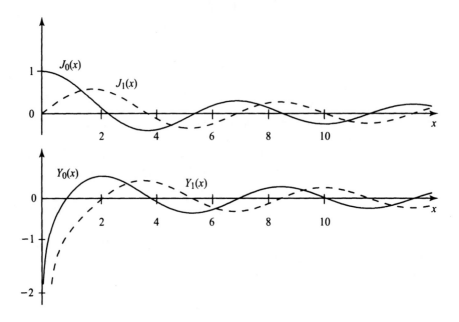

Figure 8.6.1 Graphs of $J_\nu(x)$ and $Y_\nu(x)$.

For $|x| \gg 1$ and $|x| \gg \nu$, the asymptotic expansion of $J_\nu(x)$ is

$$
J_{\nu-1}(x) \sim \sqrt{\frac{2}{\pi x}} \left[\left\{ 1 - \frac{\left(4\nu^2 - 1^2\right)\left(4\nu^2 - 3^2\right)}{2!\,(8x)^2} \right. \right.
$$
$$
\left. + \frac{\left(4\nu^2 - 1^2\right)\left(4\nu^2 - 3^2\right)\left(4\nu^2 - 5^2\right)\left(4\nu^2 - 7^2\right)}{4!\,(8x)^4} - \ldots \right\} \cos\phi
$$
$$
\left. - \left\{ \frac{\left(4\nu^2 - 1^2\right)}{8x} - \frac{\left(4\nu^2 - 1^2\right)\left(4\nu^2 - 3^2\right)\left(4\nu^2 - 5^2\right)}{3!\,(8x)^3} + \ldots \right\} \sin\phi \right]
$$
$$
(8.6.24)
$$

where

$$
\phi = x - \left(\nu + \frac{1}{2}\right)\frac{\pi}{2}.
$$

For $|x| \gg 1$ and $|x| \gg \nu$, the asymptotic expansion of $Y_\nu(x)$ is

$$Y_\nu(x) \sim \sqrt{\frac{2}{\pi x}} \left[\left\{ 1 - \frac{\left(4\nu^2 - 1^2\right)\left(4\nu^2 - 3^2\right)}{2!\,(8x)^2} \right. \right.$$

$$+ \frac{\left(4\nu^2 - 1^2\right)\left(4\nu^2 - 3^2\right)\left(4\nu^2 - 5^2\right)\left(4\nu^2 - 7^2\right)}{4!\,(8x)^4} - \cdots \left. \right\} \sin\phi$$

$$+ \left\{ \frac{\left(4\nu^2 - 1^2\right)}{8x} - \frac{\left(4\nu^2 - 1^2\right)\left(4\nu^2 - 3^2\right)\left(4\nu^2 - 5^2\right)}{3!\,(8x)^3} + \cdots \right\} \cos\phi \left. \right].$$

$$(8.6.25)$$

When $\nu = \pm(1/2)$, Bessel's function may be expressed in the form

$$J_{\frac{1}{2}}(x) = \sqrt{\frac{2}{\pi x}} \sin x, \qquad (8.6.26)$$

$$J_{-\frac{1}{2}}(x) = \sqrt{\frac{2}{\pi x}} \cos x. \qquad (8.6.27)$$

The Bessel functions which satisfy the condition

$$J_\nu(ak_m) + h J_\nu'(ak_m) = 0, \quad h,\ a = \text{constant}, \qquad (8.6.28)$$

are orthogonal to each other with respect to the weight function x, that is, for the nonnegative integer ν, the orthogonal relation is

$$\int_0^a x J_\nu(xk_n) J_\nu(xk_m)\,dx = 0, \quad n \neq m. \qquad (8.6.29)$$

When $n = m$, we have the norm

$$\|J_\nu(xk_m)\|^2 = \int_0^a x\left[J_\nu(xk_m)\right]^2 dx$$

$$= \frac{1}{2k_m^2}\left\{ a^2 k_m^2 \left[J_\nu'(ak_m)\right]^2 + \left(a^2 k_m^2 - \nu^2\right)\left[J_\nu(ak_m)\right]^2 \right\},$$

$$(8.6.30)$$

where k_m are the roots of (8.6.28).

We now give a particular example of the eigenfunction expansion theorem discussed in Sections 8.4 and 8.5. Assume a formal expansion of the function $f(x)$ defined in $0 \leq x \leq a$ in the form

$$f(x) = \sum_{m=1}^{\infty} a_m J_\nu(xk_m), \qquad (8.6.31)$$

where the summation is taken over all the positive roots $k_1, k_2, k_3, \ldots$, of equation (8.6.28). Multiplying (8.6.31) by $x J_\nu(xk_n)$, integrating, and utilizing (8.6.30), we obtain

$$\int_0^a x f(x) J_\nu(xk_m)\, dx = a_m \int_0^a x \left[J_\nu(xk_m) \right]^2 dx$$

$$= \frac{a_m}{2k_m^2} \left\{ a^2 k_m^2 \left[J_\nu'(ak_m) \right]^2 \right.$$

$$\left. + \left(a^2 k_m^2 - \nu^2 \right) \left[J_\nu(ak_m) \right]^2 \right\}. \quad (8.6.32)$$

Thus, we have the following theorem:

Theorem 8.6.1. *If*

$$b_m = \int_0^a x f(x) J_\nu(xk_m)\, dx \qquad (8.6.33)$$

then the expansion (8.6.31) of $f(x)$ takes the form

$$f(x) = \sum_{m=1}^\infty \frac{2k_m^2 b_m J_\nu(xk_m)}{k_m^2 \left[J_\nu'(ak_m) \right]^2 + \left(a^2 k_m^2 - \nu^2 \right) \left[J_\nu(ak_m) \right]^2}. \qquad (8.6.34)$$

In particular, when $h = 0$ in (8.6.28), that is, when k_m are the positive roots of $J_\nu(ak_m) = 0$, then (8.6.34) becomes

$$f(x) = \frac{2}{a^2} \sum_{m=1}^\infty \frac{b_m J_\nu(xk_m)}{k_m^2 \left[J_\nu'(ak_m) \right]^2} = \frac{2}{a^2} \sum_{m=1}^\infty \frac{b_m J_\nu(xk_m)}{\left[J_{\nu+1}(ak_m) \right]^2}. \qquad (8.6.35)$$

These expansions are known as the Bessel–Fourier series for $f(x)$. They are generated by Sturm–Liouville problems involving the Bessel equation, and arise from problems associated with partial differential equations.

Closely related to Bessel's functions are *Hankel's functions of the first and second kind*, defined by

$$H_\nu^{(1)}(x) = J_\nu(x) + i Y_\nu(x), \qquad H_\nu^{(2)}(x) = J_\nu(x) - i Y_\nu(x), \quad (8.6.36)$$

respectively, where $i = \sqrt{-1}$.

Other closely related functions are the modified Bessel functions. Consider Bessel's equation containing a parameter λ, namely,

$$x^2 y'' + x y' + \left(\lambda^2 x^2 - \nu^2 \right) y = 0. \qquad (8.6.37)$$

The general solution of this equation is

$$y(x) = c_1 J_\nu(\lambda x) + c_2 Y_\nu(\lambda x).$$

If $\lambda = i$, then

$$y(x) = c_1 J_\nu(ix) + c_2 Y_\nu(ix).$$

We write

$$J_\nu (ix) = \sum_{k=0}^{\infty} \frac{(-1)^k (ix)^{2k+\nu}}{2^{2k+\nu} k! \, \Gamma (\nu + k + 1)}$$

$$= i^\nu I_\nu (x),$$

where

$$I_\nu (x) = \sum_{k=0}^{\infty} \frac{x^{2k+\nu}}{2^{2k+\nu} k! \, \Gamma (\nu + k + 1)}, \tag{8.6.38}$$

$I_\nu (x)$ is called *the modified Bessel function of the first kind of order ν.* As in the case of J_ν and $J_{-\nu}$, I_ν and $I_{-\nu}$ (which is defined in a similar manner) are linearly independent solutions except when ν is an integer. Consequently, we define *the modified Bessel function of the second kind of order ν* by

$$K_\nu (x) = \frac{\pi}{2} \left(\frac{I_{-\nu} (x) - I_\nu (x)}{\sin \nu \pi} \right). \tag{8.6.39}$$

Thus, we obtain the general solution of the modified Bessel equation

$$x^2 y'' + x y' - \left(x^2 + \nu^2 \right) y = 0 \tag{8.6.40}$$

in the form

$$y (x) = c_1 I_\nu (x) + c_2 K_\nu (x). \tag{8.6.41}$$

We should note that

$$I_\nu (0) = \begin{cases} 1, & \nu = 0 \\ 0, & \nu > 0 \end{cases} \tag{8.6.42}$$

and that K_ν approaches infinity as $x \to 0$.

For a detailed treatment of Bessel and related functions, refer to Watson's (1966) *Theory of Bessel Functions.*

The eigenvalue problems which involve Bessel's functions will be described in Section 8.8 on singular Sturm–Liouville systems.

8.7 Adjoint Forms and Lagrange Identity

Self-adjoint equations play a very important role in many areas of applied mathematics and mathematical physics. Here we will give a brief account of self-adjoint operators and the Lagrange identity.

We consider the equation

$$L [y] = a_0 (x) y'' + a_1 (x) y' + a_2 (x) y = 0$$

defined on an interval I. Integrating $z(x) L[y]$ by parts from a to x, we have

$$\int_a^x zL[y]\,dx = \left[(za_0)\,y' - (za_0)'\,y + (za_1)\,y\right]_a^x$$
$$+ \int_a^x \left[(za_0)'' - (za_1)' + (za_2)\right] y\,dx. \quad (8.7.1)$$

Now, if we define the second-order operator L^* by

$$L^*[z] = (za_0)'' - (za_1)' + (za_2) = a_0\,z'' \,(2a_0' - a_1)\,z' + (a_0'' - a_1' + a_2)\,z$$

the relation (8.7.1) takes the form

$$\int_a^x (zL[y] - yL^*[z])\,dx = \left[a_0\,(y'z - yz') + (a_1 - a_0')\,yz\right]_a^x. \quad (8.7.2)$$

The operator L^* is called the *adjoint operator* corresponding to the operator L. It can be readily verified that the adjoint of L^* is L itself. If L and L^* are the same, L is said to be *self-adjoint*. The necessary and sufficient condition for this is that

$$a_1 = 2a_0' - a_1, \qquad a_2 = a_0'' - a_1' + a_2,$$

which is satisfied if

$$a_1 = a_0'.$$

Thus, if L is self-adjoint, we have

$$L(y) = a_0 y'' + a_0' y' + a_2 y$$
$$= (a_0 y')' + a_2(x)\,y. \quad (8.7.3)$$

In general, $L[y]$ is not self-adjoint. But if we let

$$h(x) = \frac{1}{a_0} \exp\left\{ \int^x \frac{a_1(t)}{a_0(t)}\,dt \right\} \quad (8.7.4)$$

then $h(x) L[y]$ is self-adjoint. Thus, any second-order linear differential equation

$$a_0(x)\,y'' + a_1(x)\,y' + a_2(x)\,y = 0 \quad (8.7.5)$$

can be made self-adjoint. Multiplying by $h(x)$ given by equation (8.7.4), equation (8.7.5) is transformed into the self-adjoint form

$$\frac{d}{dx}\left[p(x)\,\frac{dy}{dx}\right] + q(x)\,y = 0, \quad (8.7.6)$$

where

$$p(x) = \exp\left\{\int^x \frac{a_1(t)}{a_0(t)} dt\right\}, \qquad q(x) = \left(\frac{a_2}{a_0}\right) \exp\left\{\int^x \frac{a_1(t)}{a_0(t)} dt\right\}. \quad (8.7.7)$$

For example, the self-adjoint form of the Legendre equation

$$\left(1 - x^2\right) y'' - 2xy' + n(n+1) y = 0$$

is

$$\frac{d}{dx}\left[\left(1 - x^2\right)\frac{dy}{dx}\right] + n(n+1) y = 0, \qquad (8.7.8)$$

and the self-adjoint form of the Bessel equation

$$x^2 y'' + xy' + \left(x^2 - \nu^2\right) y = 0$$

is

$$\frac{d}{dx}\left(x\frac{dy}{dx}\right) + \left(x - \frac{\nu^2}{x}\right) y = 0. \qquad (8.7.9)$$

Now, if we differentiate both sides of equation (8.7.2), we obtain

$$z L[y] - y L^*[z] = \frac{d}{dx}\left[a_0\left(y'z - yz'\right) + \left(a_1 - a_0'\right) yz\right] \qquad (8.7.10)$$

which is known as the *Lagrange identity* for the operator L.

If we consider the integral from a to b of equation (8.7.2), we obtain *Green's identity*

$$\int_a^b \left(z L[y] - y L^*[z]\right) dx = \left[a_0\left(y'z - yz'\right) + \left(a_1 - a_0'\right) yz\right]_a^b. \qquad (8.7.11)$$

When L is self-adjoint, this relation becomes

$$\int_a^b \left(z L[y] - y L[z]\right) dx = \left[a_0\left(y'z - yz'\right)\right]_a^b. \qquad (8.7.12)$$

8.8 Singular Sturm–Liouville Systems

A Sturm–Liouville equation is called *singular* when it is given on a semi-infinite or infinite interval, or when the coefficient $p(x)$ or $s(x)$ vanishes, or when one of the coefficients becomes infinite at one end or both ends of a finite interval. A singular Sturm–Liouville equation together with appropriate linear homogeneous end conditions is called a *singular Sturm–Liouville (SSL) system*. The conditions imposed in this case are not like the separated boundary end conditions in the regular Sturm–Liouville system. The condition that is often necessary to prescribe is the boundedness of the function $y(x)$ at the singular end point. To exhibit this, let us consider a problem with a singularity at the end point $x = a$. By the relation (8.7.12), for any twice continuously differentiable functions $y(x)$ and $z(x)$, we have on (a, b)

$$\int_{a+\varepsilon}^{b} \{zL[y] - yL[z]\} \, dx = p(b)[y'(b)z(b) - y(b)z'(b)]$$
$$-p(a+\varepsilon)[y'(a+\varepsilon)z(a+\varepsilon) - y(a+\varepsilon)z'(a+\varepsilon)],$$

where ε is a small positive number. If the conditions

$$\lim_{x \to a+} p(x)[y'(x)z(x) - y(x)z'(x)] = 0, \qquad (8.8.1)$$
$$p(b)[y'(b)z(b) - y(b)z'(b)] = 0, \qquad (8.8.2)$$

are imposed on y and z, it follows that

$$\int_{a}^{b} \{zL[y] - yL[z]\} \, dx = 0. \qquad (8.8.3)$$

For example, when $p(a) = 0$, the relations (8.8.1) and (8.8.2) are replaced by the conditions
1. $y(x)$ and $y'(x)$ are finite as $x \to a$
2. $b_1 y(b) + b_2 y'(b) = 0$.

Thus, we say that the singular Sturm–Liouville system is *self-adjoint*, if any functions $y(x)$ and $z(x)$ that satisfy the end conditions satisfy

$$\int_{a}^{b} \{zL[y] - yL[z]\} \, dx = 0.$$

Example 8.8.1. Consider the singular Sturm–Liouville system involving Legendre's equation

$$\frac{d}{dx}\left[(1 - x^2)\frac{dy}{dx}\right] + \lambda y = 0, \quad -1 < x < 1,$$

with the conditions that y and y' are finite as $x \to \pm 1$.

In this case, $p(x) = 1 - x^2$ and $s(x) = 1$, and $p(x)$ vanishes at $x = \pm 1$. The Legendre functions of the first kind, $P_n(x)$, $n = 0, 1, 2, \ldots$, are the eigenfunctions which are finite as $x \to \pm 1$. The corresponding eigenvalues are $\lambda_n = n(n+1)$ for $n = 0, 1, 2, \ldots$. We observe here that the singular Sturm–Liouville system has infinitely many real eigenvalues, and the eigenfunctions $P_n(x)$ are orthogonal to each other.

Example 8.8.2. Another example of a singular Sturm–Liouville system is the Bessel equation for fixed ν

$$\frac{d}{dx}\left(x\frac{dy}{dx}\right) + \left(\lambda x - \frac{\nu^2}{x}\right)y = 0, \qquad 0 < x < a,$$

with the end conditions that $y(a) = 0$ and y and y' are finite as $x \to 0+$.

Here $p(x) = x$, $q(x) = -\frac{\nu^2}{x}$, and $s(x) = x$. Now $p(0) = 0$, $q(x)$ becomes infinite as $x \to 0+$, and $s(0) = 0$; therefore, the system is singular. If $\lambda = k^2$, the eigenfunctions of the system are the Bessel functions of the first kind of order ν, namely $J_\nu(k_n x)$, $n = 1, 2, 3, \ldots$, where $k_n a$ is the nth zero of J_ν. The Bessel function J_ν and its derivative are both finite as $x \to 0+$. The eigenvalues are $\lambda_n = k_n^2$. Thus, the system has infinitely many eigenvalues, and the eigenfunctions are orthogonal to each other with respect to the weight function $s(x) = x$.

In the preceding examples, we have seen that the eigenfunctions are orthogonal with respect to the weight function $s(x)$. In general, the eigenfunctions of a singular Sturm–Liouville system are orthogonal if they are square-integrable with respect to the weight function $s(x) = x$.

Theorem 8.8.1. *The square-integrable eigenfunctions corresponding to distinct eigenvalues of a singular Sturm–Liouville system are orthogonal with respect to the weight function $s(x)$.*

Proof. Proceeding as in Theorem 8.2.1, we arrive at

$$(\lambda_j - \lambda_k)\int_a^b s\,\phi_j\,\phi_k\,dx = p(b)\left[\phi_j(b)\,\phi_k'(b) - \phi_j'(b)\,\phi_k(b)\right]$$
$$- p(a)\left[\phi_j(a)\,\phi_k'(a) - \phi_j'(a)\,\phi_k(a)\right].$$

Suppose the boundary term vanishes, as in the case mentioned earlier, where $p(a) = 0$, y and y' are finite as $x \to a$, and at the other end $b_1 y(b) + b_2 y'(b) = 0$. Then, we have

$$(\lambda_j - \lambda_k)\int_a^b s\,\phi_j\,\phi_k\,dx = 0.$$

This integral exists by virtue of (8.3.2). Thus, for distinct eigenvalues $\lambda_j \neq \lambda_k$, the square-integrable functions ϕ_j and ϕ_k are orthogonal with respect to the weight function $s(x)$.

Example 8.8.3. Consider the singular Sturm–Liouville system involving the *Hermite equation*

$$u'' - 2xu' + \lambda u = 0, \qquad -\infty < x < \infty, \qquad (8.8.4)$$

which is not self-adjoint.

If we let $y(x) = e^{-x^2/2}u(x)$, the Hermite equation takes the self-adjoint form

$$y'' + \left[(1 - x^2) + \lambda\right]y = 0, \qquad -\infty < x < \infty.$$

Here $p = 1$, $q(x) = 1 - x^2$, $s = 1$. The eigenvalues are $\lambda_n = 2n$ for nonnegative integers n, and the corresponding eigenfunctions are $\phi_n(x) = e^{-x^2/2} H_n(x)$, where $H_n(x)$ are the Hermite polynomials which are solutions of the Hermite equation (8.8.4) (See Magnus and Oberhettinger (1949)).

Now, we impose the end conditions that y tends to zero as $x \to \pm\infty$. This is satisfied because $H_n(x)$ are polynomials in x and in fact $x^n e^{-x^2/2} \to 0$ as $x \to \pm\infty$. Since $\phi_n(x)$ are square-integrable, we have

$$\int_{-\infty}^{\infty} H_m(x) H_n(x) e^{-x^2} dx = 0, \qquad m \neq n.$$

Example 8.8.4. Consider the problem of the transverse vibration of a thin elastic circular membrane

$$u_{tt} = c^2 \left(u_{rr} + \frac{1}{r} u_r \right), \qquad r < 1, \quad t > 0,$$

$$u(r,0) = f(r), \qquad u_t(r,0) = 0, \qquad 0 \le r \le 1, \qquad (8.8.5)$$

$$u(1,t) = 0, \qquad \lim_{r \to 0} u(r,t) < \infty, \qquad t \ge 0.$$

We seek a nontrivial separable solution in the form

$$u(r,t) = R(r) T(t).$$

Substituting this in the wave equation yields

$$\frac{R'' + (1/r) R'}{R} = \frac{1}{c^2} \frac{T''}{T} = -\alpha^2,$$

where α is a positive constant. The negative sign in front of α^2 is chosen to obtain the solution periodic in time. Thus, we have

$$rR'' + R' + \alpha^2 rR = 0, \qquad T'' + \alpha^2 c^2 T = 0.$$

The solution $T(t)$ is therefore given by

$$T(t) = A\cos(\alpha ct) + B\sin(\alpha ct).$$

Next, it is required to determine the solution $R(r)$ of the following singular Sturm–Liouville system

$$\frac{d}{dr} \left[r \frac{dR}{dr} \right] + \alpha^2 rR = 0, \qquad (8.8.6)$$

$$R(1) = 0, \qquad \lim_{r \to 0} R(r) < \infty. \qquad (8.8.7)$$

We note that in this case, $p = r$ which vanishes at $r = 0$. The condition on the boundedness of the function $R(r)$ is obtained from the fact that

$$\lim_{r \to 0} u(r,t) = \lim_{r \to 0} R(r) T(t) < \infty$$

which implies that

$$\lim_{r \to 0} R(r) < \infty \qquad (8.8.8)$$

for arbitrary $T(t)$. Equation (8.8.6) is Bessel's equation of order zero, the solution of which is given by

$$R(r) = C J_0(\alpha r) + D Y_0(\alpha r), \qquad (8.8.9)$$

where J_0 and Y_0 are Bessel's functions of the first and second kinds respectively of order zero. The condition (8.8.7) requires that $D = 0$ since $Y_0(\alpha r) \to -\infty$ as $r \to 0$. Hence,

$$R(r) = C J_0(\alpha r).$$

The remaining condition $R(1) = 0$ yields $J_0(\alpha) = 0$.

This transcendental equation has infinitely many positive zeros

$$\alpha_1 < \alpha_2 < \alpha_3 < \dots.$$

Thus, the solution of problem (8.8.5) is given by

$$u_n(r,t) = J_0(\alpha_n r)(A_n \cos \alpha_n ct + B_n \sin \alpha_n ct), \qquad n = 1, 2, 3, \dots.$$

Since the Bessel equation is linear and homogeneous, the linear superposition principle gives

$$u(r,t) = \sum_{n=1}^{\infty} J_0(\alpha_n r)(A_n \cos \alpha_n ct + B_n \sin \alpha_n ct), \qquad (8.8.10)$$

is also a solution, provided the series converges and is sufficiently differentiable with respect to r and t. Differentiating (8.8.10) formally with respect to t, we obtain

$$u_t(r,t) = \sum_{n=1}^{\infty} J_0(\alpha_n r)(-A_n \alpha_n c \sin \alpha_n ct + B_n \alpha_n c \cos \alpha_n ct).$$

Application of the initial condition $u_t(r,0) = 0$ yields $B_n = 0$. Consequently, we have

$$u(r,t) = \sum_{n=1}^{\infty} A_n J_0(\alpha_n r) \cos(\alpha_n ct). \qquad (8.8.11)$$

It now remains to show that $u(r,t)$ satisfies the initial condition $u(r,0) = f(r)$. For this, we have

$$u(r,0) = f(r) \sim \sum_{n=1}^{\infty} A_n J_0(\alpha_n r).$$

If $f(r)$ is piecewise smooth on $[0, 1]$, then the eigenfunctions $J_0(\alpha_n r)$ form a complete orthogonal system with respect to the weight function r over the interval $(0, 1)$. Hence, we can formally expand $f(r)$ in terms of the eigenfunctions. Thus,

$$f(r) = \sum_{n=1}^{\infty} A_n J_0(\alpha_n r), \qquad (8.8.12)$$

where the coefficient A_n is represented by

$$A_n = \int_0^1 rf(r) J_0(\alpha_n r)\, dr \Big/ \int_0^1 r\, [J_0(\alpha_n r)]^2\, dr. \qquad (8.8.13)$$

The solution of the problem (8.8.5) is therefore given by (8.8.11) with the coefficients A_n given by (8.8.13).

8.9 Legendre's Equation and Legendre's Function

The Legendre equation is

$$(1 - x^2)\, y'' - 2xy' + \nu(\nu + 1)\, y = 0, \qquad (8.9.1)$$

where ν is a real number. This equation arises in problems with spherical symmetry in mathematical physics. Its coefficients are analytic at $x = 0$. Thus, if we expand near the point $x = 0$, the coefficients are

$$p(x) = -\frac{2x}{1 - x^2} = -2x \sum_{m=0}^{\infty} x^{2m} = \sum_{m=0}^{\infty} (-2)\, x^{2m+1},$$

and

$$q(x) = \frac{\nu(\nu + 1)}{1 - x^2} = \nu(\nu + 1) \sum_{m=0}^{\infty} x^{2m} = \sum_{m=0}^{\infty} \nu(\nu + 1)\, x^{2m}.$$

We see that these series converge for $|x| < 1$. Thus, the Legendre equation on $|x| < 1$ has convergent power series solution at $x = 0$.

Now to find the solution near the ordinary point $x = 0$, we assume

$$y(x) = \sum_{m=0}^{\infty} a_m x^m.$$

Substituting y, y', and y'' in the Legendre equation, we obtain

$$(1 - x^2) \sum_{m=0}^{\infty} m(m-1)\, a_m x^{m-2} - 2x \sum_{m=0}^{\infty} m a_m x^{m-1}$$

$$+ \nu(\nu + 1) \sum_{m=0}^{\infty} a_m x^m = 0.$$

Simplification gives

$$\sum_{m=0}^{\infty} \left[(m+1)(m+2) a_{m+2} + (\nu - m)(\nu + m + 1) a_m\right] x^m = 0.$$

Therefore the coefficients in the power series must satisfy the recurrence relation

$$a_{m+2} = -\frac{(\nu - m)(\nu + m + 1)}{(m+1)(m+2)} a_m, \qquad m \geq 0. \qquad (8.9.2)$$

This relation determines $a_2, a_4, a_6, \ldots$ in terms of a_0, and $a_3, a_5, a_7, \ldots$ in terms of a_1. It can easily be verified that a_{2k} and a_{2k+1} can be expressed in terms of a_0 and a_1 respectively as

$$a_{2k} = \frac{(-1)^k \nu (\nu - 2) \ldots (\nu - 2k + 2)(\nu + 1)(\nu + 3) \ldots (\nu + 2k - 1)}{(2k)!} a_0$$

and

$$a_{2k+1} = \frac{(-1)^k (\nu - 1)(\nu - 3) \ldots (\nu - 2k + 1)(\nu + 2)(\nu + 4) \ldots (\nu + 2k)}{(2k+1)!} a_1.$$

Hence, the solution of the Legendre equation is

$$y(x) = a_0 \left[1 \right.$$
$$\left. + \sum_{k=1}^{\infty} \frac{(-1)^k \nu (\nu - 2) \ldots (\nu - 2k + 2)(\nu + 1)(\nu + 3) \ldots (\nu + 2k - 1) x^{2k}}{(2k)!} \right]$$
$$+ a_1 \left[x \right.$$
$$\left. + \sum_{k=1}^{\infty} \frac{(-1)^k (\nu - 1)(\nu - 3) \ldots (\nu - 2k + 1)(\nu + 2)(\nu + 4) \ldots (\nu + 2k) x^{2k+1}}{(2k+1)!} \right]$$
$$= a_0 \phi_\nu(x) + a_1 \psi_\nu(x). \qquad (8.9.3)$$

It can easily be proved that the functions $\phi_\nu(x)$ and $\psi_\nu(x)$ converge for $x < 1$ and are linearly independent.

Now consider the case in which $\nu = n$, with n a nonnegative integer. It is then evident from the recurrence relation (8.9.2) that, when $m = n$,

$$a_{n+2} = a_{n+4} = \ldots = 0.$$

Consequently, when n is even, the series $\phi_n(x)$ terminates with x^n, whereas the series for $\psi_n(x)$ does not terminate. When n is odd, it is the series for $\psi_n(x)$ which terminates with x^n, while that for $\phi_n(x)$ does not terminate.

In the first case (n even), $\phi_n(x)$ is a polynomial of degree n; the same is true for $\psi_n(x)$ in the second case (n odd).

Thus, for any nonnegative integer n, either $\phi_n(x)$ or $\psi_n(x)$, but not both, is a polynomial of degree n. Consequently, the general solution of the Legendre equation contains a polynomial solution $P_n(x)$ and an infinite series solution $Q_n(x)$ for a nonnegative integer n. To find the polynomial solution $P_n(x)$, it is convenient to choose a_n so that $P_n(1) = 1$. Let this a_n be

$$a_n = \frac{(2n)!}{2^n (n!)^2}. \tag{8.9.4}$$

Rewriting the recurrence relation (8.9.2), we have

$$a_{n-2} = -\frac{(n-1)n}{2(2n-1)} a_n.$$

Substituting a_n from (8.9.4) into this relation, we obtain

$$a_{n-2} = -\frac{(2n-2)!}{2^n (n-1)!(n-2)!},$$

and

$$a_{n-4} = \frac{(2n-4)!}{2^n 2! (n-2)!(n-4)!}.$$

It follows by induction that

$$a_{n-2k} = \frac{(-1)^k (2n-2k)!}{2^n k! (n-k)!(n-2k)!}.$$

Hence, we may write $P_n(x)$ in the form

$$P_n(x) = \sum_{k=0}^{N} \frac{(-1)^k (2n-2k)!}{2^n k! (n-k)!(n-2k)!} x^{n-2k}, \tag{8.9.5}$$

where $N = (n/2)$ when n is even, and $N = (n-1)/2$ when n is odd. This polynomial $P_n(x)$ is called the *Legendre function of the first kind of order* n. It is also known as the *Legendre polynomial of degree* n.

The first few Legendre polynomials are

$$P_0(x) = 1,$$
$$P_1(x) = x,$$
$$P_2(x) = \frac{1}{2}\left(3x^2 - 1\right),$$
$$P_3(x) = \frac{1}{2}\left(5x^3 - 3x\right),$$
$$P_4(x) = \frac{1}{8}\left(35x^4 - 30x^2 + 3\right).$$

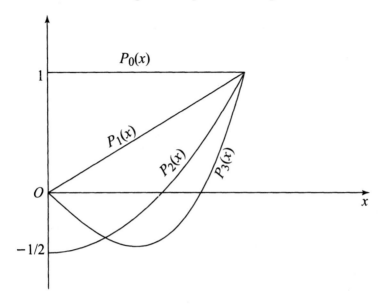

Figure 8.9.1 The first four Legendre's polynomials.

These polynomials are plotted in Figure 8.9.1 for small values of x.

Recall that for a given nonnegative integer n, only one of the two solutions $\phi_n(x)$ and $\psi_n(x)$ of Legendre's equation is a polynomial, while the other in an infinite series. This infinite series, when appropriately normalized, is called the *Legendre function of the second kind*. It is defined for $|x| < 1$ by

$$Q_n(x) = \begin{cases} \phi_n(1)\,\psi_n(x) & \text{for } n \text{ even} \\ -\psi_n(1)\,\phi_n(x) & \text{for } n \text{ odd} \end{cases}. \tag{8.9.6}$$

Thus, when n is a nonnegative integer, the general solution of the Legendre equation is given by

$$y(x) = c_1 P_n(x) + c_2 Q_n(x). \tag{8.9.7}$$

The Legendre polynomial may also be expressed in the form

$$P_n(x) = \frac{1}{2^n n!} \frac{d^n}{dx^n} (x^2 - 1)^n. \tag{8.9.8}$$

This expression is known as the *Rodriguez formula*.

Like Bessel's functions, Legendre polynomials satisfy certain recurrence relations. Some of the important relations are

$$(n + 1) P_{n+1} (x) - (2n + 1) x P_n (x) + n P_{n-1} (x) = 0, \qquad n \geq 1, \qquad (8.9.9)$$
$$(x^2 - 1) P_n' (x) = n x P_n (x) - n P_{n-1} (x), \qquad n \geq 1, \qquad (8.9.10)$$
$$n P_n (x) + P_{n-1}' (x) - x P_n' (x) = 0, \qquad n \geq 1, \qquad (8.9.11)$$
$$P_{n+1}' (x) = x P_n' (x) + (n + 1) P_n (x), \qquad n \geq 0. \qquad (8.9.12)$$

In addition,

$$P_{2n} (-x) = P_{2n} (x), \qquad (8.9.13)$$
$$P_{2n+1} (-x) = -P_{2n+1} (x). \qquad (8.9.14)$$

These indicate that $P_n (x)$ is an even function for even n, and an odd function for odd n.

It can easily be shown that the Legendre polynomials form a sequence of orthogonal functions on the interval $[-1, 1]$. Thus, we have

$$\int_{-1}^{1} P_n (x) P_m (x) \, dx = 0, \qquad \text{for} \quad n \neq m. \qquad (8.9.15)$$

The norm of the function $P_n (x)$ is given by

$$\| P_n (x) \|^2 = \int_{-1}^{1} P_n^2 (x) \, dx = \frac{2}{2n + 1}. \qquad (8.9.16)$$

Another important equation in mathematical physics, one which is closely related to the Legendre equation (8.9.1), is *Legendre's associated equation*:

$$(1 - x^2) y'' - 2xy' + \left[n (n + 1) - \frac{m^2}{1 - x^2} \right] y = 0, \qquad (8.9.17)$$

where m is an integer. Although this equation is independent of the algebraic sign of the integer m, it is often convenient to have the solutions for negative m differ somewhat from those for positive m.

We consider first the case for a nonnegative integer m. Introducing the change of variable

$$y = (1 - x^2)^{m/2} u, \qquad |x| < 1,$$

Legendre's associated equation becomes

$$(1 - x^2) u'' - 2 (m + 1) xu' + (n - m) (n + m + 1) u = 0.$$

But this is the same as the equation obtained by differentiating the Legendre equation (8.9.1) m times. Thus, the general solution of (8.9.17) is given by

$$y (x) = (1 - x^2)^{m/2} \frac{d^m Y (x)}{dx^m}, \qquad (8.9.18)$$

and

$$Y(x) = c_1 P_n(x) + c_2 Q_n(x) \tag{8.9.19}$$

is the general solution of (8.9.1). Hence, we have the linearly independent solutions of (8.9.17), known as the *associated Legendre functions of the first and second kind*, respectively given by

$$P_n^m(x) = \left(1 - x^2\right)^{m/2} \frac{d^m P_n(x)}{dx^m}, \tag{8.9.20}$$

and

$$Q_n^m(x) = \left(1 - x^2\right)^{m/2} \frac{d^m Q_n(x)}{dx^m}. \tag{8.9.21}$$

We observe that

$$P_n^0(x) = P_n(x), \qquad Q_n^0(x) = Q_n(x),$$

and that $P_n^m(x)$ vanishes for $m > n$.

The functions $P_n^{-m}(x)$ and $Q_n^{-m}(x)$ are defined by

$$P_n^{-m}(x) = (-1)^m \frac{(n-m)!}{(n+m)!} P_n^m(x), \qquad m = 0, 1, 2, \ldots, n, \tag{8.9.22}$$

$$Q_n^{-m}(x) = (-1)^m \frac{(n-m)!}{(n+m)!} Q_n^m(x), \qquad m = 0, 1, 2, \ldots, n. \tag{8.9.23}$$

The first few associated Legendre functions are

$$P_1^1(x) = \left(1 - x^2\right)^{\frac{1}{2}},$$
$$P_2^1(x) = 3x \left(1 - x^2\right)^{\frac{1}{2}},$$
$$P_2^2(x) = 3 \left(1 - x^2\right).$$

The associated Legendre functions of the first kind also form a sequence of orthogonal functions in the interval $[-1, 1]$. Their orthogonality, as well as their norm, is expressed by the equation

$$\int_{-1}^{1} P_n^m(x) P_k^m(x)\, dx = \frac{2(n-m)!}{(2n+1)(n+m)!} \delta_{nk}. \tag{8.9.24}$$

Note that (8.9.15) and (8.9.16) are special cases of (8.9.24), corresponding to the choice $m = 0$.

We finally observe that $P_n^m(x)$ is bounded everywhere in the interval $[-1, 1]$, whereas $Q_n^m(x)$ is unbounded at the end points $x = \pm 1$.

Problems in which Legendre's polynomials arise will be treated in Chapter 10.

8.10 Boundary-Value Problems Involving Ordinary Differential Equations

A *boundary-value problem* consists in finding an unknown solution which satisfies an ordinary differential equation and appropriate boundary conditions at two or more points. This is in contrast to an initial-value problem for which a unique solution exists for an equation satisfying prescribed initial conditions at one point.

The linear two-point boundary-value problem, in general, may be written in the form

$$L[y] = f(x), \qquad a < x < b,$$

$$U_i[y] = \alpha_i, \qquad 1 \leq i \leq n,$$

(8.10.1)

where L is a linear operator of order n and U_i is the boundary operator defined by

$$U_i[y] = \sum_{j=1}^{n} a_{ij} \, y^{(j-1)}(a) + \sum_{j=1}^{n} b_{ij} \, y^{(j-1)}(b). \qquad (8.10.2)$$

Here a_{ij}, b_{ij}, and α_i are constants. The treatment of this problem can be found in Coddington and Levinson (1955). More complicated boundary conditions occur in practice. Treating a general differential system is rather complicated and difficult.

A large class of boundary-value problems that occur often in the physical sciences consists of the second-order equations of the type

$$y'' = f(x, y, y'), \qquad a < x < b$$

with the boundary conditions

$$U_1[y] = a_1 y(a) + a_2 y'(a) = \alpha$$
$$U_2[y] = b_1 y(b) + b_2 y'(b) = \beta$$

where a_1, a_2, b_1, b_2, α, and β are constants. The existence and uniqueness of solutions to this problem are treated by Keller (1968). Here we are interested in considering a special case where the linear boundary-value problem consists of the differential equation

$$L[y] = y'' + p(x) y' + q(x) y = f(x) \qquad (8.10.3)$$

and the boundary conditions

$$U_1[y] = a_1 y(a) + a_2 y'(a) = \alpha,$$
$$U_2[y] = b_1 y(b) + b_2 y'(b) = \beta, \qquad (8.10.4)$$

where the constants a_1 and a_2, and likewise b_1 and b_2, are not both zero, and α, and β are constants.

In general, a boundary-value problem may not possess a solution, and if it does, the solution may not be unique. We illustrate this with a simple problem.

Example 8.10.1. We first consider the boundary-value problem

$$y'' + y = 1,$$
$$y(0) = 0, \quad y\left(\frac{\pi}{2}\right) = 0.$$

By the method of variation of parameters, we find a unique solution

$$y(x) = 1 - \cos x - \sin x.$$

We observe that the solution of the associated homogeneous boundary-value problem

$$y'' + y = 0,$$
$$y(0) = 0, \quad y\left(\frac{\pi}{2}\right) = 0,$$

is trivial.

Next we consider the boundary-value problem

$$y'' + y = 1,$$
$$y(0) = 0, \quad y(\pi) = 0.$$

The general solution is

$$y(x) = c_1 \cos x + c_2 \sin x + 1.$$

Applying the boundary conditions, we see that

$$y(0) = c_1 + 1 = 0, \quad y(\pi) = -c_1 + 1 = 0.$$

This is not possible, and hence, the boundary-value problem has no solution. However, if we consider its associated homogeneous boundary-value problem

$$y'' + y = 0,$$
$$y(0) = 0, \quad y(\pi) = 0,$$

we can easily determine that solutions exist and are given by

$$y(x) = c_2 \sin x,$$

where c_2 is an arbitrary constant.

This leads to the following alternative theorem:

Theorem 8.10.1. *Let $p(x)$, $q(x)$, and $f(x)$ be continuous on $[a, b]$. Then either the boundary-value problem*

$$L[y] = f$$

$$(8.10.5)$$

$$U_1[y] = \alpha, \quad U_2[y] = \beta,$$

has a unique solution for any given constants α and β, or else the associated homogeneous boundary-value problem

$$L[y] = 0$$

$$(8.10.6)$$

$$U_1[y] = 0, \quad U_2[y] = 0,$$

has a nontrivial solution.

Proof of this theorem can be found in Myint-U (1978).

8.11 Green's Functions for Ordinary Differential Equations

In the present section, we will introduce Green's functions. Let us consider the linear homogeneous ordinary differential equation of second order:

$$L[y] = -f(x), \quad a < x < b, \quad (8.11.1)$$

where

$$L \equiv \frac{d}{dr}\left[p(x)\frac{d}{dx}\right] + q(x),$$

with the homogeneous boundary conditions

$$a_1 y(a) + a_2 y'(a) = 0, \quad (8.11.2)$$
$$b_1 y(b) + b_2 y'(b) = 0, \quad (8.11.3)$$

where the constants a_1 and a_2, and likewise b_1 and b_2, are not both zero. We shall assume that f and q are continuous and that p is continuously differentiable and does not vanish in the interval $[a, b]$.

According to the theory of ordinary differential equations, the general solution of (8.11.1) is given by

$$y(x) = A y_1(x) + B y_2(x) + y_p(x) \quad (8.11.4)$$

where A and B are arbitrary constants, $y_1 = y_1(x)$ and $y_2 = y_2(x)$ are two linearly independent solutions of the corresponding homogeneous equation $L[y] = 0$, and $y_p(x)$ is a particular integral of (8.11.1).

Using the method of variation of parameters, the particular integral will be sought by replacing the constants A and B by arbitrary functions of x, $u_1(x)$ and $u_2(x)$, to get

$$y_p(x) = u_1(x) y_1(x) + u_2(x) y_2(x). \qquad (8.11.5)$$

These arbitrary functions are to be determined so that (8.11.5) satisfies equation (8.11.1). The substitution of the above trial form for $y_p(x)$ into (8.11.1) imposes one condition that (8.11.5) must satisfy. However, there are two arbitrary functions, and hence, two conditions are needed to determine them. If follows that another condition is available in solving the problem.

The first task is to substitute the trial solution into equation (8.11.1). Differentiating $y_p(x)$, we obtain

$$y_p' = u_1 y_1' + u_2 y_2' + u_1' y_1 + u_2' y_2.$$

It is convenient to set the second condition noted above to require that

$$u_1' y_1 + u_2' y_2 = 0, \qquad (8.11.6)$$

leaving

$$y_p' = u_1 y_1' + u_2 y_2'. \qquad (8.11.7)$$

A second differentiation of y_p gives

$$y_p'' = u_1 y_1'' + u_2 y_2'' + u_1' y_1' + u_2' y_2'.$$

Putting these results into equation (8.11.1) and grouping yields

$$u_1 [p(x) y_1'' + p' y_1' + q y_1] + u_2 [p(x) y_2'' + p' y_2' + q y_2]$$
$$+ p(x) \{u_1' y_1' + u_2' y_2'\} = -f(x).$$

Since $y_1(x)$ and $y_2(x)$ are solutions of the homogeneous equations, both the square brackets in the above expression vanish. The result is that

$$u_1' y_1' + u_2' y_2' = -\frac{f(x)}{p(x)}. \qquad (8.11.8)$$

Thus, two equations ((8.11.6) and (8.11.8)) determine $u_1(x)$ and $u_2(x)$. Solving these equations algebraically for u_1' and u_2' produces

$$u_1'(x) = \frac{f(x) y_2(x)}{p(x) W(x)}, \quad u_2'(x) = -\frac{f(x) y_1(x)}{p(x) W(x)},$$

where $W(x) = (y_1 y_2' - y_2 y_1')$ is the non-zero Wronskian of the solutions y_1 and y_2. Integration of these results yields

$$u_1(x) = + \int \frac{f(x) y_2(x) \, dx}{p(x) W(x)}, \quad u_2(x) = - \int \frac{f(x) y_1(x) \, dx}{p(x) W(x)}.$$

The substitution of these results into (8.11.5) gives the solution in (a, b)

$$y_p(x) = y_1(x) \int_b^x \frac{f(\xi) y_2(\xi)}{p(\xi) W(\xi)} d\xi - y_2(x) \int_a^x \frac{f(\xi) y_1(\xi)}{p(\xi) W(\xi)} d\xi$$

$$= - \int_a^x \frac{y_2(x) y_1(\xi)}{p(\xi) W(\xi)} f(\xi) \, d\xi - \int_x^b \frac{y_1(x) y_2(\xi)}{p(\xi) W(\xi)} f(\xi) \, d\xi. \quad (8.11.9)$$

This form suggests the definition

$$G(x, \xi) = \begin{cases} -\frac{y_2(x) y_1(\xi)}{p(\xi) W(\xi)}, & a \le \xi < x, \\ -\frac{y_1(x) y_2(\xi)}{p(\xi) W(\xi)}, & x < \xi \le b. \end{cases} \quad (8.11.10)$$

This is called *Green's function.* Thus, the solution becomes

$$y_p(x) = \int_a^b G(x, \xi) f(\xi) \, d\xi \quad (8.11.11)$$

provided $G(x, \xi)$ is continuous and $f(\xi)$ is at least piecewise continuous in (a, b). The existence of Green's function is evident from equations (8.11.10) provided $W(\xi) \ne 0$.

In order to obtain a deeper insight into the role of Green's function, certain properties are important to note. These are

(i) $G(x, \xi)$ is continuous at $x = \xi$, and consequently, throughout the interval (a, b). This follows from the fact that $G(x, \xi)$ is constructed from the solutions of the homogeneous equation which are continuous in the intervals $a \le \xi < x$ and $x < \xi \le b$, $W(\xi) \ne 0$.
(ii) Its first and second derivatives are continuous for all $x \ne \xi$ in $a \le x$, $\xi \le b$.
(iii) $G(x, \xi)$ is symmetric in x and ξ, that is, if x and ξ are interchanged in (8.11.10), the definition is not changed.
(iv) The first derivative of $G(x, \xi)$ has a finite discontinuity at $x = \xi$

$$\lim_{x \to \xi+} \left(\frac{\partial G}{\partial x} \right) - \lim_{x \to \xi-} \left(\frac{\partial G}{\partial x} \right) = -\frac{1}{p(\xi)}. \quad (8.11.12)$$

(v) $G(x, \xi)$ is a solution of the homogeneous equation throughout the interval except at $x = \xi$. This point can be pressed further by substitution of (8.11.11) and (8.11.1), note that $y_p(x)$ is a solution of (8.11.1). This leads to

$$\int_a^b \left[\frac{d}{dx} \left\{ p(x) \frac{dG(x,\xi)}{dx} \right\} + q(x) G(x,\xi) \right] f(\xi) \, d\xi = -f(x),$$

where the quantity in the square bracket is zero except at $x = \xi$. If follows that, in order for this result to hold,

$$\frac{d}{dx} \left[p(x) \frac{dG(x,\xi)}{dx} \right] + q(x) G(x,\xi) = -\delta(\xi - x), \quad (8.11.13)$$

where $\delta(\xi - x)$ is the Dirac function which has the following properties:

$$\delta(\xi - x) = 0, \quad \text{except at } x = \xi,$$

$$\int_a^b \delta(\xi - x) \, d\xi = 1, \quad a < x < b.$$

For any continuous function $f(\xi)$ in (a, b) and for $a \leq x \leq b$,

$$\int_a^b f(\xi) \delta(\xi - x) \, d\xi = f(x).$$

(vi) For fixed ξ, $G(x, \xi)$ satisfies the given boundary conditions (8.11.2)–(8.11.3).

Differentiating (8.11.11) with respect to x by Leibniz's rule, we obtain

$$y_p'(x) = \int_a^b G'(x,\xi) f(\xi) \, d\xi + G(x, x-) f(x)$$

$$+ \int_a^b G'(x,\xi) f(\xi) \, d\xi - G(x, x+) f(x)$$

$$= \int_a^b G'(x,\xi) f(\xi) \, d\xi,$$

since $G(x, \xi)$ is continuous in ξ, that is, $G(x, x-) = G(x, x+)$. Because $G(x, \xi)$ satisfies the boundary conditions, it follows that

$$a_1 y_p(a) + a_2 y_p'(a) = \int_a^b [a_1 G(a, \xi) + a_2 G'(a, \xi)] f(\xi) \, d\xi = 0.$$

Similarly,

$$b_1 y_p(b) + b_2 y_p'(b) = 0.$$

The result embodied in equation (8.11.13) permits a meaningful physical interpretation of the role of Green's function. If we assume that the differential equation (8.11.1) represents a vibrating mechanical system driven by a uniformly distributed force $-f(x)$ in the interval (a, b), where the system is defined, then $G(x, \xi)$ governed by (8.11.13) represents the displacement of the system at a point x resulting from a unit impulse of force at $x = \xi$. The displacement at the point x due to uniformly distributed force $f(\xi)$ per unit length over an elementary interval $(\xi, \xi + d\xi)$ is given by $f(\xi)\, G(x, \xi)\, d\xi$. Finally, the total displacement of the system at the point x results from superposition (addition) of these contributions so that the total solution over the entire interval (a, b) is given by (8.11.1), that is,

$$y_p(x) = \int_a^b G(x, \xi) f(\xi) \, d\xi.$$

Combining all above results, we state the fundamental theorem for Green's function:

Theorem 8.11.1. *If $f(x)$ is continuous on $[a, b]$, then the function*

$$y(x) = \int_a^b G(x, \xi) f(\xi) \, d\xi$$

is a solution of the boundary-value problem

$$L[y] = -f(x),$$
$$a_1 y(a) + a_2 y'(a) = 0, \qquad b_1(b) + b_2 y'(b) = 0.$$

Example 8.11.1. Consider the problem

$$y'' = -x, \qquad y(0) = 0, \qquad y(1) = 0. \tag{8.11.14}$$

For a fixed value of ξ, Green's function $G(x, \xi)$ satisfies the associated homogeneous equation

$$G'' = 0$$

in $0 < x < \xi$, $\xi < x < 1$, and the boundary conditions

$$G(0, \xi) = 0, \qquad G(1, \xi) = 0.$$

In addition, it satisfies

$$\frac{dG}{dx}(x, \xi)\Big|_{x=\xi-}^{x=\xi+} = -\frac{1}{p(\xi)}.$$

Now choose $G(x, \xi)$ such that

$$G(x, \xi) = \begin{cases} G_1(x, \xi) = (1 - \xi)x, & \text{for } 0 \le x \le \xi \\ G_2(x, \xi) = (1 - x)\xi, & \text{for } \xi \le x \le 1. \end{cases}$$

It can be seen that $G'' = 0$ over the intervals $0 < x < \xi$, $\xi < x < 1$. Also

$$G_1 (0, \xi) = 0, \qquad G_2 (1, \xi) = 0.$$

Moreover,

$$G_2' (x, \xi) - G_1' (x, \xi) = -\xi - (1 - \xi) = -1$$

which is the value of the jump $-1/p (\xi)$, because in this case $p = 1$. Hence, by Theorem 8.11.1, keeping in mind that ξ is the variable in $G (x, \xi)$, the solution of (8.11.14) is

$$
\begin{aligned}
y (x) &= \int_0^x G (x, \xi) f (\xi) \, d\xi + \int_x^1 G (x, \xi) f (\xi) \, d\xi \\
&= \int_0^x (1 - x) \, \xi^2 d\xi + \int_x^1 x (1 - \xi) \, \xi d\xi = \frac{x}{6} \left(1 - x^2\right).
\end{aligned}
$$

8.12 Construction of Green's Functions

In the above example, we see that the solution was obtained immediately as soon as Green's function was obtained properly. Thus, the real problem is not that of finding the solution but that of determining Green's function for the problem. We will now show by construction that there exists a Green's function for $L [y]$ satisfying the prescribed boundary conditions.

We first assume that the associated homogeneous equation satisfying the conditions (8.11.2) and (8.11.3) has the trivial solution only, as in Example 8.11.1. We construct the solution $y_1 (x)$ of the equation

$$L [y] = 0$$

satisfying $a_1 y (a) + a_2 y' (a) = 0$. We see that $c_1 y_1 (x)$ is the most general such solution, where c_1 is an arbitrary constant.

In a similar manner, we let $c_2 y_2 (x)$, with c_2 is an arbitrary constant, be the most general solution of

$$L [y] = 0$$

satisfying $b_1 y (b) + b_2 y' (b) = 0$. Thus, y_1 and y_2 exist in the interval (a, b) and are linearly independent. For, if they were linearly dependent, then $y_1 = c \, y_2$, which shows that y_1 would satisfy both the boundary conditions at $x = a$ and $x = b$. This contradicts our assumption about the trivial solution. Consequently, Green's function can take the form

$$G (x, \xi) = \begin{cases} c_1 (\xi) y_1 (x), & \text{for } x < \xi \\[2mm] c_2 (\xi) y_2 (x), & \text{for } x > \xi \end{cases} \qquad (8.12.1)$$

Since $G (x, \xi)$ is continuous at $x = \xi$, we have

$$c_2\left(\xi\right)y_2\left(\xi\right)-c_1\left(\xi\right)y_1\left(\xi\right)=0. \tag{8.12.2}$$

The discontinuity in the derivative of G at the point requires that

$$\left.\frac{dG}{dx}\left(x,\xi\right)\right|_{x=\xi-}^{x=\xi+}=c_2\left(\xi\right)y_2'\left(\xi\right)-c_1\left(\xi\right)y_1'\left(\xi\right)=-\frac{1}{p\left(\xi\right)}. \tag{8.12.3}$$

Solving equations (8.12.2) and (8.12.3) for c_1 and c_2, we find

$$c_1\left(\xi\right)=\frac{-y_2\left(\xi\right)}{p\left(\xi\right)W\left(y_1,y_2;\xi\right)},\quad c_2\left(\xi\right)=\frac{-y_1\left(\xi\right)}{p\left(\xi\right)W\left(y_1,y_2;\xi\right)}, \tag{8.12.4}$$

where $W\left(y_1,y_2;\xi\right)$ is the Wronskian given by $W\left(y_1,y_2;\xi\right)=y_1\left(\xi\right)y_2'\left(\xi\right)-y_2\left(\xi\right)y_1'\left(\xi\right)$. Since the two solutions are linearly independent, the Wronskian differs from zero.

From Theorem 8.2.6, with $\lambda=0$, we have

$$pW=\text{constant}=C. \tag{8.12.5}$$

Hence, Green's function is given by

$$G\left(x,\xi\right)=\begin{cases} -y_1\left(x\right)y_2\left(\xi\right)/C, & \text{for}\quad x\leq\xi \\ \\ -y_2\left(x\right)y_1\left(\xi\right)/C, & \text{for}\quad x\geq\xi. \end{cases} \tag{8.12.6}$$

Thus, we state the following theorem:

Theorem 8.12.1. *If the associated homogeneous boundary-value problem of (8.11.1)–(8.11.3) has the trivial solution only, then Green's function exists and is unique.*

Proof. The proof for uniqueness of Green's function is left as an exercise for the reader.

Example 8.12.1. Consider the problem

$$y''+y=-1,\qquad y\left(0\right)=0,\qquad y\left(\frac{\pi}{2}\right)=0. \tag{8.12.7}$$

The solution of $L\left[y\right]=\left(dy'/dx\right)+y=0$ satisfying $y\left(0\right)=0$ is

$$y_1\left(x\right)=\sin x,\qquad 0\leq x<\xi$$

and the solution of $L\left[y\right]=0$ satisfying $y\left(\pi/2\right)=0$ is

$$y_2\left(x\right)=\cos x,\qquad \xi<x\leq\frac{\pi}{2}.$$

The Wronskian of y_1 and y_2 is then given by

$$W\left(\xi\right)=y_1\left(\xi\right)y_2'\left(\xi\right)-y_2\left(\xi\right)y_1'\left(\xi\right)=-1.$$

Since in this case $p = 1$, (8.12.6) becomes

$$G(x, \xi) = \begin{cases} \sin x \cos \xi, & \text{for } x \leq \xi \\ \cos x \sin \xi, & \text{for } x \geq \xi. \end{cases}$$

Therefore, the solution of (8.12.7) is

$$y(x) = \int_0^x G(x, \xi) f(\xi) d\xi + \int_x^{\pi/2} G(x, \xi) f(\xi) d\xi$$

$$= \int_0^x \cos x \sin \xi \, d\xi + \int_x^{\pi/2} \sin x \cos \xi \, d\xi$$

$$= -1 + \sin x + \cos x.$$

It can be seen in the formula (8.12.6) that Green's function is symmetric in x and ξ.

Example 8.12.2. Construct the Green's function for the two-point boundary-value problem

$$y''(x) + \omega^2 y = f(x), \qquad y(a) = y(b) = 0.$$

This describes the forced oscillation of an elastic string with fixed ends at $x = a$ and $x = b$.

It is easy to check that $\sin \omega x$ and $\cos \omega x$ are two functions which satisfy the homogeneous equation $y'' + \omega^2 y = 0$. These are used to construct two functions $y_1(x)$ and $y_2(x)$ which satisfy the boundary conditions $y_1(a) = y_2(b) = 0$. Accordingly, $y_1(x) = A \sin \omega x + B \cos \omega x$ and $y_2(x) = C \sin \omega x + D \cos \omega x$, and the resulting functions are

$$y_1(x) = \sin \omega (x - a), \qquad y_2(x) = \sin \omega (x - b).$$

The corresponding Wronskian is $W = -\omega \sin \omega (a - b)$. Substituting these results into (8.11.10) yields

$$G(x, \xi) = \begin{cases} \frac{\sin \omega(\xi - a) \sin \omega(x - b)}{-\omega \sin \omega(a - b)}, & a \leq \xi < x \\ \frac{\sin \omega(x - a) \sin \omega(\xi - b)}{-\omega \sin \omega(a - b)}, & x \leq \xi \leq b \end{cases}$$

provided $\sin \omega(a - b) \neq 0$.

8.13 The Schrödinger Equation and Linear Harmonic Oscillator

The quantum mechanical motion of the harmonic oscillator is described by the one-dimensional Schrödinger equation

$$H\psi\left(x\right) = E\psi\left(x\right),\qquad(8.13.1)$$

where the Hamiltonian H is given by

$$H = -\left(\frac{\hbar^2}{2M}\right)\frac{d^2}{dx^2} + V\left(x\right),\qquad V\left(x\right) = \frac{1}{2}M\omega^2 x^2,\qquad(8.13.2)$$

and E is the energy, $V\left(x\right)$ is the potential, $h = 2\pi\hbar$ is the Planck constant, M is the mass of the particle, and ω is the classical frequency of the oscillator.

We solve equation (8.13.1) subject to the requirement that the solution be bounded as $|x| \to \infty$. The solution of (8.13.1) is facilitated by the first solving the equation for large x. In terms of the constants

$$\beta = \frac{2ME}{\hbar^2},\qquad \alpha = \frac{M\omega}{\hbar} > 0,$$

the equation (8.13.1) takes the form

$$\frac{d^2\psi}{dx^2} + \left(\beta - \alpha^2 x^2\right)\psi = 0.\qquad(8.13.3)$$

For small β and large x, $\beta - \alpha^2 x^2 \sim -\alpha^2 x^2$ so that the equation becomes

$$\frac{d^2\psi}{dx^2} - \alpha^2 x^2 \psi = 0.$$

As $|x| \to \infty$, $\psi\left(x\right) = x^n \exp\left(\pm\frac{\alpha x^2}{2}\right)$ satisfies (8.13.3) for a finite n so far as leading terms $\left(\sim -\alpha^2 x^2\right)$ are concerned. The positive exponential factor is unacceptable because of the boundary conditions, so the asymptotic solution $\psi\left(x\right) = x^n \exp\left(-\frac{\alpha x^2}{2}\right)$ suggests the possibility of the exact solution in the form $\psi\left(x\right) = v\left(x\right)\exp\left(-\frac{\alpha x^2}{2}\right)$ where $v\left(x\right)$ is to be determined. Substituting this result into (8.13.3), we obtain

$$\frac{d^2 v}{dx^2} - 2\alpha x\frac{dv}{dx} + \left(\beta - \alpha\right)v = 0.\qquad(8.13.4)$$

In terms of a new independent variable $\zeta = x\sqrt{\alpha}$, this equation reduces to the form

$$\frac{d^2 v}{d\zeta^2} - 2\zeta\frac{dv}{d\zeta} + \left(\frac{\beta}{\alpha} - 1\right)v = 0.\qquad(8.13.5)$$

We seek a power series solution

$$v\left(\zeta\right) = \sum_{n=0}^{\infty} a_n\zeta^n.\qquad(8.13.6)$$

Substituting this series into equation (8.13.5) and equating the coefficients of ζ^n to zero, we obtain the recurrence relation

$$a_{n+2} = \frac{(2n + 1 - \beta/\alpha)}{(n + 1)(n + 2)} a_n \qquad (8.13.7)$$

which gives

$$\frac{a_{n+2}}{a_n} \sim \frac{2}{n} \quad \text{as} \quad n \to \infty. \qquad (8.13.8)$$

This ratio is the same as that of the series for $\zeta^n \exp\left(\zeta^2\right) \left(\sim x^n e^{\alpha x^2}\right)$ with finite n. This leads to the fact that $\psi(x) = v(x) e^{-\alpha x^2/2} \sim x^n e^{\alpha x^2/2}$ which does not satisfy the basic requirement for $|x| \to \infty$. This unacceptable result can only be avoided if n is an integer and the series terminates so that it becomes a polynomial of degree n. This means that $a_{n+2} = 0$ but $a_n \neq 0$ so that

$$2n + 1 - \frac{\beta}{\alpha} = 0, \qquad (8.13.9)$$

or

$$\frac{\beta}{\alpha} = (2n + 1).$$

Substituting the values for α and β, it turns out that

$$E \equiv E_n = \left(n + \frac{1}{2}\right)\omega\hbar, \qquad n = 0, 1, 2, \ldots. \qquad (8.13.10)$$

This represents a discrete set of energies. Thus, in quantum mechanics, a stationary state of the harmonic oscillator can assume only one of the values from the set E_n. The energy is thus quantized, and forms a discrete spectrum. According to the classical theory, the energy forms a continuous spectrum, that is, all non-negative numbers are allowed for the energy of a harmonic oscillator. This shows a remarkable contrast between the results of the classical and quantum theory.

The number n which characterizes the energy eigenvalues and eigenfunctions is called the *quantum number*. The value of $n = 0$ corresponds to the minimum value of the quantum number with the energy

$$E_0 = \frac{1}{2}\omega\hbar. \qquad (8.13.11)$$

This is called the *lowest* (or *ground*) *state energy* which never vanishes as the lowest possible classical energy would. E_0 is proportional to $\hbar$, representing a quantum phenomenon. The discrete energy spectrum is in perfect agreement with the quantization rules of the quantum theory.

To determine the eigenfunctions for the harmonic oscillator associated with the eigenvalues E_n, we obtain the solution of equation (8.13.5) which has the form

$$\frac{d^2v}{d\zeta^2} - 2\zeta\frac{dv}{d\zeta} + 2nv = 0. \tag{8.13.12}$$

This is a well-known differential equation for the *Hermite polynomials* $H_n(\zeta)$ of degree n. Thus, the complete eigenfunctions can be expressed in terms of $H_n(\zeta)$ as

$$\psi_n(x) = A_n H_n(x\sqrt{\alpha}) \exp\left(-\frac{\alpha x^2}{2}\right), \tag{8.13.13}$$

where A_n are arbitrary constants.

The Hermite polynomials $H_n(x)$ are usually defined by

$$H_n(x) = (-1)^n e^{x^2} D^n\left(e^{-x^2}\right), \qquad D \equiv \frac{d}{dx}. \tag{8.13.14}$$

They form an orthogonal system in $(-\infty, \infty)$ with the weight function $\exp\left(-x^2\right)$.

The orthogonal relation for these polynomials is

$$\int_{-\infty}^{\infty} e^{-x^2} H_m(x) H_n(x)\, dx = \begin{cases} 0, & n \neq m \\ 2^n n! \sqrt{\pi}, & n = m. \end{cases}$$

The Hermite polynomials $H_n(x)$ for $n = 0, 1, 2, 3, 4$ are

$$H_0(x) = 1$$
$$H_1(x) = 2x$$
$$H_2(x) = -2 + 4x^2$$
$$H_3(x) = -12x + 8x^3$$
$$H_4(x) = 12 - 48x^2 + 16x^4.$$

Finally, the eigenfunctions ψ_n of the linear harmonic oscillator for the quantum number $n = 0, 1, 2, 3$ are given in Figure 8.13.1.

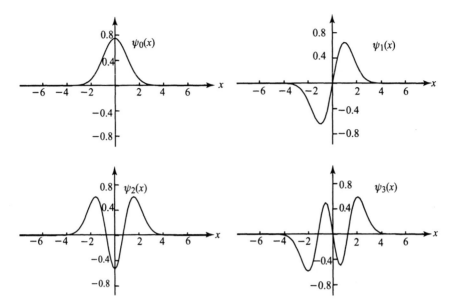

Figure 8.13.1 Eigenfunctions ψ_n for $n = 0, 1, 2, 3$.

8.14 Exercises

1. Determine the eigenvalues and eigenfunctions of the following regular Sturm–Liouville systems:

(a) $y'' + \lambda y = 0$,

$\quad y(0) = 0, \ y(\pi) = 0$.

(b) $y'' + \lambda y = 0$,

$\quad y(0) = 0, \ y'(1) = 0$.

(c) $y'' + \lambda y = 0$,

$\quad y'(0) = 0, \ y'(\pi) = 0$.

(d) $y'' + \lambda y = 0$,

$\quad y(1) = 0, \ y(0) + y'(0) = 0$.

2. Find the eigenvalues and eigenfunctions of the following periodic Sturm–Liouville systems:

(a) $y'' + \lambda y = 0$,

$$y(-1) = y(-1), \ y'(-1) = y'(1).$$

(b) $y'' + \lambda y = 0$,

$$y(0) = y(2\pi), \ y'(0) = y'(2\pi).$$

(c) $y'' + \lambda y = 0$,

$$y(0) = y(\pi), \ y'(0) = y'(\pi).$$

3. Obtain the eigenvalues and eigenfunctions of the following Sturm–Liouville systems:

(a) $y'' + y' + (1 + \lambda) y = 0$,

$$y(0) = 0, \ y(1) = 0.$$

(b) $y'' + 2y' + (1 - \lambda) y = 0$,

$$y(0) = 0, \ y'(1) = 0.$$

(c) $y'' - 3y' + 3(1 + \lambda) y = 0$,

$$y'(0) = 0, \ y'(\pi) = 0.$$

4. Find the eigenvalues and eigenfunctions of the following regular Sturm–Liouville systems:

(a) $x^2 y'' + 3xy' + \lambda y = 0, \ 1 \leq x \leq e$,

$$y(1) = 0, \ y(e) = 0.$$

(b) $\frac{d}{dx}\left[(2 + x)^2 y'\right] + \lambda y = 0, \ -1 \leq x \leq 1$,

$$y(-1) = 0, \ y(1) = 0.$$

(c) $(1 + x)^2 y'' + 2(1 + x) y' + 3\lambda y = 0, \ 0 \leq x \leq 1$,

$$y(0) = 0, \ y(1) = 0.$$

5. Determine all eigenvalues and eigenfunctions of the Sturm–Liouville systems:

(a) $x^2 y'' + x y' + \lambda y = 0$,

$y(1) = 0$, y, y' are bounded at $x = 0$.

(b) $y'' + \lambda y = 0$,

$y(0) = 0$, y, y' are bounded at infinity.

6. Expand the function

$$f(x) = \sin x, \quad 0 \le x \le \pi$$

in terms of the eigenfunctions of the Sturm–Liouville problem

$$y'' + \lambda y = 0,$$
$$y(0) = 0, \; y(\pi) + y'(\pi) = 0.$$

7. Find the expansion of

$$f(x) = x, \qquad 0 \le x \le \pi$$

in a series of eigenfunctions of the Sturm–Liouville system

$$y'' + \lambda y = 0,$$
$$y'(0) = 0, \; y'(\pi) = 0.$$

8. Transform each of the following equations into the equivalent self-adjoint form:
(a) The Laguerre equation

$$x y'' + (1 - x) y' + n y = 0, \qquad n = 0, 1, 2, \dots.$$

(b) The Hermite equation

$$y'' - 2x y' + 2n y = 0, \qquad n = 0, 1, 2, \dots.$$

(c) The Tchebycheff equation

$$(1 - x^2) y'' - x y' + n^2 y = 0, \qquad n = 0, 1, 2, \dots.$$

9. If $q(x)$ and $s(x)$ are continuous and $p(x)$ is twice continuously differentiable in $[a, b]$, show that the solutions of the fourth-order Sturm–Liouville system

$$[p(x)y'']'' + [q(x) + \lambda s(x)]y = 0,$$

$$\left[a_1 y + a_2 (py'')'\right]_{x=a} = 0, \left[b_1 y + b_2 (py'')'\right]_{x=b} = 0,$$

$$[c_1 y' + c_2 (py'')]_{x=a} = 0, \ [d_1 y' + d_2 (py'')]_{x=b} = 0,$$

where $a_1^2 + a_2^2 \neq 0$, $b_1^2 + b_2^2 \neq 0$, $c_1^2 + c_2^2 \neq 0$, $d_1^2 + d_2^2 \neq 0$,
are orthogonal with respect to $s(x)$ in $[a, b]$.

10. If the eigenfunctions of the problem

$$\frac{1}{r}\frac{d}{dr}(ry') + \lambda y = 0, \ 0 < r < a,$$

$$c_1 y(a) + c_2 y'(a) = 0,$$

$$\lim_{r \to 0+} y(r) < \infty,$$

satisfy

$$\lim_{r \to 0+} ry'(r) = 0,$$

show that all the eigenvalues are real for real c_1 and c_2.

11. Find the Green's function for each of the following problems:

(a) $L[y] = y'' = 0$,

$y(0) = 0, \ y'(1) = 0.$

(b) $L[y] = (1 - x^2)y'' - 2xy' = 0$,

$y(0) = 0, \ y'(1) = 0.$

(c) $L[y] = y'' + a^2 y = 0, \ a = $ constant,

$y(0) = 0, \ y(1) = 0.$

12. Determine the solution of each of the following boundary-value problems:

(a) $y'' + y = 1$,

$y(0) = 0, \ y(1) = 0.$

(b) $y'' + 4y = e^x$,

$y(0) = 0, \ y'(1) = 0.$

(c) $y'' = \sin x$,

$y(0) = 0, \ y(1) + 2y'(1) = 0.$

(d) $y'' + 4y = -2$,

$y(0) = 0, \ y\left(\frac{\pi}{4}\right) = 0.$

(e) $y'' = -x$,

$y(0) = 2, \ y(1) + y'(1) = 4.$

(f) $y'' = -x^2$,

$y(0) + y'(0) = 4, \ y'(1) = 2.$

(g) $y'' = -x$,

$y(0) = 1, \ y'(1) = 2.$

13. Determine the solution of the following boundary-value problems:

(a) $y'' = -f(x), \quad y(0) = 0, \quad y'(1) = 0.$

(b) $y'' = -f(x), \quad y(-1) = 0, \quad y(1) = 0.$

14. Find the solution of the following boundary-value problems:

(a) $y'' - y = -f(x), \quad y(0) = y(1) = 0.$

(b) $y'' - y = -f(x), \quad y'(0) = y'(1) = 0.$

15. Show that the Green's function $G(t, \xi)$ for the forced harmonic oscillator described by initial-value problem

$$\ddot{x} + \omega^2 x = \left(\frac{F}{m}\right) \sin \Omega t,$$
$$x(0) = a, \quad \dot{x}(0) = 0,$$

is

$$G\left(t,\xi\right) = \frac{1}{\omega}\sin\omega\left(t-\xi\right).$$

Hence, the particular solution is

$$x_p\left(t\right) = \frac{F}{m\omega}\int_0^t \sin\omega\left(t-\xi\right)\sin\left(\Omega\xi\right)d\xi.$$

16. Determine the Green's function for the boundary-value problem

$$xy'' + y' = -f\left(x\right),$$

$$y\left(1\right) = 0, \quad \lim_{x\to 0}\left|y\left(x\right)\right| < \infty.$$

17. Determine the Green's function for the boundary-value problem

$$xy'' + y' - \frac{n^2}{x}y = -f\left(x\right),$$

$$y\left(1\right) = 0, \quad \lim_{x\to 0}\left|y\left(x\right)\right| < \infty.$$

18. Determine the Green's function for the boundary-value problem

$$\left[\left(1-x^2\right)y'\right]' - \frac{h^2}{\left(1-x^2\right)}y = -f\left(x\right), \quad h = 1,2,3,\ldots,$$

$$\lim_{r\to\pm 1}\left|y\left(x\right)\right| < \infty.$$

19. Prove the uniqueness of the Green's function for the boundary-value problem

$$L\left[y\right] = -f\left(x\right),$$

$$a_1 y\left(a\right) + a_2 y'\left(a\right) = 0,$$

$$b_1 y\left(b\right) + b_2 y'\left(b\right) = 0.$$

20. Find the Green's function for the boundary-value problem

$$L\left[y\right] = y^{(iv)} = -f\left(x\right),$$

$$y\left(0\right) = y\left(1\right) = y'\left(0\right) = y'\left(1\right) = 0.$$

Prove that the homogeneous problem has a trivial solution only, and prove that the nonhomogeneous problem has a unique solution.

21. Determine the Green's function for the boundary-value problem

$$y'' = -f(x), \ y(-1) = y(1), \ y'(-1) = y'(1).$$

22. Consider the nonself-adjoint boundary-value problem

$$L[y] = y'' + 3y' + 2y = -f(x),$$

$$2y(0) - y(1) = 0, \qquad y'(1) = 2.$$

By direct integration of $GL[y]$ from 0 to 1, show that

$$y(x) = -2G(1,x) - \int_0^1 G(x,\xi) f(\xi) \, d\xi$$

is the solution of the boundary-value problem, if G satisfies the system

$$G_{\xi\xi} - 3G_\xi + 2G = 0, \quad \xi \neq x,$$
$$G(0,x) = 0,$$
$$6G(1,x) - 2G_\xi(1,x) + G_\xi(0,x) = 0.$$

Find the Green's function $G(x,\xi)$.

23. Show that

$$\frac{dG(x,\xi)}{dx}\bigg|_{\xi=x-}^{\xi=x+} = \frac{1}{p(x)}$$

is equivalent to

$$\frac{dG(x,\xi)}{dx}\bigg|_{x=\xi-}^{x=\xi+} = -\frac{1}{p(\xi)}.$$

24. (a) Apply the Prüfer transformation

$$R^2 = y^2 + p^2 (y')^2, \qquad \theta = \tan^{-1}\left(\frac{y}{py'}\right)$$

to transform the Sturm–Liouville equation (8.1.3) into the first order nonlinear equation in the form

$$\frac{dR}{dx} = \frac{1}{2}r\left(\frac{1}{p} - q - \lambda r\right)\sin 2\theta, \qquad \frac{d\theta}{dx} = (q + \lambda r)\sin^2\theta + \frac{1}{p}\cos^2\theta,$$

where $a < x < b$.
(b) Draw the direction field $\left(\frac{d\theta}{dx}\right)$ in the (θ, x) plane with $p = x$, $q = -\frac{1}{x}$ and $r = x$. Hence draw the solution curves of the θ-equation for different λ with data $\theta(a) = \alpha$ and $\theta(b)$ is arbitrary.

9

Boundary-Value Problems and Applications

"The enormous usefulness of mathematics in the natural sciences is something bordering on the mysterious and there is no rational explanation for it. It is not at all natural that "laws of nature" exist, much less that man is able to discover them. The miracle of the appropriateness of the language of mathematics for the formulation of the laws of physics is a wonderful gift which we neither understand nor deserve."

Eugene Wigner

9.1 Boundary-Value Problems

In the preceding chapters, we have treated the initial-value and initial boundary-value problems. In this chapter, we shall be concerned with boundary-value problems. Mathematically, a boundary-value problem is finding a function which satisfies a given partial differential equation and particular boundary conditions. Physically speaking, the problem is independent of time, involving only space coordinates. Just as initial-value problems are associated with hyperbolic partial differential equations, boundary-value problems are associated with partial differential equations of elliptic type. In marked contrast to initial-value problems, boundary-value problems are considerably more difficult to solve. This is due to the physical requirement that solutions must hold in the large unlike the case of initial-value problems, where solutions in the small, say over a short interval of time, may still be of physical interest.

The second-order partial differential equation of the elliptic type in n independent variables x_1, x_2, ..., x_n is of the form

$$\nabla^2 u = F\left(x_1, x_2 \ldots, x_n, u_{x_1}, u_{x_2}, \ldots, u_{x_n}\right), \qquad (9.1.1)$$

where

$$\nabla^2 u = \sum_{i=1}^{n} u_{x_i x_i}.$$

Some well-known elliptic equations include
A. Laplace equation

$$\nabla^2 u = 0. \tag{9.1.2}$$

B. Poisson equation

$$\nabla^2 u = g(x), \tag{9.1.3}$$

where

$$g(x) = g(x_1, x_2, \ldots, x_n).$$

C. Helmholtz equation

$$\nabla^2 u + \lambda u = 0, \tag{9.1.4}$$

where λ is a positive constant
D. Schrödinger equation (time independent)

$$\nabla^2 u + [\lambda - q(x)] u = 0. \tag{9.1.5}$$

We shall not attempt to treat general elliptic partial differential equations. Instead, we shall begin by presenting the simplest boundary-value problems for the Laplace equation in two dimensions.

We first define a harmonic function. A function is said to be *harmonic* in a domain D if it satisfies the Laplace equation and if it and its first two derivatives are continuous in D.

We may note here that, since the Laplace equation is linear and homogeneous, a linear combination of harmonic functions is harmonic.

1. The First Boundary-Value Problem

The Dirichlet Problem: Find a function $u(x, y)$, harmonic in D, which satisfies

$$u = f(s) \quad \text{on} \quad B, \tag{9.1.6}$$

where $f(s)$ is a prescribed continuous function on the boundary B of the domain D. D is the interior of a simple closed piecewise smooth curve B.

We may physically interpret the solution u of the Dirichlet problem as the steady-state temperature distribution in a body containing no sources or sinks of heat, with the temperature prescribed at all points on the boundary.

2. The Second Boundary-Value Problem

The Neumann Problem: Find a function $u(x, y)$, harmonic in D, which satisfies

$$\frac{\partial u}{\partial n} = f(s) \qquad \text{on } B, \tag{9.1.7}$$

with

$$\int_B f(s)\, ds = 0. \tag{9.1.8}$$

The symbol $\partial u/\partial n$ denotes the directional derivative of u along the outward normal to the boundary B. The last condition (9.1.8) is known as *the compatibility condition*, since it is a consequence of (9.1.7) and the equation $\nabla^2 u = 0$. Here the solution u may be interpreted as the steady-state temperature distribution in a body containing no heat sources or heat sinks when the heat flux across the boundary is prescribed.

The compatibility condition, in this case, may be interpreted physically as the heat requirement that the net heat flux across the boundary be zero.

3. The Third Boundary-Value Problem

Find a function $u(x, y)$ harmonic in D which satisfies

$$\frac{\partial u}{\partial n} + h(s)\, u = f(s) \qquad \text{on } B, \tag{9.1.9}$$

where h and f are given continuous functions. In this problem, the solution u may be interpreted as the steady-state temperature distribution in a body, from the boundary of which the heat radiates freely into the surrounding medium of prescribed temperature.

4. The Fourth Boundary-Value Problem

The Robin Problem: Find a function $u(x, y)$, harmonic in D, which satisfies boundary conditions of different types on different portions of the boundary B. An example involving such boundary conditions is

$$u = f_1(s) \qquad \text{on } B_1, \tag{9.1.10}$$
$$\frac{\partial u}{\partial n} = f_2(s) \qquad \text{on } B_2,$$

where $B = B_1 \cup B_2$.

Problems 1 through 4 are called *interior boundary-value problems*. These differ from *exterior boundary-value problems* in two respects:

i. For problems of the latter variety, part of the boundary is at infinity.

ii. Solutions of exterior problems must satisfy an additional requirement, namely, that of boundedness at infinity.

9.2 Maximum and Minimum Principles

Before we prove the uniqueness and continuity theorems for the interior Dirichlet problem for the two-dimensional Laplace equation, we first prove the maximum and minimum principles.

Theorem 9.2.1. *(The Maximum Principle)* *Suppose that $u(x,y)$ is harmonic in a bounded domain D and continuous in $\mathbf{D} = D \cup B$. Then u attains its maximum on the boundary B of D.*

Physically, we may interpret this as meaning that the temperature of a body which was neither a source nor a sink of heat acquires its largest (and smallest) values on the surface of the body, and the electrostatic potential in a region which does not contain any free charge attains its maximum (and minimum) values on the boundary of the region.

Proof. Let the maximum of u on B be M. Let us now suppose that the maximum of u in $\mathbf{D}$ is not attained at any point of B. Then it must be attained at some point $P_0(x_0, y_0)$ in D. If $M_0 = u(x_0, y_0)$ denotes the maximum of u in D, then M_0 must also be the maximum of u in $\mathbf{D}$.

Consider the function

$$v(x, y) = u(x, y) + \frac{M_0 - M}{4R^2}\left[(x - x_0)^2 + (y - y_0)^2\right], \qquad (9.2.1)$$

where the point $P(x, y)$ is in D and where R is the radius of a circle containing D. Note that

$$v(x_0, y_0) = u(x_0, y_0) = M_0.$$

We have $v(x, y) \leq M + (M_0 - M)/2 = \frac{1}{2}(M + M_0) < M_0$ on B. Thus, $v(x, y)$ like $u(x, y)$ must attain its maximum at a point in D. It follows from the definition of v that

$$v_{xx} + v_{yy} = u_{xx} + u_{yy} + \frac{(M_0 - M)}{R^2} = \frac{(M_0 - M)}{R^2} > 0. \qquad (9.2.2)$$

But for v to be a maximum in D,

$$v_{xx} \leq 0, \qquad v_{yy} \leq 0.$$

Thus,

$$v_{xx} + v_{yy} \leq 0$$

which contradicts equation (9.2.2). Hence, the maximum of u must be attained on B.

Theorem 9.2.2. *(The Minimum Principle)* *If $u(x, y)$ is harmonic in a bounded domain D and continuous in $\mathbf{D} = D \cup B$, then u attains its minimum on the boundary B of D.*

Proof. The proof follows directly by applying the preceding theorem to the harmonic function $-u(x, y)$.

As a result of the above theorems, we see that $u =$ constant which is evidently harmonic attains the same value in the domain D as on the boundary B.

9.3 Uniqueness and Continuity Theorems

Theorem 9.3.1. *(**Uniqueness Theorem**) The solution of the Dirichlet problem, if it exists, is unique.*

Proof. Let $u_1(x, y)$ and $u_2(x, y)$ be two solutions of the Dirichlet problem. Then u_1 and u_2 satisfy

$$\nabla^2 u_1 = 0, \qquad \nabla^2 u_2 = 0 \qquad \text{in } D,$$
$$u_1 = f, \qquad u_2 = f \qquad \text{on } B.$$

Since u_1 and u_2 are harmonic in D, $(u_1 - u_2)$ is also harmonic in D. But

$$u_1 - u_2 = 0 \qquad \text{on } B.$$

The maximum-minimum principle gives

$$u_1 - u_2 = 0$$

at all interior points of D. Thus, we have

$$u_1 = u_2.$$

Therefore, the solution is unique.

Theorem 9.3.2. *(**Continuity Theorem**) The solution of the Dirichlet problem depends continuously on the boundary data.*

Proof. Let u_1 and u_2 be the solutions of

$$\nabla^2 u_1 = 0 \qquad \text{in } D,$$
$$u_1 = f_1 \qquad \text{on } B,$$

and

$$\nabla^2 u_2 = 0 \qquad \text{in } D,$$
$$u_2 = f_2 \qquad \text{on } B.$$

If $v = u_1 - u_2$, then v satisfies

$$\nabla^2 v = 0 \qquad \text{in } D,$$
$$v = f_1 - f_2 \qquad \text{on } B.$$

By maximum and minimum principles, $f_1 - f_2$ attains the maximum and minimum of v on B. Thus, if $|f_1 - f_2| < \varepsilon$, then

$$-\varepsilon < v_{min} \le v_{max} < \varepsilon \qquad \text{on } B.$$

Thus, at any interior point in D, we have

$$-\varepsilon < v_{min} \le v \le v_{max} < \varepsilon.$$

Therefore, $|v| < \varepsilon$ in D. Hence,

$$|u_1 - u_2| < \varepsilon.$$

Theorem 9.3.3. *Let $\{u_n\}$ be a sequence of functions harmonic in D and continuous in $\mathbf{D}$. Let f_i be the values of u_i on B. If a sequence $\{u_n\}$ converges uniformly on B, then it converges uniformly in $\mathbf{D}$.*

Proof. By hypothesis, $\{f_n\}$ converges uniformly on B. Thus, for $\varepsilon > 0$, there exists an integer N such that everywhere on B

$$|f_n - f_m| < \varepsilon \quad \text{for} \quad n, m > N.$$

It follows from the continuity theorem that for all $n, m > N$

$$|u_n - u_m| < \varepsilon$$

in D, and hence, the theorem is proved.

9.4 Dirichlet Problem for a Circle

1. Interior Problem

We shall now establish the existence of the solution of the Dirichlet problem for a circle.

The Dirichlet problem is

$$\nabla^2 u = u_{rr} + \frac{1}{r} u_r + \frac{1}{r^2} u_{\theta\theta} = 0, \quad 0 \le r < a, \quad 0 < \theta \le 2\pi, \tag{9.4.1}$$
$$u(a, \theta) = f(\theta) \text{ for all } \theta \text{ in } [0, 2\pi]. \tag{9.4.2}$$

By the method of separation of variables, we seek a solution in the form

$$u(r, \theta) = R(r) \Theta(\theta) \ne 0.$$

Substitution of this in equation (9.4.1) yields

$$r^2 \frac{R^{''}}{R} + r \frac{R^{'}}{R} = -\frac{\Theta^{''}}{\Theta} = \lambda.$$

Hence,

$$r^2 R^{''} + r R^{'} - \lambda R = 0, \tag{9.4.3}$$

$$\Theta^{''} + \lambda \Theta = 0. \tag{9.4.4}$$

Because of the periodicity conditions $\Theta(0) = \Theta(2\pi)$ and $\Theta^{'}(0) = \Theta^{'}(2\pi)$ which ensure that the function Θ is single-valued, the case $\lambda < 0$ does not yield an acceptable solution. When $\lambda = 0$, we have

$$u(r, \theta) = (A + B \log r)(C\theta + D).$$

Since $\log r \to -\infty$ as $r \to 0+$ (note that $r = 0$ is a singular point of equation (9.4.1)), B must vanish in order for u to be finite at $r = 0$. C must also vanish in order for u to be periodic with period 2π. Hence, the solution for $\lambda = 0$ is $u = $ constant. When $\lambda > 0$, the solution of equation (9.4.4) is

$$\Theta(\theta) = A \cos \sqrt{\lambda}\,\theta + B \sin \sqrt{\lambda}\,\theta.$$

The periodicity conditions imply

$$\sqrt{\lambda} = n \quad \text{for} \quad n = 1, 2, 3, \ldots .$$

Equation (9.4.3) is the Euler equation and therefore, the general solution is

$$R(r) = Cr^n + Dr^{-n}.$$

Since $r^{-n} \to \infty$ as $r \to 0$, D must vanish for u to be continuous at $r = 0$.

Thus, the solution is

$$u(r, 0) = Cr^n (A \cos n\,\theta + B \sin n\,\theta) \quad \text{for} \quad n = 1, 2, \ldots .$$

Hence, the general solution of equation (9.4.1) may be written in the form

$$u(r, \theta) = \frac{a_0}{2} + \sum_{n=1}^{\infty} \left(\frac{r}{a}\right)^n (a_n \cos n\theta + b_n \sin n\theta), \tag{9.4.5}$$

where the constant term $(a_0/2)$ represents the solution for $\lambda = 0$, and a_n and b_n are constants. Letting $\rho = r/a$, we have

$$u(\rho, \theta) = \frac{a_0}{2} + \sum_{n=1}^{\infty} \rho^n (a_n \cos n\theta + b_n \sin n\theta). \tag{9.4.6}$$

Our next task is to show that $u(r, \theta)$ is harmonic in $0 \leq r < a$ and continuous in $0 \leq r \leq a$. We must also show that u satisfies the boundary condition (9.4.2).

We assume that a_n and b_n are the Fourier coefficients of $f(\theta)$, that is,

$$a_n = \frac{1}{\pi} \int_0^{2\pi} f(\theta) \cos n\theta \, d\theta, \qquad n = 0, 1, 2, 3, \ldots,$$

(9.4.7)

$$b_n = \frac{1}{\pi} \int_0^{2\pi} f(\theta) \sin n\theta \, d\theta, \qquad n = 1, 2, 3, \ldots.$$

Thus, from their very definitions, a_n and b_n are bounded, that is, there exists some number $M > 0$ such that

$$|a_0| < M, \qquad |a_n| < M, \qquad |b_n| < M, \qquad n = 1, 2, 3, \ldots.$$

Thus, if we consider the sequence of functions $\{u_n\}$ defined by

$$u_n(\rho, \theta) = \rho^n (a_n \cos n\theta + b_n \sin n\theta),$$

(9.4.8)

we see that

$$|u_n| < 2\rho_0^n M, \qquad 0 \leq \rho \leq \rho_0 < 1.$$

Hence, in any closed circular region, series (9.4.6) converges uniformly. Next, differentiate u_n with respect to r. Then, for $0 \leq \rho \leq \rho_0 < 1$,

$$\left| \frac{\partial u_n}{\partial r} \right| = \left| \frac{n}{a} \rho^{n-1} (a_n \cos n\theta + b_n \sin n\theta) \right| < 2 \left(\frac{n}{a} \right) \rho_0^{n-1} M.$$

Thus, the series obtained by differentiating series (9.4.6) term by term with respect to r converges uniformly. In a similar manner, we can prove that the series obtained by twice differentiating series (9.4.6) term by term with respect to r and θ converge uniformly. Consequently,

$$\nabla^2 u = u_{rr} + \frac{1}{r} u_r + \frac{1}{r^2} u_{\theta\theta}$$

$$= \sum_{n=1}^{\infty} \frac{\rho^{n-2}}{a^2} (a_n \cos n\theta + b_n \sin n\theta) \left[n(n-1) + n - n^2 \right]$$

$$= 0, \qquad 0 \leq \rho \leq \rho_0 < 1.$$

Since each term of series (9.4.6) is a harmonic function, and since the series converges uniformly, $u(r, \theta)$ is harmonic at any interior point of the region $0 \leq \rho < 1$. It now remains to show that u satisfies the boundary data $f(\theta)$.

Substitution of the Fourier coefficients a_n and b_n into equation (9.4.6) yields

$$u(\rho, \theta) = \frac{1}{2\pi} \int_0^{2\pi} f(\theta) \, d\theta + \frac{1}{\pi} \sum_{n=1}^{\infty} \rho^n \int_0^{2\pi} f(\tau)$$

$$\times (\cos n\tau \cos n\theta + \sin n\tau \sin n\theta) \, d\tau$$

$$= \frac{1}{2\pi} \int_0^{2\pi} \left[1 + 2 \sum_{n=1}^{\infty} \rho^n \cos n(\theta - \tau) \right] f(\tau) \, d\tau. \tag{9.4.9}$$

The interchange of summation and integration is permitted due to the uniform convergence of the series. For $0 \le \rho \le 1$

$$1 + 2 \sum_{n=1}^{\infty} [\rho^n \cos n(\theta - \tau)] = 1 + \sum_{n=1}^{\infty} \left[\rho^n e^{in(\theta-\tau)} + \rho^n e^{-in(\theta-\tau)} \right]$$

$$= 1 + \frac{\rho \, e^{i(\theta-\tau)}}{1 - \rho \, e^{i(\theta-\tau)}} + \frac{\rho \, e^{-i(\theta-\tau)}}{1 - \rho \, e^{-i(\theta-\tau)}}$$

$$= \frac{1 - \rho^2}{1 - \rho \, e^{i(\theta-\tau)} - \rho \, e^{-i(\theta-\tau)} + \rho^2}$$

$$= \frac{1 - \rho^2}{1 - 2\rho \cos(\theta - \tau) + \rho^2}.$$

Hence,

$$u(\rho, \theta) = \frac{1}{2\pi} \int_0^{2\pi} \frac{1 - \rho^2}{1 - 2\rho \cos(\theta - \tau) + \rho^2} f(\tau) \, d\tau. \tag{9.4.10}$$

The integral on the right side of (9.4.10) is called the *Poisson integral formula for a circle*.

Now if $f(\theta) = 1$, then, according to series (9.4.9), $u(r, \theta) = 1$ for $0 \le \rho \le 1$. Thus, equation (9.4.10) gives

$$1 = \frac{1}{2\pi} \int_0^{2\pi} \frac{1 - \rho^2}{1 - 2\rho \cos(\theta - \tau) + \rho^2} \, d\tau.$$

Hence,

$$f(\theta) = \frac{1}{2\pi} \int_0^{2\pi} \frac{1 - \rho^2}{1 - 2\rho \cos(\theta - \tau) + \rho^2} f(\theta) \, d\tau, \qquad 0 \le \rho < 1.$$

Therefore,

$$u(\rho, \theta) - f(\theta) = \frac{1}{2\pi} \int_0^{2\pi} \frac{(1 - \rho^2)[f(\tau) - f(\theta)]}{1 - 2\rho \cos(\theta - \tau) + \rho^2} \, d\tau. \tag{9.4.11}$$

Since $f(\theta)$ is uniformly continuous on $[0, 2\pi]$, for given $\varepsilon > 0$, there exists a positive number $\delta(\varepsilon)$ such that $|\theta - \tau| < \delta$ implies $|f(\theta) - f(\tau)| < \varepsilon$. If $|\theta - \tau| \ge \delta$ so that $\theta - \tau \ne 2n\pi$ for $n = 0, 1, 2, \ldots$, then

$$\lim_{\rho \to 1-} \frac{1 - \rho^2}{1 - 2\rho \cos (\theta - \tau) + \rho^2} = 0.$$

In other words, there exists ρ_0 such that if $|\theta - \tau| \geq \delta$, then

$$\frac{1 - \rho^2}{1 - 2\rho \cos (\theta - \tau) + \rho^2} < \varepsilon,$$

for $0 \leq \rho \leq \rho_0 < 1$. Hence, equation (9.4.10) yields

$$
\begin{aligned}
|u(r,\theta)| - f(\theta)| &\leq \frac{1}{2\pi} \int_{|\theta - \tau| \geq \delta}^{2\pi} \frac{(1 - \rho^2) |f(\tau) - f(\theta)|}{1 - 2\rho \cos (\theta - \tau) + \rho^2} \, d\tau \\
&\quad + \frac{1}{2\pi} \int_{|\theta - \tau| < \delta}^{2\pi} \frac{(1 - \rho^2) |f(\theta) - f(\tau)|}{1 - 2\rho \cos (\theta - \tau) + \rho^2} \, d\tau \\
&\leq \frac{1}{2\pi} (2\pi\varepsilon) \left[2 \max_{0 \leq \theta \leq 2\pi} |f(\theta)| \right] + \frac{\varepsilon}{2\pi} \cdot 2\pi \\
&= \varepsilon \left[1 + 2 \left(\max_{0 \leq \theta \leq 2\pi} |f(\theta)| \right) \right]
\end{aligned}
$$

which implies that

$$\lim_{\rho \to 1-} u(r,\theta) = f(\theta)$$

uniformly in θ. Therefore, we state the following theorem:

Theorem 9.4.1. *There exists one and only one harmonic function $u(r,\theta)$ which satisfies the continuous boundary data $f(\theta)$. This function is either given by*

$$u(r,\theta) = \frac{1}{2\pi} \int_0^{2\pi} \frac{a^2 - r^2}{a^2 - 2ar \cos(\theta - \tau) + r^2} f(\tau) \, d\tau, \qquad (9.4.12)$$

or

$$u(r,\theta) = \frac{a_0}{2} + \sum_{n=1}^{\infty} \frac{r^n}{a^n} (a_n \cos n\theta + b_n \sin n\theta), \qquad (9.4.13)$$

where a_n and b_n are the Fourier coefficients of $f(\theta)$.

For $\rho = 0$, the Poisson integral formula (9.4.10) becomes

$$u(0,\theta) = u(0) = \frac{1}{2\pi} \int_0^{2\pi} f(\tau) \, d\tau. \qquad (9.4.14)$$

This result may be stated as follows:

Theorem 9.4.2. (Mean Value Theorem) *If u is harmonic in a circle, then the value of u at the center is equal to the mean value of u on the boundary of the circle.*

Several comments are in order. First, the Continuity Theorem 9.3.2 for the Dirichlet problem for the Laplace equation is a special example of the general result that the Dirichlet problems for all elliptic equations are well-posed. Second, the formula (9.4.12) represents the unique continuous solution of the Laplace equation in $0 \le r < a$ even when $f(\theta)$ is discontinuous. This means that, for Laplace's equation, discontinuities in boundary conditions are smoothed out in the interior of the domain. This is a remarkable contrast to linear hyperbolic equations where any discontinuity in the data propagates along the characteristics. Third, the integral solution (9.4.12) can be written as

$$u(r, \theta) = \int_{-\pi}^{\pi} P(r, \tau - \theta) f(\tau) \, d\tau,$$

where $P(r, \tau - \theta)$ is called the *Poisson kernel* given by

$$P(r, \tau - \theta) = \frac{1}{2\pi} \frac{(a^2 - r^2)}{[a^2 - 2ar \cos(\tau - \theta) + r^2]}.$$

Clearly, $P(a, \tau - \theta) = 0$ except at $\tau = \theta$. Also

$$f(\theta) = \lim_{r \to a^-} u(r, \theta) = \int_{-\pi}^{\pi} \lim_{r \to a^-} P(r, \tau - \theta) f(\tau) \, d\tau.$$

This implies that

$$\lim_{r \to a^-} P(r, \tau - \theta) = \delta(\tau - \theta),$$

where $\delta(x)$ is the Dirac delta function.

As in the preceding section, the exterior Dirichlet problem for a circle can readily be solved. For the exterior problem u must be bounded as $r \to \infty$. The general solution, therefore, is

$$u(r, \theta) = \frac{a_0}{2} + \sum_{n=1}^{\infty} \left(\frac{r^{-n}}{a^{-n}} \right) (a_n \cos n\theta + b_n \sin n\theta). \qquad (9.4.15)$$

Applying the boundary condition $u(a, \theta) = f(\theta)$, we obtain

$$f(\theta) = \frac{a_0}{2} + \sum_{n=1}^{\infty} (a_n \cos n\theta + b_n \sin n\theta).$$

Hence, we find

$$a_n = \frac{1}{\pi} \int_0^{2\pi} f(\tau) \cos n\tau \, d\tau, \qquad n = 0, 1, 2, \ldots, \qquad (9.4.16)$$

$$b_n = \frac{1}{\pi} \int_0^{2\pi} f(\tau) \sin n\tau \, d\tau, \qquad n = 0, 1, 2, \ldots. \qquad (9.4.17)$$

Substituting a_n and b_n into equation (9.4.15) yields

$$u(r, \theta) = \frac{1}{2\pi} \int_0^{2\pi} \left[1 + 2 \sum_{n=1}^{\infty} \left(\frac{a}{r} \right)^n \cos n(\theta - \tau) \right] f(\tau) \, d\tau.$$

Comparing this equation with (9.4.9), we see that the only difference between the exterior and interior problem is that ρ^n is replaced by ρ^{-n}. Therefore, the final result takes the form

$$u(\rho, \theta) = \frac{1}{2\pi} \int_0^{2\pi} \frac{\rho^2 - 1}{1 - 2\rho \cos(\theta - \tau) + \rho^2} f(\tau) \, d\tau, \text{for } \rho > 1. (9.4.18)$$

9.5 Dirichlet Problem for a Circular Annulus

The natural extension of the Dirichlet problem for a circle is the Dirichlet problem for a circular annulus, that is

$$\nabla^2 u = 0, \qquad r_2 < r < r_1, \tag{9.5.1}$$
$$u(r_1, \theta) = f(\theta), \quad u(r_2, \theta) = g(\theta). \tag{9.5.2}$$

In addition, $u(r, \theta)$ must satisfy the periodicity condition. Accordingly, $f(\theta)$ and $g(\theta)$ must also be periodic with period 2π.

Proceeding as in the case of the Dirichlet problem for a circle, we obtain for $\lambda = 0$

$$u(r, \theta) = (A + B \log r)(C\theta + D).$$

The periodicity condition on u requires that $C = 0$. Then, $u(r, \theta)$ becomes

$$u(r, \theta) = \frac{a_0}{2} + \frac{b_0}{2} \log r,$$

where $a_0 = 2AD$ and $b_0 = 2BD$.

The solution for the case $\lambda > 0$ is

$$u(r, \theta) = \left(Cr^{\sqrt{\lambda}} + Dr^{-\sqrt{\lambda}} \right) \left(A \cos \sqrt{\lambda} \, \theta + B \sin \sqrt{\lambda} \, \theta \right),$$

where $\sqrt{\lambda} = n = 1, 2, 3, \ldots$. Thus, the general solution is

$$u(r, \theta) = \frac{1}{2} (a_0 + b_0 \log r) + \sum_{n=1}^{\infty} \left[\left(a_n r^n + b_n r^{-n} \right) \cos n\theta \right.$$
$$\left. + \left(c_n r^n + d_n r^{-n} \right) \sin n\theta \right], \tag{9.5.3}$$

where a_n, b_n, c_n, and d_n are constants.

Applying the boundary conditions (9.5.2), we find that the coefficients are given by

$$a_0 + b_0 \log r_1 = \frac{1}{\pi} \int_0^{2\pi} f(\tau) \, d\tau,$$

$$a_n r_1^n + b_n r_1^{-n} = \frac{1}{\pi} \int_0^{2\pi} f(\tau) \cos n\tau \, d\tau,$$

$$c_n r_1^n + d_n r_1^{-n} = \frac{1}{\pi} \int_0^{2\pi} f(\tau) \sin n\tau \, d\tau,$$

and

$$a_0 + b_0 \log r_2 = \frac{1}{\pi} \int_0^{2\pi} g(\tau) \, d\tau,$$

$$a_n r_2^n + b_n r_2^{-n} = \frac{1}{\pi} \int_0^{2\pi} g(\tau) \cos n\tau \, d\tau,$$

$$c_n r_2^n + d_n r_2^{-n} = \frac{1}{\pi} \int_0^{2\pi} g(\tau) \sin n\tau \, d\tau.$$

The constants a_0, b_0, a_n, b_n, c_n, d_n for $n = 1, 2, 3, \ldots$ can then be determined. Hence, the solution of the Dirichlet problem for an annulus is given by (9.5.3).

9.6 Neumann Problem for a Circle

Let u be a solution of the Neumann problem

$$\nabla^2 u = 0 \qquad \text{in } D,$$

$$\frac{\partial u}{\partial n} = f \qquad \text{on } B.$$

It is evident that $u +$ constant is also a solution. Thus, we see that the solution of the Neumann problem is not unique, and it differs from another by a constant.

Consider the interior Neumann problem

$$\nabla^2 u = 0, \qquad r < R, \tag{9.6.1}$$

$$\frac{\partial u}{\partial n} = \frac{\partial u}{\partial r} = f(\theta), \qquad r = R. \tag{9.6.2}$$

Before we determine a solution of the Neumann problem, a necessary condition for the existence of a solution will be established.

In Green's second formula

$$\int\int_D \left(v \nabla^2 u - u \nabla^2 v \right) dS = \int_B \left(v \frac{\partial u}{\partial n} - u \frac{\partial v}{\partial n} \right) ds, \tag{9.6.3}$$

we put $v = 1$, so that $\nabla^2 v = 0$ in D and $\partial v / \partial n = 0$ on B. Then, the result is

$$\int_D \int \nabla^2 u \, dS = \int_B \frac{\partial u}{\partial n} \, ds. \qquad (9.6.4)$$

Substituting of (9.6.1) and (9.6.2) into equation (9.6.4) yields

$$\int_B f \, ds = 0 \qquad (9.6.5)$$

which may also be written in the form

$$R \int_0^{2\pi} f(\theta) \, d\theta = 0. \qquad (9.6.6)$$

As in the case of the interior Dirichlet problem for a circle, the solution of the Laplace equation is

$$u(r, \theta) = \frac{a_0}{2} + \sum_{k=1}^{\infty} r^k (a_k \cos k\theta + b_k \sin k\theta). \qquad (9.6.7)$$

Differentiating this with respect to r and applying the boundary condition (9.6.2), we obtain

$$\frac{\partial u}{\partial r}(R, \theta) = \sum_{k=1}^{\infty} k R^{k-1} (a_k \cos k\theta + b_k \sin k\theta) = f(\theta). \qquad (9.6.8)$$

Hence, the coefficients are given by

$$a_k = \frac{1}{k\pi R^{k-1}} \int_0^{2\pi} f(\tau) \cos k\tau \, d\tau, \qquad k = 1, 2, 3, \ldots,$$

$$\qquad (9.6.9)$$

$$b_k = \frac{1}{k\pi R^{k-1}} \int_0^{2\pi} f(\tau) \sin k\tau \, d\tau, \qquad k = 1, 2, 3, \ldots.$$

Note that the expansion of $f(\theta)$ in a series of the form (9.6.8) is possible only by virtue of the compatibility condition (9.6.6) since

$$a_0 = \frac{1}{\pi} \int_0^{2\pi} f(\tau) \, d\tau = 0.$$

Inserting a_k and b_k in equation (9.6.7), we obtain

$$u(r, \theta) = \frac{a_0}{2} + \frac{R}{\pi} \int_0^{2\pi} \left[\sum_{k=1}^{\infty} \left(\frac{r}{R}\right)^k \cos k(\theta - \tau) \right] f(\tau) \, d\tau.$$

Using the identity

$$-\frac{1}{2} \log \left[1 + \rho^2 - 2\rho \cos(\theta - \tau) \right] = \sum_{k=1}^{\infty} \frac{1}{k} \rho^k \cos \{k(\theta - \tau)\},$$

with $\rho = (r/R)$, we find that

$$u\left(r,\theta\right) = \frac{a_0}{2} - \frac{R}{2\pi} \int_0^{2\pi} \log\left[R^2 - 2rR\cos\left(\theta - \tau\right) + r^2\right] f\left(\tau\right) d\tau. \quad (9.6.10)$$

in which a constant factor R^2 in the argument of the logarithm was eliminated by virtue of equation (9.6.6).

In a similar manner, for the exterior Neumann problem, we can readily find the solution in the form

$$u\left(r,\theta\right) = \frac{a_0}{2} + \frac{R}{2\pi} \int_0^{2\pi} \log\left[R^2 - 2rR\cos\left(\theta - \tau\right) + r^2\right] f\left(\tau\right) d\tau. \quad (9.6.11)$$

9.7 Dirichlet Problem for a Rectangle

We first consider the boundary-value problem

$$\begin{array}{lll} \nabla^2 u = u_{xx} + u_{yy} = 0, & 0 < x < a, \quad 0 < y < b, & (9.7.1) \\ u\left(x,0\right) = f\left(x\right), & u\left(x,b\right) = 0, \quad 0 \le x \le a, & (9.7.2) \\ u\left(0,y\right) = 0, \quad u\left(a,y\right) = 0, & 0 \le y \le b. & (9.7.3) \end{array}$$

We seek a nontrivial separable solution in the form

$$u\left(x,y\right) = X\left(x\right)Y\left(y\right)$$

Substituting $u\left(x,y\right)$ in the Laplace equation, we obtain

$$\begin{aligned} X'' - \lambda X &= 0, & (9.7.4) \\ Y'' + \lambda Y &= 0, & (9.7.5) \end{aligned}$$

where λ is a separation constant. Since the boundary conditions are homogeneous for $x = 0$ and $x = a$, we choose $\lambda = -\alpha^2$ with $\alpha > 0$ in order to obtain nontrivial solutions of the eigenvalue problem

$$X'' + \alpha^2 X = 0,$$
$$X\left(0\right) = X\left(a\right) = 0.$$

It is easily found that the eigenvalues are

$$\alpha = \frac{n\pi}{a}, \qquad n = 1, 2, 3, \ldots.$$

and the corresponding eigenfunctions are $\sin\left(n\pi x/a\right)$. Hence

$$X_n\left(x\right) = B_n \sin\left(\frac{n\pi x}{a}\right).$$

The solution of equation (9.7.5) is $Y(y) = C \cosh \alpha y + D \sinh \alpha y$, which may also be written in the form

$$Y(y) = E \sinh \alpha (y + F),$$

where $E = \left(D^2 - C^2\right)^{\frac{1}{2}}$ and $F = (1/\alpha) \tanh^{-1}(C/D)$. Applying the remaining homogeneous boundary condition

$$u(x, b) = X(x) Y(b) = 0,$$

we obtain

$$Y(b) = E \sinh \alpha (b + F) = 0,$$

and hence,

$$F = -b, \qquad E \neq 0$$

for a nontrivial solution $u(x, y)$. Thus, we have

$$Y_n(y) = E_n \sinh \left\{ \frac{n\pi}{a}(y - b) \right\}.$$

Because of linearity, the solution becomes

$$u(x, y) = \sum_{n=1}^{\infty} a_n \sin \left(\frac{n\pi x}{a} \right) \sinh \left\{ \frac{n\pi}{a}(y - b) \right\},$$

where $a_n = B_n E_n$. Now, we apply the nonhomogeneous boundary condition to obtain

$$u(x, 0) = f(x) = \sum_{n=1}^{\infty} a_n \sinh \left(\frac{-n\pi b}{a} \right) \sin \left(\frac{n\pi x}{a} \right).$$

This is a Fourier sine series and hence,

$$a_n = \frac{-2}{a \sinh \left(\frac{n\pi b}{a} \right)} \int_0^a f(x) \sin \left(\frac{n\pi x}{a} \right) dx.$$

Thus, the formal solution is given by

$$u(x, y) = \sum_{n=1}^{\infty} a_n^* \frac{\sinh \left\{ \frac{n\pi}{a}(b - y) \right\}}{\sinh \left(\frac{n\pi b}{a} \right)} \sin \left(\frac{n\pi x}{a} \right), \qquad (9.7.6)$$

where

$$a_n^* = \frac{2}{a} \int_0^a f(x) \sin \left(\frac{n\pi x}{a} \right) dx.$$

To prove the existence of solution (9.7.6), we first note that

$$\frac{\sinh \frac{n\pi}{a}(b-y)}{\sinh \frac{n\pi b}{a}} = e^{-n\pi y/a}\left[\frac{1-e^{-(2n\pi/a)(b-y)}}{1-e^{-2n\pi b/a}}\right] \leq C_1 e^{-n\pi y/a},$$

where C_1 is a constant. Since $f(x)$ is bounded, we have

$$|a_n^*| \leq \frac{2}{a}\int_0^a |f(x)|\,dx = C_2.$$

Thus, the series for $u(x,y)$ is dominated by the series

$$\sum_{n=1}^\infty M e^{-n\pi y_0/a} \quad \text{for} \quad y \geq y_0 > 0, \quad M = \text{constant},$$

and hence, $u(x,y)$ converges uniformly in x and y whenever $0 \leq x \leq a$, $y \geq y_0 > 0$. Consequently, $u(x,y)$ is continuous in this region and satisfies the boundary values $u(0,y) = u(a,y) = u(x,b) = 0$.

Now differentiating u twice with respect to x, we obtain

$$u_{xx}(x,y) = \sum_{n=1}^\infty -a_n^*\left(\frac{n\pi}{a}\right)^2 \frac{\sinh \frac{n\pi}{a}(b-y)}{\sinh \frac{n\pi b}{a}} \sin\left(\frac{n\pi x}{a}\right)$$

and differentiating u twice with respect to y, we obtain

$$u_{yy}(x,y) = \sum_{n=1}^\infty a_n^*\left(\frac{n\pi}{a}\right)^2 \frac{\sinh \frac{n\pi}{a}(b-y)}{\sinh \frac{n\pi b}{a}} \sin\left(\frac{n\pi x}{a}\right).$$

It is evident that the series for u_{xx} and u_{yy} are both dominated by

$$\sum_{n=1}^\infty M^* n^2 e^{-n\pi y_0/a}$$

and hence, converge uniformly for any $0 < y_0 < b$. It follows that u_{xx} and u_{yy} exist, and hence, u satisfies the Laplace equation.

It now remains to show that $u(x,0) = f(x)$. Let $f(x)$ be a continuous function and let $f'(x)$ be piecewise continuous on $[0,a]$. If, in addition, $f(0) = f(a) = 0$, then, the Fourier series for $f(x)$ converges uniformly. Putting $y = 0$ in the series for $u(x,y)$, we obtain

$$u(x,0) = \sum_{n=1}^\infty a_n^* \sin\left(\frac{n\pi x}{a}\right).$$

Since $u(x,0)$ converges uniformly to $f(x)$, we write, for $\varepsilon > 0$,

$$|s_m(x,0) - s_n(x,0)| < \varepsilon \quad \text{for} \quad m,n > N_\varepsilon,$$

where

$$s_m(x,y) = \sum_{n=1}^{\infty} a_n^* \sin\left(\frac{n\pi x}{a}\right).$$

We also know that $s_m(x,y) - s_n(x,y)$ satisfies the Laplace equation and the boundary conditions at $x = 0$, $x = a$ and $y = b$. Then, by the maximum principle,

$$|s_m(x,y) - s_n(x,y)| < \varepsilon \quad \text{for} \quad m, n > N_\varepsilon$$

in the region $0 \le x \le a$, $0 \le y \le b$. Thus, the series for $u(x,y)$ converges uniformly, and as a consequence, $u(x,y)$ is continuous in the region $0 \le x \le a$, $0 \le y \le b$. Hence, we obtain

$$u(x,0) = \sum_{n=1}^{\infty} a_n^* \sin\left(\frac{n\pi x}{a}\right) = f(x).$$

Thus, the solution (9.7.6) is established.

The general Dirichlet problem

$$\nabla^2 u = 0, \qquad 0 < x < a, \qquad\qquad 0 < y < b,$$
$$u(x,0) = f_1(x), \quad u(x,a) = f_2(x), \quad 0 \le x \le a,$$
$$u(0,y) = f_3(y), \quad u(b,y) = f_4(y), \quad 0 \le y \le b$$

can be solved by separating it into four problems, each of which has one nonhomogeneous boundary condition and the rest zero. Thus, determining each solution as in the preceding problem and then adding the four solutions, the solution of the Dirichlet problem for a rectangle can be obtained.

9.8 Dirichlet Problem Involving the Poisson Equation

The solution of the Dirichlet problem involving the Poisson equation can be obtained for simple regions when the solution of the corresponding Dirichlet problem for the Laplace equation is known.

Consider the Poisson equation

$$\nabla^2 u = u_{xx} + u_{uu} = f(x,y) \quad \text{in } D,$$

with the condition

$$u = g(x,y) \quad \text{on } B.$$

Assume that the solution can be written in the form

$$u = v + w,$$

where v is a particular solution of the Poisson equation and w is the solution of the associated homogeneous equation, that is,

$$\nabla^2 w = 0,$$
$$\nabla^2 v = f.$$

As soon as v is ascertained, the solution of the Dirichlet problem

$$\nabla^2 w = 0 \qquad \text{in } D,$$
$$w = -v + g(x,y) \quad \text{on } B$$

can be determined. The usual method of finding a particular solution for the case in which $f(x,y)$ is a polynomial of degree n is to seek a solution in the form of a polynomial of degree $(n+2)$ with undetermined coefficients.

As an example, we consider the torsion problem

$$\nabla^2 u = -2, \qquad 0 < x < a, \qquad 0 < y < b,$$
$$u(0,y) = 0, \quad u(a,y) = 0; \quad u(x,0) = 0, \quad u(x,b) = 0.$$

We let $u = v + w$. Now assume v to be the form

$$v(x,y) = A + Bx + Cy + Dx^2 + Exy + Fy^2.$$

Substituting this in the Poisson equation, we obtain

$$2D + 2F = -2.$$

The simplest way of satisfying this equation is to choose

$$D = -1 \quad \text{and} \quad F = 0.$$

The remaining coefficients are arbitrary. Thus, we take

$$v(x,y) = ax - x^2$$

so that v reduces to zero on the sides $x = 0$ and $x = a$. Next, we find w from

$$\nabla^2 w = 0, \qquad 0 < x < a, \qquad 0 < y < b,$$
$$w(0,y) = -v(0,y) = 0,$$
$$w(a,y) = -v(a,0) = 0,$$
$$w(x,0) = -v(x,0) = -\left(ax - x^2\right),$$
$$w(x,b) = -v(x,b) = -\left(ax - x^2\right).$$

As in the Dirichlet problem (Section 9.7), the solution is found to be

$$w(x,y) = \sum_{n=1}^{\infty} \left(a_n \cosh \frac{n\pi y}{a} + b_n \sinh \frac{n\pi y}{a}\right) \sin\left(\frac{n\pi x}{a}\right).$$

Application of the nonhomogeneous boundary conditions yields

$$w\left(x,0\right) = -\left(ax - x^2\right) = \sum_{n=1}^{\infty} a_n \sin\left(\frac{n\pi x}{a}\right),$$

$$w\left(x,b\right) = -\left(ax - x^2\right) = \sum_{n=1}^{\infty} \left(a_n \cosh\frac{n\pi b}{a} + b_n \sinh\frac{n\pi b}{a}\right)\sin\left(\frac{n\pi x}{a}\right),$$

from which we find

$$a_n = \frac{2}{a}\int_0^a \left(x^2 - ax\right)\sin\left(\frac{n\pi x}{a}\right)dx$$

$$= \begin{cases} 0, & \text{if } n \text{ is even} \\ \frac{-8a^2}{\pi^3 n^3} & \text{if } n \text{ is odd} \end{cases}$$

and

$$\left(a_n \cosh\frac{n\pi b}{a} + b_n \sinh\frac{n\pi b}{a}\right) = \frac{2}{a}\int_0^a \left(x^2 - ax\right)\sin\left(\frac{n\pi x}{a}\right)dx.$$

Thus, we have

$$b_n = \frac{\left(1 - \cosh\frac{n\pi b}{a}\right)a_n}{\sinh\left(\frac{n\pi b}{a}\right)}.$$

Hence, the solution of the Dirichlet problem for the Poisson equation is given by

$$u\left(x,y\right) = \left(a - x\right)x$$

$$- \frac{8a^2}{\pi^3}\sum_{n=1}^{\infty} \frac{\left[\sinh\left(2n-1\right)\frac{\pi(b-y)}{a} + \sinh\left(2n-1\right)\frac{\pi y}{a}\right]}{\sinh\left(2n-1\right)\frac{\pi b}{a}}\frac{\sin\left(2n-1\right)\frac{\pi x}{a}}{\left(2n-1\right)^3}.$$

9.9 The Neumann Problem for a Rectangle

Consider the Neumann problem

$$\nabla^2 u = 0, \qquad 0 < x < a, \qquad\qquad 0 < y < b, \qquad (9.9.1)$$
$$u_x\left(0,y\right) = f_1\left(y\right), \quad u_x\left(a,y\right) = f_2\left(y\right), \quad 0 \le y \le b, \qquad (9.9.2)$$
$$u_y\left(x,0\right) = g_1\left(x\right), \quad u_y\left(x,b\right) = g_2\left(x\right), \quad 0 \le x \le a. \qquad (9.9.3)$$

The compatibility condition that must be fulfilled in this case is

$$\int_0^a \left[g_1\left(x\right) - g_2\left(x\right)\right]dx + \int_0^b \left[f_1\left(y\right) - f_2\left(y\right)\right]dy = 0. \qquad (9.9.4)$$

We assume a solution in the form

$$u\left(x,y\right) = u_1\left(x,y\right) + u_2\left(x,y\right), \tag{9.9.5}$$

where $u_1\left(x,y\right)$ is a solution of

$$\nabla^2 u_1 = 0,$$
$$\frac{\partial u_1}{\partial x}\left(0,y\right) = 0,$$
$$\frac{\partial u_1}{\partial x}\left(a,y\right) = 0, \tag{9.9.6}$$
$$\frac{\partial u_1}{\partial x}\left(x,0\right) = g_1\left(x\right),$$
$$\frac{\partial u_1}{\partial x}\left(x,b\right) = g_2\left(x\right),$$

and where g_1 and g_2 satisfy the compatibility condition

$$\int_0^a \left[g_1\left(x\right) - g_2\left(x\right)\right]\,dx = 0. \tag{9.9.7}$$

The function $u_2\left(x,y\right)$ is a solution of

$$\nabla^2 u_2 = 0,$$
$$\frac{\partial u_2}{\partial x}\left(0,y\right) = f_1\left(y\right)$$
$$\frac{\partial u_2}{\partial x}\left(a,y\right) = f_2\left(y\right) \tag{9.9.8}$$
$$\frac{\partial u_2}{\partial y}\left(x,0\right) = 0,$$
$$\frac{\partial u_2}{\partial y}\left(x,b\right) = 0,$$

where f_1 and f_2 satisfy the compatibility condition

$$\int_0^b \left[f_1\left(y\right) - f_2\left(y\right)\right]\,dy = 0. \tag{9.9.9}$$

Hence, $u_1\left(x,y\right)$ and $u_2\left(x,y\right)$ can be determined. Conditions (9.9.7) and (9.9.9) ensure that condition (9.9.4) is fulfilled. Thus, the problem is solved.

However, the solution obtained in this manner is rather restrictive. In general, condition (9.9.4) does not imply conditions (9.9.7) and (9.9.9). Thus, generally speaking, it is *not* possible to obtain a solution of the Neumann problem for a rectangle by the method described above.

To obtain a general solution, Grunberg (1946) proposed the following method. Suppose we assume a solution in the form

$$u\left(x,y\right) = \frac{Y_0}{2}\left(y\right) + \sum_{n=1}^{\infty} X_n\left(x\right) Y_n\left(y\right), \tag{9.9.10}$$

where $X_n(x) = \cos(n\pi x/a)$ is an eigenfunction of the eigenvalue problem

$$X'' + \lambda X = 0,$$
$$X'(0) = X'(a) = 0,$$

corresponding to the eigenvalue $\lambda_n = (n\pi/a)^2$. Then, from equation (9.9.10), we see that

$$Y_n(y) = \frac{2}{a} \int_0^a u(x, y) X_n(x)\, dx,$$

$$= \frac{2}{a} \int_0^a u(x, y) \cos\left(\frac{n\pi x}{a}\right) dx. \qquad (9.9.11)$$

Multiplying both sides of equation (9.9.1) by $2\cos(n\pi x/a)$ and integrating with respect to x from 0 to a, we obtain

$$\frac{2}{a} \int_0^a (u_{xx} + u_{yy}) \cos\left(\frac{n\pi x}{a}\right) dx = 0,$$

or,

$$Y_n'' + \frac{2}{a} \int_0^a u_{xx} \cos\left(\frac{n\pi x}{a}\right) dx = 0.$$

Integrating the second term by parts and applying the boundary conditions (9.9.2), we obtain

$$Y_n''(y) - \left(\frac{n\pi}{a}\right)^2 Y_n(y) = F_n(y), \qquad (9.9.12)$$

where $F_n(y) = 2[f_1(y) - (-1)^n f_2(y)]/a$. This is an ordinary differential equation whose solution may be written in the form

$$Y_n(y) = A_n \cosh\left(\frac{n\pi y}{a}\right) + B_n \sinh\left(\frac{n\pi y}{a}\right)$$

$$+ \frac{2}{\pi n} \int_0^y F_n(\tau) \sinh\left\{\frac{n\pi}{a}(y - \tau)\right\} d\tau. \quad (9.9.13)$$

The coefficients A_n and B_n are determined from the boundary conditions

$$Y_n'(0) = \frac{2}{a} \int_0^a u_y(x, 0) \cos\left(\frac{n\pi x}{a}\right) dx$$

$$= \frac{2}{a} \int_0^a g_1(x) \cos\left(\frac{n\pi x}{a}\right) dx \qquad (9.9.14)$$

and

$$Y_n'(b) = \frac{2}{a} \int_0^a g_2(x) \cos\left(\frac{n\pi x}{a}\right) dx. \qquad (9.9.15)$$

For $n = 0$, equation (9.9.12) takes the form

$$Y_0'' (y) = \frac{2}{a} [f_1 (y) - f_2 (y)]$$

and hence,

$$Y_0' (y) = \frac{2}{a} \int_0^y [f_1 (\tau) - f_2 (\tau)] \, d\tau + C,$$

where C is an integration constant. Employing the condition (9.9.14) for $n = 0$, we find

$$C = \frac{2}{a} \int_0^a g_1 (x) \, dx.$$

Thus, we have

$$Y_0' (y) = \frac{2}{a} \left\{ \int_0^y [f_1 (\tau) - f_2 (\tau)] \, d\tau + \int_0^a g_1 (x) \, dx \right\}.$$

Consequently,

$$Y_0' (b) = \frac{2}{a} \left\{ \int_0^b [f_1 (\tau) - f_2 (\tau)] \, d\tau + \int_0^a g_1 (x) \, dx \right\}.$$

Also from equation (9.9.14), we have

$$Y_0' (b) = \frac{2}{a} \int_0^a g_2 (x) \, dx.$$

It follows from these two expressions for $Y_0' (b)$ that

$$\int_0^b [f_1 (y) - f_2 (y)] \, dy + \int_0^a [g_1 (x) - g_2 (x)] \, dx = 0.$$

which is the necessary condition for the existence of a solution to the Neumann problem for a rectangle.

9.10 Exercises

1. Reduce the Neumann problem to the Dirichlet problem in the two-dimensional case.

2. Reduce the wave equation

$$u_n = c^2 (u_{xx} + u_{yy} + u_{zz})$$

to the Laplace equation

$$u_{xx} + u_{yy} + u_{zz} + u_{\tau\tau} = 0$$

by letting $\tau = ict$ where $i = \sqrt{-1}$. Obtain the solution of the wave equation in cylindrical coordinates via the solution of the Laplace equation. Assume that $u(r, \theta, z, \tau)$ is independent of z.

3. Prove that, if $u(x, t)$ satisfies

$$u_t = k\, u_{xx}$$

for $0 \leq x \leq 1$, $0 \leq t \leq t_0$, then the maximum value of u is attained either at $t = 0$ or at the end points $x = 0$ or $x = 1$ for $0 \leq t \leq t_0$. This is called the *maximum principle* for the heat equation.

4. Prove that a function which is harmonic everywhere on a plane and is bounded either above or below is a constant. This is called the *Liouville theorem*.

5. Show that the compatibility condition for the Neumann problem

$$\nabla^2 u = f \quad \text{in } D$$
$$\frac{\partial u}{\partial n} = g \quad \text{on } B$$

is

$$\int_D f\, dS + \int_B g\, ds = 0,$$

where B is the boundary of domain D.

6. Show that the second degree polynomial

$$P = Ax^2 + Bxy + Cy^2 + Dyz + Fz^2 + Exz$$

is harmonic if

$$E = -(A + C)$$

and obtain

$$P = A\left(x^2 - z^2\right) + Bxy + C\left(y^2 - z^2\right) + Dyz + Fxz.$$

7. Prove that a solution of the Neumann problem

$$\nabla^2 u = f \quad \text{in } D$$
$$u = g \quad \text{on } B$$

differs from another solution by at most a constant.

8. Determine the solution of each of the following problems:

(a) $\nabla^2 u = 0$, $1 < r < 2$, $0 < \theta < \pi$,

   $u(1,\theta) = \sin\theta$, $u(2,\theta) = 0$, $0 \le \theta \le \pi$,

   $u(r,0) = 0$, $u(r,\pi) = 0$, $1 \le r \le 2$.

(b) $\nabla^2 u = 0$, $1 < r < 2$, $0 < \theta < \pi$,

   $u(1,\theta) = 0$, $u(2,\theta) = \theta(\theta - \pi)$, $0 \le \theta \le \pi$,

   $u(r,0) = 0$, $u(r,\pi) = 0$, $1 \le r \le 2$.

(c) $\nabla^2 u = 0$, $1 < r < 3$, $0 < \theta < \pi/2$,

   $u(1,\theta) = 0$, $u(3,\theta) = 0$, $0 \le \theta \le \pi/2$,

   $u(r,0) = (r-1)(r-3)$, $u\left(r,\frac{\pi}{2}\right) = 0$, $1 \le r \le 3$.

(d) $\nabla^2 u = 0$, $1 < r < 3$, $0 < \theta < \pi/2$,

   $u(1,\theta) = 0$, $u(3,\theta) = 0$, $0 \le \theta \le \pi$,

   $u(r,0) = 0$, $u\left(r,\frac{\pi}{2}\right) = f(r)$, $1 \le r \le 3$.

9. Solve the boundary-value problem

$$\nabla^2 u = 0, \qquad a < r < b, \qquad 0 < \theta < \alpha,$$

$$u(a,\theta) = f(\theta), \quad u(b,\theta) = 0, \qquad 0 \le \theta \le \alpha,$$

$$u(r,\alpha) = 0, \qquad u(r,0) = f(r), \quad a \le r \le b.$$

10. Verify directly that the Poisson integral is a solution of the Laplace equation.

11. Solve

$$\nabla^2 u = 0, \qquad 0 < r < a, \qquad 0 < \theta < \pi,$$

$$u(r,0) = 0, \qquad u(r,\pi) = 0,$$

$$u(a,\theta) = \theta(\pi - \theta), \quad 0 \le \theta \le \pi,$$

$$u(0,\theta) \text{ is bounded.}$$

12. Solve

$$\nabla^2 u + u = 0, \qquad 0 < r < a, \qquad 0 < \theta < \alpha,$$

$$u(r,0) = 0, \qquad u(r,\alpha) = 0,$$

$$u(a,\theta) = f(\theta), \quad 0 \le \theta \le \alpha,$$

$$u(0,\theta) \text{ is bounded.}$$

13. Find the solution of the Dirichlet problem

$$\nabla^2 u = -2, \qquad r < a, \qquad 0 < \theta < 2\pi,$$

$$u(a,0) = 0.$$

14. Solve the following problems:

(a) $\nabla^2 u = 0,$ $\qquad\qquad 1 < r < 2, \qquad 0 < \theta < 2\pi,$

$\qquad u_r(1,\theta) = \sin\theta, \qquad u_r(2,\theta) = 0, \qquad 0 \le \theta \le 2\pi,$

(b) $\nabla^2 u = 0,$ $\qquad\qquad 1 < r < 2, \qquad 0 < \theta < 2\pi,$

$\qquad u_r(1,\theta) = 0, \qquad u_r(2,\theta) = \theta - \pi, \qquad 0 \le \theta \le 2\pi.$

15. Solve

$$\nabla^2 u = 0, \qquad a < r < b, \qquad 0 < \theta < 2\pi,$$

$$u_r(a,\theta) = f(\theta), \quad u_r(b,\theta) = g(\theta), \quad 0 \le \theta \le 2\pi,$$

where

$$\int_{r=a} f\,ds + \int_{r=b} g\,ds = 0.$$

16. Solve the Robin problem for a semicircular disk

$$\nabla^2 u = 0, \qquad r < R, \qquad 0 < \theta < \pi,$$

$$u_r(R,\theta) = \sin\theta, \quad 0 \le \theta \le \pi,$$

$$u(r,0) = 0, \qquad u(r,\pi) = 0.$$

17. Solve

$$\nabla^2 u = 0, \qquad a < r < b, \qquad\qquad 0 < \theta < \alpha,$$

$$u_r(a, \theta) = 0, \qquad u_r(b, \theta) = f(\theta), \qquad 0 \le \theta \le \alpha,$$

$$u(r, 0) = 0, \qquad u(r, \alpha) = 0, \qquad a < r < b.$$

18. Determine the solution of the mixed boundary-value problem

$$\nabla^2 u = 0, \quad r < R, \qquad 0 < \theta < 2\pi,$$

$$u_r(R, \theta) + hu(R, \theta) = f(\theta), \quad h = \text{constant}.$$

19. Solve

$$\nabla^2 u = 0, \qquad a < r < b, \qquad 0 < \theta < 2\pi,$$

$$u_r(a, \theta) + hu(a, \theta) = f(\theta), \quad u_r(b, \theta) + hu(b, \theta) = g(\theta).$$

20. Find a solution of the Neumann problem

$$\nabla^2 u = -r^2 \sin 2\theta, \qquad r_1 < r < r_2, \qquad 0 < \theta < 2\pi,$$

$$u_r(r_1, \theta) = 0, \qquad u_r(r_2, \theta) = 0, \quad 0 \le \theta \le 2\pi.$$

21. Solve the Robin problem

$$\nabla^2 u = -r^2 \sin 2\theta,$$

$$u(r_1, \theta) = 0, \qquad u_r(r_2, \theta) = 0.$$

22. Solve the following Dirichlet problems:
   (a) $\nabla^2 u = 0,$ $0 < x < 1,$ $0 < y < 1,$

   $$u(x, 0) = x(x - 1), \qquad u(x, 1) = 0, \qquad 0 \le x \le 1,$$

   $$u(0, y) = 0, \qquad u(1, y) = 0, \qquad 0 \le y \le 1.$$

   (b) $\nabla^2 u = 0,$ $0 < x < 1,$ $0 < y < 1,$

   $$u(x, 0) = 0, \qquad u(x, 1) = \sin(\pi x), \qquad 0 \le x \le 1,$$

   $$u(0, y) = 0, \qquad u(1, y) = 0, \qquad 0 \le y \le 1.$$

(c) $\nabla^2 u = 0,$ $0 < x < 1,$ $0 < y < 1,$

   $u(x, 0) = 0,$ $u(x, 1) = 0,$ $0 \leq x \leq 1,$

   $u(0, y) = \left(\cos \frac{\pi y}{2} - 1\right) \cos \left(\frac{\pi y}{2}\right), u(1, y) = 0,$ $0 \leq y \leq 1.$

(d) $\nabla^2 u = 0,$ $0 < x < 1,$ $0 < y < 1,$

   $u(x, 0) = 0,$ $u(x, 1) = 0,$ $0 \leq x \leq 1,$

   $u(0, y) = 0,$ $u(1, y) = \sin \pi y \cos \pi y,$ $0 \leq y \leq 1.$

23. Solve the following Neumann problems:

(a) $\nabla^2 u = 0,$ $0 < x < \pi,$ $0 < y < \pi,$

   $u_x(0, y) = \left(y - \frac{\pi}{2}\right),$ $u_x(\pi, y) = 0,$ $0 \leq y \leq \pi,$

   $u_y(x, 0) = x,$ $u_y(x, \pi) = x,$ $0 \leq x \leq \pi.$

(b) $\nabla^2 u = 0,$ $0 < x < \pi,$ $0 < y < \pi,$

   $u_x(0, y) = 0,$ $u_x(\pi, y) = 2\cos y,$ $0 \leq y \leq \pi,$

   $u_y(x, 0) = 0,$ $u_y(x, \pi) = 0,$ $0 \leq x \leq \pi.$

(c) $\nabla^2 u = 0,$ $0 < x < \pi,$ $0 < y < \pi,$

   $u_x(0, y) = 0,$ $u_x(\pi, y) = 0,$ $0 \leq y \leq \pi,$

   $u_y(x, 0) = \cos x,$ $u_y(x, \pi) = 0,$ $0 \leq x \leq \pi.$

(d) $\nabla^2 u = 0,$ $0 < x < \pi,$ $0 < y < \pi,$

   $u_x(0, y) = y,$ $u_x(\pi, y) = y,$ $0 \leq y \leq \pi,$

   $u_y(x, 0) = x,$ $u_y(x, \pi) = x.$ $0 \leq x \leq \pi.$

24. The steady-state temperature distribution in a rectangular plate of
length a and width b is described by

$$\nabla^2 u = 0, \qquad 0 < x < a, \qquad 0 < y < b.$$

At $x = 0$, the temperature is kept at zero degrees, while at $x = a$, the
plate is insulated. The temperature is prescribed at $y = 0$. At $y = b$,

heat is allowed to radiate freely into the surrounding medium of zero degree temperature. That is, the boundary conditions are

$$u\,(0, y) = 0, \qquad u_x\,(a, y) = 0, \qquad\qquad 0 \le y \le b,$$

$$u\,(x, 0) = f\,(x), \quad u_y\,(x, b) + h\,u\,(x, b) = 0, \quad 0 \le x \le a.$$

Determine the temperature distribution.

25. Solve the Dirichlet problem

$$\nabla^2 u = -2y, \qquad 0 < x < 1, \qquad 0 < y < 1,$$

$$u\,(0, y) = 0, \qquad u\,(1, y) = 0, \qquad 0 \le y \le 1,$$

$$u\,(x, 0) = 0, \qquad u\,(x, 1) = 0, \qquad 0 \le x \le 1.$$

26. Find the harmonic function which vanishes on the hypotenuse and has prescribed values on the other two sides of an isosceles right-angled triangle formed by $x = 0$, $y = 0$, and $y = a - x$, where a is a constant.

27. Find a solution of the Neumann problem

$$\nabla^2 u = \left(x^2 - y^2\right), \qquad 0 < x < a, \qquad 0 < y < a,$$

$$u_x\,(0, y) = 0, \qquad\qquad u_x\,(a, y) = 0, \qquad 0 \le y \le a,$$

$$u_y\,(x, 0) = 0, \qquad\qquad u_y\,(x, a) = 0, \qquad 0 \le x \le a.$$

28. Solve the third boundary-value problem

$$\nabla^2 u = 0, \qquad 0 < x < 1, \qquad 0 < y < 1,$$

$$u_x\,(0, y) + h\,u\,(0, y) = 0, \qquad 0 \le y \le 1, \qquad h = \text{constant}$$

$$u_x\,(1, y) + h\,u\,(1, y) = 0, \qquad 0 \le y \le 1,$$

$$u_y\,(x, 0) + h\,u\,(x, 0) = 0, \qquad 0 \le x \le 1,$$

$$u_y\,(x, 1) + h\,u\,(x, 1) = f\,(x), \quad 0 \le x \le 1.$$

29. Determine the solution of the boundary-value problem

$$\nabla^2 u = 1, \qquad 0 < x < \pi, \qquad 0 < y < \pi,$$

$$u(0, y) = 0, \qquad u_x(\pi, y) = 0, \quad 0 \le y \le \pi,$$

$$u_y(x, 0) = 0, \qquad u_y(x, \pi) + h\, u(x, \pi) = f(x), \quad 0 \le x \le \pi.$$

30. Obtain the integral representation of the Neumann problem

$$\nabla^2 u = f \quad \text{in } D,$$

$$\frac{\partial u}{\partial n} = g \quad \text{on boundary } B \text{ of } D.$$

31. Find the solution in terms of the Green's function of the Poisson problem

$$\nabla^2 u = f \quad \text{in } D,$$

$$\frac{\partial u}{\partial n} + hu = g \quad \text{on boundary } B \text{ of } D.$$

32. Find the steady-state temperature distribution inside a circular annular region governed by the boundary-value problem

$$u_{rr} + \frac{1}{r} u_r + \frac{1}{r^2} u_{\theta\theta} = 0, \qquad 1 < r < 2, \qquad -\pi < 0 < \pi,$$

$$u(1, \theta) = \sin^2 \theta, \qquad u_r(2, \theta) = 0, \qquad -\pi < \theta < \pi.$$

33. Consider a radially symmetric steady-state problem in a solid homogeneous cylinder of radius unity and height h. The temperature distribution function $u(r, z)$ satisfies the equation

$$\nabla^2 u = u_{rr} + \frac{1}{r} u_r + u_{zz} = 0, \qquad 0 < r < 1, \qquad 0 < z < h.$$

Solve the Dirichlet boundary-value problem with the following boundary conditions:

(a) $u(1, z) = f(z),$ $u(r, 0) = 0,$ $u(r, h) = 0,$

(b) $u(1, z) = 0,$ $u(r, 0) = g(r),$ $u(r, h) = 0,$

(c) $u(1, z) = 0,$ $u(r, 0) = 0,$ $u(r, h) = w(r),$

(d) $u(1, z) = a \sin\left(\frac{3\pi z}{h}\right),$ $u(r, 0) = 0 = u(r, h).$

34. Show that the solution of the problem 33(c) is given by

$$u\left(r, z\right) = \sum_{n=1}^{\infty} a_n J_0\left(k_n r\right) \frac{\sinh k_n z}{\sinh k_n h},$$

where $J_0\left(k_n\right) = 0$, $n = 1, 2, 3, \ldots$.

35. Solve the following boundary-value problem:

(a) $\nabla^2 u = 0$, $0 < r < 1$, $0 < z < 1$,

     $u\left(1, z\right) = 0 = u\left(r, 0\right)$, $u\left(r, 1\right) = 1 - r^2$, $0 \leq r \leq 1$.

(b) $\nabla^2 u = 0$, $0 < r < 1$, $0 < z < \pi$,

     $u\left(1, z\right) = A = \text{constant}$, $u\left(r, 0\right) = 0$, $u_z\left(r, \pi\right) = 0$.

(c) $\nabla^2 u = 0$, $0 < r < a$, $0 < z < h$,

     $u\left(a, z\right) = f\left(z\right)$, $u\left(r, 0\right) = 0 = u\left(r, h\right)$, $0 \leq r \leq a$.

(d) $\nabla^2 u = 0$, $0 < r < a$, $0 < z < h$,

     $u\left(a, z\right) = z\left(h - z\right)$, $u\left(r, 0\right) = 0 = u\left(r, h\right)$.

In each of the above problems (a)–(d),

$$\nabla^2 u = u_{rr} + \frac{1}{r} u_r + u_{zz}.$$

10

Higher-Dimensional Boundary-Value Problems

"As long as a branch of knowledge offers an abundance of problems, it is full of vitality."

David Hilbert

10.1 Introduction

The treatment of problems in more than two space variables is much more involved than problems in two space variables. Here a number of multi-dimensional problems involving the Laplace equation, wave and heat equations with various boundary conditions will be presented. Included are the Dirichlet problem for a cube, for a cylinder and for a sphere, wave and heat equations in three dimensional rectangular, cylindrical polar and spherical polar coordinates. The solution of the three-dimensional Schrödinger equation in a central field with applications to hydrogen and helium atoms is discussed. We also consider the forced vibration of a rectangular membrane described by the three-dimensional, nonhomogeneous wave equation with moving boundaries.

10.2 Dirichlet Problem for a Cube

The steady-state temperature distribution in a cube is described by the Laplace equation

$$\nabla^2 u = u_{xx} + u_{yy} + u_{zz} = 0 \qquad (10.2.1)$$

for $0 < x < \pi$, $0 < y < \pi$, $0 < z < \pi$. The faces are kept at zero degree temperature except for the face $z = 0$, that is,

$$u\left(0,y,z\right) = u\left(\pi,y,z\right) = 0$$
$$u\left(x,0,z\right) = u\left(x,\pi,z\right) = 0$$
$$u\left(x,y,\pi\right) = 0,\ u\left(x,y,0\right) = f\left(x,y\right),\quad 0 \le x \le \pi,\quad 0 \le y \le \pi.$$

We assume a nontrivial separable solution in the form

$$u\left(x,y,z\right) = X\left(x\right)Y\left(y\right)Z\left(z\right).$$

Substituting this in the Laplace equation, we obtain

$$X''YZ + XY''Z + XYZ'' = 0.$$

Division by XYZ yields

$$\frac{X''}{X} + \frac{Y''}{Y} = -\frac{Z''}{Z}.$$

Since the right side depends only on z and the left side is independent of z, both terms must be equal to a constant. Thus, we have

$$\frac{X''}{X} + \frac{Y''}{Y} = -\frac{Z''}{Z} = \lambda.$$

By the same reasoning, we have

$$\frac{X''}{X} = \lambda - \frac{Y''}{Y} = \mu.$$

Hence, we obtain the three ordinary differential equations

$$X'' - \mu X = 0,$$
$$Y'' - \left(\lambda - \mu\right)Y = 0,$$
$$Z'' + \lambda Z = 0.$$

Using the boundary conditions, the eigenvalue problem for X,

$$X'' - \mu X = 0,$$
$$X\left(0\right) = X\left(\pi\right) = 0,$$

yields the eigenvalues $\mu = -m^2$ for $m = 1, 2, 3, \ldots$ and the corresponding eigenfunctions $\sin mx$.

Similarly, the eigenvalue problem for Y

$$Y'' - \left(\lambda - \mu\right)Y = 0,$$
$$Y\left(0\right) = Y\left(\pi\right) = 0,$$

gives the eigenvalues $\lambda - \mu = -n^2$ where $n = 1, 2, 3, \ldots$ and the corresponding eigenfunctions $\sin ny$.

Since λ is given by $-\left(m^2 + n^2\right)$, it follows that the solution of $Z'' + \lambda Z = 0$ satisfying the condition $Z(\pi) = 0$ is

$$Z(z) = C \sinh\left[\sqrt{m^2 + n^2}\,(\pi - z)\right].$$

Thus, the solution of the Laplace equation satisfying the homogeneous boundary conditions takes the form

$$u(x, y, z) = \sum_{m=1}^{\infty}\sum_{n=1}^{\infty} a_{mn} \sinh\left(\sqrt{m^2 + n^2}\,(\pi - z)\right) \sin mx \sin ny.$$

Applying the nonhomogeneous boundary condition, we formally obtain

$$f(x, y) = \sum_{m=1}^{\infty}\sum_{n=1}^{\infty} a_{mn} \sinh\left(\sqrt{m^2 + n^2}\,\pi\right) \sin mx \sin ny.$$

The coefficient of the double Fourier series is thus given by

$$a_{mn} \sinh\left(\sqrt{m^2 + n^2}\,\pi\right) = \frac{4}{\pi^2}\int_0^\pi \int_0^\pi f(x, y) \sin mx \sin ny \, dx \, dy.$$

Therefore, the formal solution to the Dirichlet problem for a cube may be written in the form

$$u(x, y, z) = \sum_{m=1}^{\infty}\sum_{n=1}^{\infty} b_{mn} \frac{\sinh\left(\sqrt{m^2 + n^2}\,(\pi - z)\right)}{\sinh\left(\sqrt{m^2 + n^2}\,\pi\right)} \sin mx \sin ny, \quad (10.2.2)$$

where

$$b_{mn} = a_{mn} \sinh\left(\sqrt{m^2 + n^2}\,\pi\right).$$

10.3 Dirichlet Problem for a Cylinder

Example 10.3.1. We consider the problem of determining the electric potential u inside a charge-free cylinder. The potential u satisfies the Laplace equation in cylindrical polar coordinates

$$\nabla^2 u = u_{rr} + \frac{1}{r} u_r + \frac{1}{r^2} u_{\theta\theta} + u_{zz} = 0, \quad (10.3.1)$$

for $0 \le r < a$, $0 < z < l$. Let the lateral surface $r = a$ and the top $z = l$ be grounded, that is, zero potential. Let the potential on the base $z = 0$ be given by

$$u(r, \theta, 0) = f(r, \theta), \quad (10.3.2)$$

where $f(a, \theta) = 0$.

We assume a nontrivial separable solution in the form

$$u(r, \theta, z) = R(r)\,\Theta(\theta)\,Z(z).$$

Substituting this in the Laplace equation yields

$$\frac{R'' + \frac{1}{r}R'}{R} + \frac{1}{r^2}\frac{\Theta''}{\Theta} = -\frac{Z''}{Z} = \lambda.$$

It follows that

$$\frac{r^2 R'' + r R'}{R} - r^2 \lambda = -\frac{\Theta''}{\Theta} = \mu.$$

Thus, we obtain the equations

$$r^2 R'' + r R' - \left(\lambda r^2 + \mu\right) R = 0, \qquad (10.3.3)$$
$$\Theta'' + \mu\Theta = 0, \qquad (10.3.4)$$
$$Z'' + \lambda Z = 0. \qquad (10.3.5)$$

Using the periodicity conditions, the eigenvalue problem for $\Theta(\theta)$,

$$\Theta'' + \mu\Theta = 0,$$
$$\Theta(0) = \Theta(2\pi), \quad \Theta'(0) = \Theta'(2\pi),$$

yields the eigenvalues $\mu = n^2$ for $n = 0, 1, 2, \ldots$ with the corresponding eigenfunctions $\sin n\theta$, $\cos n\theta$. Thus,

$$\Theta(\theta) = A\cos n\theta + B\sin n\theta. \qquad (10.3.6)$$

Suppose λ is real and negative and let $\lambda = -\beta^2$ where $\beta > 0$. If the condition $Z(l) = 0$ is imposed, then the solution of equation (10.3.5) can be written in the form

$$Z(z) = C\sinh\beta(l - z). \qquad (10.3.7)$$

Next we introduce the new independent variable $\xi = \beta r$. Equation (10.3.3) transforms into

$$\xi^2 \frac{d^2 R}{d\xi^2} + \xi \frac{dR}{d\xi} + \left(\xi^2 - n^2\right) R = 0$$

which is the Bessel equation of order n. The general solution is

$$R_n(\xi) = D J_n(\xi) + E Y_n(\xi)$$

where J_n and Y_n are the Bessel functions of the first and second kind respectively. In terms of the original variable, we have

$$R_n(r) = DJ_n(\beta r) + EY_n(\beta r).$$

Since $Y_n(\beta r)$ is unbounded at $r = 0$, we choose $E = 0$. The condition $R(a) = 0$ requires that

$$J_n(\beta a) = 0.$$

For each $n \geq 0$, there exist positive zeros. Arranging these in an infinite increasing sequence, we have

$$0 < \alpha_{n1} < \alpha_{n2} < \ldots < \alpha_{nm} < \ldots.$$

Thus, we obtain

$$\beta_{nm} = (\alpha_{nm}/a).$$

Consequently,

$$R_n(r) = DJ_n(\alpha_{nm}r/a).$$

The solution u then finally takes the form

$$u(r, \theta, z) = \sum_{n=0}^{\infty} \sum_{m=1}^{\infty} J_n\left(\frac{r}{a}\alpha_{nm}\right)(a_{nm}\cos n\theta + b_{nm}\sin n\theta)$$
$$\times \sinh\left[\frac{(l-z)}{a}\alpha_{nm}\right].$$

To satisfy the nonhomogeneous boundary condition, it is required that

$$f(r, \theta) = \sum_{n=0}^{\infty} \sum_{m=1}^{\infty} J_n\left(\frac{r}{a}\alpha_{nm}\right)(a_{nm}\cos n\theta + b_{nm}\sin n\theta)\sinh\left(\frac{l}{a}\alpha_{nm}\right).$$

The coefficients a_{nm} and b_{nm} are given by

$$a_{0m} = \frac{1}{\pi a^2 \sinh\left(\frac{1}{a}\alpha_{0m}\right)[J_1(\alpha_{0m})]^2}\int_0^a \int_0^{2\pi} f(r, \theta) J_0\left(\frac{r}{a}\alpha_{0m}\right) r\,dr\,d\theta,$$

$$a_{nm} = \frac{2}{\pi a^2 \sinh\left(\frac{1}{a}\alpha_{nm}\right)[J_{n+1}(\alpha_{nm})]^2}\int_0^a \int_0^{2\pi} f(r, \theta) J_n\left(\frac{r}{a}\alpha_{nm}\right)$$
$$\times \cos n\theta\, r\,dr\,d\theta,$$

$$b_{nm} = \frac{2}{\pi a^2 \sinh\left(\frac{1}{a}\alpha_{nm}\right)[J_{n+1}(\alpha_{nm})]^2}\int_0^a \int_0^{2\pi} f(r, \theta) J_n\left(\frac{r}{a}\alpha_{nm}\right)$$
$$\times \sin n\theta\, r\,dr\,d\theta.$$

Example 10.3.2. We shall illustrate the same problem with different boundary conditions. Consider the problem

$$\nabla^2 u = 0, \quad 0 \le r < a, \quad 0 < z < \pi,$$
$$u(r, \theta, 0) = 0, \quad u(r, \theta, \pi) = 0,$$
$$u(a, \theta, z) = f(\theta, z).$$

As before, by the separation of variables, we obtain

$$r^2 R'' + r R' - \left(\lambda r^2 + \mu\right) R = 0,$$
$$\Theta'' + \mu \Theta = 0,$$
$$Z'' + \lambda Z = 0.$$

By the periodicity conditions, again as in the previous example, the Θ equation yields the eigenvalues $\mu = n^2$ with $n = 0, 1, 2, \ldots$; the corresponding eigenfunctions are $\sin n\theta$, $\cos n\theta$. Thus, we have

$$\Theta(\theta) = A_n \cos n \cos \theta + B_n \sin n\theta.$$

Now let $\lambda = \beta^2$ with $\beta > 0$. Then, the boundary value problem

$$Z'' + \beta^2 Z = 0$$
$$Z(0) = 0, \quad Z(\pi) = 0,$$

has the solution

$$Z(z) = C_m \sin mz, \qquad m = 1, 2, 3, \ldots.$$

Finally, we have

$$r^2 R'' + r R' - \left(m^2 r^2 + n^2\right) R = 0,$$

or

$$R'' + \frac{1}{r} R' - \left(m^2 + \frac{n^2}{r^2}\right) R = 0,$$

the general solution of which is

$$R(r) = D I_n(mr) + E K_n(mr),$$

where I_n and K_n are the modified Bessel functions of the first and second kind, respectively.

Since R must remain finite at $r = 0$, we set $E = 0$. Then R takes the form

$$R(r) = D I_n(mr).$$

Applying the nonhomogeneous condition, we find the solution

$$u\left(r,\theta,z\right) = \sum_{m=1}^{\infty} \left(\frac{a_{m0}}{2}\right) \frac{I_0\left(mr\right)}{I_0\left(ma\right)} \sin mz$$

$$+ \sum_{m=1}^{\infty}\sum_{n=1}^{\infty} \left(a_{mn}\cos n\theta + b_{mn}\sin n\theta\right) \frac{I_n\left(mr\right)}{I_n\left(ma\right)} \sin mz,$$

where

$$a_{mn} = \frac{2}{\pi^2} \int_0^\pi \int_0^{2\pi} f\left(\theta,z\right) \sin mz \cos n\theta \, d\theta \, dz,$$

$$b_{mn} = \frac{2}{\pi^2} \int_0^\pi \int_0^{2\pi} f\left(\theta,z\right) \sin mz \sin n\theta \, d\theta \, dz.$$

10.4 Dirichlet Problem for a Sphere

Example 10.4.3. To determine the potential in a sphere, we transform the Laplace equation into spherical coordinates. It has the form

$$\nabla^2 u = u_{rr} + \frac{2}{r} u_r + \frac{1}{r^2} u_{\theta\theta} + \frac{\cot\theta}{r^2} u_\theta + \frac{1}{r^2 \sin^2\theta} u_{\varphi\varphi}, \quad (10.4.1)$$

where $0 \le r < a$, $0 < \theta < \pi$, and $0 < \varphi < 2\pi$.

Let the prescribed potential on the sphere be

$$u\left(a,\theta,\varphi\right) = f\left(\theta,\varphi\right). \quad (10.4.2)$$

We assume a nontrivial separable solution in the form

$$u\left(r,\theta,\varphi\right) = R\left(r\right)\Theta\left(\theta\right)\Phi\left(\varphi\right).$$

Substitution of u in the Laplace equation yields

$$r^2 R'' + 2r R' - \lambda R = 0, \quad (10.4.3)$$

$$\sin^2\theta\,\Theta'' + \sin\theta\cos\theta\,\Theta' + \left(\lambda\sin^2\theta - \mu\right)\Theta = 0, \quad (10.4.4)$$

$$\Phi'' + \mu\Phi = 0. \quad (10.4.5)$$

The general solution of equation (10.4.5) is

$$\Phi\left(\varphi\right) = A\cos\sqrt{\mu}\,\varphi + B_n \sin\sqrt{\mu}\,\varphi. \quad (10.4.6)$$

The periodicity condition requires that

$$\sqrt{\mu} = m, \qquad m = 0,1,2,\ldots.$$

Since equation (10.4.3) is of Euler type, the solution is of the form

$$R\left(r\right) = r^\beta.$$

Inserting this is equation (10.4.3), we obtain

$$\beta^2 + \beta - \lambda = 0.$$

The roots are $\beta = \left(-1 + \sqrt{1 + 4\lambda}\right)/2$ and $-(1 + \beta)$. Hence, the general solution of equation (10.4.3) is

$$R(r) = C r^\beta + D r^{-(1+\beta)}. \qquad (10.4.7)$$

The variable $\xi = \cos\theta$ transforms equation (10.4.4) into the form

$$\left(1 - \xi^2\right)\Theta'' - 2\xi\Theta' + \left[\beta(\beta + 1) - \frac{m^2}{1 - \xi^2}\right]\Theta = 0 \qquad (10.4.8)$$

which is Legendre's associated equation. The general solution with $\beta = n$ for $n = 0, 1, 2, \ldots$ is

$$\Theta(\theta) = E P_n^m(\cos\theta) + F Q_n^m(\cos\theta).$$

Continuity of $\Theta(\theta)$ at $\theta = 0$, π corresponds to continuity of $\Theta(\xi)$ at $\xi = \pm 1$. Since $Q_n^m(\xi)$ has a logarithmic singularity at $\xi = 1$, we choose $F = 0$. Thus, the solution of equation (10.4.8) becomes

$$\Theta(\theta) = E P_n^m(\cos\theta).$$

Consequently, the solution of the Laplace equation in spherical coordinates is

$$u(r, \theta, \varphi) = \sum_{n=0}^{\infty} \sum_{m=0}^{n} r^n P_n^m(\cos\theta)(a_{nm}\cos m\varphi + b_{nm}\sin m\varphi).$$

In order for u to satisfy the prescribed function on the boundary, it is necessary that

$$f(\theta, \varphi) = \sum_{n=0}^{\infty} \sum_{m=0}^{n} a^n P_n^m(\cos\theta)(a_{nm}\cos m\varphi + b_{nm}\sin m\varphi)$$

for $0 \le \theta \le \pi$, $0 \le \varphi \le 2\pi$. By the orthogonal properties of the functions $P_n^m(\cos\theta)\cos m\varphi$ and $P_n^m(\cos\theta)\sin m\varphi$, the coefficients are given by

$$a_{nm} = \frac{(2n+1)}{2\pi a^n}\frac{(n-m)!}{(n+m)!}\int_0^{2\pi}\int_0^{\pi} f(\theta, \varphi) P_n^m(\cos\theta)\cos m\varphi \sin\theta \, d\theta \, d\varphi,$$

$$b_{nm} = \frac{(2n+1)}{2\pi a^n}\frac{(n-m)!}{(n+m)!}\int_0^{2\pi}\int_0^{\pi} f(\theta, \varphi) P_n^m(\cos\theta)\sin m\varphi \sin\theta \, d\theta \, d\varphi,$$

for $m = 1, 2, \ldots$, and $n = 1, 2, \ldots$, and

$$a_{n0} = \frac{(2n+1)}{4\pi a^n}\int_0^{2\pi}\int_0^{\pi} f(\theta, \varphi) P_n(\cos\theta)\sin\theta \, d\theta \, d\varphi,$$

for $n = 0, 1, 2, \ldots$.

Example 10.4.4. Determine the potential of a grounded conducting sphere in a uniform field that satisfies the problem

$$\nabla^2 u = 0, \qquad 0 \le r < a, \ 0 < \theta < \pi, \ 0 < \phi < 2\pi,$$
$$u(a, \theta) = 0, \qquad u \to -E_0 \, r \cos \theta, \quad \text{as} \quad r \to \infty.$$

Let the field be in the z direction so that the potential u will be independent of ϕ. Then, the Laplace equation takes the form

$$u_{rr} + \frac{2}{r} u_r + \frac{1}{r^2} u_{\theta\theta} + \frac{\cot \theta}{r^2} u_\theta = 0.$$

We assume a nontrivial separable solution in the form

$$u(r, \theta) = R(r) \, \Theta(\theta).$$

Substitution of this in the Laplace equation yields

$$r^2 R'' + 2r R' - \lambda R = 0,$$
$$\sin^2 \theta \, \Theta'' + \sin \theta \cos \theta \, \Theta' + \lambda \sin^2 \theta \, \Theta = 0.$$

If we set $\lambda = n(n+1)$ with $n = 0, 1, 2, \ldots$, then the second equation is the Legendre equation. The general solution of this equation is

$$\Theta(\theta) = A_n \, P_n(\cos \theta) + B_n \, Q_n(\cos \theta),$$

where P_n and Q_n are the Legendre functions of the first and second kind respectively. In order for the solution not to be singular at $\theta = 0$ and $\theta = \pi$, we set $B_n = 0$. Thus, $\Theta(\theta)$ becomes

$$\Theta(\theta) = A_n \, P_n(\cos \theta).$$

The solution of the R-equation is obtained in the form

$$R(r) = C_n \, r^n + D_n \, r^{-(n+1)}.$$

Thus, the potential function is

$$u(r, \theta) = \sum_{n=0}^{\infty} \left(a_n \, r^n + b_n \, r^{-(n+1)} \right) P_n(\cos \theta).$$

To satisfy the condition at infinity, we must have

$$a_1 = -E_0, \quad \text{and} \quad a_n = 0, \quad \text{for} \quad n \ge 2$$

and hence,

$$u(r, \theta) = -E_0 \, r \cos \theta + \sum_{n=1}^{\infty} \frac{b_n}{r^{n+1}} P_n(\cos \theta).$$

The condition $u(a, \theta) = 0$ yields

$$0 = -E_0 a \cos \theta + \sum_{n=1}^{\infty} \frac{b_n}{a^{n+1}} P_n (\cos \theta).$$

Using the orthogonality of the Legendre functions, we find that b_n are given by

$$b_n = \frac{(2n + 1)}{2} E_0 a^{n+2} \int_{-\pi}^{\pi} \cos \theta\, P_n (\cos \theta)\, d(\cos \theta) = E_0 a^3 \delta_{n1},$$

since the integral vanishes for all n except $n = 1$. Hence, the potential function is given by

$$u(r, \theta) = -E_0 r \cos \theta + E_0 \frac{a^3}{r^2} \cos \theta.$$

Example 10.4.5. A dielectric sphere of radius a is placed in a uniform electric field E_0. Determine the potentials inside and outside the sphere.

The problem is to find potentials u_1 and u_2 that satisfy

$$\nabla^2 u_1 = \nabla^2 u_2 = 0,$$
$$K \frac{\partial u_1}{\partial r} = \frac{\partial u_2}{\partial r}, \quad u_1 = u_2, \quad \text{on} \quad r = a,$$
$$u_2 \to -E_0 r \cos \theta \quad \text{as} \quad r \to \infty,$$

where u_1 and u_2 are the potentials inside and outside the sphere, respectively, and K is the dielectric constant.

As in the preceeding example, the potential function is

$$u(r, \theta) = \sum_{n=0}^{\infty} \left(a_n r^n + b_n r^{-(n+1)} \right) P_n (\cos \theta). \qquad (10.4.9)$$

Since u_1 must be finite at the origin, we take

$$u_1 (r, \theta) = \sum_{n=0}^{\infty} a_n r^n P_n (\cos \theta) \quad \text{for} \quad r \leq a. \qquad (10.4.10)$$

For u_2, which must approach infinity in the prescribed manner, we choose

$$u_2 (r, \theta) = -E_0 r \cos \theta + \sum_{n=0}^{\infty} b_n r^{-(n+1)} P_n (\cos \theta). \qquad (10.4.11)$$

From the two continuity conditions at $r = a$, we obtain

$$a_1 = -E_0 + \frac{b_1}{a^3}, \qquad K a_1 = -E_0 - \frac{2b_1}{a^3},$$
$$a_n = b_n = 0, \qquad n \geq 2.$$

The coefficients a_1 and b_1 are then found to be

$$a_1 = -\frac{3E_0}{K+2}, \qquad b_1 = E_0\, a^3 \frac{(K-1)}{(K+2)}.$$

Hence, the potential for $r \le a$ is given by

$$u_1(r, \theta) = -\frac{3E_0}{K+2}\, r \cos\theta,$$

and the potential for $r \ge a$ is given by

$$u_2(r, \theta) = -E_0\, r \cos\theta + E_0\, a^3 \frac{(K-1)}{(K+2)}\, r^{-2} \cos\theta.$$

Example 10.4.6. Determine the potential between concentric spheres held at different constant potentials.

Here we need to solve

$$\nabla^2 u = 0, \qquad a < r < b,$$
$$u = A \quad \text{on} \quad r = a,$$
$$u = B \quad \text{on} \quad r = b.$$

In this case, the potential depends only on the radial distance. Hence, we have

$$\frac{1}{r^2} \frac{\partial}{\partial r}\left(r^2 \frac{\partial u}{\partial r}\right) = 0.$$

By elementary integration, we obtain

$$u(r) = c_1 + \frac{c_2}{r},$$

where c_1 and c_2 are arbitrary constants.

Applying the boundary conditions, we obtain

$$c_1 = \frac{Bb - Aa}{b-a}, \qquad c_2 = (A-B)\frac{ab}{b-a}.$$

Thus, the solution is

$$u(r) = \frac{Bb - Aa}{(b-a)} + \frac{(A-B)\, ab}{(b-a)\, r}$$
$$= \left(\frac{Bb}{r}\right)\left(\frac{r-a}{b-a}\right) + \left(\frac{Aa}{r}\right)\left(\frac{b-r}{b-a}\right).$$

10.5 Three-Dimensional Wave and Heat Equations

The wave equation in three space variables may be written as

$$u_{tt} = c^2 \nabla^2 u, \tag{10.5.1}$$

where ∇^2 is the three-dimensional Laplace operator.

We assume a nontrivial separable solution in the form

$$u(x, y, z, t) = U(x, y, z) T(t).$$

Substituting this into equation (10.5.1), we obtain

$$T'' + \lambda c^2 T = 0, \tag{10.5.2}$$
$$\nabla^2 U + \lambda U = 0, \tag{10.5.3}$$

where $-\lambda$ is a separation constant. The variables are separated and the solutions of equations (10.5.2) and (10.5.3) are to be determined.

Next we consider the heat equation

$$u_t = k \nabla^2 u. \tag{10.5.4}$$

As before, we seek a nontrivial separable solution in the form

$$u(x, y, z, t) = U(x, y, z) T(t).$$

Substituting this into equation (10.5.4), we obtain

$$T' + \lambda k T = 0,$$
$$\nabla^2 U + \lambda U = 0.$$

Thus, we see that the problem here, as in the previous case, is essentially that of solving the Helmholtz equation

$$\nabla^2 U + \lambda U = 0.$$

10.6 Vibrating Membrane

As a specific example of the higher-dimensional wave equation, let us determine the solution of the problem of the vibrating membrane of length a and width b. The initial boundary-value problem for the displacement function $u(x, y, t)$ is

$$u_{tt} = c^2 (u_{xx} + u_{yy}), \qquad 0 < x < a,\ 0 < y < b,\ t > 0, \tag{10.6.1}$$
$$u(x, y, 0) = f(x, y), \qquad 0 \le x \le a, \quad 0 \le y \le b, \tag{10.6.2}$$
$$u_t(x, y, 0) = g(x, y), \qquad 0 \le x \le a, \quad 0 \le y \le b, \tag{10.6.3}$$
$$u(0, y, t) = 0, \qquad u(a, y, t) = 0, \tag{10.6.4}$$
$$u(x, 0, t) = 0, \qquad u(x, b, t) = 0. \tag{10.6.5}$$

We have just shown that the separated equations for the wave equation are

$$T'' + \lambda c^2 T = 0, \qquad\qquad (10.6.6)$$
$$\nabla^2 U + \lambda U = 0, \qquad\qquad (10.6.7)$$

where, in this case, $\nabla^2 U = U_{xx} + U_{yy}$. Let $\lambda = \alpha^2$. Then the solution of equation (10.6.6) is

$$T(t) = A \cos \alpha c t + B \sin \alpha c t.$$

Now we look for a nontrivial solution of equation (10.6.7) in the form

$$U(x, y) = X(x) Y(y).$$

Substituting this into equation (10.6.7) yields

$$X'' - \mu X = 0,$$
$$Y'' + (\lambda + \mu) Y = 0.$$

If we let $\mu = -\beta^2$, then the solutions of these equations take the form

$$X(x) = C \cos \beta x + D \sin \beta x.$$
$$Y(y) = E \cos \gamma y + F \sin \gamma y,$$

where

$$\gamma^2 = (\lambda + \mu) = \alpha^2 - \beta^2.$$

The homogeneous boundary conditions in x require that $C = 0$ and

$$D \sin \beta a = 0$$

which implies that $\beta = (m\pi/a)$ with $D \neq 0$. Similarly, the homogeneous boundary conditions in y require that $E = 0$ and

$$F \sin \gamma b = 0$$

which implies that $\gamma = (n\pi/b)$ with $F \neq 0$. Noting that m and n are independent integers, we obtain the displacement function in the form

$$u(x, y, t) = \sum_{m=1}^{\infty} \sum_{n=1}^{\infty} (a_{mn} \cos \alpha_{mn} ct + b_{mn} \sin \alpha_{mn} ct) \sin \left(\frac{m\pi x}{a}\right) \sin \left(\frac{n\pi y}{b}\right),$$

$$(10.6.8)$$

where $\alpha_{mn} = (m^2 \pi^2 / a^2) + (n^2 \pi^2 / b^2)$, a_{mn} and b_{mn} are constants.
Now applying the nonhomogeneous initial conditions, we have

$$u\left(x,y,0\right) = f\left(x,y\right) = \sum_{m=1}^{\infty} \sum_{n=1}^{\infty} a_{mn} \sin\left(\frac{m\pi x}{a}\right) \sin\left(\frac{n\pi y}{b}\right),$$

and thus,

$$a_{mn} = \frac{4}{ab} \int_0^a \int_0^b f\left(x,y\right) \sin\left(\frac{m\pi x}{a}\right) \sin\left(\frac{n\pi y}{b}\right) dx\, dy. \quad (10.6.9)$$

In a similar manner, the initial condition on u_t implies

$$u_t\left(x,y,0\right) = g\left(x,y\right) = \sum_{m=1}^{\infty} \sum_{n=1}^{\infty} b_{mn}\, \alpha_{mn}\, c \sin\left(\frac{m\pi x}{a}\right) \sin\left(\frac{n\pi y}{b}\right),$$

from which it follows that

$$b_{mn} = \frac{4}{\alpha_{mn}\, abc} \int_0^a \int_0^b g\left(x,y\right) \sin\left(\frac{m\pi x}{a}\right) \sin\left(\frac{n\pi y}{b}\right) dx\, dy. \quad (10.6.10)$$

The solution of the rectangular membrane problem is, therefore, given by equation (10.6.8).

Example 10.6.1. (**Vibration of a Circular Membrane**). For a circular elastic membrane that is stretched over a circular frame of radius a, the motion of the membrane can be described by a function $u\left(r,\theta,t\right)$ that satisfies the partial differential equation

$$\frac{1}{c^2} u_{tt} = u_{rr} + \frac{1}{r} u_r + \frac{1}{r^2} u_{\theta\theta}, \quad (10.6.11)$$

where $c^2 = \left(T/\rho\right)$, T is the tension in the membrane and ρ is its mass density.

We consider the *synchronous vibrations* of the vibration of the membrane defined by the separable solution

$$u\left(r,\theta,t\right) = v\left(r,\theta,t\right) \cos\left(\omega ct\right). \quad (10.6.12)$$

Substituting (10.6.12) into (10.6.11) gives

$$v_{rr} + \frac{1}{r} v_r + \frac{1}{r^2} u_{\theta\theta} + \omega^2 v = 0. \quad (10.6.13)$$

We seek a nontrivial separable solution

$$v\left(r,\theta\right) = R\left(r\right)\Theta\left(\theta\right)$$

of equation (10.6.13) so that

$$\frac{r^2 R'' + r\, R'}{R} + \omega^2 r^2 = -\frac{\Theta''}{\Theta} = \lambda^2. \quad (10.6.14)$$

This must hold for all points of the membrane, $0 < r < a$ and $0 \leq \theta \leq 2\pi$. Consequently,

$$r^2 R'' + r R' + \left(\omega^2 r^2 - \lambda^2\right) R = 0, \qquad 0 < r < a, \qquad (10.6.15)$$

$$\Theta'' + \lambda^2 \Theta = 0, \qquad 0 \leq \theta \leq 2\pi. \qquad (10.6.16)$$

The general solution of (10.6.16) is

$$\Theta(\theta) = A \cos \lambda\theta + B \sin \lambda\theta. \qquad (10.6.17)$$

This represents a single-valued solution at all points of the disk only if $\lambda = n$ is an integer. Thus,

$$\Theta(\theta) = \Theta_n(\theta) = A_n \cos n\theta + B_n \sin n\theta. \qquad (10.6.18)$$

With $\lambda = n$, the radial equation (10.6.15) becomes

$$r^2 R'' + r R' + \left(\omega^2 r^2 - n^2\right) R = 0. \qquad (10.6.19)$$

The parameter ω can be eliminated by defining

$$x = \omega r, \quad \text{and} \quad y(x) = y(\omega r) = R(r).$$

Substituting these into (10.6.19) gives

$$x^2 y'' + x y' + \left(x^2 - n^2\right) y = 0. \qquad (10.6.20)$$

Or, equivalently,

$$y'' + \frac{1}{x} y' + \left(1 - \frac{n^2}{x^2}\right) y = 0. \qquad (10.6.21)$$

This is the well-known *Bessel equation of order n*, which has been discussed in Section 8.6.

10.7 Heat Flow in a Rectangular Plate

Another example of a two-dimensional problem is the conduction of heat in a thin rectangular plate. Let the plate of length a and width b be perfectly insulated at the faces $x = 0$ and $x = a$. Let the two other sides be maintained at zero temperature. Let the initial temperature distribution be $f(x, y)$. Then, we seek the solution of the initial boundary-value problem

$$u_t = k \nabla^2 u, \qquad 0 < x < a, \quad 0 < y < b, \quad t > 0, \qquad (10.7.1)$$

$$u(x, y, 0) = f(x, y), \qquad 0 \leq x \leq a, \quad 0 \leq y \leq b, \qquad (10.7.2)$$

$$u_x\left(0, y, t\right) = 0, \qquad u_x\left(a, y, t\right) = 0, \tag{10.7.3ab}$$
$$u\left(x, 0, t\right) = 0, \qquad u\left(x, b, t\right) = 0. \tag{10.7.4ab}$$

As shown earlier, the separated equations for this problem are found to be

$$T' + \lambda k T = 0, \tag{10.7.5}$$
$$\nabla^2 U + \lambda U = 0. \tag{10.7.6}$$

We assume a nontrivial separable solution in the form

$$U\left(x, y\right) = X\left(x\right) Y\left(y\right).$$

Inserting this in equation (10.7.6), we obtain

$$X'' - \mu X = 0, \tag{10.7.7}$$
$$Y'' + \left(\lambda + \mu\right) Y = 0. \tag{10.7.8}$$

Because the conditions in x are homogeneous, we choose $\mu = -\alpha^2$ so that

$$X\left(x\right) = A \cos \alpha x + B \sin \alpha x.$$

Since $X'\left(0\right) = 0$, $B = 0$ and since $X'\left(a\right) = 0$,

$$\sin \alpha a = 0, \qquad A \neq 0$$

which gives

$$\alpha = \left(m\pi/a\right), \qquad m = 1, 2, 3, \ldots.$$

We note that $\mu = 0$ is also an eigenvalue. Consequently,

$$X_m\left(x\right) = A_m \cos\left(m\pi x/a\right), \qquad m = 0, 1, 2, \ldots.$$

Similarly, for nontrivial solution Y, we select $\beta^2 = \lambda + \mu = \lambda - \alpha^2$ so that the solution of equation (10.7.8) is

$$Y\left(y\right) = C \cos \beta y + D \sin \beta y.$$

Applying the homogeneous conditions, we find $C = 0$ and

$$\sin \beta b = 0, \qquad D \neq 0.$$

Thus, we obtain

$$\beta = \left(n\pi/b\right); \qquad n = 1, 2, 3, \ldots,$$

and

$$Y_n (y) = D_n \sin (n\pi y/b).$$

Recalling that $\lambda = \alpha^2 + \beta^2$, the solution of equation (10.7.5) may be written in the form

$$T_{mn}(t) = E_{mn}\, e^{-\left(m^2/a^2 + n^2/b^2\right)\pi^2 kt}.$$

Thus, the solution of the heat equation satisfying the prescribed boundary conditions may be written as

$$u(x,y,t) = \sum_{m=0}^{\infty} \sum_{n=1}^{\infty} a_{mn}\, e^{-\left(m^2/a^2 + n^2/b^2\right)\pi^2 kt} \cos\left(\frac{m\pi x}{a}\right) \sin\left(\frac{n\pi y}{b}\right),$$

$$(10.7.9)$$

where $a_{mn} = A_m D_m E_{mn}$ are arbitrary constants.

Applying the initial condition, we obtain

$$u(x,y,0) = f(x,y) = \sum_{m=0}^{\infty} \sum_{n=1}^{\infty} a_{mn} \cos\left(\frac{m\pi x}{a}\right) \sin\left(\frac{n\pi y}{b}\right). \quad (10.7.10)$$

This is a double Fourier series, and the coefficients are given by

$$a_{0n} = \left(\frac{2}{ab}\right) \int_0^a \int_0^b f(x,y) \sin\left(\frac{n\pi y}{b}\right) dx\, dy,$$

and for $m \geq 1$

$$a_{mn} = \left(\frac{4}{ab}\right) \int_0^a \int_0^b f(x,y) \cos\left(\frac{m\pi x}{a}\right) \sin\left(\frac{n\pi y}{b}\right) dx\, dy.$$

The solution of the heat equation is thus given by equation (10.7.9).

Example 10.7.1. (**Steady-state temperature in a Circular Disk**). We next consider the steady-state temperature distribution $u(r,\theta)$ in a circular disk of radius $r = a$ that satisfies the Laplace equation

$$u_{rr} + \frac{1}{r} u_r + \frac{1}{r^2} u_{\theta\theta} = 0, \qquad 0 \leq r \leq a, \quad 0 \leq \theta \leq 2\pi, \ (10.7.11)$$

$$u(r,\theta) = f(\theta), \quad \text{on} \quad r = a \quad \text{for all} \quad \theta, \quad (10.7.12)$$

where $f(\theta)$ is a given function of θ.

This is exactly the Dirichlet problem for a circle that was already solved in Section 9.4.

We also consider the steady-state temperature distribution $u(r,\theta,\phi)$ in a sphere of radius a where $0 \leq r < a$, $0 < \theta < \pi$ and $0 < \phi < 2\pi$. For simplicity, we assume only steady temperature distribution which depends on r and θ. Thus, u is independent of the longitudinal coordinate

ϕ, and hence, the steady-state temperature distribution $u(r, \theta)$ satisfies the Laplace equation in spherical polar coordinates in the form

$$\frac{\partial}{\partial r}\left(r^2 \frac{\partial u}{\partial r}\right) + \frac{1}{\sin\theta}\frac{\partial}{\partial\theta}\left(\sin\theta \frac{\partial u}{\partial\theta}\right) = 0. \qquad (10.7.13)$$

We seek a separable solution of (10.7.13) in the form $u(r, \theta) = R(r)\Theta(\theta)$ so that (10.7.13) leads to

$$\frac{1}{R}\frac{d}{dr}\left(r^2 \frac{dR}{dr}\right) = -\frac{1}{\Theta(\theta)\sin\theta}\frac{d}{d\theta}\left(\sin\theta \frac{d\Theta}{d\theta}\right) = 0. \qquad (10.7.14)$$

This must hold for $0 < r < a$ and $0 < \theta < \pi$. Consequently,

$$\frac{1}{R}\frac{d}{dr}\left(r^2 \frac{dR}{dr}\right) = -\frac{1}{\Theta(\theta)\sin\theta}\frac{d}{d\theta}\left(\sin\theta \frac{d\Theta}{d\theta}\right) = \lambda, \qquad (10.7.15)$$

or

$$\frac{d}{dr}\left(r^2 \frac{dR}{dr}\right) - \lambda R = 0, \qquad 0 < r < a, \qquad (10.7.16)$$

$$\frac{d}{d\theta}\left(\sin\theta \frac{d\Theta}{d\theta}\right) + \lambda \sin\theta \, \Theta(\theta) = 0, \qquad 0 < \theta < \pi. \qquad (10.7.17)$$

Equation for $R(r)$ can also be written as

$$R'' + \frac{2}{r}R' - \frac{\lambda}{r^2}R = 0. \qquad (10.7.18)$$

We simplify equation (10.7.17) by the change of variable.

$$x = \cos\theta, \qquad y(x) = \Theta(\theta).$$

Using the chain rule we obtain

$$\frac{d\Theta}{d\theta} = \frac{dy}{dx}\frac{dx}{d\theta} = -(\sin\theta)\frac{dy}{dx}$$

and hence,

$$\begin{aligned}
\frac{d}{d\theta}\left(\sin\theta \frac{d\Theta}{d\theta}\right) &= -\frac{d}{d\theta}\left(\sin^2\theta \frac{dy}{dx}\right) \\
&= -\frac{d}{dx}\left[(1 - x^2)\frac{dy}{dx}\right]\frac{dx}{d\theta} \\
&= \sin\theta \frac{d}{dx}\left[(1 - x^2)\frac{dy}{dx}\right].
\end{aligned}$$

Combining this result with (10.7.17) leads to the Legendre equation

$$\frac{d}{dx}\left[\left(1-x^2\right)\frac{dy}{dx}\right] + \lambda y = 0, \qquad -1 \le x \le 1, \qquad (10.7.19)$$

or, equivalently,

$$\left(1-x^2\right)\frac{d^2y}{dx^2} - 2x\,\frac{dy}{dx} + \lambda\,y = 0. \qquad (10.7.20)$$

This equation was completely solved in Section 8.9. Equation (10.7.19) is the well-known Sturm–Liouville equation with $y\,(-1)$ and $y\,(+1)$ finite. The results are

$$\lambda = \lambda_n = n\,(n+1), \qquad y\,(x) = P_n\,(x), \qquad n = 0, 1, 2, 3, \ldots,$$

where $P_n\,(x)$ is the Legendre polynomial of degree n.

10.8 Waves in Three Dimensions

The propagation of waves due to an initial disturbance in a rectangular volume is best described by the solution of the initial boundary-value problem

$$u_{tt} = c^2\nabla^2 u, \qquad 0 < x < a, \quad 0 < y < b, \quad 0 < z < d, \quad t > 0, \tag{10.8.1}$$

$$u\,(x,y,z,0) = f\,(x,y,z), \quad 0 \le x \le a, \quad 0 \le y \le b, \quad 0 \le z \le d, \tag{10.8.2}$$

$$u_t\,(x,y,z,0) = g\,(x,y,z), \quad 0 \le x \le a, \quad 0 \le y \le b, \quad 0 \le z \le d, \tag{10.8.3}$$

$$u\,(0,y,z,t) = 0, \quad u\,(a,y,z,t) = 0, \tag{10.8.4}$$

$$u\,(x,0,z,t) = 0, \quad u\,(x,b,z,t) = 0, \tag{10.8.5}$$

$$u\,(x,y,0,t) = 0, \quad u\,(x,y,d,t) = 0. \tag{10.8.6}$$

We assume a nontrivial separable solution in the form

$$u\,(x,y,z,t) = U\,(x,y,z)\,T\,(t)\,.$$

The separated equations are given by

$$T'' + \lambda c^2 T = 0, \tag{10.8.7}$$

$$\nabla^2 U + \lambda U = 0. \tag{10.8.8}$$

We assume that U has the nontrivial separable solution in the form

$$U\,(x,y,z) = X\,(x)\,Y\,(y)\,Z\,(z)\,.$$

Substitution of this into equation (10.8.8) yields

$$X'' - \mu X = 0, \tag{10.8.9}$$

$$Y'' - \nu Y = 0, \tag{10.8.10}$$

$$Z'' + (\lambda + \mu + \nu)\,Z = 0. \tag{10.8.11}$$

Because of the homogeneous conditions in x, we let $\mu = -\alpha^2$ so that

$$X(x) = A\cos\alpha x + B\sin\alpha x.$$

As in the preceding examples, we obtain

$$X_l(x) = B_l \sin\left(\frac{l\pi x}{a}\right), \qquad l = 1, 2, 3, \dots.$$

In a similar manner, we let $\nu = -\beta^2$ to obtain

$$Y(y) = C\cos\beta y + D\sin\beta y$$

and accordingly,

$$Y_m(y) = D_m \sin\left(\frac{m\pi y}{b}\right), \qquad m = 1, 2, 3, \dots.$$

We again choose $\gamma^2 = \lambda + \mu + \nu = \lambda - \alpha^2 - \beta^2$ so that

$$Z(z) = E\cos(\gamma z) + F\sin(\gamma z).$$

Applying the homogeneous conditions in z, we obtain

$$Z_n(z) = F_n \sin\left(\frac{n\pi z}{d}\right).$$

Since the solution of equation (10.8.7) is

$$T(t) = G\cos\left(\sqrt{\lambda}\,ct\right) + H\sin\left(\sqrt{\lambda}\,ct\right),$$

the solution of the wave equation has the form

$$u(x, y, z, t) = \sum_{l=1}^{\infty}\sum_{m=1}^{\infty}\sum_{n=1}^{\infty}\left(a_{lmn}\cos\sqrt{\lambda}\,ct + b_{lmn}\sin\sqrt{\lambda}\,ct\right)$$

$$\times \sin\left(\frac{l\pi x}{a}\right)\sin\left(\frac{m\pi y}{b}\right)\sin\left(\frac{n\pi z}{d}\right)$$

where a_{lmn} and b_{lmn} are arbitrary constants. The coefficients a_{lmn} are determined from the initial condition $u(x, y, z, 0) = f(x, y, z)$ and are found to be

$$a_{lmn} = \frac{8}{abd}\int_0^a\int_0^b\int_0^d f(x, y, z)\sin\left(\frac{l\pi x}{a}\right)\sin\left(\frac{m\pi y}{b}\right)\sin\left(\frac{n\pi z}{d}\right)dx\,dy\,dz.$$

Similarly the coefficients b_{lmn} are determined from the initial condition $u(x, y, z, 0) = g(x, y, z)$ and are found to be

$$b_{lmn}$$

$$= \frac{8}{\sqrt{\lambda}\,acbd}\int_0^a\int_0^b\int_0^d g(x, y, z)\sin\left(\frac{l\pi x}{a}\right)\sin\left(\frac{m\pi y}{b}\right)\sin\left(\frac{n\pi z}{d}\right)dx\,dy\,dz,$$

where

$$\lambda = \left(\frac{l^2}{a^2} + \frac{m^2}{b^2} + \frac{n^2}{d^2}\right)\pi^2.$$

10.9 Heat Conduction in a Rectangular Volume

As in the case of the wave equation, the solution of the heat equation in three spaces variables can be determined. Consider the problem of heat distribution in a rectangular volume. The faces are maintained at zero degree temperature. The solid is initially heated so that the problem may be written as

$$u_t = k\nabla^2 u, \qquad 0 < x < a, \quad 0 < y < b, \quad 0 < z < d, \quad t > 0,$$
$$u(x, y, z, 0) = f(x, y, z), \quad 0 \le x \le a, \quad 0 \le y \le b, \quad 0 \le z \le d,$$
$$u(0, y, z, t) = 0, \qquad u(a, y, z, t) = 0,$$
$$u(x, 0, z, t) = 0, \qquad u(x, b, z, t) = 0,$$
$$u(x, y, 0, t) = 0, \qquad u(x, y, d, t) = 0.$$

As before, the separable equations are

$$T' + \lambda k T = 0, \tag{10.9.1}$$
$$\nabla^2 U + \lambda U = 0. \tag{10.9.2}$$

If we assume the solution U to be of the form

$$U(x, y, z) = X(x) Y(y) Z(z),$$

then the solution of the Helmholtz equation is

$$U_{lmn}(x, y, z) = B_l D_m F_n \sin\left(\frac{l\pi x}{a}\right) \sin\left(\frac{m\pi y}{b}\right) \sin\left(\frac{n\pi z}{d}\right).$$

Since the solution of equation (10.9.1) is

$$T(t) = G e^{-\lambda k t},$$

the solution of the heat equation takes the form

$$u(x, y, z, t) = \sum_{l=1}^{\infty} \sum_{m=1}^{\infty} \sum_{n=1}^{\infty} a_{lmn} e^{-\lambda k t} \sin\left(\frac{l\pi x}{a}\right) \sin\left(\frac{m\pi y}{b}\right) \sin\left(\frac{n\pi z}{d}\right),$$

where $\lambda = \left[(l^2/a^2) + (m^2/b^2) + (n^2/d^2)\right]\pi^2$ and a_{lmn} are constants.

Application of the initial condition yields

$$a_{lmn} = \left(\frac{8}{abd}\right) \int_0^a \int_0^b \int_0^d f(x, y, z) \sin\left(\frac{l\pi x}{a}\right) \sin\left(\frac{m\pi y}{b}\right)$$
$$\sin\left(\frac{n\pi z}{d}\right) dx\, dy\, dz.$$

10.10 The Schrödinger Equation and the Hydrogen Atom

In quantum mechanics, the Hamiltonian (or energy operator) is usually denoted by H and is defined by

$$H = \frac{\mathbf{p}^2}{2M} + V(\mathbf{r}) \tag{10.10.1}$$

where $\mathbf{p} = (\hbar/i)\nabla = -i\hbar\nabla$ is the momentum of a particle of mass M, $h = 2\pi\hbar$ is the Planck constant, and $V(\mathbf{r})$ is the potential energy.

The physical state of a particle at time t is described as fully as possible by the wave function $\Psi(\mathbf{r}, t)$. The probability of finding the particle at position $\mathbf{r} = (x, y, z)$ within a finite volume $dV = dx\,dy\,dz$ is

$$\iiint |\Psi|^2\, dx\, dy\, dz.$$

The particle must always be somewhere in the space, so the probability of finding the particle within the whole space is one, that is,

$$\int_{-\infty}^{\infty}\int_{-\infty}^{\infty}\int_{-\infty}^{\infty} |\Psi|^2\, dx\, dy\, dz = 1.$$

The time dependent Schrödinger equation for the function $\Psi(\mathbf{r}, t)$ is

$$i\hbar\Psi_t = H\Psi, \tag{10.10.2}$$

where H is explicitly given by

$$H = -\frac{\hbar^2}{2M}\nabla^2 + V(\mathbf{r}). \tag{10.10.3}$$

Given the potential $V(\mathbf{r})$, the fundamental problem of quantum mechanics is to obtain a solution of (10.10.2) which agrees with a given initial state $\Psi(\mathbf{r}, 0)$.

For the stationary state solutions, we seek a solution of the form

$$\Psi(\mathbf{r}, t) = f(t)\psi(\mathbf{r}).$$

Substituting this into (10.10.2) gives

$$\frac{df}{dt} + \frac{iE}{\hbar}f = 0, \tag{10.10.4}$$

$$H\psi(\mathbf{r}) = E\psi(\mathbf{r}), \tag{10.10.5}$$

where E is a separation constant and has the dimension of energy. Integration of (10.10.4) gives

$$f(t) = A \exp\left(-\frac{iEt}{\hbar}\right),$$ (10.10.6)

where A is an arbitrary constant.

Equation (10.10.5) is called the *time independent Schrödinger equation*. The great importance of this equation follows from the fact that the separation of variables gives not just some particular solution of (10.10.5), but generally yields all solutions of physical interest. If $\psi_E(\mathbf{r})$ represents one particular solution of (10.10.5), then most general solutions of (10.10.2) can be obtained by the principle of superposition of such particular solutions. In fact, the general solution is given by

$$\psi(\mathbf{r}, t) = \sum_E A_E \exp\left(-\frac{iEt}{\hbar}\right) \psi_E(\mathbf{r}),$$ (10.10.7)

where the summation is taken over all admissible values of E, and A_E is an arbitrary constant to be determined from the initial conditions.

We now solve the eigenvalue problem for the Schrödinger equation for the spherically symmetric potential so that $V(\mathbf{r}) = V(r)$. The equation for the wave function $\psi(r)$ is

$$\nabla^2 \psi + \frac{2M}{\hbar^2}[E - V(r)]\psi = 0,$$ (10.10.8)

where ∇^2 is the three-dimensional Laplacian.

To determine the wave function ψ, it is convenient to introduce spherical polar coordinates (r, θ, ϕ) so that equation (10.10.8) takes the form

$$\frac{1}{r^2} \frac{\partial}{\partial r}\left(r^2 \frac{\partial \psi}{\partial r}\right) + \frac{1}{r^2 \sin\theta} \frac{\partial}{\partial \theta}\left(\sin\theta \frac{\partial \psi}{\partial \theta}\right) + \frac{1}{r^2 \sin^2\theta} \frac{\partial^2 \psi}{\partial \phi^2}$$
$$+ K[E - V(r)]\psi = 0, \quad (10.10.9)$$

where $K = (2M/\hbar^2)$, $\psi \equiv \psi(r, \theta, \phi)$, $0 \le r < \infty$, $0 \le \theta \le \pi$, and $0 \le \phi \le 2\pi$.

We seek a nontrivial separable solution of the form

$$\psi = R(r) Y(\theta, \phi)$$

and then substitute into (10.10.9) to obtain the following equations

$$\frac{d}{dr}\left(r^2 \frac{dR}{dr}\right) + [K(E - V)r^2 - \lambda]R = 0, \quad (10.10.10)$$

$$\left[\frac{1}{\sin\theta} \frac{\partial}{\partial \theta}\left(\sin\theta \frac{\partial}{\partial \theta}\right) + \frac{1}{\sin^2\theta} \frac{\partial^2}{\partial \phi^2}\right]Y + \lambda Y = 0, \quad (10.10.11)$$

where λ is a separation constant.

We first solve (10.10.11) by separation of variables through $Y = \Theta(\theta)\Phi(\phi)$ so that the equation becomes

$$\sin\theta \frac{d}{d\theta}\left(\sin\theta \frac{d\Theta}{d\theta}\right) + \left(\lambda\sin^2\theta - m^2\right)\Theta = 0, \qquad (10.10.12)$$

$$\frac{d^2\Phi}{d\phi^2} + m^2\Phi = 0, \qquad (10.10.13)$$

where m^2 is a separation constant.

The general solution of (10.10.13) is

$$\Phi = A e^{im\phi} + B e^{-im\phi},$$

where A and B are arbitrary constants to be determined by the boundary conditions on $\psi(r,\theta,\phi) = R(r)\Theta(\theta)\Phi(\phi)$ which will now be formulated.

According to the fundamental postulate of quantum mechanics, the wave function for a particle without spin must have a definite value at every point in space. Hence, we assume that ψ is a single-valued function of position. In particular, ψ must have the same value whether the azimuthal coordinate ϕ is given by ϕ or $\phi + 2\pi$, that is, $\Phi(\phi) = \Phi(\phi + 2\pi)$. Consequently, the solution for Φ has the form

$$\Phi = C e^{im\Phi}, \qquad m = 0, \pm 1, \pm 2, \ldots, \qquad (10.10.14)$$

where C is an arbitrary constant.

In order to solve (10.10.12), it is convenient to change the variable $x = \cos\theta$, $\Theta(\theta) = u(x)$, $-1 \le x \le 1$ so that this equation becomes

$$\frac{d}{dx}\left[(1-x^2)\frac{du}{dx}\right] + \left(\lambda - \frac{m^2}{1-x^2}\right)u = 0. \qquad (10.10.15)$$

For the particular case $m = 0$, this equation becomes

$$\frac{d}{dx}\left[(1-x^2)\frac{du}{dx}\right] + \lambda u = 0. \qquad (10.10.16)$$

This is known as the *Legendre equation*, which gives the Legendre polynomials $P_l(x)$ of degree l as solutions provided $\lambda = l(l+1)$ where l is a positive integer or zero.

When $m \ne 0$, equation (10.10.15) with $\lambda = l(l+1)$ admits solutions which are well known as *associated Legendre functions*, $P_l^m(x)$ of degree l and order m defined by

$$P_l^m(x) = (1-x^2)^{m/2}\frac{d^m}{dx^m}P_l^m(x), \qquad x = \cos\theta.$$

Clearly, $P_l^m(x)$ vanishes when $m > l$. As for the negative integral values of m, it can be readily shown that

$$P_l^{-m}(x) = (-1)^m \frac{(l-m)!}{(l+m)!} P_l^m(x).$$

Hence, the functions $P_l^{-m}(x)$ differ from $P_l^m(x)$ by a constant factor, and as a consequence, m is restricted to a positive integer or zero. Thus, the associated Legendre functions $P_l^m(x)$ with $|m| \le l$ are the only nonsingular and physically acceptable solutions of (10.10.15). Since $|m| \le l$, when $l = 0$, $m = 0$; when $l = 1$, $m = -1$, 0, $+1$; when $l = 2$, $m = -2$, -1, 0, 1, 2, etc. This means that, given l, there are exactly $(2l+1)$ different values of $m = -l, \ldots, -1, 0, 1, \ldots, l$. The numbers l and m are called the *orbital quantum member* and the *magnetic quantum number* respectively.

It is convenient to write down the solutions of (10.10.11) as functions which are normalized with respect to an integration over the whole solid angle. They are called *spherical harmonics* and are given by, for $m \ge 0$,

$$Y_l^m(\theta, \phi) = \left[\frac{(2l+1)}{4\pi} \frac{(l-m)!}{(l+m)!} \right]^{\frac{1}{2}} (-1)^m e^{im\phi} P_l^m(\cos\theta). \quad (10.10.17)$$

Spherical harmonics with negative m and with $|m| \le l$ are defined by

$$Y_l^m(\theta, \phi) = (-1)^m \overline{Y_l^{-m}(\theta, \phi)}. \quad (10.10.18)$$

We now return to a general discussion of the radial equation (10.10.10) which becomes, under the transformation $R(r) = P(r)/r$,

$$\frac{d^2 P}{dr^2} + \left[K(E - V) - \frac{\lambda}{r^2} \right] P(r) = 0. \quad (10.10.19)$$

Almost all cases of physical interest require $V(r)$ to be finite everywhere except at the origin $r = 0$. Also, $V(r) \to 0$ as $r \to \infty$. The Coulomb and square well potentials are typical examples of this kind. In the neighborhood of $r = 0$, $V(r)$ can be neglected compared to the centrifugal term $\left(\sim 1/r^2 \right)$ so that equation (10.10.19) takes the form

$$\frac{d^2 P}{dr^2} - \frac{l(l+1)}{r^2} P(r) = 0 \quad (10.10.20)$$

for all states with $l \ne 0$. The general solution of this equation is

$$P(r) = A r^{l+1} + B r^{-l}, \quad (10.10.21)$$

where A and B are arbitrary constants. With the boundary condition $P(0) = 0$, $B = 0$ so that the solution is proportional to r^{l+1}.

On the other hand, in view of the assumption that $V(r) \to 0$ as $r \to \infty$, the radial equation (10.10.19) reduces to

$$\frac{d^2 P}{dr^2} + KE\, P(r) = 0. \quad (10.10.22)$$

The general solution of this equation is

$$P(r) = C e^{ir\sqrt{KE}} + D e^{-ir\sqrt{KE}}. \tag{10.10.23}$$

The solution is oscillatory for $E > 0$, and exponential in nature for $E < 0$. The oscillatory solutions are not physically acceptable because the wave function does not tend to zero as $r \to \infty$. When $E < 0$, the second term in (10.10.23) tends to infinity as $r \to \infty$. Consequently, the only physically acceptable solutions for $E > 0$, have the asymptotic form

$$P(r) = C e^{-\alpha r/2}, \tag{10.10.24}$$

where $KE = -(\alpha^2/4)$.

Thus, the general solution of (10.10.19) can be written as

$$P(\mathbf{r}) = f(r) e^{-(\alpha/2)r},$$

so that $f(r)$ satisfies the ordinary differential equation

$$\frac{d^2 f}{dr^2} - \alpha \frac{df}{dr} - \left[KV + \frac{l(l+1)}{r^2} \right] f = 0. \tag{10.10.25}$$

Note that this general solution is physically acceptable because the wave function tends to zero as $r \to 0$ and as $r \to \infty$.

We now specify the form of the potential $V(r)$. One of the most common potentials is the Coulomb potential $V(r) = -Ze^2/r$ representing the attraction between an atomic nucleus of charge $+Ze$ and a moving electron of charge $-e$. For the hydrogen atom $Z = 1$. It is a two particle system consisting of a negatively charged electron interacting with a positively charged proton. On the other hand, a helium atom consists of two protons and two neutrons. There are two electrons in orbit around the nucleus of a helium atom. For the singly charged helium ion $Z = 2$, where Z represents the number of unit charges of the nucleus. Consequently, equation (10.10.25) reduces to

$$\frac{d^2 f}{dr^2} - \alpha \frac{df}{dr} + \left[\frac{KZe^2}{r} - \frac{l(l+1)}{r^2} \right] f(r) = 0. \tag{10.10.26}$$

We seek a power series solution of this equation in the form

$$f(r) = r^k \sum_{s=1}^{\infty} a_s r^s, \qquad k \neq 0. \tag{10.10.27}$$

Substituting this series into (10.10.26), we obtain

$$r^k \sum_{s=1}^{\infty} [(s+k)(s+k-1) - l(l+1)] a_s r^{s+k-1}$$

$$+ \sum_{s=1}^{\infty} [Zke^2 - \alpha(s+k)] a_s r^{s+k-1} = 0.$$

Clearly, the lowest power of r is $(k-1)$, so that

$$[k(k+1) - l(l+1)] a_1 = 0.$$

This implies that $k = l$ or $-(l+1)$ provided $a_1 \neq 0$. The negative root of k is not acceptable because it leads to an unbounded solution. Equating the coefficient of r^{s+k-1}, we get the recurrence relation for the coefficients as

$$a_{s+1} = \frac{\alpha(s+l) - ZKe^2}{s(s+2l+1)} a_s, \qquad s = 1, 2, 3, \ldots . \qquad (10.10.28)$$

The asymptotic nature of this result is

$$\frac{a_{s+1}}{a_s} \sim \frac{\alpha}{s} \qquad \text{as} \qquad s \to \infty.$$

This ratio is the same as that of the series for $e^{\alpha r}$. This means that $R(r)$ is unbounded as $r \to \infty$, which is physically unacceptable. Hence, the series for $f(r)$ must terminate, and $f(r)$ must be a polynomial so that $a_{s+1} = 0$, but $a_s \neq 0$. Hence

$$\alpha(s+l) - ZKe^2 = 0, \qquad s = 1, 2, 3, \ldots,$$

or,

$$\frac{\alpha^2}{4} = \frac{Z^2 K^2 e^4}{4(s+l)^2} = -KE. \qquad (10.10.29)$$

Putting $K = (2M/\hbar^2)$, the energy levels are given by

$$E = E_n = -\frac{Z^2 K^2 e^4}{4n^2 K} = -\frac{M Z^2 e^4}{2\hbar^2 n^2}, \qquad (10.10.30)$$

where $n = (s+l)$ is called the *principal quantum number* and $n = 1, 2, 3, \ldots$.
   Thus, it turns out that the complete solution of the Schrödinger equation is given by

$$\psi_{n,l,m}(r, \theta, \phi) = R_{n,l}(r) Y_l^m(\theta, \phi),$$

where the radial part is the solution of the radial equation (10.10.10), and it depends on the principle quantum number n (energy levels) and the orbital quantum number l. However, it does not depend on the magnetic quantum number m. Of course, there are $(2l+1)$ states with the same l value but with different m values. Each of these states has the same energy, and therefore, such systems have a $(2l+1)$-fold degeneracy, as a result of rotational symmetry.
   For the hydrogen atom, $Z = 1$, the discrete energy spectrum is

$$E_n = -\frac{M e^4}{2\hbar^2 n^2} = -\frac{e^2}{2an^2}, \qquad (10.10.31)$$

where $a = (\hbar^2/e^2 M)$ is called the *Bohr radius of the hydrogen atom* of mass M and charge of the electron, $-e$. This discrete energy spectrum depends only on the principle quantum number n (but not on m) and has an excellent agreement with experimental prediction of spectral lines.

For a given n, there are n sets of l and s

$$n = 1, \quad \{l = 0, \quad s = 1\}; \qquad n = 2, \quad \begin{Bmatrix} l = 0, \; s = 2 \\ l = 1, \; s = 1 \end{Bmatrix};$$

$$n = 3, \quad \begin{Bmatrix} l = 0, \; s = 3 \\ l = 1, \; s = 2 \\ l = 2, \; s = 1 \end{Bmatrix}; \qquad \text{etc.}$$

Given n, there are exactly n values of l $(l = 0, 1, 2, \ldots, n - 1)$ and the highest value of l is $n - 1$.

Thus, the three numbers n, l, m, determine a unique eigenfunction, $\psi_{n,l,m}(r, \theta, \phi) = R_{n,l}(r) Y_l^m(\theta, \phi)$. Since the energy levels depend only on the principle quantum number n, there are, in general, several linearly independent eigenfunctions of the Schrödinger equation for the hydrogen atom corresponding to each energy level, so the energy levels are said to be degenerate. There are $(2l + 1)$ different eigenfunctions of the same energy obtained by varying the magnetic quantum number m from $-l$ to l. In general, the total number of degenerate energy states E_n for the hydrogen atom is then

$$\sum_{l=0}^{n-1} (2l + 1) = 2 \frac{n(n - 1)}{2} + n = n^2. \tag{10.10.32}$$

The energy levels of the hydrogen atom (10.10.31) can be expressed in terms the Rydberg, Ry, as

$$E_n = -\frac{Ry}{n^2}, \tag{10.10.33}$$

where Ry represents the Rydberg given by

$$Ry = \frac{Me^4}{2\hbar^2} = \frac{Mc^2 e^4}{2(\hbar c)^2} = \frac{Mc^2}{2} \times \left(\frac{e^2}{\hbar c}\right)^2$$

$$\simeq \frac{5 \times 10^5}{2} eV \times \left(\frac{1}{137}\right)^2 \simeq 13.3\, eV.$$

Consequently,

$$E_n = -\frac{13.3}{n^2} eV. \tag{10.10.34}$$

Thus, the ground state of the hydrogen atom, which is the most tightly bound, has an energy $-13.3\,eV$ (more accurately $-13.6\,eV$) and therefore, it would take $13.6\,eV$ to release the electron from its ground state. Therefore, this is called the *binding energy* of the hydrogen atom.

Finally, the electron is treated here as a nonrelativistic particle. However, in reality, small relativistic effects can be calculated. These are known as *fine structure corrections*. Thus, the nonrelativistic Schrödinger equation describes the hydrogen atom extremely well.

Example 10.10.1. (**Infinite Square well potential $V_0 \to \infty$**). We consider the one-dimensional Schrödinger equation (10.10.8) in the form

$$\left[-\left(\frac{\hbar^2}{2M}\right)\frac{d^2}{dx^2} + V(x)\right]\psi(x) = E\psi(x), \qquad (10.10.35)$$

where the potential $V(x)$ is given by

$$V(x) = \begin{cases} V_0, & x \leq -a, \quad x \geq a, \\ 0, & -a \leq x \leq a \end{cases} \qquad (10.10.36)$$

and take the limit $V_0 \to +\infty$ as shown in Figure 10.10.1.

It is noted that the potential is zero inside the square well and E is the kinetic energy of the particle in this region $(-a \leq x \leq a)$ which must be positive, $E > 0$. It is convenient to fix the origin at the center of the well so that $V(x)$ is an even function of x. A case of special interest is that $V_0 > E \geq 0$ and eventually, $V_0 \to \infty$.

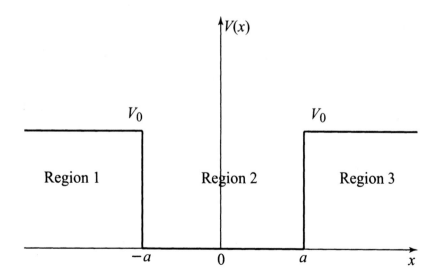

Figure 10.10.1 Square well with potential $V_0 \to \infty$.

The given potential is different in different regions, we solve (10.10.35) separately in three regions.

Region 1. $V = V_0$ in this region $x \leq -a$.

The Schrödinger equation (10.10.35) in this region is

$$-\frac{\hbar^2}{2M}\psi_{xx} + V_0\psi = E\psi.$$

Or, equivalently,

$$\psi_{xx} = \left(\frac{2M}{\hbar^2}\right)(V_0 - E)\,\psi, \qquad V_0 > E > 0. \qquad (10.10.37)$$

The general solution of (10.10.37) is

$$\psi_1(x) = A\,e^{kx} + B\,e^{-kx}, \qquad (10.10.38)$$

where A and B are constants and

$$k = \left[\frac{2M}{\hbar^2}(V_0 - E)\right]^{\frac{1}{2}}. \qquad (10.10.39)$$

The wave function $\psi_1(x)$ must be bounded as $x \to -\infty$ to retain its probabilistic interpretation, hence $B = 0$, and the solution in $x \leq -a$ is

$$\psi_1(x) = A\,e^{kx}.$$

As $V_0 \to \infty$, $k \to \infty$, and, in this limit, the solution must vanish, that is,

$$\psi_1(x) = 0, \quad \text{for} \quad x \leq -a. \qquad (10.10.40)$$

Region 2. $V = 0$ in $-a \leq x \leq a$.
    In this case, the equation takes the form

$$\psi_{xx} + k^2\psi = 0, \qquad (10.10.41)$$

where

$$k^2 = \frac{2ME}{\hbar^2}. \qquad (10.10.42)$$

The general solution of (10.10.41) is given by

$$\psi_2(x) = C\sin kx + D\cos kx, \qquad (10.10.43)$$

where C and D are arbitrary constants.
    Region 3. $V = V_0$ in this region $x \geq a$.
    An argument similar to region 1 leads to zero solution, that is,

$$\psi_3\left(x\right) = 0, \quad \text{for} \quad x \geq a. \tag{10.10.44}$$

From a physical point of view, the solution of the Schrödinger equation must be continuous everywhere including at the boundaries. Thus, matching of solutions at $x = \pm a$ is required so that

$$\psi_2\left(a\right) = C\sin ak + D\cos ak = 0 = \psi_3\left(a\right), \tag{10.10.45}$$
$$\psi_2\left(-a\right) = -C\sin ak + D\cos ak = 0 = \psi_1\left(-a\right). \tag{10.10.46}$$

This system of linear homogeneous equations has nontrivial solutions for C and D *only* if the determinant of the coefficient matrix vanishes. This means that

$$\sin ak\cos ak = 0. \tag{10.10.47}$$

There are two possible nontrivial solutions for the set of conditions (10.10.47).

Case 1. Even solution: $\cos ak = 0$.

In this case, it follows from (10.10.45)–(10.10.46) that $C = 0$. Hence,

$$ak = (2n+1)\,\frac{\pi}{2}, \quad n \quad \text{is an integer},$$

or,

$$k^2 = k_n^2 = \left[(2n+1)\,\frac{\pi}{2a}\right]^2. \tag{10.10.48}$$

Consequently, (10.10.42) gives the energy levels $E = E_n$ as

$$E_n = \frac{(2n+1)^2\,\pi^2\hbar^2}{8Ma^2}. \tag{10.10.49}$$

In this case, the nontrivial solution in region 2 takes the form

$$\psi_2\left(x\right) = D\cos kx, \quad -a \leq x \leq a. \tag{10.10.50}$$

Case 2. Odd solution: $\sin ak = 0$.

It follows from (10.10.45)–(10.10.46) that $D = 0$ and $\sin ak = 0$ holds. Consequently,

$$ak = n\pi, \quad n \quad \text{is an integer}, \ n \neq 0,$$

or,

$$k_n^2 = \left(\frac{n\pi}{a}\right)^2. \tag{10.10.51}$$

Thus, the energy levels are given by

$$E_n = \frac{\hbar^2 k_n^2}{2M} = \frac{(n\pi\hbar)^2}{2Ma^2}. \tag{10.10.52}$$

The nontrivial solution in region 2 is

$$\psi_2(x) = C \sin kx, \qquad -a \le x \le a. \tag{10.10.53}$$

Thus, it turns out that, corresponding to every value of E_n given by (10.10.49) or (10.10.52), there exists a physically acceptable solution. Hence, the general solution of the Schrödinger equation is obtained from (10.10.7) in the form

$$\psi(x,t) = \sum_n C_n \psi_n(x) \exp\left(-\frac{itE_n}{\hbar}\right), \tag{10.10.54}$$

where C_n are constants.

In classical mechanics, the motion of the particle is allowed for $E > 0$. In quantum mechanics, it follows from (10.10.49) or (10.10.52) that particle motion is allowed for *discrete values of energy*, that is, the energy for this system is quantized. This is a remarkable contrast between the results of the classical mechanics and quantum mechanics.

Finally, it follows from the above analysis is that

$$\lim_{|x|\to\infty} \psi(x) = 0. \tag{10.10.55}$$

Such a system, where the wave function vanishes beyond range or asymptotically, is called a *bound state,* and energy is quantized. A very common example is the hydrogen atom which was discussed in this section. In the present system $\psi(x) = 0$ for $x^2 \ge a^2$. Therefore, this system is also referred to as a particle in a box of length $2a$. The probability for finding the particle outside this region is zero.

10.11 Method of Eigenfunctions and Vibration of Membrane

Consider the nonhomogeneous initial boundary-value problem

$$L[u] = \rho u_{tt} - G \qquad \text{in} \quad D \tag{10.11.1}$$

with prescribed homogeneous boundary conditions on the boundary B of D, and the initial conditions

$$u(x_1, x_2, \ldots, x_n, 0) = f(x_1, x_2, \ldots, x_n), \tag{10.11.2}$$

$$u_t(x_1, x_2, \ldots, x_n, 0) = g(x_1, x_2, \ldots, x_n). \tag{10.11.3}$$

Here $\rho \equiv \rho(x_1, x_2, \ldots, x_n)$ is a real-valued positive continuous function and $G \equiv G(x_1, x_2, \ldots, x_n)$ is a real-valued continuous function.

We assume that the only solution of the associated homogeneous problem

$$L[u] = \rho u_{tt} \tag{10.11.4}$$

with the prescribed boundary conditions is the trivial solution. Then, if there exists a solution of the given problem (10.11.1)–(10.11.3), it can be represented by a series of eigenfunctions of the associated eigenvalue problem

$$L[\varphi] + \lambda \rho \varphi = 0 \tag{10.11.5}$$

with φ satisfying the boundary conditions given for u. For problems with one space variable, see Section 7.8.

As a specific example, we shall determine the solution of the problem of forced vibration of a rectangular membrane of length a and width b. The problem is

$$\begin{align}
u_{tt} - c^2 \nabla^2 u &= F(x, y, t) \quad \text{in} \quad D && \text{(10.11.6)} \\
u(x, y, 0) &= f(x, y), \quad 0 \le x \le a, \quad 0 \le y \le b, && \text{(10.11.7)} \\
u_t(x, y, 0) &= g(x, y), \quad 0 \le x \le a, \quad 0 \le y \le b, && \text{(10.11.8)} \\
u(0, y, t) &= 0, \quad u(a, y, t) = 0, && \text{(10.11.9)} \\
u(x, 0, t) &= 0, \quad u(x, b, t) = 0. && \text{(10.11.10)}
\end{align}$$

The associated eigenvalue problem is

$$\begin{align}
\nabla^2 \varphi + \lambda \varphi &= 0 \quad \text{in} \quad D, \\
\varphi &= 0 \quad \text{on the boundary} \quad B \text{ of } D.
\end{align}$$

The eigenvalues for this problem according to Section 10.6 are given by

$$\alpha_{mn} = \left(\frac{m^2 \pi^2}{a^2} + \frac{n^2 \pi^2}{b^2} \right), \quad m, n = 1, 2, 3 \ldots$$

and the corresponding eigenfunctions are

$$\varphi_{mn}(x, y) = \sin\left(\frac{m \pi x}{a} \right) \sin\left(\frac{n \pi y}{b} \right).$$

Thus, we assume the solution

$$u(x, y, t) = \sum_{m=1}^{\infty} \sum_{n=1}^{\infty} u_{mn}(t) \sin\left(\frac{m \pi x}{a} \right) \sin\left(\frac{n \pi y}{b} \right)$$

and the forcing function

$$F(x, y, t) = \sum_{m=1}^{\infty} \sum_{n=1}^{\infty} F_{mn}(t) \sin\left(\frac{m\pi x}{a}\right) \sin\left(\frac{n\pi y}{b}\right).$$

Here $F_{mn}(t)$ are given by

$$F_{mn}(t) = \frac{4}{ab} \int_0^a \int_0^b F(x, y, t) \sin\left(\frac{m\pi x}{a}\right) \sin\left(\frac{n\pi y}{b}\right) dx \, dy.$$

Note that u automatically satisfies the homogeneous boundary conditions. Now inserting $u(x, y, t)$ and $F(x, y, t)$ in equation (10.11.6), we obtain

$$\ddot{u}_{mn} + c^2 \alpha_{mn}^2 u_{mn} = F_{mn},$$

where $\alpha_{mn}^2 = (m\pi/a)^2 + (n\pi/b)^2$. We have assumed that u is twice continuously differentiable with respect to t. Thus, the solution of the preceding ordinary differential equation takes the form

$$u_{mn}(t) = A_{mn} \cos(\alpha_{mn} ct) + B_{mn} \sin(\alpha_{mn} ct)$$

$$+ \frac{1}{(\alpha_{mn} c)} \int_0^t F_{mn}(\tau) \sin[\alpha_{mn} c(t - \tau)] \, d\tau.$$

The first initial condition gives

$$u(x, y, 0) = f(x, y) = \sum_{m=1}^{\infty} \sum_{n=1}^{\infty} A_{mn} \sin\left(\frac{m\pi x}{a}\right) \sin\left(\frac{n\pi y}{b}\right).$$

Assuming that $f(x, y)$ is continuous in x and y, the coefficient A_{mn} of the double Fourier series is given by

$$A_{mn} = \frac{4}{ab} \int_0^a \int_0^b f(x, y) \sin\left(\frac{m\pi x}{a}\right) \sin\left(\frac{n\pi y}{b}\right) dx \, dy.$$

Similarly, from the remaining initial condition, we have

$$u_t(x, y, 0) = g(x, y) = \sum_{m=1}^{\infty} \sum_{n=1}^{\infty} B_{mn} (\alpha_{mn} c) \sin\left(\frac{m\pi x}{a}\right) \sin\left(\frac{n\pi y}{b}\right),$$

and hence, for continuous $g(x, y)$,

$$B_{mn} = \frac{4}{(ab\,\alpha_{mn} c)} \int_0^a \int_0^b g(x, y) \sin\left(\frac{m\pi x}{a}\right) \sin\left(\frac{n\pi y}{b}\right) dx \, dy.$$

The solution of the given initial boundary-value problem is therefore given by

$$u(x, y, t) = \sum_{m=1}^{\infty} \sum_{n=1}^{\infty} u_{mn}(t) \sin\left(\frac{m\pi x}{a}\right) \sin\left(\frac{n\pi y}{b}\right),$$

provided the series for u and its first and second derivatives converge uniformly.

If $F(x, y, t) = e^{x+y} \cos \omega t$, then

$$F_{mn}(t) = \frac{4mn\pi^2}{(m^2\pi^2 + a^2)(n^2\pi^2 + b^2)} \left[1 + (-1)^{m+1} e^a\right]$$
$$\times \left[1 + (-1)^{n+1} e^b\right] \cos \omega t$$
$$= C_{mn} \cos \omega t.$$

Hence, we have

$$u_{mn}(t) = \frac{1}{(\alpha_{mn}c)} \int_0^t C_{mn} \cos \omega t \sin\left[\alpha_{mn}c\left(t - \tau\right)\right] d\tau$$
$$= \frac{C_{mn}}{(\alpha_{mn}^2 c^2 - \omega^2)} (\cos \omega t - \cos \alpha_{mn} ct)$$

provided $\omega \neq (\alpha_{mn}c)$. Thus, the solution may be written in the form

$$u(x, y, t) = \sum_{m=1}^{\infty} \sum_{n=1}^{\infty} \frac{C_{mn}}{(\alpha_{mn}^2 c^2 - \omega^2)} (\cos \omega t - \cos \alpha_{mn} ct)$$
$$\times \sin\left(\frac{m\pi x}{a}\right) \sin\left(\frac{n\pi y}{b}\right).$$

10.12 Time-Dependent Boundary-Value Problems

The preceding chapters have been devoted to problems with homogeneous boundary conditions. Due to the frequent occurrence of problems with time dependent boundary conditions in practice, we consider the forced vibration of a rectangular membrane with moving boundaries. The problem here is to determine the displacement function u which satisfies

$$\begin{array}{llll} u_{tt} - c^2 \nabla^2 u = F(x, y, t), & 0 < x < a, & 0 < y < b, & (10.12.1) \\ u(x, y, 0) = f(x, y), & 0 \le x \le a, & 0 \le y \le b, & (10.12.2) \\ u_t(x, y, 0) = g(x, y), & 0 \le x \le a, & 0 \le y \le b, & (10.12.3) \\ u(0, y, t) = p_1(y, t), & 0 \le y \le b, & t \ge 0, & (10.12.4) \\ u(a, y, t) = p_2(y, t), & 0 \le y \le b, & t \ge 0, & (10.12.5) \\ u(x, 0, t) = q_1(x, t), & 0 \le x \le a, & t \ge 0, & (10.12.6) \\ u(x, b, t) = q_2(x, t), & 0 \le x \le a, & t \ge 0. & (10.12.7) \end{array}$$

For such problems, we seek a solution in the form

$$u(x, y, t) = U(x, y, t) + v(x, y, t), \qquad (10.12.8)$$

where v is the new dependent variable to be determined. Before finding v, we must first determine U. If we substitute equation (10.12.8) into equations (10.12.1)–(10.12.7), we respectively obtain

$$v_{tt} - c^2 \left(v_{xx} + v_{yy} \right) = F - U_{tt} + c^2 \left(U_{xx} + U_{yy} \right) = \tilde{F} \left(x, y, t \right)$$

and

$$
\begin{aligned}
v \left(x, y, 0 \right) &= f \left(x, y \right) - U \left(x, y, 0 \right) = \tilde{f} \left(x, y \right), \\
v_t \left(x, y, 0 \right) &= g \left(x, y \right) - U_t \left(x, y, 0 \right) = \tilde{g} \left(x, y \right), \\
v \left(0, y, t \right) &= p_1 \left(y, t \right) - U \left(0, y, t \right) = \tilde{p}_1 \left(y, t \right), \\
v \left(a, y, t \right) &= p_2 \left(y, t \right) - U \left(a, y, t \right) = \tilde{p}_2 \left(y, t \right), \\
v \left(x, 0, t \right) &= q_1 \left(x, t \right) - U \left(x, 0, t \right) = \tilde{q}_1 \left(x, t \right), \\
v \left(x, b, t \right) &= q_2 \left(x, t \right) - U \left(x, b, t \right) = \tilde{q}_2 \left(x, t \right).
\end{aligned}
$$

In order to make the conditions on v homogeneous, we set

$$\tilde{p}_1 = \tilde{p}_2 = \tilde{q}_1 = \tilde{q}_2 = 0,$$

so that

$$
\begin{array}{llr}
U \left(0, y, t \right) = p_1 \left(y, t \right), & U \left(a, y, t \right) = p_2 \left(y, t \right), & (10.12.9) \\
U \left(x, 0, t \right) = q_1 \left(x, t \right), & U \left(x, b, t \right) = q_2 \left(x, t \right). & (10.12.10)
\end{array}
$$

In order that the boundary conditions be compatible, we assume that the prescribed functions take the forms

$$
\begin{array}{ll}
p_1 \left(y, t \right) = \varphi \left(y \right) p_1^* \left(y, t \right), & p_2 \left(y, t \right) = \varphi \left(y \right) p_2^* \left(y, t \right), \\
q_1 \left(x, t \right) = \psi \left(x \right) q_1^* \left(x, t \right), & q_2 \left(x, t \right) = \psi \left(x \right) q_2^* \left(x, t \right),
\end{array}
$$

where the function φ must vanish at the end points $y = 0$, $y = b$ and the function ψ must vanish at $x = 0$, $x = a$. Thus, $U \left(x, y, t \right)$ which satisfies equations (10.12.9)–(10.12.10) takes the form

$$U \left(x, y, t \right) = \varphi \left(y \right) \left[p_1^* + \frac{x}{a} \left(p_2^* + p_1^* \right) \right] + \psi \left(x \right) \left[q_1^* + \frac{y}{b} \left(q_2^* + q_1^* \right) \right].$$

The problem then is to find the function $v \left(x, y, t \right)$ which satisfies

$$
\begin{aligned}
v_{tt} - c^2 \left(v_{xx} + v_{yy} \right) &= \tilde{F} \left(x, y, t \right), \\
v \left(x, y, 0 \right) = \tilde{f} \left(x, y \right), & \qquad v_t \left(x, y, 0 \right) = \tilde{g} \left(x, y \right), \\
v \left(0, y, t \right) = 0, & \qquad v \left(a, y, t \right) = 0, \\
v \left(x, 0, t \right) = 0, & \qquad v \left(x, b, t \right) = 0.
\end{aligned}
$$

This is an initial boundary-value problem with homogeneous boundary conditions, which has already been solved.

As a particular case, consider the following problem

$$u_{tt} - c^2 \left(u_{xx} + u_{yy} \right) = 0,$$

$$u \left(x, y, 0 \right) = 0, \qquad u_t \left(x, y, 0 \right) = \frac{y}{b} \sin \left(\frac{\pi x}{a} \right),$$

$$u \left(0, y, t \right) = 0, \qquad u \left(a, y, t \right) = 0,$$

$$u \left(x, 0, t \right) = 0, \qquad u \left(x, b, t \right) = \sin \left(\frac{\pi x}{a} \right) \sin t.$$

We assume a solution in the form

$$u \left(x, y, t \right) = v \left(x, y, t \right) + U \left(x, y, t \right).$$

The function $U \left(x, y, t \right)$ which satisfies

$$U \left(0, y, t \right) = 0, \qquad U \left(a, y, t \right) = 0,$$

$$U \left(x, 0, t \right) = 0, \qquad U \left(x, b, t \right) = \sin \left(\frac{\pi x}{a} \right) \sin t$$

is

$$U \left(x, y, t \right) = \sin \left(\frac{\pi x}{a} \right) \left(\frac{y}{b} \sin t \right).$$

Thus, the new problem to be solved is

$$v_{tt} - c^2 \left(v_{xx} + v_{yy} \right) = \left(1 - \frac{c^2 \pi^2}{a^2} \right) \frac{y}{b} \sin \left(\frac{\pi x}{a} \right) \sin t,$$

$$v \left(x, y, 0 \right) = 0, \qquad v_t \left(x, y, 0 \right) = 0,$$

$$v \left(0, y, t \right) = 0, \qquad v \left(a, y, t \right) = 0,$$

$$v \left(x, 0, t \right) = 0, \qquad v \left(x, b, t \right) = 0.$$

Then, we find F_{mn} from

$$F_{mn} \left(t \right) = \frac{4}{ab} \int_0^a \int_0^b F \left(x, y, t \right) \sin \left(\frac{m\pi x}{a} \right) \sin \left(\frac{n\pi y}{b} \right) dx \, dy,$$

where

$$F \left(x, y, t \right) = \left(1 - \frac{c^2 \pi^2}{a^2} \right) \frac{y}{b} \sin \left(\frac{\pi x}{a} \right) \sin t,$$

and obtain

$$F_{mn} \left(t \right) = \frac{2 \left(-1 \right)^{n+1}}{an} \left(1 - \frac{c^2 \pi^2}{a^2} \right) \sin t.$$

Now we determine $v_{mn} \left(t \right)$ which are given by

$$v_{mn}(t) = A_{mn} \cos(\alpha_{mn}ct) + B_{mn} \sin(\alpha_{mn}ct)$$

$$+ \frac{1}{\alpha_{mn}c} \int_0^t F_{mn}(\tau) \sin[\alpha_{mn}c(t - \tau)] \, d\tau.$$

Since $v(x, y, 0) = 0$, $A_{mn} = 0$, but

$$B_{mn} = \frac{4}{ab\,\alpha_{mn}c} \int_0^a \int_0^b \left(-\frac{y}{b} \sin\frac{\pi x}{a}\right) \sin\left(\frac{m\pi x}{a}\right) \sin\left(\frac{n\pi y}{b}\right) dx \, dy$$

$$= \frac{2(-1)^n}{\alpha_{mn}nac}.$$

Thus, we have

$$v_{mn}(t) = \frac{2(-1)^n}{\alpha_{mn}nac} \sin(\alpha_{mn}ct)$$

$$+ \frac{2(-1)^n}{\alpha_{mn}ca^3 n\,(1 - \alpha^2 c^2)} \left(a^2 - c^2\pi^2\right) (\sin\alpha_{mn}ct - \alpha c \sin t).$$

The solution is therefore given by

$$u(x, y, t) = \frac{y}{b} \sin\left(\frac{\pi x}{a}\right) \sin t + \sum_{m=1}^{\infty} \sum_{n=1}^{\infty} v_{mn}(t) \sin\left(\frac{m\pi x}{a}\right) \sin\left(\frac{n\pi y}{b}\right).$$

10.13 Exercises

1. Solve the Dirichlet problem

$$\nabla^2 u = 0, \quad 0 < x < a, \quad 0 < y < b, \quad 0 < z < c,$$

$$u(0, y, z) = \sin\left(\frac{\pi y}{b}\right) \sin\left(\frac{\pi z}{c}\right), \quad u(a, y, z) = 0,$$

$$u(x, 0, z) = 0, \quad u(x, b, z) = 0,$$

$$u(x, y, 0) = 0, \quad u(x, y, c) = 0.$$

2. Solve the Neumann problem

$$\nabla^2 u = 0, \quad 0 < x < 1, \quad 0 < y < 1, \quad 0 < z < 1,$$

$$u_x(0, y, z) = 0, \quad u_x(1, y, z) = 0,$$

$$u_y(x, 0, z) = 0, \quad u_y(x, 1, z) = 0,$$

$$u_z(x, y, 0) = \cos\pi x \cos\pi y, \quad u_z(x, y, 1) = 0.$$

3. Solve the Robin boundary-value problem

$$\nabla^2 u = 0, \quad 0 < x < \pi, \quad 0 < y < \pi, \quad 0 < z < \pi,$$

$$u(0, y, z) = f(y, z), \quad u(\pi, y, z) = 0,$$

$$u_y(x, 0, z) = 0, \quad u_y(x, \pi, z) = 0,$$

$$\left.\begin{array}{l} u_z(x, y, 0) + h\,u(x, y, 0) = 0, \\[2mm] u_z(x, y, \pi) + h\,u(x, y, \pi) = 0, \end{array}\right\} \quad h = \text{constant}.$$

4. Determine the solution of each of the following problems for a cylinder:

(a)
$$\nabla^2 u = 0, \quad r < a, \quad 0 < \theta < 2\pi, \quad 0 < z < l,$$

$$u\,(a, \theta, z) = 0, \quad u\,(r, \theta, l) = 0, \quad u\,(r, \theta, 0) = f\,(r, \theta).$$

(b)
$$\nabla^2 u = 0, \quad r < a, \quad 0 < \theta < 2\pi, \quad 0 < z < l,$$

$$u\,(a, \theta, z) = f\,(\theta, z), \quad u_z\,(r, \theta, 0) = 0, \quad u_z\,(r, \theta, l) = 0.$$

5. Find the solution of the Dirichlet problem for a sphere

$$\nabla^2 u = 0, \quad r < a, \quad 0 < \theta < \pi, \quad 0 < \varphi < 2\pi,$$

$$u\,(a, \theta, \varphi) = \cos^2 \theta.$$

6. Solve the Dirichlet problem in a region bounded by two concentric spheres

$$\nabla^2 u = 0, \quad a < r < b, \quad 0 < \theta < \pi, \quad 0 < \phi < 2\pi,$$

$$u\,(a, \theta, \phi) = f\,(\theta, \phi), \quad u\,(b, \theta, \phi) = g\,(\theta, \phi).$$

7. Find the steady-state temperature distribution in a cylinder of radius a if a constant flow of heat T is supplied at the end $z = 0$, and the surface $r = a$ and the end $z = l$ are maintained at zero temperature.

8. Find the potential of the electrostatic field inside a cylinder of length l and radius a, if each end of the cylinder is grounded, and the surface is charged to a potential u_0.

9. Determine the potential of the electric field inside a sphere of radius a, if the upper half of the sphere is charged to a potential u_1 and the lower half to a potential u_2.

10. Solve the Dirichlet problem for a half cylinder

$$\nabla^2 u = 0, \quad r < 1, \quad 0 < \theta < \pi, \quad 0 < z < 1,$$

$$u\,(1, \theta, z) = 0, \quad u\,(r, 0, z) = 0, \quad u\,(r, \pi, z) = 0,$$

$$u\,(r, \theta, 0) = 0, \quad u\,(r, \theta, 1) = f\,(r, \theta).$$

11. Solve the Neumann problem for a sphere

$$\nabla^2 u = 0, \quad r < 1, \quad 0 < \theta < \pi, \quad 0 < \varphi < 2\pi,$$

$$u_r(1, \theta, \varphi) = f(\theta, \varphi),$$

where

$$\int_0^{2\pi} \int_0^\pi f(\theta, \varphi) \sin\theta \, d\theta \, d\varphi = 0.$$

12. Find the solution of the initial boundary-value problem

$$u_{tt} = c^2 \nabla^2 u, \quad 0 < x < 1, \quad 0 < y < 1, \quad t > 0,$$

$$u(x, y, 0) = \sin^2 \pi x \, \sin \pi y, \quad u_t(x, y, 0) = 0, \quad 0 \le x \le 1, \quad 0 \le y \le 1,$$

$$u(0, y, t) = 0, \qquad u(1, y, t) = 0, \quad 0 \le y \le 1, \quad t > 0,$$

$$u(x, 0, t) = 0, \qquad u(x, 1, t) = 0, \quad 0 \le x \le 1, \quad t > 0.$$

13. Obtain the solution of the problem

$$u_{tt} = c^2 \nabla^2 u, \quad r < a, \quad 0 < \theta < 2\pi, \quad t > 0,$$

$$u(r, \theta, 0) = f(r, \theta), \qquad u_t(r, \theta, 0) = g(r, \theta), \quad u(a, \theta, t) = 0.$$

14. Determine the temperature distribution in a rectangular plate with radiation from its surface. The temperature distribution is described by

$$u_t = k(u_{xx} + u_{yy}) - h(u - u_0), \quad 0 < x < a, 0 < y < b, t > 0,$$
$$u(x, y, 0) = f(x, y),$$
$$u(0, y, t) = 0, \qquad u(a, y, t) = 0,$$
$$u(x, 0, t) = 0, \qquad u(x, b, t) = 0,$$

where k, h and u_0 are constants.

15. Solve the heat conduction problem in a circular plate

$$u_t = k\left(u_{rr} + \frac{1}{r} u_r + \frac{1}{r^2} u_{\theta\theta}\right), \quad r < 1, \quad 0 < \theta < 2\pi, \quad t > 0,$$
$$u(r, \theta, 0) = f(r, \theta), \qquad u(1, \theta, t) = 0.$$

16. Solve the initial boundary-value problem

$$u_{tt} = c^2 \nabla^2 u, \quad 0 < x < 1, \quad 0 < y < 1, \quad 0 < z < 1, \quad t > 0,$$

$$u(x, y, z, 0) = \sin \pi x \, \sin \pi y \, \sin \pi z,$$

$$u_t\,(x, y, z, 0) = 0,$$

$$u\,(0, y, z, t) = u\,(1, y, z, t) = 0,$$

$$u\,(x, 0, z, t) = u\,(x, 1, z, t) = 0,$$

$$u\,(x, y, 0, t) = u\,(x, y, 1, t) = 0.$$

17. Solve

$$u_{tt} + k\,u_t = c^2 \nabla^2 u, \quad 0 < x < a, \quad 0 < y < b, \quad 0 < z < d, \quad t > 0,$$

$$u\,(x, y, z, 0) = f\,(x, y, z), \qquad u_t\,(x, y, z, 0) = g\,(x, y, z),$$

$$u\,(0, y, z, t) = u\,(a, y, z, t) = 0,$$

$$u\,(x, 0, z, t) = u\,(x, b, z, t) = 0,$$

$$u\,(x, y, 0, t) = u\,(x, y, d, t) = 0.$$

18. Obtain the solution of the problem for $t > 0$,

$$u_{tt} = c^2 \left(u_{rr} + \frac{1}{r} u_r + \frac{1}{r^2} u_{\theta\theta} + u_{zz} \right), \quad r < a,\ 0 < \theta < 2\pi,\ 0 < z < l,$$

$$u\,(r, \theta, z, 0) = f\,(r, \theta, z), \qquad u_t\,(r, \theta, z, 0) = g\,(r, \theta, z),$$

$$u\,(a, \theta, z, t) = 0, \qquad u\,(r, \theta, 0, t) = u\,(r, \theta, l, t) = 0.$$

19. Determine the solution of the heat conduction problem

$$u_t = k \nabla^2 u, \quad 0 < x < a, \quad 0 < y < b, \quad 0 < z < c, \quad t > 0,$$

$$u\,(x, y, z, 0) = f\,(x, y, z),$$

$$u_x\,(0, y, z, t) = u_x\,(a, y, z, t) = 0,$$

$$u_y\,(x, 0, z, t) = u_y\,(x, b, z, t) = 0,$$

$$u_z\,(x, y, 0, t) = u_z\,(x, y, c, t) = 0.$$

20. Solve the problem

$$u_t = k \nabla^2 u, \quad r < a, \quad 0 < \theta < 2\pi, \quad 0 < z < l, \quad t > 0,$$

$$u(r, \theta, z, 0) = f(r, \theta, z),$$

$$u_r(a, \theta, z, t) = 0,$$

$$u(r, \theta, 0, t) = u(r, \theta, l, t) = 0.$$

21. Find the temperature distribution in the section of a sphere cut out by the cone $\theta = \theta_0$. The surface temperature is zero while the initial temperature is given by $f(r, \theta, \varphi)$.

22. Solve the initial boundary-value problem

$$u_{tt} = c^2 \nabla^2 u + F(x, y, t), \quad 0 < x < a, \quad 0 < y < b, \quad t > 0,$$

$$u(x, y, 0) = f(x, y), \qquad u_t(x, y, 0) = g(x, y),$$

$$u_x(0, y, t) = u_x(a, y, t) = 0 \quad \text{for all } t > 0,$$

$$u_y(x, 0, t) = u_y(x, b, t) = 0 \quad \text{for all } t > 0.$$

23. Solve the problem

$$u_{tt} = c^2 \nabla^2 u + xy \sin t, \quad 0 < x < \pi, \quad 0 < y < \pi, \quad t > 0,$$

$$u(x, y, 0) = 0, \qquad u_t(x, y, 0) = 0,$$

$$u(0, y, t) = u(\pi, y, t) = 0,$$

$$u(x, 0, t) = u(x, \pi, t) = 0.$$

24. Solve

$$u_t = k \nabla^2 u + F(x, y, z, t), \quad 0 < x < a, \ 0 < y < b, \ 0 < z < c,$$
$$t > 0,$$

$$u(x, y, z, 0) = f(x, y, z),$$

$$u(0, y, z, t) = u(a, y, z, t) = 0,$$

$$u\left(x,0,z,t\right)=u\left(x,b,z,t\right)=0,$$

$$u_z\left(x,y,0,t\right)=u_z\left(x,y,c,t\right)=0.$$

25. Solve the nonhomogeneous diffusion problem

$$u_t=k\,\nabla^2u+A,\quad 0<x<\pi,\quad 0<y<\pi,\quad t>0,$$

$$u\left(x,y,0\right)=0,$$

$$u\left(0,y,t\right)=u\left(\pi,y,t\right)=0,$$

$$u_y\left(x,0,t\right)+u\left(x,0,t\right)=0,$$

$$u_y\left(x,\pi,t\right)+u\left(x,\pi,t\right)=0,$$

where k and A are constants.

26. Find the temperature distribution of the composite cylinder consisting of an inner cylinder $0\le r\le r_0$ and an outer cylindrical tube $r_0\le r\le a$. The surface temperature is maintained at zero degrees, and the initial temperature distribution is given by $f\left(r,\theta,z\right)$.

27. Solve the initial boundary-value problem

$$u_t-c^2\nabla^2u=0,\quad 0<x<\pi,\quad 0<y<\pi,\quad t>0,$$

$$u\left(x,y,0\right)=0,$$

$$u\left(0,y,t\right)=u\left(\pi,y,t\right)=0,$$

$$u\left(x,0,t\right)=x\left(x-\pi\right)\sin t,\quad u\left(x,\pi,t\right)=0,\quad 0\le x\le\pi,\quad t\ge0.$$

28. Solve the problem

$$u_{tt}=c^2\nabla^2u,\quad r<a,\quad 0<\theta<2\pi,\quad t>0,$$

$$u\left(r,\theta,0\right)=f\left(r,\theta\right),$$

$$u_t\left(r,\theta,0\right)=g\left(r,\theta\right),$$

$$u\left(a,\theta,t\right)=p\left(\theta,t\right).$$

29. Solve

$$u_t = c^2 \nabla^2 u, \qquad r < a, \quad t > 0,$$

$$u(r, \theta, 0) = f(r, \theta), \qquad u_t(a, \theta, t) = g(\theta, t), \qquad 0 < \theta < \pi.$$

30. Determine the solution of the biharmonic equation

$$\nabla^4 u = q/D$$

with the boundary conditions

$$u(x, 0) = u(x, b) = 0,$$

$$u\left(-\frac{a}{2}, y\right) = u\left(\frac{a}{2}, y\right) = 0,$$

$$u_{xx}\left(-\frac{a}{2}, y\right) = u_{xx}\left(\frac{a}{2}, y\right) = 0,$$

$$u_{yy}(x, 0) = u_{yy}(x, b) = 0,$$

where q is the load per unit area and D is the flexural rigidity of the plate. This is the problem of the deflection of a uniformly loaded plate, the sides of which are simply supported.

31. (a) Show that the solution of the one-dimensional Schrödinger equation for a free particle of mass M

$$\psi_t = \left(\frac{i\hbar}{2M}\right)\psi_{xx}$$

is

$$\psi(x, t) = \left(\frac{N}{b}\right)\exp\left(-\frac{x^2}{2b^2}\right), \qquad b = \left(a^2 + \frac{i\hbar t}{M}\right)^{\frac{1}{2}},$$

where a is an integrating constant that can be determined from the initial value of the wave function $\psi(x, t)$, and N is also a constant that can be determined from the normalization of the probability (wave function) of finding the particle.

(b) Show that the Gaussian probability density is

$$|\psi|^2 = \psi\psi^* = \frac{|N|^2}{ac}\exp\left(-\frac{x^2}{c^2}\right),$$

and its mean width is

$$\delta = \frac{c}{\sqrt{2}}, \qquad C = \left(a^2 + \frac{\hbar^2 t^2}{M^2 a^2}\right)^{\frac{1}{2}}.$$

32. Analogous to Example 10.10.1, solve the problem for a finite square well potential (see Figure 10.10.1) with a finite value for the height of the potential given as

$$V(x) = \begin{cases} 0, & \text{for} \quad -a \leq x \leq a \\ V_0, & \text{for } x \leq -a, \quad x \geq a. \end{cases}$$

33. Consider the quantum mechanical problem described by the one-dimensional Schrödinger equation

$$\psi_{xx} + k^2 \psi = 0$$

where the wavenumber $k = \frac{1}{\hbar}\sqrt{2M(E-V)}$ in the rectangular potential barrier of height V_0 and width $2a$, and

$$V(x) = V_0 H(a - |x|),$$

where H is the Heaviside unit step function. The particle is free for $x < -a$ and $x > a$, and $V(x)$ is an even function; the case $V_0 > E$ is of great interest here.

Show that the general solution of the Schrödinger equation for $V_0 > E$ is

$$\psi(x) = \begin{cases} A\,e^{ikx} + B\,e^{-ikx}, & x \leq -a, \\ C\,e^{-\kappa x} + D\,e^{+\kappa x}, & -a \leq x \leq a, \\ F\,e^{ikx} + G\,e^{-ikx}, & x \geq a, \end{cases}$$

where $\hbar k = \sqrt{2ME}$ and $\hbar\kappa = \sqrt{2M(V_0 - E)}$.

Matching the boundary conditions at $x = -a$, show that

$$\begin{bmatrix} A \\ B \end{bmatrix} = \frac{1}{2} \begin{bmatrix} \left(1 + \frac{i\kappa}{k}\right)\exp(\kappa a + ika) & \left(1 - \frac{i\kappa}{k}\right)\exp(-\kappa a + ika) \\ \left(1 - \frac{i\kappa}{k}\right)\exp(\kappa a - ika) & \left(1 + \frac{i\kappa}{k}\right)\exp(-\kappa a - ika) \end{bmatrix} \begin{bmatrix} C \\ D \end{bmatrix}$$

where [ ] denotes a matrix.

Using the matching conditions at $x = a$, show that

$$\begin{bmatrix} C \\ D \end{bmatrix} = \frac{1}{2} \begin{bmatrix} \left(1 - \frac{ik}{\kappa}\right)\exp(a\kappa + iak) & \left(1 + \frac{ik}{\kappa}\right)\exp(a\kappa - iak) \\ \left(1 + \frac{ik}{\kappa}\right)\exp(-a\kappa + iak) & \left(1 - \frac{ik}{\kappa}\right)\exp(-a\kappa - ika) \end{bmatrix} \begin{bmatrix} F \\ G \end{bmatrix}.$$

Hence, deduce

$$\begin{bmatrix} A \\ B \end{bmatrix} = \begin{bmatrix} \left(\cosh 2a\kappa + \frac{i\varepsilon}{2}\sinh 2a\kappa\right)e^{2iak} & \frac{1}{2}(i\eta)\sinh 2a\kappa \\ -\frac{1}{2}(i\eta)\sinh 2a\kappa & \left(\cosh 2a\kappa - \frac{1}{2}i\varepsilon \sinh 2a\kappa\right)e^{-2ika} \end{bmatrix} \begin{bmatrix} F \\ G \end{bmatrix}$$

where $\varepsilon = \left(\frac{\kappa}{k} - \frac{k}{\kappa}\right)$ and $\eta = \left(\frac{\kappa}{k} + \frac{k}{\kappa}\right)$.

11

Green's Functions and Boundary-Value Problems

"Potential theory has developed out of the vector analysis created by Gauss, Green, and Kelvin for the mathematical theories of gravitational attraction, of electrostatics and of the hydrodynamics of perfect fluids (i.e., incompressible and inviscid fluids). The first stage of abstraction was the study of harmonic functions, i.e., potential functions in space free from masses, charges, sources, or sinks. This led to the inspired intuition of Dirichlet and the early attempts to justify his 'principle'."

George Temple

11.1 Introduction

Boundary-value problems associated with either ordinary or partial differential equations arise most frequently in mathematics, mathematical physics and engineering science. The linear superposition principle is one of the most elegant and effective methods to represent solutions of boundary-value problems in terms of an auxiliary function known as *Green's function*. Such a function was first introduced by George Green as early as 1828. Subsequently, the method of Green's functions became a very useful analytical method in mathematics and in many of the applied sciences.

In previous chapters, it has been shown that the eigenfunction method can effectively be used to express the solutions of differential equations as infinite series. On the other hand, solutions of differential equations can be obtained as an integral superposition in terms of Green's functions. So the method of Green's functions offers several advantages over eigenfunction expansions. First, an integral representation of solutions provides a direct way of describing the general analytical structure of a solution that may be obscured by an infinite series representation. Second, from an analytical point of view, the evaluation of a solution from an integral representation may prove simpler than finding the sum of an infinite series, particularly near rapidly-varying features of a function, where the convergence of an eigen-function expansion may be slow. Third, in view of the Gibbs phenomenon discussed in Chapter 6, the integral representation seems to impose less stringent requirements on the functions that describe the values that the solution must assume on a given boundary than the expansion based on eigenfunctions.

Many physical problems are described by second-order nonhomogeneous differential equations with homogeneous boundary conditions or by second-order homogeneous equations with nonhomogeneous boundary conditions. Such problems can be solved by the method of Green's functions.

We consider a nonhomogeneous partial differential equation of the form

$$L_{\mathbf{x}} u(\mathbf{x}) = f(\mathbf{x}), \tag{11.1.1}$$

where $\mathbf{x} = (x, y, z)$ is a vector in three (or higher) dimensions, $L_{\mathbf{x}}$ is a linear partial differential operator in three or more independent variables with constant coefficients, and $u(\mathbf{x})$ and $f(\mathbf{x})$ are functions of three or more independent variables. The Green's function $G(\mathbf{x}, \xi)$ of this problem satisfies the equation

$$L_{\mathbf{x}} G(\mathbf{x}, \xi) = \delta(\mathbf{x} - \xi) \tag{11.1.2}$$

and represents the effect at the point $\mathbf{x}$ of the Dirac delta function source at the point $\xi = (\xi, \eta, \zeta)$.

Multiplying (11.1.2) by $f(\xi)$ and integrating over the volume V of the ξ space, so that $dV = d\xi\, d\eta\, d\zeta$, we obtain

$$\int_V L_{\mathbf{x}} G(\mathbf{x}, \xi) f(\xi)\, d\xi = \int_V \delta(\mathbf{x} - \xi) f(\xi)\, d\xi = f(\mathbf{x}). \tag{11.1.3}$$

Interchanging the order of the operator $L_{\mathbf{x}}$ and integral sign in (11.1.3) gives

$$L_{\mathbf{x}} \left[\int_V G(\mathbf{x}, \xi) f(\xi)\, d\xi \right] = f(\mathbf{x}). \tag{11.1.4}$$

A simple comparison of (11.1.4) with (11.1.1) leads to the solution of (11.1.1) in the form

$$u(\mathbf{x}) = \int_V G(\mathbf{x}, \xi) f(\xi)\, d\xi. \tag{11.1.5}$$

Clearly, (11.1.5) is valid for any finite number of components of $\mathbf{x}$. Accordingly, the Green's function method can be applied, in general, to any linear, constant coefficient, nonhomogeneous partial differential equation in any number of independent variables.

Another way to approach the problem is by looking for the inverse operator $L_{\mathbf{x}}^{-1}$. If it is possible to find $L_{\mathbf{x}}^{-1}$, then the solution of (11.1.1) can be obtained as $u(\mathbf{x}) = L_{\mathbf{x}}^{-1}(f(\mathbf{x}))$. It turns out that in many important cases it is possible, and the inverse operator can be expressed as an integral operator of the form

$$u(\mathbf{x}) = L_{\mathbf{x}}^{-1}(f(\xi)) = \int_V G(\mathbf{x}, \xi) f(\xi) \, d\xi. \qquad (11.1.6)$$

The kernel $G(\mathbf{x}, \xi)$ is called the *Green's function* which is, in fact, the characteristic of the operator $L_{\mathbf{x}}$ for any finite number of independent variables.

In our study of partial differential equations with the aid of Green's functions, special attention will be given to those three partial differential equations which occur most frequently in mathematics, mathematical physics and engineering science; the wave equation

$$u_{tt} - c^2 \nabla^2 u = f(\mathbf{x}), \qquad (11.1.7)$$

the heat or diffusion equation

$$u_t - \kappa \nabla^2 u = f(\mathbf{x}), \qquad (11.1.8)$$

and the potential or the Laplace equation

$$\nabla^2 u = f(\mathbf{x}), \qquad (11.1.9)$$

where the Laplacian ∇^2 in an n-dimensional Euclidean space is given by

$$\nabla^2 \equiv \frac{\partial^2}{\partial x_1^2} + \frac{\partial^2}{\partial x_2^2} + \ldots + \frac{\partial^2}{\partial x_n^2}, \qquad (11.1.10)$$

and $\mathbf{x} = (x_1, x_2, \ldots, x_n)$.

Clearly, the solutions of the wave and heat equations are functions of $(n+1)$ coordinates consisting of n space dimensions, $\mathbf{x} = (x_1, x_2, \ldots, x_n)$ and one time dimension t, whereas the solutions of the Laplace equation are functions of n space dimensions.

This chapter deals with the basic idea and properties of Green's functions and how to construct such functions for finding solutions of partial differential equations. Some examples of applications are provided in this chapter and in the next chapter.

11.2 The Dirac Delta Function

The application of Green's functions to boundary-value problems in ordinary differential equations was described earlier in Chapter 8. The Green's function method is applied here to boundary-value problems in partial differential equations. The method provides solutions in integral form and is applicable to a wide class of problems in applied mathematics and mathematical physics.

Before developing the method of Green's functions, we will first define the Dirac delta function $\delta\left(x - \xi, y - \eta\right)$ in two dimensions by

(a) $\delta\left(x - \xi, y - \eta\right) = 0, \qquad x \neq \xi, \qquad y \neq \eta,$ \hfill (11.2.1)

(b) $\displaystyle\iint_{R_\varepsilon} \delta\left(x - \xi, y - \eta\right) dx\, dy = 1, \qquad R_\varepsilon : \left(x - \xi\right)^2 + \left(y - \eta\right)^2 < \varepsilon^2,$

\hfill (11.2.2)

(c) $\displaystyle\iint_R F\left(x, y\right) \delta\left(x - \xi, y - \eta\right) dx\, dy = F\left(\xi, \eta\right),$ \hfill (11.2.3)

for arbitrary continuous function F in the region R_ε.

The delta function is not a function in the ordinary sense. For an elegant treatment of the delta function as a generalized function, see L. Schwartz, *Théorie des Distributions* (1950, 1951). It is a symbolic function, and is often viewed as the limit of a distribution.

If $\delta\left(x - \xi\right)$ and $\delta\left(y - \eta\right)$ are one-dimensional delta functions, we have

$$\iint_R F\left(x, y\right) \delta\left(x - \xi\right) \delta\left(y - \eta\right) dx\, dy = F\left(\xi, \eta\right).$$ \hfill (11.2.4)

Since (11.2.3) and (11.2.4) hold for an arbitrary continuous function F, we conclude that

$$\delta\left(x - \xi, y - \eta\right) = \delta\left(x - \xi\right) \delta\left(y - \eta\right).$$ \hfill (11.2.5)

Thus, we may state that the two-dimensional delta function is the product of two one-dimensional delta functions.

Higher dimensional delta functions can be defined in a similar manner.

$$\delta\left(x_1, x_2, \ldots, x_n\right) = \delta\left(x_1\right) \delta\left(x_2\right) \ldots \delta\left(x_n\right).$$ \hfill (11.2.6)

The expression for the δ-function become much more complicated when we introduce curvilinear coordinates. However, for simplicity, we transform the two-dimensional delta function from Cartesian coordinates x, y to curvilinear coordinates α, β by means of the transformation

$$x = u\left(\alpha, \beta\right) \quad \text{and} \quad y = v\left(\alpha, \beta\right),$$ \hfill (11.2.7)

where u and v are single-valued continuous and differentiable functions of their arguments. We assume that under this transformation $\alpha = \alpha_1$ and $\beta = \beta_1$ correspond to $x = \xi$ and $y = \eta$ respectively. Changing the coordinates according to (11.2.7), we reduce equation (11.2.4) to

$$\iint F\left(u, v\right) \delta\left(u - \xi\right) \delta\left(v - \eta\right) |J|\, d\alpha\, d\beta = F\left(\xi, \eta\right),$$ \hfill (11.2.8)

where J is the Jacobian of the transformation defined by

$$J = \frac{\partial (u, v)}{\partial (\alpha, \beta)} = \begin{vmatrix} u_\alpha & u_\beta \\ v_\alpha & v_\beta \end{vmatrix} \neq 0. \tag{11.2.9}$$

Consequently, we can write

$$\delta (u - \xi) \, \delta (v - \eta) \, |J| = \delta (\alpha - \alpha_1) \, \delta (\beta - \beta_1). \tag{11.2.10}$$

In particular, the transformation from rectangular Cartesian coordinates (x, y) to polar coordinates (r, θ) is defined by

$$x = r \cos \theta, \qquad y = r \sin \theta, \tag{11.2.11}$$

so that the Jacobian J is

$$J = \begin{vmatrix} x_r & y_r \\ x_\theta & y_\theta \end{vmatrix} = x_r y_\theta - y_r r_\theta = r. \tag{11.2.12}$$

In this case, J vanishes at the origin and the transformation is singular at $r = 0$ for any θ. Hence, θ can be ignored and

$$\delta (x) \, \delta (y) = \frac{\delta (r)}{|J_1|} = \frac{1}{2\pi} \frac{\delta (r)}{r}, \tag{11.2.13}$$

where

$$J_1 = \int_0^\pi J \, d\theta = 2\pi r.$$

Similarly, the transformation from three-dimensional rectangular Cartesian coordinates (x, y, z) to spherical polar coordinates (r, θ, ϕ) is given by

$$x = r \sin \theta \cos \phi, \qquad y = r \sin \theta \sin \phi, \qquad z = r \cos \theta, \tag{11.2.14}$$

where $0 \leq r < \infty$, $0 \leq \theta \leq \pi$, and $0 \leq \phi \leq 2\pi$.

The Jacobian of the transformation is

$$J = r^2 \sin \theta.$$

This Jacobian vanishes for all points on the z-axis, that is, for $\theta = 0$, and hence, the coordinate ϕ may be ignored. Also, J vanishes at the origin $(r = 0)$ in which case both θ and ϕ may be ignored. Consequently,

$$\delta (x) \, \delta (y) \, \delta (z) = \frac{\delta (r)}{|J_2|} = \frac{\delta (r)}{4\pi r^2}, \tag{11.2.15}$$

where

$$J_2 = \int_0^\pi \int_0^{2\pi} J \, d\theta \, d\phi = \int_0^\pi \int_0^{2\pi} r^2 \sin \theta \, d\phi = 4\pi r^2.$$

11.3 Properties of Green's Functions

The solution of the Dirichlet problem in a domain D with boundary B

$$\nabla^2 u = h(x, y) \quad \text{in } D$$
$$u = f(x, y) \quad \text{on } B \tag{11.3.1}$$

is given in Section 11.5 and has the form

$$u(x, y) = \iint_D G(x, y; \xi, \eta) h(\xi, \eta) \, d\xi \, d\eta + \int_B f \frac{\partial G}{\partial n} \, ds, \tag{11.3.2}$$

where G is the Green's function and n denotes the outward normal to the boundary B of the region D. It is rather obvious then that the solution $u(x, y)$ can be determined as soon as the Green's function G is ascertained, so the problem in this technique is really to find the Green's function.

First, we shall define the Green's function for the Dirichlet problem involving the Laplace operator. Then, the Green's function for the Dirichlet problem involving the Helmholtz operator may be defined in a completely analogous manner.

The Green's function for the Dirichlet problem involving the Laplace operator is the function which satisfies

(a)
$$\nabla^2 G = \delta(x - \xi, y - \eta) \quad \text{in } D, \tag{11.3.3}$$
$$G = 0 \quad \text{on } B. \tag{11.3.4}$$

(b) G is symmetric, that is,

$$G(x, y; \xi, \eta) = G(\xi, \eta; x, y), \tag{11.3.5}$$

(c) G is continuous in x, y, ξ, η, but $(\partial G/\partial n)$ has a discontinuity at the point (ξ, η) which is specified by the equation

$$\lim_{\varepsilon \to 0} \int_{C_\varepsilon} \frac{\partial G}{\partial n} \, ds = 1, \tag{11.3.6}$$

where n is the outward normal to the circle

$$C_\varepsilon : (x - \xi)^2 + (y - \eta)^2 = \varepsilon^2.$$

The Green's function G may be interpreted as the response of the system at a field point (x, y) due to a δ function input at the source point (ξ, η). G is continuous everywhere in D, and its first and second derivatives are continuous in D except at (ξ, η). Thus, property (a) essentially states that $\nabla^2 G = 0$ everywhere except at the source point (ξ, η).

We will now prove property (b).

Theorem 11.3.1. *The Green's function is symmetric.*

Proof. Applying the Green's second formula

$$\iint_D \left(\phi \nabla^2 \psi - \psi \nabla^2 \phi\right) dS = \int_B \left(\phi \frac{\partial \psi}{\partial n} - \psi \frac{\partial \phi}{\partial n}\right) ds, \qquad (11.3.7)$$

to the functions $\phi = G\left(x, y; \xi, \eta\right)$ and $\psi = G\left(x, y; \xi^*, \eta^*\right)$, we obtain

$$\iint_D \left[G\left(x, y; \xi, \eta\right) \nabla^2 G\left(x, y; \xi^*, \eta^*\right) - G\left(x, y; \xi^*, \eta^*\right) \nabla^2 G\left(x, y; \xi, \eta\right)\right] dx\, dy$$

$$= \int_B \left[G\left(x, y; \xi, \eta\right) \frac{\partial G}{\partial n}\left(x, y; \xi^*, \eta^*\right) - G\left(x, y; \xi^*, \eta^*\right) \frac{\partial G}{\partial n}\left(x, y; \xi, \eta\right)\right] ds.$$

Since $G\left(x, y; \xi, \eta\right)$ and hence, $G\left(x, y; \xi^*, \eta^*\right)$ must vanish on B, we have

$$\iint_D \left[G\left(x, y; \xi, \eta\right) \nabla^2 G\left(x, y; \xi^*, \eta^*\right)\right.$$

$$\left. - G\left(x, y; \xi^*, \eta^*\right) \nabla^2 G\left(x, y; \xi, \eta\right)\right] dx\, dy = 0.$$

But

$$\nabla^2 G\left(x, y; \xi, \eta\right) = \delta\left(x - \xi, y - \eta\right),$$

and

$$\nabla^2 G\left(x, y; \xi^*, \eta^*\right) = \delta\left(x - \xi^*, y - \eta^*\right).$$

Since

$$\iint_D G\left(x, y; \xi, \eta\right) \delta\left(x - \xi^*, y - \eta^*\right) dx\, dy = G\left(\xi^*, \eta^*; \xi, \eta\right),$$

and

$$\iint_D G\left(x, y; \xi^*, \eta^*\right) \delta\left(x - \xi, y - \eta\right) dx\, dy = G\left(\xi, \eta; \xi^*, \eta^*\right),$$

we obtain

$$G\left(\xi, \eta; \xi^*, \eta^*\right) = G\left(\xi^*, \eta^*; \xi, \eta\right).$$

Theorem 11.3.2. $\partial G / \partial n$ *is discontinuous at* (ξ, η); *in particular*

$$\lim_{\varepsilon \to 0} \int_{C_\varepsilon} \frac{\partial G}{\partial n} ds = 1, \qquad C_\varepsilon : (x - \xi)^2 + (y - \eta)^2 = \varepsilon^2.$$

Proof. Let R_ε be the region bounded by C_ε. Then, integrating both sides of equation (11.3.3), we obtain

$$\iint_{R_\varepsilon} \nabla^2 G\, dx\, dy = \iint_R \delta\left(x - \xi, y - \eta\right) dx\, dy = 1.$$

It therefore follows that

$$\lim_{\varepsilon \to 0} \iint_{R_\varepsilon} \nabla^2 G \, dx \, dy = 1. \tag{11.3.8}$$

Thus, by the Divergence theorem of calculus,

$$\lim_{\varepsilon \to 0} \iint_{C_\varepsilon} \frac{\partial G}{\partial n} \, ds = 1.$$

11.4 Method of Green's Functions

It is often convenient to seek G as the sum of a particular integral of the nonhomogeneous equation and the solution of the associated homogeneous equation. That is, G may assume the form

$$G\left(\xi, \eta; x, y\right) = F\left(\xi, \eta; x, y\right) + g\left(\xi, \eta; x, y\right), \tag{11.4.1}$$

where F, known as the free-space Green's function, satisfies

$$\nabla^2 F = \delta\left(\xi - x, \eta - y\right) \quad \text{in } D, \tag{11.4.2}$$

and g satisfies

$$\nabla^2 g = 0 \quad \text{in } D, \tag{11.4.3}$$

so that by superposition $G = F + g$ satisfies equation (11.3.3). Also $G = 0$ on B requires that

$$g = -F \quad \text{on } B. \tag{11.4.4}$$

Note that F need not satisfy the boundary condition. Hereafter, (x, y) will denote the source point.

Before we determine the solution of a particular problem, let us first find F for the Laplace and Helmholtz operators.

(1) Laplace Operator

In this case, F must satisfy the equation

$$\nabla^2 F = \delta\left(\xi - x, \eta - y\right) \quad \text{in } D.$$

Then, for $r = \left[(\xi - x)^2 + (\eta - y)^2\right]^{\frac{1}{2}} > 0$, that is, for $\xi \neq x, \eta \neq y$, we have with (x, y) as the center

$$\nabla^2 F = \frac{1}{r} \frac{\partial}{\partial r} \left(r \frac{\partial F}{\partial r}\right) = 0,$$

since F is independent of θ. Therefore, the solution is

$$F = A + B \log r.$$

Applying condition (11.3.6), it follows directly from equation (11.3.8) with $\nabla^2 g = 0$, that

$$\lim_{\varepsilon \to 0} \int_{C_\varepsilon} \frac{\partial F}{\partial n} \, ds = \lim_{\varepsilon \to 0} \int_0^{2\pi} \frac{B}{r} r \, d\theta = 1.$$

Thus, $B = 1/2\pi$ and A is arbitrary. For simplicity, we choose $A = 0$. Then F takes the form

$$F = \frac{1}{2\pi} \log r. \tag{11.4.5}$$

(2) Helmholtz Operator

Here F is required to satisfy

$$\nabla^2 F + \kappa^2 F = \delta \left(x - \xi, y - \eta \right).$$

Again for $r > 0$, we find

$$\frac{1}{r} \frac{\partial}{\partial r} \left(r \frac{\partial F}{\partial r} \right) + \kappa^2 F = 0,$$

or,

$$r^2 F_{rr} + r F_r + \kappa^2 r^2 F = 0.$$

This is the Bessel equation of order zero, the solution of which is

$$F \left(\kappa r \right) = A J_0 \left(\kappa r \right) + B Y_0 \left(\kappa r \right).$$

Since the behavior of J_0 at $r = 0$ is not singular, we set $A = 0$. Thus, we have

$$F \left(\kappa r \right) = B Y_0 \left(\kappa r \right).$$

But, for very small r,

$$Y_0 \left(\kappa r \right) \sim \frac{2}{\pi} \log r.$$

Applying condition (11.3.6), we obtain

$$1 = \lim_{\varepsilon \to 0} \int_{C_\varepsilon} \frac{\partial F}{\partial n} \, ds = \lim_{\varepsilon \to 0} \int_{C_\varepsilon} B \frac{\partial Y_0}{\partial r} \, ds = B \cdot \frac{2}{\pi r} \cdot 2\pi r$$

and hence, $B = 1/4$. Thus, $F \left(\kappa r \right)$ becomes

$$F(\kappa r) = \frac{1}{4} Y_0 (\kappa r).$$ (11.4.6)

We may point out that, since

$$\left(\nabla^2 + \kappa^2\right) \text{ approaches } \nabla^2 \text{ as } \kappa \to 0,$$

it should (and does) follow that

$$\frac{1}{4} Y_0 (\kappa r) \to \frac{1}{2} \log r \text{ as } \kappa \to 0+.$$

11.5 Dirichlet's Problem for the Laplace Operator

We are now in a position to determine the solution of the Dirichlet problem

$$\nabla^2 u = h \quad \text{in } D,$$

$$u = f \quad \text{on } B,$$ (11.5.1)

by the method of Green's function.

By putting $\phi(\xi, \eta) = G(\xi, \eta; x, y)$ and $\psi(\xi, \eta) = u(\xi, \eta)$ in equation (11.3.7), we obtain

$$\iint_D \left[G(\xi, \eta; x, y) \nabla^2 u - u(\xi, \eta) \nabla^2 G \right] d\xi \, d\eta$$

$$= \int_B \left[G(\xi, \eta; x, y) \frac{\partial u}{\partial n} - u(\xi, \eta) \frac{\partial G}{\partial n} \right] ds.$$

But

$$\nabla^2 u = h(\xi, \eta) \quad \text{in } D,$$

and

$$\nabla^2 G = \delta(\xi - x, \eta - y) \quad \text{in } D.$$

Thus, we have

$$\iint_D \left[G(\xi, \eta; x, y) h(\xi, \eta) - u(\xi, \eta) \delta(\xi - x, \eta - y) \right] d\xi \, d\eta$$

$$= \int_B \left[G(\xi, \eta; x, y) \frac{\partial u}{\partial n} - u(\xi, \eta) \frac{\partial G}{\partial n} \right] ds.$$ (11.5.2)

Since $G = 0$ and $u = f$ on B, and since G is symmetric, it follows that

$$u(x, y) = \iint_D G(x, y; \xi, \eta) h(\xi, \eta) \, d\xi \, d\eta + \int_B f \frac{\partial G}{\partial n} \, ds$$ (11.5.3)

which is the solution given by (11.3.2).

As a specific example, consider the Dirichlet problem for a unit circle. Then

$$\nabla^2 g = g_{\xi\xi} + g_{\eta\eta} = 0 \quad \text{in } D,$$

$$g = -F \qquad \qquad \text{on } B.$$

(11.5.4)

But we already have from equation (11.4.5) that $F = (1/2\pi) \log r$.

If we introduce the polar coordinates (see Figure 11.5.1) ρ, θ, σ, β by means of the equations

$$x = \rho \cos \theta, \qquad \xi = \sigma \cos \beta,$$

$$y = \rho \sin \theta, \qquad \eta = \sigma \sin \beta,$$

(11.5.5)

then the solution of equation (11.5.4) is [see Section 9.4]

$$g(\sigma, \beta) = \frac{a_0}{2} + \sum_{n=1}^{\infty} \sigma^n (a_n \cos n\beta + b_n \sin n\beta),$$

where

$$g = -\frac{1}{4\pi} \log \left[1 + \rho^2 - 2\rho \cos (\beta - \theta)\right] \quad \text{on } B.$$

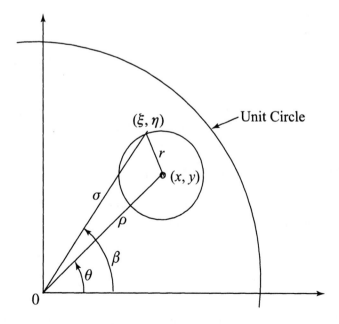

Figure 11.5.1 Image point.

By using the relation

$$\log\left[1 + \rho^2 - 2\rho\cos\left(\beta - \theta\right)\right] = -2\sum_{n=1}^{\infty}\frac{\rho^n\cos n\left(\beta - \theta\right)}{n},$$

and equating the coefficients of $\sin n\beta$ and $\cos n\beta$ to determine a_n and b_n, we find

$$a_n = \frac{\rho^n}{2\pi n}\cos n\theta, \qquad b_n = \frac{\rho^n}{2\pi n}\sin n\theta.$$

It therefore follows that

$$g\left(\rho, \theta; \sigma, \beta\right) = \frac{1}{2\pi}\sum_{n=1}^{\infty}\frac{\left(\sigma\rho\right)^n}{n}\cos n\left(\beta - \theta\right)$$

$$= -\frac{1}{4\pi}\log\left[1 + \left(\sigma\rho\right)^2 - 2\left(\sigma\rho\right)\cos\left(\beta - \theta\right)\right].$$

Hence, the Green's function for the problem is

$$G\left(\rho, \theta; \sigma, \beta\right) = \frac{1}{4\pi}\log\left[\sigma^2 + \rho^2 - 2\sigma\rho\cos\left(\beta - \theta\right)\right]$$

$$- \frac{1}{4\pi}\log\left[1 + \left(\sigma\rho\right)^2 - 2\sigma\rho\cos\left(\beta - \theta\right)\right], \quad (11.5.6)$$

from which we find

$$\left.\frac{\partial G}{\partial n}\right|_{\text{on } B} = \left(\frac{\partial G}{\partial \sigma}\right)_{\sigma=1} = \frac{1}{2\pi}\frac{1 - \rho^2}{\left[1 + \rho^2 - 2\rho\cos\left(\beta - \theta\right)\right]}.$$

If $h = 0$, then solution (11.5.3) reduces to the Poisson integral formula similar to (9.4.10) and assumes the form

$$u\left(\rho, \theta\right) = \frac{1}{2\pi}\int_0^{2\pi}\frac{1 - \rho^2}{1 + \rho^2 - 2\rho\cos\left(\beta - \theta\right)}f\left(\beta\right)d\beta.$$

11.6 Dirichlet's Problem for the Helmholtz Operator

We will now determine the Green's function solution of the Dirichlet problem involving the Helmholtz operator, namely,

$$\nabla^2 u + \kappa^2 u = h \quad \text{in } D,$$

$$(11.6.1)$$

$$u = f \quad \text{on } B,$$

where D is a circular domain of unit radius with boundary B. Then, the Green's function must satisfy

$$\nabla^2 G + \kappa^2 G = \delta\left(\xi - x, \eta - y\right) \quad \text{in } D,$$

$$(11.6.2)$$

$$G = 0 \qquad \text{on } B.$$

Again, we seek the solution in the form

$$G\left(\xi, \eta; x, y\right) = F\left(\xi, \eta; x, y\right) + g\left(\xi, \eta; x, y\right). \qquad (11.6.3)$$

From equation (11.4.6), we have

$$F = \frac{1}{4} Y_0\left(\kappa r\right), \qquad (11.6.4)$$

where $r = \left[\left(\xi - x\right)^2 + \left(\eta - y\right)^2\right]^{\frac{1}{2}}$. The function g must satisfy

$$\nabla^2 g + \kappa^2 g = 0 \qquad \text{in } D,$$

$$(11.6.5)$$

$$g = -\frac{1}{4} Y_0\left(\kappa r\right) \quad \text{on } B.$$

This solution can be determined easily by the method of separation of variables. Thus, the solution in the polar coordinates defined by equation (11.5.5) may be written in the form

$$g\left(\rho, \theta, \sigma, \beta\right) = \sum_{n=0}^{\infty} J_n\left(\kappa\sigma\right)\left[a_n \cos n\beta + b_n \sin n\beta\right], \qquad (11.6.6)$$

where

$$a_0 = -\frac{1}{8\pi J_0\left(\kappa\right)} \int_{-\pi}^{\pi} Y_0\left[\kappa\sqrt{1 + \rho^2 - 2\rho\cos\left(\beta - \theta\right)}\right] d\beta,$$

$$a_n = -\frac{1}{4\pi J_n\left(\kappa\right)} \int_{-\pi}^{\pi} Y_0\left[\kappa\sqrt{1 + \rho^2 - 2\rho\cos\left(\beta - \theta\right)}\right] \cos n\beta\, d\beta$$

$$b_n = -\frac{1}{4\pi J_n\left(\kappa\right)} \int_{-\pi}^{\pi} Y_0\left[\kappa\sqrt{1 + \rho^2 - 2\rho\cos\left(\beta - \theta\right)}\right] \sin n\beta\, d\beta \left.\right\} \quad n = 1, 2, \ldots.$$

To find the solution of the Dirichlet problem, we multiply both sides of the first equation of equation (11.6.1) by G and integrate. Thus, we have

$$\iint_D \left(\nabla^2 u + \kappa^2 u\right) G\left(\xi, \eta; x, y\right) d\xi\, d\eta = \iint_D h\left(\xi, \eta\right) G\left(\xi, \eta; x, y\right) d\xi\, d\eta.$$

We then apply Green's theorem on the left side of the preceding equation and obtain

$$\iint_D h\left(\xi, \eta\right) G\left(\xi, \eta; x, y\right) d\xi\, d\eta - \iint_D u\left(\nabla^2 G + \kappa^2 G\right) d\xi\, d\eta$$

$$= \int_B \left(G\, u_n - u\, G_n\right) ds.$$

But $\nabla^2 G + \kappa^2 G = \delta\,(\xi - x, \eta - y)$ in D and $G = 0$ on B. We, therefore, have

$$u\,(x, y) = \iint_D h\,(\xi, \eta)\,G\,(\xi, \eta; x, y)\,d\xi\,d\eta + \int_B f\,(\xi, \eta)\,G_n ds, \quad (11.6.7)$$

where G is given by equation (11.6.3) with equations (11.6.4) and (11.6.6).

11.7 Method of Images

We shall describe another method of obtaining Green's functions. This method, called the *method of images*, is based essentially on the construction of Green's function for a finite domain from that of an infinite domain. The disadvantage of this method is that it can be applied only to problems with simple boundary geometry.

As an illustration, we consider the same Dirichlet problem solved in Section 11.5.

Let $P\,(\xi, \eta)$be a point in the unit circle D, and let $Q\,(x, y)$ be the source point also in D. The distance between P and Q is r. Let Q' be the *image* which lies outside of D on the ray from the origin opposite to the source point Q (as shown in Figure 11.7.1) such that $OQ/\sigma = \sigma/OQ'$, where σ is the radius of the circle passing through P centered at the origin.

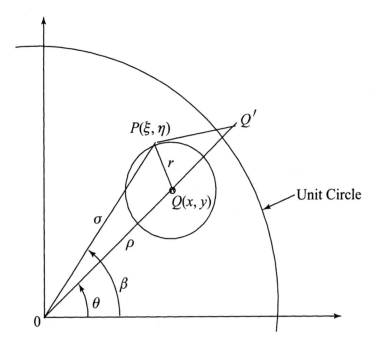

Figure 11.7.1 Image point.

Since the two triangles OPQ and OPQ' are similar by virtue of the hypothesis $(OQ)(OQ') = \sigma^2$ and by possessing a common angle at O, we have

$$\frac{r'}{r} = \frac{\sigma}{\rho}, \qquad (11.7.1)$$

where $r' = PQ'$ and $\rho = OQ$.

If $\sigma = 1$, equation (11.7.1) becomes

$$\left(\frac{r}{r'}\right)\frac{1}{\rho} = 1.$$

Then, we clearly see that the quantity

$$\frac{1}{2\pi}\log\left(\frac{r}{r'}\frac{1}{\rho}\right) = \frac{1}{2\pi}\log r - \frac{1}{2\pi}\log r' + \frac{1}{2\pi}\log\frac{1}{\rho} \qquad (11.7.2)$$

which vanishes on the boundary $\sigma = 1$, is harmonic in D except at Q, and satisfies equation (11.3.3). (Note the $\log r'$ is harmonic everywhere except at Q', which is outside the domain D.) This suggests that we should choose the Green's function

$$G = \frac{1}{2\pi}\log r - \frac{1}{2\pi}\log r' + \frac{1}{2\pi}\log\frac{1}{\rho}. \qquad (11.7.3)$$

Noting that Q' is at $(1/\rho, \theta)$, the function G in polar coordinates takes the form

$$G(\rho, \theta, \sigma, \beta) = \frac{1}{4\pi}\log\left[\sigma^2 + \rho^2 - 2\sigma\rho\cos(\beta - \theta)\right]$$
$$-\frac{1}{4\pi}\log\left[\frac{1}{\sigma^2} + \rho^2 - 2\frac{\rho}{\sigma}\cos(\beta - \theta)\right] + \frac{1}{2\pi}\log\frac{1}{\sigma} \quad (11.7.4)$$

which is the same as G given by (11.5.6).

It is quite interesting to observe the physical interpretation of the Green's function (11.7.3) and (11.7.4). The first term represents the potential due to a unit line charge at the source point, whereas the second term represents the potential due to a negative unit charge at the image point. The third term represents a uniform potential. The sum of these potentials makes up the total potential field.

Example 11.7.1. To illustrate an obvious and simple case, consider the semi-infinite plane $\eta > 0$. The problem is to solve

$$\nabla^2 u = h \quad \text{in } \eta > 0,$$

$$u = f \quad \text{on } \eta = 0.$$

The image point should be obvious by inspection. Thus, if we construct

$$G = \frac{1}{4\pi} \log \left[(\xi - x)^2 + (\eta - y)^2 \right] - \frac{1}{4\pi} \left[(\xi - x)^2 + (\eta + y)^2 \right], \quad (11.7.5)$$

the condition that $G = 0$ on $\eta = 0$ is clearly satisfied. It is also evident that G is harmonic in $\eta > 0$ except at the source point, and that G satisfies equation (11.3.3).

With $G_n|_B = [-G_\eta]_{\eta=0}$, the solution (11.5.3) is given by

$$u(x, y) = \frac{y}{\pi} \int_{-\infty}^{\infty} \frac{f(\xi)\, d\xi}{(\xi - x)^2 + y^2}$$

$$+ \frac{1}{4\pi} \int_0^{\infty} \int_{-\infty}^{\infty} \log \left[\frac{(\xi - x)^2 + (\eta - y)^2}{(\xi - x)^2 + (\eta + y)^2} \right] h(\xi, \eta)\, d\xi\, d\eta. \quad (11.7.6)$$

Example 11.7.2. Another example that illustrates the method of images well is the Robin's problem on the quarter infinite plane, that is,

$$\begin{aligned}
\nabla^2 u &= h(\xi, \eta) & &\text{in } \xi > 0, \quad \eta > 0, \\
u &= f(\eta) & &\text{on } \xi = 0, & &(11.7.7) \\
u_n &= g(\xi) & &\text{on } \eta = 0.
\end{aligned}$$

This illustrated in Figure 11.7.2.

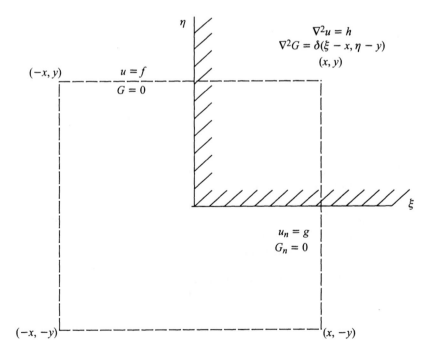

Figure 11.7.2 Images in the Robin problem.

Let $(-x, y)$, $(-x, -y)$, and $(x, -y)$ be the three image points of the source point (x, y). Then, by inspection, we can immediately construct Green's function

$$G = \frac{1}{4\pi} \log \frac{\left[(\xi - x)^2 + (\eta - y)^2\right]\left[(\xi - x)^2 + (\eta + y)^2\right]}{\left[(\xi + x)^2 + (\eta - y)^2\right]\left[(\xi + x)^2 + (\eta + y)^2\right]}. \quad (11.7.8)$$

This function satisfies $\nabla^2 G = 0$ except at the source point, and $G = 0$ on $\xi = 0$ and $G_\eta = 0$ on $\eta = 0$.

The solution from equation (11.3.3) is thus given by

$$u(x, y) = \iint_D G h \, d\xi \, d\eta + \int_B (G u_n - u G_n) \, ds,$$

$$= \int_0^\infty \int_0^\infty G h \, d\xi \, d\eta + \int_0^\infty g(\xi) G(\xi, 0; x, y) \, d\xi,$$

$$+ \int_0^\infty f(\eta) G_\xi (0, \eta; x, y) \, d\xi.$$

11.8 Method of Eigenfunctions

In this section, we will apply the method of eigenfunctions, described in Chapter 10, to obtain the Green's function.

We consider the boundary-value problem

$$\nabla^2 u = h \quad \text{in } D,$$

$$u = f \quad \text{on } B. \quad (11.8.1)$$

For this problem, G must satisfy

$$\nabla^2 G = \delta(\xi - x, \eta - y) \quad \text{in } D,$$

$$G = 0 \quad \text{on } B, \quad (11.8.2)$$

and hence, the associated eigenvalue problem is

$$\nabla^2 \phi + \lambda \phi = 0 \quad \text{in } D,$$

$$\phi = 0 \quad \text{on } B. \quad (11.8.3)$$

Let ϕ_{mn} be the eigenfunctions and λ_{mn} be the corresponding eigenvalues. We then expand G and δ in terms of the eigenfunctions ϕ_{mn}. Consequently, we write

$$G\left(\xi, \eta; x, y\right) = \sum_m \sum_n a_{mn}\left(x, y\right) \phi_{mn}\left(\xi, \eta\right), \qquad (11.8.4)$$

$$\delta\left(\xi - x, \eta - y\right) = \sum_m \sum_n b_{mn}\left(x, y\right) \phi_{mn}\left(\xi, \eta\right), \qquad (11.8.5)$$

where

$$b_{mn} = \frac{1}{\|\phi_{mn}\|^2} \iint_D \delta\left(\xi - x, \eta - y\right) \phi_{mn}\left(\xi, \eta\right) d\xi\, d\eta$$

$$= \frac{\phi_{mn}\left(x, y\right)}{\|\phi_{mn}\|^2} \qquad (11.8.6)$$

in which

$$\|\phi_{mn}\|^2 = \iint_D \phi_{mn}^2 d\xi\, d\eta.$$

Now substituting equations (11.8.4) and (11.8.5) into equation (11.8.2) and using the relation from equation (11.8.3) that

$$\nabla^2 \phi_{mn} + \lambda_{mn} \phi_{mn} = 0,$$

we obtain

$$-\sum_m \sum_n \lambda_{mn} a_{mn}\left(x, y\right) \phi_{mn}\left(\xi, \eta\right) = \sum_m \sum_n \frac{\phi_{mn}\left(x, y\right) \phi_{mn}\left(\xi, \eta\right)}{\|\phi_{mn}\|^2}.$$

Hence,

$$a_{mn}\left(x, y\right) = -\frac{\phi_{mn}\left(x, y\right)}{\lambda_{mn} \|\phi_{mn}\|^2}, \qquad (11.8.7)$$

and the Green's function is therefore given by

$$G\left(\xi, \eta; x, y\right) = -\sum_m \sum_n \frac{\phi_{mn}\left(x, y\right) \phi_{mn}\left(\xi, \eta\right)}{\lambda_{mn} \|\phi_{mn}\|^2}. \qquad (11.8.8)$$

Example 11.8.1. As a particular example, consider the Dirichlet problem in a rectangular domain

$$\nabla^2 u = h \quad \text{in } D\ \{0 < x < a, \quad 0 < y < b\},$$
$$u = 0 \quad \text{on } B.$$

The eigenfunctions can be obtained explicitly by the method of separation of variables. We assume a nontrivial solution in the form

$$u\left(\xi, \eta\right) = X\left(\xi\right) Y\left(\eta\right).$$

Substituting this in the following system

$$\nabla^2 u + \lambda u = 0 \quad \text{in } D,$$
$$u = 0 \quad \text{on } B,$$

yields, with α^2 as separation constant,

$$X'' + \alpha^2 X = 0,$$
$$Y'' + \left(\lambda - \alpha^2\right) Y = 0.$$

With the homogeneous boundary conditions $X(0) = X(a) = 0$ and $Y(0) = Y(b) = 0$, functions X and Y are found to be

$$X_m(\xi) = A_m \sin\left(\frac{m\pi\xi}{a}\right),$$
$$Y_n(\eta) = B_n \sin\left(\frac{n\pi\eta}{b}\right).$$

We then have

$$\lambda_{mn} = \pi^2 \left(\frac{m^2}{a^2} + \frac{n^2}{b^2}\right).$$

Thus, we obtain the eigenfunctions

$$\phi_{mn}(\xi, \eta) = \sin\left(\frac{m\pi\xi}{a}\right)\sin\left(\frac{n\pi\eta}{b}\right).$$

Knowing ϕ_{mn}, we compute $\|\phi_{mn}\|$ and obtain

$$\|\phi_{mn}\|^2 = \int_0^a \int_0^b \sin^2\left(\frac{m\pi\xi}{a}\right)\sin^2\left(\frac{n\pi\eta}{b}\right) d\xi\, d\eta = \left(\frac{ab}{4}\right).$$

We then obtain from equation (11.8.8) the Green's function

$$G(\xi, \eta; x, y) = -\frac{4ab}{\pi^2} \sum_{m=1}^{\infty} \sum_{n=1}^{\infty} \frac{\sin\left(\frac{m\pi x}{a}\right)\sin\left(\frac{n\pi y}{b}\right)\sin\left(\frac{m\pi\xi}{a}\right)\sin\left(\frac{n\pi\eta}{b}\right)}{(m^2 b^2 + n^2 a^2)}.$$

11.9 Higher-Dimensional Problems

The Green's function method can be easily extended for applications in three and more dimensions. Since most of the problems encountered in the physical sciences are in three dimensions, we will illustrate the method with some examples suitable for practical application.

We first extend our definition of the Green's function in three dimensions.

The Green's function for the Dirichlet problem involving the Laplace operator is the function that satisfies

(a) $\nabla^2 G = \delta(x - \xi, y - \eta, z - \zeta)$ in R,   (11.9.1)

   $G = 0$ on S.   (11.9.2)

(b) $G(x, y, z; \xi, \eta, \zeta) = G(\xi, \eta, \zeta; x, y, z)$.   (11.9.3)

(c) $\displaystyle \lim_{\varepsilon \to 0} \iint_{S_\varepsilon} \frac{\partial G}{\partial n}\, ds = 1$,   (11.9.4)

where n is the outward unit normal to the surface

$$S_\varepsilon : (x - \xi)^2 + (y - \eta)^2 + (z - \zeta)^2 = \varepsilon^2.$$

Proceeding as in the two-dimensional case, the solution of the Dirichlet problem

$$\nabla^2 u = h \quad \text{in } R,$$

$$u = f \quad \text{on } S,$$
(11.9.5)

is

$$u(x, y, z) = \iiint_R h\, G\, dR + \iint_S f\, G_n\, dS.$$
(11.9.6)

Again we let

$$G(\xi, \eta, \zeta; x, y, z) = F(\xi, \eta, \zeta; x, y, z) + g(\xi, \eta, \zeta; x, y, z),$$

where

$$\nabla^2 F = \delta(x - \xi, y - \eta, z - \zeta) \quad \text{in } R,$$

and

$$\nabla^2 g = 0 \quad \text{in } R,$$

$$u = -F \quad \text{on } S.$$

Example 11.9.1. We consider a spherical domain with radius a. We must have

$$\nabla^2 F = 0$$

except at the source point. For

$$r = \left[(\xi - x)^2 + (\eta - y)^2 + (\zeta - z)^2\right]^{\frac{1}{2}} > 0$$

with (x, y, z) as the origin, we have

$$\nabla^2 F = \frac{1}{r^2} \frac{d}{dr}\left(r^2 \frac{dF}{dr}\right) = 0.$$

Integration then yields

$$F = A + \frac{B}{r} \quad \text{for} \quad r > 0.$$

Applying condition (11.9.4) we obtain

$$\lim_{\varepsilon \to 0} \iint_{S_\varepsilon} G_n \, dS = \lim_{\varepsilon \to 0} \iint_{S_\varepsilon} F_r \, dS = 1.$$

Consequently, $B = -(1/4\pi)$ and A is arbitrary. If we set the boundedness condition at infinity for exterior problems so that $A = 0$, we have

$$F = -\frac{1}{4\pi r}. \tag{11.9.7}$$

We apply the method of images to obtain the Green's function. If we draw a three-dimensional diagram analogous to Figure 11.7.1, we will have a relation similar to (11.7.1), namely,

$$r' = \left(\frac{a}{\rho}\right) r, \tag{11.9.8}$$

where r' and ρ are measured in three-dimensional space. Thus, we seek Green's function

$$G = \frac{-1}{4\pi r} + \frac{a/\rho}{4\pi r'} \tag{11.9.9}$$

which is harmonic everywhere in r except at the source point, and is zero on the surface S.

In terms of spherical coordinates

$$\begin{aligned} \xi &= \tau \cos \psi \sin \alpha, & x &= \rho \cos \phi \sin \theta, \\ \eta &= \tau \sin \psi \sin \alpha, & y &= \rho \sin \phi \sin \theta, \\ \zeta &= \tau \cos \alpha, & z &= \rho \cos \theta, \end{aligned}$$

the Green's function G can be written in the form

$$G = \frac{-1}{4\pi \left(\tau^2 + \rho^2 - 2\tau\rho \cos \gamma\right)^{\frac{1}{2}}} + \frac{1}{4\pi \left[\frac{\tau^2 \rho^2}{a^2} + a^2 - 2\tau\rho \cos \gamma\right]^{\frac{1}{2}}}, \tag{11.9.10}$$

where γ is the angle between r and r'. Now differentiating G, we have

$$\left[\frac{\partial G}{\partial \tau}\right]_{\tau=a} = \frac{a^2 - \rho^2}{4\pi a \left(a^2 + \rho^2 - 2a\rho \cos \gamma\right)^{\frac{1}{2}}}.$$

Thus, the solution of the Dirichlet problem for $h = 0$ is

$$u\left(\rho, \theta, \phi\right) = \frac{a\left(a^2 - \rho^2\right)}{4\pi} \int_0^{2\pi} \int_0^\pi \frac{f\left(\alpha, \psi\right) \sin \alpha \, d\alpha \, d\psi}{\left(a^2 + \rho^2 - 2a\rho \cos \gamma\right)^{\frac{3}{2}}}, \quad (11.9.11)$$

where $\cos \gamma = \cos \alpha \cos \theta + \sin \alpha \sin \theta \cos \left(\psi - \phi\right)$. This integral is called the *three-dimensional Poisson integral formula*.

For the exterior problem where the outward normal is radially inward towards the origin, the solution can be simply obtained by replacing $\left(a^2 - \rho^2\right)$ by $\left(\rho^2 - a^2\right)$ in equation (11.9.11).

Example 11.9.2. An example involving the Helmholtz operator is the three-dimensional radiation problem

$$\nabla^2 u + \kappa^2 u = 0, \quad (11.9.12)$$

$$\lim_{r \to \infty} r\left(u_r + i\kappa u\right) = 0,$$

where $i = \sqrt{-1}$; the limit condition is called the *radiation condition*, and r is the field point distance.

In this case, the Green's function must satisfy

$$\nabla^2 G + \kappa^2 G = \delta\left(\xi - x, \eta - y, \zeta - z\right). \quad (11.9.13)$$

Since the point source solution is dependent only on r, we write the Helmholtz equation

$$G_{rr} + \frac{2}{r} G_r + \kappa^2 G = 0 \quad \text{for} \quad r > 0.$$

Note that the source point is taken as the origin. If we write the above equation in the form

$$(Gr)_{rr} + \kappa^2\left(Gr\right) = 0 \quad \text{for} \quad r > 0 \quad (11.9.14)$$

then the solution can easily be seen to be

$$Gr = Ae^{i\kappa r} + Be^{-i\kappa r},$$

or, equivalently,

$$G = A\frac{e^{i\kappa r}}{r} + B\frac{e^{-i\kappa r}}{r}. \quad (11.9.15)$$

In order for G to satisfy the radiation condition

$$\lim_{r \to \infty} r\left(G_r + i\kappa G\right) = 0,$$

the constant $A = 0$, and G then takes the form

$$G = B\frac{e^{-i\kappa r}}{r}.$$

To determine B, we have

$$\lim_{\varepsilon \to 0} \iint_{S_\varepsilon} \frac{\partial G}{\partial n} dS = -\lim_{\varepsilon \to 0} \iint_{S_\varepsilon} B \frac{e^{-i\kappa r}}{r} \left(\frac{1}{r} + i\kappa\right) dS = 1$$

from which we obtain $B = -1/4\pi$, and consequently,

$$G = -\frac{e^{-i\kappa r}}{4\pi r}. \qquad (11.9.16)$$

Note that this reduces to $(1/4\pi r)$ when $\kappa = 0$.

Example 11.9.3. Show that the solution of the Poisson equation

$$-\nabla^2 u = f(x, y, z), \qquad (11.9.17)$$

is

$$u(x, y, z) = \iiint G(r) f(\xi, \eta, \zeta) \, d\xi \, d\eta \, d\zeta, \qquad (11.9.18)$$

where the Green's function $G(r)$ is

$$G(r) = \frac{1}{4\pi r} = \frac{1}{4\pi} \left\{ (x - \xi)^2 + (y - \eta)^2 + (z - \zeta)^2 \right\}^{-\frac{1}{2}}. \ (11.9.19)$$

The Green's function G satisfies the equation

$$-\nabla^2 G = \delta(x - \xi) \, \delta(y - \eta) \, \delta(z - \zeta). \qquad (11.9.20)$$

It is noted that everywhere except at $(x, y, z) = (\xi, \eta, \zeta)$, equation (11.9.20) is a homogeneous equation that can be solved by the method of separation of variables. However, at the point (ξ, η, ζ) this equation is no longer homogeneous. Usually, this point (ξ, η, ζ) represents a *source point* or a *source point singularity* in electrostatics or fluid mechanics. In order to solve (11.9.17), it is necessary to take into account the source point at (ξ, η, ζ). Without loss of generality, it is convenient to transform the frame of reference so that the source point is at the origin. This can be done by the transformation $x_1 = x - \xi$, $y_1 = y - \eta$, and $z_1 = z - \zeta$. Consequently, equation (11.9.20) becomes

$$\nabla^2 G = -\delta(x_1) \, \delta(y_1) \, \delta(z_1), \qquad (11.9.21)$$

where ∇^2 is the Laplacian in terms of x_1, y_1, and z_1.
    Introducing the spherical polar coordinates

$$x_1 = r \sin\theta \cos\phi, \qquad y_1 = r \sin\theta \sin\phi, \qquad z_1 = r \cos\theta,$$

equation (11.9.21) reduces to the form

$$\nabla^2 G = -\frac{\delta(r)}{4\pi r^2}, \tag{11.9.22}$$

where

$$\nabla^2 G \equiv \frac{1}{r^2}\frac{\partial}{\partial r}\left(r^2\frac{\partial G}{\partial r}\right) + \frac{1}{r^2\sin\theta}\frac{\partial}{\partial\theta}\left(\sin\theta\frac{\partial G}{\partial\theta}\right) + \frac{1}{r^2\sin^2\theta}\frac{\partial^2 G}{\partial\phi^2}. \tag{11.9.23}$$

Since the right hand side of (11.9.22) is a function of r alone, and hence, G must be a function of r alone, we write (11.9.22) with (11.9.23) as

$$\frac{1}{r^2}\frac{\partial}{\partial r}\left(r^2\frac{\partial G}{\partial r}\right) = -\frac{\delta(r)}{4\pi r^2}. \tag{11.9.24}$$

We assume that G tends to zero as $r \to \infty$.

The solution of the corresponding homogeneous equation of (11.9.24) is

$$G(r) = \frac{a}{r} + b, \tag{11.9.25}$$

where a and b are constants of integration. Since $G \to 0$ as $r \to \infty$, $b = 0$ and we set $a = \frac{1}{4\pi}$. Consequently, the solution for G is

$$G(r) = \frac{1}{4\pi r}. \tag{11.9.26}$$

This solution can be interpreted as the potential produced by a point charge at the point (ξ, η, ζ).

Finally, the solution of (11.9.17) is then given by

$$u(x, y, z) = -\iiint_{-\infty}^{\infty} G(r) f(\xi, \eta, \zeta) \, d\xi \, d\eta \, d\zeta$$

$$= \frac{1}{4\pi}\iiint_{-\infty}^{\infty}\left[(x-\xi)^2 + (y-\eta)^2 + (z-\zeta)^2\right]^{-\frac{1}{2}}$$

$$\times f(\xi, \eta, \zeta) \, d\xi \, d\eta \, d\zeta. \tag{11.9.27}$$

Physically, this solution of the Poisson equation represents the potential $u(x, y, z)$ produced by a charge distribution of volume density $f(x, y, z)$.

11.10 Neumann Problem

We have noted in the chapter on boundary-value problems that the Neumann problem requires more attention than Dirichlet's problem because an additional condition is necessary for the existence of a solution of the Neumann problem.

We now consider the Neumann problem

$$\nabla^2 u + \kappa^2 u = h \quad \text{in } R,$$
$$\frac{\partial u}{\partial n} = 0 \quad \text{on } S.$$

By the divergence theorem of calculus, we have

$$\iiint_R \nabla^2 u \, dR = \iint_S \frac{\partial u}{\partial n} \, dS.$$

Thus, if we integrate the Helmholtz equation and use the preceding result, we obtain

$$\kappa^2 \iiint_R u \, dR = \iiint_R h \, dR.$$

In the case of Poisson's equation where $\kappa = 0$, this relation is satisfied only when

$$\iiint_R h \, dR = 0.$$

If we consider a heat conduction problem, this condition may be interpreted as the requirement that the net heat generation be zero. This is physically reasonable since the boundary is insulated in such a way that the net flux across it is zero.

If we define Green's function G, in this case, by

$$\nabla^2 G + \kappa^2 G = \delta \left(\xi - x, \eta - y, \zeta - z \right) \quad \text{in } R,$$
$$\frac{\partial G}{\partial n} = 0 \qquad\qquad\qquad\qquad \text{on } S.$$

Then we must have

$$\kappa^2 \iiint_R G \, dR = 1$$

which cannot be satisfied for $\kappa = 0$. But, we know from a physical point of view that a solution exists if

$$\iiint_R h \, dR = 0.$$

Hence, we will modify the definition of Green's function so that

$$\frac{\partial G}{\partial n} = C \quad \text{on } S,$$

where C is a constant. Integrating $\nabla^2 G = \delta$ over R, we obtain

$$C \iint_S dS = 1.$$

It is not difficult to show that G remains symmetric if

$$\iint_S G\, dS = 0.$$

Thus, under this condition, if we take C to be reciprocal of the surface area, the solution of the Neumann problem for Poisson's equation is

$$u\,(x,y,z) = C^* + \iiint_R G\,(x,y,z;\xi,\eta,\zeta)\,h\,(\xi,\eta,\zeta)\,d\xi\,d\eta\,d\zeta,$$

where C^* is a constant.

We should remark here that the method of Green's functions provides the solution in integral form. This is made possible by replacing a problem involving nonhomogeneous boundary conditions with a problem of finding Green's function G with homogeneous boundary conditions.

Regardless of method employed, the Green's function of a problem with nonhomogeneous equation and homogeneous boundary conditions is the same as the Green's function of a problem with homogeneous equation and nonhomogeneous boundary conditions, since one problem can be transferred to the other without difficulty. To illustrate, we consider the problem

$$Lu = f \quad \text{in } R,$$
$$u = 0 \quad \text{on } \partial R,$$

where ∂R denotes the boundary of R.

If we let $v = w - u$, where w satisfies $Lw = f$ in R, then the problem becomes

$$Lv = 0 \quad \text{in } R,$$
$$v = w \quad \text{on } \partial R.$$

Conversely, if we consider the problem

$$Lu = 0 \quad \text{in } R,$$
$$u = g \quad \text{on } \partial R,$$

we can easily transform this problem into

$$Lv = Lw \equiv w^* \quad \text{in } R,$$
$$v = 0 \qquad\qquad \text{on } \partial R,$$

by putting $v = w - u$ and finding w that satisfies $w = g$ on ∂R.

In fact, if we have

$$Lu = f \quad \text{in } R,$$
$$u = g \quad \text{on } \partial R,$$

we can transform this problem into either one of the above problems.

11.11 Exercises

1. If L denotes the partial differential operator

$$Lu = Au_{xx} + Bu_{xy} + Cu_{yy} + Du_x + Eu_y + Fu,$$

and if M denotes the adjoint operator

$$Mv = (Av)_{xx} + (Bv)_{xy} + (Cv)_{yy} - (Dv)_x - (Ev)_y + Fv,$$

show that

$$\iint_R (vLu - uMv)\, dx\, dy = \int_{\partial R} [U \cos(n, x) + V \cos(n, y)]\, ds,$$

where

$$U = Avu_x - u(Av)_x - u(Bv)_y + Duv,$$
$$V = Bvu_x + Cvu_y - u(Cv)_y + Euv,$$

and ∂R is the boundary of a region R.

2. Prove that the Green's function for a problem, if it exists, is unique.
3. Determine the Green's function for the exterior Dirichlet problem for a unit circle

$$\nabla^2 u = 0 \quad \text{in } r > 1,$$
$$u = f \quad \text{on } r = 1.$$

4. Prove that for $x = x(\xi, \eta)$ and $y = y(\xi, \eta)$

$$\delta(x - x_0)\,\delta(y - y_0) = \frac{1}{|J|}\,\delta(\xi - \xi_0)\,\delta(\eta - \eta_0),$$

where J is the Jacobian and (x_0, y_0) corresponds to (ξ_0, η_0). Hence, show that for polar coordinates

$$\delta(x - x_0)\,\delta(y - y_0) = \frac{1}{r}\,\delta(r - r_0)\,\delta(\theta - \theta_0).$$

5. Determine, for an infinite wedge, the Green's function that satisfies

$$\nabla^2 G + \kappa^2 G = \frac{1}{r}\,\delta(r - r_0, \theta - \theta_0),$$
$$G = 0, \quad \theta = 0, \quad \text{and} \quad \theta = \alpha.$$

6. Determine, for the Poisson's equation, the Green's function which vanishes on the boundary of a semicircular domain of radius R.

7. Find the solution of the Dirichlet problem

$$\nabla^2 u = 0, \quad 0 < x < a, \quad 0 < y < b,$$
$$u(0, y) = u(a, y) = u(x, b) = 0,$$
$$u(x, 0) = f(x).$$

8. Determine the solution of Dirichlet's problem

$$\nabla^2 u = f(r, \theta) \quad \text{in } D,$$
$$u = 0, \qquad \text{on } \partial D,$$

where ∂D is the boundary of a circle D of radius R.

9. Determine the Green's function for the semi-infinite region $\zeta > 0$ for

$$\nabla^2 G + \kappa^2 G = \delta(\xi - x, \eta - y, \zeta - z),$$
$$G = 0, \quad \text{on} \quad \zeta = 0.$$

10. Determine the Green's function for the semi-infinite region $\zeta > 0$ for

$$\nabla^2 G + \kappa^2 G = \delta(\xi - x, \eta - y, \zeta - z),$$
$$\frac{\partial G}{\partial n} = 0, \quad \text{on} \quad \zeta = 0.$$

11. Find the Green's function in the quarter plane $\xi > 0$, $\eta > 0$ which satisfies

$$\nabla^2 G = \delta(\xi - x, \eta - y),$$
$$G = 0, \quad \text{on} \quad \xi = 0 \quad \text{and} \quad \eta = 0.$$

12. Find the Green's function in the quarter plane $\xi > 0$, $\eta > 0$ which satisfies

$$\nabla^2 G = \delta(\xi - x, \eta - y),$$
$$G_\xi(0, \eta) = 0, \qquad G(\xi, 0) = 0.$$

13. Find the Green's function in the half plane $0 < x < \infty$, $-\infty < y < \infty$ for the problem

$$\nabla^2 u = f \quad \text{in } R,$$
$$u = 0, \quad \text{on } x = 0.$$

14. Determine the Green's function that satisfies

$$\nabla^2 G = \delta(x - \xi, y - \eta) \quad \text{in} \quad D : 0 < x < a, \quad 0 < y < \infty,$$
$$G = 0, \quad \text{on} \quad \partial D : x = 0, \quad x = a, \quad y = 0,$$
$$G \text{ is bounded at infinity.}$$

15. Find the Green's function that satisfies

$$\nabla^2 G = \frac{1}{r}\delta\left(r - \rho, \theta - \beta\right), \quad 0 < \theta < \frac{\pi}{3}, \quad 0 < r < 1,$$

$$G = 0, \quad \text{on} \quad \theta = 0, \quad \text{and} \quad \theta < \frac{\pi}{3},$$

$$\frac{\partial G}{\partial n} = 0, \quad \text{on} \quad r = 1.$$

16. Solve the boundary-value problem

$$\frac{1}{r}\frac{\partial G}{\partial r}\left(r\frac{\partial u}{\partial r}\right) + \frac{\partial^2 u}{\partial z^2} + \kappa^2 u = 0, \quad r \geq 0, \quad z > 0,$$

$$\frac{\partial u}{\partial z} = \begin{cases} 0, & r > a, \quad z = 0, \\ C, & r < a, \quad z = 0, \end{cases} \quad C = \text{constant.}$$

17. Obtain the solution of the Laplace equation

$$\nabla^2 u = 0, \quad 0 < r < \infty, \quad 0 < \theta < 2\pi,$$
$$u\left(r, 0+\right) = u\left(r, 2\pi-\right) = 0.$$

18. Determine the Green's function for the equation

$$\nabla^2 u - \kappa^2 u = 0,$$

vanishing on all sides of the rectangle $0 \leq x \leq a$, $0 \leq y \leq b$.

19. Determine the Green's function of the Helmholtz equation

$$\nabla^2 u + \kappa^2 u = 0, \quad 0 < x < a, \quad -\infty < y < \infty,$$

vanishing on $x = 0$ and $x = a$.

20. Solve the exterior Dirichlet problem

$$\nabla^2 u = 0, \quad \text{in} \quad r > 1,$$
$$u\left(1, \theta, \phi\right) = f\left(\theta, \phi\right).$$

21. By the method of images, determine the potential due to a point charge q near a conducting sphere of radius R with potential V.

22. By the method of images, show that the potential due to a conducting sphere of radius R in a uniform electric field E_0 is given by

$$U = -E_0\left(r - \frac{R^2}{r^2}\right)\cos\theta,$$

where r, θ are polar coordinates with origin at the center of the sphere.

23. Determine the potential in a cylinder of radius R and length l. The potential on the ends is zero, while the potential on the cylindrical surface is prescribed to be $f(\theta, z)$.

24. Consider the fundamental solution of the Fokker–Planck equation defined by

$$\left[\frac{\partial}{\partial t} - \frac{\partial}{\partial x}\left(\frac{\partial}{\partial x} + x\right)\right] G(x, x'; t, t') = \delta(x - x')\,\delta(t - t').$$

Using the transformation of variables employed in Example 7.8.4, show that the above equation becomes

$$\left[\frac{\partial}{\partial \tau} - \frac{\partial}{\partial \xi^2}\right] G(\xi, \xi'; \tau, \tau') = \delta(\xi - \xi')\,\delta(\tau - \tau').$$

Show that
(a) the fundamental solution of the Fokker–Planck equation (see Reif (1965)) is

$$G(x, x'; t, t') = [2\pi\{1 - \exp[-2(t - t')]\}]^{-\frac{1}{2}}$$
$$\times \exp\left[-\frac{1}{2}\frac{[x - x'\exp\{-(t - t')\}]^2}{1 - \exp\{-2(t - t')\}}\right],$$

(b) $$\lim_{t \to \infty} G(x, x'; t, t') = \frac{1}{\sqrt{2\pi}}\exp\left(-\frac{1}{2}x^2\right),$$

(c) $$\lim_{t \to \infty} u(x, t) = \frac{1}{\sqrt{2\pi}}\exp\left(-\frac{1}{2}x^2\right)\int_{-\infty}^{\infty} f(x')\,dx',$$

where

$$u(x, 0) = f(x).$$

Give an interpretation of this asymptotic solution $u(x, t)$ as $t \to \infty$.

25. (a) Use the transformation $u = ve^{-t}$ to show that the telegraph equation

$$u_{tt} - c^2 u_{xx} + 2u_t = 0,$$

can be reduced to the form

$$v_{tt} - c^2 v_{xx} + v = 0.$$

(b) Show that the fundamental solution of the transformed telegraph equation is given by

$$G(x - x', t - t') = \frac{1}{2} I_0\left[\sqrt{(t - t')^2 - (x - x')^2}\right]$$
$$\times H[(t - t') - (x - x')]\,H[(t - t') + (x - x')].$$

(c) If the initial data for the telegraph equation are

$$u(x,0) = f(x), \quad u_t(x,0) = g(x),$$

show that the solution of the telegraph equation is given by

$$u(x,t) = \left\{ \frac{1}{2} \int_{x-t}^{x+t} g(\xi) I_0 \left[\sqrt{t^2 - (x-\xi)^2} \right] d\xi \right.$$

$$\left. + \frac{1}{2} \frac{\partial}{\partial t} \left[\int_{x-t}^{x+t} f(\xi) I_0 \left[\sqrt{t^2 - (x-\xi)^2} \right] d\xi \right] \right\} e^{-t},$$

which is, by evaluating the second term,

$$= e^{-t} \left\{ \frac{1}{2} [f(x-t) + f(x+t)] + \frac{1}{2} \int_{x-t}^{x+t} g(\xi) I_0 \left[\sqrt{t^2 - (x-\xi)^2} \right] d\xi \right.$$

$$\left. + \frac{t}{2} \int_{x-t}^{x+t} f(\xi) \left[t^2 - (x-\xi)^2 \right]^{-\frac{1}{2}} I_1 \left[\sqrt{t^2 - (x-\xi)^2} \right] d\xi \right\}.$$

12

Integral Transform Methods with Applications

"The theory of Fourier series and integrals has always had major difficulties and necessitated a large mathematical apparatus in dealing with questions of convergence. It engendered the development of methods of summation, although these did not lead to a completely satisfactory solution of the problem.... For the Fourier transform, the introduction of distribution (hence the space S) is inevitable either in an explicit or hidden form.... As a result one may obtain all that is desired from the point of view of the continuity and inversion of the Fourier transform."

L. Schwartz

"In every mathematical investigation, the question will arise whether we can apply our mathematical results to the real world."

V. I. Arnold

12.1 Introduction

The linear superposition principle is one of the most effective and elegant methods to represent solutions of partial differential equations in terms of eigenfunctions or Green's functions. More precisely, the eigenfunction expansion method expresses the solution as an infinite series, whereas the integral solution can be obtained by integral superposition or by using Green's functions with initial and boundary conditions. The latter offers several advantages over eigenfunction expansion. First, an integral representation provides a direct way of describing the general analytical structure of a solution that may be obscured by an infinite series representation. Second, from a practical point of view, the evaluation of a solution from an integral representation may prove simpler than finding the sum of an infinite series, particularly near rapidly-varying features of a function, where the convergence of an eigenfunction expansion is expected to be slow. Third, in view of the Gibbs phenomenon discussed in Chapter 6, the integral representation seems to be less stringent requirements on the functions that describe the initial conditions or the values of a solution are required to assume on a given boundary than expansions based on eigenfunctions.

Integral transform methods are found to be very useful for finding solutions of initial and/or boundary-value problems governed by partial differential equations for the following reason. The differential equations can readily be replaced by algebraic equations that are inverted by the inverse transform so that the solution of the differential equations can then be obtained in terms of the original variables. The aim of this chapter is to provide an introduction to the use of integral transform methods for students of applied mathematics, physics, and engineering. Since our major interest is the application of integral transforms, no attempt will be made to discuss the basic results and theorems relating to transforms in their general forms. The present treatment is restricted to classes of functions which usually occur in physical and engineering applications.

12.2 Fourier Transforms

We first give a formal definition of the Fourier transform by using the complex Fourier integral formula (6.13.10).

Definition 12.2.1. *If $u(x,t)$ is a continuous, piecewise smooth, and absolutely integrable function, then the Fourier transform of $u(x,t)$ with respect to $x \in \mathbb{R}$ is denoted by $U(k,t)$ and is defined by*

$$\mathcal{F}\{u(x,t)\} = U(k,t) = \frac{1}{\sqrt{2\pi}} \int_{-\infty}^{\infty} e^{-ikx} u(x,t)\, dx, \qquad (12.2.1)$$

where k is called the Fourier transform variable and $\exp(-ikx)$ is called the kernel of the transform.

Then, for all $x \in \mathbb{R}$, the inverse Fourier transform of $U(k,t)$ is defined by

$$\mathcal{F}^{-1}\{U(k,t)\} = u(x,t) = \frac{1}{\sqrt{2\pi}} \int_{-\infty}^{\infty} e^{ikx} U(k,t)\, dk. \qquad (12.2.2)$$

We may note that the factor $(1/2\pi)$ in the Fourier integral formula (6.13.9) has been split and placed in front of the integrals (12.2.1) and (12.2.2). Often the factor $(1/2\pi)$ can be placed in only one of the relations (12.2.1) and (12.2.2). It is not uncommon to adopt the kernel $\exp(ikx)$ in (12.2.1) instead of $\exp(-ikx)$, and as a consequence, $\exp(-ikx)$ would be replaced by $\exp(ikx)$ in (12.2.2).

Example 12.2.1. Show that

(a) $\mathcal{F}\left\{\exp\left(-ax^2\right)\right\} = \dfrac{1}{\sqrt{2a}}\exp\left(-\dfrac{k^2}{4a}\right),$ $a > 0,$ (12.2.3)

(b) $\mathcal{F}\left\{\exp\left(-a\,|x|\right)\right\} = \sqrt{\dfrac{2}{\pi}}\,\dfrac{a}{(a^2 + k^2)},$ $a > 0,$ (12.2.4)

(c) $\mathcal{F}\left\{\chi_{[-a,a]}\left(x\right)\right\} = \sqrt{\dfrac{2}{\pi}}\left(\dfrac{\sin ak}{k}\right),$ (12.2.5)

where $\chi_{[-a,a]}\left(x\right) = H\left(a - |x|\right) = \begin{Bmatrix} 1, & |x| < a \\ 0, & |x| > a \end{Bmatrix}.$ (12.2.6)

Proof. We have, by definition (12.2.1),

$$\mathcal{F}\left\{\exp\left(-ax^2\right)\right\} = \dfrac{1}{\sqrt{2\pi}}\int_{-\infty}^{\infty} e^{-ikx-ax^2}\,dx$$

$$= \dfrac{1}{\sqrt{2\pi}}\int_{-\infty}^{\infty}\exp\left[-a\left(x + \dfrac{ik}{2a}\right)^2 - \dfrac{k^2}{4a}\right]\,dx$$

$$= \dfrac{1}{\sqrt{2\pi}}\exp\left(-\dfrac{k^2}{4a}\right)\int_{-\infty}^{\infty} e^{-ay^2}\,dy$$

$$= \dfrac{1}{\sqrt{2a}}\exp\left(-\dfrac{k^2}{4a}\right),$$

in which the change of variable $y = \left(x + \frac{ik}{2a}\right)$ is used. The above result is correct, and the change of variable can be justified by methods of complex analysis because $(ik/2a)$ is complex. If $a = \frac{1}{2}$, then

$$\mathcal{F}\left\{\exp\left(-\dfrac{1}{2}x^2\right)\right\} = \exp\left(-\dfrac{1}{2}k^2\right).$$ (12.2.7)

This shows that $\mathcal{F}\left\{f\left(x\right)\right\} = f\left(k\right)$. Such a function is said to be *self-reciprocal* under the Fourier transformation.

The graphs of $f\left(x\right) = e^{-ax^2}$ and $F\left(k\right) = \mathcal{F}\left\{f\left(x\right)\right\}$ are shown in Figure 12.2.1 for $a = 1$.

To prove (b), we write

$$\mathcal{F}\left\{\exp\left(-a\,|x|\right)\right\} = \dfrac{1}{\sqrt{2\pi}}\int_{-\infty}^{\infty}\exp\left(-a\,|x| - ikx\right)\,dx$$

$$= \dfrac{1}{\sqrt{2\pi}}\left[\int_{-\infty}^{0}\exp\left\{(a - ik)\,x\right\}\,dx\right.$$

$$\left. + \int_{0}^{\infty}\exp\left\{-(a + ik)\,x\right\}\,dx\right]$$

$$= \dfrac{1}{\sqrt{2\pi}}\left[\dfrac{1}{a - ik} + \dfrac{1}{a + ik}\right] = \sqrt{\dfrac{2}{\pi}}\,\dfrac{a}{a^2 + k^2}.$$

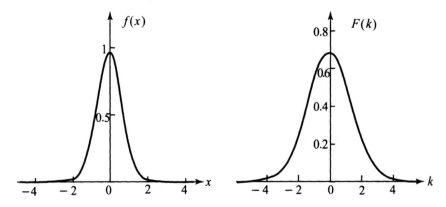

Figure 12.2.1 Graphs of $f(x) = \exp(-ax^2)$ and $F(k)$.

It is noted that $f(x) = \exp(-a|x|)$ decreases rapidly at infinity, and it is not differentiable at $x = 0$. The graphs of $f(x)$ and its Fourier transform $F(k)$ are shown in Figure 12.2.2.

To prove (c), we have

$$F_a(k) = \mathcal{F}\{\chi_{[-a,a]}(x)\} = \frac{1}{\sqrt{2\pi}} \int_{-\infty}^{\infty} e^{-ikx}\chi_{[-a,a]}(x)\,dx$$

$$= \frac{1}{\sqrt{2\pi}} \int_{-a}^{a} e^{-ikx}\,dx = \sqrt{\frac{2}{\pi}}\left(\frac{\sin ak}{a}\right).$$

The graphs of $\chi_{[-a,a]}(x)$ and $F_a(k)$ are shown in Figure 12.2.3 with $a = 1$.

Analogous to the Fourier cosine and sine series, there are Fourier cosine and sine integral transforms for odd and even functions respectively.

Definition 12.2.2. *Let $f(x)$ be defined for $0 \le x < \infty$, and extended as an even function in $(-\infty, \infty)$ satisfying the conditions of Fourier Inte-*

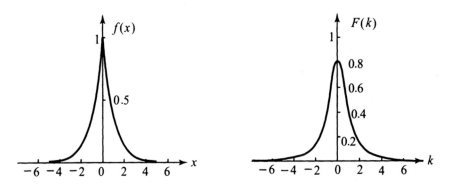

Figure 12.2.2 Graphs of $f(x) = \exp(-a|x|)$ and $F(k)$.

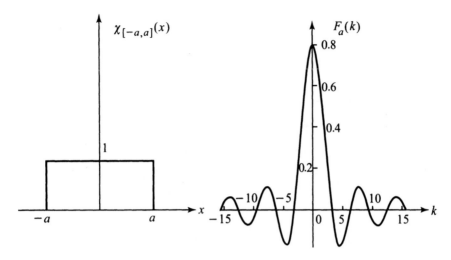

Figure 12.2.3 Graphs of $\chi_{[-a,a]}(x)$ and $F_a(k)$.

gral formula (6.13.9). Then, at the points of continuity, the Fourier cosine transform of $f(x)$ and its inverse transform are defined by

$$\mathcal{F}_c\{f(x)\} = F_c(k) = \sqrt{\frac{2}{\pi}} \int_0^\infty \cos kx \, f(x) \, dx, \qquad (12.2.8)$$

$$\mathcal{F}_c^{-1}\{F_c(k)\} = f(x) = \sqrt{\frac{2}{\pi}} \int_0^\infty \cos kx \, F_c(k) \, dk, \qquad (12.2.9)$$

where $\mathcal{F}_c$ is the Fourier cosine transformation and $\mathcal{F}_c^{-1}$ is its inverse transformation respectively.

Definition 12.2.3. *Similarly, the Fourier sine integral formula (6.13.3) leads to the Fourier sine transform and its inverse defined by*

$$\mathcal{F}_s\{f(x)\} = F_s(k) = \sqrt{\frac{2}{\pi}} \int_0^\infty \sin kx \, f(x) \, dx, \qquad (12.2.10)$$

$$\mathcal{F}_s^{-1}\{F_s(k)\} = f(x) = \sqrt{\frac{2}{\pi}} \int_0^\infty \sin kx \, F_s(k) \, dk, \qquad (12.2.11)$$

where $\mathcal{F}_s$ is called the Fourier sine transformation and $\mathcal{F}_s^{-1}$ is its inverse.

Example 12.2.2. Show that

(a) $\qquad \mathcal{F}_c\{e^{-ax}\} = \sqrt{\frac{2}{\pi}} \frac{a}{(a^2 + k^2)}, \qquad a > 0, \qquad (12.2.12)$

(b) $\qquad \mathcal{F}_s\{e^{-ax}\} = \sqrt{\frac{2}{\pi}} \frac{k}{(a^2 + k^2)}, \qquad a > 0, \qquad (12.2.13)$

(c) $\qquad \mathcal{F}_s^{-1}\left\{\frac{1}{k} e^{-sk}\right\} = \sqrt{\frac{2}{\pi}} \tan^{-1}\left(\frac{x}{s}\right). \qquad (12.2.14)$

We have, by definition,

$$\mathcal{F}_c\left\{e^{-ax}\right\} = \sqrt{\frac{2}{\pi}} \int_0^\infty e^{-ax} \cos kx \, dk,$$

$$= \frac{1}{2}\sqrt{\frac{2}{\pi}} \int_0^\infty \left[e^{-(a-ik)x} + e^{-(a+ik)x}\right] dx,$$

$$= \frac{1}{2}\sqrt{\frac{2}{\pi}} \left[\frac{1}{a-ik} + \frac{1}{a+ik}\right] = \sqrt{\frac{2}{\pi}} \frac{a}{(a^2+k^2)}.$$

The proof of (b) is similar and is left to the reader as an exercise.

To prove (c), we use the standard definite integral

$$\sqrt{\frac{\pi}{2}} \mathcal{F}_s^{-1}\left\{e^{-sk}\right\} = \int_0^\infty e^{-sk} \sin kx \, dk = \frac{x}{s^2+x^2}.$$

Integrating both sides with respect to s from s to ∞ gives

$$\int_0^\infty \frac{e^{-sk}}{k} \sin kx \, dk = \int_s^\infty \frac{x \, ds}{x^2+s^2} = \left[\tan^{-1}\frac{s}{x}\right] = (\pi/2) - \tan^{-1}\left(\frac{s}{x}\right).$$

Consequently,

$$\mathcal{F}_s^{-1}\left\{\frac{1}{k}e^{-sk}\right\} = \sqrt{\frac{2}{\pi}} \int_0^\infty \frac{1}{k}e^{-sk} \sin kx \, dk = \sqrt{\frac{2}{\pi}} \tan^{-1}\left(\frac{x}{s}\right).$$

12.3 Properties of Fourier Transforms

Theorem 12.3.1. *(Linearity).* The Fourier transformation $\mathcal{F}$ is linear.

Proof. We have

$$\mathcal{F}[f(x)] = \frac{1}{\sqrt{2\pi}} \int_{-\infty}^\infty e^{-ikx} f(x) \, dx.$$

Then, for any constants a and b,

$$\mathcal{F}[af(x) + bg(x)] = \frac{1}{\sqrt{2\pi}} \int_{-\infty}^\infty [af(x) + bg(x)] e^{-ikx} dx,$$

$$= \frac{a}{\sqrt{2\pi}} \int_{-\infty}^\infty f(x) e^{-ikx} dx + \frac{b}{\sqrt{2\pi}} \int_{-\infty}^\infty g(x) e^{-ikx} dx,$$

$$= a\mathcal{F}[f(x)] + b\mathcal{F}[g(x)].$$

Theorem 12.3.2. *(Shifting).* Let $\mathcal{F}[f(x)]$ be a Fourier transform of $f(x)$. Then

$$\mathcal{F}[f(x-c)] = e^{-ixc}\mathcal{F}[f(x)],$$

where c is a real constant.

Proof. From the definition, we have, for $c > 0$,

$$\mathcal{F}\left[f\left(x - c\right)\right] = \frac{1}{\sqrt{2\pi}} \int_{-\infty}^{\infty} f\left(x - c\right) e^{-ikx} dx,$$

$$= \frac{1}{\sqrt{2\pi}} \int_{-\infty}^{\infty} f\left(\xi\right) e^{-ik(\xi + c)} d\xi, \quad \text{where} \quad \xi = x - c$$

$$= e^{-ikc} \mathcal{F}\left[f\left(x\right)\right].$$

Theorem 12.3.3. *(Scaling).* If $\mathcal{F}$ is the Fourier transform of f, then

$$\mathcal{F}\left[f\left(cx\right)\right] = \left(1/\left|c\right|\right) F\left(k/c\right),$$

where c is a real nonzero constant.

Proof. For $c \neq 0$,

$$\mathcal{F}\left[f\left(cx\right)\right] = \frac{1}{\sqrt{2\pi}} \int_{-\infty}^{\infty} f\left(cx\right) e^{-ikx} dx.$$

If we let $\xi = cx$, then

$$\mathcal{F}\left[f\left(cx\right)\right] = \frac{1}{\left|c\right|} \frac{1}{\sqrt{2\pi}} \int_{-\infty}^{\infty} f\left(\xi\right) e^{-i(k/c)\xi} d\xi$$

$$= \left(1/\left|c\right|\right) F\left(k/c\right).$$

Theorem 12.3.4. *(Differentiation).* Let f be continuous and piecewise smooth in $(-\infty, \infty)$. Let $f\left(x\right)$ approach zero as $\left|x\right| \to \infty$. If f and f' are absolutely integrable, then

$$\mathcal{F}\left[f'\left(x\right)\right] = ik\mathcal{F}\left[f\left(x\right)\right] = ikF\left(k\right).$$

Proof.

$$\mathcal{F}\left[f'\left(x\right)\right] = \frac{1}{\sqrt{2\pi}} \int_{-\infty}^{\infty} f'\left(x\right) e^{-ikx} dx,$$

$$= \frac{1}{\sqrt{2\pi}} \left[f\left(x\right) e^{-ikx} \Big|_{-\infty}^{\infty} + \frac{ik}{\sqrt{2\pi}} \int_{-\infty}^{\infty} f\left(x\right) e^{-ikx} dx\right],$$

$$= ik\mathcal{F}\left[f\left(x\right)\right] = ikF\left(k\right).$$

This result can be easily extended. If f and its first $(n - 1)$ derivatives are continuous, and if its nth derivative is piecewise continuous, then

$$\mathcal{F}\left[f^{(n)}\left(x\right)\right] = (ik)^n \mathcal{F}\left[f\left(x\right)\right] = (ik)^n F\left(k\right), \quad n = 0, 1, 2, \ldots \quad (12.3.1)$$

provided f and its derivatives are absolutely integrable. In addition, we assume that f and its first $(n - 1)$ derivatives tend to zero as $\left|x\right|$ tends to infinity.

If $u(x,t) \to 0$ as $|x| \to \infty$, then

$$\mathcal{F}\left\{\frac{\partial u}{\partial x}\right\} = \frac{1}{\sqrt{2\pi}} \int_{-\infty}^{\infty} e^{-ikx} \left(\frac{\partial u}{\partial x}\right) dx,$$

which is, integrating by parts,

$$= \frac{1}{\sqrt{2\pi}} \left[e^{-ikx} u(x,t)\right]_{-\infty}^{\infty} + \frac{ik}{\sqrt{2\pi}} \int_{-\infty}^{\infty} e^{-ikx} u(x,t)\, dx,$$

$$= ik\, \mathcal{F}\{u(x,t)\} = ik\, U(k,t). \tag{12.3.2}$$

Similarly, if $u(x,t)$ is continuously n times differentiable, and $\frac{\partial^m m}{\partial x^m} \to 0$ as $|x| \to \infty$ for $m = 1, 2, 3, \ldots, (n-1)$ then

$$\mathcal{F}\left\{\frac{\partial^n u}{\partial x^n}\right\} = (ik)^n\, \mathcal{F}\{u(x,t)\} = (ik)^n\, U(k,t). \tag{12.3.3}$$

It also follows from the definition (12.2.1) that

$$\mathcal{F}\left\{\frac{\partial u}{\partial t}\right\} = \frac{dU}{dt}, \quad \mathcal{F}\left\{\frac{\partial^2 u}{\partial t^2}\right\} = \frac{d^2 U}{dt^2}, \quad \ldots, \quad \mathcal{F}\left\{\frac{\partial^n u}{\partial t^n}\right\} = \frac{d^n U}{dt^n}. \tag{12.3.4}$$

The definition of the Fourier transform (12.2.1) shows that a sufficient condition for $u(x,t)$ to have a Fourier transform is that $u(x,t)$ is absolutely integrable in $-\infty < x < \infty$. This existence condition is too strong for many practical applications. Many simple functions, such as a constant function, $\sin \omega x$, and $x^n H(x)$, do not have Fourier transforms even though they occur frequently in applications.

The above definition of the Fourier transform has been extended for a more general class of functions to include the above and other functions. We simply state the fact that there is a sense, useful in practical applications, in which the above stated functions and many others do have Fourier transforms. The following are examples of such functions and their Fourier transforms (see Lighthill, 1964):

$$\mathcal{F}\{H(a - |x|)\} = \sqrt{\frac{2}{\pi}} \left(\frac{\sin ak}{k}\right), \tag{12.3.5}$$

where $H(x)$ is the Heaviside unit step function,

$$\mathcal{F}\{\delta(x - a)\} = \frac{1}{\sqrt{2\pi}} \exp(-iak), \tag{12.3.6}$$

where $\delta(x - a)$ is the Dirac delta function, and

$$\mathcal{F}\{H(x - a)\} = \sqrt{\frac{\pi}{2}} \left[\frac{1}{i\pi k} + \delta(k)\right] \exp(-iak). \tag{12.3.7}$$

Example 12.3.1. Find the solution of the Dirichlet problem in the half-plane $y > 0$

$$u_{xx} + u_{yy} = 0, \qquad -\infty < x < \infty, \quad y > 0,$$
$$u(x, 0) = f(x), \qquad -\infty < x < \infty,$$

u and u_x vanish as $|x| \to \infty$, and u is bounded as $y \to \infty$.

Let $U(k, y)$ be the Fourier transform of $u(x, y)$ with respect to x. Then

$$U(k, y) = \frac{1}{\sqrt{2\pi}} \int_{-\infty}^{\infty} u(x, y) e^{-ikx} dx.$$

Application of the Fourier transform with respect to x gives

$$U_{yy} - k^2 U = 0, \tag{12.3.8}$$
$$U(k, 0) = F(k) \quad \text{and} \quad U(k, y) \to 0 \quad \text{as} \quad y \to \infty. \tag{12.3.9}$$

The solution of this transformed system is

$$U(k, y) = F(k) e^{-|k|y}.$$

The inverse Fourier transform of $U(k, y)$ gives the solution in the form

$$u(x, y) = \frac{1}{\sqrt{2\pi}} \int_{-\infty}^{\infty} \left[\frac{1}{\sqrt{2\pi}} \int_{-\infty}^{\infty} f(\xi) e^{-|k|y} e^{-ik\xi} d\xi \right] e^{ikx} dk,$$
$$= \frac{1}{2\pi} \int_{-\infty}^{\infty} f(\xi) d\xi \int_{-\infty}^{\infty} e^{-k[i(\xi-x)]-|k|y} dk.$$

It follows from the proof of Example 12.2.1 (b) that

$$\int_{-\infty}^{\infty} e^{-k[i(\xi-x)]-|k|y} dk = \frac{2y}{(\xi - x)^2 + y^2}.$$

Hence, the solution of the Dirichlet problem in the half-plane $y > 0$ is

$$u(x, y) = \frac{y}{\pi} \int_{-\infty}^{\infty} \frac{f(\xi)}{(\xi - x)^2 + y^2} d\xi.$$

From this solution, we can readily deduce a solution of the Neumann problem in the half-plane $y > 0$.

Example 12.3.2. Find the solution of Neumann's problem in the half-plane $y > 0$

$$u_{xx} + u_{yy} = 0, \qquad -\infty < x < \infty, \quad y > 0,$$
$$u_y(x, 0) = g(x), \qquad -\infty < x < \infty,$$

u is bounded as $y \to \infty$, u and u_x vanish as $|x| \to \infty$.

Let $v(x,y) = u_y(x,y)$. Then

$$u(x,y) = \int^y v(x,\eta)\,d\eta$$

and the Neumann problem becomes

$$\frac{\partial^2 v}{\partial x^2} + \frac{\partial^2 v}{\partial y^2} = \frac{\partial^2 u_y}{\partial x^2} + \frac{\partial^2 u_y}{\partial y^2} = \frac{\partial}{\partial y}\left(u_{xx} + u_{yy}\right) = 0.$$

$$v(x,0) = u_y(x,0) = g(x).$$

This is the Dirichlet problem for $v(x,y)$, and its solution is given by

$$v(x,y) = \frac{y}{\pi}\int_{-\infty}^{\infty} \frac{g(\xi)\,d\xi}{(\xi - x)^2 + y^2}.$$

Thus, we have

$$\begin{aligned}
u(x,y) &= \frac{1}{\pi}\int^y \eta \int_{-\infty}^{\infty} \frac{g(\xi)\,d\xi}{(\xi - x)^2 + \eta^2}\,d\eta, \\
&= \frac{1}{2\pi}\int_{-\infty}^{\infty} g(\xi)\,d\xi \int^y \frac{2\eta\,d\eta}{(\xi - x)^2 + \eta^2}, \\
&= \frac{1}{2\pi}\int_{-\infty}^{\infty} g(\xi)\log\left[(x - \xi)^2 + y^2\right]d\xi,
\end{aligned}$$

where an arbitrary constant can be added to this solution. In other words, the solution of any Neumann's problem is uniquely determined up to an arbitrary constant.

12.4 Convolution Theorem of the Fourier Transform

The function

$$(f * g)(x) = \frac{1}{\sqrt{2\pi}}\int_{-\infty}^{\infty} f(x - \xi)g(\xi)\,d\xi \qquad (12.4.1)$$

is called the *convolution* of the functions f and g over the interval $(-\infty, \infty)$.

Theorem 12.4.1. *(Convolution Theorem)*. If $F(k)$ and $G(k)$ are the Fourier transforms of $f(x)$ and $g(x)$ respectively, then the Fourier transform of the convolution $(f * g)$ is the product $F(k)G(k)$. That is,

$$\mathcal{F}\{f(x) * g(x)\} = F(k)G(k). \qquad (12.4.2)$$

Or, equivalently,

$$\mathcal{F}^{-1}\left\{F\left(k\right)G\left(k\right)\right\}=f\left(x\right)*g\left(x\right).\qquad(12.4.3)$$

More explicitly,

$$\frac{1}{\sqrt{2\pi}}\int_{-\infty}^{\infty}F\left(k\right)G\left(k\right)e^{ikx}dk=\left(f*g\right)\left(x\right)$$

$$=\frac{1}{\sqrt{2\pi}}\int_{-\infty}^{\infty}f\left(x-\xi\right)g\left(\xi\right)d\xi.\qquad(12.4.4)$$

Proof. By definition, we have

$$\mathcal{F}\left[\left(f*g\right)\left(x\right)\right]=\frac{1}{2\pi}\int_{-\infty}^{\infty}e^{-ikx}dx\int_{-\infty}^{\infty}f\left(x-\xi\right)g\left(\xi\right)d\xi,$$

$$=\frac{1}{2\pi}\int_{-\infty}^{\infty}g\left(\xi\right)e^{-ik\xi}d\xi\int_{-\infty}^{\infty}f\left(x-\xi\right)e^{-ik(x-\xi)}dx.$$

With the change of variable $\eta=x-\xi$, we have

$$\mathcal{F}\left[\left(f*g\right)\left(x\right)\right]=\frac{1}{\sqrt{2\pi}}\int_{-\infty}^{\infty}g\left(\xi\right)e^{-ik\xi}d\xi\,\frac{1}{\sqrt{2\pi}}\int_{-\infty}^{\infty}f\left(\eta\right)e^{-ik\eta}d\eta$$

$$=F\left(k\right)G\left(k\right).$$

The convolution satisfies the following properties:

1. $f*g=g*f$ (commutative).
2. $f*\left(g*h\right)=\left(f*g\right)*h$ (associative).
3. $f*\left(ag+bh\right)=a\left(f*g\right)+b\left(f*h\right),$ (distributive),
   where a and b are constants.

Theorem 12.4.2. *(Parseval's formula).*

$$\int_{-\infty}^{\infty}\left|f\left(x\right)\right|^{2}dx=\int_{-\infty}^{\infty}\left|F\left(k\right)\right|^{2}dk.\qquad(12.4.5)$$

Proof. The convolution formula (12.4.4) gives

$$\int_{-\infty}^{\infty}f\left(x\right)g\left(\xi-x\right)dx=\int_{-\infty}^{\infty}F\left(k\right)G\left(k\right)e^{ik\xi}dk$$

which is, by putting $\xi=0$,

$$\int_{-\infty}^{\infty}f\left(x\right)g\left(-x\right)dx=\int_{-\infty}^{\infty}F\left(k\right)G\left(k\right)dk.\qquad(12.4.6)$$

Putting $g\left(-x\right)=\overline{f\left(x\right)}$,

$$G\left(k\right)=\frac{1}{\sqrt{2\pi}}\int_{-\infty}^{\infty}g\left(x\right)e^{-ikx}dx=\frac{1}{\sqrt{2\pi}}\int_{-\infty}^{\infty}\overline{f\left(-x\right)}e^{-ikx}dx,$$

$$=\overline{\frac{1}{\sqrt{2\pi}}\int_{-\infty}^{\infty}f\left(x\right)e^{-ikx}dx}=\overline{F\left(k\right)},$$

where the bar denotes the complex conjugate.

Thus, result (12.4.6) becomes

$$\int_{-\infty}^{\infty} f(x)\,\overline{f(x)}\,dx = \int_{-\infty}^{\infty} F(k)\,\overline{F(k)}\,dk,$$

or,

$$\int_{-\infty}^{\infty} |f(x)|^2\,dx = \int_{-\infty}^{\infty} |F(k)|^2\,dk.$$

In terms of the notation of the norm, this is

$$\|f\| = \|F\|.$$

For physical systems, the quantity $|f|^2$ is a measure of energy, and $|F|^2$ represents the power spectrum of $f(x)$.

Example 12.4.3. Obtain the solution of the initial-value problem of heat conduction in an infinite rod

$$u_t = \kappa\,u_{xx}, \qquad -\infty < x < \infty, \quad t > 0, \qquad (12.4.7)$$
$$u(x,0) = f(x), \qquad -\infty < x < \infty, \qquad (12.4.8)$$
$$u(x,t) \to 0, \qquad \text{as} \quad |x| \to \infty,$$

where $u(x,t)$ represents the temperature distribution and is bounded, and κ is a constant of diffusivity.

The Fourier transform of $u(x,t)$ with respect to x is defined by

$$U(k,t) = \frac{1}{\sqrt{2\pi}} \int_{-\infty}^{\infty} e^{-ikx} u(x,t)\,dx.$$

In view of this transformation, equations (12.4.7)–(12.4.8) become

$$U_t + \kappa\,k^2\,U = 0, \qquad (12.4.9)$$
$$U(k,0) = F(k). \qquad (12.4.10)$$

The solution of the transformed system is

$$U(k,t) = F(k)\,e^{-k^2 \kappa t}.$$

The inverse Fourier transformation gives the solution

$$u(x,t) = \frac{1}{\sqrt{2\pi}} \int_{-\infty}^{\infty} F(k)\,e^{-k^2 \kappa t} e^{ikx}\,dk$$

which is, by the convolution theorem 12.4.1,

$$= \frac{1}{\sqrt{2\pi}} \int_{-\infty}^{\infty} f(\xi) g(x - \xi) d\xi,$$

where $g(x)$ is the inverse transform of $G(k) = e^{-k^2 \kappa t}$ and has the form

$$g(x) = \mathcal{F}^{-1}\left\{ e^{-\kappa k^2 t} \right\} = \frac{1}{\sqrt{2\pi}} \int_{-\infty}^{\infty} e^{-k^2 \kappa t + ikx} dk = \frac{1}{\sqrt{2\kappa t}} e^{-x^2/4\kappa t}.$$

Consequently, the final solution is

$$u(x, t) = \frac{1}{\sqrt{4\pi\kappa t}} \int_{-\infty}^{\infty} f(\xi) \exp\left[-\frac{(x - \xi)^2}{4\kappa t} \right] d\xi, \qquad (12.4.11)$$

$$= \int_{-\infty}^{\infty} f(\xi) G(x - \xi, t) d\xi, \qquad (12.4.12)$$

where

$$G(x - \xi, t) = \frac{1}{\sqrt{4\pi\kappa t}} \exp\left[-\frac{(x - \xi)^2}{4\kappa t} \right], \qquad (12.4.13)$$

is called the *Green's function* (or the *fundamental solution*) of the diffusion equation.

This means that temperature at any point x and any time t is represented by the definite integral (12.4.12) that is made up of the contribution due to the initial source $f(\xi)$ and the Green's function $G(x - \xi, t)$.

Solution (12.4.12) represents the temperature response along the rod at time t due to an initial unit impulse of heat at $x = \xi$. The physical meaning of the solution is that the initial temperature distribution $f(x)$ is decomposed into a spectrum of impulses of magnitude $f(\xi)$ at each point $x = \xi$ to form the resulting temperature $f(\xi) G(x - \xi, t)$. Thus, the resulting temperature is integrated to find the solution (12.4.11). This is the so-called *principle of superposition*.

Using the change of variable

$$\frac{\xi - x}{2\sqrt{\kappa t}} = \zeta, \qquad d\zeta = \frac{d\xi}{2\sqrt{\kappa t}},$$

we obtain

$$u(x, t) = \frac{1}{\sqrt{\pi}} \int_{-\infty}^{\infty} f\left(x + 2\sqrt{\kappa t}\, \zeta \right) e^{-\zeta^2} d\zeta. \qquad (12.4.14)$$

Integral (12.4.11) or (12.4.14) is called the *Poisson integral representation* of the temperature distribution. This integral is convergent for $t > 0$, and integrals obtained from it by differentiation under the integral sign with respect to x and t are uniformly convergent in the neighborhood of the point (x, t). Hence, $u(x, t)$ and its derivatives of all orders exist for $t > 0$.

In the limit $t \to 0+$, solution (12.4.12) becomes formally

$$u(x, 0) = f(x) = \int_{-\infty}^{\infty} f(\xi) \lim_{t \to 0+} G(x - \xi, t) \, d\xi.$$

This limit represents the Dirac delta function

$$\delta(x - \xi) = \lim_{t \to 0+} \frac{1}{\sqrt{4\pi\kappa t}} e^{-(x-\xi)^2/4\kappa t}. \qquad (12.4.15)$$

Consider a special case where

$$f(x) = \left\{ \begin{array}{l} 0, \ x < 0 \\ a, \ x > 0 \end{array} \right\} = a H(x).$$

Then, the solution (12.4.11) gives

$$u(x, t) = \frac{a}{2\sqrt{\pi\kappa t}} \int_0^{\infty} \exp\left[-\frac{(x - \xi)^2}{4\kappa t} \right] d\xi.$$

If we introduce a change of variable

$$\eta = \frac{\xi - x}{2\sqrt{\kappa t}}$$

then the above solution becomes

$$\begin{aligned} u(x, t) &= \frac{a}{\sqrt{\pi}} \int_{-x/2\sqrt{\kappa t}}^{\infty} e^{-\eta^2} d\eta \\ &= \frac{a}{\sqrt{\pi}} \left[\int_{-x/2\sqrt{\kappa t}}^{0} e^{-\eta^2} d\eta + \int_0^{\infty} e^{-\eta^2} d\eta \right] \\ &= \frac{a}{\sqrt{\pi}} \left[\int_0^{x/2\sqrt{\kappa t}} e^{-\eta^2} d\eta + \frac{\sqrt{\pi}}{2} \right] \\ &= \frac{a}{2} \left[1 + \operatorname{erf}\left(\frac{x}{2\sqrt{\kappa t}} \right) \right], \end{aligned}$$

where $\operatorname{erf}(x)$ is called the *error function* and is defined by

$$\operatorname{erf}(x) = \frac{2}{\sqrt{\pi}} \int_0^x e^{-\eta^2} d\eta. \qquad (12.4.16)$$

This is a widely used and tabulated function.

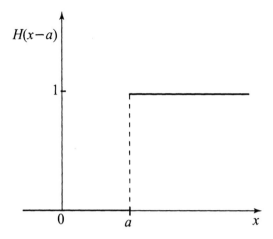

Figure 12.5.1 The Heaviside unit step function.

12.5 The Fourier Transforms of Step and Impulse Functions

In this section, we shall determine the Fourier transforms of the step function and the impulse function, functions which occur frequently in applied mathematics and mathematical physics.

The Heaviside unit step function is defined by

$$H\left(x - a\right) = \begin{cases} 0, & x < a \\ \\ 1, & x \ge a \end{cases} \qquad a \ge 0, \qquad (12.5.1)$$

as shown in Figure 12.5.1.

The Fourier transform of the Heaviside unit step function can be easily determined. We consider first

$$\mathcal{F}\left[H\left(x - a\right)\right] = \frac{1}{\sqrt{2\pi}} \int_{-\infty}^{\infty} H\left(x - a\right) e^{-ikx} dx,$$

$$= \frac{1}{\sqrt{2\pi}} \int_{a}^{\infty} e^{-ikx} dx.$$

This integral does not exist. However, we can prove the existence of this integral by defining a new function

$$H\left(x - a\right) e^{-\alpha x} = \begin{cases} 0, & x < a \\ \\ e^{-\alpha x}, & x \ge a. \end{cases}$$

This is evidently the unit step function as $\alpha \to 0$. Thus, we find the Fourier transform of the unit step function as

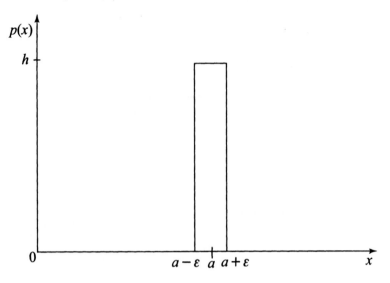

Figure 12.5.2 Impulse function $p(x)$.

$$\mathcal{F}[H(x-a)] = \lim_{\alpha \to 0} \mathcal{F}\left[H(x-a)e^{-\alpha x}\right]$$

$$= \lim_{\alpha \to 0} \frac{1}{\sqrt{2\pi}} \int_{-\infty}^{\infty} H(x-a)e^{-\alpha x}e^{-ikx}dx$$

$$= \lim_{\alpha \to 0} \frac{1}{\sqrt{2\pi}} \int_{a}^{\infty} e^{-(\alpha+ik)x}dx$$

$$= \frac{e^{-iak}}{\sqrt{2\pi}\,ik}. \qquad (12.5.2)$$

For $a = 0$,

$$\mathcal{F}[H(x)] = \left(\sqrt{2\pi}\,ik\right)^{-1}. \qquad (12.5.3)$$

An *impulse function* is defined by

$$p(x) = \begin{cases} h, & a - \varepsilon < x < a + \varepsilon \\ \\ 0, & x \le a - \varepsilon \quad \text{or} \quad x \ge a + \varepsilon \end{cases}$$

where h is large and positive, $a > 0$, and ε is a small positive constant, as shown in Figure 12.5.2. This type of function appears in practical applications; for instance, a force of large magnitude may act over a very short period of time.

The Fourier transform of the impulse function is

$$\mathcal{F}\left[p\left(x\right)\right] = \frac{1}{\sqrt{2\pi}} \int_{-\infty}^{\infty} p\left(x\right) e^{-ikx} dx$$

$$= \frac{1}{\sqrt{2\pi}} \int_{a-\varepsilon}^{a+\varepsilon} h\, e^{-ikx} dx$$

$$= \frac{h}{\sqrt{2\pi}} \frac{e^{-iak}}{ik} \left(e^{ik\varepsilon} - e^{-ik\varepsilon}\right)$$

$$= \frac{2h\varepsilon}{\sqrt{2\pi}} e^{-iak} \left(\frac{\sin k\varepsilon}{k\varepsilon}\right).$$

Now if we choose the value of h to be $(1/2\varepsilon)$, then the impulse defined by

$$I\left(\varepsilon\right) = \int_{-\infty}^{\infty} p\left(x\right) dx$$

becomes

$$I\left(\varepsilon\right) = \int_{a-\varepsilon}^{a+\varepsilon} \frac{1}{2\varepsilon} dx = 1$$

which is a constant independent of ε. In the limit as $\varepsilon \to 0$, this particular function $p_\varepsilon\left(x\right)$ with $h = (1/2\varepsilon)$ satisfies

$$\lim_{\varepsilon \to 0} p_\varepsilon\left(x\right) = 0, \qquad x \neq a,$$

$$\lim_{\varepsilon \to 0} I\left(\varepsilon\right) = 1.$$

Thus, we arrive at the result

$$\delta\left(x-a\right) = 0, \qquad x \neq a,$$

$$\int_{-\infty}^{\infty} \delta\left(x-a\right) dx = 1. \tag{12.5.4}$$

This is the Dirac delta function which was defined earlier in Section 8.11.

We now define the Fourier transform of $\delta\left(x\right)$ as the limit of the transform of $p_\varepsilon\left(x\right)$. We then consider

$$\mathcal{F}\left[\delta\left(x-a\right)\right] = \lim_{\varepsilon \to 0} \mathcal{F}\left[p_\varepsilon\left(x\right)\right]$$

$$= \lim_{\varepsilon \to 0} \frac{e^{-iak}}{\sqrt{2\pi}} \left(\frac{\sin k\varepsilon}{k\varepsilon}\right)$$

$$= \frac{e^{-iak}}{\sqrt{2\pi}} \tag{12.5.5}$$

in which we note that, by L'Hospital's rule, $\lim_{\varepsilon \to 0} (\sin k\varepsilon / k\varepsilon) = 1$. When $a = 0$, we obtain

$$\mathcal{F}\left[\delta\left(x\right)\right] = \left(1/\sqrt{2\pi}\right). \tag{12.5.6}$$

Example 12.5.1. Slowing-down of Neutrons (see Sneddon (1951), p. 215). Consider the following physical problem

$$u_t = u_{xx} + \delta(x)\,\delta(t), \qquad (12.5.7)$$

$$u(x,0) = \delta(x), \qquad (12.5.8)$$

$$\lim_{|x|\to\infty} u(x,t) = 0. \qquad (12.5.9)$$

This is the problem of an infinite medium which slows neutrons, in which a source of neutrons is located. Here $u(x,t)$ represents the number of neutrons per unit volume per unit time and $\delta(x)\,\delta(t)$ represents the source function.

Let $U(k,t)$ be the Fourier transform of $u(x,t)$. Then the Fourier transformation of equation (12.5.7) yields

$$\frac{dU}{dt} + k^2 U = \frac{1}{\sqrt{2\pi}}\,\delta(t).$$

The solution of this, after applying the condition $U(k,0) = (1/\sqrt{2\pi})$, is

$$U(k,t) = \frac{1}{\sqrt{2\pi}}\,e^{-k^2 t}.$$

Hence, the inverse Fourier transform gives the solution of the problem

$$u(x,t) = \frac{1}{\sqrt{2\pi}} \int_{-\infty}^{\infty} e^{-k^2 t + ikx}\,dk,$$

$$= \frac{1}{\sqrt{4\pi t}}\,e^{-x^2/4t}.$$

12.6 Fourier Sine and Cosine Transforms

For semi-infinite regions, the Fourier sine and cosine transforms determined in Section 12.2 are particularly appropriate in solving boundary-value problems. Before we illustrate their applications, we must first prove the differentiation theorem.

Theorem 12.6.1. *Let $f(x)$ and its first derivative vanish as $x \to \infty$. If $F_c(k)$ is the Fourier cosine transform, then*

$$\mathcal{F}_c[f''(x)] = -k^2 F_c(k) - \sqrt{\frac{2}{\pi}}\,f'(0). \qquad (12.6.1)$$

Proof.

$$\mathcal{F}_c[f''(x)] = \sqrt{\frac{2}{\pi}} \int_0^{\infty} f''(x)\cos kx\,dx$$

$$= \sqrt{\frac{2}{\pi}} \left[f'(x) \cos kx \right]_0^\infty + \sqrt{\frac{2}{\pi}} k \int_0^\infty f'(x) \sin kx \, dx$$

$$= -\sqrt{\frac{2}{\pi}} f'(0) + \sqrt{\frac{2}{\pi}} k \left[f(x) \sin kx \right]_0^\infty$$

$$- \sqrt{\frac{2}{\pi}} k^2 \int_0^\infty f(x) \cos kx \, dx$$

$$= -\sqrt{\frac{2}{\pi}} f'(0) - k^2 F_c(k).$$

In a similar manner, the Fourier cosine transforms of higher-order derivatives of $f(x)$ can be obtained.

Theorem 12.6.2. *Let $f(x)$ and its first derivative vanish as $x \to \infty$. If $F_s(k)$ is the Fourier sine transform, then*

$$\mathcal{F}_s[f''(x)] = \sqrt{\frac{2}{\pi}} k f(0) - k^2 F_s(k). \tag{12.6.2}$$

The proof is left to the reader.

Example 12.6.2. Find the temperature distribution in a semi-infinite rod for the following cases with zero initial temperature distribution:

(a) The heat supplied at the end $x = 0$ at the rate $g(t)$;

(b) The end $x = 0$ is kept at a constant temperature T_0.

The problem here is to solve the heat conduction equation

$$u_t = \kappa u_{xx}, \qquad x > 0, \quad t > 0,$$
$$u(x,0) = 0, \qquad x > 0.$$

(a) $u_x(0,t) = g(t)$ and (b) $u(0,t) = T_0$, $t \geq 0$. Here we assume that $u(x,t)$ and $u_x(x,t)$ vanish as $x \to \infty$.

For case (a), let $U(k,t)$ be the Fourier cosine transform of $u(x,t)$. Then the transformation of the heat conduction equation yields

$$U_t + \kappa k^2 U = -\sqrt{\frac{2}{\pi}} g(t) \kappa.$$

The solution of this equation with $U(k,0) = 0$ is

$$u(x,t) = \sqrt{\frac{2}{\pi}} \int_0^\infty U(k,t) \cos kx \, dk$$

$$= -\frac{2\kappa}{\pi} \int_0^t g(\tau) \, d\tau \int_0^\infty e^{-\kappa k^2 (t-\tau)} \cos kx \, dk.$$

The inner integral is given by (see Problem 6, Exercises 12.18)

$$\int_0^\infty e^{-k^2\kappa(t-\tau)}\cos kx\, dk = \frac{1}{2}\sqrt{\frac{\pi}{\kappa(t-\tau)}}\exp\left[-\frac{x^2}{4\kappa(t-\tau)}\right].$$

The solution, therefore, is

$$u(x,t) = -\sqrt{\frac{\kappa}{\pi}}\int_0^t \frac{g(\tau)}{\sqrt{t-\tau}}e^{-x^2/4\kappa(t-\tau)}\,d\tau. \tag{12.6.3}$$

For case (b), we apply the Fourier sine transform $U(k,t)$ of $u(x,t)$ to obtain the transformed equation

$$U_t + \kappa k^2 U = \sqrt{\frac{2}{\pi}}k\,T_0\,\kappa.$$

The solution of this equation with zero initial condition is

$$U(k,t) = T_0\sqrt{\frac{2}{\pi}}\frac{\left(1-e^{-\kappa t k^2}\right)}{k}.$$

Then the inverse Fourier sine transformation gives

$$u(x,t) = \frac{2T_0}{\pi}\int_0^\infty \frac{\sin kx}{k}\left(1-e^{-\kappa t k^2}\right)dk.$$

Making use of the integral

$$\int_0^\infty e^{-a^2 x^2}\left(\frac{\sin kx}{k}\right)dk = \frac{\pi}{2}\mathrm{erf}\left(\frac{x}{2a}\right),$$

the solution is found to be

$$u(x,t) = \frac{2T_0}{\pi}\left[\frac{\pi}{2}-\frac{\pi}{2}\mathrm{erf}\left(\frac{x}{2\sqrt{\kappa t}}\right)\right]$$

$$= T_0\,\mathrm{erfc}\left(\frac{x}{2\sqrt{\kappa t}}\right), \tag{12.6.4}$$

where $\mathrm{erfc}(x) = 1 - \mathrm{erf}(x)$ is the complementary error function defined by

$$\mathrm{erfc}(x) = \frac{2}{\sqrt{\pi}}\int_x^\infty e^{-\alpha^2}d\alpha.$$

12.7 Asymptotic Approximation of Integrals by Stationary Phase Method

Although definite integrals represent exact solutions for many physical problems, the physical meaning of the solutions is often difficult to determine. In many cases the exact evaluation of the integrals is a formidable task. It is then necessary to resort to asymptotic methods.

We consider the typical integral solution

$$u\left(x,t\right) = \int_a^b F\left(k\right) e^{it\theta(k)} dk, \qquad (12.7.1)$$

where $F\left(k\right)$ is called the *spectral function* determined by the initial or boundary data in $a < k < b$, and $\theta\left(k\right)$, known as the *phase function*, is given by

$$\theta\left(k\right) \equiv k\frac{x}{t} - \omega\left(k\right), \qquad x > 0. \qquad (12.7.2)$$

We examine the asymptotic behavior of (12.7.1) for both large x and large t; one of the interesting limits is $t \to \infty$ with (x/t) held fixed. Integral (12.7.1) can be evaluated by the Kelvin stationary phase method for large t. As $t \to \infty$, the integrand of (12.7.1) oscillates very rapidly; consequently, the contributions to $u\left(x,t\right)$ from adjacent parts of the integrand cancel one another except in the neighborhood of the points, if any, at which the phase function $\theta\left(k\right)$ is stationary, that is, $\theta'\left(k\right) = 0$. Thus, the main contribution to the integral for large t comes from the neighborhood of the point $k = k_1$ which determined by the solution of

$$\theta'\left(k_1\right) = \frac{x}{t} - \omega'\left(k_1\right) = 0, \qquad a < k_1 < b. \qquad (12.7.3)$$

The point $k = k_1$ known as the *point of stationary phase*, or simply, *stationary point*.

We expand both $F\left(k\right)$ and $\theta\left(k\right)$ in Taylor series about $k = k_1$ so that

$$u\left(x,t\right) = \int_a^b \left[F\left(k_1\right) + \left(k - k_1\right) F'\left(k_1\right) + \frac{1}{2}\left(k - k_1\right)^2 F''\left(k_1\right) + \ldots \right]$$

$$\times \exp\left\{ it\left[\theta\left(k_1\right) + \frac{1}{2}\left(k - k_1\right)^2 \theta''\left(k_1\right)\right.\right.$$

$$\left.\left. + \frac{1}{6}\left(k - k_1\right)^3 \theta'''\left(k_1\right) + \ldots \right] \right\} dk \qquad (12.7.4)$$

provided that $\theta''\left(k_1\right) \neq 0$.

Introducing the change of variable $k - k_1 = \varepsilon\alpha$, where

$$\varepsilon\left(t\right) = \left\{ \frac{2}{t\left|\theta''\left(k_1\right)\right|} \right\}^{\frac{1}{2}}, \qquad (12.7.5)$$

we find that the significant contribution to integral (12.7.4) is

$$u\left(x,t\right) \sim \varepsilon \int_{-(k_1-a)/\varepsilon}^{(b-k_1)/\varepsilon} \left[F\left(k_1\right) + \varepsilon\alpha F'\left(k_1\right) + \frac{1}{2}\varepsilon^2\alpha^2 F''\left(k_1\right) + \ldots \right]$$

$$\times \exp\left\{ i\left[t\theta\left(k_1\right) + \alpha^2 \operatorname{sgn}\theta''\left(k_1\right) + \frac{1}{3}\varepsilon\left(\frac{\theta'''\left(k_1\right)}{\left|\theta''\left(k_1\right)\right|}\right)\alpha^3 + \ldots \right] \right\} d\alpha,$$

$$(12.7.6)$$

where sgn x denotes the *signum function* defined by sgn $x = 1$, $x > 0$ and sgn $x = -1$, $x < 0$.

We then proceed to the limit as $\varepsilon \to 0$ $(t \to \infty)$ and use the standard integral

$$\int_{-\infty}^{\infty} \exp\left(\pm i\alpha^2\right) d\alpha = \sqrt{\pi} \exp\left(\pm \frac{i\pi}{4}\right) \tag{12.7.7}$$

to obtain the asymptotic approximation as $t \to \infty$,

$$u(x,t) \sim F(k_1) \left[\frac{2\pi}{t\,|\theta''(k_1)|}\right]^{\frac{1}{2}} \exp\left\{i\left[t\,\theta(k_1) + \frac{\pi}{4}\,\text{sgn}\,\theta''(k_1)\right]\right\} + O\left(\varepsilon^2\right), \tag{12.7.8}$$

where $O\left(\varepsilon^2\right)$ means that a function tends to zero like $\varepsilon^2\,(t)$ as $t \to \infty$. If there is more than one stationary point, each one contributes a term similar to (12.7.8) and we obtain, for n stationary points $k = k_r$, $r = 1, 2, \ldots n$;

$$u(x,t) \sim \sum_{r=1}^{n} F(k_r) \left\{\frac{2\pi}{t\,|\theta''(k_r)|}\right\}^{\frac{1}{2}} \exp\left\{i\left[t\,\theta(k_r) + \frac{\pi}{4}\,\text{sgn}\,\theta''(k_r)\right]\right\}, \quad t \to \infty. \tag{12.7.9}$$

If $\theta''(k_1) = 0$, but $\theta'''(k_1) \neq 0$, then asymptotic approximation (12.7.8) fails. This important special case can be handled in a similar fashion. The asymptotic approximation of (12.7.1) is then given by

$$u(x,t) = F(k_1) \exp\left\{it\theta(k_1)\right\} \int_{-\infty}^{\infty} \exp\left[\frac{i}{6}\,t\,\theta'''(k_1)\,(k - k_1)^3\right] dk$$

$$\sim \Gamma\left(\frac{4}{3}\right) \left[\frac{6}{t\,|\theta'''(k_1)|}\right]^{\frac{1}{3}} F(k_1) \exp\left[it\theta(k_1) + \frac{\pi i}{6}\right] + O\left(t^{-\frac{2}{3}}\right) \quad \text{as } t \to \infty. \tag{12.7.10}$$

For an elaborate treatment of the stationary phase method, see Copson (1965).

12.8 Laplace Transforms

Because of their simplicity, Laplace transforms are frequently used to solve a wide class of partial differential equations. Like other transforms, Laplace transforms are used to determine particular solutions. In solving partial differential equations, the general solutions are difficult, if not impossible, to obtain. The transform technique sometimes offers a useful tool for finding particular solutions.

The Laplace transform is closely related to the complex Fourier transform, so the Fourier integral formula (6.13.10) can be used to define the Laplace transform and its inverse. We replace $f(x)$ in (6.13.10) by $H(x) e^{-cx} f(x)$ for $x > 0$ to obtain

$$f(x) H(x) e^{-cx} = \frac{1}{2\pi} \int_{-\infty}^{\infty} e^{ikx} dk \int_0^{\infty} f(t) e^{-t(c+ik)} dt$$

or

$$f(x) H(x) = \frac{1}{2\pi} \int_{-\infty}^{\infty} e^{x(c+ik)} dk \int_0^{\infty} f(t) e^{-t(c+ik)} dt.$$

Substituting $s = c + ik$ so that $ds = idk$, we obtain, for $x > 0$,

$$f(x) H(x) = \frac{1}{2\pi i} \int_{c-i\infty}^{c+i\infty} e^{xs} ds \int_0^{\infty} f(t) e^{-st} dt. \qquad (12.8.1)$$

Thus, we give the following definition of the Laplace transform: If $f(t)$ is defined for all values of $t > 0$, then the *Laplace transform* of $f(t)$ is denoted by $\bar{f}(s)$ or $\mathcal{L}\{f(t)\}$ and is defined by the integral

$$\bar{f}(s) = \mathcal{L}\{f(t)\} = \int_0^{\infty} e^{-st} f(t) dt, \qquad (12.8.2)$$

where s is a positive real number or a complex number with a positive real part so that the integral is convergent.

Hence, (12.8.1) gives

$$f(x) = \mathcal{L}^{-1}\{\bar{f}(s)\} = \frac{1}{2\pi i} \int_{c-i\infty}^{c+i\infty} e^{xs} \bar{f}(s) ds, \quad c > 0, \quad (12.8.3)$$

for $x > 0$ and zero for $x < 0$. This complex integral is used to define the *inverse Laplace transform* which is denoted by $\mathcal{L}^{-1}\{\bar{f}(s)\} = f(t)$. It can be verified easily that both $\mathcal{L}$ and $\mathcal{L}^{-1}$ are linear integral operators.

We now find the Laplace transforms of some elementary functions.

1. Let $f(t) = c$, c is a constant.

$$\mathcal{L}[c] = \int_0^{\infty} e^{-st} c \, dt$$

$$= \left[-\frac{ce^{-st}}{s} \right]_0^{\infty} = \frac{c}{s}.$$

2. Let $f(t) = e^{at}$, a is a constant.

$$\mathcal{L}[e^{at}] = \int_0^{\infty} e^{-st} e^{at} dt$$

$$= \left[-\frac{e^{-(s-a)t}}{(s-a)} \right]_0^{\infty} = \frac{1}{s-a}, \quad s \geq a.$$

3. Let $f(t) = t^2$. Then

$$\mathcal{L}\left[t^2\right] = \int_0^\infty e^{-st}\, t^2 \, dt.$$

Integration by parts yields

$$\mathcal{L}\{t^2\} = \left[-\frac{t^2 e^{-st}}{s}\right]_0^\infty + \int_0^\infty \frac{e^{-st}}{s}\, 2t \, dt.$$

Since $t^2 e^{-st} \to 0$ as $t \to \infty$, we have, integrating by parts again,

$$\mathcal{L}\left[t^2\right] = \frac{2}{s}\left[-\frac{e^{-st}}{s}t\right] + \frac{2}{s}\int_0^\infty \frac{e^{-st}}{s}\, dt = \frac{2}{s^3}.$$

4. Let $f(t) = \sin \omega t$.

$$\bar{f}(s) = \mathcal{L}\left[\sin \omega t\right] = \int_0^\infty e^{-st} \sin \omega t \, dt$$

$$= \left[-\frac{e^{-st}}{s}\sin \omega t\right]_0^\infty + \int_0^\infty \frac{e^{-st}}{s}\, \omega \cos \omega t \, dt$$

$$= \frac{\omega}{s}\left[-\frac{e^{-st}}{s}\cos \omega t\right] - \frac{\omega}{s}\int_0^\infty \frac{e^{-st}}{s}\, \omega \sin \omega t \, dt$$

$$\bar{f}(s) = \frac{\omega}{s^2} - \frac{\omega^2}{s^2}\bar{f}(s).$$

Thus, solving for $\bar{f}(s)$, we obtain

$$\mathcal{L}\left[\sin \omega t\right] = \omega / \left(s^2 + \omega^2\right).$$

A function $f(t)$ is said to be of *exponential order* as $t \to \infty$ if there exist real constants M and a such that $|f(t)| \le Me^{at}$ for $0 \le t < \infty$.

Theorem 12.8.1. *Let f be piecewise continuous in the interval $[0, T]$ for every positive T, and let f be of exponential order, that is, $f(t) = O(e^{at})$ as $t \to \infty$ for some $a > 0$. Then, the Laplace transform of $f(t)$ exists for $\mathrm{Re}\, s > a$.*

Proof. Since f is piecewise continuous and of exponential order, we have

$$|\mathcal{L}(f(t))| = \left|\int_0^\infty e^{-st} f(t)\, dt\right|$$

$$\le \int_0^\infty e^{-st} |f(t)|\, dt$$

$$\le \int_0^\infty e^{-st} Me^{at} dt$$

$$= M \int_0^\infty e^{-(s-a)t} dt = M/(s-a), \quad \mathrm{Re}\, s > a.$$

Thus,

$$\int_0^\infty e^{-st} f(t)\, dt$$

exists for $\mathrm{Re}\, s > a$.

12.9 Properties of Laplace Transforms

Theorem 12.9.1. (**Linearity**) *If $\mathcal{L}[f(t)]$ and $\mathcal{L}[g(t)]$ are Laplace trans-forms of $f(t)$ and $g(t)$ respectively, then*

$$\mathcal{L}[af(t) + bg(t)] = a\mathcal{L}[f(t)] + b\mathcal{L}[g(t)]$$

where a and b are constants.

Proof.

$$\mathcal{L}[af(t) + bg(t)] = \int_0^\infty [af(t) + bg(t)] e^{-st} dt$$

$$= a\int_0^\infty f(t) e^{-st} dt + b\int_0^\infty g(t) e^{-st} dt$$

$$= a\mathcal{L}[f(t)] + b\mathcal{L}[g(t)].$$

This shows that $\mathcal{L}$ is a linear operator.

Theorem 12.9.2. (**Shifting**) *If $\bar{f}(s)$ is the Laplace transform of $f(t)$, then the Laplace transform of $e^{at} f(t)$ is $\bar{f}(s-a)$.*

Proof. By definition, we have

$$\mathcal{L}\left[e^{at} f(t)\right] = \int_0^\infty e^{-st} e^{at} f(t) \, dt$$

$$= \int_0^\infty e^{-(s-a)t} f(t) \, dt$$

$$= \bar{f}(s-a).$$

Example 12.9.1.

(a) If $\mathcal{L}[t^2] = 2/s^3$, then $\mathcal{L}[t^s e^t] = 2/(s-1)^3$.

(b) If $\mathcal{L}[\sin \omega t] = \omega/(s^2 + \omega^2)$, then $\mathcal{L}[e^{at} \sin \omega t] = \omega/\left[(s-1)^2 + \omega^2\right]$.

(c) If $\mathcal{L}\{\cos \omega t\} = \frac{s}{s^2+\omega^2}$, then $\mathcal{L}\{e^{at} \cos \omega t\} = \frac{s-a}{(s-a)^2+\omega^2}$.

(d) If $\mathcal{L}\{t^n\} = \frac{n!}{s^{n+1}}$, then $\mathcal{L}\{e^{at} t^n\} = \frac{n!}{(s-a)^{n+1}}$.

Theorem 12.9.3. (*Scaling*) *If the Laplace transform of $f(t)$ is $\bar{f}(s)$, then the Laplace transform of $f(ct)$ with $c > 0$ is $(1/c)\bar{f}(s/c)$.*

Proof. By definition, we have

$$\mathcal{L}[f(ct)] = \int_0^\infty e^{-st} f(ct) \, dt$$

$$= \int_0^\infty \frac{1}{c} e^{-(s\xi/c)} f(\xi) \, d\xi \quad \text{(substituting } \xi = ct)$$

$$= (1/c)\bar{f}(s/c).$$

Example 12.9.2.

(a) If $\frac{s}{s^2+1} = \mathcal{L}[\cos t]$, then

$$\frac{1}{\omega} \frac{s/\omega}{(s/\omega)^2 + 1} = \frac{s}{s^2 + \omega^2} = \mathcal{L}[\cos \omega t].$$

(b) If $\frac{1}{s-1} = \mathcal{L}[e^t]$, then

$$\frac{1}{a} \frac{1}{\left(\frac{s}{a} - 1\right)} = \mathcal{L}\left[e^{at}\right],$$

or

$$\mathcal{L}\left[e^{at}\right] = \frac{1}{s-a}.$$

Theorem 12.9.4. (*Differentiation*) *Let f be continuous and f' piecewise continuous, in $0 \le t \le T$ for all $T > 0$. Let f also be of exponential order as $t \to \infty$. Then, the Laplace transform of $f'(t)$ exists and is given by*

$$\mathcal{L}[f'(t)] = s\mathcal{L}[f(t)] - f(0) = s\bar{f}(s) - f(0).$$

Proof. Consider the definite integral

$$\int_0^T e^{-st} f'(t)\, dt = \left[e^{-st} f(t)\right]_0^T + \int_0^T s e^{-st} f(t)\, dt$$

$$= e^{-sT} f(T) - f(0) + s \int_0^T e^{-st} f(t)\, dt.$$

Since $|f(t)| \le Me^{at}$ for large t, with $a > 0$ and $M > 0$,

$$\left|e^{-sT} f(T)\right| \le Me^{-(s-a)T}.$$

In the limit as $T \to \infty$, $e^{-sT} f(T) \to 0$ whenever $s > a$. Hence,

$$\mathcal{L}[f'(t)] = s\mathcal{L}[f(t)] - f(0) = s\bar{f}(s) - f(0).$$

If f' and f'' satisfy the same conditions imposed on f and f' respectively, then, the Laplace transform of $f''(t)$ can be obtained immediately by applying the preceding theorem; that is

$$\mathcal{L}[f''(t)] = s\mathcal{L}[f'(t)] - f'(0)$$
$$= s\{s\mathcal{L}[f(t)] - f(0)\} - f'(0)$$
$$= s^2\mathcal{L}[f(t)] - sf(0) - f'(0)$$
$$= s^2\bar{f}(s) - sf(0) - f'(0).$$

Clearly, the Laplace transform of $f^{(n)}(t)$ can be obtained in a similar manner by successive application of Theorem 12.9.4. The result may be written as

$$\mathcal{L}\left[f^{(n)}(t)\right] = s^n\mathcal{L}[f(t)] - s^{n-1}f(0) - \ldots - sf^{(n-2)}(0) - f^{(n-1)}(0).$$

Theorem 12.9.5. *(Integration)* *If $\bar{f}(s)$ is the Laplace transform of $f(t)$,* then

$$\mathcal{L}\left[\int_0^t f(\tau)\,d\tau\right] = \bar{f}(s)/s.$$

Proof.

$$\mathcal{L}\left[\int_0^t f(\tau)\,d\tau\right] = \int_0^\infty \left[\int_0^t f(\tau)\,d\tau\right] e^{-st}\,dt$$

$$= \left[-\frac{e^{-st}}{s}\int_0^t f(\tau)\,d\tau\right]_0^\infty + \frac{1}{s}\int_0^\infty f(t)\,e^{-st}\,dt$$

$$= \bar{f}(s)/s$$

since $\displaystyle\int_0^t f(\tau)\,d\tau$ is of exponential order.

In solving problems by the Laplace transform method, the difficulty arises in finding inverse transforms. Although the inversion formula exists, its evaluation requires a knowledge of functions of complex variables. However, for some problems of mathematical physics, we need not use this inversion formula. We can avoid its use by expanding a given transform by the method of partial fractions in terms of simple fractions in the transform variables. With these simple functions, we refer to the table of Laplace transforms given in the end of the book and obtain the inverse transforms. Here, we should note that we use the assumption that there is essentially a one-to-one correspondence between functions and their Laplace transforms. This may be stated as follows:

Theorem 12.9.6. *(Lerch)* *Let f and g be piecewise continuous functions of exponential order. If there exists a constant s_0, such that $\mathcal{L}[f] = \mathcal{L}[g]$ for all $s > s_0$, then $f(t) = g(t)$ for all $t > 0$ except possibly at the points of discontinuity.*

For a proof, the reader is referred to Kreider et al. (1966).

In order to find a solution of linear partial differential equations, the following formulas and results are useful.
If $\mathcal{L}\{u(x,t)\} = \bar{u}(x,s)$, then

$$\mathcal{L}\left\{\frac{\partial u}{\partial t}\right\} = s\,\bar{u}(x,s) - u(x,0),$$

$$\mathcal{L}\left\{\frac{\partial^2 u}{\partial t^2}\right\} = s^2\,\bar{u}(x,s) - s\,u(x,0) - u_t(x,0),$$

and so on.
Similarly, it is easy to show that

$$\mathcal{L}\left\{\frac{\partial u}{\partial x}\right\} = \frac{d\bar{u}}{dx}, \quad \mathcal{L}\left\{\frac{\partial^2 u}{\partial x^2}\right\} = \frac{d^2\bar{u}}{dx^2}, \dots, \mathcal{L}\left\{\frac{\partial^n u}{\partial x^n}\right\} = \frac{d^n\bar{u}}{dx^n}.$$

The following results are useful for applications:

$$\mathcal{L}\left\{\mathrm{erfc}\left(\frac{a}{2\sqrt{t}}\right)\right\} = \frac{1}{s}\exp\left(-a\sqrt{s}\right), \qquad a \geq 0, \qquad (12.9.1)$$

$$\mathcal{L}\left\{\exp\left(at\right)\mathrm{erf}\left(\sqrt{at}\right)\right\} = \frac{\sqrt{a}}{\sqrt{s}\,(s-a)}, \qquad a > 0. \qquad (12.9.2)$$

Example 12.9.3. Consider the motion of a semi-infinite string with an external force $f(t)$ acting on it. One end is kept fixed while the other end is allowed to move freely in the vertical direction. If the string is initially at rest, the motion of the string is governed by

$$u_{tt} = c^2 u_{xx} + f(t), \qquad 0 < x < \infty, \quad t > 0,$$
$$u(x,0) = 0, \qquad u_t(x,0) = 0,$$
$$u(0,t) = 0, \qquad u_x(x,t) \to 0, \quad \text{as} \quad x \to \infty.$$

Let $\bar{u}(x,s)$ be the Laplace transform of $u(x,t)$. Transforming the equation of motion and using the initial conditions, we obtain

$$\bar{u}_{xx} - \left(s^2/c^2\right)\bar{u} = -\bar{f}(s)/c^2.$$

The solution of this ordinary differential equation is

$$\bar{u}(x,s) = Ae^{sx/c} + Be^{-sx/c} + \left[\bar{f}(s)/s^2\right].$$

The transformed boundary conditions are given by

$$\bar{u}(0,s) = 0, \quad \text{and} \quad \lim_{x\to\infty} \bar{u}_x(x,s) = 0.$$

In view of the second condition, we have $A = 0$. Now applying the first condition, we obtain

$$\bar{u}(0,s) = B + \left[\bar{f}(s)/s^2\right] = 0.$$

Hence

$$\bar{u}(x,s) = \left[\bar{f}(s)/s^2\right]\left[1 - e^{-sx/c}\right].$$

(a) When $f(t) = f_0$, a constant, then

$$\bar{u}(x,s) = f_0\left(\frac{1}{s^3} - \frac{1}{s^3}e^{-sx/c}\right).$$

The inverse Laplace transform gives the solution

$$u(x,t) = \frac{f_0}{2}\left[t^2 - \left(t - \frac{x}{c}\right)^2\right] \quad \text{when} \quad t \geq x/c,$$

$$= (f_0/2)\, t^2 \quad\quad\quad\quad \text{when} \quad t \leq x/c.$$

(b) When $f(t) = \cos\omega t$, where ω is a constant, then

$$\bar{f}(s) = \int_0^\infty e^{-st} \cos\omega t\, dt = s/\left(\omega^2 + s^2\right).$$

Thus, we have

$$\bar{u}(x,s) = \frac{1}{s\left(\omega^2 + s^2\right)}\left(1 - e^{-sx/c}\right). \tag{12.9.3}$$

By the method of partial fractions, we write

$$\frac{1}{s\left(s^2 + \omega^2\right)} = \frac{1}{\omega^2}\left[\frac{1}{s} - \frac{1}{s^2 + \omega^2}\right].$$

Hence

$$\mathcal{L}^{-1}\left[\frac{1}{s\left(s^2 + \omega^2\right)}\right] = \frac{1}{\omega^2}(1 - \cos\omega t) = \frac{2}{\omega^2}\sin^2\left(\frac{\omega t}{2}\right).$$

If we denote

$$\psi(t) = \sin^2\left(\frac{\omega t}{2}\right),$$

then the Laplace inverse of equation (12.9.3) may be written in the form

$$u(x,t) = \frac{2}{\omega^2}\left[\psi(t) - \psi\left(t - \frac{x}{c}\right)\right] \quad \text{when} \quad t \geq x/c,$$

$$= \frac{2}{\omega^2}\psi(t) \quad\quad\quad\quad\quad \text{when} \quad t \leq x/c.$$

12.10 Convolution Theorem of the Laplace Transform

The function

$$(f * g)(t) = \int_0^t f(t - \xi) g(\xi)\, d\xi \tag{12.10.1}$$

is called the *convolution* of the functions f and g.

Theorem 12.10.1. (*Convolution*) If $\bar{f}(s)$ and $\bar{g}(s)$ are the Laplace transforms of $f(t)$ and $g(t)$ respectively, then the Laplace transform of the convolution $(f * g)(t)$ is the product $\bar{f}(s)\bar{g}(s)$.

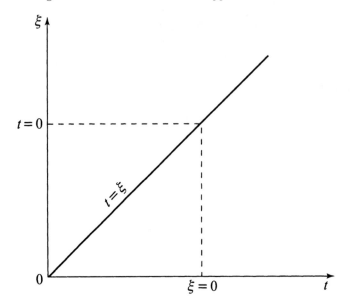

Figure 12.10.1 Region of integration.

Proof. By definition, we have

$$\mathcal{L}\left[(f * g)(t)\right] = \int_0^\infty e^{-st} \int_0^t f(t - \xi)\, g(\xi)\, d\xi\, dt$$

$$= \int_0^\infty \int_0^t e^{-st} f(t - \xi)\, g(\xi)\, d\xi\, dt.$$

The region of integration is shown in Figure 12.10.1. By reversing the order of integration, we have

$$\mathcal{L}\left[(f * g)(t)\right] = \int_0^\infty \int_\xi^\infty e^{-st} f(t - \xi)\, g(\xi)\, dt\, d\xi$$

$$= \int_0^\infty g(\xi) \int_\xi^t e^{-st} f(t - \xi)\, dt\, d\xi.$$

If we introduce the new variable $\eta = (t - \xi)$ in the inner integral, we obtain

$$\mathcal{L}\left[(f * g)(t)\right] = \int_0^\infty g(\xi) \int_0^\infty e^{-s(\xi + \eta)} f(\eta)\, d\eta\, d\xi$$

$$= \int_0^\infty g(\xi)\, e^{-s\xi}\, d\xi \int_0^\infty e^{-s\eta} f(\eta)\, d\eta$$

$$= \bar{f}(s)\, \bar{g}(s). \tag{12.10.2}$$

The convolution satisfies the following properties:

1. $f * g = g * f$ (commutative).
2. $f * (g * h) = (f * g) * h$ (associative).
3. $f * (\alpha g + \beta h) = \alpha (f * g) + \beta (f * h)$,   (distributive),
   where α and β are constants.

Example 12.10.1. Find the temperature distribution in a semi-infinite radiating rod. The temperature is kept constant at $x = 0$, while the other end is kept at zero temperature. If the initial temperature distribution is zero, the problem is governed by

$$u_t = k u_{xx} - hu, \quad 0 < x < \infty, \quad t > 0, \quad h = \text{constant},$$
$$u(x,0) = 0, \quad\quad\quad u(0,t) = u_0, \quad t > 0, \quad u_0 = \text{constant},$$
$$u(x,t) \to 0, \quad\quad\quad \text{as} \quad x \to \infty.$$

Let $\bar{u}(x,s)$ be the Laplace transform of $u(x,t)$. Then the transformation with respect to t yields

$$\bar{u}_{xx} - \left(\frac{s+h}{k}\right)\bar{u} = 0,$$

$$\bar{u}(0,s) = u_0/s, \quad \lim_{x\to\infty} \bar{u}(x,s) = 0.$$

The solution of this equation is

$$\bar{u}(x,s) = A\, e^{x\sqrt{(s+h)/k}} + B\, e^{-x\sqrt{(s+h)/k}}.$$

The boundary condition at infinity requires that $A = 0$. Applying the other boundary condition gives

$$\bar{u}(0,s) = B = u_0/s.$$

Hence, the solution takes the form

$$\bar{u}(x,s) = (u_0/s) \exp\left[-x\sqrt{(s+h)/k}\right].$$

We find (by using the Table of Laplace Transforms) that

$$\mathcal{L}^{-1}\left[\frac{u_0}{s}\right] = u_0,$$

and

$$\mathcal{L}^{-1}\left[\exp\left\{-x\sqrt{(s+h)/k}\right\}\right] = \frac{x \exp\left[-ht - (x^2/4kt)\right]}{2\sqrt{\pi k t^3}}.$$

Thus, the inverse Laplace transform of $\bar{u}(x,s)$ is

$$u(x,t) = \mathcal{L}^{-1}\left[\frac{u_0}{s}\exp\left\{-x\sqrt{(s+h)/k}\right\}\right].$$

By the Integration Theorem 12.9.5, we have

$$u(x,t) = \int_0^t \frac{u_0 \, x \exp\left[-h\tau - \left(x^2/4k\tau\right)\right]}{2\sqrt{\pi k} \, \tau^{\frac{3}{2}}} \, d\tau.$$

Substituting the new variable $\eta = \left(x/2\sqrt{k\tau}\right)$ yields

$$u(x,t) = \frac{2\,u_0}{\sqrt{\pi}} \int_{x/2\sqrt{kt}}^{\infty} \exp\left[-\eta^2 + \left(hx^2/4k\eta^2\right)\right] d\eta.$$

For the case $h = 0$, the solution $u(x,t)$ becomes

$$u(x,t) = \frac{2\,u_0}{\sqrt{\pi}} \int_{x/2\sqrt{kt}}^{\infty} e^{-\eta^2} d\eta$$

$$= \frac{2\,u_0}{\sqrt{\pi}} \int_0^{\infty} e^{-\eta^2} d\eta - \frac{2\,u_0}{\sqrt{\pi}} \int_0^{x/2\sqrt{kt}} e^{-\eta^2} d\eta$$

$$= u_0 \left[1 - \operatorname{erf}\left(\frac{x}{2\sqrt{kt}}\right)\right] = u_0 \operatorname{erfc}\left(\frac{x}{2\sqrt{kt}}\right).$$

12.11 Laplace Transforms of the Heaviside and Dirac Delta Functions

We have defined the Heaviside unit step function. Now, we will find its Laplace transform.

$$\mathcal{L}\left[H(t-a)\right] = \int_0^{\infty} e^{-st} H(t-a) \, dt$$

$$= \int_a^{\infty} e^{-st} dt = \left(\frac{1}{s}\right) e^{-as}, \quad s > 0. \qquad (12.11.1)$$

Theorem 12.11.1. *(Second Shifting)* *If $\bar{f}(s)$ and $\bar{g}(s)$ are the Laplace transforms of $f(t)$ and $g(t)$ respectively, then*

(a) $\mathcal{L}\left[H(t-a) f(t-a)\right] = e^{-as} \bar{f}(s) = e^{-as} \mathcal{L}\{f(t)\}.$

(b) $\mathcal{L}\{H(t-a) g(t)\} = e^{-as} \mathcal{L}\{g(t+a)\}.$

Proof. (a) By definition

$$\mathcal{L}\left[H(t-a) f(t-a)\right] = \int_0^{\infty} e^{-st} H(t-a) f(t-a) \, dt$$

$$= \int_a^{\infty} e^{-st} f(t-a) \, dt.$$

Introducing the new variable $\xi = t - a$, we obtain

$$\mathcal{L}\left[H\left(t-a\right)f\left(t-a\right)\right] = \int_0^\infty e^{-(\xi+a)s}f\left(\xi\right)d\xi$$

$$= e^{-as}\int_0^\infty e^{-\xi s}f\left(\xi\right)d\xi = e^{-as}\bar{f}\left(s\right).$$

To prove (b), we write

$$\mathcal{L}\left\{H\left(t-a\right)g\left(t\right)\right\} = \int_a^\infty e^{-st}g\left(t\right)dt \qquad (t-a=\tau)$$

$$= \int_0^\infty e^{-s(a+\tau)}g\left(a+\tau\right)d\tau = e^{-sa}\mathcal{L}\left\{g\left(t+a\right)\right\}.$$

Example 12.11.1. (a) Given that

$$f\left(t\right) = \left\{\begin{matrix} 0, & t<2 \\ t-2, & t\geq 2 \end{matrix}\right\} = (t-2)H\left(t-2\right),$$

find the Laplace transform of $f\left(t\right)$.

We have

$$\mathcal{L}\left[f\left(t\right)\right] = \mathcal{L}\left[H\left(t-2\right)\left(t-2\right)\right] = e^{-2s}\mathcal{L}\left[t\right] = \left(\frac{1}{s^2}\right)e^{-2s}.$$

(b) Find the inverse Laplace transform of

$$\bar{f}\left(s\right) = \frac{1+e^{-2s}}{s^2}.$$

$$\mathcal{L}^{-1}\left[\bar{f}\left(s\right)\right] = \mathcal{L}^{-1}\left(\frac{1}{s^2} + \frac{e^{-2s}}{s^2}\right)$$

$$= \mathcal{L}^{-1}\left[\frac{1}{s^2}\right] + \mathcal{L}^{-1}\left[\frac{e^{-2s}}{s^2}\right] = t + H\left(t-2\right)\left(t-2\right)$$

$$= \left\{\begin{matrix} t, & 0\leq t<2, \\ 2\left(t-1\right), & t\geq 2. \end{matrix}\right.$$

The Laplace transform of the impulse function $p\left(t\right)$ is given by

$$\mathcal{L}\left[p\left(t\right)\right] = \int_0^\infty e^{-st}p\left(t\right)dt$$

$$= \int_{a-\varepsilon}^{a+\varepsilon} h\,e^{-st}dt$$

$$= h\left[-\frac{e^{-st}}{s}\right]_{a-\varepsilon}^{a+\varepsilon}$$

$$= \frac{h\,e^{-as}}{s}\left(e^{\varepsilon s} - e^{-\varepsilon s}\right)$$

$$= 2\frac{h\,e^{-as}}{s}\sinh\left(\varepsilon s\right). \qquad (12.11.2)$$

If we choose the value of h to be $(1/2\varepsilon)$, then the impulse is given by

$$I(\varepsilon) = \int_{-\infty}^{\infty} p(t)\, dt = \int_{a-\varepsilon}^{a+\varepsilon} \frac{1}{2\varepsilon}\, dt = 1.$$

Thus, in the limit as $\varepsilon \to 0$, this particular impulse function satisfies

$$\lim_{\varepsilon \to 0} p_\varepsilon(t) = 0, \qquad t \neq a,$$

$$\lim_{\varepsilon \to 0} I(\varepsilon) = 1.$$

From this result, we obtain the Dirac delta function which satisfies

$$\delta(t-a) = 0, \qquad t \neq a,$$

$$\int_{-\infty}^{\infty} \delta(t-a)\, dt = 1. \tag{12.11.3}$$

Thus, we may define the Laplace transform of $\delta(t)$ as the limit of the transform of $p_\varepsilon(t)$.

$$\mathcal{L}[\delta(t-a)] = \lim_{\varepsilon \to 0} \mathcal{L}[p_\varepsilon(t)],$$

$$= \lim_{\varepsilon \to 0} e^{-as} \frac{\sinh(\varepsilon s)}{\varepsilon s} \tag{12.11.4}$$

$$= e^{-as}.$$

If $a = 0$, we have

$$\mathcal{L}[\delta(t)] = 1. \tag{12.11.5}$$

One very useful result that can be derived is the integral of the product of the delta function and any continuous function $f(t)$.

$$\int_{-\infty}^{\infty} \delta(t-a) f(t)\, dt = \lim_{\varepsilon \to 0} \int_{-\infty}^{\infty} p_\varepsilon(t) f(t)\, dt,$$

$$= \lim_{\varepsilon \to 0} \int_{a-\varepsilon}^{a+\varepsilon} \frac{f(t)}{2\varepsilon}\, dt,$$

$$= \lim_{\varepsilon \to 0} \frac{1}{2\varepsilon} \cdot 2\varepsilon\, f(t^*), \qquad a-\varepsilon < t^* < a+\varepsilon$$

$$= f(a). \tag{12.11.6}$$

Suppose that $f(t)$ is a periodic function with period T. Let f be piecewise continuous on $[0, T]$. Then, the Laplace transform of $f(t)$ is

$$\mathcal{L}[f(t)] = \int_0^{\infty} e^{-st} f(t)\, dt,$$

$$= \sum_{n=0}^{\infty} \int_{nT}^{(n+1)T} e^{-st} f(t)\, dt.$$

If we introduce a new variable $\xi = t - nT$, then

$$\mathcal{L}[f(t)] = \sum_{n=0}^{\infty} e^{-nTs} \int_0^T e^{-s\xi} f(\xi)\, d\xi,$$

$$= \sum_{n=0}^{\infty} e^{-nTs} \overline{f}_1(s),$$

where $\overline{f}_1(s) = \int_0^T e^{-s\xi} f(\xi)\, d\xi$ is the transform of the function f over the first period. Since the series is a geometric series, we obtain for the transform of a periodic function

$$\mathcal{L}[f(t)] = \frac{\overline{f}_1(s)}{(1 - e^{-Ts})}. \tag{12.11.7}$$

Example 12.11.2. Find the Laplace transform of the square wave function with period $2c$ given by

$$f(t) = \begin{cases} h, & 0 < t < c \\ \\ -h, & c < t < 2c \end{cases} \quad \text{with} \quad f(t + 2c) = f(t),$$

as shown in Figure 12.11.1.

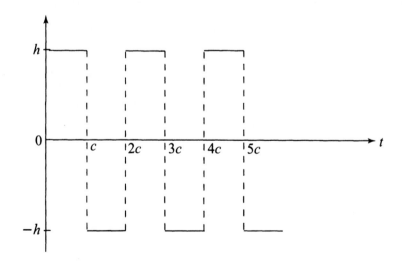

Figure 12.11.1 Square wave function.

$$\overline{f}_1(s) = \int_0^{2c} e^{-s\xi} f(\xi)\, d\xi,$$

$$= \int_0^c e^{-s\xi} h\, d\xi + \int_c^{2c} e^{-s\xi} (-h)\, d\xi,$$

$$= \frac{h}{s} \left(1 - e^{-cs}\right)^2.$$

Thus, the Laplace transform of $f(t)$ is, by (12.11.7),

$$\mathcal{L}[f(t)] = \frac{\overline{f}_1(s)}{1 - e^{-2cs}} = \frac{h\left(1 - e^{-cs}\right)^2}{s\left(1 - e^{-2cs}\right)}$$

$$= \frac{h\left(1 - e^{-cs}\right)}{s\left(1 + e^{-cs}\right)} = \frac{h}{s} \tanh\left(\frac{cs}{2}\right).$$

Example 12.11.3. A uniform bar of length l is fixed at one end. Let the force

$$f(t) = \begin{cases} f_0, & t > 0 \\ 0, & t < 0 \end{cases}$$

be suddenly applied at the end $x = l$. If the bar is initially at rest, find the longitudinal displacement for $t > 0$.

The motion of the bar is governed by the differential system

$$u_{tt} = a^2 u_{xx}, \quad 0 < x < l, \quad t > 0, \quad a = \text{constant},$$
$$u(x,0) = 0, \quad u_t(x,0) = 0,$$
$$u(0,t) = 0, \quad u_x(l,t) = (f_0/E), \quad \text{where } E \text{ is a constant and } t > 0.$$

Let $\overline{u}(x,s)$ be the Laplace transform of $u(x,t)$. Then, $\overline{u}(x,s)$ satisfies the system

$$\overline{u}_{xx} - \frac{s^2}{a^2} \overline{u} = 0,$$
$$\overline{u}(0,s) = 0, \quad \overline{u}_x(l,s) = (f_0/Es).$$

The solution of this differential equation is

$$\overline{u}(x,s) = A e^{xs/a} + B e^{-xs/a}.$$

Applying the boundary conditions, we have

$$A + B = 0, \quad \left(\frac{s}{a} e^{ls/a}\right) A + \left(-\frac{s}{a} e^{-ls/a}\right) B = f_0/Es.$$

Solving for A and B, we obtain

$$A = -B = \frac{af_0}{Es^2 \left(e^{ls/a} + e^{-ls/a}\right)}.$$

Hence, the transform of the displacement function is given by

$$\bar{u}(x,s) = \frac{af_0 \left(e^{xs/a} - e^{-xs/a}\right)}{Es^2 \left(e^{ls/a} + e^{-ls/a}\right)}.$$

Before finding the inverse transform of $\bar{u}(x,s)$, multiply the numerator and denominator by $\left(e^{-ls/a} - e^{-3ls/a}\right)$. Thus, we have

$$\bar{u}(x,s) = \left(\frac{af_0}{Es^2}\right)\left[e^{-(l-x)s/a} - e^{-(l+x)s/a} - e^{-(3l-x)s/a} + e^{-(3l+x)s/a}\right]$$

$$\times \frac{1}{\left(1 - e^{-4ls/a}\right)}.$$

Since the denominator has the term $\left(1 - e^{-4ls/a}\right)$, the inverse transform $u(x,t)$ is periodic with period $(4l/a)$. Hence, the final solution may be written in the form

$$u(x,t) = \begin{cases} 0, & 0 < t < \frac{l-x}{a}, \\[2mm] \frac{af_0}{E}\left(t - \frac{l-x}{a}\right), & \frac{l-x}{a} < t < \frac{l+x}{a}, \\[2mm] \frac{af_0}{E}\left[\left(t - \frac{l-x}{a}\right) - \left(t - \frac{l+x}{a}\right)\right], & \frac{l+x}{a} < t < \frac{3l-x}{a}, \\[2mm] \frac{af_0}{E}\left[\left(t - \frac{l-x}{a}\right) - \left(t - \frac{l+x}{a}\right) - \left(t - \frac{3l-x}{a}\right)\right], & \frac{3l-x}{a} < t < \frac{3l+x}{a}, \\[2mm] \frac{af_0}{E}\left[\left(t - \frac{l-x}{a}\right) - \left(t - \frac{l+x}{a}\right) - \left(t - \frac{3l-x}{a}\right) + \left(t - \frac{3l+x}{a}\right)\right], & \\[2mm] 0, & \frac{3l+x}{a} < t < \frac{4l}{a}, \end{cases}$$

which may be simplified to obtain

$$u(x,t) = \begin{cases} 0, & 0 < t < (l-x)/a, \\[2mm] \frac{af_0}{E}\left(t - \frac{l-x}{a}\right), & (l-x)/a < t < (l+x)/a, \\[2mm] \frac{af_0}{E}\left(\frac{2x}{a}\right), & (l+x)/a < t < (3l-x)/a, \\[2mm] \frac{af_0}{E}\left(-t + \frac{3l+x}{a}\right), & (3l-x)/a < t < (3l+x)/a, \\[2mm] 0, & (3l+x)/a < t < 4l/a. \end{cases}$$

This result can clearly be seen in Figure 12.11.2.

Example 12.11.4. Consider a semi-infinite string fixed at the end $x = 0$. The string is initially at rest. Let there be an external force

$$f(x,t) = -f_0 \, \delta \left(t - \frac{x}{v} \right),$$

acting on the string. This is a concentrated force f_0 acting at the point $x = vt$.

The motion of the string is governed by the initial boundary-value problem

$$u_{tt} = c^2 u_{xx} - f_0 \, \delta \left(t - \frac{x}{v} \right),$$
$$u(x,0) = 0, \quad u_t(x,0) = 0,$$
$$u(0,t) = 0, \quad u(x,t) \text{ is bounded as } x \to \infty.$$

Let $\bar{u}(x,s)$ be the Laplace transform of $u(x,t)$. Transforming the wave equation and using the initial conditions, we obtain

$$\bar{u}_{xx} - \frac{s^2}{c^2} \bar{u} = \frac{f_0}{c^2} \exp(-xs/v).$$

The solution of this equation is

$$\bar{u}(x,s) = \left(A e^{sx/c} + B e^{-sx/c} \right) + \begin{cases} \dfrac{f_0 v^2 e^{-sx/v}}{(c^2 - v^2) s^2} & \text{for} \quad v \neq c, \\[2ex] -\dfrac{f_0 x e^{-sx/v}}{2cs} & \text{for} \quad v = c. \end{cases}$$

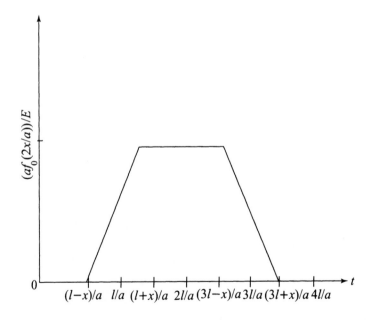

Figure 12.11.2 Graph of $u(x,t)$.

The condition that $u(x, t)$ must be bounded at infinity requires that $A = 0$. Application of the condition

$$\overline{u}(0, s) = 0,$$

yields

$$B = \begin{cases} \frac{-f_0 v^2}{(c^2 - v^2)s^2} & \text{for} \quad v \neq c, \\ 0 & \text{for} \quad v = c. \end{cases}$$

Hence, the Laplace transform is given by

$$\overline{u}(x, s) = \begin{cases} \frac{f_0 v^2 \left(e^{-xs/v} - e^{-xs/c}\right)}{(c^2 - v^2)s^2} & \text{for} \quad v \neq c, \\ -\frac{f_0 x e^{-xs/c}}{2cs} & \text{for} \quad v = c. \end{cases}$$

The inverse Laplace transform is therefore given by

$$u(x, t) = \begin{cases} \frac{f_0 v^2}{(c^2 - v^2)} \left[\left(t - \frac{x}{v}\right)u\left(t - \frac{x}{v}\right) - \left(t - \frac{x}{c}\right)u\left(t - \frac{x}{c}\right)\right], & \text{for} \quad v \neq c, \\ -\left(\frac{f_0 x}{2c}\right)u\left(t - \frac{x}{c}\right), & \text{for} \quad v = c. \end{cases}$$

Example 12.11.5. (The Stokes Problem and the Rayleigh Problem in fluid dynamics). Solve the Stokes problem which is concerned with the unsteady boundary layer flows induced in a semi-infinite viscous fluid bounded by an infinite horizontal disk at $z = 0$ due to oscillations of the disk in its own plane with a given frequency ω.

We solve the boundary layer equation for the velocity $u(z, t)$

$$u_t = \nu u_{zz}, \quad z > 0, \quad t > 0,$$

with the boundary and initial conditions

$$\begin{aligned} u(z, t) &= U_0 e^{i\omega t}, & z &= 0, & t &> 0, \\ u(z, t) &\to 0, & \text{as} \quad z &\to \infty, & t &> 0, \\ u(z, 0) &= 0, & \text{at} \quad t &\leq 0 \quad \text{for all} \quad z &> 0, \end{aligned}$$

where $u(z, t)$ is the velocity of fluid of kinematic viscosity ν and U_0 is a constant.

The Laplace transform solution of the equation with the transformed boundary conditions is

$$\overline{u}(z, s) = \frac{U_0}{(s - i\omega)} \exp\left(-z\sqrt{\frac{s}{\nu}}\right).$$

Using a standard table of inverse Laplace transforms, we obtain the solution

$$u(z,t) = \frac{U_0}{2} e^{i\omega t} \left[\exp\left(-\lambda z\right) \operatorname{erfc}\left(\zeta - \sqrt{i\omega t}\right) \right.$$
$$\left. + \exp\left(\lambda z\right) \operatorname{erfc}\left(\zeta + \sqrt{i\omega t}\right) \right],$$

where $\zeta = \left(z/2\sqrt{\nu t}\right)$ is called the *similarity variable* of the viscous boundary layer theory, and $\lambda = (i\omega/\nu)^{\frac{1}{2}}$. This result describes the unsteady boundary layer flow.

In view of the asymptotic formula for the complementary error function

$$\operatorname{erfc}\left(\zeta \mp \sqrt{i\omega t}\right) \sim (2,0) \quad \text{as} \quad t \to \infty,$$

the above solution for $u(z,t)$ has the asymptotic representation

$$u(z,t) \sim U_0 \exp\left(i\omega t - \lambda z\right)$$
$$= U_0 \exp\left[i\omega t - \left(\frac{\omega}{2\nu}\right)^{\frac{1}{2}} (1+i)\, z\right]. \qquad (12.11.8)$$

This is called the *Stokes steady-state solution*. This represents the propagation of shear waves which spread out from the oscillating disk with velocity $\omega/k = \sqrt{2\nu\omega}$ $\left(k = (\omega/2\nu)^{\frac{1}{2}}\right)$ and exponentially decaying amplitude. The boundary layer associated with the solution has thickness of the order $(\nu/\omega)^{\frac{1}{2}}$ in which the shear oscillations imposed by the disk decay exponentially with distance z from the disk. This boundary layer is called the *Stokes layer*. In other words, the thickness of the Stokes layer is equal to the depth of penetration of vorticity which is essentially confined to the immediate vicinity of the disk for high frequency ω.

The Stokes problem with $\omega = 0$ becomes the Rayleigh problem. In other words, the motion is generated in the fluid from rest by moving the disk impulsively in its own plane with constant velocity U_0. In this case, the Laplace transform solution is

$$\overline{u}(z,s) = \frac{U_0}{s} \exp\left(-z\sqrt{\frac{s}{\nu}}\right),$$

so that the inversion gives the Rayleigh solution

$$u(z,t) = U_0 \operatorname{erfc}\left(\frac{z}{2\sqrt{\nu t}}\right). \qquad (12.11.9)$$

This describes the growth of a boundary layer adjacent to the disk. The associated boundary layer is called the *Rayleigh layer* of thickness of the order $\delta \sim \sqrt{\nu t}$ which grows with increasing time t. The rate of growth is of the order $d\delta/dt \sim \sqrt{\nu/t}$, which diminishes with increasing time.

The vorticity of the unsteady flow is given by

$$\frac{\partial u}{\partial z} = \frac{U_0}{\sqrt{\pi \nu t}} \exp\left(-\zeta^2\right) \tag{12.11.10}$$

which decays exponentially to zero as $z \gg \delta$.

Note that the vorticity is everywhere zero at $t = 0$ except at $z = 0$. This implies that it is generated at the disk and diffuses outward within the Rayleigh layer. The total viscous diffusion time is $T_d \sim \delta^2/\nu$.

Another physical quantity related to the Stokes and Rayleigh problems is the *skin friction* on the disk defined by

$$\tau_0 = \mu \left(\frac{\partial u}{\partial z}\right)_{z=0}, \tag{12.11.11}$$

where $\mu = \nu\rho$ is the dynamic viscosity and ρ is the density of the fluid. The skin friction can readily be calculated from the flow field given by (12.11.8) or (12.11.9).

Example 12.11.6. (The Nonhomogeneous Cauchy Problem for the Wave Equation). We consider the nonhomogeneous Cauchy problem

$$u_{tt} - c^2 u_{xx} = q\left(x,t\right), \quad x \in R, \quad t > 0, \tag{12.11.12}$$
$$u\left(x,0\right) = f\left(x\right), \quad u_t\left(x,0\right) = g\left(x\right) \text{ for all } x \in R, \tag{12.11.13}$$

where $q\left(x,t\right)$ is a given function representing a source term.

We use the joint Laplace and Fourier transform of $u\left(x,t\right)$

$$\overline{U}\left(k,s\right) = \mathcal{L}\left[\mathcal{F}\left\{u\left(x,t\right)\right\}\right] = \frac{1}{\sqrt{2\pi}} \int_{-\infty}^{\infty} e^{-ikx} dx \int_{0}^{\infty} e^{-st} u\left(x,t\right) dt. \tag{12.11.14}$$

Application of the joint transform leads to the solution of the transformed Cauchy problem in the form

$$\overline{U}\left(k,s\right) = \frac{s\,F\left(k\right) + G\left(k\right) + \overline{Q}\left(k,s\right)}{\left(s^2 + c^2 k^2\right)}. \tag{12.11.15}$$

The inverse Laplace transform of (12.11.15) gives

$$U\left(k,t\right) = F\left(k\right)\cos\left(ckt\right) + \frac{1}{ck}G\left(k\right)\sin\left(ckt\right) + \frac{1}{ck}\mathcal{L}^{-1}\left\{\frac{ck}{s^2 + c^2 k^2} \cdot \overline{Q}\left(k,s\right)\right\}$$
$$= F\left(k\right)\cos\left(ckt\right) + \frac{G\left(k\right)}{ck}\sin\left(ckt\right) + \frac{1}{ck}\int_{0}^{t}\sin ck\left(t-\tau\right)Q\left(k,\tau\right)d\tau. \tag{12.11.16}$$

The inverse Fourier transform leads to the exact integral solution

$$u(x,t) = \frac{1}{2\sqrt{2\pi}} \int_{-\infty}^{\infty} \left(e^{ickt} + e^{-ickt}\right) e^{ikx} F(k) \, dk$$

$$+ \frac{1}{2\sqrt{2\pi}} \int_{-\infty}^{\infty} \left(e^{ickt} - e^{-ickt}\right) e^{ikx} \cdot \frac{G(k)}{ick} \, dk$$

$$+ \frac{1}{\sqrt{2\pi}} \cdot \frac{1}{2c} \int_0^t d\tau \int_{-\infty}^{\infty} \frac{Q(k,\tau)}{ik} \left[e^{ick(t-\tau)} - e^{-ick(t-\tau)}\right] e^{ikx} \, dk$$

$$= \frac{1}{2} \left[f(x+ct) + f(x-ct)\right] + \frac{1}{2c} \int_{x-ct}^{x+ct} g(\xi) \, d\xi$$

$$+ \frac{1}{2c} \int_0^t d\tau \frac{1}{\sqrt{2\pi}} \int_{-\infty}^{\infty} Q(k,\tau) \, dk \int_{x-c(t-\tau)}^{x+c(t-\tau)} e^{ik\xi} \, d\xi$$

$$= \frac{1}{2} \left[f(x-ct) + f(x+ct)\right] + \frac{1}{2c} \int_{x-ct}^{x+ct} g(\xi) \, d\xi$$

$$+ \frac{1}{2c} \int_0^t d\tau \int_{x-c(t-\tau)}^{x+c(t-\tau)} q(\xi,\tau) \, d\xi. \quad (12.11.17)$$

In the case of the homogeneous Cauchy problem, $q(x,t) \equiv 0$, the solution of (12.11.17) reduces to the famous d'Alembert solution (5.3.8).

Example 12.11.7. (The Heat Conduction Equation in a Semi-Infinite Medium and Fractional Derivatives). Solve the one-dimensional diffusion equation

$$u_t = \kappa \, u_{xx}, \qquad x > 0, \quad t > 0, \qquad\qquad (12.11.18)$$

with the initial and boundary conditions

$$u(x,0) = 0, \qquad x > 0, \qquad\qquad\qquad (12.11.19)$$
$$u(0,t) = f(t), \qquad t > 0, \qquad\qquad\qquad (12.11.20)$$
$$u(x,t) \to 0, \qquad \text{as } x \to \infty, \quad t > 0. \qquad (12.11.21)$$

Application of the Laplace transform with respect to t to (12.11.18) gives

$$\frac{d^2\bar{u}}{dx^2} - \frac{s}{\kappa}\bar{u} = 0. \qquad\qquad\qquad (12.11.22)$$

The general solution of this equation is

$$\bar{u}(x,s) = A \exp\left(-x\sqrt{\frac{s}{\kappa}}\right) + B \exp\left(x\sqrt{\frac{s}{\kappa}}\right),$$

where A and B are integrating constants. For bounded solutions, $B \equiv 0$, and using $\bar{u}(0,s) = \bar{f}(s)$, we obtain the solution

$$\bar{u}(x,s) = \bar{f}(s) \exp\left(-x\sqrt{\frac{s}{\kappa}}\right). \qquad\qquad (12.11.23)$$

The Laplace inversion theorem gives the solution

$$u\left(x,t\right) = \frac{x}{2\sqrt{\pi\kappa}} \int_0^t f\left(t - \tau\right) \tau^{-\frac{3}{2}} \exp\left(-\frac{x^2}{4\kappa\tau}\right) d\tau, \quad (12.11.24)$$

which, by setting $\lambda = \frac{x}{2\sqrt{\kappa\tau}}$, or $d\lambda = -\frac{x}{4\sqrt{\kappa}}\tau^{-\frac{3}{2}}d\tau$,

$$= \frac{2}{\sqrt{\pi}} \int_{\frac{x}{2\sqrt{\kappa t}}}^{\infty} f\left(t - \frac{x^2}{4\kappa\lambda^2}\right) e^{-\lambda^2} d\lambda. \quad (12.11.25)$$

This is the formal solution of the heat conduction problem.

In particular, if $f\left(t\right) = T_0 = $ constant, solution (12.11.25) becomes

$$u\left(x,t\right) = \frac{2T_0}{\sqrt{\pi}} \int_{\frac{x}{2\sqrt{\kappa t}}}^{\infty} e^{-\lambda^2} d\lambda = T_0 \operatorname{erfc}\left(\frac{x}{2\sqrt{\kappa t}}\right). \quad (12.11.26)$$

Clearly, the temperature distribution tends asymptotically to the constant value T_0, as $t \to \infty$.

Alternatively, solution (12.11.23) can be written as

$$\overline{u}\left(x,s\right) = \overline{f}\left(s\right) s\,\overline{u}_0\left(x,s\right), \quad (12.11.27)$$

where

$$s\,\overline{u}_0\left(x,s\right) = \exp\left(-x\sqrt{\frac{s}{\kappa}}\right). \quad (12.11.28)$$

Consequently, the inversion of (12.11.27) gives a new representation

$$u\left(x,t\right) = \int_0^t f\left(t - \tau\right) \left(\frac{\partial u_0}{\partial \tau}\right) d\tau. \quad (12.11.29)$$

This is called the *Duhamel formula* for the diffusion equation.

We consider another physical problem: determining the temperature distribution of a semi-infinite solid when the rate of flow of heat is prescribed at the end $x = 0$. Thus, the problem is to solve diffusion equation (12.11.18) subject to conditions (12.11.19), (12.11.21), and

$$-k\left(\frac{\partial u}{\partial x}\right) = g\left(t\right) \quad \text{at } x = 0, \quad t > 0, \quad (12.11.30)$$

where k is a constant called *thermal conductivity*.

Application of the Laplace transform gives the solution of the transformed problem

$$\overline{u}\left(x,s\right) = \frac{1}{k}\sqrt{\frac{\kappa}{s}}\,\overline{g}\left(s\right) \exp\left(-x\sqrt{\frac{s}{\kappa}}\right). \quad (12.11.31)$$

The inverse Laplace transform yields the solution

$$u(x,t) = \frac{1}{k}\sqrt{\frac{\kappa}{\pi}} \int_0^t g(t-\tau)\,\tau^{-\frac{1}{2}} \exp\left(-\frac{x^2}{4\kappa\tau}\right) d\tau, \quad (12.11.32)$$

which is, by the change of variable $\lambda = \frac{x}{2\sqrt{\kappa\tau}}$,

$$= \frac{x}{k\sqrt{\pi}} \int_{\frac{x}{2\sqrt{\kappa t}}}^{\infty} g\left(t - \frac{x^2}{4\kappa\lambda^2}\right) \lambda^{-2} e^{-\lambda^2} d\lambda. \quad (12.11.33)$$

In particular, if $g(t) = T_0 = $ constant, this solution becomes

$$u(x,t) = \frac{T_0\, x}{k\sqrt{\pi}} \int_{\frac{x}{\sqrt{4\kappa t}}}^{\infty} \lambda^{-2} e^{-\lambda^2} d\lambda.$$

Integrating this result by parts gives

$$u(x,t) = \frac{T_0}{k}\left[2\sqrt{\frac{\kappa t}{\pi}} \exp\left(-\frac{x^2}{4\kappa t}\right) - x\,\mathrm{erfc}\left(\frac{x}{2\sqrt{\kappa t}}\right) \right]. \quad (12.11.34)$$

Alternatively, the heat conduction problem (12.11.18)–(12.11.21) can be solved by using fractional derivatives (Debnath 1995). We recall (12.11.23) and rewrite it as

$$\frac{\partial \bar{u}}{\partial x} = -\sqrt{\frac{s}{\kappa}}\,\bar{u}. \quad (12.11.35)$$

This can be expressed in terms of a fractional derivative of order $\frac{1}{2}$ as

$$\frac{\partial u}{\partial x} = -\frac{1}{\sqrt{\kappa}} \mathcal{L}^{-1}\left\{\sqrt{s}\,\bar{u}(x,s)\right\} = -\frac{1}{\sqrt{\kappa}}\,{_0}D_t^{\frac{1}{2}} u(x,t). \quad (12.11.36)$$

Thus, the heat flux is expressed in terms of the fractional derivative. In particular, when $u(0,t) = $ constant $= T_0$, then the heat flux at the surface is given by

$$-k\left(\frac{\partial u}{\partial x}\right)_{x=0} = \frac{k}{\sqrt{\kappa}} D_t^{\frac{1}{2}} T_0 = \frac{kT_0}{\sqrt{\pi\kappa t}}. \quad (12.11.37)$$

Example 12.11.8. (*Diffusion Equation in a Finite Medium*). Solve the diffusion equation

$$u_t = \kappa\, u_{xx}, \qquad 0 < x < a,\ t > 0, \quad (12.11.38)$$

with the initial and boundary conditions

$$u(x,0) = 0, \quad 0 < x < a, \quad (12.11.39)$$
$$u(0,t) = U, \quad t > 0, \quad (12.11.40)$$
$$u_x(a,t) = 0, \quad t > 0, \quad (12.11.41)$$

where U is a constant.

We introduce the Laplace transform of $u(x,t)$ with respect to t to obtain

$$\frac{d^2\bar{u}}{dx^2} - \frac{s}{\kappa}\bar{u} = 0, \qquad 0 < x < a, \qquad (12.11.42)$$

$$\bar{u}(0, s) = \frac{U}{s}, \qquad \left(\frac{d\bar{u}}{dx}\right)_{x=a} = 0. \qquad (12.11.43)$$

The general solution of (12.11.42) is

$$\bar{u}(x, s) = A \cosh\left(x\sqrt{\frac{s}{\kappa}}\right) + B \sinh\left(x\sqrt{\frac{s}{\kappa}}\right), \qquad (12.11.44)$$

where A and B are constants of integration. Using (12.11.43), we obtain the values of A and B, so that the solution (12.11.44) becomes

$$\bar{u}(x, s) = \frac{U}{s} \cdot \frac{\cosh\left[(a-x)\sqrt{\frac{s}{\kappa}}\right]}{\cosh\left(a\sqrt{\frac{s}{\kappa}}\right)}. \qquad (12.11.45)$$

The inverse Laplace transform gives the solution

$$u(x, t) = U\mathcal{L}^{-1}\left\{\frac{\cosh(a-x)\sqrt{\frac{s}{\kappa}}}{s\cosh\sqrt{a\frac{s}{\kappa}}}\right\}. \qquad (12.11.46)$$

The inversion can be carried out by the Cauchy residue theorem to obtain the solution

$$u(x, t) = U\left[1 + \frac{4}{\pi}\sum_{n=1}^{\infty}\frac{(-1)^n}{(2n-1)}\cos\left\{\frac{(2n-1)(a-x)\pi}{2a}\right\}\right.$$
$$\left. \times \exp\left\{-(2n-1)^2\left(\frac{\pi}{2a}\right)^2\kappa t\right\}\right]. \qquad (12.11.47)$$

By expanding the cosine term, this becomes

$$u(x, t) = U\left[1 - \frac{4}{\pi}\sum_{n=1}^{\infty}\frac{1}{(2n-1)}\sin\left\{\left(\frac{2n-1}{2a}\right)\pi x\right\}\right.$$
$$\left. \times \exp\left\{-(2n-1)^2\left(\frac{\pi}{2a}\right)^2\kappa t\right\}\right]. \qquad (12.11.48)$$

This result can be obtained by solving the problem by the method of separation of variables.

Example 12.11.9. (Diffusion in a Finite Medium). Solve the one-dimensional diffusion equation in a finite medium $0 < z < a$, where the concentration function $C(z, t)$ satisfies the equation

$$C_t = \kappa \, C_{zz}, \qquad 0 < z < a, \ t > 0, \tag{12.11.49}$$

and the initial and boundary data

$$C(z,0) = 0 \qquad \text{for} \quad 0 < z < a, \tag{12.11.50}$$

$$C(z,t) = C_0 \qquad \text{for} \quad z = a, \ t > 0, \tag{12.11.51}$$

$$\frac{\partial C}{\partial z} = 0 \qquad \text{for} \quad z = 0, \ t > 0, \tag{12.11.52}$$

where C_0 is a constant.

Application of the Laplace transform of $C(z,t)$ with respect to t gives

$$\frac{d^2 \overline{C}}{dz^2} - \left(\frac{s}{\kappa}\right) \overline{C} = 0, \quad 0 < z < a,$$

$$\overline{C}(a,s) = \frac{C_0}{s}, \qquad \left(\frac{d\overline{C}}{dz}\right)_{z=0} = 0.$$

The solution of this differential equation system is

$$\overline{C}(z,s) = \frac{C_0 \cosh\left(z\sqrt{\frac{s}{\kappa}}\right)}{s \cosh\left(a\sqrt{\frac{s}{\kappa}}\right)}, \tag{12.11.53}$$

which, by writing $\alpha = \sqrt{\frac{s}{\kappa}}$,

$$= \frac{C_0 \left(e^{\alpha z} + e^{-\alpha z}\right)}{s \left(e^{\alpha a} + e^{-\alpha a}\right)}$$

$$= \frac{C_0}{s} \left[\exp\left\{-\alpha(a-z)\right\} + \exp\left\{-\alpha(a+z)\right\}\right] \sum_{n=0}^{\infty} (-1)^n \exp(-2n\alpha a)$$

$$= \frac{C_0}{s} \left[\sum_{n=0}^{\infty} (-1)^n \exp\left[-\alpha\left\{(2n+1)a - z\right\}\right]\right.$$

$$\left. + \sum_{n=0}^{\infty} (-1)^n \exp\left[-\alpha\left\{(2n+1)a + z\right\}\right]\right]. \tag{12.11.54}$$

Using the result (12.9.1), we obtain the final solution

$$C(z,t) = C_0 \left\{\sum_{n=0}^{\infty} (-1)^n \left[\operatorname{erfc}\left\{\frac{(2n+1)a - z}{2\sqrt{\kappa t}}\right\}\right.\right.$$

$$\left.\left. + \operatorname{erfc}\left\{\frac{(2n+1)a + z}{2\sqrt{\kappa t}}\right\}\right]\right\}. \tag{12.11.55}$$

This solution represents an infinite series of complementary error functions. The successive terms of this series are, in fact, the concentrations at depth $a - z$, $a + z$, $3a - z$, $3a + z$, $\ldots$ in the medium. The series converges rapidly for all except large values of $\left(\frac{\kappa t}{a^2}\right)$.

Example 12.11.10. (*The Wave Equation for the Transverse Vibration of a Semi-Infinite String*). Find the displacement of a semi-infinite string, which is initially at rest in its equilibrium position. At time $t = 0$, the end $x = 0$ is constrained to move so that the displacement is $u(0, t) = A f(t)$ for $t \geq 0$, where A is a constant. The problem is to solve the one-dimensional wave equation

$$u_{tt} = c^2 u_{xx}, \qquad 0 \leq x < \infty, \quad t > 0, \qquad (12.11.56)$$

with the boundary and initial conditions

$$u(x, t) = A f(t) \qquad \text{at} \quad x = 0, \quad t \geq 0, \qquad (12.11.57)$$

$$u(x, t) \to 0 \qquad \text{as} \quad x \to \infty, \quad t \geq 0, \qquad (12.11.58)$$

$$u(x, t) = 0 = \frac{\partial u}{\partial t} \qquad \text{at} \quad t = 0 \quad \text{for} \quad 0 < x < \infty. \qquad (12.11.59)$$

Application of the Laplace transform of $u(x, t)$ with respect to t gives

$$\frac{d^2 \bar{u}}{dx^2} - \frac{s^2}{c^2} \bar{u} = 0, \qquad \text{for} \quad 0 \leq x < \infty,$$

$$\bar{u}(x, s) = A \bar{f}(s) \qquad \text{at} \quad x = 0,$$

$$\bar{u}(x, s) \to 0 \qquad \text{as} \quad x \to \infty.$$

The solution of this differential equation system is

$$\bar{u}(x, s) = A \bar{f}(s) \exp\left(-\frac{xs}{c}\right). \qquad (12.11.60)$$

Inversion gives the solution

$$u(x, t) = A f\left(t - \frac{x}{c}\right) H\left(t - \frac{x}{c}\right). \qquad (12.11.61)$$

In other words, the solution is

$$u(x, t) = \begin{cases} A f\left(t - \frac{x}{c}\right), & t > \frac{x}{c} \\ 0, & t < \frac{x}{c}. \end{cases} \qquad (12.11.62)$$

This solution represents a wave propagating at a velocity c with the characteristic $x = ct$.

Example 12.11.11. (*The Cauchy–Poisson Wave Problem in Fluid Dynamics*). We consider the two-dimensional Cauchy–Poisson problem (Debnath 1994) for an inviscid liquid of infinite depth with a horizontal free surface. We assume that the liquid has constant density ρ and negligible surface tension. Waves are generated on the free surface of liquid initially at rest for time $t < 0$ by the prescribed free surface displacement at $t = 0$.

In terms of the velocity potential $\phi(x, z, t)$ and the free surface elevation $\eta(x, t)$, the linearized surface wave motion in Cartesian coordinates (x, y, z) is governed by the following equation and free surface and boundary conditions:

$$\nabla^2 \phi = \phi_{xx} + \phi_{zz} = 0, \quad -\infty < z \leq 0, \quad x \in R, \quad t < 0, \quad (12.11.63)$$

$$\left. \begin{array}{l} \phi_z - \eta_t = 0 \\ \phi_t + g\eta = 0 \end{array} \right\} \text{ on } z = 0, \ t > 0, \quad (12.11.64)$$

$$\phi_z \to 0 \quad \text{as} \quad z \to -\infty. \quad (12.11.65)$$

The initial conditions are

$$\phi(x, 0, 0) = 0 \quad \text{and} \quad \eta(x, 0) = \eta_0(x), \quad (12.11.66)$$

where $\eta_0(x)$ is a given initial elevation with compact support.

We introduce the Laplace transform with respect to t and the Fourier transform with respect to x defined by

$$\left[\bar{\tilde{\phi}}(k, z, s), \bar{\tilde{\eta}}(k, s) \right] = \frac{1}{\sqrt{2\pi}} \int_{-\infty}^{\infty} e^{-ikx} dx \int_0^{\infty} e^{-st} [\phi, \eta] \, dt. \quad (12.11.67)$$

Application of the joint transform method to the above system gives

$$\bar{\tilde{\phi}}_{zz} - k^2 \bar{\tilde{\phi}} = 0, \quad -\infty < z \leq 0, \quad (12.11.68)$$

$$\left. \begin{array}{l} \bar{\tilde{\phi}} = s\bar{\tilde{\eta}} - \tilde{\eta}_0(k) \\[2mm] s\bar{\tilde{\phi}} + g\bar{\tilde{\eta}} = 0 \end{array} \right\} \text{ on } z = 0, \quad (12.11.69)$$

$$\bar{\tilde{\phi}}_z \to 0 \quad \text{as} \quad z \to -\infty, \quad (12.11.70)$$

where

$$\tilde{\eta}_0(k) = \mathcal{F}\{\eta_0(x)\}.$$

The bounded solution of equation (12.11.68) is

$$\bar{\tilde{\phi}}(k, s) = \overline{A} \exp(|k| z), \quad (12.11.71)$$

where $\overline{A} = \overline{A}(s)$ is an arbitrary function of s.

Substituting (12.11.71) into (12.11.69) and eliminating $\bar{\tilde{\eta}}$ from the resulting equations gives $\overline{A}$. Hence, the solutions for $\bar{\tilde{\phi}}$ and $\bar{\tilde{\eta}}$ are

$$\left[\bar{\tilde{\phi}}, \bar{\tilde{\eta}} \right] = \left[-\frac{g \, \tilde{\eta}_0 \exp(|k| z)}{s^2 + \omega^2}, \frac{s \, \tilde{\eta}_0}{s^2 + \omega^2} \right], \quad (12.11.72)$$

and the associated the dispersion relation is

$$\omega^2 = g\,|k|\,. \qquad (12.11.73)$$

The inverse Laplace and Fourier transforms give the solutions

$$\phi(x,z,t) = -\frac{g}{\sqrt{2\pi}} \int_{-\infty}^{\infty} \frac{\sin\omega t}{\omega} \exp\left(ikx + |k|\,z\right) \tilde{\eta}_0(k)\,dk, \qquad (12.11.74)$$

$$\eta(x,t) = \frac{1}{\sqrt{2\pi}} \int_{-\infty}^{\infty} \tilde{\eta}_0(k)\cos\omega t\, e^{ikx} dk,$$

$$= \frac{1}{\sqrt{2\pi}} \int_{0}^{\infty} \tilde{\eta}_0(k) \left[e^{i(kx-\omega t)} + e^{i(kx+\omega t)} \right] dk, \qquad (12.11.75)$$

in which $\tilde{\eta}_0(-k) = \tilde{\eta}_0(k)$ is assumed.

Physically, the first and second integrals of (12.11.75) represent waves traveling in the positive and negative directions of x, respectively, with phase velocity $\frac{\omega}{k}$. These integrals describe superposition of all such waves over the wavenumber spectrum $0 < k < \infty$.

For the classical Cauchy–Poisson wave problem, $\eta_0(x) = a\delta(x)$, where $\delta(x)$ is the Dirac delta function, so that $\tilde{\eta}_0(k) = \left(a/\sqrt{2\pi}\right)$. Thus, solution (12.11.75) becomes

$$\eta(x,t) = \frac{a}{2\pi} \int_{0}^{\infty} \left[e^{i(kx-\omega t)} + e^{i(kx+\omega t)} \right] dk. \qquad (12.11.76)$$

The wave integrals (12.11.74) and (12.11.75) represent the exact solution for the velocity potential ϕ and the free surface elevation η for all x and $t > 0$. However, they do not lend any physical interpretations. In general, the exact evaluation of these integrals is a formidable task. So it is necessary to resort to asymptotic methods. It would be sufficient for the determination of the principal features of the wave motions to investigate (12.11.75) or (12.11.76) asymptotically for large time t and large distance x with (x,t) held fixed. The asymptotic solution for this kind of problem is available in many standard books; (for example, see Debnath 1994, p. 85). We use the stationary phase approximation of a typical wave integral (12.7.1), for $t \to \infty$, given by (12.7.8)

$$\eta(x,t) = \int_{a}^{b} F(k) \exp[it\theta(k)]dk \qquad (12.11.77)$$

$$\sim f(k_1) \left[\frac{2\pi}{t\,|\theta''(k_1)|} \right]^{\frac{1}{2}} \exp\left[i\left\{ t\theta(k_1) + \frac{\pi}{4}\,\text{sgn}\,\theta''(k_1) \right\} \right], \qquad (12.11.78)$$

where $\theta(k) = \frac{kx}{t} - \omega(k)$, $x > 0$, and $k = k_1$ is a stationary point that satisfies the equation

$$\theta'(k_1) = \frac{x}{t} - \omega'(k_1) = 0, \qquad a < k_1 < b. \qquad (12.11.79)$$

Application of (12.11.78) to (12.11.75) shows that only the first integral in (12.11.75) has a stationary point for $x > 0$. Hence, the stationary phase approximation (12.11.78) gives the asymptotic solution, as $t \to \infty$, $x > 0$,

$$\eta(x,t) \sim \left[\frac{1}{t \left| \omega''(k_1) \right|} \right]^{\frac{1}{2}} \tilde{\eta}_0(k_1) \exp\left[i\{k_1 x - t\omega(k_1)\} + \frac{i\pi}{4} \operatorname{sgn}\{-\omega''(k_1)\} \right],$$

(12.11.80)

where $k_1 = \left(gt^2/4x^2 \right)$ is the root of the equation $\omega'(k) = \frac{x}{t}$.

On the other hand, when $x < 0$, only the second integral of (12.11.75) has a stationary point $k_1 = \left(gt^2/4x^2 \right)$, and hence, the same result (12.11.78) can be used to obtain the asymptotic solution for $t \to \infty$ and $x < 0$ as

$$\eta(x,t) \sim \left[\frac{1}{t \left| \omega''(k_1) \right|} \right]^{\frac{1}{2}} \tilde{\eta}_0(k_1) \exp\left[i\{t\omega(k_1) - k_1 |x|\} + \frac{i\pi}{4} \operatorname{sgn}\omega''(k_1) \right].$$

(12.11.81)

In particular, for the classical Cauchy–Poisson solution (12.11.76), the asymptotic representation for $\eta(x,t)$ follows from (12.11.81) in the form

$$\eta(x,t) \sim \frac{at}{2\sqrt{2\pi}} \frac{\sqrt{g}}{x^{3/2}} \cos\left(\frac{gt^2}{4x} \right), \qquad gt^2 \gg 4x \qquad (12.11.82)$$

and gives a similar result for $\eta(x,t)$, when $x < 0$ and $t \to \infty$.

12.12 Hankel Transforms

We introduce the definition of the Hankel transform from the two-dimensional Fourier transform and its inverse given by

$$\mathcal{F}\{f(x,y)\} = F(k,l) = \frac{1}{2\pi} \int_{-\infty}^{\infty} \int_{-\infty}^{\infty} \exp\{-i(\boldsymbol{\kappa} \cdot \mathbf{r})\} f(x,y) \, dx \, dy,$$

(12.12.1)

$$\mathcal{F}^{-1}\{F(k,l)\} = f(x,y) = \frac{1}{2\pi} \int_{-\infty}^{\infty} \int_{-\infty}^{\infty} \exp\{i(\boldsymbol{\kappa} \cdot \mathbf{r})\} F(k,l) \, dk \, dl,$$

(12.12.2)

where $\mathbf{r} = (x,y)$ and $\boldsymbol{\kappa} = (k,l)$. Introducing polar coordinates $(x,y) = r(\cos\theta, \sin\theta)$ and $(k,l) = \kappa(\cos\phi, \sin\phi)$, we find $\boldsymbol{\kappa} \cdot \mathbf{r} = \kappa r \cos(\theta - \phi)$ and then

$$F(\kappa,\phi) = \frac{1}{2\pi} \int_{0}^{\infty} r\,dr \int_{0}^{2\pi} \exp\left[-i\kappa r \cos(\theta - \phi)\right] f(r,\theta) \, d\theta.$$

(12.12.3)

We next assume $f(r, \theta) = \exp(in\theta) f(r)$, which is not a very severe restriction, and make a change of variable $\theta - \phi = \alpha - \frac{\pi}{2}$ to reduce (12.12.3) to the form

$$F(\kappa, \phi) = \frac{1}{2\pi} \int_0^\infty rf(r) \, dr$$

$$\times \int_{\phi_0}^{2\pi+\phi_0} \exp\left[in\left(\phi - \frac{\pi}{2}\right) + i(n\alpha - \kappa r \sin\alpha)\right] d\alpha, \quad (12.12.4)$$

where $\phi_0 = \frac{\pi}{2} - \phi$.

We use the integral representation of the Bessel function of order n

$$J_n(\kappa r) = \frac{1}{2\pi} \int_{\phi_0}^{2\pi+\phi_0} \exp\left[i(n\alpha - \kappa r \sin\alpha)\right] d\alpha \qquad (12.12.5)$$

so that integral (12.12.4) becomes

$$F(\kappa, \phi) = \exp\left[in\left(\phi - \frac{\pi}{2}\right)\right] \int_0^\infty r J_n(\kappa r) f(r) \, dr \qquad (12.12.6)$$

$$= \exp\left[in\left(\phi - \frac{\pi}{2}\right)\right] \tilde{f}_n(\kappa), \qquad (12.12.7)$$

where $\tilde{f}_n(\kappa)$ is called the *Hankel transform* of $f(r)$ and is defined formally by

$$\mathcal{H}_n\{f(r)\} = \tilde{f}_n(\kappa) = \int_0^\infty r J_n(\kappa r) f(r) \, dr. \qquad (12.12.8)$$

Similarly, in terms of the polar variables with the assumption $f(x, y) = f(r, \theta) = e^{in\theta} f(r)$ and with result (12.12.7), the inverse Fourier transform (12.12.2) becomes

$$e^{in\theta} f(r) = \frac{1}{2\pi} \int_0^\infty \kappa \, d\kappa \int_0^{2\pi} \exp\left[i\kappa r \cos(\theta - \phi)\right] F(\kappa, \phi) \, d\phi$$

$$= \frac{1}{2\pi} \int_0^\infty \kappa \tilde{f}_n(\kappa) \, d\kappa \int_0^{2\pi} \exp\left[in\left(\phi - \frac{\pi}{2}\right) + i\kappa r \cos(\theta - \phi)\right] d\phi,$$

which is, by the change of variables $\theta - \phi = -\left(\alpha + \frac{\pi}{2}\right)$ and $\theta_0 = -\left(\theta + \frac{\pi}{2}\right)$,

$$= \frac{1}{2\pi} \int_0^\infty \kappa \tilde{f}_n(\kappa) \, d\kappa \int_{\theta_0}^{2\pi+\theta_0} \exp\left[in(\theta + \alpha) - i\kappa r \sin\alpha\right] d\alpha$$

$$= e^{in\theta} \int_0^\infty \kappa J_n(\kappa r) \tilde{f}_n(\kappa) \, d\kappa, \qquad \text{by (12.12.5).} \qquad (12.12.9)$$

Thus, the *inverse Hankel transform* is defined by

$$\mathcal{H}_n^{-1}\left[\tilde{f}_n(\kappa)\right] = f(r) = \int_0^\infty \kappa J_n(\kappa r) \tilde{f}_n(\kappa) \, d\kappa. \qquad (12.12.10)$$

Instead of $\tilde{f}_n(\kappa)$, we often simply write $\tilde{f}(\kappa)$ for the Hankel transform specifying the order. Integrals (12.12.8) and (12.12.10) exist for certain large classes of functions, which usually occur in physical applications.

Alternatively, the famous Hankel integral formula (Watson, 1966, p 453)

$$f(r) = \int_0^\infty \kappa J_n(\kappa r)\, d\kappa \int_0^\infty p\, J_n(\kappa p)\, f(p)\, dp, \qquad (12.12.11)$$

can be used to define the Hankel transform (12.12.8) and its inverse (12.12.10).

In particular, the Hankel transforms of zero order ($n = 0$) and of order one ($n = 1$) are often useful for the solution of problems involving Laplace's equation in an axisymmetric cylindrical geometry.

Example 12.12.1. Obtain the zero-order Hankel transforms of

$$\text{(a) } r^{-1}\exp(-ar), \qquad \text{(b) } \tfrac{\delta(r)}{r}, \qquad \text{(c) } H(a-r),$$

where $H(r)$ is the Heaviside unit step function.

(a) $\tilde{f}_0(\kappa) = \mathcal{H}_0\left\{\tfrac{1}{r}\exp(-ar)\right\} = \int_0^\infty \exp(-ar)\, J_0(\kappa r)\, dr = \frac{1}{\sqrt{\kappa^2+a^2}}.$

(b) $\tilde{f}_0(\kappa) = \mathcal{H}_0\left\{\tfrac{\delta(r)}{r}\right\} = \int_0^\infty \delta(r)\, J_0(\kappa r)\, dr = 1.$

(c) $\tilde{f}_0(\kappa) = \mathcal{H}_0\{H(a-r)\} = \int_0^a r\, J_0(\kappa r)\, dr = \tfrac{1}{\kappa^2}\int_0^{a\kappa} p\, J_0(p)\, dp$

$$= \tfrac{1}{\kappa^2}\left[p\, J_1(p)\right]_0^{a\kappa} = \tfrac{a}{\kappa}\, J_1(a\kappa).$$

Example 12.12.2. Find the first-order Hankel transform of the following functions:

$$\text{(a) } f(r) = e^{-ar}, \qquad \text{(b) } f(r) = \frac{1}{r}\, e^{-ar}.$$

(a) $\tilde{f}(\kappa) = \mathcal{H}_1\{e^{-ar}\} = \int_0^\infty r\, e^{-ar} J_1(\kappa r)\, dr = \dfrac{\kappa}{(a^2+\kappa^2)^{\frac{3}{2}}}.$

(b) $\tilde{f}(\kappa) = \mathcal{H}_1\left\{\tfrac{e^{-ar}}{r}\right\} = \int_0^\infty e^{-ar} J_1(\kappa r)\, dr = \tfrac{1}{\kappa}\left[1 - a\left(\kappa^2 + a^2\right)^{-\frac{1}{2}}\right].$

Example 12.12.3. Find the nth-order Hankel transforms of

$$\text{(a) } f(r) = r^n H(a-r), \qquad \text{(b) } f(r) = r^n \exp(-ar^2).$$

(a) $\tilde{f}(\kappa) = \mathcal{H}_n\left[r^n H(a-r)\right] = \int_0^a r^{n+1} J_n(\kappa r)\, dr = \frac{a^{n+1}}{\kappa} J_{n+1}(a\kappa)$.

(b) $\tilde{f}(\kappa) = \mathcal{H}_n\left[r^n \exp\left(-ar^2\right)\right]$

$$= \int_0^\infty r^{n+1} J_n(\kappa r) \exp\left(-ar^2\right) dr = \frac{\kappa^n}{(2a)^{n+1}} \exp\left(-\frac{\kappa^2}{4a}\right).$$

12.13 Properties of Hankel Transforms and Applications

We state the following properties of the Hankel transforms:

(i) The Hankel transform operator, $\mathcal{H}_n$ is a linear integral operator, that is,

$$\mathcal{H}_n\left\{af(r) + bg(r)\right\} = a\mathcal{H}_n\left\{f(r)\right\} + b\mathcal{H}_n\left\{g(r)\right\}$$

for any constants a and b.

(ii) The Hankel transform satisfies the *Parseval relation*

$$\int_0^\infty rf(r)g(r)\, dr = \int_0^\infty k\tilde{f}(k)\tilde{g}(k)\, dk \qquad (12.13.1)$$

where $\tilde{f}(k)$ and $\tilde{g}(k)$ are Hankel transforms of $f(r)$ and $g(r)$ respectively.

To prove (12.13.1), we proceed formally to obtain

$$\int_0^\infty k\tilde{f}(k)\tilde{g}(k)\, dk = \int_0^\infty k\tilde{f}(k)\, dk \int_0^\infty rJ_n(kr)g(r)\, dr$$

$$= \int_0^\infty rg(r)\, dr \int_0^\infty kJ_n(kr)\tilde{f}(k)\, dk$$

$$= \int_0^\infty rf(r)g(r)\, dr.$$

(iii) $\mathcal{H}_n\left\{f'(r)\right\} = \frac{k}{2n}\left[(n-1)\tilde{f}_{n+1}(k) - (n+1)\tilde{f}_{n-1}(k)\right]$
provided $rf(r)$ vanishes as $r \to 0$ and as $r \to \infty$.

(iv)

$$\mathcal{H}_n\left\{\frac{1}{r}\frac{d}{dr}\left(r\frac{df}{dr}\right) - \frac{n^2}{r^2}f(r)\right\} = -k^2\tilde{f}_n(k) \qquad (12.13.2)$$

provided both $\left(r\frac{df}{dr}\right)$ and $rf(r)$ vanish as $r \to 0$ and as $r \to \infty$.

We have, by definition,

$$\mathcal{H}_n \left\{ \frac{1}{r} \frac{d}{dr} \left(r \frac{df}{dr} \right) - \frac{n^2}{r^2} f(r) \right\}$$

$$= \int_0^\infty \frac{d}{dr} \left(r \frac{df}{dr} \right) J_n(kr) \, dr - \int_0^\infty \frac{n^2}{r^2} r f(r) J_n(kr) \, dr$$

$$= \left[r \frac{df}{dr} J_n(kr) \right]_0^\infty - \int_0^\infty k J_n'(kr) \, r \frac{df}{dr} dr$$

$$\quad - \int_0^\infty \frac{n^2}{r^2} [rf(r)] J_n(kr) \, dr, \qquad \text{by partial integration}$$

$$= - [f(r) \, kr J_n'(kr)]_0^\infty + \int_0^\infty \frac{d}{dr} [k r J_n'(kr)] f(r) \, dr$$

$$\quad - \int_0^\infty \frac{n^2}{r^2} r f(r) J_n(kr) \, dr, \qquad \text{by partial integration}$$

which is, by the given assumption and Bessel's differential equation (8.6.1),

$$= - \int_0^\infty \left(k^2 - \frac{n^2}{r^2} \right) r f(r) J_n(kr) \, dr - \int_0^\infty \frac{n^2}{r^2} [rf(r)] J_n(kr) \, dr$$

$$= -k^2 \int_0^\infty r f(r) J_n(kr) \, dr$$

$$= -k^2 \mathcal{H}_n \{ f(r) \} = -k^2 \tilde{f}_n(k).$$

(v) (*Scaling*). If $\mathcal{H}_n \{ f(r) \} = \tilde{f}_n(\kappa)$, then

$$\mathcal{H}_n \{ f(ar) \} = \frac{1}{a^2} \tilde{f}_n \left(\frac{\kappa}{a} \right), \qquad a > 0. \qquad (12.13.3)$$

Proof. We have, by definition,

$$\mathcal{H}_n \{ f(ar) \} = \int_0^\infty r J_n(\kappa r) f(ar) \, dr$$

$$= \frac{1}{a^2} \int_0^\infty s J_n \left(\frac{\kappa}{a} s \right) f(s) \, ds = \frac{1}{a^2} \tilde{f}_n \left(\frac{\kappa}{a} \right).$$

These results are used very widely in solving partial differential equations in the axisymmetric cylindrical configurations. We illustrate this point by considering the following examples of applications.

Example 12.13.1. Obtain the solution of the free vibration of a large circular membrane governed by the initial-value problem

$$\frac{\partial^2 u}{\partial r^2} + \frac{1}{r} \frac{\partial u}{\partial r} = \frac{1}{c^2} \frac{\partial^2 u}{\partial t^2}, \qquad 0 < r < \infty, \qquad t > 0, \qquad (12.13.4)$$

$$u(r,0) = f(r), \qquad u_t(r,0) = g(r), \qquad 0 \le r < \infty, \quad (12.13.5)$$

where $c^2 = (T/\rho) = \text{constant}$, T is the tension in the membrane, and ρ is the surface density of the membrane.

Application of the Hankel transform of order zero

$$\tilde{u}(k,t) = \int_0^\infty r u(r,t) J_0(kr) \, dr$$

to the vibration problem gives

$$\frac{d^2\tilde{u}}{dt^2} + k^2 c^2 \tilde{u} = 0$$

$$\tilde{u}(k,0) = \tilde{f}(k), \qquad \tilde{u}_t(k,0) = \tilde{g}(k).$$

The general solution of this transformed system is

$$\tilde{u}(k,t) = \tilde{f}(k)\cos(ckt) + \frac{\tilde{g}(k)}{ck}\sin(ckt).$$

The inverse Hankel transformation gives

$$u(r,t) = \int_0^\infty k\tilde{f}(k)\cos(ckt) J_0(kr) \, dk$$

$$+ \frac{1}{c}\int_0^\infty \tilde{g}(k)\sin(ckt) J_0(kr) \, dr. \quad (12.13.6)$$

This is the desired solution.

In particular, we consider the following initial conditions

$$u(r,0) = f(r) = \frac{A}{\left(1 + \frac{r^2}{a^2}\right)^{\frac{1}{2}}}, \qquad u_t(r,0) = g(r) = 0$$

so that $\tilde{g}(k) = 0$ and

$$\tilde{f}(k) = Aa \int_0^\infty \frac{r J_0(kr) \, dr}{(a^2 + r^2)^{\frac{1}{2}}} = \frac{Aa}{k} e^{-ak}$$

by means of Example 12.12.1(a).

Thus, solution (12.13.6) becomes

$$u(r,t) = Aa \int_0^\infty e^{-ak} J_0(kr)\cos(ckt) \, dk$$

$$= Aa \operatorname{Re} \int_0^\infty e^{-k(a+ict)} J_0(kr) \, dk$$

$$= Aa \operatorname{Re} \left\{ r^2 + (a + ict)^2 \right\}^{-\frac{1}{2}}. \quad (12.13.7)$$

Example 12.13.2. Obtain the steady-state solution of the axisymmetric acoustic radiation problem governed by the wave equation in cylindrical polar coordinates (r, θ, z):

$$c^2 \nabla^2 u = u_{tt}, \qquad 0 < r < \infty, \qquad z > 0, \qquad t > 0 \qquad (12.13.8)$$

$$u_z = f(r, t) \quad \text{on} \quad z = 0, \qquad\qquad\qquad (12.13.9)$$

where $f(r, t)$ is a given function and c is a constant. We also assume that the solution is bounded and behaves as outgoing spherical waves. This is referred to as the *Sommerfeld radiation condition.*

We seek a solution of the acoustic radiation potential $u = e^{i\omega t} \phi(r, z)$ so that ϕ satisfies the Helmholtz equation

$$\phi_{rr} + \frac{1}{r} \phi_r + \phi_{zz} + \frac{\omega^2}{c^2} \phi = 0, \quad 0 < r < \infty, \quad z > 0 \quad (12.13.10)$$

with the boundary condition representing the normal velocity prescribed on the $z = 0$ plane

$$\phi_z = f(r) \quad \text{on} \quad z = 0, \qquad\qquad\qquad (12.13.11)$$

where $f(r)$ is a known function of r.

We solve the problem by means of the zero-order Hankel transformation

$$\tilde{\phi}(k, z) = \int_0^\infty r J_0(kr) \phi(r, z) \, dr$$

so that the given differential system becomes

$$\tilde{\phi}_{zz} = \kappa^2 \tilde{\phi}, \qquad z > 0, \qquad \tilde{\phi}_z = \tilde{f}(k) \quad \text{on} \quad z = 0$$

where $\kappa = \left[k^2 - \left(\omega^2/c^2 \right) \right]^{\frac{1}{2}}$.

The solution of this system is

$$\tilde{\phi}(k, z) = -\kappa^{-1} \tilde{f}(k) e^{-\kappa z}, \qquad\qquad\qquad (12.13.12)$$

where κ is real and positive for $k > \omega/c$, and purely imaginary for $k < \omega/c$.

The inverse transformation yields the solution

$$\phi(r, z) = - \int_0^\infty \kappa^{-1} \tilde{f}(k) k J_0(kr) e^{-\kappa z} dk. \qquad (12.13.13)$$

Since the exact evaluation of this integral is difficult, we choose a simple form of $f(r)$ as

$$f(r) = A H(a - r), \qquad\qquad\qquad (12.13.14)$$

where A is a constant and $H(x)$ is the Heaviside unit step function so that

$$\tilde{f}(k) = \int_0^a k J_0(ak)\, dk = \frac{a}{k} J_1(ak).$$

Then the solution for this special case is given by

$$\phi(r, z) = -Aa \int_0^\infty \kappa^{-1} J_1(ak) J_0(kr) e^{-\kappa z} dk. \qquad (12.13.15)$$

For an asymptotic evaluation of this integral, we express it in terms of the spherical polar coordinates (R, θ, ϕ), $(x = R \sin\theta \cos\phi, \quad y = R \sin\theta \sin\phi,$ $z = R \cos\theta)$, combined with the asymptotic result

$$J_0(kr) \sim \left(\frac{2}{\pi k r}\right)^{\frac{1}{2}} \cos\left(kr - \frac{\pi}{4}\right) \qquad \text{as} \quad r \to \infty$$

so that the acoustic potential $u = e^{i\omega t} \phi$ is

$$u \sim -\frac{Aa\sqrt{2}\, e^{i\omega t}}{\sqrt{\pi R \sin\theta}} \int_0^\infty J_1(ka) \cos\left(kR\sin\theta - \frac{\pi}{4}\right) e^{-kz} dk,$$

where $z = R \cos\theta$.

This integral can be evaluated asymptotically for $R \to \infty$ by using the stationary phase approximation formula (12.7.8) to obtain

$$u \sim -\frac{Aac}{\omega R \sin\theta} J_1(k_1 a)\, e^{i(\omega t - \omega R/c)}, \qquad (12.13.16)$$

where $k_1 = \omega/c \sin\theta$ is the stationary point. This solution represents the outgoing spherical waves with constant velocity c and decaying amplitude as $R \to \infty$.

12.14 Mellin Transforms and their Operational Properties

If $f(t)$ is not necessarily zero for $t < 0$, it is possible to define the *two-sided* (or *bilateral*) Laplace transform

$$\overline{f}(p) = \int_{-\infty}^\infty e^{-pt} f(t)\, dt. \qquad (12.14.1)$$

Then replacing $f(x)$ with $e^{-cx} f(x)$ in Fourier integral formula (6.13.9), we obtain

$$e^{-cx} f(x) = \frac{1}{2\pi} \int_{-\infty}^\infty e^{-ikx} dk \int_{-\infty}^\infty f(t)\, e^{-t(c-ik)} dt,$$

or

$$f(x) = \frac{1}{2\pi} \int_{-\infty}^{\infty} e^{x(c-ik)} dk \int_{-\infty}^{\infty} f(t) e^{-t(c-ik)} dt.$$

Making a change of variable $p = c - ik$ and using definition (12.14.1), we obtain the formal inverse transform after replacing x by t as

$$f(t) = \frac{1}{2\pi i} \int_{c-i\infty}^{c+i\infty} e^{pt} \overline{f}(p) \, dp, \qquad c > 0. \qquad (12.14.2)$$

If we put $e^{-t} = x$ into (12.14.1) with $f(-\log x) = g(x)$ and $\overline{f}(p) \equiv G(p)$, then (12.14.1)–(12.14.2) become

$$G(p) = \mathcal{M}\{g(x)\} = \int_0^{\infty} x^{p-1} g(x) \, dx, \qquad (12.14.3)$$

$$g(x) = \mathcal{M}^{-1}\{G(p)\} = \frac{1}{2\pi i} \int_{c-i\infty}^{c+i\infty} x^{-p} G(p) \, dp. \qquad (12.14.4)$$

The function $G(p)$ is called the *Mellin transform* of $g(x)$ defined by (12.14.3). The *inverse Mellin transformation* is given by (12.14.4).

We state the following operational properties of the Mellin transforms:

(i) Both $\mathcal{M}$ and $\mathcal{M}^{-1}$ are linear integral operators,

(ii) $\mathcal{M}[f(ax)] = a^{-p} F(p)$,

(iii) $\mathcal{M}[x^a f(x)] = F(p+a)$,

(iv) $\mathcal{M}[f'(x)] = -(p-1) F(p-1)$, provided that $[f(x) x^{p-1}]_0^{\infty} = 0$,

$\mathcal{M}[f''(x)] = (p-1)(p-2) F(p-2)$,

$\cdots \quad \cdots \quad \cdots \quad \cdots \quad \cdots \quad \cdots$,

$\mathcal{M}[f^{(n)}(x)] = \frac{(-1)^n \Gamma(p)}{\Gamma(p-n)} F(p-n)$,

provided $\lim_{x \to 0} x^{p-r-1} f^{(r)}(x) = 0$, $r = 0, 1, 2, \ldots, (n-1)$,

(v) $\mathcal{M}\{xf'(x)\} = -p\mathcal{M}\{f(x)\} = -pF(p)$, provided that $[x^p f(x)]_0^{\infty} = 0$,

$\mathcal{M}\{x^2 f''(x)\} = (-1)^2 (p+p^2) F(p)$,

$\cdots \quad \cdots \quad \cdots \quad \cdots \quad \cdots \quad \cdots$,

$\mathcal{M}\{x^n f^{(n)}(x)\} = (-1)^n \frac{\Gamma(p+n)}{\Gamma(p)} F(p)$.

(vi) $\mathcal{M}\left\{\left(x\frac{d}{dx}\right)^n f(x)\right\} = (-1)^n p^n F(p)$, $n = 1, 2, \ldots$.

(vii) Convolution Property

$$\mathcal{M}\left[\int_0^\infty f(x\xi)\, g(\xi)\, d\xi\right] = F(p)\, G(1-p),$$

$$\mathcal{M}\left[\int_0^\infty f\left(\frac{x}{\xi}\right) g(\xi)\, \frac{d\xi}{\xi}\right] = F(p)\, G(p).$$

(viii) If $F(p) = \mathcal{M}(f(x))$ and $G(p) = \mathcal{M}(g(x))$, then, the following convolution result holds:

$$\mathcal{M}[f(x)\, g(x)] = \frac{1}{2\pi i}\int_{c-i\infty}^{c+i\infty} F(s)\, G(p-s)\, ds.$$

In particular, when $p = 1$, we obtain the Parseval formula

$$\int_0^\infty f(x)\, g(x)\, dx = \frac{1}{2\pi i}\int_{c-i\infty}^{c+i\infty} F(s)\, G(1-s)\, ds.$$

The reader is referred to Debnath (1995) for other properties of the Mellin transform.

Example 12.14.1. Show that the Mellin transform of $(1+x)^{-1}$ is $\pi\,\mathrm{cosec}\,\pi p$, $0 < \mathrm{Re}\,p < 1$.

We consider the standard definite integral

$$\int_0^1 (1-t)^{m-1}\, t^{p-1}\, dt = \frac{\Gamma(m)\,\Gamma(p)}{\Gamma(m+p)}, \qquad \mathrm{Re}\,p > 0, \quad \mathrm{Re}\,m > 0,$$

and then change the variable $t = \frac{x}{1+x}$ to obtain

$$\int_0^\infty \frac{x^{p-1}\, dx}{(1+x)^{m+p}} = \frac{\Gamma(m)\,\Gamma(p)}{\Gamma(m+p)}.$$

If we replace $m + p$ by α, this gives

$$\mathcal{M}\left[(1+x)^{-\alpha}\right] = \frac{\Gamma(p)\,\Gamma(\alpha-p)}{\Gamma(\alpha)}.$$

Setting $\alpha = 1$ and using the result

$$\Gamma(p)\,\Gamma(1-p) = \pi\,\mathrm{cosec}\,\pi p, \qquad 0 < \mathrm{Re}\,p < 1,$$

we obtain

$$\mathcal{M}\left[(1+x)^{-1}\right] = \pi\,\mathrm{cosec}\,\pi p, \qquad 0 < \mathrm{Re}\,p < 1.$$

Example 12.14.2. Obtain the solution of the boundary-value problem

$$x^2 u_{xx} + x u_x + u_{yy} = 0, \qquad 0 \le x < \infty, \quad 0 < y < 1,$$

$$u(x,0) = 0, \qquad \text{and} \quad u(x,1) = \begin{cases} A, & 0 \le x \le 1 \\ 0, & x > 1 \end{cases},$$

where A is constant.

We apply the Mellin transform

$$U(p,y) = \int_0^\infty x^{p-1} u(x,y)\, dx$$

to reduce the system to the form

$$U_{yy} + p^2 U = 0, \qquad 0 < y < 1,$$

$$U(p,0) = 0, \quad \text{and} \quad U(p,1) = A \int_0^1 x^{p-1} dx = \frac{A}{p}.$$

The solution of this differential system is

$$U(p,y) = \frac{A}{p} \frac{\sin(py)}{\sin p}, \qquad 0 < \operatorname{Re} p < 1.$$

The inverse Mellin transform gives

$$u(x,y) = \frac{A}{2\pi i} \int_{c-i\infty}^{c+i\infty} \frac{x^{-p}}{p} \frac{\sin(py)}{\sin p}\, dp,$$

where $U(p,y)$ is analytic in a vertical strip $0 < \operatorname{Re} p < \pi$ and hence, $0 < c < \pi$. The integrand has simple poles at $p = r\pi$, $r = 1, 2, 3, \ldots$ which lie inside a semi-circular contour in the right half-plane. Application of the theory of residues gives the solution for $x > 1$

$$u(x,y) = \frac{A}{\pi} \sum_{r=1}^\infty \frac{(-1)^r x^{-r\pi}}{r} \sin(r\pi y).$$

Example 12.14.3. Find the Mellin transform of the *Weyl fractional integral*

$$w(x,\alpha) = W_\alpha\left[f(\xi)\right] = \frac{1}{\Gamma(\alpha)} \int_x^\infty f(\xi)(\xi - x)^{\alpha-1}\, d\xi.$$

We rewrite the Weyl integral by setting

$$k(x) = x^\alpha f(x), \qquad g(x) = \frac{1}{\Gamma(\alpha)}(1-x)^{\alpha-1} H(1-x),$$

so that

$$\omega\left(x,\alpha\right)=\int_{0}^{\infty}k\left(\xi\right)g\left(\frac{x}{\xi}\right)\frac{d\xi}{\xi}.$$

The Mellin transform of this result is obtained by the convolution property (vii):

$$\Omega\left(p,a\right)=K\left(p\right)G\left(p\right),$$

where $K\left(p\right)=\mathcal{M}\left[k\left(x\right)\right]=\mathcal{M}\left[x^{\alpha}f\left(x\right)\right]=F\left(p+\alpha\right)$ and

$$G\left(p\right)=\frac{1}{\Gamma\left(\alpha\right)}\int_{0}^{1}\left(1-x\right)^{\alpha-1}x^{p-1}dx=\frac{\Gamma\left(p\right)}{\Gamma\left(p+\alpha\right)}.$$

Thus,

$$\Omega\left(p,a\right)=\mathcal{M}\left[W_{\alpha}f\left(\xi\right)\right]=\frac{\Gamma\left(p\right)}{\Gamma\left(p+\alpha\right)}F\left(p+\alpha\right).\qquad(12.14.5)$$

If α is complex with a positive real part such that $n-1<\mathrm{Re}\,\alpha<n$ where n is a positive integer, the *fractional derivative of order* α of a function $f\left(x\right)$ is defined by the formula

$$D_{\infty}^{\alpha}f\left(x\right)=\frac{d^{n}}{dx^{n}}W_{n-\alpha}f\left(x\right)=\frac{1}{\Gamma\left(n-\alpha\right)}\frac{d^{n}}{dx^{n}}\int_{x}^{\infty}f\left(\xi\right)\left(\xi-x\right)^{n-\alpha-1}d\xi.$$

$$(12.14.6)$$

The Mellin transform of this fractional derivative can be given by using operational property (iv) and (12.14.5)

$$\mathcal{M}\left[D_{\infty}^{\alpha}f\left(x\right)\right]=\frac{\left(-1\right)^{n}\Gamma\left(p\right)}{\Gamma\left(p-n\right)}\Omega\left(p-n,n-\alpha\right),$$

$$=\frac{\left(-1\right)^{n}\Gamma\left(p\right)}{\Gamma\left(p-a\right)}F\left(p-\alpha\right).$$

This is an obvious generalization of the third result listed under (iv).

12.15 Finite Fourier Transforms and Applications

The finite Fourier transforms are often used in determining solutions of nonhomogeneous problems. These finite transforms, namely the sine and cosine transforms, follow immediately from the theory of Fourier series.

Let $f\left(x\right)$ be a piecewise continuous function in a finite interval, say, $\left(0,\pi\right)$. This interval is introduced for convenience, and the change of interval can be made without difficulty.

The finite Fourier sine transform denoted by $F_{s}\left(n\right)$ of the function $f\left(x\right)$ may be defined by

$$F_s(n) = \mathcal{F}_s[f(x)] = \frac{2}{\pi} \int_0^\pi f(x) \sin nx \, dx, \qquad n = 1, 2, 3, \ldots, \quad (12.15.1)$$

and the inverse of the transform follows at once from the Fourier sine series; that is

$$f(x) = \sum_{n=1}^\infty F_s(n) \sin nx. \qquad (12.15.2)$$

The finite Fourier cosine transform $F_c(n)$ of $f(x)$ may be defined by

$$F_c(n) = \mathcal{F}_c[f(x)] = \frac{2}{\pi} \int_0^\pi f(x) \cos nx \, dx, \qquad n = 0, 1, 2, \ldots. \quad (12.15.3)$$

The inverse of the Fourier cosine transform is given by

$$f(x) = \frac{F_c(0)}{2} + \sum_{n=1}^\infty F_c(n) \cos nx. \qquad (12.15.4)$$

Theorem 12.15.1. *Let $f'(x)$ be continuous and $f''(x)$ be piecewise continuous in $[0, \pi]$. If $F_s(n)$ is the finite Fourier sine transform of $f(x)$, then*

$$\mathcal{F}_s[f''(x)] = \frac{2n}{\pi}[f(0) - (-1)^n f(\pi)] - n^2 F_s(n). \qquad (12.15.5)$$

Proof. By definition

$$\mathcal{F}_s[f''(x)] = \frac{2}{\pi} \int_0^\pi f''(x) \sin nx \, dx$$

$$= \frac{2}{\pi}[f'(x) \sin nx]_0^\pi - \frac{2n}{\pi} \int_0^\pi f'(x) \cos nx \, dx$$

$$= -\frac{2n}{\pi}[f(x) \cos nx]_0^\pi - \frac{2n^2}{\pi} \int_0^\pi f(x) \sin nx \, dx$$

$$= -\frac{2n}{\pi}[f(\pi)(-1)^n - f(0)] - n^2 F_s(n).$$

The transforms of higher-order derivatives can be derived in a similar manner.

Theorem 12.15.2. *Let $f'(x)$ be continuous and $f''(x)$ be piecewise continuous in $[0, \pi]$. If $F_c(n)$ is the finite Fourier cosine transform of $f(x)$, then*

$$\mathcal{F}_c[f''(x)] = \frac{2}{\pi}[(-1)^n f'(\pi) - f'(0)] - n^2 F_c(n). \qquad (12.15.6)$$

The proof is left to the reader.

Example 12.15.1. Consider the motion of a string of length π due to a force acting on it. Let the string be fixed at both ends. The motion is thus governed by

$$u_{tt} = c^2 u_{xx} + f(x,t), \qquad 0 < x < \pi, \quad t > 0,$$
$$u(x,0) = 0, \qquad u_t(x,0) = 0, \qquad 0 < x < \pi, \qquad (12.15.7)$$
$$u(0,t) = 0, \qquad u(\pi,t) = 0, \qquad t > 0.$$

Applying the finite Fourier sine transform to the equation of motion with respect to x gives

$$\mathcal{F}_s\left[u_{tt} - c^2 u_{xx} - f(x,t)\right] = 0.$$

Due to its linearity (see Problem 52, 12.18 Exercises), this can be written in the form

$$\mathcal{F}_s[u_{tt}] - c^2 \mathcal{F}_s[u_{xx}] = \mathcal{F}_s[f(x,t)]. \qquad (12.15.8)$$

Let $U(n,t)$ be the finite Fourier sine transform of $u(x,t)$. Then we have

$$\mathcal{F}_s[u_{tt}] = \frac{2}{\pi} \int_0^\pi u_{tt} \sin nx \, dx$$
$$= \frac{d^2}{dt^2}\left[\frac{2}{\pi}\int_0^\pi u(x,t) \sin nx \, dx\right] = \frac{d^2 U_s(n,t)}{dt^2}.$$

We also have, from Theorem 12.15.1,

$$\mathcal{F}_s[u_{xx}] = \frac{2n}{\pi}\left[u(0,t) - (-1)^n u(\pi,t)\right] - n^2 U_s(n,t).$$

Because of the boundary conditions

$$u(0,t) = u(\pi,t) = 0,$$

$\mathcal{F}_s[u_{xx}]$ becomes

$$\mathcal{F}_s[u_{xx}] = -n^2 U_s(n,t).$$

If we denote the finite Fourier sine transform of $f(x,t)$ by

$$F_s(n,t) = \frac{2}{\pi}\int_0^\pi f(x,t) \sin nx \, dx,$$

then, equation (12.15.8) takes the form

$$\frac{d^2 U_s}{dt^2} + n^2 c^2 U_s = F_s(n,t).$$

This is a second-order ordinary differential equation, the solution of which is given by

$$U_s(n,t) = A\cos nct + B\sin nct + \frac{1}{nc}\int_0^t F_s(n,\tau)\sin nc\,(t-\tau)\,d\tau.$$

Applying the initial conditions

$$\mathcal{F}_s[u(x,0)] = \frac{2}{\pi}\int_0^\pi u(x,0)\sin nx\,dx = U_s(n,0) = 0,$$

and

$$\mathcal{F}_s[u_t(x,0)] = \frac{d}{dt}U_s(n,0) = 0,$$

we have

$$U_s(n,t) = \frac{1}{nc}\int_0^t F_s(n,\tau)\sin nc\,(t-\tau)\,d\tau.$$

Thus, the inverse transform of $U_s(n,t)$ is

$$u(x,t) = \sum_{n=1}^\infty U_s(n,t)\sin nx$$

$$= \sum_{n=1}^\infty \left[\frac{1}{nc}\int_0^t F_s(n,\tau)\sin nc\,(t-\tau)\,d\tau\right]\sin nx.$$

In the case when $f(x,t) = h$ which is a constant, then

$$\mathcal{F}_s[h] = \frac{2}{\pi}\int_0^\pi h\sin nx\,dx = \frac{2h}{n\pi}\left[1-(-1)^n\right].$$

Now, we evaluate

$$U_s(n,t) = \frac{1}{nc}\int_0^t \frac{2h}{n\pi}\left[1-(-1)^n\right]\sin nc\,(t-\tau)\,d\tau$$

$$= \frac{2h}{n^3\pi c^2}\left[1-(-1)^n\right](1-\cos nct).$$

Hence, the solution is given by

$$u(x,t) = \frac{2h}{\pi c^2}\sum_{n=1}^\infty \frac{[1-(-1)^n]}{n^3}(1-\cos nct)\sin nx.$$

Example 12.15.2. Find the temperature distribution in a rod of length π. The heat is generated in the rod at the rate $g(x,t)$ per unit time. The ends are insulated. The initial temperature distribution is given by $f(x)$.

The problem is to find the temperature function $u(x,t)$ of the system

$$u_t = u_{xx} + g(x,t), \quad 0 < x < \pi, \quad t > 0,$$
$$u(x,0) = f(x), \quad 0 \le x \le \pi,$$
$$u_x(0,t) = 0, \quad u_x(\pi,t) = 0, \quad t \ge 0.$$

Let $U_s(n,t)$ be the finite Fourier cosine transform of $u(x,t)$. As before, transformation of the heat equation with respect to x, using the boundary conditions, yields

$$\frac{dU_s}{dt} = -n^2 U_s + G_s(n,t),$$

where

$$G_s(n,t) = \frac{2}{\pi} \int_0^\pi g(x,t) \cos nx\, dx.$$

Rewriting this equation, we obtain

$$\frac{d}{dt}\left(e^{n^2 t} U_s\right) = G_s\, e^{n^2 t}.$$

Thus, the solution is

$$U_s(n,t) = \int_0^t e^{-n^2(t-\tau)} G_s(n,\tau)\, d\tau + A\, e^{-n^2 t}.$$

Transformation of the initial condition gives

$$U_s(n,0) = \frac{2}{\pi} \int_0^\pi u(x,0) \cos nx\, dx = \frac{2}{\pi} \int_0^\pi f(x) \cos nx\, dx.$$

Hence, $U_s(n,t)$ takes the form

$$U_s(n,t) = \int_0^t e^{-n^2(t-\tau)} G_s(n,\tau)\, d\tau + U_s(n,0)\, e^{-n^2 t}.$$

The solution $u(x,t)$, therefore, is given by

$$u(x,t) = \frac{U_s(0,0)}{2} + \sum_{n=1}^\infty U_s(n,t) \cos nx.$$

Example 12.15.3. A rod with diffusion constant κ contains a fuel which produces neutrons by fission. The ends of the rod are perfectly reflecting. If the initial neutron distribution is $f(x)$, find the neutron distribution $u(x,t)$ at any subsequent time t.

The problem is governed by

$$u_t = \kappa u_{xx} + bu,$$
$$u(x,0) = f(x), \qquad\qquad 0 < x < l, \qquad t > 0,$$
$$u_x(0,t) = u_x(l,t) = 0.$$

If $U(n,t)$ is the finite Fourier cosine transform of $u(x,t)$, then by transforming the equation and using the boundary conditions, we obtain

$$U_t + \left(\kappa n^2 - b \right) U = 0.$$

The solution of this equation is

$$U(n,t) = C e^{-\left(\kappa n^2 - b \right) t}$$

where C is a constant. Then applying the initial condition, we obtain

$$U(n,t) = U(n,0)\, e^{-\left(\kappa n^2 - b \right) t},$$

where

$$U(n,0) = \frac{2}{l} \int_0^l f(x) \cos nx \, dx.$$

Thus, the solution takes the form

$$u(x,t) = \frac{U(0,0)}{2} + \sum_{n=1}^{\infty} U(n,t) \cos nx.$$

If for instance $f(x) = x$ in $0 < x < \pi$, then, $U(0,0) = \pi$, and

$$U(n,0) = \frac{2}{n^2 \pi} \left[(-1)^n - 1 \right], \qquad n = 1,2,3,\ldots$$

the solution is given by

$$u(x,t) = \frac{\pi}{2} + \sum_{n=1}^{\infty} \frac{2}{n^2 \pi} \left[(-1)^n - 1 \right] \exp \left\{ - \left(\kappa n^2 - b \right) t \right\} \cos nx.$$

12.16 Finite Hankel Transforms and Applications

The Fourier–Bessel series representation of a function $f(r)$ defined in $0 \le r \le a$ can be stated in the following theorem:

Theorem 12.16.1. *If $f(r)$ is defined in $0 \le r \le a$ and*

$$F_n(k_i) = \mathcal{H}_n \{ f(r) \} = \int_0^a r f(r) J_n(r k_i) \, dr, \qquad (12.16.1)$$

then

$$f(r) = \mathcal{H}_n^{-1} \{ F_n(k_i) \} = \frac{2}{a^2} \sum_{i=1}^{\infty} F_n(k_i) \frac{J_n(r k_i)}{J_{n+1}^2(a k_i)}, \qquad (12.16.2)$$

where k_i $(0 < k_1 < k_2 < \ldots)$ are the roots of the equation

$$J_n(a k_i) = 0.$$

The function $F_n(k_i)$ defined by (12.16.1) is called the nth-order *finite Hankel transform* of $f(r)$, and the *inverse Hankel transform* is defined by (12.16.2). In particular, when $n = 0$, the finite Hankel transform of order zero and its inverse are defined by the integral and series respectively

$$F(k_i) = \mathcal{H}_0\{f(r)\} = \int_0^a rf(r) J_0(rk_i)\, dr, \qquad (12.16.3)$$

$$f(r) = \mathcal{H}_0^{-1}\{F(k_i)\} = \frac{2}{a^2} \sum_{i=1}^{\infty} F(k_i) \frac{J_0(rk_i)}{J_1^2(ak_i)}. \qquad (12.16.4)$$

Example 12.16.1. Find the nth-order finite Hankel transform of $f(r) = r^n$.
We have the following result for the Bessel function

$$\int_0^a r^{n+1} J_n(k_i r)\, dr = \frac{a^{n+1}}{k_i} J_{n+1}(ak_i),$$

so that

$$\mathcal{H}_n\{r^n\} = \frac{a^{n+1}}{k_i} J_{n+1}(ak_i).$$

In particular, when $n = 0$

$$\mathcal{H}_0\{1\} = \frac{a}{k_i} J_1(ak_i)$$

or, equivalently,

$$1 = \mathcal{H}_0^{-1}\left\{\frac{a}{k_i} J_1(ak_i)\right\} = \frac{2}{a} \sum_{i=1}^{\infty} \frac{J_0(rk_i)}{k_i J_1(ak_i)}.$$

Example 12.16.2. Find $\mathcal{H}_0\{(a^2 - r^2)\}$.
We have by definition

$$\mathcal{H}_0\{(a^2 - r^2)\} = \int_0^a (a^2 - r^2) r J_0(k_i r)\, dr$$

$$= \frac{4a}{k_i^3} J_1(ak_i) - \frac{2a^2}{k_i^2} J_0(ak_i),$$

where k_i is a root of $J_0(ax) = 0$. Hence,

$$\mathcal{H}_0\{(a^2 - r^2)\} = \frac{4a}{k_i^3} J_1(ak_i).$$

We state the following operational properties of the finite Hankel transform:

(i) $\mathcal{H}_n \left\{ \dfrac{df}{dr} \right\} = \dfrac{k_i}{2n} \left[(n-1) H_{n+1} \{ f(r) \} - (n+1) H_{n-1} \{ f(r) \} \right],$

(ii) $\mathcal{H}_1 \left\{ \dfrac{df}{dr} \right\} = -k_i H_0 \{ f(r) \},$

(iii) $\mathcal{H}_n \left[\dfrac{1}{r} \dfrac{d}{dr} \{ r f'(r) \} - \dfrac{n^2}{r^2} f(r) \right] = -ak_i f(a) J_n'(ak_i) - k_i^2 H_n \{ f(r) \},$

$$(12.16.5)$$

(iv) $\mathcal{H}_0 \left[f''(r) + \dfrac{1}{r} f'(r) \right] = ak_i f(a) J_1(ak_i) - k_i^2 H_0 \{ f(r) \}.$ $(12.16.6)$

Example 12.16.3. Find the solution of the axisymmetric heat conduction equation

$$u_t = \kappa \left(u_{rr} + \frac{1}{r} u_r \right), \qquad 0 < r < a, \quad t > 0,$$

with the boundary and initial conditions

$$u(r,t) = f(t) \quad \text{on} \quad r = a, \quad t \ge 0,$$
$$u(r,0) = 0, \qquad 0 \le r \le a,$$

where $u(r,t)$ represents the temperature distribution.

We apply the finite Hankel transform defined by

$$U(k,t) = \mathcal{H}_0 \{ u(r,t) \} = \int_0^a r J_0(k_i r) u(r,t) \, dr$$

so that the given equation with the boundary condition becomes

$$\frac{dU}{dt} + \kappa k_i^2 U = \kappa a k_i J_1(ak_i) f(t).$$

The solution of this equation with the transformed initial condition is

$$U(k,t) = a\kappa k_i J_1(ak_i) \int_0^t f(\tau) e^{-\kappa k_i^2 (t-\tau)} d\tau.$$

The inverse transformation gives the solution as

$$u(r,t) = \left(\frac{2\kappa}{a} \right) \sum_{i=1}^{\infty} \frac{k_i J_0(rk_i)}{J_1(ak_i)} \int_0^t f(\tau) e^{-\kappa k_i^2 (t-\tau)} d\tau. \quad (12.16.7)$$

In particular, if $f(t) = T_0 = \text{constant}$, then this solution becomes

$$u(r,t) = \left(\frac{2T_0}{a} \right) \sum_{i=1}^{\infty} \frac{J_0(rk_i)}{k_i J_1(ak_i)} \left(1 - e^{-\kappa k_i^2 t} \right). \quad (12.16.8)$$

In view of Example 12.16.1, result (12.16.8) becomes

$$u\left(r,t\right) = T_0 \left[1 - \frac{2}{a} \sum_{i=1}^{\infty} \frac{J_0\left(rk_i\right)}{k_i J_1\left(ak_i\right)} e^{-\kappa k_i^2 t}\right]. \qquad (12.16.9)$$

This solution consists of the steady-state term, and the transient term which tends to zero as $t \to \infty$. Consequently, the steady-state is attained in the limit $t \to \infty$.

Example 12.16.4. (*Unsteady Viscous Flow in a Rotating Cylinder*). The axisymmetric unsteady motion of a viscous fluid in an infinitely long circular cylinder of radius a is governed by

$$v_t = \nu\left(v_{rr} + \frac{1}{r}v_r - \frac{v}{r^2}\right), \qquad 0 \le r \le a, \quad t > 0,$$

where $v = v\left(r,t\right)$ is the tangential fluid velocity and ν is the kinematic viscosity of the fluid.

The cylinder is at rest until at $t = 0+$ it is caused to rotate, so that the boundary and initial conditions are

$$v\left(r,t\right) = a\Omega f\left(t\right) H\left(t\right) \quad \text{on} \quad r = a,$$
$$v\left(r,t\right) = 0, \quad \text{at} \quad t = 0 \quad \text{for} \quad r < a,$$

where $f\left(t\right)$ is a physically realistic function of t.

We solve the problem by using the joint Laplace and finite Hankel transforms of order one defined by

$$\overline{V}\left(k_m, s\right) = \int_0^a r J_1\left(rk_m\right) dr \int_0^{\infty} e^{-st} v\left(r,t\right) dt,$$

where $\overline{V}\left(k_m, s\right)$ is the Laplace transform of $V\left(k_m, t\right)$, and k_m are the roots of equation $J_1\left(ak_m\right) = 0$.

Application of the transform yields

$$\left(\frac{s}{\nu}\right)\overline{V}\left(k_m, s\right) = -ak_m \overline{V}\left(a, s\right) J_1'\left(ak_m\right) - k_m^2 \overline{V}\left(k_m, s\right),$$
$$\overline{V}\left(a, s\right) = a\,\Omega\overline{f}\left(s\right),$$

where $\overline{f}\left(s\right)$ is the Laplace transform of $f\left(t\right)$.

The solution of this system is

$$\overline{V}\left(k_m, s\right) = -\frac{a^2 \nu k_m \Omega\, \overline{f}\left(s\right) J_1'\left(ak_m\right)}{\left(s + \nu k_m^2\right)}.$$

The joint inverse transformation gives

$$v(r,t) = \frac{2}{a^2} \sum_{m=1}^{\infty} \frac{J_1(rk_m)}{[J_1'(ak_m)]^2} \frac{1}{2\pi i} \int_{c-i\infty}^{c+i\infty} e^{st}\,\overline{V}(k_m,s)\,ds$$

$$= -2\nu\Omega \sum_{m=1}^{\infty} \frac{k_m J_1(rk_m)}{J_1'(ak_m)} \frac{1}{2\pi i} \int_{c-i\infty}^{c+i\infty} \frac{e^{st}\,\overline{f}(s)}{(s+\nu k_m^2)}\,ds$$

$$= -2\nu\Omega \sum_{m=1}^{\infty} \frac{k_m J_1(rk_m)}{J_1'(ak_m)} \int_0^t f(\tau)\exp\left[-\nu k_m^2(t-\tau)\right]d\tau,$$

by the Convolution Theorem of the Laplace transform.

In particular, when $f(t) = \cos\omega t$, the velocity field becomes

$$v(r,t) = -2\nu\Omega \sum_{m=1}^{\infty} \frac{k_m J_1(rk_m)}{J_1'(ak_m)} \int_0^t \cos\omega\tau\,\exp\left[-\nu k_m^2(t-\tau)\right]d\tau$$

$$= 2\nu\Omega \sum_{m=1}^{\infty} \frac{k_m J_1(rk_m)}{J_1'(ak_m)}$$

$$\times \left[\frac{\nu k_m^2 \exp\left(-\nu t k_m^2\right) - \left(\omega\sin\omega t + \nu k_m^2 \cos\omega t\right)}{(\omega^2 + \nu^2 k_m^4)}\right]$$

$$= v_{st}(s,t) + v_{tr}(r,t),\qquad\qquad (12.16.10)$$

where the steady-state flow field v_{st} and the transient flow field v_{tr} are given by

$$v_{st}(r,t) = -2\nu\Omega \sum_{m=1}^{\infty} \frac{k_m J_1(rk_m)\left(\omega\sin\omega t + \nu k_m^2 \cos\omega t\right)}{J_1'(ak_m)(\omega^2 + \nu^2 k_m^4)},\quad (12.16.11)$$

$$v_{tr}(r,t) = 2\nu^2\Omega \sum_{m=1}^{\infty} \frac{J_1(rk_m)\,k_m^3\,e^{-\nu t k_m^2}}{J_1'(ak_m)(\omega^2 + \nu^2 k_m^4)}.\qquad (12.16.12)$$

Thus, the solution consists of the steady-state and transient components. In the limit $t \to \infty$, the latter decays to zero, and the ultimate steady-state is attained and is given by (12.16.11), which has the form

$$v_{st}(r,t) = -2\nu\Omega \sum_{m=1}^{\infty} \frac{k_m J_1(rk_m)\cos(\omega t - \alpha)}{J_1'(ak_m)(\omega^2 + \nu^2 k_m^4)^{\frac{1}{2}}},\qquad (12.16.13)$$

where $\tan\alpha = \left(\omega/\nu k_m^2\right)$. Thus, we see that the steady solution suffers from a phase change of $\alpha + \pi$. The amplitude of the motion remains bounded for all values of ω.

The frictional couple exerted on the fluid by unit length of the cylinder of radius $r = a$ is given by

$$C = \int_0^{2\pi} [P_{r\theta}]_{r=a}\,a^2\,d\theta = 2\pi a^2\,[P_{r\theta}]_{r=a},$$

where $P_{r\theta} = \mu r \left(d/dr \right) \left(v/r \right)$ with $\mu = \nu \rho$ calculated from (12.16.10).
Thus,

$$
C =
$$
$$
4\pi\mu\Omega \left[-a \sum_{m=1}^{\infty} \frac{\nu k_m J_1 \left(ak_m \right) \left[\nu k_m^2 \exp \left(-\nu t k_m^2 \right) - \left(\nu k_m^2 \cos \omega t + \omega \sin \omega t \right) \right]}{\left(\omega^2 + \nu^2 k_m^4 \right) J_1' \left(ak_m \right)} \right.
$$
$$
\left. + a^2 \sum_{m=1}^{\infty} \frac{\nu k_m^2 \left[\nu k_m^2 \exp \left(-\nu t k_m^2 \right) - \left(\nu k_m^2 \cos \omega t + \omega \sin \omega t \right) \right]}{\left(\omega^2 + \nu^2 k_m^4 \right) J_1' \left(ak_m \right)} \right]. \qquad (12.16.14)
$$

A particular case corresponding to $\omega = 0$ is of special interest. The solution assumes the form

$$
v \left(r, t \right) = -2\Omega \sum_{m=1}^{\infty} \frac{J_1 \left(rk_m \right) \left(1 - e^{-\nu t k_m^2} \right)}{k_m J_1' \left(ak_m \right)} = v_{st} + v_{tr}, \qquad (12.16.15)
$$

where v_{st} and v_{tr} represent the steady-state and the transient flow fields respectively given by

$$
v_{st} \left(r, t \right) = -2\Omega \sum_{m=1}^{\infty} \frac{J_1 \left(rk_m \right)}{k_m J_1' \left(ak_m \right)}, \qquad (12.16.16)
$$

$$
v_{tr} \left(r, t \right) = 2\Omega \sum_{m=1}^{\infty} \frac{J_1 \left(rk_m \right)}{k_m J_1' \left(ak_m \right)} e^{-\nu t k_m^2}. \qquad (12.16.17)
$$

It follows from (12.16.16) that

$$
v_{st} \left(r, t \right) = 2\Omega \sum_{m=1}^{\infty} \frac{J_1 \left(rk_m \right)}{k_m J_2 \left(ak_m \right)}
$$
$$
= \frac{2\Omega}{a^2} \sum_{m=1}^{\infty} \frac{a^2}{k_m} J_2 \left(ak_m \right) \cdot \frac{J_1 \left(rk_m \right)}{J_2^2 \left(ak_m \right)}
$$
$$
= \Omega \mathcal{H}_1^{-1} \left\{ \frac{a^2}{k_m} J_2 \left(ak_m \right) \right\} = \Omega r, \quad \text{by Example 12.16.1.}
$$

Thus, the steady-state solution has the closed form

$$
v_{st} \left(r, t \right) = r\Omega. \qquad (12.16.18)
$$

This represents the rigid body rotation of the fluid inside the cylinder. Thus, the final form of (12.16.15) is given by

$$
v \left(r, t \right) = r\Omega - 2\Omega \sum_{m=1}^{\infty} \frac{J_1 \left(rk_m \right) e^{-\nu t k_m^2}}{k_m J_2 \left(ak_m \right)}. \qquad (12.16.19)
$$

In the limit $t \to \infty$, the transients die out and the ultimate steady-state is attained as the rigid body rotation about the axis of the cylinder.

12.17 Solution of Fractional Partial Differential Equations

(a) Fractional Diffusion Equation
The fractional diffusion equation is given by

$$\frac{\partial^\alpha u}{\partial t^\alpha} = \kappa \, \frac{\partial^2 u}{\partial x^2}, \quad x \in R, \; t > 0, \tag{12.17.1}$$

with the boundary and initial conditions

$$u(x,t) \to 0 \text{ as } |x| \to \infty, \tag{12.17.2}$$

$$\left[{}_0D_t^{\alpha-1} u(x,t)\right]_{t=0} = f(x) \quad \text{for } x \in R, \tag{12.17.3}$$

where κ is a diffusivity constant and $0 < \alpha \le 1$.

Application of the Fourier transform to (12.17.1) with respect to x and using the boundary conditions (12.17.2) and (12.17.3) yields

$$ {}_0D_t^\alpha \, \tilde{u}(k,t) = -\kappa \, k^2 \, \tilde{u}, \tag{12.17.4}$$

$$\left[{}_0D_t^{\alpha-1}\tilde{u}(k,t)\right]_{t=0} = \tilde{f}(k), \tag{12.17.5}$$

where $\tilde{u}(k,t)$ is the Fourier transform of $u(x,t)$ defined by (12.2.1).
The Laplace transform solution of (12.17.4) and (12.17.5) yields

$$\bar{\tilde{u}}(k,s) = \frac{\tilde{f}(k)}{(s^\alpha + \kappa \, k^2)}. \tag{12.17.6}$$

The inverse Laplace transform of (12.17.6) gives

$$\tilde{u}(k,t) = \tilde{f}(k) \, t^{\alpha-1} E_{\alpha,\alpha}\left(-\kappa \, k^2 t^\alpha\right), \tag{12.17.7}$$

where $E_{\alpha,\beta}$ is the Mittag-Leffler function defined by

$$E_{\alpha,\beta}(z) = \sum_{m=0}^{\infty} \frac{z^m}{\Gamma(\alpha m + \beta)}, \quad \alpha > 0, \; \beta > 0. \tag{12.17.8}$$

Finally, the inverse Fourier transform leads to the solution

$$u(x,t) = \int_{-\infty}^{\infty} G(x - \xi, \, t) \, f(\xi) \, d\xi, \tag{12.17.9}$$

where

$$G(x,t) = \frac{1}{\pi} \int_{-\infty}^{\infty} t^{\alpha-1} E_{\alpha,\alpha}\left(-\kappa \, k^2 t^\alpha\right) \cos kx \, dk. \tag{12.17.10}$$

This integral can be evaluated by using the Laplace transform of $G(x, t)$ as

$$\overline{G}(x, s) = \frac{1}{\pi} \int_{-\infty}^{\infty} \frac{\cos kx \, dk}{s^{\alpha} + \kappa k^2} = \frac{1}{\sqrt{4\kappa}} s^{-\alpha/2} \exp\left(-\frac{|x|}{\sqrt{\kappa}} s^{\alpha/2}\right), \quad (12.17.11)$$

where

$$\mathcal{L}\left[t^{m\alpha + \beta - 1} E_{\alpha, \beta}^{(m)}(\pm a t^{\alpha})\right] = \frac{m! \, s^{\alpha - \beta}}{(s^{\alpha} \mp a)^{m+1}}, \quad (12.17.12)$$

and

$$E_{\alpha, \beta}^{(m)}(z) = \frac{d^m}{dz^m} E_{\alpha, \beta}(z). \quad (12.17.13)$$

The inverse Laplace transform of (12.17.11) gives the explicit solution

$$G(x, t) = \frac{1}{\sqrt{4\kappa}} t^{\frac{\alpha}{2} - 1} W\left(-\xi, -\frac{\alpha}{2}, \frac{\alpha}{2}\right), \quad (12.17.14)$$

where $\xi = \frac{|x|}{\sqrt{\kappa} \, t^{\alpha/2}}$, and $W(z, \alpha, \beta)$ is the Wright function (see Erdélyi 1953, formula 18.1 (27)) defined by

$$W(z, \alpha, \beta) = \sum_{n=0}^{\infty} \frac{z^n}{n! \, \Gamma(\alpha n + \beta)}. \quad (12.17.15)$$

It is important to note that when $\alpha = 1$, the initial-value problem (12.17.1)–(12.17.3) reduces to the classical diffusion problem and solution (12.17.9) reduces to the classical solution because

$$G(x, t) = \frac{1}{\sqrt{4\kappa t}} W\left(-\frac{x}{\sqrt{\kappa t}}, -\frac{1}{2}, \frac{1}{2}\right) = \frac{1}{\sqrt{4\pi\kappa t}} \exp\left(-\frac{x^2}{4\kappa t}\right). \quad (12.17.16)$$

The fractional diffusion equation (12.17.1) has also been solved by other authors including Schneider and Wyss (1989), Mainardi (1994, 1995), Debnath (2003) and Nigmatullin (1986) with a physical realistic initial condition

$$u(x, 0) = f(x), \quad x \in R. \quad (12.17.17)$$

The solutions obtained by these authors are in total agreement with (12.17.9).

It is noted that the order α of the derivative with respect to time t in equation (12.17.1) can be of arbitrary real order including $\alpha = 2$ so that equation (12.17.1) may be called the *fractional diffusion-wave equation*. For $\alpha = 2$, it becomes the classical wave equation. The equation (12.17.1) with $1 < \alpha \leq 2$ will be solved next in some detail.

(b) Fractional Nonhomogeneous Wave Equation

The fractional nonhomogeneous wave equation is given by

$$\frac{\partial^\alpha u}{\partial t^\alpha} - c^2 \frac{\partial^2 u}{\partial x^2} = q(x, t), \quad x \in R, \quad t > 0 \qquad (12.17.18)$$

with the initial condition

$$u(x, 0) = f(x), \quad u_t(x, 0) = g(x), \quad x \in R, \qquad (12.17.19)$$

where c is a constant and $1 < \alpha \le 2$.

Application of the joint Laplace transform with respect to t and the Fourier transform with respect to x gives the transform solution

$$\tilde{\bar{u}}(k, s) = \frac{\tilde{f}(k) s^{\alpha-1}}{s^\alpha + c^2 k^2} + \frac{\tilde{g}(k) s^{\alpha-2}}{s^\alpha + c^2 k^2} + \frac{\tilde{\bar{q}}(k, s)}{s^\alpha + c^2 k^2}, \qquad (12.17.20)$$

where k is the Fourier transform variable and s is the Laplace transform variable.

The inverse Laplace transform produces the following result:

$$\tilde{u}(k, t) = \tilde{f}(k) \mathcal{L}^{-1}\left\{\frac{s^{\alpha-1}}{s^\alpha + c^2 k^2}\right\} + \tilde{g}(k) \mathcal{L}^{-1}\left\{\frac{s^{\alpha-2}}{s^\alpha + c^2 k^2}\right\}$$

$$+ \mathcal{L}^{-1}\left\{\frac{\tilde{\bar{q}}(k, s)}{s^\alpha + c^2 k^2}\right\}, \qquad (12.17.21)$$

which, by (12.17.12),

$$= \tilde{f}(k) E_{\alpha,1}\left(-c^2 k^2 t^\alpha\right) + \tilde{g}(k) t E_{\alpha,2}\left(-c^2 k^2 t^\alpha\right)$$

$$+ \int_0^t \tilde{q}(k, t - \tau) \tau^{\alpha-1} E_{\alpha,\alpha}\left(-c^2 k^2 \tau^\alpha\right) d\tau. \quad (12.17.22)$$

Finally, the inverse Fourier transform gives the formal solution

$$u(x, t) = \frac{1}{\sqrt{2\pi}} \int_{-\infty}^\infty \tilde{f}(k) E_{\alpha,1}\left(-c^2 k^2 t^\alpha\right) e^{ikx} dk$$

$$+ \frac{1}{\sqrt{2\pi}} \int_{-\infty}^\infty t \tilde{g}(k) E_{\alpha,2}\left(-c^2 k^2 \tau^\alpha\right) e^{ikx} dk$$

$$+ \frac{1}{\sqrt{2\pi}} \int_0^t \tau^{\alpha-1} d\tau \int_{-\infty}^\infty \tilde{q}(k, t - \tau) E_{\alpha,\alpha}\left(-c^2 k^2 \tau^\alpha\right) e^{ikx} dk.$$

$$(12.17.23)$$

In particular, when $\alpha = 2$, the fractional wave equation (12.17.18) reduces to the classical nonhomogeneous wave equation. In this particular case, we use

$$E_{2,1}\left(-c^2k^2t^2\right) = \cosh\left(ickt\right) = \cos\left(ckt\right), \tag{12.17.24}$$

$$tE_{2,2}\left(-c^2k^2t^2\right) = t \cdot \frac{\sinh\left(ickt\right)}{ickt} = \frac{1}{ck}\sin ckt. \tag{12.17.25}$$

Consequently, solution (12.17.23) reduces to solution (12.11.17) for $\alpha = 2$ as

$$u\left(x,t\right) = \frac{1}{\sqrt{2\pi}} \int_{-\infty}^{\infty} \tilde{f}\left(k\right) \cos\left(ckt\right) e^{ikx} dk + \frac{1}{\sqrt{2\pi}} \int_{-\infty}^{\infty} \tilde{g}\left(k\right) \frac{\sin\left(ckt\right)}{ck} e^{ikx} dk$$

$$+ \frac{1}{\sqrt{2\pi}\,c} \int_0^t d\tau \int_{-\infty}^{\infty} \tilde{q}\left(k,\tau\right) \frac{\sin ck\left(t-\tau\right)}{k} e^{ikx} dk \quad (12.17.26)$$

$$= \frac{1}{2}\left[f\left(x-ct\right) + f\left(x+ct\right)\right] + \frac{1}{2c} \int_{x-ct}^{x+ct} g\left(\xi\right) d\xi$$

$$+ \frac{1}{2c} \int_0^t d\tau \int_{x-c(t-\tau)}^{x+c(t-\tau)} q\left(\xi,\tau\right) d\xi. \tag{12.17.27}$$

We now derive the solution of the nonhomogeneous fractional diffusion equation (12.17.18) with $c^2 = \kappa$ and $g\left(x\right) = 0$. In this case, the joint transform solution (12.17.20) becomes

$$\tilde{\tilde{u}}\left(k,s\right) = \frac{\tilde{f}\left(k\right)s^{\alpha-1}}{\left(s^\alpha + \kappa k^2\right)} + \frac{\tilde{\tilde{q}}\left(k,s\right)}{\left(s^\alpha + \kappa k^2\right)} \tag{12.17.28}$$

which is inverted by using (12.17.12) to obtain $\tilde{u}\left(k,t\right)$ in the form

$$\tilde{u}\left(k,t\right)$$

$$= \tilde{f}\left(k\right) E_{\alpha,1}\left(-\kappa\,k^2t^\alpha\right) + \int_0^t \left(t-\tau\right)^{\alpha-1} E_{\alpha,\alpha}\left\{-\kappa\,k^2\left(t-\tau\right)^\alpha\right\} \tilde{q}\left(k,\tau\right) d\tau. \tag{12.17.29}$$

Finally, the inverse Fourier transform gives the exact solution

$$u\left(x,t\right) = \frac{1}{\sqrt{2\pi}} \int_{-\infty}^{\infty} \tilde{f}\left(k\right) E_{\alpha,1}\left(-\kappa\,k^2t^\alpha\right) e^{ikx} dk$$

$$+ \frac{1}{\sqrt{2\pi}} \int_0^t d\tau \int_{-\infty}^{\infty} \left(t-\tau\right)^{\alpha-1} E_{\alpha,\alpha}\left\{-\kappa\,k^2\left(t-\tau\right)^\alpha\right\} \tilde{q}\left(k,\tau\right) e^{ikx} dk. \tag{12.17.30}$$

Application of the Convolution Theorem of the Fourier transform gives the final solution in the form

$$u\left(x,t\right) = \int_{-\infty}^{\infty} G_1\left(x-\xi,\,t\right) f\left(\xi\right) d\xi$$

$$+ \int_0^t \left(t-\tau\right)^{\alpha-1} d\tau \int_{-\infty}^{\infty} G_2\left(x-\xi,\,t-\tau\right) q\left(\xi,\tau\right) d\xi, \tag{12.17.31}$$

where

$$G_1(x,t) = \frac{1}{2\pi} \int_{-\infty}^{\infty} e^{ikx} E_{\alpha,1}\left(-\kappa\, k^2 t^\alpha\right) dk, \qquad (12.17.32)$$

and

$$G_2(x,t) = \frac{1}{2\pi} \int_{-\infty}^{\infty} e^{ikx} E_{\alpha,\alpha}\left(-\kappa\, k^2 t^\alpha\right) dk. \qquad (12.17.33)$$

In particular, when $\alpha = 1$, the classical solution of the nonhomogeneous diffusion equation (12.17.18) is obtained in the form

$$u(x,t) = \int_{-\infty}^{\infty} G_1(x - \xi,\, t) f(\xi)\, d\xi$$

$$+ \int_0^t d\tau \int_{-\infty}^{\infty} G_2(x - \xi,\, t - \tau) q(\xi,\tau)\, d\xi, \qquad (12.17.34)$$

where

$$G_1(x,t) = G_2(x,t) = \frac{1}{\sqrt{4\pi\kappa t}} \exp\left(-\frac{x^2}{4\kappa t}\right). \qquad (12.17.35)$$

In the case of classical homogeneous diffusion equation (12.17.18), solutions (12.17.30) and (12.17.34) are in perfect agreement with those of Mainardi (1996), who obtained the solution by using the Laplace transform method together with complicated evaluation of the Laplace inversion integral and the auxiliary function $M(z,\alpha)$. He obtained the solution in terms of $M\left(z,\frac{\alpha}{2}\right)$ and discussed the nature of the solution for different values of α. He made some comparisons between ordinary diffusion ($\alpha = 1$) and fractional diffusion $\left(\alpha = \frac{1}{2} \text{ and } \alpha = \frac{2}{3}\right)$. For cases $\alpha = \frac{4}{3}$ and $\alpha = \frac{3}{2}$, the solution exhibits a striking difference from ordinary diffusion with a transition from the Gaussian function centered at $z = 0$ (ordinary diffusion) to the Dirac delta function centered at $z = 1$ (wave propagation). This indicates a possibility of an *intermediate process* between diffusion and wave propagation. A special difference is observed between the solutions of the fractional diffusion equation ($0 < \alpha \leq 1$) and the fractional wave equation ($1 < \alpha \leq 2$). In addition, the solution exhibits a slow process for the case with $0 < \alpha \leq 1$ and an intermediate process for $1 < \alpha \leq 2$.

(c) Fractional-Order Diffusion Equation in Semi-Infinite Medium

We consider the fractional-order diffusion equation in a semi-infinite medium $x > 0$, when the boundary is kept at a temperature $u_0 f(t)$ and the initial temperature is zero in the whole medium. Thus, the initial boundary-value problem is governed by the equation

$$\frac{\partial^\alpha u}{\partial t^\alpha} = \kappa\, \frac{\partial^2 u}{\partial x^2}, \quad 0 < x < \infty,\ t > 0, \qquad (12.17.36)$$

with

$$u(x,0) = 0, \qquad x > 0, \tag{12.17.37}$$

$$u(0,t) = u_0 f(t), \quad t > 0, \text{ and } u(x,t) \to 0 \text{ as } x \to \infty. \tag{12.17.38}$$

Application of the Laplace transform with respect to t gives

$$\frac{d^2\bar{u}}{dx^2} - \left(\frac{s^\alpha}{\kappa}\right)\bar{u}(x,s) = 0, \qquad x > 0, \tag{12.17.39}$$

$$\bar{u}(0,s) = u_0\bar{f}(s), \quad \bar{u}(x,s) \to 0 \text{ as } x \to \infty. \tag{12.17.40}$$

Evidently, the solution of this transformed boundary-value problem is

$$\bar{u}(x,s) = u_0\,\bar{f}(s)\exp(-ax), \tag{12.17.41}$$

where $a = (s^\alpha/\kappa)^{\frac{1}{2}}$. Thus, the solution (12.17.41) is given by

$$u(x,t) = u_0 \int_0^t f(t-\tau)\,g(x,\tau)\,d\tau = u_0 f(t) * g(x,t), \tag{12.17.42}$$

where

$$g(x,t) = \mathcal{L}^{-1}\{\exp(-ax)\}.$$

When $\alpha = 1$ and $f(t) = 1$, solution (12.17.41) becomes

$$\bar{u}(x,s) = \frac{u_0}{s}\exp\left(-x\sqrt{\frac{s}{\kappa}}\right), \tag{12.17.43}$$

which yields the classical solution in terms of the complementary error function (see Debnath 1995)

$$u(x,t) = u_0\,\text{erfc}\left(\frac{x}{2\sqrt{\kappa t}}\right). \tag{12.17.44}$$

In the classical case ($\alpha = 1$), the more general solution is given by

$$u(x,t) = u_0 \int_0^t f(t-\tau)\,g(x,\tau)\,d\tau = u_0 f(t) * g(x,t), \tag{12.17.45}$$

where

$$g(x,t) = \mathcal{L}^{-1}\left\{\exp\left(-x\sqrt{\frac{s}{\kappa}}\right)\right\} = \frac{x}{2\sqrt{\pi\kappa t^3}}\exp\left(-\frac{x^2}{4\kappa t}\right). \tag{12.17.46}$$

(d) The Fractional Stokes and Rayleigh Problems in Fluid Dynamics

The classical Stokes problem (see Debnath 1995) deals with the unsteady boundary layer flows induced in a semi-infinite viscous fluid bounded by an infinite horizontal disk at $z = 0$ due to nontorsional oscillations of the disk in its own plane with a given frequency ω. When $\omega = 0$, the Stokes problem reduces to the classical Rayleigh problem where the unsteady boundary layer flow is generated in the fluid from rest by moving the disk impulsively in its own plane with constant velocity U.

We consider the unsteady fractional boundary layer equation for the fluid velocity $u(z, t)$ that satisfies the equation

$$\frac{\partial^\alpha u}{\partial t^\alpha} = \nu \frac{\partial^2 u}{\partial z^2}, \quad 0 < z < \infty, \ t > 0, \tag{12.17.47}$$

with the given boundary and initial conditions

$$u(0, t) = U f(t), \quad u(z, t) \to 0 \quad \text{as} \quad z \to \infty, \ t > 0, \tag{12.17.48}$$
$$u(z, 0) = 0 \quad \text{for all } z > 0, \tag{12.17.49}$$

where ν is the kinematic viscosity, U is a constant velocity, and $f(t)$ is an arbitrary function of time t.

Application of the Laplace transform with respect to t gives

$$s^\alpha \bar{u}(z, s) = \nu \frac{d^2 \bar{u}}{dz^2}, \quad 0 < z < \infty, \tag{12.17.50}$$

$$\bar{u}(0, s) = U \bar{f}(s), \quad \bar{u}(z, s) \to 0 \quad \text{as} \quad z \to \infty. \tag{12.17.51}$$

Use of the Fourier sine transform (see Debnath 1995) with respect to z yields

$$\bar{U}_s(k, s) = \left(\sqrt{\frac{2}{\pi}} \, \nu U \right) \frac{k \bar{f}(s)}{(s^\alpha + \nu k^2)}. \tag{12.17.52}$$

The inverse Fourier sine transform of (12.17.52) leads to the solution

$$\bar{u}(z, s) = \left(\frac{2}{\pi} \nu U \right) \bar{f}(s) \int_0^\infty \frac{k \sin kz}{(s^\alpha + \nu k^2)} dk, \tag{12.17.53}$$

and the inverse Laplace transform gives the solution for the velocity

$$u(z, t) = \left(\frac{2}{\pi} \nu U \right) \int_0^\infty k \sin kz \, dk \int_0^t f(t - \tau) \tau^{\alpha-1} E_{\alpha,\alpha} \left(-\nu k^2 \tau^\alpha \right) d\tau. \tag{12.17.54}$$

When $f(t) = \exp(i\omega t)$, the solution of the fractional Stokes problem is

$$u(z, t) = \left(\frac{2\nu U}{\pi} \right) e^{i\omega t} \int_0^\infty k \sin kz \, dk \int_0^t e^{-i\omega \tau} \tau^{\alpha-1} E_{\alpha,\alpha} \left(-\nu k^2 \tau^\alpha \right) d\tau. \tag{12.17.55}$$

When $\alpha = 1$, solution (12.17.55) reduces to the classical Stokes solution

$$u\left(z,t\right) = \left(\frac{2\nu U}{\pi}\right) \int_0^\infty \left(1 - e^{-\nu t k^2}\right) \frac{k \sin kz}{\left(i\omega + \nu k^2\right)} dk. \quad (12.17.56)$$

For the fractional Rayleigh problem, $f\left(t\right) = 1$ and the solution follows from (12.17.54) in the form

$$u\left(z,t\right) = \left(\frac{2\nu U}{\pi}\right) \int_0^\infty k \sin kz \, dk \int_0^t \tau^{\alpha-1} E_{\alpha,\alpha}\left(-\nu k^2 \tau^\alpha\right) d\tau. \quad (12.17.57)$$

This solution reduces to the classical Rayleigh solution when $\alpha = 1$ as

$$u\left(z,t\right) = \left(\frac{2\nu U}{\pi}\right) \int_0^\infty k \sin kz \, dk \int_0^t E_{1,1}\left(-\nu \tau k^2\right) d\tau$$

$$= \left(\frac{2\nu U}{\pi}\right) \int_0^\infty k \sin kz \, dk \int_0^t \exp\left(-\nu \tau k^2\right) d\tau$$

$$= \left(\frac{2U}{\pi}\right) \int_0^\infty \left(1 - e^{-\nu t k^2}\right) \frac{\sin kz}{k} dk,$$

which is (by (2.10.10) of Debnath 1995),

$$= \left(\frac{2U}{\pi}\right) \left[\frac{\pi}{2} - \frac{\pi}{2}\operatorname{erf}\left(\frac{z}{2\sqrt{\nu t}}\right)\right] = U \operatorname{erfc}\left(\frac{z}{2\sqrt{\nu t}}\right). \quad (12.17.58)$$

The above analysis is in full agreement with the classical solutions of the Stokes and Rayleigh problems (see Debnath 1995).

(e) The Fractional Unsteady Couette Flow

We consider the unsteady viscous fluid flow between the plate at $z = 0$ at rest and the plate $z = h$ in motion parallel to itself with a variable velocity $U\left(t\right)$ in the x-direction. The fluid velocity $u\left(z,t\right)$ satisfies the fractional equation of motion

$$\frac{\partial^\alpha u}{\partial t^\alpha} = P\left(t\right) + \nu \frac{\partial^2 u}{\partial z^2}, \quad 0 \leq z \leq h, \, t > 0, \quad (12.17.59)$$

with the boundary and initial conditions

$$u\left(0,t\right) = 0 \text{ and } u\left(h,t\right) = U\left(t\right), \quad t > 0, \quad (12.17.60)$$

$$u\left(z,t\right) = 0 \text{ at } t \leq 0 \quad \text{for} \quad 0 \leq z \leq h, \quad (12.17.61)$$

where $-\frac{1}{\rho} p_x = P\left(t\right)$ and ν is the kinematic viscosity of the fluid.

We apply the joint Laplace transform with respect to t and the finite Fourier sine transform with respect to z defined by

$$\bar{\tilde{u}}_s\left(n,s\right) = \int_0^\infty e^{-st} dt \int_0^h u\left(z,t\right) \sin\left(\frac{n\pi z}{h}\right) dz \quad (12.17.62)$$

to the system (12.17.59)–(12.17.61) so that the transform solution is

$$\bar{\tilde{u}}_s(n, s) = \frac{\overline{P}(s)\frac{1}{a}[1 - (-1)^n]}{(s^\alpha + \nu a^2)} + \frac{\nu a (-1)^{n+1}\overline{U}(s)}{(s^\alpha + \nu a^2)}, \quad (12.17.63)$$

where $a = \left(\frac{n\pi}{h}\right)$ and n is the finite Fourier sine transform variable.

Thus, the inverse Laplace transform yields

$$\tilde{u}_s(n, t) = \frac{1}{a}[1 - (-1)^n]\int_0^t P(t - \tau)\tau^{\alpha-1}E_{\alpha,\alpha}\left(-\nu a^2\tau^\alpha\right)d\tau$$

$$+\nu a (-1)^{n+1}\int_0^t U(t - \tau)\tau^{\alpha-1}E_{\alpha,\alpha}\left(-\nu a^2\tau^\alpha\right)d\tau. \quad (12.17.64)$$

Finally, the inverse finite Fourier sine transform leads to the solution

$$u(z, t) = \frac{2}{h}\sum_{n=1}^\infty \tilde{u}_s(n, t)\sin\left(\frac{n\pi z}{h}\right). \quad (12.17.65)$$

In particular, when $\alpha = 1$, $P(t) = $ constant, and $U(t) = $ constant, then solution (12.17.65) reduces to the solution of the generalized Couette flow (see p. 277 Debnath 1995).

(f) Fractional Axisymmetric Wave-Diffusion Equation

The fractional axisymmetric wave-diffusion equation in an infinite domain

$$\frac{\partial^\alpha u}{\partial t^\alpha} = a\left(\frac{\partial^2 u}{\partial r^2} + \frac{1}{r}\frac{\partial u}{\partial r}\right), \quad 0 < r < \infty, \ t > 0, \quad (12.17.66)$$

is called the *diffusion* or *wave equation* accordingly as $a = \kappa$ or $a = c^2$.

For the fractional diffusion equation, we prescribe the initial condition

$$u(r, 0) = f(r), \quad 0 < r < \infty. \quad (12.17.67)$$

Application of the joint Laplace transform with respect to t and the Hankel transform of zero order (see Section 12.12) with respect to r to (12.17.66) and (12.17.67) gives the transform solution

$$\bar{\tilde{u}}(k, s) = \frac{s^{\alpha-1}\tilde{f}(k)}{(s^\alpha + \kappa k^2)}, \quad (12.17.68)$$

where k, s are the Hankel and Laplace transform variables respectively.

The joint inverse transform leads to the solution

$$u(r, t) = \int_0^\infty k J_0(kr)\tilde{f}(k)E_{\alpha,1}\left(-\kappa k^2 t^\alpha\right)dk, \quad (12.17.69)$$

where $J_0(kr)$ is the Bessel function of the first kind of order zero and $\tilde{f}(k)$ is the zero-order Hankel transform of $f(r)$.

When $\alpha = 1$, solution (12.17.69) reduces to the classical solution that was obtained by Debnath (see p 66, Debnath 2005).

On the other hand, we can solve the wave equation (12.17.66) with $a = c^2$ and the initial conditions

$$u(r,0) = f(r), \, u_t(r,0) = g(r) \text{ for } 0 < r < \infty, \qquad (12.17.70)$$

provided the Hankel transforms of $f(r)$ and $g(r)$ exist.

Application of the joint Laplace and Hankel transform leads to the transform solution

$$\bar{\tilde{u}}(k,s) = \frac{s^{\alpha-1}\tilde{f}(k)}{(s^\alpha + c^2 k^2)} + \frac{s^{\alpha-2}\tilde{g}(k)}{(s^\alpha + c^2 k^2)}. \qquad (12.17.71)$$

The joint inverse transformation gives the solution

$$u(r,t) = \int_0^\infty k J_0(k,r)\,\tilde{f}(k)\, E_{\alpha,1}\left(-c^2 k^2 t^\alpha\right) dk$$

$$+ \int_0^\infty k J_0(k,r)\,\tilde{g}(k)\, t E_{\alpha,2}\left(-c^2 k^2 t^\alpha\right) dk. \quad (12.17.72)$$

When $\alpha = 2$, (12.17.72) reduces to the classical solution (12.13.6).

In a finite domain $0 \le r \le a$, the fractional diffusion equation (12.17.66) has the boundary and initial data

$$u(r,t) = f(t) \text{ on } r = a, \quad t > 0, \qquad (12.17.73)$$
$$u(r,0) = 0 \qquad \text{for all } r \text{ in } (0,a). \qquad (12.17.74)$$

Application of the joint Laplace and finite Hankel transform of zero order (see pp. 317, 318, Debnath 1995) yields the solution

$$u(r,t) = \frac{2}{a^2} \sum_{i=1}^\infty \tilde{u}(k_i,t)\, \frac{J_0(rk_i)}{J_1^2(ak_i)}, \qquad (12.17.75)$$

where

$$\tilde{u}(k_i,t) = (a\kappa\,k_i)\, J_1(ak_i) \int_0^t f(t-\tau)\, \tau^{\alpha-1} E_{\alpha,\alpha}\left(-\kappa\,k_i^2\,\tau^\alpha\right) d\tau. \quad (12.17.76)$$

When $\alpha = 1$, (12.17.75) reduces to (12.16.7).

Similarly, the fractional wave equation (12.17.66) with $a = c^2$ in a finite domain $0 \le r \le a$ with the boundary and initial conditions

$$u(r,t) = 0 \text{ on } r = a, \, t > 0, \qquad (12.17.77)$$
$$u(r,0) = f(r) \text{ and } u_t(r,0) = g(r) \text{ for } 0 < r < a, \qquad (12.17.78)$$

can be solved by means of the joint Laplace and finite Hankel transforms. The solution of this problem is

$$u\left(r,t\right) = \frac{2}{a^2} \sum_{i=1}^{\infty} \tilde{u}\left(k_i, t\right) \frac{J_0\left(rk_i\right)}{J_1^2\left(ak_i\right)}, \qquad (12.17.79)$$

where

$$\tilde{u}\left(k_i, t\right) = \tilde{f}\left(k_i\right) E_{\alpha,1}\left(-c^2 k_i^2 t^\alpha\right) + \tilde{g}\left(k_i\right) t E_{\alpha,2}\left(-c^2 k_i^2 t^\alpha\right). \quad (12.17.80)$$

When $\alpha = 2$, solution (12.17.79) reduces to the solution (11.4.26) obtained by Debnath (1995).

(g) The Fractional Schrödinger Equation in Quantum Mechanics

The one-dimensional fractional Schrödinger equation for a free particle of mass m is

$$i\hbar \frac{\partial^\alpha \psi}{\partial t^\alpha} = -\frac{\hbar^2}{2m} \frac{\partial^2 \psi}{\partial x^2}, \qquad -\infty < x < \infty, \ t > 0, \quad (12.17.81)$$

$$\psi\left(x, 0\right) = \psi_0\left(x\right), \qquad -\infty < x < \infty, \qquad (12.17.82)$$

$$\psi\left(x, t\right) \to 0 \text{ as } |x| \to \infty, \qquad (12.17.83)$$

where $\psi\left(x, t\right)$ is the wave function, $h = 2\pi\hbar = 6.625 \times 10^{-27}\text{erg sec} = 4.14 \times 10^{-21}\text{MeV sec}$ is the Planck constant, and $\psi_0\left(x\right)$ is an arbitrary function.

Application of the joint Laplace and Fourier transform to (12.17.81)–(12.17.83) gives the solution in the transform space in the form

$$\overline{\Psi}\left(k, s\right) = \frac{s^{\alpha-1}\Psi_0\left(k\right)}{s^\alpha + ak^2}, \qquad a = \frac{i\hbar}{2m}, \qquad (12.17.84)$$

where k, s represent the Fourier and the Laplace transforms variables.

The use of the joint inverse transform yields the solution

$$\psi\left(x, t\right) = \frac{1}{\sqrt{2\pi}} \int_{-\infty}^{\infty} e^{ikx} \tilde{\psi}_0\left(k\right) E_{\alpha,1}\left(-ak^2 t^\alpha\right) dk. \quad (12.17.85)$$

$$= \mathcal{F}^{-1}\left\{ \tilde{\psi}_0\left(k\right) E_{\alpha,1}\left(-ak^2 t^\alpha\right) \right\}, \qquad (12.17.86)$$

which is, by Theorem 12.4.1, the Convolution Theorem of the Fourier transform

$$= \int_{-\infty}^{\infty} G\left(x - \xi, t\right) \psi_0\left(\xi\right) d\xi, \qquad (12.17.87)$$

where

$$G\left(x, t\right) = \frac{1}{\sqrt{2\pi}} \mathcal{F}^{-1}\left\{ E_{\alpha,1}\left(-ak^2 t^\alpha\right) \right\}$$

$$= \frac{1}{2\pi} \int_{-\infty}^{\infty} e^{ikx} E_{\alpha,1}\left(-ak^2 t^\alpha\right) dk. \qquad (12.17.88)$$

When $\alpha = 1$, solution (12.17.87) becomes

$$\psi(x,t) = \int_{-\infty}^{\infty} G(x-\xi, t)\,\psi_0(\xi)\,d\xi, \tag{12.17.89}$$

where the Green's function $G(x,t)$ is given by

$$\begin{aligned} G(x,t) &= \frac{1}{2\pi} \int_{-\infty}^{\infty} e^{ikx} E_{1,1}\left(-ak^2t\right) dk \\ &= \frac{1}{2\pi} \int_{-\infty}^{\infty} \exp\left(ikx - atk^2\right) dk \\ &= \frac{1}{\sqrt{4\pi at}} \exp\left(-\frac{x^2}{4at}\right). \end{aligned} \tag{12.17.90}$$

This solution (12.17.89) is in perfect agreement with the classical solution obtained by Debnath (1995).

12.18 Exercises

1. Find the Fourier transform of

   (a) $f(x) = \exp\left(-ax^2\right)$,   (b) $f(x) = \exp\left(-a\,|x|\right)$,

   where a is a constant.

2. Find the Fourier transform of the gate function

$$f_a(x) = \begin{cases} 1, & |x| < a, \quad a \text{ is a positive constant.} \\ 0, & |x| \geq a. \end{cases}$$

3. Find the Fourier transform of

   (a) $f(x) = \frac{1}{|x|}$,   (b) $f(x) = \chi_{[-a,a]}(x) = \begin{cases} 1, & -a < x < a \\ 0, & \text{otherwise,} \end{cases}$

   (c) $f(x) = \begin{cases} 1 - \frac{|x|}{a}, & |x| \leq a \\ 0, & |x| \geq a, \end{cases}$ (d) $f(x) = \frac{1}{(x^2+a^2)}$.

4. Find the Fourier transform of

   (a) $f(x) = \sin\left(x^2\right)$,   (b) $f(x) = \cos\left(x^2\right)$.

5. Show that

$$I = \int_0^\infty e^{-a^2 x^2} \, dx = \sqrt{\pi}/2a, \qquad a > 0,$$

by noting that

$$I^2 = \int_0^\infty \int_0^\infty e^{-a^2(x^2 + y^2)} \, dx \, dy = \int_0^{\pi/2} \int_0^\infty e^{-a^2 r^2} r \, dr \, d\theta.$$

6. Show that

$$\int_0^\infty e^{-a^2 x^2} \cos bx \, dx = \left(\sqrt{\pi}/2a\right) e^{-b^2/4a^2}, \qquad a > 0.$$

7. Prove that

(a) $f(x) = \frac{1}{\sqrt{2\pi}} \int_{-\infty}^\infty e^{ikx} F(k) \, dk = \mathcal{F}^{-1} \{F(k)\},$

(b) $\mathcal{F}[f(ax - b)] = \frac{1}{|a|} e^{ikb/a} F(k/a).$

8. Prove the following properties of the Fourier convolution:

(a) $f(x) * g(x) = g(x) * f(x),$ \qquad\qquad (b) $f * (g * h) = (f * g) * h,$

(c) $f * (ag + bh) = a(f * g) + b(f * h),$ where a and b are constants,

(d) $f * 0 = 0 * f = 0,$ \qquad\qquad\qquad (e) $f * 1 \neq f,$

(f) $f * \sqrt{2\pi}\, \delta = f = \sqrt{2\pi}\, \delta * f,$

(g) $\mathcal{F}\{f(x) g(x)\} = (F * G)(k) = \frac{1}{\sqrt{2\pi}} \int_{-\infty}^\infty F(k - \xi) G(\xi) \, d\xi,$

9. Prove the following properties of the Fourier convolution:

(a) $\frac{d}{dx}\{f(x) * g(x)\} = f'(x) * g(x) = f(x) * g'(x),$

(b) $\frac{d^2}{dx^2}[(f * g)(x)] = (f' * g')(x) = (f'' * g)(x),$

(c) $(f * g)^{(m+n)}(x) = \left[f^{(m)} * g^{(n)}\right](x),$

(d) $\int_{-\infty}^\infty (f * g)(x) \, dx = \int_{-\infty}^\infty f(u) \, du \int_{-\infty}^\infty g(v) \, dv.$

(e) If $g(x) = \frac{1}{2a} H(a - x),$ then

$(f * g)(x)$ is the average value of $f(x)$ in $[x - a, \, x + a].$

(f) If $G_t(x) = \frac{1}{\sqrt{4\pi kt}} \exp\left(-\frac{x^2}{4\kappa t}\right)$, then $(G_t * G_s)(x) = G_{t+s}(x)$.

10. Prove the following results:

(a) $\frac{1}{\sqrt{2\pi}} \int_{-\infty}^{\infty} e^{-k^2 t - ikx}\, dk = \frac{1}{\sqrt{2t}} e^{-x^2/4t}$,

(b) $\int_{-\infty}^{\infty} F(k)\, g(k)\, e^{ikx}\, dk = \int_{-\infty}^{\infty} f(y)\, G(y-x)\, dy$,

(c) $\int_{-\infty}^{\infty} F(k)\, g(k)\, dk = \int_{-\infty}^{\infty} f(y)\, G(y)\, dy$,

(d) $\sin x * e^{-a|x|} = \sqrt{\frac{2}{\pi}} \frac{a \sin x}{(1+a^2)}$,   (e) $e^{ax} * \chi_{[0,\infty)}(x) = \frac{1}{a} \frac{e^{ax}}{\sqrt{2\pi}}$, $a > 0$,

(f) $\frac{1}{\sqrt{2a}} \exp\left(-\frac{x^2}{4a}\right) * \frac{1}{\sqrt{2b}} \exp\left(-\frac{x^2}{4b}\right) = \frac{1}{\sqrt{2(a+b)}} \exp\left(-\frac{x^2}{4(a+b)}\right)$.

11. Determine the solution of the initial-value problem

$$u_{tt} = c^2 u_{xx}, \qquad -\infty < x < \infty, \qquad t > 0,$$
$$u(x,0) = f(x), \qquad u_t(x,0) = g(x), \qquad -\infty < x < \infty.$$

12. Solve

$$u_t = u_{xx}, \qquad x > 0, \qquad t > 0,$$
$$u(x,0) = f(x), \qquad u(0,t) = 0.$$

13. Solve

$$u_{tt} = c^2 u_{xxxx} = 0, \qquad -\infty < x < \infty, \qquad t > 0,$$
$$u(x,0) = f(x), \qquad u_t(x,0) = 0, \qquad -\infty < x < \infty.$$

14. Solve

$$u_{tt} + c^2 u_{xxxx} = 0, \qquad\qquad x > 0, \qquad t > 0,$$
$$u(x,0) = 0, \qquad\qquad u_t(x,0) = 0, \qquad x > 0,$$
$$u(0,t) = g(t), \qquad u_{xx}(0,t) = 0, \qquad t > 0.$$

15. Solve

$$\phi_{xx} + \phi_{yy} = 0, \qquad -a < x < a, \qquad 0 < y < \infty,$$
$$\phi_y(x,0) = \begin{cases} \delta_0, & 0 < |x| < a, \\ \\ 0, & |x| > a. \end{cases}$$
$$\phi(x,y) \to 0 \text{ uniformly in } x \text{ as } y \to \infty.$$

16. Solve

$$u_t = u_{xx} + t\,u, \quad -\infty < x < \infty, \quad t > 0,$$
$$u(x,0) = f(x), \quad u(x,t) \text{ is bounded}, \quad -\infty < x < \infty.$$

17. Solve

$$u_t - u_{xx} + hu = \delta(x)\,\delta(t), \quad -\infty < x < \infty, \quad t > 0,$$
$$u(x,0) = 0, \quad u(x,t) \to 0 \text{ uniformly in } t \text{ as } |x| \to \infty.$$

18. Solve

$$u_t - u_{xx} + h(t)\,u_x = \delta(x)\,\delta(t), \quad 0 < x < \infty, \quad t > 0,$$
$$u(x,0) = 0, \quad u_x(0,t) = 0,$$
$$u(x,t) \to 0 \text{ uniformly in } t \text{ as } x \to \infty.$$

19. Solve

$$u_{xx} + u_{yy} = 0, \quad 0 < x < \infty, \quad 0 < y < \infty,$$
$$u(x,0) = f(x), \quad 0 \le x < \infty,$$
$$u_x(0,y) = g(y), \quad 0 \le y < \infty,$$
$$u(x,y) \to 0 \text{ uniformly in } x \text{ as } x \to \infty \text{ and uniformly in } y \text{ as } x \to \infty.$$

20. Solve

$$u_{xx} + u_{yy} = 0, \quad -\infty < x < \infty, \quad 0 < y < a,$$
$$u(x,0) = f(x), \quad u(x,a) = 0, \quad -\infty < x < \infty,$$
$$u(x,y) \to 0 \text{ uniformly in } y \text{ as } |x| \to \infty.$$

21. Solve

$$u_t = u_{xx}, \quad x > 0, \quad t > 0,$$
$$u(x,0) = 0, \quad x > 0, \quad u(0,t) = f(t), \quad t > 0,$$
$$u(x,t) \text{ is bounded for all } x \text{ and } t.$$

22. Solve

$$u_{xx} + u_{yy} = 0, \quad x > 0, \quad 0 < y < 1,$$
$$u(x,0) = f(x), \quad u(x,1) = 0, \quad x > 0,$$
$$u(0,y) = 0, \quad u(x,y) \to 0 \text{ uniformly in } y \text{ as } x \to \infty.$$

23. Find the Laplace transform of each of the following functions:

(a) t^n,

(b) $\cos \omega t$,

(c) $\sinh kt$,

(d) $\cosh kt$,

(e) te^{at},

(f) $e^{at} \sin \omega t$,

(g) $e^{at} \cos \omega t$,

(h) $t \sinh kt$,

(i) $t \cosh kt$,

(j) $\sqrt{\frac{1}{t}}$,

(k) $\sqrt{t}$,

(l) $\frac{\sin at}{t}$.

24. Find the inverse transform of each of the following functions:

(a) $\frac{s}{(s^2+a^2)(s^2+b^2)}$,

(b) $\frac{1}{(s^2+a^2)(s^2+b^2)}$,

(c) $\frac{1}{(s-a)(s-b)}$,

(d) $\frac{1}{s(s+a)^2}$,

(e) $\frac{1}{s(s+a)}$,

(f) $\frac{s^2-a^2}{(s^2+a^2)^2}$.

25. The velocity potential $\phi(x, z, t)$ and the free-surface evaluation $\eta(x, t)$ for surface waves in water of infinite depth satisfy the Laplace equation

$$\phi_{xx} + \phi_{zz} = 0, \qquad -\infty < x < \infty, \qquad -\infty < z \leq 0, \qquad t > 0,$$

with the free-surface, boundary, and initial conditions

$$\phi_z = \eta_t \quad \text{on} \quad z = 0, \qquad t > 0,$$
$$\phi_t + g\eta = 0 \quad \text{on} \quad z = 0, \qquad t > 0,$$
$$\phi_z \to 0 \quad \text{as} \quad z \to -\infty,$$
$$\phi(x, 0, 0) = 0 \quad \text{and} \quad \eta(x, 0) = f(x), \qquad -\infty < x < \infty,$$

where g is the constant acceleration due to gravity.
Show that

$$\phi(x, z, t) = -\frac{\sqrt{g}}{\sqrt{2\pi}} \int_{-\infty}^{\infty} k^{-\frac{1}{2}} F(k) e^{|k|z - ikx} \sin\left(\sqrt{g|k|}\, t\right) dk,$$

$$\eta(x, t) = \frac{1}{\sqrt{2\pi}} \int_{-\infty}^{\infty} F(k) e^{-ikx} \cos\left(\sqrt{g|k|}\, t\right) dk,$$

where k represents the Fourier transform variable.
Find the asymptotic solution for $\eta(x, t)$ as $t \to \infty$.

26. Use the Fourier transform method to show that the solution of the one-dimensional Schrödinger equation for a free particle of mass m,

$$i\hbar\psi_t = -\frac{\hbar^2}{2m}\psi_{xx}, \qquad -\infty < x < \infty, \qquad t > 0,$$
$$\psi(x, 0) = f(x), \qquad -\infty < x < \infty,$$

where ψ and ψ_x tend to zero as $|x| \to \infty$, and $h = 2\pi\hbar$ is the Planck constant, is given by

$$\psi(x,t) = \frac{1}{\sqrt{2\pi}} \int_{-\infty}^{\infty} f(\xi) G(x - \xi) \, d\xi,$$

where $G(x,t) = \frac{(1-i)}{2\sqrt{\gamma t}} \exp\left[-\frac{x^2}{4i\gamma t}\right]$ is the Green's function and $\gamma = \frac{\hbar}{2m}$.

27. Prove the following properties of the Laplace convolution:

(a) $f * g = g * f$,   (b) $f * (g * h) = (f * g) * h$,

(c) $f * (\alpha g + \beta h) = \alpha (f * g) + \beta (f * h)$, α and β are constants,

(d) $f * 0 = 0 * f$,

(e) $\frac{d}{dt} [(f * g)(t)] = f'(t) * g(t) + f(0) g(t)$,

(f) $\frac{d^2}{dt^2} [(f * g)(t)] = f''(t) * g(t) + f'(0) g(t) + f(0) g'(t)$,

(g) $\frac{d^n}{dt^n} [(f * g)(t)] = f^{(n)}(t) * g(t) + \sum_{k=0}^{n-1} f^{(k)}(0) g^{(n-k-1)}(t)$.

28. Obtain the solution of the problem

$$u_{tt} = c^2 u_{xx}, \quad 0 < x < \infty, \quad t > 0,$$
$$u(x,0) = f(x), \quad u_t(x,0) = 0,$$
$$u(0,t) = 0, \quad u(x,t) \to 0 \text{ uniformly in } t \text{ as } x \to \infty.$$

29. Solve

$$u_{tt} = c^2 u_{xx}, \quad 0 < x < l, \quad t > 0,$$
$$u(x,0) = 0, \quad u_t(x,0) = 0,$$
$$u(0,t) = f(t), \quad u(l,t) = 0, \quad t \geq 0.$$

30. Solve

$$u_t = \kappa u_{xx}, \quad 0 < x < \infty, \quad t > 0,$$
$$u(x,0) = f_0, \quad 0 < x < \infty,$$
$$u(0,t) = f_1, \quad u(x,t) \to f_0 \text{ uniformly in } t \text{ as } x \to \infty, \quad t > 0.$$

31. Solve

$$u_t = \kappa u_{xx}, \quad 0 < x < \infty, \quad t > 0,$$
$$u(x,0) = x, \quad x > 0,$$
$$u(0,t) = 0, \quad u(x,t) \to x \text{ uniformly in } t \text{ as } x \to \infty, \quad t > 0.$$

32. Solve

$$u_t = \kappa u_{xx}, \quad 0 < x < \infty, \qquad t > 0,$$
$$u(x,0) = 0, \qquad 0 < x < \infty,$$
$$u(0,t) = t^2, \qquad u(x,t) \to 0 \text{ uniformly in } t \text{ as } x \to \infty, \quad t \geq 0.$$

33. Solve

$$u_t = \kappa u_{xx} - hu, \quad 0 < x < \infty, \quad t > 0, \quad h = \text{constant},$$
$$u(x,0) = f_0, \qquad x > 0,$$
$$u(0,t) = 0, \qquad u_x(0,t) \to 0 \text{ uniformly in } t \text{ as } x \to \infty, \quad t > 0.$$

34. Solve

$$u_t = \kappa u_{xx}, \quad 0 < x < \infty, \quad t > 0,$$
$$u(x,0) = 0, \qquad 0 < x < \infty,$$
$$u(0,t) = f_0, \qquad u(x,t) \to 0 \text{ uniformly in } t \text{ as } x \to \infty, \quad t > 0.$$

35. Solve

$$u_{tt} = c^2 u_{xx}, \quad 0 < x < \infty, \quad t > 0,$$
$$u(x,0) = 0, \qquad u_t(x,0) = f_0, \quad 0 < x < \infty,$$
$$u(0,t) = 0, \qquad u_x(x,t) \to 0 \text{ uniformly in } t \text{ as } x \to \infty, \quad t > 0.$$

36. Solve

$$u_{tt} = c^2 u_{xx}, \quad 0 < x < \infty, \quad t > 0,$$
$$u(x,0) = f(x), \quad u_t(x,0) = 0, \quad 0 < x < \infty,$$
$$u(0,t) = 0, \qquad u_x(x,t) \to 0 \text{ uniformly in } t \text{ as } x \to \infty, \quad t > 0.$$

37. A semi-infinite lossless transmission line has no initial current or potential. A time dependent EMF, $V_0(t) H(t)$ is applied at the end $x = 0$. Find the potential $V(x,t)$. Then determine the potential for cases: (i) $V_0(t) = V_0 = \text{constant}$, and (ii) $V_0(t) = V_0 \cos \omega t$.

38. Solve the Blasius problem of an unsteady boundary layer flow in a semi-infinite body of viscous fluid enclosed by an infinite horizontal disk at $z = 0$. The governing equation, boundary, and initial conditions are

$$\frac{\partial u}{\partial t} = \nu \frac{\partial^2 u}{\partial z^2}, \quad z > 0, \qquad t > 0,$$
$$u(z,t) = Ut \quad \text{on } z = 0, \qquad t > 0,$$
$$u(z,t) \to 0 \quad \text{as } z \to \infty, \qquad t > 0,$$
$$u(z,t) = 0 \quad \text{at } t \leq 0, \qquad z > 0.$$

Explain the implication of the solution.

39. The stress-strain relation and equation of motion for a viscoelastic rod in the absence of external force are

$$\frac{\partial e}{\partial t} = \frac{1}{E}\frac{\partial \sigma}{\partial t} + \frac{\sigma}{\eta}, \qquad \frac{\partial \sigma}{\partial x} = \rho \frac{\partial^2 u}{\partial t^2},$$

where e is the strain, η is the coefficient of viscosity, and the displacement $u(x,t)$ is related to the strain by $e = \partial u/\partial x$. Prove that the stress $\sigma(x,t)$ satisfies the modified wave equation

$$\frac{\partial^2 \sigma}{\partial x^2} - \frac{\rho}{\eta}\frac{\partial \sigma}{\partial t} = \frac{1}{c^2}\frac{\partial^2 \sigma}{\partial t^2}, \qquad c^2 = E/\rho.$$

Show that the stress distribution in a semi-infinite viscoelastic rod subject to the boundary and initial conditions,

$$\dot{u}(0,t) = U\,H(t), \qquad \sigma(x,t) \to 0 \quad \text{as} \quad x \to \infty, \qquad t > 0,$$
$$\sigma(x,0) = 0, \qquad \dot{u}(x,0) = 0,$$

is given by

$$\sigma(x,t) = -U\rho c \exp\left(-\frac{Et}{2\eta}\right) I_0\left[\frac{Et}{2\eta}\left(t^2 - \frac{x^2}{c^2}\right)^{\frac{1}{2}}\right] H\left(t - \frac{x}{c}\right).$$

40. An elastic string is stretched between $x = 0$ and $x = l$ and is initially at rest in the equilibrium position. Show that the Laplace transform solution for the displacement field subject to the boundary conditions $y(0,t) = f(t)$ and $y(l,t) = 0$, $t > 0$ is

$$\bar{y}(x,s) = \bar{f}(s)\,\frac{\sinh\left\{\frac{s}{c}(l-x)\right\}}{\sinh\frac{sl}{c}}.$$

41. The end $x = 0$ of a semi-infinite submarine cable is maintained at a potential $V_0 H(t)$. If the cable has no initial current and potential, determine the potential $V(x,t)$ at point x and at time t.

42. Obtain the solution of the Stokes–Ekman problem (see Debnath, 1995) of an unsteady boundary layer flow in a semi-infinite body of viscous fluid bounded by an infinite horizontal disk at $z = 0$, when both the fluid and the disk rotate with a uniform angular velocity Ω about the z-axis. The governing boundary layer equation, the boundary conditions, and the initial conditions are

$$\frac{\partial q}{\partial t} + 2\Omega i q = \nu \frac{\partial^2 q}{\partial z^2}, \qquad\qquad z > 0, \qquad t > 0,$$
$$q(z,t) = ae^{i\omega t} + be^{-i\omega t} \quad \text{on} \quad z = 0, \qquad t > 0,$$
$$q(z,t) \to 0 \quad \text{as} \quad z \to \infty, \qquad\qquad\qquad t > 0,$$
$$q(z,t) = 0 \quad \text{at} \quad t \le 0, \quad \text{for all} \quad z > 0,$$

where $q = u + iv$, is the complex velocity field, w is the frequency of oscillations of the disk, and a and b are complex constants. Hence, deduce the steady-state solution, and determine the structure of the associated boundary layers.

43. Show that, when $w = 0$ in the Stokes–Ekman problem 42, the steady flow field is given by

$$q(z,t) \sim (a + b) \exp \left\{ - \left(\frac{2i\Omega}{\nu} \right)^{\frac{1}{2}} z \right\}.$$

Hence determine the thickness of the Ekman layer.

44. For problem 14 (e) (iii) in 3.9 Exercises, show that the potential $V(x,t)$ and the current $I(x,t)$ satisfy the partial differential equation

$$\left(\frac{\partial^2}{\partial t^2} + 2k\frac{\partial}{\partial t} + k^2 \right) (V, I) = c^2 \frac{\partial^2}{\partial x^2} (V, I).$$

Find the solution for $V(x,t)$ with the boundary and initial data

$$V(x,t) = V_0(t) \quad \text{at} \quad x = 0, \qquad t > 0,$$
$$V(x,t) \to 0 \quad \text{as} \quad x \to \infty, \qquad t > 0,$$
$$V(x,0) = V_t(x,0) = 0 \quad \text{for} \quad 0 \le x < \infty.$$

45. Use the Laplace transform to solve the Abel integral equation

$$g(t) = \int_0^t f'(\tau)(t - \tau)^{-\alpha} \, d\tau, \qquad 0 < \alpha < 1.$$

46. Solve Abel's problem of tautochronous motion described in problem 17 of 14.11 Exercises.

47. The velocity potential $\phi(r, z, t)$ and the free-surface elevation $\eta(r, t)$ for axisymmetric surface waves in water of infinite depth satisfy the equation

$$\phi_{rr} + \frac{1}{r}\phi_r + \phi_{zz} = 0, \quad 0 \le r < \infty, \quad -\infty < z \le 0, \qquad t > 0,$$

with the free-surface, boundary, and initial conditions

$$\phi_z = \eta_z \quad \text{on} \quad z = 0, \qquad t > 0,$$
$$\phi_t + g\eta = 0, \quad \text{on} \quad z = 0, \qquad t > 0,$$
$$\phi_z \to 0, \quad z \to -\infty,$$
$$\phi(r, 0, 0) = 0, \quad \text{and} \quad \eta(r, 0) = f(r), \quad 0 \le r < \infty,$$

where g is the acceleration due to gravity and $f(r)$ represents the initial elevation.
Show that

$$\phi\left(r,z,t\right)=-\sqrt{g}\int_{0}^{\infty}\sqrt{k}\,\tilde{f}\left(k\right)J_{0}\left(kr\right)e^{kz}\sin\left(\sqrt{gk}\,t\right)dk,$$

$$\eta\left(r,t\right)=\int_{0}^{\infty}k\,\tilde{f}\left(k\right)J_{0}\left(kr\right)\cos\left(\sqrt{gk}\,t\right)dk,$$

where $\tilde{f}\left(k\right)$ is the zero-order Hankel transform of $f\left(r\right)$.
Derive the asymptotic solution

$$\eta\left(r,t\right)\sim\frac{gt^{2}}{2^{\frac{3}{2}}r^{3}}\,\tilde{f}\left(\frac{gt^{2}}{4r^{2}}\right)\cos\left(\frac{gt^{2}}{4r}\right)\quad\text{as}\quad t\to\infty.$$

48. Write the solution for the Cauchy–Poisson problem where the initial
    elevation is concentrated in the neighborhood of the origin, that is,
    $f\left(r\right)=\left(a/2\pi r\right)\delta\left(r\right)$, where a is the total volume of the fluid displaced.

49. The steady temperature distribution $u\left(r,z\right)$ in a semi-infinite solid with
    $z\geq0$ is governed by the system

    $$u_{rr}+\frac{1}{r}u_{r}+u_{zz}=-Aq\left(r\right),\qquad0<r<\infty,\qquad z>0,$$
    $$u\left(r,0\right)=0,$$

    where A is a constant and $q\left(r\right)$ represents the steady heat source. Show
    that the solution is given by

    $$u\left(r,z\right)=A\int_{0}^{\infty}\tilde{q}\left(k\right)J_{0}\left(kr\right)k^{-1}\left(1-e^{-kz}\right)dk,$$

    where $\tilde{q}\left(k\right)$ is the zero-order Hankel transform of $q\left(r\right)$.

50. Find the solution for the small deflection $u\left(r\right)$ of an elastic membrane
    subjected to a concentrated loading distribution which is governed by

    $$u_{rr}+\frac{1}{r}u_{r}-\kappa^{2}u=\frac{1}{2\pi}\frac{\delta\left(r\right)}{r},\qquad0\leq r<\infty,$$

    where u and its derivatives vanish as $r\to\infty$.

51. Obtain the solution for the potential $v\left(r,z\right)$ due to a flat electrified disk
    of radius unity with the center of the disk at the origin and the axis
    along the z-axis. The function $v\left(r,z\right)$ satisfies the Laplace equation

    $$v_{rr}+\frac{1}{r}v_{r}-v_{zz}=0,\qquad0<r<\infty,\qquad z>0,$$

    with the boundary conditions

    $$v\left(r,0\right)=v_{0},\qquad0\leq r<1,$$
    $$v_{z}\left(r,0\right)=0,\qquad r>1.$$

52. Prove that the Fourier sine and cosine transforms are linear.

53. If $\mathcal{F}_s(n)$ is the Fourier sine transform of $f(x)$ on $0 \le x \le l$, show that

$$\mathcal{F}_s[f''(x)] = \frac{2n\pi}{l^2}[f(0) - (-1)^n f(l)] - \left(\frac{n\pi}{l}\right)^2 \mathcal{F}_s(n).$$

54. If $\mathcal{F}_c(n)$ is the Fourier cosine transform of $f(x)$ on $0 \le x \le l$, show that

$$\mathcal{F}_c[f''(x)] = \frac{2}{l}[(-1)^n f'(l) - f'(0)] - \left(\frac{n\pi}{l}\right)^2 \mathcal{F}_c(n).$$

When $l = \pi$, show that

$$\mathcal{F}_c[f''(x)] = \frac{2}{\pi}[(-1)^n f'(\pi) - f'(0)] - n^2 \mathcal{F}_c(n).$$

55. By the transform method, solve

$$\begin{aligned}
u_t &= u_{xx} + g(x,t), & 0 < x < \pi, & \quad t > 0, \\
u(x,0) &= f(x), & 0 \le x \le \pi, & \\
u(0,t) &= 0, & u(\pi,t) \to 0 & \quad t > 0.
\end{aligned}$$

56. By the transform method, solve

$$\begin{aligned}
u_t &= u_{xx} + g(x,t), & 0 < x < \pi, & \quad t > 0, \\
u(x,0) &= 0, & 0 < x < \pi, & \\
u(0,t) &= 0, & u_x(\pi,t) + hu(\pi,t) = 0, & \quad t > 0.
\end{aligned}$$

57. By the transform method, solve

$$\begin{aligned}
u_t &= u_{xx} + g(x,t), & 0 < x < \pi, & \quad t > 0, \\
u(x,0) &= 0, & 0 < x < \pi, & \\
u(0,t) &= 0, & u_x(\pi,t) = 0, & \quad t > 0.
\end{aligned}$$

58. By the transform method, solve

$$\begin{aligned}
u_t &= u_{xx} - hu, & 0 < x < \pi, & \quad t > 0, \\
u(x,0) &= \sin x, & 0 \le x \le \pi, & \\
u(0,t) &= 0, & u(\pi,t) = 0, & \quad t > 0.
\end{aligned}$$

59. By the transform method, solve

$$\begin{aligned}
u_{tt} &= u_{xx} + h, & 0 < x < \pi, & \quad t > 0, & h = \text{constant}, \\
u(x,0) &= 0, & u_t(x,0) = 0, & \quad 0 < x < \pi, \\
u_x(0,t) &= 0, & u_x(\pi,t) = 0, & \quad t > 0.
\end{aligned}$$

60. By the transform method, solve

$$u_{tt} = u_{xx} + g(x), \qquad 0 < x < \pi, \qquad t > 0,$$
$$u(x, 0) = 0, \qquad u_t(x, 0) = 0, \qquad 0 < x < \pi,$$
$$u(0, t) = 0, \qquad u(\pi, t) = 0, \qquad t > 0.$$

61. By the transform method, solve

$$u_{tt} + c^2 u_{xxxx} = 0, \qquad 0 < x < \pi, \qquad t > 0,$$
$$u(x, 0) = 0, \qquad u_t(x, 0) = 0, \qquad 0 < x < \pi,$$
$$u(0, t) = 0, \qquad u(\pi, t) = 0, \qquad t > 0.$$
$$u_{xx}(0, t) = 0, \qquad u_{xx}(\pi, t) = \sin t, \qquad t \geq 0.$$

62. Find the temperature distribution $u(r, t)$ in a long cylinder of radius a when the initial temperature is constant, u_0, and radiation occurs at the surface into a medium with zero temperature. Here $u(r, t)$ satisfies the initial boundary-problem

$$u_t = \kappa \left(u_{rr} + \frac{1}{r} u_r \right), \qquad 0 \leq r < a, \qquad t > 0,$$
$$u_r + \alpha u = 0 \quad \text{at} \quad r = a, \qquad t > 0,$$
$$u(r, 0) = u_0 \quad \text{for} \quad 0 \leq r < a,$$

where κ and α are constants.

63. Apply the finite Fourier sine transform to solve the longitudinal displacement field in a uniform bar of length l and cross section A subjected to an external force FA applied at the end $x = l$. The governing equation and boundary and initial conditions are

$$c^2 u_{xx} = u_{tt}, \qquad \left(c^2 = \frac{E}{\rho} \right), \qquad 0 < x < l, \qquad t > 0,$$
$$u(0, t) = 0 \qquad E u(l, t) = F, \qquad t > 0,$$
$$u(x, 0) = u_t(x, 0) = 0, \qquad 0 < x < l,$$

where E is the constant Young's modulus, ρ is the density, and F is constant.

64. Use the finite Fourier cosine transform to solve the heat conduction problem

$$u_t = \kappa u_{xx}, \qquad 0 < x < l, \qquad t > 0,$$
$$u_x(x, t) = 0 \quad \text{at} \quad x = 0 \quad \text{and} \quad x = l, \qquad t > 0,$$
$$u(x, 0) = u_0 \quad \text{for} \quad 0 < x < l,$$

where u_0 and κ are constant.

65. Use the Mellin transform to find the solution of the integral equation

$$\int_{-\infty}^{\infty} f(x) k(xt) \, dx = g(t), \qquad t > 0.$$

66. Use the Mellin transform to show the following results:

(a) $\displaystyle\sum_{n=1}^{\infty} f(n) = \frac{1}{2\pi i} \int_{c-i\infty}^{c+i\infty} \zeta(p) F(p) \, dp,$

(b) $\displaystyle\sum_{n=1}^{\infty} f(nx) = \mathcal{M}^{-1}\left[\zeta(p) F(p)\right],$

where $\zeta(s)$ is the Riemann zeta function defined by (6.7.13).

67. Show that the solution of the boundary-value problem

$$u_{rr} + \frac{1}{r} u_r + u_{zz} = 0, \qquad r \geq 0, \quad z > 0,$$
$$u(r,0) = u_0 \quad \text{for} \quad 0 \leq r \leq a,$$
$$u(r,z) \to 0 \quad \text{as} \quad z \to \infty,$$

is

$$u(r,z) = au_0 \int_0^{\infty} J_1(ak) J_0(kr) e^{-kz} dk.$$

68. Show that the asymptotic representation of the Bessel function $J_n(kr)$ for large kr is

$$J_n(kr) = \frac{1}{\pi} \int_0^{\pi} \cos(n\theta - kr\sin\theta) \, d\theta \sim \left(\frac{2}{\pi kr}\right)^{\frac{1}{2}} \cos\left(kr - \frac{n\pi}{2} - \frac{\pi}{4}\right).$$

69. (a) Use the Laplace transform to solve the heat conduction problem

$$u_t = \kappa u_{xx}, \qquad 0 < x < \infty, \quad t > 0,$$
$$u(x,0) = 0, \quad x > 0,$$
$$u(0,t) = f(t), \quad u(x,t) \to 0 \quad \text{as} \quad x \to \infty, \quad t > 0.$$

(b) Derive Duhamel's formula

$$u(x,t) = \int_0^t f(t-\tau) \left(\frac{\partial u_0}{\partial \tau}\right) d\tau,$$

where

$$\left(\frac{\partial u_0}{\partial t}\right) = \frac{x}{\sqrt{4\pi\kappa}} t^{-3/2} \exp\left(-\frac{x^2}{4\kappa t}\right).$$

13

Nonlinear Partial Differential Equations with Applications

"True Laws of Nature cannot be linear."

Albert Einstein

"... the progress of physics will to a large extent depend on the progress of nonlinear mathematics, of methods to solve nonlinear equations ... and therefore we can learn by comparing different nonlinear problems."

Werner Heisenberg

13.1 Introduction

The three-dimensional linear wave equation

$$u_{tt} = c^2 \, \nabla^2 u, \tag{13.1.1}$$

arises in the areas of elasticity, fluid dynamics, acoustics, magnetohydrodynamics, and electromagnetism.

The general solution of the one-dimensional equation (13.1.1) is

$$u(x,t) = \phi(x - ct) + \psi(x + ct), \tag{13.1.2}$$

where ϕ and ψ are determined by the initial or boundary conditions. Physically, ϕ and ψ represent waves moving with constant speed c and without change of shape, along the positive and the negative directions of x respectively.

The solutions ϕ and ψ correspond to the two factors when the one-dimensional equation (13.1.1) is written in the form

$$\left(\frac{\partial}{\partial t} + c \frac{\partial}{\partial x} \right) \left(\frac{\partial}{\partial t} - c \frac{\partial}{\partial x} \right) u = 0. \tag{13.1.3}$$

Obviously, the simplest linear wave equation is

$$u_t + c\,u_x = 0, \tag{13.1.4}$$

and its solution $u = \phi\,(x - ct)$ represents a wave moving with a constant velocity c in the positive x-direction without change of shape.

13.2 One-Dimensional Wave Equation and Method of Characteristics

The simplest first-order nonlinear wave equation is given by

$$u_t + c\,(u)\,u_x = 0, \qquad -\infty < x < \infty, \qquad t > 0, \tag{13.2.1}$$

where $c\,(u)$ is a given function of u.

We solve this nonlinear equation subject to the initial condition

$$u\,(x, 0) = f\,(x), \qquad -\infty < x < \infty. \tag{13.2.2}$$

Before we discuss the method of solution, the following comments are in order. First, unlike linear differential equations, the principle of superposition cannot be applied to find the general solution of nonlinear partial differential equations. Second, the effect of nonlinearity can change the entire nature of the solution. Third, a study of the above initial-value problem reveals most of the important ideas for nonlinear hyperbolic waves. Finally, a large number of physical and engineering problems are governed by the above nonlinear system or an extension of it.

Although the nonlinear system governed by (13.2.1)–(13.2.2) looks simple, it poses nontrivial problems in applied mathematics, and it leads surprisingly to new phenomena. We solve the system by the method of characteristics.

In order to construct continuous solutions, we consider the total differential du given by

$$du = \frac{\partial u}{\partial t}dt + \frac{\partial u}{\partial x}dx, \tag{13.2.3}$$

so that the points (x, t) are assumed to lie on a curve Γ. Then, dx/dt represents the slope of the curve Γ at any point P on Γ. Thus, equation (13.2.3) becomes

$$\frac{du}{dt} = u_t + \left(\frac{dx}{dt}\right)u_x. \tag{13.2.4}$$

It follows from this result that (13.2.1) can be regarded as the ordinary differential equation

$$\frac{du}{dt} = 0, \tag{13.2.5}$$

along any member of the family of curves Γ which are the solution curves of

$$\frac{dx}{dt} = c\,(u).$$

(13.2.6)

These curves Γ are called the *characteristic curves* of the main equation (13.2.1). Thus, the solution of (13.2.1) has been reduced to the solution of a pair of simultaneous ordinary differential equations (13.2.5) and (13.2.6). Clearly, both the characteristic speed and the characteristics depend on the solution u.

Equation (13.2.5) implies that $u = $ constant along each characteristic curve Γ, and each $c\,(u)$ remains constant on Γ. Therefore, (13.2.6) shows that the characteristic curves of (13.2.1) form a family of straight lines in the (x, t)-plane with slope $c\,(u)$. This indicates that the general solution of (13.2.1) depends on finding the family of lines. Also, each line with slope $c\,(u)$ corresponds to the value of u on it. If the initial point on the characteristic curve Γ is denoted by ξ and if one of the curves Γ intersects $t = 0$ at $x = \xi$, then $u\,(\xi, 0) = f\,(\xi)$ on the whole of that curve Γ as shown in Figure 13.2.1.

Thus, we have the following characteristic form on Γ:

$$\frac{dx}{dt} = c\,(u), \qquad x\,(0) = \xi,$$

(13.2.7)

$$\frac{du}{dt} = 0, \qquad u\,(\xi, 0) = f\,(\xi).$$

(13.2.8)

These constitute a pair of coupled ordinary differential equations on Γ.

Equation (13.2.7) cannot be solved independently because c is a function of u. However, (13.2.8) can readily be solved to obtain $u = $ constant and

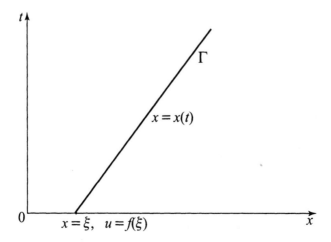

Figure 13.2.1 A characteristic curve.

hence, $u = f(\xi)$ on the whole of Γ. Thus, (13.2.7) leads to

$$\frac{dx}{dt} = F(\xi), \qquad x(0) = \xi, \qquad (13.2.9)$$

where

$$F(\xi) = c(f(\xi)). \qquad (13.2.10)$$

Integrating equation (13.2.9) gives

$$x = tF(\xi) + \xi. \qquad (13.2.11)$$

This represents the characteristic curve which is a straight line whose slope is not a constant, but depends on ξ.

Combining these results, we obtain the solution of the initial-value problem in parametric form

$$\left. \begin{array}{rcl} u(x,t) &=& f(\xi) \\ x &=& \xi + tF(\xi) \end{array} \right\}, \qquad (13.2.12)$$

where

$$F(\xi) = c(f(\xi)). $$

We next verify that this final form represents an analytic expression of the solution. Differentiating (13.2.12) with respect to x and t, we obtain

$$u_x = f'(\xi)\,\xi_x, \qquad u_t = f'(\xi)\,\xi_t,$$
$$1 = \{1 + tF'(\xi)\}\,\xi_x,$$
$$0 = F(\xi) + \{1 + tF'(\xi)\}\,\xi_t.$$

Elimination of ξ_x and ξ_t gives

$$u_x = \frac{f'(\xi)}{1 + tF'(\xi)}, \qquad u_t = -\frac{F(\xi)\,f'(\xi)}{1 + tF'(\xi)}. \qquad (13.2.13)$$

Since $F(\xi) = c(f(\xi))$, equation (13.2.1) is satisfied provided $1 + tF'(\xi) \neq 0$.

The solution (13.2.12) also satisfies the initial condition at $t = 0$, since $\xi = x$, and the solution (13.2.12) is unique.

Suppose that $u(x,t)$ and $v(x,t)$ are two solutions. Then, on $x = \xi + tF(\xi)$,

$$u(x,t) = u(\xi,0) = f(\xi) = v(x,t).$$

Thus, we have proved the following:

Theorem 13.2.1. *The nonlinear initial-value problem*

$$u_t + c(u) u_x = 0, \qquad -\infty < x < \infty, \qquad t > 0,$$
$$u(x, t) = f(x), \quad at \quad t = 0, \qquad -\infty < x < \infty,$$

has a unique solution provided $1 + t F'(\xi) \neq 0$, f *and* c *are* $C^1(R)$ *functions where* $F(\xi) = c(f(\xi))$.

The solution is given in the parametric form:

$$u(x, t) = f(\xi),$$
$$x = \xi + t F(\xi).$$

Remark: When $c(u) = \text{constant} = c > 0$, equation (13.2.1) becomes the linear wave equation (13.1.4). The characteristic curves are $x = ct + \xi$ and the solution u is given by

$$u(x, t) = f(\xi) = f(x - ct).$$

Physical Significance of (13.2.12).

We assume $c(u) > 0$. The graph of u at $t = 0$ is the graph of f. In view of the fact

$$u(x, t) = u(\xi + t F(\xi), t) = f(\xi)$$

the point $(\xi, f(\xi))$ moves parallel to the x-axis in the positive direction through a distance $t F(\xi) = ct$, and the distance moved $(x = \xi + ct)$ depends on ξ. This is a typical nonlinear phenomenon. In the linear case, the curve moves parallel to the x-axis with constant velocity c, and the solution represents waves travelling without change of shape. Thus, there is a striking difference between the linear and the nonlinear solution.

Theorem 13.2.1 asserts that the solution of the nonlinear initial-value problem exists provided

$$1 + t F'(\xi) \neq 0, \qquad x = \xi + t F'(\xi). \tag{13.2.14}$$

However, the former condition is always satisfied for sufficiently small time t. By a solution of the problem, we mean a differentiable function $u(x, t)$. It follows from results (13.2.13) that both u_x and u_t tend to infinity as $1 + t F'(\xi) \to 0$. This means that the solution develops a singularity (discontinuity) when $1 + t F'(\xi) = 0$. We consider a point $(x, t) = (\xi, 0)$ so that this condition is satisfied on the characteristics through the point $(\xi, 0)$ at a time t such that

$$t = -\frac{1}{F'(\xi)} \tag{13.2.15}$$

which is positive provided $F'(\xi) = c'(f) f'(\xi) < 0$. If we assume $c'(f) > 0$, the above inequality implies that $f'(\xi) < 0$. Hence, the solution (13.2.12) ceases to exist for all time if the initial data is such that $f'(\xi) < 0$ for some value of ξ. Suppose $t = \tau$ is the time when the solution first develops a singularity (discontinuity) for some value of ξ. Then

$$\tau = -\frac{1}{\min_{-\infty < \xi < \infty} \{c'(f) f'(\xi)\}} > 0.$$

We draw the graphs of the nonlinear solution $u(x, t) = f(\xi)$ below for different values of $t = 0, \tau, 2\tau, \ldots$ The shape of the initial curve for $u(x, t)$ changes with increasing values of t, and the solution becomes multiple-valued for $t \geq \tau$. Therefore, the solution breaks down when $F'(\xi) < 0$ for some ξ, and such breaking is a typical nonlinear phenomenon. In linear theory, such breaking will never occur.

More precisely, the development of a singularity in the solution for $t \geq \tau$ can be seen by the following consideration. If $f'(\xi) < 0$, we can find two values of $\xi = \xi_1, \xi_2$ ($\xi_1 < \xi_2$) on the initial line such that the characteristics through them have different slopes $1/c(u_1)$ and $1/c(u_2)$ where $u_1 = f(\xi_1)$ and $u_2 = f(\xi_2)$ and $c(u_2) < c(u_1)$. Thus, these two characteristics will intersect at a point in the (x, t)-plane for some $t > 0$. Since the characteristics carry constant values of u, the solution ceases to be single-valued at their point of intersection. Figure 13.2.2 shows that the wave profile progressively distorts itself, and at any instant of time there exists an interval on the x-axis, where u assumes three values for a given x. The end result is the development of a nonunique solution, and this leads to breaking.

Therefore, when conditions (13.2.14) are violated the solution develops a discontinuity known as a *shock*. The analysis of shock involves extension of a solution to allow for discontinuities. Also, it is necessary to impose on the solution certain restrictions to be satisfied across its discontinuity. This point will be discussed further in a subsequent section.

13.3 Linear Dispersive Waves

We consider a single linear partial differential equation with constant coefficients in the form

$$P\left(\frac{\partial}{\partial t}, \frac{\partial}{\partial x}, \frac{\partial}{\partial y}, \frac{\partial}{\partial z}\right) u(\mathbf{x}, t) = 0, \tag{13.3.1}$$

where P is a polynomial in partial derivatives and $\mathbf{x} = (x, y, z)$.

We seek an elementary plane wave solution of (13.3.1) in the form

$$u(\mathbf{x}, t) = a\, e^{i(\boldsymbol{\kappa} \cdot \mathbf{x} - \omega t)}, \tag{13.3.2}$$

where a is the amplitude, $\boldsymbol{\kappa} = (k, l, m)$ is the wavenumber vector, ω is the frequency and a, $\boldsymbol{\kappa}$, ω are constants. When this plane wave solution is

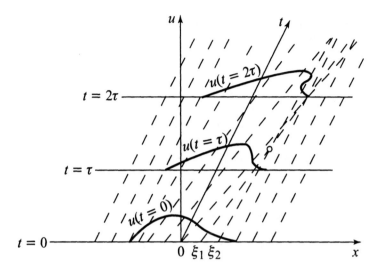

Figure 13.2.2 The solution $u(x,t)$ for different times $t = 0$, τ and 2τ; the characteristics are shown by the dotted lines; two of them from $x = \xi_1$ and $x = \xi_2$ intersect at $t > \tau$.

substituted in the equation, $\partial/\partial t$, $\partial/\partial x$, $\partial/\partial y$, and $\partial/\partial z$ produce factors $-i\omega$, ik, il, and im respectively, and the solution exists provided ω and κ are related by an equation

$$P(-i\omega, ik, il, im) = 0. \qquad (13.3.3)$$

This equation is known as the *dispersion relation*. Evidently, we have a direct correspondence between equation (13.3.1) and the dispersion relation (13.3.3) through the correspondence

$$\frac{\partial}{\partial t} \leftrightarrow -i\omega, \qquad \left(\frac{\partial}{\partial x}, \frac{\partial}{\partial y}, \frac{\partial}{\partial z}\right) \leftrightarrow i(k, l, m). \qquad (13.3.4)$$

Equation (13.3.1) and the corresponding dispersion relation (13.3.3) indicate that the former can be derived from the latter and vice-versa by using (13.3.4). The dispersion relation characterizes the plane wave motion. In many problems, the dispersion relation can be written in the explicit form

$$\omega = W(k, l, m). \qquad (13.3.5)$$

The *phase* and *the group velocities* of the waves are defined by

$$\mathbf{C}_p(\boldsymbol{\kappa}) = \frac{\omega}{\kappa}\,\hat{\boldsymbol{\kappa}}, \qquad (13.3.6)$$

$$\mathbf{C}_g(\boldsymbol{\kappa}) = \nabla_{\boldsymbol{\kappa}}\omega, \qquad (13.3.7)$$

where $\widehat{\boldsymbol{\kappa}}$ is the unit vector in the direction of wave vector $\boldsymbol{\kappa}$.

In the one-dimensional case, (13.3.5)–(13.3.7) become

$$\omega = W(k), \qquad C_p = \frac{\omega}{k}, \qquad C_g = \frac{d\omega}{dk}. \qquad (13.3.8)$$

The one-dimensional waves given by (13.3.2) are called *dispersive* if the group velocity $C_g \equiv \omega'(k)$ is not constant, that is, $\omega''(k) \neq 0$. Physically, as time progresses, the different waves disperse in the medium with the result that a single hump breaks into wavetrains.

Example 13.3.1.

(i) Linearized one-dimensional wave equation

$$u_{tt} - c^2 u_{xx} = 0, \qquad \omega = \pm\, ck. \qquad (13.3.9)$$

(ii) Linearized Korteweg and de Vries (KdV) equation for long water waves

$$u_t + \alpha u_x + \beta u_{xxx} = 0, \qquad \omega = \alpha k - \beta k^3. \qquad (13.3.10)$$

(iii) Klein–Gordon equation

$$u_{tt} - c^2 u_{xx} + \alpha^2 u = 0, \qquad \omega = \pm \left(c^2 k^2 + \alpha^2\right)^{\frac{1}{2}}. \qquad (13.3.11)$$

(iv) Schrödinger equation in quantum mechanics and de Broglie waves

$$i\hbar\, \psi_t - \left(V - \frac{\hbar^2}{2m}\nabla^2\right)\psi = 0, \qquad \hbar\omega = \frac{\hbar^2\kappa^2}{2m} + V, \qquad (13.3.12)$$

where V is a constant potential energy, and $h = 2\pi\hbar$ is the Planck constant.

The group velocity of de Broglie wave is $(\hbar\boldsymbol{\kappa}/m)$, and through the correspondence principle, $\hbar\omega$ is to be interpreted as the total energy, $\left(\hbar^2\kappa^2/2m\right)$ as the kinetic energy, and $\hbar\boldsymbol{\kappa}$ as the particle momentum. Hence, the group velocity is the classical particle velocity.

(v) Equation for vibration of a beam

$$u_{tt} + \alpha^2 u_{xxxx} = 0, \qquad \omega = \pm\, \alpha k^2. \qquad (13.3.13)$$

(vi) The dispersion relation for water waves in an ocean of depth h

$$\omega^2 = gk \tanh kh, \qquad (13.3.14)$$

where g is the acceleration due to gravity.

(vii) The Boussinesq equation

$$u_{tt} - \alpha^2 \nabla^2 u - \beta^2 \nabla^2 u_{tt} = 0, \qquad \omega = \pm \frac{\alpha\kappa}{\sqrt{1 + \beta^2 \kappa^2}}. \qquad (13.3.15)$$

This equation arises in elasticity for longitudinal waves in bars, long water waves, and plasma waves.

(viii) Electromagnetic waves in dielectrics

$$\left(u_{tt} + \omega_0^2 u\right)\left(u_{tt} - c_0^2 u_{xx}\right) - \omega_p^2 u_{tt} = 0,$$
$$\left(\omega^2 - \omega_0^2\right)\left(\omega^2 - c_0^2 k^2\right) - \omega_p^2 \omega^2 = 0, \qquad (13.3.16)$$

where ω_0 is the natural frequency of the oscillator, c_0 is the speed of light, and ω_p is the plasma frequency.

In view of the superposition principle, the general solution can be obtained from (13.3.2) with the dispersion solution (13.3.3). For the one-dimensional case, the general solution has the Fourier integral representation

$$u\left(x, t\right) = \int_{-\infty}^{\infty} F\left(k\right) e^{i[kx - tW(k)]} dk, \qquad (13.3.17)$$

where $F\left(k\right)$ is chosen to satisfy the initial or boundary data provided the data are physically realistic enough to have Fourier transforms.

In many cases, as cited in Example 13.3.1, there are two modes $\omega = \pm W\left(k\right)$ so that the solution (13.3.17) has the form

$$u\left(x, t\right) = \int_{-\infty}^{\infty} F_1\left(k\right) e^{i[kx - tW(k)]} dk + \int_{-\infty}^{\infty} F_2\left(k\right) e^{i[kx - tW(k)]} dk, \qquad (13.3.18)$$

with the initial data at $t = 0$

$$u\left(x, t\right) = \phi\left(x\right), \qquad u_t\left(x, t\right) = \psi\left(x\right). \qquad (13.3.19)$$

The initial conditions give

$$\phi\left(x\right) = \int_{-\infty}^{\infty} [F_1\left(k\right) + F_2\left(k\right)] e^{ikx} dk,$$

$$\psi\left(x\right) = -i \int_{-\infty}^{\infty} [F_1\left(k\right) + F_2\left(k\right)] W\left(k\right) e^{ikx} dk.$$

Applying the Fourier inverse transformations, we have

$$F_1\left(k\right) + F_2\left(k\right) = \Phi\left(k\right) = \frac{1}{\sqrt{2\pi}} \int_{-\infty}^{\infty} \phi\left(x\right) e^{-ikx} dx,$$

$$-iW\left(k\right) [F_1\left(k\right) - F_2\left(k\right)] = \Psi\left(k\right) = \frac{1}{\sqrt{2\pi}} \int_{-\infty}^{\infty} \psi\left(x\right) e^{-ikx} dx,$$

so that

$$[F_1(k) + F_2(k)] = \frac{1}{2}\left[\Phi(k) \pm \frac{i\Psi(k)}{W(k)}\right]. \qquad (13.3.20)$$

The asymptotic behavior of $u(x,t)$ for large t with fixed (x/t) can be obtained by the Kelvin stationary phase approximation. For real $\phi(x)$, $\psi(x)$, $\Phi(-k) = \Phi^*(k)$ and $\Psi(-k) = \Psi^*(k)$, where the asterisk denotes a complex conjugate. It follows from (13.3.20) that, for $W(k)$ even

$$[F_1(-k), F_2(-k)] = [F_2^*(k), F_1^*(k)], \qquad (13.3.21)$$

and for $W(k)$ odd,

$$[F_1(-k), F_2(-k)] = [F_1^*(k), F_2^*(k)]. \qquad (13.3.22)$$

In particular, when $\phi(x) = \delta(x)$ and $\psi(x) \equiv 0$, then $F_1(k) = F_2(k) = 1/\sqrt{8\pi}$, and the solution (13.3.18) reduces to the form

$$u(x,t) = \sqrt{\frac{2}{\pi}} \int_0^\infty \cos kx \cos\{tW(k)\}\, dk. \qquad (13.3.23)$$

In order to obtain the asymptotic approximation by the Kelvin stationary phase method (see Section 12.7) for $t \to \infty$, we consider both cases when $W(k)$ is even ($W'(k)$ is odd) and when $W(k)$ is odd ($W'(k)$ is even) and make an extra reasonable assumption that $W'(k)$ is monotonic and positive for $k > 0$. It turns out that the asymptotic solution for $t \to \infty$ is

$$u(x,t) \sim 2\operatorname{Re}\left\{F_1(k)\left\{\frac{2\pi}{t\,|W''(k)|}\right\}^{\frac{1}{2}} \exp\left[i\left\{\theta(x,t) - \frac{\pi}{4}\operatorname{sgn}W''(k)\right\}\right]\right\}$$

$$+ O\left(\frac{1}{t}\right),$$

$$= \operatorname{Re}\left[a(x,t)\, e^{i\theta(x,t)}\right], \qquad (13.3.24)$$

where $k(x,t)$ is the positive root of the equation

$$W'(k) = \frac{x}{t}, \qquad \omega = W(k), \qquad \frac{x}{t} > 0, \qquad (13.3.25ab)$$

$$\theta(x,t) = x\,k(x,t) - t\,\omega(x,t), \qquad (13.3.26)$$

and

$$a(x,t) = 2F_1(k)\left\{\frac{2\pi}{t\,|W''(k)|}\right\}^{\frac{1}{2}} \exp\left\{-\frac{i\pi}{4}\operatorname{sgn}W''(k)\right\}. \qquad (13.3.27)$$

It is important to point out that solution (13.3.24) has a form similar to that of the elementary plane wave solution, but k, ω, and a are no longer constants; they are functions of space variable x and time t. The solution still represents an oscillatory wavetrain with the phase function $\theta(x,t)$ describing the variations between local maxima and minima. Unlike the elementary plane wavetrain, the present asymptotic result (13.3.24) represents a nonuniform wavetrain in the sense that the amplitude, the distance, and the time between successive maxima are not constant.

It also follows from (13.3.25a) that

$$\frac{k_t}{k} = -\frac{W'(k)}{kW''(k)t}\frac{1}{t} \sim O\left(\frac{1}{t}\right), \qquad (13.3.28)$$

$$\frac{k_x}{k} = -\frac{1}{kW''(k)}\frac{1}{t} \sim O\left(\frac{1}{t}\right). \qquad (13.3.29)$$

These results indicate the $k(x,t)$ is a slowly varying function of x and t as $t \to \infty$. Applying a similar argument to ω and a, we conclude that $k(x,t)$, $\omega(x,t)$, and $a(x,t)$ are slowly varying functions of x and t as $t \to \infty$.

Finally, all these results seem to provide an important clue for natural generalization of the concept of nonlinear and nonuniform wavetrains.

13.4 Nonlinear Dispersive Waves and Whitham's Equations

To describe a slowly varying nonlinear and nonuniform oscillatory wavetrain in a medium (see Whitham, 1974), we assume the existence of a solution in the form (13.3.24) so that

$$u(x,t) = a(x,t)e^{i\theta(x,t)} + \text{c.c.}, \qquad (13.4.1)$$

where c.c. stands for the complex conjugate, $a(x,t)$ is the complex amplitude given by (13.3.27), and the phase function $\theta(x,t)$ is

$$\theta(x,t) = xk(x,t) - t\omega(x,t), \qquad (13.4.2)$$

and k, ω, and a are slowly varying function of x and t.

Due to slow variations of k and ω, it is reasonable to assume that these quantities still satisfy the dispersion relation

$$\omega = W(k). \qquad (13.4.3)$$

Differentiating (13.4.2) with respect to x and t respectively, we obtain

$$\theta_x = k + \{x - tW'(k)\}k_x, \qquad (13.4.4)$$
$$\theta_t = -W(k) + \{x - tW'(k)\}k_t. \qquad (13.4.5)$$

In the neighborhood of stationary points defined by (13.3.25a), these results become

$$\theta_x = k\left(x, t\right), \qquad \theta_t = -\omega\left(x, t\right). \tag{13.4.6}$$

These results can be used as a definition of local wavenumber and local frequency of the slowly varying nonlinear wavetrain.

In view of (13.4.6), relation (13.4.3) gives a nonlinear partial differential equation for the phase θ in the form

$$\frac{\partial \theta}{\partial t} + W\left(\frac{\partial \theta}{\partial x}\right) = 0. \tag{13.4.7}$$

The solution of this equation determines the geometry of the wave pattern.

However, it is convenient to eliminate θ from (13.4.6) to obtain

$$\frac{\partial k}{\partial t} + \frac{\partial \omega}{\partial t} = 0. \tag{13.4.8}$$

This is known as the *Whitham equation* for the conservation of waves, where k represents the density of waves and ω is the flux of waves.

Using the dispersion relation (13.4.3), we obtain

$$\frac{\partial k}{\partial t} + C_g\left(k\right)\frac{\partial k}{\partial x} = 0, \tag{13.4.9}$$

where $C_g\left(k\right) = W'\left(k\right)$ is the group velocity. This represents the simplest nonlinear wave (hyperbolic) equation for the propagation of k with the group velocity $C_g\left(k\right)$.

Since equation (13.4.9) is similar to (13.2.1), we can use the analysis of Section 13.2 to find the general solution of (13.4.9) with the initial condition $k\left(x, 0\right) = f\left(x\right)$ at $t = 0$. In this case, the solution has the form

$$k\left(x, t\right) = f\left(\xi\right), \qquad x = \xi + t F\left(\xi\right), \tag{13.4.10}$$

where $F\left(\xi\right) = C_g\left(f\left(\xi\right)\right)$. This further confirms the propagation of k with the velocity C_g. Some physical interpretations of this kind of solution have already been discussed in Section 13.2.

Equations (13.4.9) and (13.4.3) reveal that ω also satisfies the nonlinear wave (hyperbolic) equation

$$\frac{\partial \omega}{\partial t} + W'\left(k\right)\frac{\partial \omega}{\partial x} = 0. \tag{13.4.11}$$

It follows from equations (13.4.9) and (13.4.11) that both k and ω remain constant on the characteristic curves defined by

$$\frac{dx}{dt} = W'\left(k\right) = C_g\left(k\right), \tag{13.4.12}$$

in the (x, t) plane. Since k and ω is constant on each curve, the characteristic curves are straight lines with slope $C_g\left(k\right)$. The solution for k is given by (13.4.10).

Finally, it follows from the above analysis that any constant value of the phase θ propagates according to $\theta(x,t) = $ constant, and hence,

$$\theta_t + \left(\frac{dx}{dt}\right)\theta_x = 0, \tag{13.4.13}$$

which gives, by (13.4.6),

$$\frac{dx}{dt} = -\frac{\theta_t}{\theta_x} = \frac{\omega}{k} = C_p. \tag{13.4.14}$$

Thus, the phase of the waves propagates with the phase speed C_p. On the other hand, (13.4.9) ensures that the wavenumber k propagates with the group velocity $C_g(k) = (d\omega/dk) = W'(k)$.

We next investigate how the wave energy propagates in the dispersive medium. We consider the following integral involving the square of the wave amplitude (energy) given by (13.3.24) between any two points $x = x_1$ and $x = x_2 \, (0 < x_1 < x_2)$

$$Q(t) = \int_{x_1}^{x_2} |a|^2 \, dx = \int_{x_1}^{x_2} aa^* dx, \tag{13.4.15}$$

$$= 8\pi \int_{x_1}^{x_2} \frac{F_1(k) F_1^*(k)}{t\,|W''(k)|} dx, \tag{13.4.16}$$

which is, due to a change of variable $x = t\,W'(k)$,

$$= 8\pi \int_{k_1}^{k_2} F_1(k) F_1^*(k) \, dk, \tag{13.4.17}$$

where $k_r = t\,W'(k_r)$, $r = 1, 2$.

When k_r is kept fixed as t varies, $Q(t)$ remains constant so that

$$0 = \frac{dQ}{dt} = \frac{d}{dt}\int_{x_1}^{x_2} |a|^2 \, dx,$$

$$= \int_{x_1}^{x_2} \frac{\partial}{\partial t}|a|^2 \, dx + |a|_2^2 \, W'(k_2) - |a|_1^2 \, W'(k_1). \tag{13.4.18}$$

In the limit $x_2 - x_1 \to 0$, this result reduces to the partial differential equation

$$\frac{\partial}{\partial t}|a|^2 + \frac{\partial}{\partial x}\left[W'(k)|a|^2\right] = 0. \tag{13.4.19}$$

This represents the equation for the conservation of wave energy, where $|a|^2$ and $|a|^2 W'(k)$ are the *energy density* and *energy flux* respectively. It also follows that the energy propagates with the group velocity $W'(k)$. It has been shown that the wavenumber k also propagates with the group velocity. Evidently, the group velocity plays a double role.

The above analysis reveals another important fact; equations (13.4.3), (13.4.8), and (13.4.19) constitute a closed set of equations for the three quantities k, ω, and a. Indeed, these are the fundamental equations for nonlinear dispersive waves and are known as *Whitham's equations*.

13.5 Nonlinear Instability

For infinitesimal waves, the wave amplitude ($ak \ll 1$) is very small, so that nonlinear effects can be neglected altogether. However, for finite amplitude waves the terms involving a^2 cannot be neglected, and the effects of nonlinearity become important. In the theory of water waves, Stokes first obtained the connection due to inherent nonlinearity between the wave-profile and the frequency of a steady periodic wave system. According to the Stokes theory, the remarkable fact is the dependence of ω on a which couples (13.4.8) to (13.4.19). This leads to a new nonlinear phenomenon.

For finite amplitude waves, the frequency ω has the Stokes expansion

$$\omega = \omega_0\left(k\right) + a^2 \omega_2\left(k\right) + \ldots = \omega\left(k, a^2\right). \tag{13.5.1}$$

This can be regarded as the nonlinear dispersion relation which depends on both k and a^2. In the linear case, the amplitude $a \to 0$, (13.5.1) gives the linear dispersion relation (13.4.3), that is, $\omega = \omega_0\left(k\right)$.

In order to discuss nonlinear instability, we substitute (13.5.1) into (13.4.8) and retain (13.4.19) in the linear approximation to obtain the following coupled system:

$$\frac{\partial k}{\partial t} + \frac{\partial}{\partial x} + \left\{\omega_0\left(k\right) + \omega_2\left(k\right)a^2\right\} = 0, \tag{13.5.2}$$

$$\frac{\partial a^2}{\partial t} + \frac{\partial}{\partial x}\left\{\omega_0'\left(k\right)a^2\right\} = 0, \tag{13.5.3}$$

where $W\left(k\right) \equiv \omega_0\left(k\right)$.

These equations can be further approximated to obtain

$$\frac{\partial k}{\partial t} + \omega_0'\frac{\partial k}{\partial x} + \omega_2\frac{\partial a^2}{\partial x} = O\left(a^2\right), \tag{13.5.4}$$

$$\frac{\partial a^2}{\partial t} + \omega_0'\frac{\partial a^2}{\partial x} + \omega_0''a^2\frac{\partial k}{\partial x} = 0. \tag{13.5.5}$$

In matrix form, these equations read

$$\begin{pmatrix} \omega_0' & \omega_2 \\ \omega_0''a^2 & \omega_0' \end{pmatrix} \begin{pmatrix} \frac{\partial k}{\partial x} \\ \frac{\partial a^2}{\partial x} \end{pmatrix} + \begin{pmatrix} 1 & 0 \\ 0 & 1 \end{pmatrix} \begin{pmatrix} \frac{\partial k}{\partial t} \\ \frac{\partial a^2}{\partial t} \end{pmatrix} = 0. \tag{13.5.6}$$

Hence, the eigenvalues λ are the roots of the determinant equation

$$|a_{ij} - \lambda\, b_{ij}| = \begin{vmatrix} \omega_0' - \lambda & \omega_2 \\ \omega_0'' a^2 & \omega_0' - \lambda \end{vmatrix} = 0, \tag{13.5.7}$$

where a_{ij} and b_{ij} are the coefficient matrices of (13.5.6). This equation gives the characteristic velocities

$$\lambda = \frac{dx}{dt} = \omega_0' \pm (\omega_2 \omega_0'')^{\frac{1}{2}} a + O\left(a^2\right). \tag{13.5.8}$$

If $\omega_2 \omega_0'' > 0$, the characteristics are real and the system is hyperbolic. The double characteristic velocity splits into two separate real velocities. This provides a new extension of the group velocity to nonlinear problems. If the disturbance is initially finite in extent, it would eventually split into two disturbances. In general, any initial disturbance or modulating source would introduce disturbances in both families of characteristics. In the hyperbolic case, compressive modulation will progressively distort and steepen so that the question of breaking will arise. These results are remarkably different from those found in linear theory, where there is only one characteristic velocity and any hump may distort, due to the dependence of $\omega_0'(k)$ on k, but would never split.

On the other hand, if $\omega_2 \omega_0'' < 0$, the characteristics are complex and the system is elliptic. This leads to ill-posed problems. Any small perturbations in k and a will be given by the solutions of the form $\exp\left[i\alpha\left(x - \lambda t\right)\right]$ where λ is calculated from (13.5.8) for unperturbed values of k and a. In this elliptic case, λ is complex, and the perturbation will grow as $t \to \infty$. Hence, the original wavetrain will become unstable. In the linear theory, the elliptic case does not arise at all.

Example 13.5.1. For Stokes waves in deep water, the dispersion relation is

$$\omega = (gk)^{\frac{1}{2}}\left(1 + \frac{1}{2}k^2 a^2\right), \tag{13.5.9}$$

so that $\omega_0(k) = (gk)^{\frac{1}{2}}$ and $\omega_2(k) = \frac{1}{2}\sqrt{g}\, k^{\frac{3}{2}}$.

In this case, $\omega_0''(k) = -\frac{1}{4}\sqrt{g}\, k^{-\frac{3}{2}}$ so that $\omega_0'' \omega_2 = -\frac{g}{8}k < 0$. The conclusion is that Stokes waves in deep water are definitely unstable. This is one of the most remarkable results in the theory of nonlinear water waves discovered during the 1960's.

13.6 The Traffic Flow Model

We consider the flow of cars on a long highway under the assumptions that cars do not enter or leave the highway at any one of its points. We take the x-axis along the highway and assume that the traffic flows in the positive direction. Suppose $\rho(x,t)$ is the density representing the number of cars per unit length at the point x of the highway at time t, and $q(x,t)$ is the flow of cars per unit time.

We assume a conservation law which states that the change in the total amount of a physical quantity contained in any region of space must be equal to the flux of that quantity across the boundary of that region. In this case, the time rate of change of the total number of cars in any segment $x_1 \leq x \leq x_2$ of the highway is given by

$$\frac{d}{dt} \int_{x_1}^{x_2} \rho(x,t)\, dx = \int_{x_1}^{x_2} \frac{\partial \rho}{\partial t}\, dx. \qquad (13.6.1)$$

This rate of change must be equal to the net flux across x_1 and x_2 given by

$$q(x_1,t) - q(x_2,t) \qquad (13.6.2)$$

which measures the flow of cars entering the segment at x_1 minus the flow of cars leaving the segment at x_2. Thus, we have the conservation equation

$$\frac{d}{dt} \int_{x_1}^{x_2} \rho(x,t)\, dx = q(x_1,t) - q(x_2,t), \qquad (13.6.3)$$

or

$$\int_{x_1}^{x_2} \frac{\partial \rho}{\partial t}\, dx = -\int_{x_1}^{x_2} \frac{\partial q}{\partial x}\, dx,$$

or

$$\int_{x_1}^{x_2} \left(\frac{\partial \rho}{\partial t} + \frac{\partial q}{\partial x} \right) dx = 0. \qquad (13.6.4)$$

Since the integrand in (13.6.4) is continuous, and (13.6.4) holds for every segment $[x_1, x_2]$, it follows that the integrand must vanish so that we have the partial differential equation

$$\frac{\partial \rho}{\partial t} + \frac{\partial q}{\partial x} = 0. \qquad (13.6.5)$$

We now introduce an additional assumption which is supported by both theoretical and experimental findings. According to this assumption, the flow rate q depends on x and t only through the density, that is, $q = Q(\rho)$ for some function Q. This assumption seems to be reasonable in the sense that the density of cars surrounding a given car indeed controls the speed of that car. The functional relation between q and ρ depends on many factors, including speed limits, weather conditions, and road characteristics. Several specific relations are suggested by Haight (1963).

We consider here a particular relation $q = \rho v$ where v is the average local velocity of cars. We assume that v is a function of ρ to a first approximation. In view of this relation, (13.6.5) reduces to the nonlinear hyperbolic equation

$$\frac{\partial \rho}{\partial t} + c(\rho) \frac{\partial \rho}{\partial x} = 0, \tag{13.6.6}$$

where

$$c(\rho) = q'(\rho) = v + \rho v'(\rho). \tag{13.6.7}$$

In general, the local velocity $v(\rho)$ is a decreasing function of ρ so that $v(\rho)$ has a finite maximum value v_{max} at $\rho = 0$ and decreases to zero at $\rho = \rho_{max} = \rho_m$. For the value of $\rho = \rho_m$, the cars are bumper to bumper. Since $q = \rho v$, $q(\rho) = 0$ when $\rho = 0$ and $\rho = \rho_m$. This means that q is an increasing function of ρ until it attains a maximum value $q_{max} = q_M$ for some $\rho = \rho_M$ and then decreases to zero at $\rho = \rho_m$. Both $q(\rho)$ and $v(\rho)$ are shown in Figure 13.6.1.

Equation (13.6.6) is similar to (13.2.1) with the wave propagation velocity $c(\rho) = v(\rho) + \rho v'(\rho)$. Since $v'(\rho) < 0$, $c(\rho) < v(\rho)$, that is, the propagation velocity is less than the car velocity. In other words, waves propagate backwards through the stream of cars, and drivers are warned of disturbances ahead. It follows from Figure 13.6.1a that $q(\rho)$ is an increasing function in $[0, \rho_M]$, a decreasing function in $[\rho_M, \rho_m]$, and attains a maximum at ρ_M. Hence, $c(\rho) = q'(\rho)$ is positive in $[0, \rho_M]$, zero at ρ_M and negative in $[\rho_M, \rho_m]$. All these mean that waves propagate forward relative to the highway in $[0, \rho_M]$, are stationary at ρ_M, and then travel backwards in $[\rho_M, \rho_m]$.

We use Section 13.1 to solve the initial-value problem for the nonlinear equation (13.6.6) with the initial condition $\rho(x, 0) = f(x)$. The solution is

$$\rho(x, t) = f(\xi), \qquad x = \xi + t F(\xi), \tag{13.6.8}$$

where

$$F(\xi) = c(f(\xi)).$$

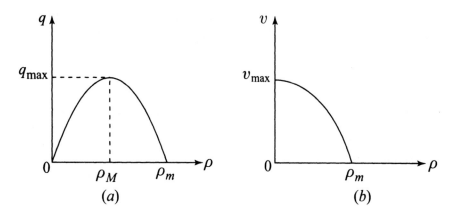

Figure 13.6.1 Graphs of $q(\rho)$ and $v(\rho)$.

Since $c'(\rho) = q''(\rho) < 0$, $q(\rho)$ is convex, and $c(\rho)$ is a decreasing function of ρ. This means that breaking occurs at the left due to formation of shock at the back. Waves propagate slower than the cars, so drivers enter such a local density increase from behind; they must decelerate rapidly through the shock but speed up slowly as they get out from the crowded area. These conclusions are in accord with observational results.

Actual observational data of traffic flow indicate that a typical result on a single lane highway is $\rho_m \approx 225$ cars per mile, $\rho_M \approx 80$ cars per mile, and $q_M \approx 1590$ cars per hour. Thus, the maximum flow rate q_M occurs at a low velocity $v = q_M/\rho_M \approx 20$ miles per hour.

13.7 Flood Waves in Rivers

We consider flood waves in a long rectangular river of constant breadth. We take the x-axis along the river which flows in the positive x-direction and assume that the disturbance is approximately the same across the breadth. In this problem, the depth $h(x,t)$ of the river plays the role of density in the traffic flow model discussed in Section 13.6. Let $q(x,t)$ be the flow per unit breadth and per unit time. According to the Conservation Law, the rate of change of the mass of the fluid in any section $x_1 \leq x \leq x_2$ must be balanced by the net flux across x_2 and x_1 so that the conservation equation becomes

$$\frac{d}{dt} \int_{x_1}^{x_2} h(x,t)\, dx + q(x_2,t) - q(x_1,t) = 0. \qquad (13.7.1)$$

An argument similar to the previous section gives

$$\frac{\partial h}{\partial t} + \frac{\partial q}{\partial x} = 0. \qquad (13.7.2)$$

Although the fluid flow is extremely complicated, we assume a simple function relation $q = Q(h)$ as a first approximation to express the increase in flow as the water level arises. Thus, equation (13.7.2) becomes

$$h_t + c(h)\, h_x = 0, \qquad (13.7.3)$$

where $c(h) = Q'(h)$ and $Q(h)$ is determined from the balance between the gravitational force and the frictional force of the river bed. This equation is similar to (13.2.1) and the method of solution has already been obtained in Section 13.2.

Here we discuss the velocity of wave propagation for some particular values of $Q(h)$. One such result is given by the Chezy result as

$$Q(h) = hv, \qquad (13.7.4)$$

where $v = \alpha\sqrt{h}$ is the velocity of fluid flow and α is a constant, so that the propagation velocity of flood waves is given by

$$c(h) = Q'(h) = \frac{3}{2}\alpha\sqrt{h} = \frac{3}{2}v. \tag{13.7.5}$$

Thus, the flood waves propagate one and a half times faster than the stream velocity.

For a general case where $v = \alpha h^n$,

$$Q(h) = hv = \alpha h^{n+1}, \tag{13.7.6}$$

so the propagation velocity of flood waves is

$$c(h) = Q'(h) = (n+1)v. \tag{13.7.7}$$

This result also indicates that flood waves propagate faster than the fluid.

13.8 Riemann's Simple Waves of Finite Amplitude

We consider a one-dimensional unsteady isentropic flow of gas of density ρ and pressure p with the direction of motion along the x-axis. Suppose $u(x,t)$ is the x-component of the velocity at time t and A is an area-element of the (y,z)-plane. The volume of the rectangular cylinder of height dx standing on the element A is then $A\,dx$ and its mass $A\rho_t\,dx\,dt$ is determined by the mass entering it, which is equal to $-A(\partial/\partial x)(\rho u)\,dx\,dt$. Its acceleration is $(u_t + uu_x)$ and the force impelling it in the positive x-direction is $-p_x A\,dx = -c^2\rho_x A\,dx$, where $p = f(\rho)$ and $c^2 = f'(\rho)$. These results lead to two coupled nonlinear partial differential equations

$$\frac{\partial\rho}{\partial t} + \frac{\partial}{\partial x}(\rho u) = 0, \tag{13.8.1}$$

$$(u_t + uu_x) + \frac{c^2}{\rho}p_x = 0. \tag{13.8.2}$$

In matrix form, this system is

$$A\frac{\partial U}{\partial x} + I\frac{\partial U}{\partial t} = 0, \tag{13.8.3}$$

where U, A and I are matrices given by

$$U = \begin{pmatrix} \rho \\ u \end{pmatrix}, \quad A = \begin{pmatrix} u & \rho \\ c^2/\rho & u \end{pmatrix} \quad \text{and} \quad I = \begin{pmatrix} 1 & 0 \\ 0 & 1 \end{pmatrix}. \tag{13.8.4}$$

The concept of characteristic curves introduced briefly in Section 13.2 requires generalization if it is to be applied to quasi-linear systems of first-order partial differential equations (13.8.1)–(13.8.2).

It is of interest to determine how a solution evolves with time t. Hence, we leave the time variable unchanged and replace the space variable x by some arbitrary curvilinear coordinate ξ so that the semi-curvilinear coordinate transformation from (x, t) to (ξ, t') can be introduced by

$$\xi = \xi(x, t), \qquad t' = t. \tag{13.8.5}$$

If the Jacobian of this transformation is nonzero, we can transform (13.8.3) by the following correspondence rule:

$$\frac{\partial}{\partial t} \equiv \frac{\partial \xi}{\partial t} \frac{\partial}{\partial \xi} + \frac{\partial t'}{\partial t} \cdot \frac{\partial}{\partial t'} = \frac{\partial \xi}{\partial t} \frac{\partial}{\partial \xi} + \frac{\partial}{\partial t'},$$

$$\frac{\partial}{\partial x} \equiv \frac{\partial \xi}{\partial x} \frac{\partial}{\partial \xi} + \frac{\partial t'}{\partial x} \frac{\partial}{\partial t'} = \frac{\partial \xi}{\partial x} \frac{\partial}{\partial \xi}.$$

This rule transforms (13.8.3) into the form

$$I \frac{\partial U}{\partial t'} + \left(\frac{\partial \xi}{\partial t} I + \frac{\partial \xi}{\partial x} A \right) \frac{\partial U}{\partial \xi} = 0. \tag{13.8.6}$$

This equation can be used to determine $\partial U / \partial \xi$ provided that the determinant of its coefficient matrix is non-zero. Obviously, this condition depends on the nature of the curvilinear coordinate curves $\xi(x, t) = \text{constant}$, which has been kept arbitrary. We assume now that the determinant vanishes for the particular choice $\xi = \eta$ so that

$$\left| \frac{\partial \eta}{\partial t} I + \frac{\partial \eta}{\partial x} A \right| = 0. \tag{13.8.7}$$

In view of this, $\partial U / \partial \eta$ will become indeterminate on the family of curves $\eta = \text{constant}$, and hence, $\partial U / \partial \eta$ may be discontinuous across the curves $\eta = \text{constant}$. This implies that each element of $\partial U / \partial \eta$ will be discontinuous across any of the curves $\eta = \text{constant}$. It is then necessary to find out how these discontinuities in the elements of $\partial U / \partial \eta$ are related across the curve $\eta = \text{constant}$. We next consider the solutions U which are everywhere continuous with discontinuous derivatives $\partial U / \partial \eta$ across the particular curve $\eta = \text{constant} = \eta_0$. Since U is continuous, elements of the matrix A are not discontinuous across $\eta = \eta_0$ so that A can be determined in the neighborhood of a point P on $\eta = \eta_0$. And since $\partial U / \partial t'$ is continuous everywhere, it is continuous across the curve $\eta = \eta_0$ at P.

In view of all of the above facts, it follows that differential equation (13.8.6) across the curve $\xi = \eta = \eta_0$ at P becomes

$$\left(\frac{\partial \eta}{\partial t} I + \frac{\partial \eta}{\partial x} A \right)_P \left[\frac{\partial U}{\partial \eta} \right]_P = 0, \tag{13.8.8}$$

where $[f]_P = f(P+) - f(P-)$ denotes the discontinuous jump in the quantity f across the curve $\eta = \eta_0$, and $f(P-)$ and $f(P+)$ represent the values to the immediate left and immediate right of the curve at P. Since P is any arbitrary point on the curve, $\partial/\partial\eta$ denotes the differentiation normal to the curves $\eta = $ constant so that equation (13.8.8) can be regarded as the compatibility condition satisfied by $\partial U/\partial\eta$ on either side of and normal to these curves in the (x, t)-plane.

Obviously, equation (13.8.8) is a homogeneous system of equations for the two jump quantities $[\partial U/\partial\eta]$. Therefore, for the existence of a non-trivial solution, the coefficient determinant must vanish, that is,

$$\left| \frac{\partial\eta}{\partial t} I + \frac{\partial\eta}{\partial x} A \right| = 0. \tag{13.8.9}$$

However, along the curves $\eta = $ constant, we have

$$0 = d\eta = \eta_t + \left(\frac{dx}{dt} \right) \eta_x, \tag{13.8.10}$$

so that these curves have the constant slope, λ

$$\frac{dx}{dt} = -\frac{\eta_t}{\eta_x} = \lambda. \tag{13.8.11}$$

Consequently, equations (13.8.9) and (13.8.8) can be expressed in terms of λ in the form

$$|A - \lambda I| = 0, \tag{13.8.12}$$

$$(A - \lambda I) \left[\frac{\partial U}{\partial \eta} \right] = 0, \tag{13.8.13}$$

where λ represents the eigenvalues of the matrix A, and $[\partial U/\partial\eta]$ is proportional to the corresponding right eigenvector of A.

Since A is a 2×2 matrix, it must have two eigenvalues. If these are real and distinct, integration of (13.8.11) leads to two distinct families of real curves Γ_1 and Γ_2 in the (x, t)-plane:

$$\Gamma_r : \frac{dx}{dt} = \lambda_r, \qquad r = 1, 2. \tag{13.8.14}$$

The families of curves Γ_r are called the *characteristic curves* of the system (13.8.3). Any one of these families of curves Γ_r may be chosen for the curvilinear coordinate curves $\eta = $ constant. The eigenvalues λ_r have the dimensions of velocity, and the λ_r associated with each family will then be the velocity of propagation of the matrix column vector $[\partial U/\partial\eta]$ along the curves Γ_r belonging to that family.

In this particular case, the eigenvalues λ of the matrix A are determined by (13.8.12), that is,

$$\begin{vmatrix} u - \lambda & \rho \\ c^2/\rho & u - \lambda \end{vmatrix} = 0, \tag{13.8.15}$$

so that

$$\lambda = \lambda_r = u \pm c, \quad r = 1, 2. \tag{13.8.16}$$

Consequently, the families of the characteristic curves Γ_r $(r = 1, 2)$ defined by (13.8.14) become

$$\Gamma_1 : \frac{dx}{dt} = u + c, \quad \text{and} \quad \Gamma_2 : \frac{dx}{dt} = u - c. \tag{13.8.17}$$

In physical terms, these results indicate that disturbances propagate with the sum of the velocities of the fluid and sound along the family of curves Γ_1. In the second family Γ_2, they propagate with the difference of the fluid velocity u and the sound velocity c.

The right eigenvectors $\mu_r \equiv \begin{pmatrix} \mu_r^{(1)} \\ \mu_r^{(2)} \end{pmatrix}$ are solutions of the equations

$$(A - \lambda_r I)\,\mu_r = 0, \quad r = 1, 2, \tag{13.8.18}$$

or,

$$\begin{pmatrix} u - \lambda_r & \rho \\ c^2/\rho & u - \lambda_r \end{pmatrix} \begin{pmatrix} \mu_r^{(1)} \\ \mu_r^{(2)} \end{pmatrix} = 0, \quad r = 1, 2. \tag{13.8.19}$$

This result combined with (13.8.13) gives

$$\begin{pmatrix} [\rho_\eta] \\ [u_\eta] \end{pmatrix} = \begin{pmatrix} \mu_r^{(1)} \\ \mu_r^{(2)} \end{pmatrix} = \alpha \begin{pmatrix} 1 \\ \pm c/\rho \end{pmatrix}, \quad r = 1, 2, \tag{13.8.20}$$

where α is a constant.

In other words, across a wavefront in the Γ_1 family of characteristic curves,

$$\frac{[\partial \rho / \partial \eta]}{1} = \frac{[\partial u / \partial \eta]}{c/\rho}, \tag{13.8.21}$$

and across a wavefront in the Γ_2 family of characteristic curves,

$$\frac{[\partial \rho / \partial \eta]}{1} = \frac{[\partial u / \partial \eta]}{-c/\rho}, \tag{13.8.22}$$

where c and ρ have values appropriate to the wavefront.

The above method of characteristics can be applied to a more general system

$$\frac{\partial U}{\partial t} + A \frac{\partial U}{\partial x} = 0, \tag{13.8.23}$$

where U is an $n \times 1$ matrix with elements $u_1, u_2, \ldots, u_n$ and A is an $n \times n$ matrix with elements a_{ij}. An argument similar to that given above leads to n eigenvalues of (13.8.13). If these eigenvalues are real and distinct, integration of equations (13.8.14) with $r = 1, 2, \ldots, n$ gives n distinct families of real curves Γ_r in the (x, t)-plane so that

$$\Gamma_r : \frac{dx}{dt} = \lambda_r, \qquad r = 1, 2, \ldots n. \tag{13.8.24}$$

When the eigenvalues λ_r of A are all real and distinct, there are n distinct linearly independent right eigenvectors μ_r of A satisfying the equation

$$A\mu_r = \lambda_r \mu_r,$$

where μ_r is an $n \times 1$ matrix with elements $\mu_r^{(1)}, \mu_r^{(2)}, \ldots, \mu_r^{(n)}$. Then across a wavefront belonging to the Γ_r family of characteristics, it turns out that

$$\frac{[\partial u_1 / \partial \eta]}{\mu_r^{(1)}} = \frac{[\partial u_2 / \partial \eta]}{\mu_r^{(2)}} = \cdots = \frac{[\partial u_n / \partial \eta]}{\mu_r^{(n)}}, \tag{13.8.25}$$

where the elements of μ_r are known on the wavefront.

In order to introduce the Riemann invariants, we form the linear combination of the eigenvectors $(\pm c/\rho, 1)$ with equations (13.8.1)–(13.8.2) to obtain

$$\pm \frac{c}{\rho} (\rho_t + \rho u_x + u\rho_x) + \left(u_t + uu_x + \frac{c^2}{\rho} \rho_x \right) = 0. \tag{13.8.26}$$

We use $\partial u / \partial \rho = \pm c/\rho$ from (13.8.21)–(13.8.22) and rewrite (13.8.26) as

$$\pm \frac{c}{\rho} [\rho_t + (u \pm c) \rho_x] + [u_t + (u \pm c) u_x] = 0. \tag{13.8.27}$$

In view of (13.8.17), equation (13.8.27) becomes

$$du \pm \frac{c}{\rho} d\rho = 0 \qquad \text{on} \quad \Gamma_r, \quad r = 1, 2, \tag{13.8.28}$$

or,

$$d[F(\rho) \pm u] = 0 \qquad \text{on} \quad \Gamma_r, \tag{13.8.29}$$

where

$$F(\rho) = \int_{\rho_0}^{\rho} \frac{c(\rho)}{\rho} d\rho. \tag{13.8.30}$$

Integration of (13.8.29) gives

$$F(\rho) + u = 2r \quad \text{on} \quad \Gamma_1 \quad \text{and} \quad F(\rho) - u = 2s \quad \text{on} \quad \Gamma_2, \qquad (13.8.31)$$

where $2r$ and $2s$ are constants of integration on Γ_1 and Γ_2, respectively.

The quantities r and s are called the *Riemann invariants*. As stated above, r is an arbitrary constant on characteristics Γ_1, and hence, in general, r will vary on each Γ_2. Similarly, s is constant on each Γ_2 but will vary on Γ_1. It is natural to introduce r and s as new curvilinear coordinates. Since r is constant on Γ_1, s can be treated as the parameter on Γ_1. Similarly, r can be regarded as the parameter on Γ_2. Then, $dx = (u \pm c)\, dt$ on Γ_r implies that

$$\frac{dx}{ds} = (u + c)\frac{dt}{ds} \quad \text{on} \quad \Gamma_1, \qquad (13.8.32)$$

$$\frac{dx}{dr} = (u - c)\frac{dt}{dr} \quad \text{on} \quad \Gamma_2. \qquad (13.8.33)$$

In fact, r is a constant on Γ_1, and s is a constant on Γ_2. Therefore, the derivatives in the two equations are really partial derivations with respect to s and r so that we can rewrite them as

$$\frac{\partial x}{\partial s} = (u + c)\frac{\partial t}{\partial s}, \qquad (13.8.34)$$

$$\frac{\partial x}{\partial r} = (u - c)\frac{\partial t}{\partial r}. \qquad (13.8.35)$$

These two first-order PDE's can, in general, be solved for $x = x(r,s)$, $t = t(r,s)$, and then, by inversion, r and s as functions x and t can be obtained. Once this is done, we use (13.8.31) to determine $u(x,t)$ and $\rho(x,t)$ in terms of r and s as

$$u(x,t) = r - s, \qquad F(\rho) = r + s. \qquad (13.8.36)$$

When one of the Riemann invariants r and s is constant throughout the flow, the corresponding solution is tremendously simplified. The solutions are known as *simple wave motions* representing simple waves in one direction only. The generating mechanisms of simple waves with their propagation laws can be illustrated by the *piston problem* in gas dynamics.

Example 13.8.1. Determine the Riemann invariants for a polytropic gas characterized by the law $p = k\rho^\gamma$, where k and γ are constants.

In this case

$$c^2 = \frac{dp}{d\rho} = k\gamma\rho^{\gamma-1}, \qquad F(\rho) = \int_0^\rho \frac{c(\rho)}{\rho} = \frac{2c}{\gamma - 1}.$$

Hence, the Riemann invariants are given by

$$\left(\frac{2c}{\gamma - 1}\right) c \pm u = (2r, 2s) \quad \text{on} \quad \Gamma_r. \qquad (13.8.37)$$

It is also possible to express the dependent variables u and c in terms of the Riemann invariants. It turns out that

$$u = r - s, \qquad c = \frac{\gamma - 1}{2}(r + s). \qquad (13.8.38)$$

Example 13.8.2. (The Piston Problem in a Polytropic Gas). The problem is to determine how a simple wave is produced by the prescribed motion of a piston in the closed end of a semi-infinite tube filled with gas.

This is a one-dimensional unsteady problem in gas dynamics. We assume that the gas is initially at rest with a uniform state $u = 0$, $\rho = \rho_0$, and $c = c_0$. The piston starts from rest at the origin and is allowed to withdraw from the tube with a variable velocity for a time t_1, after which the velocity of withdrawal remains constant. The piston path is shown by a dotted curve in Figure 13.8.1. In the (x, t)-plane, the path of the piston is given by $x = X(t)$ with $X(0) = 0$. The fluid velocity u is equal to the piston velocity $\dot{X}(t)$ on the piston $x = X(t)$, which will be used as the boundary condition for the piston.

The initial state of the gas is given by $u = u_0$, $\rho = \rho_0$, and $c = c_0$ at $t = 0$, in $x \geq 0$. The characteristic line Γ_0 that bounds it and passes through the origin is determined by the equation

$$\frac{dx}{dt} = (u + c)_{t=0} = c_0$$

so that the equation of the characteristic line Γ_0 is $x = c_0 t$.

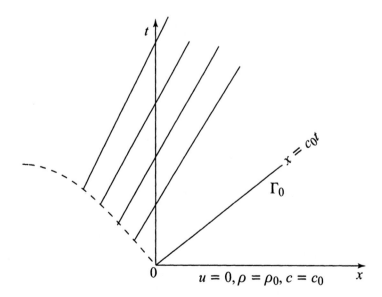

Figure 13.8.1 Simple waves generated by the motion of a piston.

In view of the uniform initial state, all the Γ_2 characteristics start on the x-axis so that the Riemann invariants s in (12.8.37b) must be constant and of the form

$$\frac{2c}{\gamma - 1} - u = \frac{2c_0}{\gamma - 1}, \tag{13.8.39}$$

or,

$$u = \frac{2\,(c - c_0)}{\gamma - 1}, \qquad c = c_0 + \frac{(\gamma - 1)}{2}u. \tag{13.8.40ab}$$

The characteristics Γ_1 meeting the piston are given by

$$\frac{2c}{\gamma - 1} + u = 2r \quad \text{on each } \Gamma_1 \text{ and } \Gamma_1 : \frac{dx}{dt} = u + c, \tag{13.8.41}$$

which is, since (13.8.40ab) holds everywhere,

$$u = \text{constant on } \Gamma_1 \text{ and } \Gamma_1 : \frac{dx}{dt} = c_0 + \frac{1}{2}\,(\gamma + 1)\,u. \tag{13.8.42}$$

Since the flow is continuous with no shocks, $u = 0$ and $c = c_0$ ahead of and on Γ_0, which separates those Γ_1 meeting the x-axis from those meeting the piston. The family of lines Γ_1 through the origin has the equation $(dx/dt) = \xi$, where ξ is a parameter with $\xi = c_0$ on Γ_0. The Γ_1 characteristics are also defined by $(dx/dt) = u + c$ so that $\xi = u + c$. Hence, elimination of c from (13.8.40b) gives

$$u = \left(\frac{2}{\gamma + 1}\right)(\xi - c_0). \tag{13.8.43}$$

Substituting this value of u in (13.8.40b), we obtain

$$c = \left(\frac{\gamma - 1}{\gamma + 1}\right)\xi + \frac{2c_0}{\gamma + 1}. \tag{13.8.44}$$

It follows from $c^2 = \gamma\,k\,\rho^{\gamma-1}$ and (13.8.40b) with the initial data, $\rho = \rho_0$, $c = c_0$ that

$$\rho = \rho_0 \left[1 + \frac{\gamma - 1}{2c_0}u\right]^{2/(\gamma-1)}. \tag{13.8.45}$$

With $\xi = (x/t)$, results (13.8.43) through (13.8.45) give the complete solution of the piston problem in terms of x and t.

Finally, the equation of the characteristic line Γ_1 is found by integrating the second equation of (13.8.42) and using the boundary condition on the piston. When a line Γ_1 intersects the piston path at time $t = \tau$, then $u = \dot{X}\,(\tau)$ along it, and the equation becomes

$$x = X\left(\tau\right) + \left\{c_0 + \frac{\gamma + 1}{2} \dot{X}\left(\tau\right)\right\}\left(t - \tau\right). \qquad (13.8.46)$$

It is noted that the family Γ_1 represents straight lines with slope dx/dt increasing with velocity u. Consequently, the characteristics are likely to overlap on the piston, that is, $\dot{X}\left(\tau\right) > 0$ for any τ. If u increases, so do c, ρ, and p so that instability develops. It shows that shocks will be formed in the compressive part of the disturbance.

13.9 Discontinuous Solutions and Shock Waves

The development of a nonunique solution of a nonlinear hyperbolic equation has already been discussed in connection with several different problems. In real physical situations, this nonuniqueness usually manifests itself in the formation of discontinuous solutions which propagate in the medium. Such discontinuous solutions across some surface are called *shock waves*. These waves are found to occur widely in high speed flows in gas dynamics.

In order to investigate the nature of discontinuous solutions, we reconsider the nonlinear conservation equation (13.6.5) that is,

$$\frac{\partial \rho}{\partial t} + \frac{\partial q}{\partial x} = 0. \qquad (13.9.1)$$

This equation has been solved under two basic assumptions: (i) There exists a functional relation between q and ρ, that is, $q = Q\left(\rho\right)$; (ii) ρ and q are continuously differentiable. In some physical situations, the solution of (13.9.1) leads to breaking phenomenon. When breaking occurs, questions arise about the validity of these assumptions. To examine the formation of discontinuities, we consider the following: (a) we assume the relation $q = Q\left(\rho\right)$ but allow jump discontinuity for ρ and q; (b) in addition to the fact that ρ and q are continuously differentiable, we assume that q is a function of ρ and ρ_x. One of the simplest forms is

$$q = Q\left(\rho\right) - \nu\rho_x, \qquad \nu > 0. \qquad (13.9.2)$$

In case (a), we assume the conservation equation (13.6.1) still holds and has the form

$$\frac{d}{dt}\int_{x_1}^{x_2} \rho\left(x, t\right) dx + q\left(x_2, t\right) - q\left(x_1, t\right) = 0. \qquad (13.9.3)$$

We now assume that there is a discontinuity at $x = s\left(t\right)$ where s is a continuously differentiable function of t, and x_1 and x_2 are chosen so that $x_2 > s\left(t\right) > x_1$, and $U\left(t\right) = \dot{s}\left(t\right)$. Equation (13.9.3) can be written as

$$\frac{d}{dt}\left[\int_{x_1}^{s^-} \rho\, dx + \int_{s^+}^{x_2} \rho\, dx\right] + q\left(x_2, t\right) - q\left(x_1, t\right) = 0,$$

which implies that

$$\int_{x_1}^{s^-} \rho_t \, dx + \dot{s}\rho\left(s^-,t\right) + \int_{s^+}^{x_2} \rho_t \, dx - \dot{s}\rho\left(s^+,t\right) + q\left(x_2,t\right) - q\left(x_1,t\right) = 0,$$

$$(13.9.4)$$

where $\rho\left(s^-,t\right)$, $\rho\left(s^+,t\right)$ are the values of $\rho\left(x,t\right)$ as $x \to s$ from below and above respectively. Since ρ_t is bounded in each of the intervals separately, the integrals tend to zero as $x_1 \to s^-$ and $x_2 \to s^+$. Thus, in the limit,

$$q\left(s^+,t\right) - q\left(s^-,t\right) = U\left\{\rho\left(s^+,t\right) - \rho\left(s^-,t\right)\right\}. \qquad (13.9.5)$$

In the conventional notation of shock dynamics, this can be written as

$$q_2 - q_1 = U\left(\rho_2 - \rho_1\right), \qquad (13.9.6)$$

or

$$-U\left[\rho\right] + \left[q\right] = 0, \qquad (13.9.7)$$

where subscripts 1 and 2 are used to denote the values behind and ahead of the shock respectively, and [ ] denotes the discontinuous jump in the quantity involved. Equation (13.9.7) is called the *shock condition*. Thus, the basic problem can be written as

$$\frac{\partial \rho}{\partial t} + \frac{\partial q}{\partial x} = 0 \quad \text{at points of continuity}, \qquad (13.9.8)$$

$$-U\left[\rho\right] + \left[q\right] = 0 \quad \text{at points of discontinuity}. \qquad (13.9.9)$$

Therefore, we have a nice correspondence

$$\frac{\partial \rho}{\partial t} \leftrightarrow -U\left[\ \right], \qquad \frac{\partial}{\partial x} \leftrightarrow \left[\ \right], \qquad (13.9.10)$$

between the differential equation and the shock condition.

It is now possible to find discontinuous solutions of (13.9.3). In any continuous part of the solution, equation (13.9.1) is still satisfied and the assumption $q = Q\left(\rho\right)$ remains valid. But we have $q_1 = Q\left(\rho_1\right)$ and $q_2 = Q\left(\rho_2\right)$ on the two sides of any shock, and the shock condition (13.9.6) has the form

$$U\left(\rho_2 - \rho_1\right) = Q\left(\rho_2\right) - Q\left(\rho_1\right). \qquad (13.9.11)$$

Example 13.9.1. The simplest example in which breaking occurs is

$$\rho_t + c\left(\rho\right)\rho_x = 0,$$

with discontinuous initial data at $t = 0$

$$\rho = \begin{cases} \rho_2, & x < 0 \\ \\ \rho_1, & x > 0 \end{cases}, \qquad (13.9.12)$$

and

$$F(x) = \begin{cases} c_2 = c_2(\rho), & x < 0 \\ \\ c_1 = c_1(\rho), & x > 0 \end{cases}, \qquad (13.9.13)$$

where

$$\rho_1 > \rho_2 \text{ and } c_2 > c_1.$$

In this case, breaking will occur immediately and this can be seen from Figure 13.9.1ab with $c'(\rho) > 0$. The multivalued region begins at the origin $\xi = 0$ and is bounded by the characteristics $x = c_1 t$ and $x = c_2 t$ with $c_1 < c_2$. This corresponds to a centered compression wave with overlapping characteristics in the (x, t)-plane.

On the other hand, if the initial condition is expansive with $c_2 < c_1$, there is a continuous solution obtained from (13.2.12) in which all values of $F(x)$ in $[c_2, c_1]$ are taken on characteristics through the origin $\xi = 0$. This corresponds to a centered fan of characteristics $x = ct$, $c_2 \le c \le c_1$ in the (x, t)-plane so that the solution has the explicit form $c = (x/t)$, $c_2 < (x/t) < c_1$. The density distribution and the expansion wave are shown in Figure 13.9.2ab.

In this case, the complete solution is given by

$$c = \begin{cases} c_2, & x \le c_2 t \\ \\ \dfrac{x}{t}, & c_2 t < x < c_1 t \\ \\ c_1, & x \ge c_1 t. \end{cases} \qquad (13.9.14)$$

13.10 Structure of Shock Waves and Burgers' Equation

In order to resolve breaking, we assumed a functional relation in ρ and q with appropriate shock conditions. Now we investigate the case when

$$q = Q(\rho) - \nu \rho_x, \qquad \nu > 0. \qquad (13.10.1)$$

Note that near breaking where ρ_x is large, (13.10.1) gives a better approximation. With (13.10.1), the basic equation (13.9.1) becomes

$$\rho_t + c(\rho) \rho_x = \nu \rho_{xx}, \qquad (13.10.2)$$

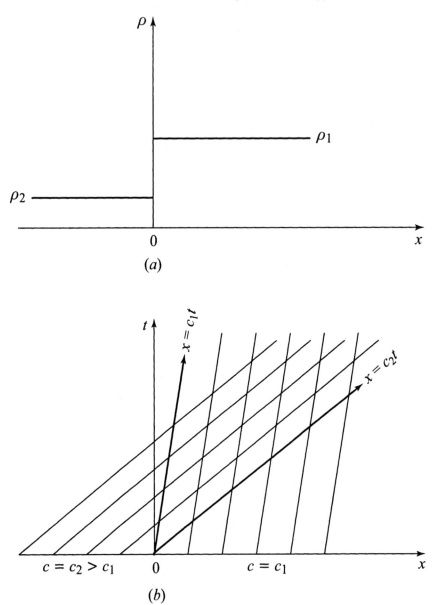

Figure 13.9.1ab Density distribution and centered compression wave with overlapping characteristics.

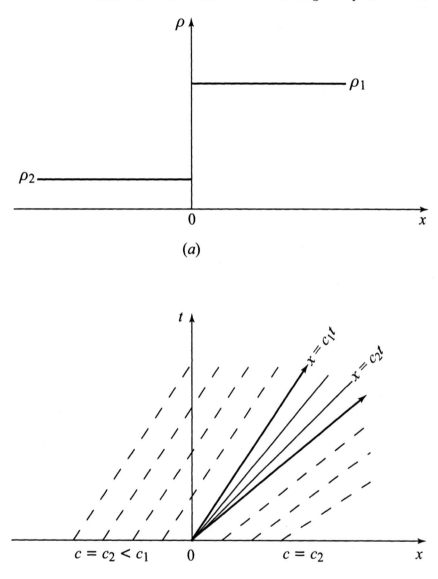

Figure 13.9.2ab Density distribution and centered expansion wave.

where $c(\rho) = Q'(\rho)$, the second and the third terms represent the effects on nonlinearity and diffusion.

We first solve (13.10.2) for two simple cases: (i) $c(\rho) = $ constant $= c$, and (ii) $c(\rho) \equiv 0$. In the first case, equation (13.10.1) becomes linear and we seek a plane wave solution

$$\rho(x, t) = a \exp\{i(kx - \omega t)\}. \tag{13.10.3}$$

Substituting this solution into the linear equation (13.10.2), we have the dispersion relation

$$\omega = ck - i\nu k^2, \tag{13.10.4}$$

where

$$\operatorname{Im}\omega = -\nu k^2 < 0, \quad \text{since} \quad \nu > 0.$$

Thus, the wave profile has the form

$$\rho(x, t) = a\, e^{-\nu k^2 t} \exp[ik(x - ct)] \tag{13.10.5}$$

which represents a diffusive wave ($\operatorname{Im}\omega < 0$) with wavenumbers k and phase velocity c whose amplitude decays exponentially with time t. The decay time is given by $t_0 = (\nu k^2)^{-1}$ which becomes smaller and smaller as k increases with fixed ν. Thus, the waves of smaller wavelengths decay faster than waves of longer wavelengths. On the other hand, for a fixed wavenumber k, t_0 decreases as ν increases so that waves of a given wavelength attenuate faster in a medium with larger ν. The quantity ν may be regarded as a measure of diffusion. Finally, after a sufficiently long time ($t \gg t_0$) only disturbances of long wavelength will survive, while all short wavelength disturbances will decay rapidly.

In the second case, (13.10.2) reduces to the linear diffusion equation

$$\rho_t = \nu \rho_{xx}. \tag{13.10.6}$$

This equation with the initial data at $t = 0$

$$\rho = \begin{cases} \rho_1, & x < 0 \\ \rho_1, & x > 0, \end{cases} \qquad \rho_1 > \rho_2, \tag{13.10.7}$$

can readily be solved, and the solution for $t > 0$ is

$$\rho(x, t) = \frac{\rho_1}{2\sqrt{\pi \nu t}} \int_{-\infty}^{0} e^{-(x-\xi)^2/4\nu t} d\xi + \frac{\rho_2}{2\sqrt{\pi \nu t}} \int_{0}^{\infty} e^{-(x-\xi)^2/4\nu t} d\xi. \tag{13.10.8}$$

After some manipulation involving changes of variables of integration $(x - \xi)/2\sqrt{\nu t} = \eta$, the solution is simplified to the form

$$u\left(x,t\right) = \frac{1}{2}\left(\rho_1 + \rho_2\right) + \left(\rho_2 - \rho_1\right)\frac{1}{\sqrt{\pi}}\int_0^{x/2\sqrt{\nu t}} e^{-\eta^2}\,d\eta, \quad (13.10.9)$$

$$= \frac{1}{2}\left(\rho_1 + \rho_2\right) + \frac{1}{2}\left(\rho_2 - \rho_1\right)\mathrm{erf}\left(\frac{x}{2\sqrt{\nu t}}\right). \qquad (13.10.10)$$

This shows that the effect of the term $\nu\rho_{xx}$ is to smooth out the initial distribution $(\nu t)^{-\frac{1}{2}}$. The solution tends to values ρ_1, ρ_2 as $x \to \mp\infty$. The absence of the term $\nu\rho_{xx}$ in (13.10.2) leads to nonlinear steepening and breaking. Indeed, equation (13.10.2) combines the two opposite effects of breaking and diffusion. The sign of ν is important; indeed, solutions are stable or unstable according as $\nu > 0$ or $\nu < 0$.

In order to investigate solutions that balance between steepening and diffusion, we seek solutions of (13.10.2) in the form

$$\rho = \rho\left(X\right), \qquad X = x - Ut, \qquad (13.10.11)$$

where U is a constant to be determined.

It follows from (13.10.2) that

$$\left[c\left(\rho\right) - U\right]\rho_X = \nu\rho_{XX}. \qquad (13.10.12)$$

Integrating this equation gives

$$Q\left(\rho\right) - U\rho + A = \nu\rho_X, \qquad (13.10.13)$$

where A is a constant of integration.

Integrating (13.10.13) with respect to X gives an implicit relation for $\rho\left(X\right)$ in the form

$$\frac{X}{\nu} = \int\frac{d\rho}{Q\left(\rho\right) - U\rho + A}. \qquad (13.10.14)$$

We would like to have a solution which tends to ρ_1, ρ_2 as $X \to \mp\infty$. If such a solution exists with $\rho_X \to 0$ as $|X| \to \infty$, the quantities U and A must satisfy

$$Q\left(\rho_1\right) - U\rho_1 + A = Q\left(\rho_2\right) - U\rho_2 + A = 0, \qquad (13.10.15)$$

which implies that

$$U = \frac{Q\left(\rho_1\right) - Q\left(\rho_2\right)}{\rho_1 - \rho_2}. \qquad (13.10.16)$$

This is exactly the same as the shock velocity obtained before.

Result (13.10.15) shows that ρ_1, ρ_2 are zeros of $Q\left(\rho\right) - U\rho + A$. In the limit $\rho \to \rho_1$ or ρ_2, the integral (13.10.14) diverges and $X \to \pm\infty$. If $c'\left(\rho\right) > 0$, then $Q\left(\rho\right) - U\rho + A \leq 0$ in $\rho_2 \leq \rho \leq \rho_1$ and then $\rho_X \leq 0$ because

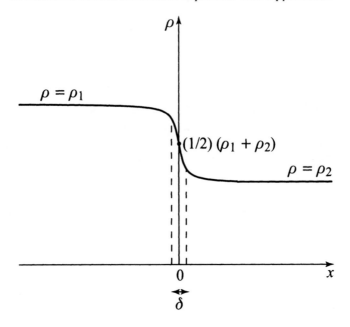

Figure 13.10.1 Shock structure and shock thickness.

of (13.10.11). Thus, ρ decreases monotonically from ρ_1 at $X = -\infty$ to ρ_2 at $X = \infty$ as shown in Figure 13.10.1.

Physically, a continuous waveform carrying an increase in ρ will progressively distort itself and eventually break forward and require a shock with $\rho_1 < \rho_2$ provided $c'(\rho) > 0$. It will break backward and require a shock with $\rho_1 > \rho_2$ and $c'(\rho) < 0$.

Example 13.10.1. Obtain the solution of (13.10.2) with the initial data (13.10.7) and $Q(\rho) = \alpha\rho^2 + \beta\rho + \gamma, \ \alpha > 0$.

We write

$$Q(\rho) - U\rho + A = -\alpha(\rho_1 - \rho)(\rho - \rho_2),$$

where

$$U = \beta + \alpha(\rho_1 + \rho_2) \quad \text{and} \quad A = \alpha\rho_1\rho_2 - \gamma.$$

Integral (13.10.14) becomes

$$\frac{X}{\nu} = -\frac{1}{\alpha}\int \frac{d\rho}{(\rho - \rho_2)(\rho_1 - \rho)} = \frac{1}{\alpha(\rho_1 - \rho_2)}\log\left(\frac{\rho_1 - \rho}{\rho - \rho_2}\right),$$

which gives the solution

$$\rho(X) = \rho_2 + (\rho_1 - \rho_2)\frac{\exp\left[\frac{\alpha X}{\nu}(\rho_2 - \rho_1)\right]}{1 + \exp\left[\frac{\alpha X}{\nu}(\rho_2 - \rho_1)\right]}. \tag{13.10.17}$$

The exponential factor in the solution indicates the existence of a transition layer of thickness δ of the order of $\nu/\left[\alpha\left(\rho_1-\rho_2\right)\right]$. This can also be referred to as the *shock thickness*. The thickness δ increases as $\rho_1 \to \rho_2$ for a fixed ν. It tends to zero as $\nu \to 0$ for a fixed ρ_1 and ρ_2.

In this case, the shock velocity (13.10.16) becomes

$$U = \alpha\left(\rho_1-\rho_2\right)+\beta = \frac{1}{2}\left(c_1+c_2\right), \qquad (13.10.18)$$

where $c\left(\rho\right)=Q'\left(\rho\right)$, $c_1=c\left(\rho_1\right)$, and $c_2=c\left(\rho_2\right)$.

We multiply (13.10.2) by $c'\left(\rho\right)$ and simplify to obtain

$$c_t+cc_x = \nu c_{xx} - \nu c''\left(\rho\right)\rho_x^2. \qquad (13.10.19)$$

Since $Q\left(\rho\right)$ is a quadratic expression in ρ, then $c\left(\rho\right)=Q'\left(\rho\right)$ becomes linear in ρ and $c''\left(\rho\right)=0$. Thus, (13.10.19) leads to *Burgers' equation* replacing c with u

$$u_t+uu_x = \nu u_{xx}. \qquad (13.10.20)$$

This equation incorporates the combined opposite effects of nonlinearity and diffusion. It is the simplest nonlinear model equation for diffusive waves in fluid dynamics. Using the Cole–Hopf transformation

$$u = -2\nu\frac{\phi_x}{\phi}. \qquad (13.10.21)$$

Burgers' equation can be solved exactly, and the opposite effects of nonlinearity and diffusion can be investigated in some detail.

We introduce the transformation in two steps. First, we write $u=\psi_x$ so that (13.10.20) can readily be integrated to obtain

$$\psi_t+\frac{1}{2}\psi_x^2 = \nu\psi_{xx}. \qquad (13.10.22)$$

The next step is to introduce $\psi=-2\nu\log\phi$ and to transform this equation into the so called *diffusion equation*

$$\phi_t = \nu\phi_{xx}. \qquad (13.10.23)$$

This equation was solved in earlier chapters. We simply quote the solution of the initial-value problem of (13.10.23) with the initial data

$$\phi\left(x,0\right)=\Phi\left(x\right), \qquad -\infty < x < \infty. \qquad (13.10.24)$$

The solution for ϕ is

$$\phi\left(x,t\right)=\frac{1}{2\sqrt{\pi\nu t}}\int_{-\infty}^{\infty}\Phi\left(\zeta\right)\exp\left[-\frac{\left(x-\zeta\right)^2}{4\nu t}\right]d\zeta, \qquad (13.10.25)$$

where $\Phi(\zeta)$ can be written in terms of the initial value $u(x,0) = F(x)$ by using (13.10.21). It turns out that, at $t = 0$,

$$\phi(x,t) = \Phi(x) = \exp\left\{-\frac{1}{2\nu}\int_0^x F(\alpha)\,d\alpha\right\}. \qquad (13.10.26)$$

It is then convenient to write down $\phi(x,t)$ in the form

$$\phi(x,t) = \frac{1}{2\sqrt{\pi\nu t}}\int_{-\infty}^{\infty}\exp\left(-\frac{f}{2\nu}\right)d\zeta, \qquad (13.10.27)$$

where

$$f(\zeta,x,t) = \int_0^\zeta F(\alpha)\,d\alpha + \frac{(x-\zeta)^2}{2t}. \qquad (13.10.28)$$

Consequently,

$$\phi_x(x,t) = -\frac{1}{4\nu\sqrt{\pi\nu t}}\int_{-\infty}^{\infty}\frac{(x-\zeta)}{t}\exp\left(-\frac{f}{2\nu}\right)d\zeta. \qquad (13.10.29)$$

Therefore, the solution for u follows from (13.10.21) and has the form

$$u(x,t) = \frac{\int_{-\infty}^{\infty}\left(\frac{x-\zeta}{t}\right)\exp\left(-\frac{f}{2\nu}\right)d\zeta}{\int_{-\infty}^{\infty}\exp\left(-\frac{f}{2\nu}\right)d\zeta}. \qquad (13.10.30)$$

Although this is the exact solution of Burgers' equation, physical interpretation can hardly be given unless a suitably simple form of $F(x)$ is specified. Even then, finding an exact evaluation of the integrals in (13.10.30) is a formidable task. It is then necessary to resort to asymptotic methods. Before we deal with asymptotic analysis, the following example may be considered for an investigation of shock formation.

Example 13.10.2. Find the solution of Burgers' equation with physical significance for the case

$$F(x) = \begin{cases} A\,\delta(x), & x < 0 \\[2mm] 0, & x > 0. \end{cases}$$

We first find

$$f(\zeta,x,t) = A\int_{0+}^\zeta \delta(\alpha)\,d\alpha + \frac{(x-\zeta)^2}{t}$$

$$= \begin{cases} \frac{(x-\zeta)^2}{2t} - A, & \zeta < 0 \\[2mm] \frac{(x-\zeta)^2}{2t}, & \zeta > 0. \end{cases}$$

Thus,

$$\int_{-\infty}^{\infty} \frac{x-\zeta}{t} \exp\left(-\frac{f}{2\nu}\right) d\zeta = \int_{-\infty}^{0} \left(\frac{x-\zeta}{t}\right) \exp\left[\frac{A}{2\nu} - \frac{(x-\zeta)^2}{4\nu t}\right] d\zeta$$

$$+ \int_{0}^{\infty} \left(\frac{x-\zeta}{t}\right) \exp\left[-\frac{(x-\zeta)^2}{4\nu t}\right] d\zeta$$

$$= 2\nu \left(e^{A/2\nu} - 1\right) \exp\left(-\frac{x^2}{4\nu t}\right),$$

which is obtained by substitution,

$$\frac{x-\zeta}{2\sqrt{\nu t}} = \alpha.$$

Similarly,

$$\int_{-\infty}^{\infty} \exp\left(-\frac{f}{2\nu}\right) d\zeta = 2\sqrt{\nu t}\left[\sqrt{\pi} + \left(e^{A/2\nu} - 1\right) \mathrm{erfc}\left(\frac{x}{2\sqrt{\nu t}}\right)\right],$$

where $\mathrm{erfc}\,(x)$ is the complementary error function defined by

$$\mathrm{erfc}\,(x) = \frac{2}{\sqrt{\pi}} \int_{x}^{\infty} e^{-\eta^2} d\eta. \tag{13.10.31}$$

Therefore, the solution for $u\,(x,t)$ is

$$u\,(x,t) = \sqrt{\frac{\nu}{t}} \frac{\left(e^{A/2\nu} - 1\right) \exp\left(-\frac{x^2}{4\nu t}\right)}{\sqrt{\pi} + \left(e^{A/2\nu} - 1\right) \left(\sqrt{\pi}/2\right) \mathrm{erfc}\left(\frac{x}{2\sqrt{\nu t}}\right)}. \tag{13.10.32}$$

In the limit as $\nu \to \infty$, the effect of diffusion would be more significant than that of nonlinearity. Since

$$\mathrm{erfc}\left(\frac{x}{2\sqrt{\nu t}}\right) \to 0, \qquad e^{A/2\nu} \sim 1 + \frac{A}{2\nu} \quad \text{as} \quad \nu \to \infty,$$

the solution (13.10.32) tends to the limiting value

$$u\,(x,t) \sim \frac{A}{2\sqrt{\pi \nu t}} \exp\left(-\frac{x^2}{4\nu t}\right). \tag{13.10.33}$$

This represents the well-known source solution of the classical linear heat equation $u_t = \nu\,u_{xx}$.

On the other hand, when $\nu \to 0$ nonlinearity would dominate over diffusion. It is expected that solution (13.10.32) tends to that of Burgers' equation as $\nu \to 0$.

We next introduce the similarity variable $\eta = x/\sqrt{2At}$ to rewrite (13.10.32) in the form

$$u(x,t) = \left(\frac{\nu}{t}\right)^{\frac{1}{2}} \frac{\left(e^{A/2\nu} - 1\right) \exp\left(-\frac{A\eta^2}{2\nu}\right)}{\sqrt{\pi} + \left(e^{A/2\nu} - 1\right)(\sqrt{\pi}/2)\operatorname{erfc}\left(\sqrt{\frac{A}{2\nu}}\,\eta\right)}, \tag{13.10.34}$$

$$\sim \left(\frac{\nu}{t}\right)^{\frac{1}{2}} \frac{\exp\left\{\frac{A}{2\nu}\left(1 - \eta^2\right)\right\}}{\sqrt{\pi} + (\sqrt{\pi}/2)\exp\left(\frac{A}{2\nu}\right)\operatorname{erfc}\left(\sqrt{\frac{A}{2\nu}}\,\eta\right)} \quad \text{as } \nu \to 0 \text{ for all } \eta,$$

$$\tag{13.10.35}$$

$$\sim 0 \quad \text{as} \quad \nu \to 0, \quad \text{for} \quad \eta < 0 \quad \text{and} \quad \eta > 1. \tag{13.10.36}$$

Invoking the asymptotic result,

$$\operatorname{erfc}(x) \sim (2/\sqrt{\pi}) \frac{e^{-x^2}}{2x} \quad \text{as} \quad x \to \infty, \tag{13.10.37}$$

the solution (13.10.34) for $0 < \eta < 1$ has the form,

$$u(x,t) \sim \left(\frac{\nu}{t}\right)^{\frac{1}{2}} \frac{2\eta\left(\frac{A}{2\nu}\right)^{\frac{1}{2}} \exp\left\{\frac{A}{2\nu}\left(1 - \eta^2\right)\right\}}{2\eta\left(\frac{A\pi}{2\nu}\right)^{\frac{1}{2}} + \exp\left\{\frac{A}{2\nu}\left(1 - \eta^2\right)\right\}}.$$

$$= \left(\frac{2A}{t}\right)^{\frac{1}{2}} \frac{\eta}{1 + 2\eta\left(\frac{A\pi}{2\nu}\right)^{\frac{1}{2}} \exp\left\{\frac{A}{2\nu}\left(\eta^2 - 1\right)\right\}}$$

$$\sim \left(\frac{2A}{t}\right)^{\frac{1}{2}} \quad \text{as} \quad \nu \to 0.$$

The final asymptotic solution as $\nu \to 0$ is

$$u(x,t) \sim \begin{cases} \frac{x}{t}, & 0 < x < (2At)^{\frac{1}{2}} \\ 0, & \text{otherwise.} \end{cases} \tag{13.10.38}$$

This result represents a shock at $x = (2At)^{\frac{1}{2}}$ with the velocity $U = (A/2t)^{\frac{1}{2}}$. This solution u has a jump from 0 to $x/t = (2A/t)^{\frac{1}{2}}$ so that the shock condition is fulfilled.

Asymptotic Behavior of Burgers' Solution as $\nu \to 0$.

We use the Kelvin stationary phase approximation method discussed in Section 12.7 to examine the asymptotic behavior of Burgers' solution (13.10.30). According to this method, the significant contribution to the integrals involved in (13.10.30) comes from points of stationary phase for fixed x and t, that is, from the roots of the equation

$$\frac{\partial f}{\partial \zeta} = F\left(\zeta\right) - \frac{\left(x - \zeta\right)}{t} = 0. \tag{13.10.39}$$

Suppose that $\zeta = \xi\left(x, t\right)$ is a root. According to result (12.7.8), integrals in (13.10.30) as $\nu \to 0$ yield

$$\int_{-\infty}^{\infty} \left(\frac{x - \xi}{t}\right) \exp\left(-\frac{f}{2\nu}\right) d\zeta \sim \frac{x - \xi}{t} \left\{\frac{4\pi\nu}{|f''\left(\xi\right)|}\right\}^{\frac{1}{2}} \exp\left\{-\frac{f\left(\xi\right)}{2\nu}\right\}$$

$$\int_{-\infty}^{\infty} \exp\left(-\frac{f}{2\nu}\right) d\zeta \sim \left\{\frac{4\pi\nu}{|f''\left(\xi\right)|}\right\}^{\frac{1}{2}} \exp\left\{-\frac{f\left(\xi\right)}{2\nu}\right\}.$$

Therefore, the final asymptotic solution is

$$u\left(x, t\right) \sim \frac{x - \xi}{t}, \tag{13.10.40}$$

where ξ satisfies (13.10.39). In other words, the solution can be rewritten in the form

$$\left.\begin{aligned} u &= F\left(\xi\right) \\ x &= \xi + t F\left(\xi\right) \end{aligned}\right\}. \tag{13.10.41}$$

This is identical with the solution (13.2.12) which was obtained in Section 13.2. In this case, the stationary point ξ represents the characteristic variable.

Although the exact solution of Burgers' equation is a single-valued and continuous function for all time t, the asymptotic solution (13.10.41) exhibits instability. It has already been shown that (13.10.41) progressively distorts itself and becomes multiple-valued after sufficiently long time. Eventually, breaking will definitely occur.

It follows from the analysis of Burgers' equation that the nonlinear and diffusion terms show opposite effects. The former introduces steepening in the solution profile, whereas the latter tends to diffuse (spread) the sharp discontinuities into a smooth profile. In view of this property, the solution represents the diffusive wave. In the context of fluid flows, ν denotes the kinematic viscosity which measures the viscous dissipation.

Finally, Burgers' equation arises in many physical problems, including one-dimensional turbulence (where this equation had its origin), sound waves in viscous media, shock waves in viscous media, waves in fluid-filled viscous elastic pipes, and magnetohydrodynamic waves in media with finite conductivity.

13.11 The Korteweg–de Vries Equation and Solitons

The celebrated dispersion relation (13.3.14) for dispersive surface waves on water of constant depth h_0 is

$$\omega = (gk \tanh kh_0)^{\frac{1}{2}}$$

$$= c_0 k \left(1 - \frac{1}{3} k^2 h_0^2 \right)^{\frac{1}{2}} \approx c_0 k \left(1 - \frac{1}{6} k^2 h_0^2 \right), \qquad (13.11.1)$$

where $c_0 = (gh_0)^{\frac{1}{2}}$ is the shallow water wave speed.

In many physical problems, wave motions with small dispersion exhibit such a k^2 term in contrast to the linearized theory value of $c_0 k$. An equation for the free surface elevation $\eta(x,t)$ with this dispersion relation is given by

$$\eta_t + c_0 \eta_x + \sigma \eta_{xxx} = 0, \qquad (13.11.2)$$

where $\sigma = \frac{1}{6} c_0 h_0^2$ is a constant for fairly long waves. This equation is called the *linearized Korteweg-de Vries (KdV) equation* for fairly long waves moving to the positive x direction only. The phase and group velocities of the waves are found from (13.11.1) and they are given by

$$C_p = \frac{\omega}{k} = c_0 - \sigma k^2, \qquad (13.11.3)$$

$$C_g = \frac{d\omega}{dk} = c_0 - 3\sigma k^2. \qquad (13.11.4)$$

It is noted that $C_p > C_g$, and the dispersion comes from the term involving k^3 in the dispersion relation (13.11.1) and hence, from the term $\sigma \eta_{xxx}$. For sufficiently long waves ($k \to 0$), $C_p = C_g = c_0$, and hence, these waves are nondispersive.

In 1895, Korteweg–de Vries derived the nonlinear equation for long water waves in a channel of depth h_0 which has the remarkable form

$$\eta_t + c_0 \left(1 + \frac{3}{2} \frac{\eta}{h_0} \right) \eta_x + \sigma \eta_{xxx} = 0. \qquad (13.11.5)$$

This is the simplest nonlinear model equation for dispersive waves, and combines nonlinearity and dispersion. The KdV equation arises in many physical problems, which include water waves of long wavelengths, plasma waves, and magnetohydrodynamics waves. Like Burgers' equation, the nonlinearity and dispersion have opposite effects on the KdV equation. The former introduces steepening of the wave profile while the latter counteracts waveform steepening. The most remarkable features is that the dispersive term in the KdV equation does allow the solitary and periodic waves which are not found in shallow water wave theory. In Burgers' equation the nonlinear term leads to steepening which produces a shock wave; on the other hand, in the KdV equation the steepening process is balanced by dispersion to give a rise to a steady solitary wave.

We now seek the traveling wave solution of the KdV equation (13.11.5) in the form

$$\eta\left(x,t\right)=h_0 f\left(X\right),\qquad X=x-Ut,\qquad\text{(13.11.6)}$$

for some function f and constant wave velocity U. We determine f and U by substitution of the form (13.11.6) into (13.11.5). This gives, with $\sigma=\frac{1}{6}c_0 h_0^2$,

$$\frac{1}{6}h_0^2 f''' + \frac{3}{2}ff' + \left(1-\frac{U}{c_0}\right)f' = 0,\qquad\text{(13.11.7)}$$

and then integration leads to

$$\frac{1}{6}h_0^2 f'' + \frac{3}{4}f^2 + \left(1-\frac{U}{c_0}\right)f + A = 0,$$

where A is an integrating constant.

We next multiply this equation by f' and integrate again to obtain

$$\frac{1}{3}h_0^2 f'^2 + f^3 + 2\left(1-\frac{U}{c_0}\right)f^2 + 4Af + B = 0,\qquad\text{(13.11.8)}$$

where B is a constant of integration.

We now seek a solitary wave solution under the boundary conditions f, f', $f'' \to 0$ as $|X| \to \infty$. Therefore, $A = B = 0$ and (13.11.8) assumes the form

$$\frac{1}{3}h_0^2 f'^2 + f^2\left(f-\alpha\right) = 0,\qquad\text{(13.11.9)}$$

where

$$\alpha = 2\left(\frac{U}{c_0}-1\right).\qquad\text{(13.11.10)}$$

Finally, we obtain

$$X = \int_0^f \frac{df}{f'} = \left(\frac{h_0^2}{3}\right)^{\frac{1}{2}}\int_0^f \frac{df}{f\sqrt{(\alpha-f)}},$$

which is, by the substitution $f = \alpha\,\mathrm{sech}^2\theta$,

$$X-X_0 = \left(\frac{4h_0^2}{3\alpha}\right)^{\frac{1}{2}}\theta,\qquad\text{(13.11.11)}$$

for some integrating constant X_0.

Therefore, the solution for $f\left(X\right)$ is

$$f\left(X\right) = \alpha\,\mathrm{sech}^2\left[\left(\frac{3\alpha}{4h_0^2}\right)^{\frac{1}{2}}\left(X-X_0\right)\right].\qquad\text{(13.11.12)}$$

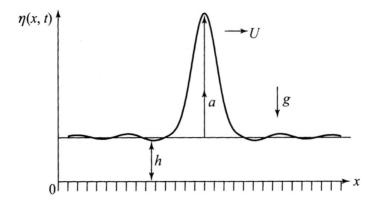

Figure 13.11.1 A soliton.

The solution $f(X)$ increases from $f = 0$ as $X \to -\infty$ so that it attains a maximum value $f = f_{\max} = \alpha$ at $X = 0$, and then decreases symmetrically to $f = 0$ as $X \to \infty$ as shown in Figure 13.11.1. These features also imply that $X_0 = 0$, so that (13.11.12) becomes

$$f(X) = \alpha \operatorname{sech}^2 \left[\left(\frac{3\alpha}{4h_0^2} \right)^{\frac{1}{2}} X \right]. \tag{13.11.13}$$

Therefore, the final solution is

$$\eta(x, t) = \eta_0 \operatorname{sech}^2 \left[\left(\frac{3\eta_0}{4h_0^3} \right)^{\frac{1}{2}} (x - Ut) \right], \tag{13.11.14}$$

where $\eta_0 = (\alpha h_0)$. This is called the *solitary wave* solution of the KdV equation for any positive constant η_0. However, it has come to be known as *soliton* since Zabusky and Kruskal coined the term in 1965. Since $\eta > 0$ for all X, the soliton is a wave of elevation which is symmetrical about $X = 0$. It propagates in the medium without change of shape with velocity

$$U = c_0 \left(1 + \frac{\alpha}{2} \right) = c_0 \left(1 + \frac{1}{2} \frac{\eta_0}{h_0} \right), \tag{13.11.15}$$

which is directly proportional to the amplitude η_0. The width, $(3\eta_0/4h_0^3)^{-\frac{1}{2}}$ is inversely proportional to $\sqrt{\eta_0}$. In other words, the solitary wave propagates to the right with a velocity U which is directly proportional to the amplitude, and has a width that is inversely proportional to the square root of the amplitude. Therefore, taller solitons travel faster and are narrower than the shorter (or slower) ones. They can overtake the shorter ones, and surprisingly, they emerge from the interaction without change of shape as shown in Figure 13.11.2. Indeed the discovery of soliton interactions confirms that solitons behave like elementary particles.

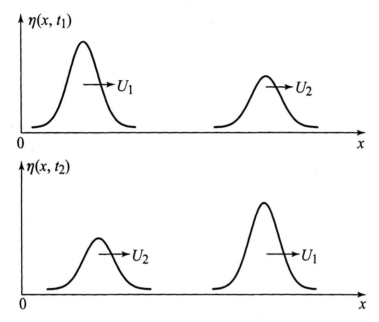

Figure 13.11.2 Interaction of two solitons ($U_1 > U_2$, $t_2 > t_1$,).

General Waves of Permanent Form.

We now consider the general case given by (13.11.8) which can be written

$$\left(\frac{h_0^2}{3}\right) f'^2 = -f^3 + \alpha f^2 - 4Af - B \equiv F\left(f\right).$$

We seek real bounded solutions for $f\left(X\right)$. Therefore, $f'^2 \geq 0$ and varies monotonically until f' is zero. Hence, the zeros of the cubic $F\left(f\right)$ are crucial. For bounded solutions, all the three zeros f_1, f_2, f_3 must be real. Without loss of generality, we choose $f_1 = 0$ and $f_2 = \alpha$. The third zero must be negative so we set $f_3 = \alpha - \beta$ with $0 < \alpha < \beta$. Therefore, the equation for $f\left(X\right)$ is

$$\frac{1}{3} h_0^2 \left(\frac{df}{dX}\right)^2 = f\left(\alpha - f\right)\left(f - \alpha + \beta\right), \qquad (13.11.16)$$

or

$$\sqrt{\frac{3}{h_0^2}} \, dX = -\frac{df}{\left[f\left(\alpha - f\right)\left(f - \alpha + \beta\right)\right]^{\frac{1}{2}}}, \qquad (13.11.17)$$

where

$$U = c_0 \left(1 + \frac{2\alpha - \beta}{2}\right). \qquad (13.11.18)$$

We put $\alpha - f = p^2$ in (13.11.17) to obtain

$$\left(\frac{3}{4h_0^2}\right)^{\frac{1}{2}} dX = \frac{dp}{[(\alpha - p^2)(\beta - p^2)]^{\frac{1}{2}}}. \tag{13.11.19}$$

We next substitute $p = \sqrt{\alpha}\, q$ into (13.11.19) to transform it into the standard form

$$\left(\frac{3\beta}{4h_0^2}\right)^{\frac{1}{2}} X = \int_0^q \frac{dq}{[(1 - q^2)(1 - m^2 q^2)]^{\frac{1}{2}}} \tag{13.11.20}$$

where $m = (\alpha/\beta)^{\frac{1}{2}}$.

The right hand side is an integral of the first kind, and hence, q can be expressed in terms of the Jacobian sn function (see Dutta and Debnath (1965))

$$q = sn\left[\left(\frac{3\beta}{4h_0^2}\right)^{\frac{1}{2}} X,\, m\right], \tag{13.11.21}$$

where m is the modulus of the Jacobian elliptic function $sn\,(z, m)$. Therefore,

$$f(X) = \alpha\left[1 - sn^2\left\{\left(\frac{3\beta}{4h_0^2}\right)^{\frac{1}{2}} X\right\}\right]$$

$$= \alpha\, cn^2\left[\left(\frac{3\beta}{4h_0^2}\right)^{\frac{1}{2}} X\right], \tag{13.11.22}$$

where $cn\,(z, m)$ is also the Jacobian elliptic function of modulus m and $cn^2\,(z) = 1 - sn^2\,(z)$.

From (13.11.20), the period P is given by

$$P = 2\left(\frac{4h_0^2}{3\beta}\right)^{\frac{1}{2}} \int_0^1 \frac{dq}{[(1 - q^2)(1 - m^2 q^2)]^{\frac{1}{2}}} \tag{13.11.23}$$

$$= \frac{4h_0}{\sqrt{3\beta}} K(m) \equiv \lambda, \tag{13.11.24}$$

where $K(m)$ is the complete elliptic integral of the first kind defined by

$$K(m) = \int_0^{\pi/2} (1 - m\sin^2\theta)^{-\frac{1}{2}} d\theta \tag{13.11.25}$$

and λ denotes the wavelength of the cnoidal wave.

It is important to note that $cn\,(z, m)$ is periodic, and hence, $\eta(X)$ represents a train of periodic waves in shallow water. Thus, these waves are called *cnoidal waves* with wavelength

Figure 13.11.3 A cnoidal wave.

$$\lambda = 2 \left(\frac{4h^3}{3b} \right)^{1/2} K(m). \tag{13.11.26}$$

The outcome of this analysis is that solution (13.11.22) represents a nonlinear wave whose shape and wavelength (or period) all depend on the amplitude of the wave. A typical cnoidal wave is shown in Figure 13.11.3. Sometimes, the cnoidal waves with slowly varying amplitude are observed in rivers. More often, wavetrains behind a weak bore (called an *undular bore*) can be regarded as cnoidal waves.

Two limiting cases are of special interest: (i) $m \to 1$ and (ii) $m \to 0$.

When $m \to 1$ ($\alpha \to \beta$), it is easy to show that $cn(z) \to \operatorname{sech} z$. Hence, the cnoidal wave solution (13.11.22) tends to the solitary wave with the wavelength λ, given by (13.11.24) which approaches infinity because $K(1) = \infty$, $K(0) = \pi/2$. The solution identically reduces to (13.11.14) with (13.11.15).

In the other limit $m \to 0$ ($\alpha \to 0$), $sn\, z \to \sin z$ and $cn\, z \to \cos z$ so that solution (13.11.22) becomes

$$f(X) = \alpha \cos^2 \left[\left(\frac{3\beta}{4h_0^2} \right)^{\frac{1}{2}} X \right], \tag{13.11.27}$$

where

$$U = c_0 \left(1 - \frac{\beta}{2} \right). \tag{13.11.28}$$

Using $\cos 2\theta = 2\cos^2 \theta - 1$, we can rewrite (13.11.27) in the form

$$f(X) = \frac{\alpha}{2} \left[1 + \cos \left(\frac{\sqrt{3\beta}}{h_0} \right) X \right]. \tag{13.11.29}$$

We next introduce $k = \sqrt{3\beta}/h_0$ (or $\beta = \frac{1}{3}k^2 h_0^2$) to simplify (13.11.29) as

$$f(X) = \frac{\alpha}{2} [1 + \cos(kx - \omega t)], \tag{13.11.30}$$

where

$$\omega = Uk = c_0 k \left(1 - \frac{1}{6}k^2 h_0^2 \right). \tag{13.11.31}$$

This corresponds to the first two terms of the series of $(gk \tanh kh_0)^{1/2}$. Thus, these results are in perfect agreement with the linearized theory.

Remark: It is important to point out that the phase velocity (13.11.3) becomes negative for $k^2 > (c_0/\sigma)$ which indicates that waves propagate in the negative x direction. This contradicts the original assumption of forward travelling waves. Moreover, the group velocity given by (13.11.4) assumes large negative values for large k so that the fine-scale features of the solution are propagated in the negative x direction. The solution of (13.11.2) involves the Airy function which shows fiercely oscillatory character for large negative arguments. This leads to a lack of continuity and a tendency to emphasize short wave components which contradicts the KdV model representing fairly long waves. In order to eliminate these physically undesirable features of the KdV equation, Benjamin, Bona, and Mahony (1972) proposed a new nonlinear model equation in the form

$$\eta_t + \eta_x + \eta\eta_x - \eta_{xxt} = 0. \tag{13.11.32}$$

This is known as the *Benjamin, Bona and Mahony (BBM) equation*. The advantage of this model over the KdV equation becomes apparent when we examine their linearized forms and the corresponding solutions. The linearized form (13.11.32) gives the dispersion relation

$$\omega = \frac{k}{1+k^2}, \tag{13.11.33}$$

which shows that both the phase velocity and the group velocity are bounded for all k, and both velocities tend to zero for large k. In other words, the model has the desirable feature of responding very insignificantly to short wave components that may be introduced into the initial wave form. Thus, the BBM model seems to be a preferable long wave model of physical interest. However, whether the BBM equation is a better model than the KdV equation has not yet been established.

Another important property of the KdV equation is that it satisfies the conservation law of the form

$$T_t + X_x = 0, \tag{13.11.34}$$

where T is called the *density* and the X is called the *flux*.

If T and X are integrable in $-\infty < x < \infty$, and $X \to 0$ as $|x| \to \infty$, then

$$\frac{d}{dt}\int_{-\infty}^{\infty} T\,dx = -\,|X|_{-\infty}^{\infty} = 0.$$

Therefore,

$$\int_{-\infty}^{\infty} T\,dx = \text{constant}$$

so that the density is conserved.

The canonical form of the KdV equation

$$u_t - 6uu_x + u_{xxx} = 0, \qquad (13.11.35)$$

can be written as

$$(u)_t + \left(-3u^2 + u_{xx}\right)_x = 0$$

so that

$$T = u \quad \text{and} \quad X = -3u^2 + u_{xx}. \qquad (13.11.36)$$

If we assume that u is periodic or that u and its derivatives decay very rapidly as $|x| \to \infty$, then

$$\frac{d}{dt} \int_{-\infty}^{\infty} u\, dx = 0.$$

This leads to the conservation of mass, that is,

$$\int_{-\infty}^{\infty} u\, dx = \text{constant}. \qquad (13.11.37)$$

This is often called the *time invariant function* of the solutions of the KdV equation.

The second conservation law for (13.11.34) can be obtained by multiplying it by u so that

$$\left(\frac{1}{2}u^2\right)_t + \left(-2u^3 + uu_x - \frac{1}{2}u_x^2\right)_x = 0. \qquad (13.11.38)$$

This gives

$$\int_{-\infty}^{\infty} \frac{1}{2}u^2 dx = \text{constant}. \qquad (13.11.39)$$

This is the principle of conservation of energy.

It is well known that the KdV equation has an infinite number of polynomial conservation laws. It is generally believed that the existence of a soliton solution to the KdV equation is closely related to the existence of an infinite number of conservation laws.

13.12 The Nonlinear Schrödinger Equation and Solitary Waves

We first derive the one-dimensional linear Schrödinger equation from the Fourier integral representation of the plane wave solution

$$\phi\left(x,t\right) = \int_{-\infty}^{\infty} F\left(k\right) \exp\left[i\left(kx - \omega t\right)\right] dk, \qquad (13.12.1)$$

where the spectrum function $F\left(k\right)$ is determined from the given initial or boundary conditions.

We assume that the wave is slowly modulated as it propagates in a dispersive medium. For such a modulated wave, most of the energy is confined in the neighborhood of $k = k_0$ so that the dispersion relation $\omega = \omega\left(k\right)$ can be expanded about the point $k = k_0$ as

$$\omega = \omega\left(k\right) = \omega_0 + \left(k - k_0\right)\omega_0' + \frac{1}{2}\left(k - k_0\right)^2 \omega_0'' + \ldots, \qquad (13.12.2)$$

where $\omega_0 \equiv \omega\left(k_0\right)$, $\omega_0' \equiv \omega'\left(k_0\right)$, $\omega_0'' \equiv \omega''\left(k_0\right)$.

In view of (13.12.2), we can rewrite (13.12.1) as

$$\phi\left(x,t\right) = \psi\left(x,t\right) \exp\left[i\left(k_0 x - \omega_0 t\right)\right], \qquad (13.12.3)$$

where the amplitude $\psi\left(x,t\right)$ is given by

$$\psi\left(x,t\right) = \int_{-\infty}^{\infty} F\left(k\right) \exp\left[i\left(k - k_0\right)x - i\left\{\left(k - k_0\right)\omega_0' \right.\right.$$
$$\left.\left. + \frac{1}{2}\left(k - k_0\right)^2 \omega_0'' \right\} t\right] dk. \quad (13.12.4)$$

Evidently, this represents the slowly varying (or modulated) part of the basic wave. A simple computation of ψ_t, ψ_x, and ψ_{xx} gives

$$\psi_t = -i\left\{\left(k - k_0\right)\omega_0' + \frac{1}{2}\left(k - k_0\right)^2 \omega_0''\right\}\psi$$
$$\psi_x = i\left(k - k_0\right)\psi$$
$$\psi_{xx} = -\left(k - k_0\right)^2 \psi$$

so that

$$i\left(\psi_t + \omega_0'\,\psi_x\right) + \frac{1}{2}\omega_0''\,\psi_{xx} = 0. \qquad (13.12.5)$$

The dispersion relation associated with this linear equation is given by

$$\omega = k\omega_0' + \frac{1}{2}k^2\omega_0''. \qquad (13.12.6)$$

If we choose a frame of reference moving with the linear group velocity, that is, $x^* = x - \omega_0' t$, $t^* = t$, the term involving ψ_x is dropped and then, ψ satisfies the linear Schrödinger equation, dropping the asterisks,

$$i\,\psi_t + \frac{1}{2}\,\omega_0''\,\psi_{xx} = 0. \tag{13.12.7}$$

We next derive the nonlinear Schrödinger equation from the nonlinear dispersion relation involving both frequency and amplitude in the most general form

$$\omega = \omega\left(k, a^2\right). \tag{13.12.8}$$

We first expand ω in a Taylor series about $k = k_0$ and $|a|^2 = 0$ in the form

$$\omega = \omega_0 + (k - k_0)\left(\frac{\partial \omega}{\partial k}\right)_{k=k_0} + \frac{1}{2}(k - k_0)^2 \left(\frac{\partial^2 \omega}{\partial k^2}\right)_{k=k_0}$$
$$+ \left(\frac{\partial \omega}{\partial |a|^2}\right)_{|a|^2=0} |a|^2, \tag{13.12.9}$$

where $\omega_0 \equiv \omega\left(k_0\right)$.

If we now replace , $\omega - \omega_0$ by $i\left(\partial/\partial t\right)$ and $k - k_0$ by $-i\left(\partial/\partial x\right)$, and assume that the resulting operators act on a, we obtain

$$i\left(a_t + \omega_0'\,a_x\right) + \frac{1}{2}\,\omega_0''\,a_{xx} + \gamma\,|a|^2\,a = 0, \tag{13.12.10}$$

where

$$\omega_0' \equiv \omega'\left(k_0\right),\ \omega_0'' \equiv \omega''\left(k_0\right),\ \text{ and }\ \gamma \equiv -\left(\frac{\partial \omega}{\partial |a|^2}\right)_{|a|^2=0} \text{ is a constant.}$$

Equation (13.12.10) is known as the *nonlinear Schrödinger (NLS) equation*. If we choose a frame of reference moving with the linear group velocity ω_0', that is, $x^* = x - \omega_0'\,t$ and $t^* = t$, the term involving a_x will drop out from (13.12.10), and the amplitude $a\left(x,t\right)$ satisfies the normalized NLS equation, dropping the asterisks,

$$i\,a_t + \frac{1}{2}\,\omega_0''\,a_{xx} + \gamma\,|a|^2\,a = 0. \tag{13.12.11}$$

The corresponding dispersion relation is given by

$$\omega = \frac{1}{2}\,\omega_0''\,k^2 - \gamma\,a^2. \tag{13.12.12}$$

According to the stability criterion established in Section 13.5, the wave modulation is stable if $\gamma\,\omega_0'' < 0$ and unstable if $\gamma\,\omega_0'' > 0$.

To study the solitary wave solution, it is convenient to use the NLS equation in the standard form

$$i\,\psi_t + \psi_{xx} + \gamma\,|\psi|^2\,\psi = 0, \qquad -\infty < x < \infty,\quad t \geq 0. \tag{13.12.13}$$

We seek waves of permanent form by assuming the solution

$$\psi = f(X) e^{i(mX - nt)}, \qquad X = x - Ut \qquad (13.12.14)$$

for some functions f and constant wave speed U to be determined, and m, n are constants.

Substitution of (13.12.14) into (13.12.13) gives

$$f'' + i(2m - U) f' + (n - m^2) f + \gamma |f|^2 f = 0. \qquad (13.12.15)$$

We eliminate f' by setting $2m - U = 0$, and then, write $n = m^2 - \alpha$ so that f can be assumed to be real. Thus, equation (13.12.15) becomes

$$f'' - \alpha f + \gamma f^3 = 0. \qquad (13.12.16)$$

Multiplying this equation by $2f'$ and integrating, we find that

$$f'^2 = A + \alpha f^2 - \frac{\gamma}{2} f^4 \equiv F(f), \qquad (13.12.17)$$

where $F(f) \equiv (\alpha_1 - \alpha_2 f^2)(\beta_1 - \beta_2 f^2)$, so that $\alpha = -(\alpha_1 \beta_2 + \alpha_2 \beta_1)$, $A = \alpha_1 \beta_1$, $\gamma = -2(\alpha_2 \beta_2)$, and $\alpha's$ and $\beta's$ are assumed to be real and distinct.

Evidently,

$$X = \int_0^f \frac{df}{\sqrt{(\alpha_1 - \alpha_2 f^2)(\beta_1 - \beta_2 f^2)}}. \qquad (13.12.18)$$

Putting $(\alpha_2 / \alpha_1)^{\frac{1}{2}} f = u$ in this integral, we deduce the following elliptic integral of the first kind (see Dutta and Debnath (1965)):

$$\sigma X = \int_0^u \frac{du}{\sqrt{(1 - u^2)(1 - \kappa^2 u^2)}}, \qquad (13.12.19)$$

where $\sigma = (\alpha_2 \beta_1)^{\frac{1}{2}}$ and $\kappa = (\alpha_1 \beta_2) / (\beta_1 \alpha_2)$.

Thus, the final solution can be expressed in terms of the Jacobian sn function

$$u = sn(\sigma X, \kappa),$$

or,

$$f(X) = \left(\frac{\alpha_1}{\alpha_2}\right)^{\frac{1}{2}} sn(\sigma X, \kappa). \qquad (13.12.20)$$

In particular, when $A = 0$, $\alpha > 0$, and $\gamma > 0$, we obtain a solitary wave solution. In this case, equation (13.12.17) can be rewritten as

$$\sqrt{\alpha}\, X = \int_0^f \frac{df}{f \left(1 - \frac{\gamma}{2\alpha} f^2\right)^{\frac{1}{2}}}. \tag{13.12.21}$$

Substitution of $(\gamma/2\alpha)^{\frac{1}{2}} f = \operatorname{sech}\theta$ in this integral gives the exact solution

$$f(X) = \left(\frac{2\alpha}{\gamma}\right)^{\frac{1}{2}} \operatorname{sech}\left[\sqrt{\alpha}\,(x - Ut)\right]. \tag{13.12.22}$$

This represents a *solitary wave* which propagates without change of shape with constant velocity U. Unlike the solution of the KdV equation, the amplitude and the velocity of the wave are independent parameters. It is noted that the solitary wave exists only for the unstable case $(\gamma > 0)$. This means that small modulations of the unstable wavetrain lead to a series of solitary waves.

The nonlinear dispersion relation for deep water waves is

$$\omega = \sqrt{gk}\left(1 + \frac{1}{2} a^2 k^2\right). \tag{13.12.23}$$

Therefore,

$$\omega_0' = \frac{\omega_0}{2k_0}, \qquad \omega_0'' = -\frac{\omega_0}{4k_0^2}, \qquad \text{and} \qquad \gamma = -\frac{1}{2}\omega_0 k_0^2, \tag{13.12.24}$$

and the NLS equation for deep water waves is obtained from (13.12.10) in the form

$$i\left(a_t + \frac{\omega_0}{2k_0} a_x\right) - \left(\frac{\omega_0}{8k_0^2}\right) a_{xx} - \frac{1}{2}\omega_0 k_0^2 |a|^2 a = 0. \tag{13.12.25}$$

The normalized form of this equation in a frame of reference moving with the linear group velocity ω_0' is

$$i\, a_t - \left(\frac{\omega_0}{8k_0^2}\right) a_{xx} = \frac{1}{2}\omega_0 k_0^2 |a|^2 a. \tag{13.12.26}$$

Since $\gamma\omega_0'' = \left(\omega_0^2/8\right) > 0$, this equation confirms the instability of deep water waves. This is one of the most remarkable recent results in the theory of water waves.

We next discuss the uniform solution and the solitary wave solution of (13.12.26). We look for solutions in the form

$$a(x,t) = A(X)\exp\left(i\gamma^2 t\right), \qquad X = x - \omega_0' t \tag{13.12.27}$$

and substitute this into equation (13.12.26) to obtain

$$A_{XX} = -\left(\frac{8k_0^2}{\omega_0}\right)\left(\gamma^2 A + \frac{1}{2}\omega_0 k_0^2 A^3\right). \tag{13.12.28}$$

We multiply this equation by $2A_X$ and then integrate to find

$$A_X^2 = -\left(A_0^4 m'^2 + \frac{8}{\omega_0} \gamma^2 k_0^2 A^2 + 2k_0^4 A^4 \right)$$
$$= \left(A_0^2 - A^2 \right) \left(A^2 - m'^2 A_0^2 \right), \qquad (13.12.29)$$

where $\left(A_0^4 m'^2 \right)$ is an integrating constant and $2k_0^4 = 1$, $m'^2 = 1 - m^2$, and $A_0^2 = 4\gamma^2/\omega_0 k_0^2 \left(m^2 - 2 \right)$ which relates A_0, γ, and m.

Finally, we rewrite equation (13.12.29) in the form

$$A_0^2 \, dX = \frac{dA}{\left[\left(1 - \frac{A^2}{A_0^2} \right) \left(\frac{A^2}{A_0^2} - m'^2 \right) \right]^{\frac{1}{2}}}, \qquad (13.12.30)$$

or,

$$A_0 \left(X - X_0 \right) = \int' \frac{ds}{\left[\left(1 - s^2 \right) \left(s^2 - m'^2 \right) \right]^{\frac{1}{2}}}, \qquad s = \left(A/A_0 \right).$$

This can readily be expressed in terms of the Jacobian dn function (see Dutta and Debnath (1965))

$$A = A_0 \, dn \left[A_0 \left(X - X_0 \right), m \right], \qquad (13.12.31)$$

where m is the modulus of the dn function.

In the limit $m \to 0$, $dn \, z \to 1$ and $\gamma^2 \to -\frac{1}{2} \omega_0 k_0^2 A_0^2$. Hence, the solution is

$$a \left(x, t \right) = A \left(t \right) = A_0 \exp \left(-\frac{1}{2} i \omega_0 k_0^2 A_0^2 t \right). \qquad (13.12.32)$$

On the other hand, when $m \to 1$, $dn \, z \to \operatorname{sech} z$ and $\gamma^2 \to -\frac{1}{4} \omega_0 k_0^2 A_0^2$. Therefore, the solitary wave solution is

$$a \left(x, t \right) = A_0 \operatorname{sech} \left[A_0 \left(x - \omega_0' t - X_0 \right) \right] \exp \left(-\frac{1}{4} \omega_0 k_0^2 A_0^2 t \right). \, (13.12.33)$$

We next use the NLS equation (13.12.26) to discuss the instability of deep water waves, which is known as the *Benjamin and Feir instability*. We consider a perturbation of (13.12.32) and write

$$a \left(X, t \right) = A \left(t \right) \left[1 + B \left(X, t \right) \right], \qquad (13.12.34)$$

where $A \left(t \right)$ is the uniform solution given by (13.12.32).

Substituting equation (13.12.34) into (13.12.26) gives

$$iA_t \left(1 + B \right) + iA \left(t \right) B_t - \left(\frac{\omega_0}{8k_0^2} \right) A \left(t \right) B_{XX}$$
$$= \frac{1}{2} \omega_0 k_0^2 A_0^2 \left[\left(1 + B \right) + BB^* \left(1 + B \right) + \left(B + B^* \right) B + \left(B + B^* \right) \right] A,$$

where B^* is the complex conjugate of B.

Neglecting squares of B, it follows that

$$i B_t - \left(\frac{\omega_0}{8k_0^2}\right) B_{XX} = \frac{1}{2}\omega_0 k_0^2 A_0^2 (B + B^*). \qquad (13.12.35)$$

We now seek a solution of the form

$$B(X,t) = B_1 e^{\Omega t + i\kappa X} + B_2 e^{\Omega t - i\kappa X}, \qquad (13.12.36)$$

where B_1, B_2 are complex constants, κ is a real wavenumber, and Ω is a growth rate (possibly complex) to be determined.

Substitution of B into (13.12.35) leads to the pair of coupled equations

$$\left(i\Omega + \frac{\omega_0\kappa^2}{8k_0^2}\right) B_1 - \frac{1}{2}\omega_0 k_0^2 A_0^2 (B_1 + B_2^*) = 0, \qquad (13.12.37)$$

$$\left(i\Omega + \frac{\omega_0\kappa^2}{8k_0^2}\right) B_2 - \frac{1}{2}\omega_0 k_0^2 A_0^2 (B_1^* + B_2) = 0. \qquad (13.12.38)$$

It is convenient to take the complex conjugate of (13.12.38) so that it assumes the form

$$\left(-i\Omega + \frac{\omega_0\kappa^2}{8k_0^2}\right) B_2^* - \frac{1}{2}\omega_0 k_0^2 A_0^2 (B_1 + B_2^*) = 0. \qquad (13.12.39)$$

The pair of linear homogeneous equations (13.12.37) and (13.12.39) for B_1 and B_2^* admits a nontrivial solution provided

$$\begin{vmatrix} i\Omega + \left(\frac{\omega_0\kappa^2}{8k_0^2}\right) - \frac{1}{2}\omega_0 k_0^2 A_0^2 & -\frac{1}{2}\omega_0 k_0^2 A_0^2 \\ -\frac{1}{2}\omega_0 k_0^2 A_0^2 & i\Omega + \left(\frac{\omega_0\kappa^2}{8k_0^2}\right) - \frac{1}{2}\omega_0 k_0^2 A_0^2 \end{vmatrix} = 0,$$

or

$$\Omega^2 = \left(\frac{\omega_0^2\kappa^2}{8k_0^2}\right)\left(k_0^2 A_0^2 - \frac{\kappa^2}{8k_0^2}\right). \qquad (13.12.40)$$

The growth rate Ω is purely imaginary or real and positive depending on whether $(\kappa^2/k_0^2) > 8k_0^2 A_0^2$ or $(\kappa^2/k_0^2) < 8k_0^2 A_0^2$. The former case corresponds to a wave (an oscillatory solution) for B, and the latter case represents the *Benjamin and Feir instability* criterion with $\tilde{\kappa} = (\kappa/k_0)$ as the non-dimensional wavenumber so that

$$\tilde{\kappa}^2 < 8k_0^2 A_0^2. \qquad (13.12.41)$$

The range of instability is given by

$$0 < \tilde{\kappa} < \tilde{\kappa}_c = 2\sqrt{2}\,(k_0 A_0). \qquad (13.12.42)$$

Since Ω is a function of $\tilde{\kappa}$, maximum instability occurs at $\tilde{\kappa} = \tilde{\kappa}_{max} = 2k_0 A_0$ with a maximum growth rate given by

$$(\operatorname{Re}\Omega)_{max} = \frac{1}{2}\omega_0 k_0^2 A_0^2. \qquad (13.12.43)$$

To establish the connection with the Benjamin–Feir instability, we have to find the velocity potential for the fundamental wave mode multiplied by $\exp(kz)$. It turns out that the term proportional to B_1 is the upper sideband, whereas that proportional to B_2 is the lower sideband. The main conclusion of the preceding analysis is that Stokes water waves are definitely unstable. In 1967, Benjamin and Feir (see Whitham (1976) or Debnath (2005)) confirmed these remarkable results both theoretically and experimentally.

Conservation Laws for the NLS Equation

Zakharov and Shabat (1972) proved that equation (13.12.13) has an infinite number of polynomial conservation laws. Each has the form of an integral, with respect to x, of a polynomial expression in terms of the function $\psi(x,t)$ and its derivatives with respect to x. These laws are somewhat similar to those already proved for the KdV equation. Therefore, the proofs of the conservation laws are based on similar assumptions used in the context of the KdV equation.

We prove here three conservation laws for the nonlinear Schrödinger equation (13.12.13):

$$\int_{-\infty}^{\infty} |\psi|^2 \, dx = \text{constant} = C_1, \qquad (13.12.44)$$

$$\int_{-\infty}^{\infty} i\left(\psi \overline{\psi}_x - \overline{\psi}\,\psi_x\right) dx = \text{constant} = C_2, \qquad (13.12.45)$$

$$\int_{-\infty}^{\infty} \left(|\psi_x|^2 - \frac{1}{2}\gamma |\psi|^4\right) dx = \text{constant} = C_3, \qquad (13.12.46)$$

where the bar denotes the complex conjugate.

We multiply (13.12.13) by $\overline{\psi}$ and its complex conjugate by ψ and subtract the latter from the former to obtain

$$i\frac{d}{dt}\left(\psi\overline{\psi}\right) + \frac{d}{dx}\left(\psi_x\overline{\psi} - \overline{\psi}_x\psi\right) = 0. \qquad (13.12.47)$$

Integration with respect to x in $-\infty < x < \infty$ gives

$$i\frac{d}{dt}\int_{-\infty}^{\infty} |\psi|^2 \, dx = 0.$$

This proves result (13.12.44).

We multiply (13.12.13) by $\overline{\psi}_x$ and its complex conjugate by ψ_x and then, add them to obtain

$$i\left(\psi_t\,\overline{\psi}_x - \overline{\psi}_t\,\psi_x\right) + \left(\psi_{xx}\,\overline{\psi}_x + \overline{\psi}_{xx}\,\psi_x\right) + \gamma\,|\psi|^2\left(\psi\,\overline{\psi}_x + \overline{\psi}_x\,\psi_x\right) = 0.$$
(13.12.48)

We differentiate (13.12.13) and its complex conjugate with respect to x, and multiply the former by $\overline{\psi}$ and the latter by ψ and then add them together. This leads to the result

$$i\left(\overline{\psi}_x\psi_{xt} - \psi\,\overline{\psi}_{xt}\right) + \left(\psi_{xxx}\,\overline{\psi} + \overline{\psi}_{xxx}\,\psi\right)$$
$$+\gamma\left[\overline{\psi}\left(|\psi|^2\,\psi\right)_x + \psi\left(|\psi|^2\,\overline{\psi}\right)_x\right] = 0. \quad (13.12.49)$$

If we subtract (13.12.49) from (13.12.48) and then simplify, we have

$$i\frac{d}{dt}\left(\overline{\psi}\,\psi_x - \psi\,\overline{\psi}_x\right)$$
$$= \frac{d}{dx}\left(\psi_x\,\overline{\psi}_x\right) + \frac{d}{dx}\left(\overline{\psi}\,\psi_{xx} + \psi\,\overline{\psi}_{xx}\right) - \frac{d}{dx}\left(\psi_x\,\overline{\psi}_x\right) + \gamma\frac{d}{dx}\left(\psi\,\overline{\psi}\right)^2$$
$$= \frac{d}{dx}\left(\overline{\psi}\,\psi_{xx} + \psi\,\overline{\psi}_{xx}\right) + \gamma\frac{d}{dx}\,|\psi|^4\,.$$

Integrating this result with respect to x, we obtain

$$\frac{d}{dt}\int_{-\infty}^{\infty} i\left(\overline{\psi}\,\psi_x - \psi\,\overline{\psi}_x\right) dx = 0.$$

This proves the second result.

We multiply (13.12.13) by $\overline{\psi}_t$ and its complex conjugate by ψ_t and add the resulting equations to derive

$$\left(\overline{\psi}_t\,\psi_{xx} + \overline{\psi}_{xx}\,\psi_t\right) + \gamma\left(\psi^2\,\overline{\psi}\,\overline{\psi}_t + \overline{\psi}^2\,\psi\,\psi_t\right) = 0,$$

or,

$$\frac{d}{dx}\left(\overline{\psi}_t\,\psi_x + \psi_t\,\overline{\psi}_x\right) - \frac{d}{dt}\left(\psi_x\,\overline{\psi}_x\right) + \frac{\gamma}{2}\frac{d}{dt}\left(\psi^2\,\overline{\psi}^2\right) = 0.$$

Integrating this with respect to x, we have

$$\frac{d}{dt}\int_{-\infty}^{\infty}\left(|\psi_x|^2 - \frac{\gamma}{2}\,|\psi|^4\right) dx = \int_{-\infty}^{\infty}\frac{d}{dx}\left(\overline{\psi}_t\,\psi_x + \psi_t\,\overline{\psi}_x\right) dx = 0. \quad (13.12.50)$$

This gives (13.12.46).

The above three conservation integrals have a simple physical meaning. In fact, the constants of motion C_1, C_2 and C_3 are related to the number of particles, the momentum, and the energy of a system governed by the nonlinear Schrödinger equation.

An analysis of this section reveals several remarkable features of the nonlinear Schrödinger equation. This equation can also be used to investigate instability phenomena in many other physical systems. Like the various forms of the KdV equation, the NLS equation arises in many physical problems, including nonlinear water waves and ocean waves, waves in plasma, propagation of heat pulses in a solid, self-trapping phenomena in nonlinear optics, nonlinear waves in a fluid filled viscoelastic tube, and various nonlinear instability phenomena in fluids and plasmas (see Debnath (2005)).

13.13 The Lax Pair and the Zakharov and Shabat Scheme

In his 1968 seminal paper, Lax developed an elegant formalism for finding isospectral potentials as solutions of a nonlinear evolution equation with all of its integrals. This work deals with some new and fundamental ideas, deeper results, and their application to the KdV model. This work subsequently paved the way to generalizations of the technique as a method for solving other nonlinear partial differential equations. Introducing the Heisenberg picture, Lax developed the method of inverse scattering based upon an abstract formulation of evolution equations and certain properties of operators on a Hilbert space, some of which are familiar in the context of quantum mechanics. His formulation has the feature of associating certain nonlinear evolution equations with linear equations that are analogs of the Schrödinger equation for the KdV equation.

To formulate Lax's method (1968), we consider two linear operators L and M. The eigenvalue equation related to the operator L corresponds to the Schrödinger equation for the KdV equation. The general form of this eigenvalue equation is

$$L\psi = \lambda\psi, \tag{13.13.1}$$

where ψ is the eigenfunction and λ is the corresponding eigenvalue. The operator M describes the change of the eigenvalues with the parameter t, which usually represents time in a nonlinear evolution equation. The general form of this evolution equation is

$$\psi_t = M\psi. \tag{13.13.2}$$

Differentiating (13.13.1) with respect to t gives

$$L_t\psi + L\psi_t = \lambda_t\psi + \lambda\psi_t. \tag{13.13.3}$$

We next eliminate ψ_t from (13.13.3) by using (13.13.2) and obtain

$$L_t\psi + LM\psi = \lambda_t\psi + \lambda M\psi = \lambda_t\psi + M\lambda\psi = \lambda_t\psi + ML\psi, \tag{13.13.4}$$

or, equivalently,

$$\frac{\partial L}{\partial t}\psi = \lambda_t \psi + (ML - LM)\psi. \tag{13.13.5}$$

Thus, eigenvalues are constant for nonzero eigenfunctions if and only if

$$\frac{\partial L}{\partial t} = -(LM - ML)\psi = -[L, M]\psi, \tag{13.13.6}$$

where $[L, M] = (LM - ML)$ is called the *commutator* of the operators L and M, and the derivative on the left-hand side of (13.13.6) is to be interpreted as the time derivative of the operator alone. Equation (13.13.6) is called the *Lax equation* and the operators L and M are called the *Lax pair*. It is the Heisenberg picture of the KdV equation. The problem, of course, is how to determine these operators for a given evolution equation. There is no systematic method of solution for this problem. For a negative integrable hierarchy, Qiao (1995) and Qiao and Strampp (2002) suggest a general approach to generate integrable equations; they also devise strategies for finding a Lax pair from a given spectral problem.

We consider the initial-value problem for $u(x, t)$ which satisfies the non-linear evolution equation system

$$u_t = N(u) \tag{13.13.7}$$
$$u(x, 0) = f(x), \tag{13.13.8}$$

where $u \in Y$ for all t, Y is a suitable function space, and $N : Y \to Y$ is a nonlinear operator that is independent of t but may involve x or derivatives with respect to x.

We must assume that the evolution equation (13.13.7) can be expressed in the Lax form

$$L_t + (LM - ML) = L_t + [L, M] = 0, \tag{13.13.9}$$

where L and M are linear operators in x on a Hilbert space H and depend on u and $L_t = u_t$ is a scalar operator. We also assume that L is self-adjoint so that $(L\phi, \psi) = (\phi, L\psi)$ for all ϕ and $\psi \in H$ with $(\cdot, \cdot)$ as an inner product.

We now formulate the eigenvalue problem for $\psi \in H$:

$$L\psi = \lambda(t)\psi, \qquad t \geq 0, \quad x \in R. \tag{13.13.10}$$

Differentiating with respect to t and making use of (13.13.9), we obtain

$$\lambda_t \psi = (L - \lambda)(\psi_t - M\psi). \tag{13.13.11}$$

The inner product of ψ with this equation yields

$$(\psi, \psi)\lambda_t = ((L - \lambda)\psi, \lambda_t - M\psi), \tag{13.13.12}$$

which, since $L - \lambda$ is self-adjoint, is given by

$$(\psi, \psi) \lambda_t = (0, \psi_t - M\psi) = 0.$$

Hence, $\lambda_t = 0$, confirming that each eigenvalue of L is a constant. Consequently, (13.13.11) becomes

$$L (\psi_t - M\psi) = \lambda (\psi_t - M\psi). \tag{13.13.13}$$

This shows that $(\psi_t - M\psi)$ is an eigenfunction of the operator L with the eigenvalue λ. It is always possible to redefine M by adding the product of the identity operator and a suitable function of t, so that the original equation (13.13.9) remains unchanged. This leads to the time evolution equation for ψ as

$$\psi_t = M\psi, \qquad t \geq 0. \tag{13.13.14}$$

Thus, we have the following.

Theorem 13.13.1. If the evolution equation (13.13.7) can be expressed as the *Lax equation*

$$L_t + [L, M] = 0 \tag{13.13.15}$$

and if (13.13.10) holds, then $\lambda_t = 0$, and ψ satisfies (13.13.14).

It is not yet clear how to find the operators L and M that satisfy the preceding conditions. To illustrate the Lax method, we choose the Schrödinger operator L in the form

$$L \equiv -\frac{\partial^2}{\partial x^2} + u, \tag{13.13.16}$$

so that $L\psi = \lambda\psi$ becomes the Sturm–Liouville problem for the self-adjoint operator L. With this given L, the problem is to find the operator M. Based on the theory of a linear unitary operator on a Hilbert space H, the linear operator M can be chosen as antisymmetric, so that $(M\phi, \psi) = -(\phi, M\psi)$ for all ϕ and $\psi \in H$. So, a suitable linear combination of odd derivatives in x is a natural choice for M. It follows from the inner product that

$$(M\phi, \psi) = \int_{-\infty}^{\infty} \frac{\partial^n \phi}{\partial x^n} \psi dx = -\int_{-\infty}^{\infty} \phi \frac{\partial^n \psi}{\partial x^n} dx = -(\phi, M\psi), \tag{13.13.17}$$

provided $M = \partial^n \phi / \partial x^n$ for odd n, and ϕ, ψ, and their derivatives with respect to x tend to zero, as $|x| \to \infty$. Moreover, we require that M has sufficient freedom in any unknown constants or functions to make $L_t + [L, M]$ a multiplicative operator, that is, of degree zero. For $n = 1$, the simplest choice for M is $M = c (\partial/\partial x)$, where c is a constant. It then follows that $[L, M] = -cu_x$, which is automatically a multiplicative operator. Thus, the Lax equation is

$$L_t + [L, M] = u_t - c u_x = 0, \qquad (13.13.18)$$

and hence, the one-dimensional wave equation

$$u_t - c u_x = 0 \qquad (13.13.19)$$

has an associated eigenvalue problem with the eigenvalues that are constants of motion.

The next natural choice is

$$M = -a \frac{\partial^3}{\partial x^3} + A \frac{\partial}{\partial x} + \frac{\partial}{\partial x} A + B, \qquad (13.13.20)$$

where a is a constant, $A = A(x, t)$, and $B = B(x, t)$, and the third term on the right-hand side of (13.13.20) can be dropped, but we retain it for convenience. It follows from an algebraic calculation that

$$[L, M] = a\, u_{xxx} - A_{xxx} - B_{xx} - 2\, u_x A$$
$$+ (3a u_{xx} - 4 A_{xx} - 2 B_x) \frac{\partial}{\partial x} + (3 a u_x - 4 A_x) \frac{\partial^2}{\partial x^2}.$$

This would be a multiplicative operator if $A = \frac{3}{4} a u$ and $B = B(t)$. Consequently, the Lax equation (13.13.15) becomes

$$u_t - \frac{3}{2} a u u_x + \frac{a}{4} u_{xxx} = 0. \qquad (13.13.21)$$

This is the standard KdV equation if $a = 4$. The operator M defined by (13.13.20) reduces to the form

$$M = -4 \frac{\partial^3}{\partial x^3} + 3 \left(u \frac{\partial}{\partial x} + \frac{\partial}{\partial x} u \right) + B(t). \qquad (13.13.22)$$

Hence, the time evolution equation for ψ can be simplified by using the Sturm–Liouville equation, $\psi_{xx} - (u - \lambda) \psi = 0$, to

$$\psi_t = 4 (\lambda \psi - u \psi)_x + 3 \psi_x + 3 (u \psi)_x + B \psi$$
$$= 2 (u + 2\lambda) \psi_x - u_x \psi + B \psi. \qquad (13.13.23)$$

We close this section by adding several comments. First, any evolution equations solvable by the inverse scattering transform (IST), like the KdV equation, can be expressed in Lax form. However, the main difficulty is that there is no completely systematic method for determining whether or not a given partial differential equation produces a Lax equation and, if so, how to find the Lax pair L and M. Indeed, Lax proved that there is an infinite number of operators, M, one associated with each odd order of $\partial/\partial x$, and hence, an infinite family of flows u_t under which the spectrum of L is preserved. Second, it is possible to study other spectral equations by choosing alternative forms for L. Third, the restriction that L and M should be limited to the class of scalar operators could be removed. In fact, L and M could be *matrix operators*. The Lax formulation has already been extended to such operators. Fourth, Zakharov and Shabat (1972, 1974) published a series of notable papers in this field extending the nonlinear Schrödinger (NLS) equation and other evolution equations. For the first time, they have generalized the Lax formalism for equations with more than one spatial variable. This extension is known as the *Zakharov and Shabat* (ZS) *scheme*, which, essentially, follows the Lax method and recasts it in a matrix form, leading to a matrix Marchenko equation.

Finally, we briefly discuss the ZS scheme for nonself-adjoint operators to obtain N-soliton solutions for the NLS equation. Zakharov and Shabat introduced an ingenious method for any nonlinear evolution equation

$$u_t = Nu, \tag{13.13.24}$$

that represents the equation in the form

$$\frac{\partial L}{\partial t} = i\,[L, M] = i\,(LM - ML)\,, \tag{13.13.25}$$

where L and M are linear differential operators that include the function u in the coefficients, and L refers to differentiating u with respect to t.

We consider the eigenvalue problem

$$L\phi = \lambda\phi. \tag{13.13.26}$$

Differentiation of (13.13.26) with respect to t gives

$$i\,\phi\left(\frac{d\lambda}{dt}\right) = (L - \lambda)\,(i\phi_t - M\phi)\,. \tag{13.13.27}$$

If ϕ satisfies (13.13.26) initially and changes in such a manner that

$$i\phi_t = M\phi, \tag{13.13.28}$$

then ϕ always satisfies (13.13.26). Equations (13.13.26) and (13.13.28) are the pair of equations coupling the function $u\,(x,t)$ in the coefficients with a scattering problem. Indeed, the nature of ϕ determines the scattering potential in (13.13.26), and the evolution of ϕ in time is given by (13.13.28).

Although this formulation is quite general, the crucial step is to factor L according to (13.13.25). Zakharov and Shabat (1972) introduce 2×2 matrices associated with (13.13.25) as follows:

$$L = i \begin{bmatrix} 1+\alpha & 0 \\ 0 & 1-\alpha \end{bmatrix} \frac{\partial}{\partial x} + \begin{bmatrix} 0 & u^* \\ u & 0 \end{bmatrix}, \tag{13.13.29}$$

$$M = -\alpha \begin{bmatrix} 1 & 0 \\ 0 & 1 \end{bmatrix} \frac{\partial^2}{\partial x^2} + \begin{bmatrix} \frac{|u|^2}{1+\alpha} & iu_x^* \\ -iu_x & \frac{-|u|^2}{1-\alpha} \end{bmatrix}, \tag{13.13.30}$$

and the NLS equation for complex $u(x,t)$ is given by

$$iu_t + u_{xx} + \gamma |u|^2 u = 0, \tag{13.13.31}$$

where

$$\gamma = 2/\left(1 - \alpha^2\right).$$

Thus, the eigenvalue problem (13.13.26) and the evolution equation (13.13.28) complete the inverse scattering problem. The initial-value problem for $u(x,t)$ can be solved for a given initial condition $u(x,0)$. It seems clear that the significant contribution would come from the point spectrum for large times $(t \to \infty)$. Physically, the disturbance tends to disintegrate into a series of solitary waves. The mathematical analysis is limited to the asymptotic solutions so that $|u| \to 0$ as $|x| \to \infty$, but a series of solitary waves is expected to be the end result of the instability of wavetrains to modulations.

13.14 Exercises

1. For the flow density relation $q = v\rho(1 - \rho/\rho_1)$, find the solution of the traffic flow problem with the initial condition $\rho(x,0) = f(x)$ for all x. Let

$$f(x) = \begin{cases} \frac{1}{3}, & x \leq 0 \\ \frac{1}{3} + \frac{5}{12}x, & 0 \leq x \leq 1 \\ \frac{3}{4}, & x \geq 1. \end{cases}$$

Show that
(i) $\rho = \frac{1}{3}\rho_0$ along the characteristic lines $ct = 3(x - x_0)$, $x_0 \leq 0$,

(ii) $\rho = \frac{3}{4}\rho_0$ along the characteristic lines $ct = 2(x_0 - x)$, $x_0 \geq 1$,

(iii) $\rho = \frac{1}{3} + \frac{5}{12}x_0\rho_0$ along $ct\left(\frac{2}{5} - x_0\right) = \frac{6}{5}(x - x_0)$, $0 \leq x_0 \leq 1$.

Discuss what happens at the intersection of the two lines $ct = 3x$ and $ct = 2(1 - x)$. Draw the characteristic lines ct versus x.

2. A mountain of height $h(x,t)$ is vulnerable to erosion if its slope h_x is very large. If h_t and h_x satisfy the functional relation $h_t = -Q(h_x)$, show that $u = h_x$ is governed by the nonlinear wave equation

$$u_t + c(u)u_x = 0, \qquad c(u) = Q'(u).$$

3. Consider the flow of water in a river carrying some particles through the solid bed. During the sedimentation process, some particles will

be deposited in the bed. Assuming that v is the constant velocity of water and that $\rho = \rho_f + \rho_b$ is the density, where ρ_f is the density of the particles carried in the fluid and ρ_b is the density of the material deposited on the solid bed, the conservation law is

$$\frac{\partial \rho}{\partial t} + \frac{\partial q}{\partial x} = 0, \qquad q = v\rho_f.$$

(a) Show that ρ_b satisfies the equation

$$\frac{\partial \rho_f}{\partial t} + c\left(\rho_f\right) \frac{\partial \rho_f}{\partial x} = 0, \qquad c\left(\rho_f\right) = \frac{v}{1 + Q'\left(\rho_f\right)},$$

where $\rho_b = Q\left(\rho_f\right)$.

(b) This problem also arises in chemical engineering with a second relation between ρ_f and ρ_b in the form

$$\frac{\partial \rho_b}{\partial t} = k_1 \left(\alpha - \rho_b\right) - k_2 \left(\beta - \rho_f\right) \rho_b,$$

where k_1, k_2 represent constant reaction rates and α, β are constant values of the saturation levels of the particles in the solid bed and fluid respectively. Show that the propagation speed c is

$$c = \frac{k_2 \beta v}{\left(k_1 \alpha + k_2 \beta\right)}$$

provided that the densities are small.

4. Show that a steady solution $u\left(x, t\right) = f\left(\zeta\right)$, $\zeta = x - ct$ of Burgers' equation with the boundary conditions $f \to u_\infty^-$ or u_∞^+ as $\zeta \to -\infty$ or $+\infty$ is

$$u\left(x, t\right) = \left[c - \frac{1}{2}\left(u_\infty^- - u_\infty^+\right) \tanh\left\{\frac{\left(u_\infty^- - u_\infty^+\right)}{4\nu}\left(x - ct\right)\right\}\right],$$

where $u_\infty^- = c + \left(c^2 + 2A\right)^{\frac{1}{2}}$, $u_\infty^+ = c - \left(c^2 - 2A\right)^{\frac{1}{2}}$ are two roots of $f^2 - 2cf - 2A = 0$ and A is a constant of integration.

5. Show that the transformations $x \to \gamma^{\frac{1}{3}}x$, $u \to -6\gamma^{\frac{1}{3}}u$, $t \to t$ reduce the KdV equation $u_t + uu_x + \gamma u_{xxx} = 0$ into the canonical form $u_t - 6uu_x + u_{xxx} = 0$.

Hence or otherwise, prove that the solution of the canonical equation with the boundary conditions that $u\left(x, t\right)$ and its derivatives tend to zero as $|x| \to \infty$ is

$$u\left(x, t\right) = -\frac{a^2}{2} \operatorname{sech}^2\left\{\frac{a}{2}\left(x - a^2 t\right)\right\}.$$

6. Verify that the Riccati transformation $u = v^2 + v_x$ transforms the KdV equation so that v satisfies the associated KdV equation

$$v_t - 6v^2 v_x + v_{xxx} = 0.$$

For a given u, show that the Riccati equation $v_x + v^2 = u$ becomes the linear Schrödinger equation, $\psi_{xx} = u\psi$ (without the energy-level term) under the transformation $v = (\psi_x/\psi)$.

7. Apply the method of characteristics to solve the pair of equations

$$\frac{\partial u}{\partial t} + \frac{\partial v}{\partial x} = 0, \qquad \frac{\partial v}{\partial t} + \frac{\partial u}{\partial x} = 0,$$

with the initial data

$$u(x,0) = e^x, \qquad v(x,0) = e^{-x},$$

Show that the Riemann invariants are

$$2r(\alpha) = 2\cosh\alpha, \qquad 2s(\beta) = 2\sinh\beta,$$

Also, show that solutions are $u = \cosh(x-t) + \sinh(x+t)$ and $v = \cosh(x-t) - \sinh(x+t)$.

8. For an isentropic flow, the Euler equations are

$$\rho_t + (\rho u)_x = 0,$$
$$u_t + uu_x + \frac{1}{\rho}p_x = 0,$$
$$S_t + uS_x = 0,$$

where ρ is the density, u is the velocity in the x direction, p is the pressure, and S is the entropy.

Show that this system has three families of characteristics.

$$\Gamma_0 : \frac{dx}{dt} = u, \qquad \Gamma_{\pm} : \frac{dx}{dt} = u \pm c,$$

where $c^2 = \left(\frac{\partial p}{\partial \rho}\right)_s = \text{constant}$.

Hence derive the following full set of characteristic equations

$$\frac{dS}{dt} = 0, \quad \text{on} \quad \Gamma_0,$$

$$\frac{dp}{dt} \pm \rho c \frac{du}{dt} = 0, \quad \text{on} \quad \Gamma_{\pm}.$$

In particular, when the flow is isentropic ($S = \text{constant}$ everywhere), show that the characteristic equations are

$$\int \frac{c(\rho)}{\rho} d\rho \pm u = \text{constant on } \Gamma_{\pm} \text{ and } \Gamma_{\pm} : \frac{dx}{dt} = u \pm c.$$

9. (a) Using equations (13.8.32)–(13.8.36), derive the second-order equation

$$t_{rs} + \phi\,(r+s)\,(t_r + t_s) = 0, \qquad \phi\,(r+s) = \frac{1}{2c}\left(1 + \frac{\rho}{c}\frac{dc}{d\rho}\right).$$

(b) For a polytropic gas, show that

$$\phi\,(r+s) = \frac{\gamma+1}{4c} = \frac{\alpha}{F\,(\rho)}, \qquad \alpha \equiv \frac{1}{2}\left(\frac{\gamma+1}{\gamma-1}\right),$$

and

$$t_{rs} + \frac{\alpha}{r+s}\,(t_r + t_s) = 0.$$

(c) With $F\,(\rho) = \frac{2c}{\gamma-1} = r+s$ and $u = r-s$, show that the differential equation in 9(b) reduces to the *Euler–Poisson–Darboux equation*

$$t_{uu} - \left(\frac{\gamma-1}{2}\right)^2\left(t_{cc} + \frac{2\alpha}{c}\,t_c\right) = 0,$$

where u and c are independent variables.

10. Show that the KdV equation

$$u_t - 6uu_x + u_{xxx} = 0$$

satisfies the conservation law in the form $U_t + V_x = 0$ when (a) $U = u$, $V = -3u^2 + u_{xx}$, and (b) $U = \frac{1}{2}u^2$, $V = -2u^3 + uu_x - \frac{1}{2}u_x^2$.

11. Show that the KdV equation

$$u_t + 6uu_x + u_{xxx} = 0$$

satisfies the conservation law

$$U_t + V_x = 0,$$

where
(a) $U = u$, $V = 3u^2 + u_{xx}$,

(b) $U = \frac{1}{2}u^2$, $V = 2u^3 + uu_{xx} - \frac{1}{2}u_x^2$,

(c) $U = u^3 - \frac{1}{2}u_x^2$, $V = \frac{9}{2}u^4 + 3u^2 u_{xx} + \frac{1}{2}u_{xx}^2 + u_x u_t$.

12. Show that the conservation laws for the associated KdV equation

$$v_t - 6v^2 v_x + v_{xxx} = 0,$$

are

(a) $(v)_t + \left(\frac{1}{3}v^3 + v_{xx}\right)_x = 0,$

(b) $\left(\frac{1}{2}v^2\right)_t + \left(\frac{1}{4}v^4 + vv_{xx} - \frac{1}{2}v_x^2\right)_x = 0.$

13. For the nonlinear Schrödinger equation (13.12.13), prove that

$$\int_{-\infty}^{\infty} T \, dx = \text{constant},$$

where (i) $T \equiv i \left(\overline{\psi} \, \psi_x - \psi \, \overline{\psi}_x \right)$ and (ii) $T \equiv |\psi_x|^2 - \frac{1}{2} \gamma \, |\psi|^4$.

14. Show that the conservation law for Burgers' equation (13.10.20) is

$$(u)_t + \left(\frac{1}{2} u^2 + \nu \, u_x \right)_x = 0.$$

15. Show that the conservation laws for the equation

$$u_t - u u_x - u_{xxt} = 0$$

are of the forms

(a) $(u)_t - \left(\frac{1}{2} u^2 + u_{xt} \right)_x = 0,$

(b) $\frac{1}{2} \left(u^2 + u_x^2 \right)_t - \left(\frac{1}{3} u^3 + u u_{xt} \right)_x = 0.$

16. Show that the linear Schrödinger system

$$i \, \psi_t + \psi_{xx} = 0, \qquad -\infty < x < \infty, \quad t > 0,$$
$$\psi \to 0, \qquad |x| \to \infty,$$
$$\psi(x, 0) = \psi(x) \quad \text{with} \quad \int_{-\infty}^{\infty} |\psi|^2 \, dx = 1.$$

has the conservation law

$$\left(i \, |\psi|^2 \right)_t + (\psi^* \psi_x - \psi \psi_x^*)_x = 0$$

and the energy integral

$$\int_{-\infty}^{\infty} |\psi|^2 \, dx = 1.$$

17. Seek a dispersive wave solution of the telegraph equation (see problem 14, 3.9 Exercises)

$$u_{tt} - c^2 u_{xx} + ac^2 u_t + bc^2 u = 0$$

in the form

$$u(x, t) = A \exp \left[i \left(kx - \omega t \right) \right].$$

(a) Show that $\omega = -\frac{1}{2} \left(iac^2 \right) \pm \frac{1}{2} \left[4c^2 k^2 + \left(4b - c^2 \right) c^2 \right]^{\frac{1}{2}}.$

(b) If $4b = a^2 c^2$, show that the solution

$$u(x, t) = A \exp \left(-\frac{1}{2} ac^2 t \right) \exp \left[ik \left(x \pm ct \right) \right]$$

represents nondispersive waves with attenuation.

14

Numerical and Approximation Methods

"The strides that have been made recently, in the theory of nonlinear partial differential equations, are as great as in the linear theory. Unlike the linear case, no wholesale liquidation of broad classes of problems has taken place; rather, it is steady progress on old fronts and on some new ones, the complete solution of some special problems, and the discovery of some brand new phenomena. The old tools – variational methods, fixed point theorems, mapping degree, and other topological tools have been augmented by some new ones. Pre-eminent for discovering new phenomena is numerical experimentation; but it is likely that in the future numerical calculations will be parts of proofs."

Peter Lax

"Almost everyone using computers has experienced instances where computational results have sparked new insights."

Norman J. Zabusky

14.1 Introduction

The preceding chapters have been devoted to the analytical treatment of linear and nonlinear partial differential equations. Several analytical methods to find the exact analytical solution of these equations within simple domains have been discussed. The boundary and initial conditions in these problems were also relatively simple, and were expressible in simple mathematical form. In dealing with many equations arising from the modelling of physical problems, the determination of such exact solutions in a simple domain is a formidable task even when the boundary and/or initial data are simple. It is then necessary to resort to numerical or approximation methods in order to deal with the problems that cannot be solved analytically. In view of the widespread accessibility of today's high speed electronic computers, numerical and approximation methods are becoming increasingly important and useful in applications.

In this chapter some of the major numerical and approximation approaches to the solution of partial differential equations are discussed in some detail. These include numerical methods based on finite difference approximations, variational methods, and the Rayleigh–Ritz, Galerkin, and Kantorovich methods of approximation. The chapter also contains a large section on analytical treatment of variational methods and the Euler–Lagrange equations and their applications. A short section on the finite element method is also included.

14.2 Finite Difference Approximations, Convergence, and Stability

The Taylor series expansion of a function $u(x, y)$ of two independent variables x and y is

$$u(x_i \pm h, y_j) = u_{i \pm 1, j} = u_{i,j} \pm h(u_x)_{i,j} + \frac{h^2}{2!}(u_{xx})_{i,j}$$

$$\pm \frac{h^3}{3!}(u_{xxx})_{i,j} + \cdots, \qquad (14.2.1ab)$$

$$u(x_i, y_j \pm k) = u_{i,j \pm 1} = u_{i,j} \pm k(u_y)_{i,j} + \frac{k^2}{2!}(u_{yy})_{i,j}$$

$$\pm \frac{k^3}{3!}(u_{yyy})_{i,j} + \cdots, \qquad (14.2.2ab)$$

where $u_{i,j} = u(x, y)$, $u_{i \pm 1, j} = u(x \pm h, y)$, and $u_{i,j \pm 1} = u(x, y \pm k)$.

We choose a set of uniformly spaced rectangles with vertices at $P_{i,j}$ with coordinates (ih, jk), where i, j, are positive or negative integers or zero, as shown in Figure 14.2.1. We denote $u(ih, jk)$ by $u_{i,j}$.

Using the above Taylor series expansion, we write approximate expressions for u_x at the vertex $P_{i,j}$ in terms of $u_{i,j}$, $u_{i \pm 1, j}$:

$$u_x = \frac{1}{h}[u(x + h, y) - u(x, y)] \sim \frac{1}{h}(u_{i+1,j} - u_{i,j}) + O(h), \qquad (14.2.3)$$

$$u_x = \frac{1}{h}[u(x, y) - u(x - h, y)] \sim \frac{1}{h}(u_{i,j} - u_{i-1,j}) + O(h), \qquad (14.2.4)$$

$$u_x = \frac{1}{2h}[u(x + h, y) - u(x - h, y)] \sim \frac{1}{2h}(u_{i+1,j} - u_{i-1,j}) + O(h^2).$$

$$(14.2.5)$$

These expressions are called the *forward first difference*, *backward first difference*, and *central first difference* of u_x, respectively. The quantity $O(h)$ or $O(h^2)$ is known as the *truncation error* in this discretization process.

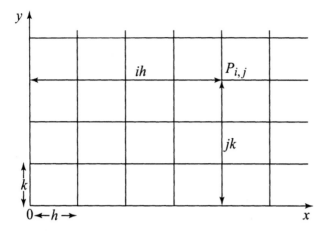

Figure 14.2.1 Uniformly spaced rectangles.

A similar approximate result for u_{xx} at the vertex $P_{i,j}$ is

$$u_{xx} = \frac{1}{h^2}\left[u\left(x+h,y\right) - 2\,u\left(x,y\right) + u\left(x-h,y\right)\right]$$

$$\sim \frac{1}{h^2}\left[u_{i+1,j} - 2\,u_{i,j} + u_{i-1,j}\right] + O\left(h^2\right). \tag{14.2.6}$$

Similarly, the approximate formulas for u_y and u_{yy} at $P_{i,j}$ are

$$u_y = \frac{1}{k}\left[u\left(x,y+k\right) - u\left(x,y\right)\right] \sim \frac{1}{k}\left(u_{i,j+1} - u_{i,j}\right) + O\left(k\right), \tag{14.2.7}$$

$$u_y = \frac{1}{k}\left[u\left(x,y\right) - u\left(x,y-k\right)\right] \sim \frac{1}{k}\left(u_{i,j} - u_{i,j-1}\right) + O\left(k\right), \tag{14.2.8}$$

$$u_y = \frac{1}{2k}\left[u\left(x,y+k\right) - u\left(x,y-k\right)\right] \sim \frac{1}{2k}\left(u_{i,j+1} - u_{i,j-1}\right) + O\left(k^2\right),$$
$$\tag{14.2.9}$$

$$u_{yy} = \frac{1}{k^2}\left[u\left(x,y+k\right) - 2\,u\left(x,y\right) + u\left(x,y-k\right)\right]$$

$$\sim \frac{1}{k^2}\left[u_{i,j+1} - 2u_{i,j} + u_{i,j-1}\right] + O\left(k^2\right). \tag{14.2.10}$$

All these difference formulas are extremely useful in finding numerical solutions of first or second order partial differential equations.

Suppose $U\left(x,y\right)$ represents the exact solution of a partial differential equation $L\left(U\right) = 0$ with independent variables x and y, and $u_{i,j}$ is the exact solution of the corresponding finite difference equation $F\left(u_{i,j}\right) = 0$. Then, the finite difference scheme is said to be *convergent* if $u_{i,j}$ tends to U as h and k tend to zero. The difference, $d_{i,j} \equiv \left(U_{i,j} - u_{i,j}\right)$ is called the *cummulative truncation* (or *discretization*) *error*.

This error can generally be minimized by decreasing the grid sizes h and k. However, this error depends not only on h and k, but also on the number of terms in the truncated series which is used to approximate each partial derivative.

Another kind of error is introduced when a partial differential equation is approximated by a finite difference equation. If the exact finite difference solution $u_{i,j}$ is replaced by the exact solution $U_{i,j}$ of the partial differential equation at the grid points $P_{i,j}$, then the value $F(U_{i,j})$ is called the *local truncation error* at $P_{i,j}$. The finite difference scheme and the partial differential equation are said to be *consistent* if $F(U_{i,j})$ tends to zero as h and k tend to zero.

In general, finite difference equations cannot be solved exactly because the numerical computation is carried out only up to a finite number of decimal places. Consequently, another kind of error is introduced in the finite difference solution during the actual process of computation. This kind of error is called the *round-off error*, and it also depends upon the type of computer used. In practice, the actual computational solution is $u_{i,j}^*$, but not $u_{i,j}$, so that the difference $r_{i,j} = \left(u_{i,j} - u_{i,j}^*\right)$ is the round-off error at the grid point $P_{i,j}$. In fact, this error is introduced into the solution of the finite difference equation by round-off errors. In reality, the round-off error depends mainly on the actual computational process and the finite difference itself. In contrast to the cummulative truncation error, the round-off error cannot be made small by allowing h and k to tend to zero.

Thus, the total error involved in the finite difference analysis at the point $P_{i,j}$ is given by

$$\left(U_{i,j} - u_{i,j}^*\right) = \left(U_{i,j} - u_{i,j}\right) + \left(u_{i,j} - u_{i,j}^*\right) = d_{i,j} - r_{i,j}. \quad (14.2.11)$$

Usually the discretization error $d_{i,j}$ is bounded when $u_{i,j}$ is bounded because the value of $U_{i,j}$ is fixed for a given partial differential equation with the prescribed boundary and initial data. This fact is used or assumed in order to introduce the concept of stability. The finite difference algorithm is said to be *stable* if the round-off errors are sufficiently small for all i as $j \to \infty$, that is, the growth of $r_{i,j}$ can be controlled. It should be pointed out again that the round-off error depends not only on the actual computational process and the type of computer used, but also on the finite difference equation itself. Lax (1954) proved a remarkable theorem which establishes the relationship between consistency, stability, and convergence for the finite difference algorithm.

Theorem 14.2.1. *(Lax's Equivalence Theorem). Given a properly posed linear initial-value problem and a finite difference approximation to it that satisfies the consistency criterion, stability is a necessary and sufficient condition for convergence.*

Von Neumann's Stability Method

This method is essentially based upon a finite Fourier series. It expresses the initial errors on the line $t = 0$ in terms of a finite Fourier series and then examines the propagation of errors as $t \to \infty$. It is convenient to denote the error function by $e_{r,s}$ instead of $e_{i,j}$ so that $e_{r,s}$ gives the initial values $e_{r,0} = e(rh) = e_r$ on the line $t = 0$ between $x = 0$ and $x = l$, where $r = 0$, 1, 2, ..., N and $Nh = l$. The finite Fourier series expansion of e_r is

$$e_r = \sum_{n=0}^{N} A_n \exp\left(in\pi x/l\right) = \sum_{n=0}^{N} A_n \exp\left(i\alpha_n rh\right), \qquad (14.2.12)$$

where $\alpha_n = (n\pi/l)$, $x = rh$, and A_n are the Fourier coefficients which are determined from the $(N+1)$ equations (14.2.12).

Since we are concerned with the linear finite difference scheme, errors form an additive system so that the total error can be found by the superposition principle. Thus, it is sufficient to consider a single term $\exp\left(i\alpha rh\right)$ in the Fourier series (14.2.12). Following the method of separation of variables commonly used for finding the analytical solution of a partial differential equation, we seek a separable solution of the finite difference equation for $e_{r,s}$ in the form

$$e_{r,s} = \exp\left(i\alpha rh + \beta sk\right) = \exp\left(i\alpha rh\right) p^s \qquad (14.2.13)$$

which reduces to $\exp\left(i\alpha rh\right)$ at $s = 0$ ($t = sk = 0$), where $p = \exp\left(\beta k\right)$, and β is a complex constant. This shows that the error is bounded as ($t \to \infty$) provided that

$$|p| \leq 1 \qquad (14.2.14)$$

is satisfied. This condition is found to be necessary and sufficient for the stability of the finite difference algorithm.

14.3 Lax–Wendroff Explicit Method

To describe this method, we consider the first-order conservation equation

$$\frac{\partial u}{\partial t} + c \frac{\partial u}{\partial x} = 0 \qquad (14.3.1)$$

where $u \equiv u(x, t)$ is some physical function of space variable x and time t. This equation occurs frequently in applied mathematics.

Lax and Wendroff use the Taylor series expansion in t in the form

$$u_{i,j+1} = u_{i,j} + k\left(u_t\right)_{i,j} + \frac{k^2}{2!}\left(u_{tt}\right)_{i,j} + \frac{k^3}{3!}\left(u_{ttt}\right)_{i,j} + \ldots, \qquad (14.3.2)$$

where $k \equiv \delta t$.

The partial derivatives in t in (14.3.2) can easily be eliminated by using $u_t = -c\,u_x$ so that (14.3.2) becomes

$$u_{i,j+1} = u_{i,j} - c\,k\,(u_x)_{i,j} + \frac{c^2 k^2}{2}\,(u_{xx})_{i,j} - \dots \qquad (14.3.3)$$

Replacing u_x, u_{xx} by the central difference formulas, (14.3.3) becomes

$$u_{i,j+1} = u_{i,j} - \left(\frac{ck}{2h}\right)(u_{i+1,j} - u_{i-1,j}) + \frac{1}{2}\left(\frac{ck}{h}\right)^2 (u_{i+1,j} - 2\,u_{i,j} + u_{i-1,j}),$$

or

$$u_{i,j+1} = \left(1 - \varepsilon^2\right) u_{i,j} + \frac{\varepsilon}{2}\,(1 + \varepsilon)\,u_{i-1,j} - \frac{\varepsilon}{2}\,(1 + \varepsilon)\,u_{i+1,j} + O\left(\varepsilon^3\right), \quad (14.3.4)$$

where $\varepsilon = (ck/h)$. This is called the *Lax–Wendroff second-order finite difference scheme*; it has been widely used to solve first-order hyperbolic equations.

Von Neumann criterion (14.2.14) can be applied to investigate the stability of the Lax–Wendroff scheme. It is noted that the error function $e_{r,s}$ given by (14.2.13) satisfies the finite difference equation (14.3.4). We then substitute (14.2.13) into (14.3.4) and cancel common factors to obtain

$$p = \left(1 - \varepsilon^2\right) + \frac{\varepsilon}{2}\left[(1 + \varepsilon)\,e^{-i\alpha h} - (1 - \varepsilon)\,e^{i\alpha h}\right]$$

$$= 1 - 2\,\varepsilon^2 \sin^2\left(\frac{\alpha h}{2}\right) - 2i\varepsilon \sin\left(\frac{\alpha h}{2}\right) \cos\left(\frac{\alpha h}{2}\right),$$

so that

$$|p|^2 = 1 - 4\varepsilon^2 \left(1 - \varepsilon^2\right) \sin^4\left(\frac{\alpha h}{2}\right). \qquad (14.3.5)$$

According to the von Neumann criterion, the Lax–Wendroff scheme (14.3.4) is stable as $t \to \infty$ if $|p| \le 1$, which gives $4\varepsilon^2\left(1 - \varepsilon^2\right) \ge 0$, that is, $0 < \varepsilon \le 1$.

The local truncation error of the Lax–Wendroff equation (14.3.4) at $P_{i,j}$ is

$$T_{i,j} = \frac{1}{k}\,(u_{i,j+1} - u_{i-1,j})$$

which is, by (14.2.2a) and (14.2.1b) with $ck = h$ ($\varepsilon = 1$),

$$= (u_t + cu_x)_{i,j} + \frac{k}{2}\,\left(u_{tt} - c^2 u_{xxx}\right)_{i,j} + \frac{k^2}{6}\,\left(u_{ttt} + c^3 u_{xxx}\right)_{i,j} + O\left((ck)^3\right).$$

$$(14.3.6)$$

The first two terms on the right side of (14.3.6) vanish by equation (14.3.1) so that the local truncation error becomes

$$T_{i,j} = \frac{1}{6}\left(k^2 u_{ttt} + c\,h^2 u_{xxx}\right)_{i,j}.$$ (14.3.7)

Another approximation to (14.3.1) with first-order accuracy is

$$\frac{1}{k}\left(u_{i,j+1} - u_{i,j}\right) + \frac{c}{h}\left(u_{i,j} - u_{i-1,j}\right) = 0.$$ (14.3.8)

A final explicit scheme for (14.3.1) is based on the central difference approximation. This scheme is called the *leap frog algorithm*. In this method, the finite difference approximation to (14.3.1) is

$$\frac{1}{2k}\left(u_{i,j+1} - u_{i,j-1}\right) + \frac{c}{2h}\left(u_{i+1,j} - u_{i-1,j}\right) = 0,$$

or,

$$u_{i,j+1} = u_{i,j-1} - \varepsilon\left(u_{i+1,j} - u_{i-1,j}\right).$$ (14.3.9)

As shown in Figure 14.3.1, this equation shows that the value of u at $P_{i,j+1}$ is computed from the previously computed values at three grid points at two previous time steps.

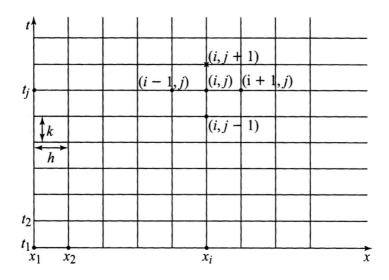

Figure 14.3.1 Grid system for the leap frog algorithm.

14.4 Explicit Finite Difference Methods

(A) Wave Equation and the Courant–Friedrichs–Lewy Convergence Criterion

The method of characteristics provides the most convenient and accurate procedure for solving Cauchy problems involving hyperbolic equations. One of the main advantages of this method is that discontinuities of the initial data propagate into the solution domain along the characteristics. However, when the initial data are discontinuous, the finite difference algorithm for the hyperbolic systems is not very convenient. Problems concerning hyperbolic equations with continuous initial data can be solved successfully by finite difference methods with rectangular grid systems.

A commonly cited problem is the propagation of a one-dimensional wave governed by the system

$$u_{tt} = c^2 \, u_{xx}, \qquad -\infty < x < \infty, \qquad t > 0, \qquad (14.4.1)$$

$$u\,(x,0) = f\,(x)\,, \qquad u_t\,(x,0) = g\,(x) \quad \text{for all } x \in \mathbb{R}. \qquad (14.4.2)$$

Using a rectangular grid system with $h = \delta x$, $k = \delta t$, $u_{i,j} = u\,(ih, jk)$, $-\infty < x < \infty$, and $0 \le j < \infty$, the central difference approximation to equation (14.4.1) is

$$\frac{1}{k^2}\,(u_{i,j+1} - 2\,u_{i,j} + u_{i,j-1}) = \frac{c^2}{h^2}\,(u_{i+1,j} - 2\,u_{i,j} + u_{i-1,j}),$$

or,

$$u_{i,j+1} = \varepsilon^2\,(u_{i+1,j} + u_{i-1,j}) + 2\,(1 - \varepsilon^2)\,u_{i,j} - u_{i,j-1}, \qquad (14.4.3)$$

where $\varepsilon \equiv (ck/h)$, and is often called the *Courant parameter*. This explicit formula allows us to determine the approximate values at the grid points on the lines $t = 2k$, $3k$, $4k$, ..., when the grid values at $t = k$ have been obtained.

The approximate values of the initial data on the line $t = 0$ are

$$u_{i,0} = f_i, \qquad \frac{1}{2k}\,(u_{i,1} - u_{i,-1}) = g_{i,0} \qquad (14.4.4)$$

so that the second result gives

$$u_{i,-1} = u_{i,1} - 2k\,g_{i,0}. \qquad (14.4.5)$$

When $j = 0$ in (14.4.3) and (14.4.5) is used, we obtain

$$u_{i,1} = \frac{1}{2}\,\varepsilon^2\,(f_{i-1} + f_{i+1}) + \left(1 - \varepsilon^2\right) f_i + k\,g_{i,0}. \qquad (14.4.6)$$

This result determines grid values on the line $t = k$.

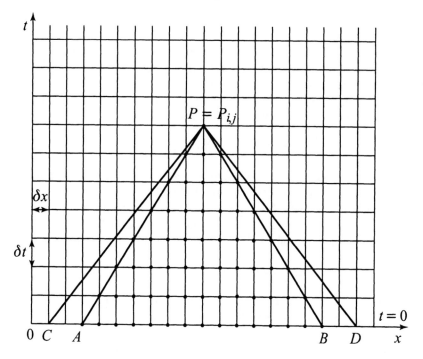

Figure 14.4.1 Computational grid systems and characteristics.

The value of u at $P_{i,j+1}$ is obtained in terms of its previously calculated values at $P_{i\pm1,j}$, $P_{i,j}$, and $P_{i,j-1}$, which are determined from previously computed values on the lines $t = (j-1)\,k$, $(j-2)\,k$, $(j-3)\,k$. Thus, the computation from the lines $t = 0$ and $t = k$ suggests that u at $P_{i,j}$ will represent a function of the values of u within the domain bounded by the lines drawn back toward $t = 0$ from P whose gradients are $(\pm\,\varepsilon)$ as shown in Figure 14.4.1. Thus the triangular regions PAB, PCD represent the domains of dependence at P of the solutions of the finite difference equation (14.4.3) and the differential equation (14.4.1). By analogy with real characteristic lines PC and PD of the differential equation, the straight lines PA and PB are called the *numerical characteristics*. Thus, it follows from Figure 14.4.1 that ΔPAB lies inside ΔPCD, which means that the solution of the finite difference system at P would remain unchanged even when the initial data along PA and PB are changed. Courant, Friedrichs and Lewy (CFL, 1928) proved that the solution of the finite difference system converges to that of the differential equation system as h and k tend to zero provided that the domain of dependence of the difference equation lies inside that of the partial differential equation. This condition for convergence is known as the *CFL condition*, which means $1/c \geq k/h$, that is, $0 < \varepsilon \leq 1$.

If the Courant parameter is $\varepsilon = 1$, equation (14.4.3) reduces to a simple form

$$u_{i,j+1} = u_{i+1,j} + u_{i-1,j} - u_{i,j-1}. \tag{14.4.7}$$

As shown in Figure 14.3.1, this equation shows that the value of u at $P_{i,j+1}$ is computed from the previously computed values at three grid points at two previous time steps. This is called the *leap frog algorithm*.

From equation (5.3.4) in Chapter 5, we know that the solution of the Cauchy problem for the wave equation has the form

$$u(x,t) = \phi(x + ct) + \psi(x - ct),$$

where the functions ϕ and ψ represent waves propagating without changing shape along the negative and positive x directions with a constant speed c. The lines of slope $(dt/dx) = \pm(1/c)$ in the $x - t$ plane, which trace the progress of the waves, are known as the *characteristics* of the wave equation.

In terms of a grid point (x_i, t_j), the above solution has the form

$$u_{i,j} = \phi(x_i + ct_j) + \psi(x_i - ct_j). \tag{14.4.8}$$

It follows from Figure 14.3.1 that $x_i = x_1 + (i-1)h$ and $t_j = t_1 + (j-1)k$ so that (14.4.8) takes the form

$$u_{i,j} = \phi(\alpha + ih + jck) + \psi(\beta + ih - jck), \tag{14.4.9}$$

where

$$\alpha = (x_1 - h) + c(t_1 - k), \qquad \beta = (x_1 - h) - c(t_1 - k).$$

Since $\varepsilon = 1$, $ck = h$, solution (14.4.9) becomes

$$u_{i,j} = \phi(\alpha + (i+j)h) + \psi(\beta + (i-j)h).$$

This satisfies equation (14.4.7). It follows that the leap frog method gives the exact solution of the partial differential equation (14.4.1).

We apply von Neumann stability analysis to investigate the stability of the above numerical method for the wave equation. We seek a separable solution of the error function $e_{r,s}$ as

$$e_{r,s} = \exp(i\alpha rh)\, p^s, \tag{14.4.10}$$

where $p = \exp(\beta k)$. This function satisfies the finite difference equation (14.4.3). Substituting (14.4.10) into (14.4.3) and cancelling the common factors, we obtain the quadratic equation for p

$$p^2 - 2bp + 1 = 0, \tag{14.4.11}$$

where $b = 1 - 2\varepsilon^2 \sin^2(\alpha h/2)$, $\varepsilon = (ck/h)$, and $b \leq 1$ for all real ε and α. This quadratic equation has two complex roots p_1 and p_2 if $b^2 < 1$. Since $p_1 \cdot p_2 = 1$, it follows that one of the roots will always have modulus greater than one unless $|p_1| = |p_2| = 1$. Thus, the scheme is unstable as $s \to \infty$ if the modulus of one of the roots exceeds unity.

On the other hand, when $-1 \leq b \leq 1$, $b^2 \leq 1$, then $|p_1| = |p_2| = 1$. Thus, the finite difference scheme is stable provided $-1 \leq b \leq 1$, which leads to the useful condition for stability as $b \geq -1$ that

$$\varepsilon^2 \leq \mathrm{cosec}^2 \left(\frac{\alpha h}{2} \right). \qquad (14.4.12)$$

This shows the dependence of the stability limit on the space-grid size h. However, this stability condition is always true if $\varepsilon^2 \leq 1$.

Example 14.4.1. Find the explicit finite difference solution of the wave equation

$$u_{tt} - u_{xx} = 0, \qquad 0 < x < 1, \quad t > 0,$$

with the boundary conditions

$$u(0,t) = u(1,t) = 0, \qquad t \geq 0,$$

and the initial conditions

$$u(x,0) = \sin \pi x, \qquad u_t(x,0) = 0, \qquad 0 \leq x \leq 1.$$

Compare the numerical solution with the analytical solution $u(x,t) = \cos \pi t \sin \pi x$ at several points.

The explicit finite difference approximation to the wave equation with $\varepsilon = (k/h) = 1$ is found from (14.4.3) in the form

$$u_{i,j+1} = u_{i-1,j} + u_{i+1,j} - u_{i,j-1}, \qquad j \geq 1.$$

The problem is symmetric with respect to $x = \frac{1}{2}$, so we need to calculate the solution only for $0 \leq x \leq \frac{1}{2}$. We take $h = k = \frac{1}{10} = 0.1$. The boundary conditions give $u_{0,j} = 0$ for $j = 0,1,2,3,4,5$. The initial condition $u_t(x,0) = 0$ yields

$$u_t(x,0) = \frac{1}{2}(u_{i,1} - u_{i,-1}) = 0,$$

or,

$$u_{i,1} = u_{i,-1}.$$

The explicit formula with $j = 0$ gives

$$u_{i,1} = \frac{1}{2}(u_{i-1,0} + u_{i+1,0}), \qquad i = 1,2,3,4,5.$$

Thus,

$$u_{1,1} = \frac{1}{2}(u_{0,0} + u_{2,0}) = \frac{1}{2}u_{2,0} = \frac{1}{2}\sin(0.2\pi) = 0.2939.$$

Similarly,

$$u_{2,1} = 0.5590, \quad u_{3,1} = 0.7695, \quad u_{4,1} = 0.9045, \quad u_{5,1} = 0.9511.$$

We next use the basic explicit formula to compute

$$u_{1,2} = u_{0,1} + u_{2,1} - u_{1,0} = 0 + 0.5590 - 0.3090 = 0.2500,$$
$$u_{2,2} = u_{1,1} + u_{3,1} - u_{2,0} = 0.2939 + 0.7695 - 0.5878 = 0.4756.$$

Similarly, we compute other values for $u_{i,j}$ which are shown in Table 14.4.1. The analytical solutions at $(x,t) = (0.1, 0.1)$ and $(0.2, 0.3)$ are given by

$$u(0.1, 0.1) = \cos(0.1\pi)\sin(0.1\pi) = (0.9511)(0.3090) = 0.2939,$$
$$u(0.2, 0.2) = \cos(0.2\pi)\sin(0.2\pi) = (0.8090)(0.5878) = 0.4577,$$
$$u(0.2, 0.3) = \cos(0.3\pi)\sin(0.2\pi) = (0.5878)(0.5878) = 0.3455.$$

Comparison of the analytical solutions with the above finite difference solutions shows that the latter results are very accurate.

(B) Parabolic Equations

As a prototype diffusion problem, we consider

$$u_t = \kappa u_{xx}, \qquad\qquad 0 < x < 1, \qquad t > 0, \qquad (14.4.13)$$
$$u(0,t) = u(1,t) = 0, \qquad \text{for all } t, \qquad\qquad (14.4.14)$$
$$u(x,0) = f(x), \qquad\qquad \text{for all } x \text{ in } (0,1), \qquad (14.4.15)$$

where $f(x)$ is a given function.

Table 14.4.1.

i	0	1	2	3	4	5
x	0.0	0.1	0.2	0.3	0.4	0.5
j t						
1 0.1	0	0.2939	0.5590	0.7695	0.9045	0.9511
2 0.2	0	0.2500	0.4577	0.6545	0.7695	0.8090
3 0.3	0	0.1817	0.3455	0.4756	0.5590	0.5878
4 0.4	0	0.9045	0.7695	0.2500	0.2939	0.3090
5 0.5	0	0	0	0	0	0

The explicit finite difference approximation to (14.4.13) is

$$\frac{1}{k}\left(u_{i,j+1} - u_{i,j}\right) = \frac{\kappa}{h^2}\left(u_{i+1,j} - 2\,u_{i,j} + u_{i-1,j}\right), \qquad (14.4.16)$$

or,

$$u_{i,j+1} = \varepsilon\left(u_{i+1,j} + u_{i-1,j}\right) + (1 - 2\,\varepsilon)\,u_{i,j}, \qquad (14.4.17)$$

where $\varepsilon = \left(\kappa k / h^2\right)$.

This explicit finite difference formula gives approximate values of u on $t = (j+1)\,k$ in terms of values on $t = jk$ with given $u_{i,0} = f_i$. Thus, $u_{i,j}$ can be obtained for all j by successive use of (14.4.17).

The problems of stability and convergence of the parabolic equation are similar to those of the wave equation. It can be shown that the solution of the finite difference equation converges to that of the differential equation system (14.4.13)–(14.4.15) as h and k tend to zero provided $\varepsilon \le \frac{1}{2}$.

In particular, when $\varepsilon = \frac{1}{2}$, equation (14.4.17) takes a simple form

$$u_{i,j+1} = \frac{1}{2}\left(u_{i+1,j} + u_{i-1,j}\right). \qquad (14.4.18)$$

This is called the *Bender–Schmidt explicit formula* which determines the solution at (x_i, t_{j+1}) as the mean of the values at the grid points $(i \pm 1, j)$. However, more accurate results can be found from (14.4.17) for $\varepsilon < \frac{1}{2}$.

To investigate the stability of the numerical scheme, we assume that the error function is

$$e_{r,s} = \exp\left(i\alpha rh\right)p^s, \qquad (14.4.19)$$

where $p = e^{\beta k}$. The error function and u_{rs} satisfy the same difference equation. Hence, we substitute (14.4.19) into (14.4.17) to obtain

$$p = 1 - 4\,\varepsilon \sin^2\left(\frac{\alpha h}{2}\right). \qquad (14.4.20)$$

Clearly, p is always less than 1 because $\varepsilon > 0$. If $p \ge 0$, the function given by (14.4.19) will decay steadily as $s = j \to \infty$. If $-1 < p < 0$, then the solution will have a decaying amplitude as $s \to \infty$. Therefore, the finite difference scheme will be stable if $p > -1$, that is, if

$$0 < \varepsilon \le \frac{1}{2}\operatorname{cosec}^2\left(\frac{\alpha h}{2}\right). \qquad (14.4.21)$$

This shows that the stability limit depends on h. However, in view of the inequality

$$\varepsilon \le \frac{1}{2} \le \frac{1}{2}\operatorname{cosec}^2\left(\frac{\alpha h}{2}\right),$$

we conclude that the stability condition is $\varepsilon \le \frac{1}{2}$. Finally, if $p < -1$, the solution oscillates with increasing amplitude as $s \to \infty$, and hence, the scheme will be unstable for $\varepsilon > \frac{1}{2}$.

Example 14.4.2. Show that the Richardson explicit finite difference scheme for (14.4.13) is unconditionally unstable.

The Richardson finite difference approximation to (14.4.13) is

$$\frac{1}{2k}\left(u_{i,j+1} - u_{i,j-1}\right) = \frac{\kappa}{h^2}\left(u_{i+1,j} - 2u_{i,j} + u_{i-1,j}\right). \quad (14.4.22)$$

To establish the instability of this equation, we use the Fourier method and assume that

$$e_{r,s} = \exp\left(i\alpha rh\right)p^s, \qquad p = e^{\beta k}.$$

This function satisfies the Richardson difference equation as does $u_{r,s}$. Consequently,

$$p - \frac{1}{p} = -8\varepsilon\sin^2\left(\frac{\alpha h}{2}\right),$$

or

$$p^2 + 8p\varepsilon\sin^2\left(\frac{\alpha h}{2}\right) - 1 = 0.$$

This quadratic equation has two roots

$$p_1, p_2 = -4\varepsilon\sin^2\left(\frac{\alpha h}{2}\right) \pm \left(1 + 16\varepsilon^2\sin^4\frac{\alpha h}{2}\right)^{\frac{1}{2}}, \qquad (14.4.23)$$

or

$$p_1, p_2 = \pm 1 - 4\varepsilon\sin^2\left(\frac{\alpha h}{2}\right)\left(1 \mp 2\varepsilon\sin^2\frac{\alpha h}{2}\right) + O\left(\varepsilon^4\right).$$

This gives $|p_1| \le 1$ and

$$|p_2| > 1 + 4\varepsilon\sin^2\left(\frac{\alpha h}{2}\right) > 1$$

for all positive ε and, consequently, the Richardson scheme is always unstable.

The unstable feature of the Richardson scheme can be eliminated by replacing $u_{i,j}$ with $\frac{1}{2}\left(u_{i,j+1} + u_{i,j-1}\right)$ in (14.4.22), which now becomes

$$(1 + 2\varepsilon)u_{i,j+1} = 2\varepsilon\left(u_{i+1,j} + u_{i-1,j}\right) + (1 - 2\varepsilon)u_{i,j-1}. \quad (14.4.24)$$

This is called the *Du Fort–Frankel explicit algorithm*, and it can be shown to be stable for all ε.

Example 14.4.3. Prove that the solution of the finite difference equation for the diffusion equation (14.4.13) in $-\infty < x < \infty$ with the initial condition $u(x,0) = e^{i\alpha x}$ converges to the exact solution of (14.4.13) as h and k tend to zero.

We obtain the exact solution of (14.4.13) by seeking a separable form

$$u(x,t) = e^{i\alpha x}\, v(t),$$

where $v(t)$ is a function of t alone which is to be determined.

Substituting this solution into (14.4.13) gives

$$\frac{dv}{dt} + \kappa \alpha^2 v = 0$$

which admits solutions

$$v(t) = A e^{-\kappa \alpha^2 t},$$

where A is an integrating constant. The initial condition $v(0) = 1$ gives $A = 1$. Hence,

$$u(x,t) = \exp\left(i\alpha x - \alpha^2 \kappa t\right). \tag{14.4.25}$$

We now solve the corresponding finite difference equation (14.4.17) by replacing i, j with r, s. We seek a separable solution of the difference equation

$$u_{r,s} = e^{i\alpha r h}\, v_s$$

with the initial condition

$$u_{r,0} = e^{i\alpha r h}\, v_0, \quad \text{so that} \quad v_0 = 1.$$

Substituting this solution into the finite difference equation (14.4.7) yields

$$v_{s+1} = \left(1 - 4\varepsilon \sin^2 \frac{\alpha h}{2}\right) v_s, \quad v_0 = 1,$$

so that the solution can be obtained by a simple inspection as

$$v_s = \left(1 - 4\varepsilon \sin^2 \frac{\alpha h}{2}\right)^s, \tag{14.4.26}$$

$$u_{r,s} = e^{i\alpha r h}\left(1 - 4\varepsilon \sin^2 \frac{\alpha h}{2}\right)^s, \tag{14.4.27}$$

where $\varepsilon = \left(\kappa k / h^2\right)$.

For small h, $1 - 4\varepsilon \sin^2(\alpha h/2) \sim 1 - \varepsilon \alpha^2 h^2$ so that $1 - \varepsilon \alpha^2 h^2 \approx \exp\left(-\varepsilon \alpha^2 h^2\right)$ for small $\varepsilon \alpha^2 h^2 = \kappa \alpha^2 k$. Consequently, the final solution becomes

$$u_{r,s} \sim e^{i\alpha rh - \kappa a^2 ks} \quad \text{as} \quad h, k \to 0. \tag{14.4.28}$$

This is identical with the exact solution of the differential equation (14.4.13) with $rh = x$ and $sk = t$. This example shows that the finite difference approximation is reasonably good.

Example 14.4.4. Calculate a finite difference solution of the initial boundary-value problem

$$u_t = u_{xx}, \qquad 0 < x < 1, \qquad t > 0,$$

with the boundary conditions

$$u(0, t) = u(1, t) = 0, \qquad t \geq 0,$$

and the initial condition

$$u(x, 0) = x(1 - x), \qquad 0 \leq x \leq 1.$$

Compare the numerical solution with the exact analytical solution at $x = 0.04$ and $t = 0.02$.

The explicit finite difference approximation to the parabolic equation is

$$u_{i,j+1} = \varepsilon u_{i-1,j} + (1 - 2\varepsilon) u_{i,j} + \varepsilon u_{i+1,j},$$

where $\varepsilon = (k/h^2)$. This gives the unknown value $u_{i,j+1}$ at the $(i, j+1)$th grid point in terms of given values of u along the jth time row.

We set $h = \frac{1}{5}$ and $k = \frac{1}{100}$ so that $\varepsilon = (k/h^2) = \frac{1}{4}$ and the above formula becomes

$$u_{i,j+1} = \frac{1}{4}(u_{i-1,j} + 2u_{i,j} + u_{i+1,j}).$$

With the notation $u_{i,0} = u(ih, 0)$, the initial condition gives

$$u_{4,0} = 0.16, \quad \text{and} \quad u_{5,0} = 0.$$

The boundary conditions yield $u_{0,j} = u(0, jk) = 0$ and $u_{5,j} = u(5h, jk) = u(1, jk) = 0$, for all $j = 0, 1, 2, \ldots$.

Using these initial and boundary data, we calculate $u_{i,j}$ as follows:

$$u_{1,1} = \frac{1}{4}(u_{0,0} + 2u_{1,0} + u_{2,0}) = 0.14, \quad u_{1,2} = \frac{1}{4}(u_{0,1} + 2u_{1,1} + u_{2,1})$$
$$= 0.125,$$

$$u_{2,1} = \frac{1}{4}(u_{1,0} + 2u_{2,0} + u_{3,0}) = 0.22, \quad u_{2,2} = \frac{1}{4}(u_{1,1} + 2u_{2,1} + u_{3,1})$$
$$= 0.200,$$

$$u_{3,1} = \frac{1}{4}(u_{2,0} + 2u_{3,0} + u_{4,0}) = 0.22, \quad u_{3,2} = \frac{1}{4}(u_{2,1} + 2u_{3,1} + u_{4,1})$$
$$= 0.200,$$

$$u_{4,1} = \frac{1}{4}(u_{3,0} + 2u_{4,0} + u_{5,0}) = 0.14, \quad u_{4,2} = \frac{1}{4}(u_{3,1} + 2u_{4,1} + u_{5,1})$$
$$= 0.125,$$

$$u_{1,3} = \frac{1}{4}(u_{0,2} + 2u_{1,2} + u_{2,2}) = 0.1125, \quad u_{1,4} = \frac{1}{4}(u_{0,3} + 2u_{1,3} + u_{2,3})$$
$$= 0.1016,$$

$$u_{2,3} = \frac{1}{4}(u_{1,2} + 2u_{2,2} + u_{3,2}) = 0.1813, \quad u_{2,4} = \frac{1}{4}(u_{1,3} + 2u_{2,3} + u_{3,3})$$
$$= 0.1641,$$

$$u_{3,3} = \frac{1}{4}(u_{2,2} + 2u_{3,2} + u_{4,2}) = 0.1813, \quad u_{3,4} = \frac{1}{4}(u_{2,3} + 2u_{3,3} + u_{4,3})$$
$$= 0.1641,$$

$$u_{4,3} = \frac{1}{4}(u_{3,2} + 2u_{4,2} + u_{5,2}) = 0.1125, \quad u_{4,4} = \frac{1}{4}(u_{3,3} + 2u_{4,3} + u_{5,3})$$
$$= 0.1016.$$

The method of separation of variables gives the analytical solution of the problem as

$$u(x,t) = \frac{8}{\pi^3} \sum_{n=0}^{\infty} \frac{1}{(2n+1)^3} \exp\left[-(2n+1)^2 \pi^2 t\right] \sin(2n+1)\pi x.$$

This exact solution $u(x,t)$ at $x = 0.4$ $(i = 2)$ and $t = 0.02$ $(j = 2)$ gives

$$u \sim \frac{8}{3}\left[\frac{1}{1^3}\exp\left(-0.02\,\pi^2\right)\sin(0.4)\,\pi + \frac{1}{3^3}\exp\left(-0.18\,\pi^2\right)\sin(1.2)\,\pi\right]$$
$$= 0.2000.$$

The analytical solution is seen to be identical with the numerical value.

Example 14.4.5. Obtain the numerical solution of the initial boundary-value problem

$$u_t = \kappa\, u_{xx}, \qquad 0 \le x \le 1, \qquad t > 0,$$
$$u(0,t) = 1, \qquad u(1,t) = 0, \qquad t \ge 0,$$
$$u(x,0) = 0, \qquad 0 \le x \le 1.$$

We use the explicit finite-difference formula (14.4.17)

$$u_{i,j+1} = \varepsilon\,(u_{i+1,j} + u_{i-1,j}) + (1 - 2\varepsilon)\,u_{i,j},$$

where $\varepsilon = (\kappa k/h^2)$.

We set $h = 0.25 = \frac{1}{4}$ and $\varepsilon = \frac{2}{5} = 0.4$ to compute $u_{i,j}$ for $i, j = 0, 1, 2, 3, 4$ as follows:

i j	0	1	2	3	4
0	1.000	0.000	0.000	0.000	0.000
1	1.000	0.400	0.000	0.000	0.000
2	1.000	0.480	0.160	0.000	0.000
3	1.000	0.560	0.224	0.064	0.000
4	1.000	0.602	0.295	0.103	0.000

(C) Elliptic Equations

As a prototype boundary-value problem, we consider the Dirichlet problem for the Laplace equation

$$\nabla^2 u \equiv u_{xx} + u_{yy} = 0, \qquad 0 \le x \le a, \quad 0 \le y \le b, \qquad (14.4.29)$$

where the value of $u(x, y)$ is prescribed everywhere on the boundary of the rectangular domain.

The rectangular grid system is the most common and convenient system for this problem. We choose the vertices of the rectangular domain as the nodal points and set $h = a/m$ and $k = b/n$ where m and n are positive integers so that the domain is divided into mn subrectangles.

The finite difference approximation to the Laplace equation (14.4.29) is

$$\frac{1}{h^2}\left(u_{i+1,j} - 2u_{i,j} + u_{i-1,j}\right) + \frac{1}{k^2}\left(u_{i,j+1} - 2u_{i,j} + u_{i,j-1}\right) = 0, (14.4.30)$$

or,

$$2\left(h^2 + k^2\right) u_{i,j} = k^2 \left(u_{i+1,j} + u_{i-1,j}\right) + h^2 \left(u_{i,j+1} + u_{i,j-1}\right), \quad (14.4.31)$$

where $1 \le i \le m - 1$ and $1 \le j \le n - 1$.

The prescribed conditions on the boundary of the rectangular domain determine the values $u_{0,j}$, $u_{m,j}$, $u_{i,0}$, and $u_{i,n}$. For a square grid system ($k = h$), equation (14.4.30) becomes

$$u_{i,j} = \frac{1}{4}\left(u_{i+1,j} + u_{i-1,j} + u_{i,j+1} + u_{i,j-1}\right). \qquad (14.4.32)$$

This means that the value of u at an interior point is equal to the average of the value of u at four adjacent points. This is the well known *mean value theorem* for harmonic functions that satisfy the Laplace equation.

As i and j vary, the present scheme reduces to a set of $(m-1)(n-1)$ linear non-homogeneous algebraic equations for $(m-1)(n-1)$ unknown values of u at interior grid points. It can be shown the solution of the finite difference equation (14.4.31) converges to the exact solution of the problem as h, $k \to 0$. The proof of the existence of a solution and its convergence to the exact solution as h and k tend to zero is essentially based on the *Maximum Modulus Principle*. It follows from the finite difference equation (14.4.30) or (14.4.31) that the value of $|u|$ at any interior grid point does not exceed its value at any of the four adjoining nodal points. In other words, the value of u at $P_{i,j}$ cannot exceed its values at the four adjoining points $P_{i \pm 1,j}$ and $P_{i,j \pm 1}$. The successive application of this argument at all interior grid points leads to the conclusion that $|u|$ at the interior grid points cannot be greater than the maximum value of $|u|$ on the boundary. This may be recognized as the finite difference analogue of the Maximum Modulus Principle discussed in Section 9.2. Thus, the success of the numerical method is directly associated with the existence of the Maximum Modulus Principle.

Clearly, the present numerical algorithm deals with a large number of algebraic equations. Even though numerical accuracy can be improved by making h and k sufficiently small, there is a major computational difficulty involved in the numerical solution of a large number of equations. It is possible to handle such a large number of algebraic equations by direct methods or by iterative methods, but it would be very difficult to obtain a numerical solution with sufficient accuracy. It is therefore necessary to develop some alternative methods of solution that can be conveniently and efficiently carried out on a computer.

In order to eliminate some of the drawbacks stated above, one of the numerical schemes, the *Liebmann's iterative method*, is useful. In this method values of u are first guessed for all interior grid points in addition to those given as the boundary points on the edges of the given domain. These values are denoted by $u_{i,j}^{(0)}$ where the superscript 0 indicates the zeroth iteration. It is convenient to choose a square grid so that the simplified finite difference equation (14.4.32) can be used. The values of u are calculated for the next iteration by using (14.4.32) at every interior point based on the values of u at the present iteration. The sequence of computation starts from the interior grid point located at the lowest left corner, proceeds upward until reaching the top, and then goes to the bottom of the next vertical line on the right. This process is repeated until the new value of u at the last interior grid point at the upper right corner has been obtained.

At the starting point, formula (14.4.32) gives

$$u_{2,2}^{(1)} = \frac{1}{4}\left(u_{3,2}^{(0)} + u_{1,2}^{(0)} + u_{2,3}^{(0)} + u_{2,1}^{(0)}\right), \qquad (14.4.33)$$

where $u_{1,2}^{(0)}$ and $u_{2,1}^{(0)}$ are boundary values which remain constant during the iteration process. They may be replaced, respectively, with $u_{1,2}^{(1)}$, $u_{2,1}^{(1)}$ in (14.4.33). The computation at the next step involves $u_{2,2}^{(0)}$. Since an improved value $u_{2,2}^{(1)}$ is available at this time, it will be utilized instead. Hence,

$$u_{2,3}^{(1)} = \frac{1}{4}\left(u_{3,3}^{(0)} + u_{1,3}^{(1)} + u_{2,4}^{(0)} + u_{2,2}^{(1)}\right), \qquad (14.4.34)$$

where $u_{1,3}^{(1)}$ is used to replace the constant boundary value $u_{1,3}^{(0)}$.

We repeat this argument to obtain a general iteration formula for computation of u at step $(n+1)$

$$u_{i,j}^{(n+1)} = \frac{1}{4}\left(u_{i+1,j}^{(n)} + u_{i-1,j}^{(n+1)} + u_{i,j+1}^{(n)} + u_{i,j-1}^{(n+1)}\right). \qquad (14.4.35)$$

This result is valid for any interior point, whether it is next to some boundary point or not. If $P_{i,j}$ is a true point, the second and fourth terms on the right side of (14.4.35) represent, respectively, the values of u at the grid points to the left of and below that point. These values have already been recomputed according to our scheme, and therefore, carry the superscript $(n+1)$. Result (14.4.35) is known as the *Liebmann* $(n+1)$ th *iteration formula*. It can be proved that $u_{i,j}^{(n)}$ converges to $u_{i,j}$ as $n \to \infty$.

Another iteration scheme similar to (14.4.35) is given by

$$u_{i,j}^{(n+1)} = \frac{1}{4}\left(u_{i+1,j}^{(n)} + u_{i-1,j}^{(n)} + u_{i,j+1}^{(n)} + u_{i,j-1}^{(n)}\right). \qquad (14.4.36)$$

This is called the *Richardson iteration formula*, and it is also useful. However, this scheme converges more slowly than that based on (14.4.35).

One of the major difficulties of the above methods is the slow rate of convergence. An improved numerical method, the *Successive Over-Relaxation (SOR) scheme* gives a faster convergence than the Liebmann or Richardson method in solving the Laplace (or the Poisson) equation. For a rectangular domain of square grids, the successive iteration scheme is given by

$$u_{i,j}^{(n+1)} = u_{i,j}^{(n)} + \frac{\omega}{4}\left(u_{i-1,j}^{(n+1)} + u_{i+1,j}^{(n)} + u_{i,j-1}^{(n+1)} + u_{i,j+1}^{(n)} - 4\,u_{i,j}^{(n)}\right), \quad (14.4.37)$$

where ω is called the *acceleration parameter* (or *relaxation factor*) to be determined. In general, ω lies in the range $1 \le \omega < 2$. The successive iterations converge fairly rapidly to the desired solution for $1 \le \omega < 2$. The most rapid rate of convergence is achieved for the optimum value of ω.

Example 14.4.6. Obtain the standard five-point formula for the Poisson equation

$$u_{xx} + u_{yy} = -f(x,y) \quad \text{in } D \subset R^2$$

with the prescribed value of $u(x,y)$ on the boundary ∂D.

We assume that the domain D is covered by a system of squares with sides of length h parallel to the x and y axes. Using the central difference approximation to the Laplace operator, we obtain

$$\frac{1}{h^2}\left(u_{i+1,j} - 2\,u_{i,j} + u_{i-1,j}\right) + \frac{1}{h^2}\left(u_{i,j+1} - 2\,u_{i,j} + u_{i,j-1}\right) = -f_{i,j},$$

or,

$$u_{i,j} = \frac{1}{4}\left(u_{i+1,j} + u_{i-1,j} + u_{i,j+1} + u_{i,j-1}\right) + \frac{1}{4}h^2 f_{i,j}$$

where $f_{i,j} = f(ih, jh)$. This is known as the *five-point formula*.

Example 14.4.7. Find the numerical solution of the torsion problem in a square beam governed by

$$\nabla^2 u = -2 \quad \text{in} \quad D = \{(x, y) : 0 \le x \le 1, 0 \le y \le 1\}$$

with $u(x, y) = 0$ on ∂D.

From the above five-point formula, we obtain

$$u_{i,j} = \frac{1}{4}\left(u_{i+1,j} + u_{i-1,j} + u_{i,j+1} + u_{i,j-1}\right) - \frac{1}{2}h^2$$

where h is the side-length of the unit square net.

We choose $h = \frac{1}{2}$, $1/2^2$, $1/2^3$, $1/2^4$ to calculate the corresponding numerical values $u_{i,j} = 0.1250, 0.1401, 0.1456, 0.1469$.

Note that the known exact analytical solution is 0.1474.

Example 14.4.8. Using the explicit finite difference method, find the solution of the Dirichlet problem

$$\begin{aligned}
u_{xx} + u_{yy} &= 0, && \text{in} \quad 0 < x < 1, \quad 0 < y < 1, \\
u(x, 0) &= x, && u(x, 1) = 0, \quad \text{on} \quad 0 \le x \le 1, \\
u(x, y) &= 0, && \text{for} \quad x = 0, \quad x = 1 \quad \text{and} \quad 0 \le y \le 1.
\end{aligned}$$

We use four interior grid points (that is, $i, j = 1, 2, 3, 4$) as shown in Figure 14.4.2 in the (x, y)-plane.

We apply the explicit finite difference formula (14.4.32) to obtain four algebraic equations

$$\begin{aligned}
-4u_{2,2} + u_{3,2} + u_{1,2} + u_{2,3} + u_{2,1} &= 0, \\
-4u_{2,3} + u_{3,3} + u_{1,3} + u_{2,4} + u_{2,2} &= 0, \\
-4u_{3,2} + u_{4,2} + u_{2,2} + u_{3,3} + u_{3,1} &= 0, \\
-4u_{3,3} + u_{4,3} + u_{2,3} + u_{3,4} + u_{3,2} &= 0.
\end{aligned}$$

The given boundary conditions imply that $u_{2,1} = u_{2,4} = u_{3,1} = u_{3,4} = u_{4,2} = u_{4,3} = 0$, $u_{1,2} = \frac{1}{3}$ and $u_{1,3} = \frac{2}{3}$ so that the above system of equations becomes

$$-4u_{2,2} + u_{3,2} + \frac{1}{3} + u_{2,3} = 0,$$

$$-4u_{2,3} + u_{3,3} + \frac{2}{3} + u_{2,2} = 0,$$

$$-4u_{3,2} + u_{2,2} + u_{3,3} = 0,$$

$$-4u_{3,3} + u_{2,3} + u_{3,2} = 0.$$

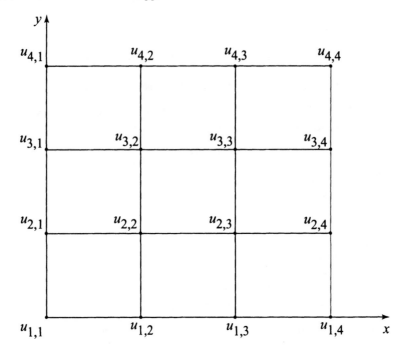

Figure 14.4.2 The square grid system.

In matrix notation, this system reads as

$$\begin{bmatrix} -4 & 1 & 1 & 0 \\ 1 & -4 & 0 & 1 \\ 1 & 0 & -4 & 1 \\ 0 & 1 & 1 & -4 \end{bmatrix} \begin{bmatrix} u_{2,2} \\ u_{2,3} \\ u_{3,2} \\ u_{3,3} \end{bmatrix} = \begin{bmatrix} -\frac{1}{3} \\ -\frac{2}{3} \\ 0 \\ 0 \end{bmatrix}.$$

The solutions of this system are

$$u_{2,2} = \frac{11}{72}, \quad u_{2,3} = \frac{16}{72}, \quad u_{3,2} = \frac{4}{72}, \quad u_{3,3} = \frac{5}{72}.$$

(D) Simultaneous First-Order Equations

We recall the wave equation (14.4.1) in $0 < x < 1$, $t > 0$. Introducing two auxiliary variables v and w by $v = u_t$ and $w = c^2 u_x$, the wave equation gives two simultaneous first-order equations

$$v_t = w_x, \qquad w_t = c^2 v_x. \tag{14.4.38}$$

The initial values of v and w are given at $t = 0$ for all x in $0 < x < 1$. The boundary condition on v and w is also prescribed on the lines $x = 0$ and $x = 1$ for $t > 0$.

The explicit finite difference method can be used to determine v and w in the triangular domain of dependence bounded by the characteristics $x - ct = 0$ and $x + ct = 1$.

The finite difference approximations to the differential equations (14.4.38) are

$$\frac{1}{k}\left(v_{i,j+1} - v_{i,j}\right) = \frac{1}{2h}\left(w_{i+1,j} - w_{i-1,j}\right), \qquad (14.4.39)$$

$$\frac{1}{k}\left(w_{i,j+1} - w_{i,j}\right) = \frac{c^2}{2h}\left(v_{i+1,j} - v_{i-1,j}\right), \qquad (14.4.40)$$

where the forward difference for v_t or w_t and the central difference for v_x or w_x are used. However, the central difference approximations to (14.4.38) can also be utilized to obtain

$$\frac{1}{2k}\left(v_{i,j+1} - v_{i,j-1}\right) = \frac{1}{2h}\left(w_{i+1,j} - w_{i-1,j}\right), \qquad (14.4.41)$$

$$\frac{1}{2k}\left(w_{i,j+1} - w_{i,j-1}\right) = \frac{c^2}{2h}\left(v_{i+1,j} - v_{i-1,j}\right). \qquad (14.4.42)$$

We examine the stability of the above two sets of finite difference formulas with $c = 1$. The von Neumann stability method is applied by replacing i and j by r and s respectively. The error function $e_{r,s}$ given by (14.4.10) is substituted in (14.4.39)–(14.4.40) to obtain the stability relations

$$A\left(p - 1\right) = \varepsilon i B \sin \alpha h, \qquad (14.4.43)$$

$$B\left(p - 1\right) = \varepsilon i A \sin \alpha h, \qquad (14.4.44)$$

where the initial perturbations in v and w along $t = 0$ are $A \exp\left(i\alpha rh\right)$ and $B \exp\left(i\alpha rh\right)$ respectively with two different constants A and B.

Elimination of A and B from the above relations gives

$$\left(p - 1\right)^2 + \varepsilon^2 \sin^2 \alpha h = 0$$

or

$$p = 1 \pm i\varepsilon \sin \alpha h,$$

and

$$|p| = \left(1 + \varepsilon^2 \sin^2 \alpha h\right)^{\frac{1}{2}} \sim 1 + \frac{1}{2}\varepsilon^2 \sin^2 \alpha h = 1 + O\left(\varepsilon^2\right). \quad (14.4.45)$$

Since $|p| > 1 + O\left(\varepsilon\right)$, the finite difference scheme for the finite time-step $t = sk$ would be unstable as the grid sizes tend to zero.

A similar stability analysis for (14.4.41)–(14.4.42) leads to the condition

$$\left(p - \frac{1}{p}\right)^2 + 4\,\varepsilon^2 \sin^2 \alpha h = 0. \tag{14.4.46}$$

This scheme is stable for $\varepsilon \leq 1$.

Another finite difference approximation to the coupled system (14.4.38) is

$$\frac{1}{2h}\,(v_{r+1,s} - v_{r-1,s}) = \frac{1}{k}\left[w_{r,s+1} - \frac{1}{2}\,(w_{r+1,s} - w_{r-1,s})\right], \tag{14.4.47}$$

$$\frac{1}{2h}\,(w_{r+1,s} - w_{r-1,s}) = \frac{1}{k}\left[v_{r,s+1} - \frac{1}{2}\,(v_{r+1,s} - v_{r-1,s})\right]. \tag{14.4.48}$$

A similar stability analysis can be carried out for these systems by substituting $v_{r,s} = Ap^s e^{i\alpha r h}$ and $w_{r,s} = Bp^s e^{i\alpha r h}$ into the equations. Elimination of A/B yields the stability equation

$$p = \cos \alpha h \pm \frac{i}{\varepsilon} \sin \alpha h,$$

or

$$|p|^2 = \cos^2 \alpha h + \frac{1}{\varepsilon^2} \sin^2 \alpha h \leq 1. \tag{14.4.49}$$

Hence, the scheme is stable provided that

$$\varepsilon \geq 1, \quad \text{that is,} \quad k \leq h. \tag{14.4.50}$$

14.5 Implicit Finite Difference Methods

From a computational point of view, the explicit finite difference algorithm is simple and convenient. However, as shown in Section 14.4(B), the major difficulty in the method for solving parabolic partial differential equations is the severe restriction on the time-step imposed by the stability condition $\varepsilon \leq \frac{1}{2}$ or $k \leq h^2/2\kappa$. This difficulty is also present in the explicit finite difference method for the solution of hyperbolic equations. In order to overcome the above difficulty, we develop implicit finite difference schemes for solving partial differential equations.

(A) Parabolic Equations

One of the successful implicit finite difference schemes is the Crank and Nicolson Method (1947), which is based on six grid points. This method eliminates the major difficulty involved in the explicit scheme. When the Crank–Nicolson implicit scheme is applied to the parabolic equation (14.4.13), u_{xx} is replaced by the mean value of the finite difference values in the jth and the $(j+1)$th row so that the finite difference approximation (14.4.13) becomes

$$\frac{1}{k}\left(u_{i,j+1} - u_{i,j}\right) = \frac{\kappa}{2h^2}\left[\left(u_{i+1,j+1} - 2u_{i,j+1} + u_{i-1,j+1}\right)\right.$$
$$\left. + \left(u_{i+1,j} - 2u_{i,j} + u_{i-1,j}\right)\right], \quad (14.5.1)$$

or

$$2\left(1+\varepsilon\right)u_{i,j+1} - \varepsilon\left(u_{i-1,j+1} + u_{i+1,j+1}\right)$$
$$= 2\left(1-\varepsilon\right)u_{i,j} + \varepsilon\left(u_{i-1,j} + u_{i+1,j}\right), \quad (14.5.2)$$

where $\varepsilon = \left(k\kappa/h^2\right)$ is a parameter.

The left side of (14.5.2) is a linear combination of three unknowns in the $(j+1)$th row, and the right side involves three known values of u in the jth row of the grid system in the (x,t)-plane. Equation (14.5.2) is called the *Crank–Nicolson implicit formula*. This formula (or its suitable modification) is widely used for solving parabolic equations. If there are n internal grid points along each jth row, then, for $j=0$ and $i=1,2,3,\ldots,$ n, the implicit formula (14.5.2) gives n simultaneous algebraic equations for n unknown values of u along the first jth row $(j=0)$ in terms of given boundary and initial data. Similarly, if $j=1$ and $i=1,2,3,\ldots,n$, equation (14.5.2) represents n unknown values of u along the second jth row $(j=1)$ and so on. This means that the method involves the solution of a system of simultaneous algebraic equations. In practice, the Crank–Nicolson scheme is convergent and unconditionally stable for all finite values of ε, and has the advantage of reducing the amount of numerical computation.

This implicit scheme can be further generalized by introducing a numerical weight factor λ in the modified version of the explicit equation (14.4.16) which is written below by approximating u_{xx} in (14.4.13) in the $(j+1)$th row instead of the jth row.

$$\frac{1}{k}\left(u_{i,j+1} - u_{i,j}\right) = \frac{\kappa}{h^2}\left(u_{i+1,j+1} - 2u_{i,j+1} + u_{i-1,j+1}\right), \quad (14.5.3)$$

or

$$u_{i,j+1} - u_{i,j} = \varepsilon\left(u_{i+1,j+1} - 2u_{i,j+1} + u_{i-1,j+1}\right). \quad (14.5.4)$$

Introducing the numerical factor λ, this can be replaced by a more general difference equation in the form

$$u_{i,j+1} - u_{i,j} = \varepsilon\left[\lambda\,\delta_x^2\,u_{i,j+1} + (1-\lambda)\,\delta_x^2\,u_{i,j}\right], \quad (14.5.5)$$

where $0 \leq \lambda \leq 1$ and δ_x^2 is the difference operator defined by

$$\delta_x^2 u_{i,j} = u_{i+1,j} - 2u_{i,j} + u_{i-1,j}. \quad (14.5.6)$$

Another equivalent form of (14.5.5) is

$$(1 + 2\varepsilon\lambda)\, u_{i,j+1} - \varepsilon\lambda\, (u_{i+1,j+1} + u_{i-1,j+1})$$
$$= \{1 - 2\varepsilon\, (1 - \lambda)\}\, u_{i,j} + \varepsilon\, (1 - \lambda)\, (u_{i+1,j} - u_{i-1,j}). \qquad (14.5.7)$$

This is a fairly general implicit formula which reduces to (14.5.4) when $\lambda = 1$. When $\lambda = \frac{1}{2}$, (14.5.7) becomes the Crank–Nicolson formula (14.5.2). Finally, if $\lambda = 0$, this implicit difference equation reduces to the explicit equation (14.4.17).

The Richardson explicit scheme was found to be unconditionally unstable in Section 14.4. This undesirable feature of the scheme can be eliminated by considering the corresponding implicit scheme. In terms of δ_x^2, the Richardson equation (14.4.22) can be expressed as

$$u_{i,j+1} = 2\varepsilon\, \delta_x^2\, u_{i,j} + u_{i,j-1}. \qquad (14.5.8)$$

To obtain the implicit Richardson formula, we replace $\delta_x^2\, u_{i,j}$ by $\frac{1}{3}\delta_x^2\, (u_{i,j+1} + u_{i,j} + u_{i,j-1})$ in (14.5.8) and we obtain

$$\left(1 - \frac{2\,\varepsilon}{3}\delta_x^2\right) u_{i,j+1} = \frac{2\varepsilon}{3}\delta_x^2 u_{i,j} + \left(1 + \frac{2\varepsilon}{3}\right) u_{i,j-1}. \qquad (14.5.9)$$

This implicit scheme can be shown to be unconditionally stable. To prove this result, we apply the von Neumann stability method with the error function (14.5.9) to obtain the equation for p as

$$(1 + a)\, p^2 + ap + (a - 1) = 0, \qquad (14.5.10)$$

where

$$a \equiv \left(\frac{8\varepsilon}{3}\right) \sin^2 \left(\frac{\alpha h}{2}\right). \qquad (14.5.11)$$

The roots of the quadratic equation are

$$p = \frac{-a \pm \left(4 - 3a^2\right)^{\frac{1}{2}}}{2\,(1 + a)}. \qquad (14.5.12)$$

This gives $|p| \leq 1$ for all values of a. Hence, the result is proved.

Example 14.5.1. Obtain the numerical solution of the following parabolic system by using the Crank–Nicolson method

$$u_t = u_{xx}, \qquad 0 < x < 1, \quad t > 0,$$
$$u\,(0, t) = u\,(1, t) = 0, \qquad t \geq 0,$$
$$u\,(x, 0) = x\,(1 - x), \qquad 0 \leq x \leq 1.$$

We recall the Crank–Nicolson equation (14.5.2) and then set $h = 0.2$ and $k = 0.01$ so that $\varepsilon = \frac{1}{4}$. The boundary and initial conditions give

$u_{0,0} = u_{5,0} = u_{0,1} = u_{5,1} = 0$ and $u_{i,0} = u(ih, 0) = ih(1 - ih)$, $i = 1, 2,$ 3, 4. Consequently, formula (14.5.2) leads to the following system of four equations:

$$-u_{0,1} - u_{2,1} + 10u_{1,1} = u_{0,0} + u_{2,0} + 6u_{1,0}$$
$$-u_{1,1} - u_{3,1} + 10u_{2,1} = u_{1,0} + u_{3,0} + 6u_{2,0}$$
$$-u_{2,1} - u_{4,1} + 10u_{3,1} = u_{2,0} + u_{4,0} + 6u_{3,0}$$
$$-u_{3,1} - u_{5,1} + 10u_{4,1} = u_{3,0} + u_{5,0} + 6u_{4,0}.$$

Using the boundary and initial conditions, the above system becomes

$$-u_{2,1} + 10u_{1,1} = 1.20$$
$$-u_{1,1} + 10u_{2,1} - u_{3,1} = 1.84$$
$$-u_{2,1} - u_{4,1} + 10u_{3,1} = 1.84$$
$$-u_{3,1} + 10u_{4,1} = 1.20.$$

These equations can be solved by direct elimination to obtain the solutions as $u_{1,1} = 0.1418$, $u_{2,1} = 0.2202$, $u_{3,1} = 0.2202$, $u_{4,1} = 0.1420$.

(B) Hyperbolic Equations

We consider an implicit finite difference scheme to solve the initial boundary-value problem consisting of the first-order hyperbolic equation

$$\frac{\partial u}{\partial t} + c\frac{\partial u}{\partial x} = 0, \qquad (c > 0), \qquad (14.5.13)$$

with the initial data $u(x, 0) = U(x)$ and the boundary condition $u(0, t) = V(t)$ where $0 \le x, t < \infty$.

The implicit finite difference approximation to (14.5.13) is

$$\frac{1}{k}(u_{i,j+1} - u_{i,j}) + \frac{c}{h}(u_{i,j+1} - u_{i-1,j+1}) = 0,$$

or

$$u_{i,j} = (1 + \varepsilon)u_{i,j+1} - \varepsilon u_{i-1,j+1}, \qquad (14.5.14)$$

where $\varepsilon = (ck/h)$.

The stability of the scheme can be examined by using the von Neumann method with the error function (14.4.10). It turns out that

$$p = [1 - \varepsilon + \varepsilon \exp(-i\alpha h)]^{-1}, \qquad (14.5.15)$$

from which it follows that $|p| \le 1$ for all h. Hence, the implicit scheme is unconditionally stable.

We next solve the wave equation $u_{tt} = c^2 u_{xx}$ by an implicit finite difference scheme. In this case, u_{tt} is replaced by the central difference formula, and u_{xx} by the mean value of the central difference values in the $(j-1)$ th and $(j+1)$ th rows. Consequently, the implicit difference approximation to the wave equation is

$$
\begin{aligned}
u_{i,j+1} - 2\,u_{i,j} + u_{i,j-1} = \frac{\varepsilon^2}{2} \big[& (u_{i+1,j+1} - 2u_{i,j+1} + u_{i-1,j+1}) \\
& + (u_{i+1,j-1} - 2\,u_{i,j-1} + u_{i-1,j-1}) \big],
\end{aligned}
$$
$$(14.5.16)$$

where $\varepsilon = (ck/h)$.

Expressing the solution for the $(j+1)$ th step in terms of the two preceding steps gives

$$
\begin{aligned}
2\left(1+\varepsilon^2\right) u_{i,j+1} & - \varepsilon^2 \left(u_{i-1,j+1} + u_{i+1,j+1}\right) \\
& = 4u_{i,j} + \varepsilon^2 \left(u_{i-1,j-1} + u_{i+1,j-1}\right) - 2\left(1+\varepsilon^2\right) u_{i,j-1}.
\end{aligned}
$$
$$(14.5.17)$$

The N grid points along the time step, $j = 0$, $i = 1, 2, 3, \ldots, N$, (14.5.17) along with the finite difference approximation to the boundary condition give N simultaneous equations for the N unknown values of u along the first time step. This constitutes a tridiagonal system of equations that can be solved by direct or iterative numerical methods.

To investigate the stability of the implicit scheme, we apply the von Neumann stability method with the error function (14.4.10). This leads to the equation

$$
p + \frac{1}{p} = 2 \left(1 + 2\,\varepsilon^2 \sin^2 \frac{\alpha h}{2}\right)^{-1},
$$

or

$$
p^2 - 2bp + 1 = 0, \tag{14.5.18}
$$

where $b = \left(1 + 2\varepsilon^2 \sin^2 \alpha h/2\right)^{-1}$ so that $0 < b \leq 1$.

Hence, the stability condition is

$$
|p| \leq 1 \tag{14.5.19}
$$

which is always satisfied provided $0 < b \leq 1$, that is, $\varepsilon < 1$ for all positive h. This confirms the unconditional stability of the scheme.

A more general implicit scheme can be introduced by replacing u_{xx} in the wave equation (14.4.1) with

$$
u_{xx} \sim \frac{1}{h^2} \left[\lambda \left(\delta_x^2\, u_{i,j+1} + \delta_x^2\, u_{i,j-1} \right) + (1+2\lambda)\, \delta_x^2\, u_{i,j} \right], \tag{14.5.20}
$$

where λ is a numerical weight (relaxation) factor and the central difference operator δ_x^2 is given by (14.5.6). This general scheme allows us to approximate the wave equation with $c = 1$ by the form

$$\delta_t^2 u_{i,j} = \varepsilon^2 \left[\lambda \left(\delta_x^2\, u_{i,j+1} + \delta_x^2\, u_{i,j-1} \right) + (1 - 2\lambda)\, \delta_x^2\, u_{i,j} \right], \quad (14.5.21)$$

where $\varepsilon = k/h$. This equation reduces to (14.5.16) when $\lambda = \frac{1}{2}$, and to the explicit finite difference result when $\lambda = 0$.

It follows from von Neumann stability analysis that the implicit scheme is unconditionally stable for $\lambda \geq \frac{1}{4}$. Von Neumann introduced another fairly general finite difference algorithm for the wave equation (14.4.1) in the form

$$\delta_t^2 u_{i,j} = \varepsilon^2 \delta_x^2 u_{i,j} + \frac{\omega}{h^2} \delta_t^2 \delta_x^2 u_{i,j}. \qquad (14.5.22)$$

This equation with appropriate boundary conditions can be solved by the tridiagonal method. Von Neumann discussed the question of stability of this implicit scheme and proved that the scheme is conditionally stable if $\omega \leq \frac{1}{4}$ and unconditionally stable if $\omega > \frac{1}{4}$.

14.6 Variational Methods and the Euler–Lagrange Equations

To describe the variational methods and Rayleigh–Ritz approximate method, it is convenient to introduce the concepts of the inner product (pre-Hilbert) and Hilbert spaces. An *inner product* space X consisting of elements u, v, w, ... over the complex number field C is a complex linear space with an inner product $\langle u, v \rangle : X \times X \to C$ such that

(i) $\langle u, v \rangle = \overline{\langle v, u \rangle}$, where the bar denotes the complex conjugate of $\langle v, u \rangle$,
(ii) $\langle \alpha u + \beta v, w \rangle = \alpha \langle u, w \rangle + \beta \langle v, w \rangle$ for any scalars α, $\beta \in C$,
(iii) $\langle u, u \rangle \geq 0$; equality holds if and only if $u = 0$.

By (i) $\langle u, u \rangle = \overline{\langle u, u \rangle}$, and so $\langle u, u \rangle$ is real. We denote $\langle u, u \rangle^{\frac{1}{2}} = \|u\|$, which is called the *norm* of u. Thus, the norm is induced by the inner product. Thus, every inner product space is a normed linear space under the norm $\|u\| = \sqrt{\langle u, u \rangle}$.

Let X be an inner product space. A sequence $\{u_n\}$ where $u_n \in X$ for every n is called a *Cauchy sequence* in X if and only if for every given $\varepsilon > 0$ (no matter how small) we can find an $N(\varepsilon)$ such that

$$\|u_n - u_m\| < \varepsilon \quad \text{for all } n, m > N(\varepsilon).$$

The space X is called *complete* if every Cauchy sequence converges to a point in X. A complete normed linear space is called a *Banach Space*. A complete linear inner product space is called a *Hilbert Space* and is usually denoted by H.

Example 14.6.1. Let C^n be the set of all n-tuples of complex numbers. Thus, C^n is an n-dimensional Hilbert space with the inner product

$$\langle x, y \rangle = \sum_{k=1}^{n} x_k \, \overline{y_k}.$$

Obviously, the set of all n-tuples of real numbers R^n is an n-dimensional Hilbert space.

Example 14.6.2. Let l_2 be the set of all sequences with entries from C such that $\sum_{k=1}^{\infty} |x_k|^2 < \infty$. This forms a Hilbert space with the inner product

$$\langle x, y \rangle = \sum_{k=1}^{\infty} x_k \, \overline{y_k}.$$

Example 14.6.3. Let $L_2([a, b])$ be the set of all square integrable functions in the Lebesgue sense in an interval $[a, b]$. $L_2([a, b])$ is a Hilbert space with the inner product

$$\langle u, v \rangle = \int_a^b u(x) \, \overline{v}(x) \, dx.$$

We next introduce the notion of an operator in a Hilbert space H. An *operator* A is a mapping from H to H (that is, $A : H \to H$). It assigns to an element u in H a new element Au in H. An operator A is called *linear* if it satisfies the property

$$A(\alpha u + \beta v) = \alpha Au + \beta Av \quad \text{for every } \alpha, \beta \in C.$$

An operator is said to be *bounded* if there exists a constant k such that $\|Au\| \leq k \|u\|$ for all $u \in H$.

We consider a bounded operator A on a Hilbert space H. For a fixed element v in H, the inner product $\langle Au, v \rangle$ in H can be regarded as a number $I(u)$ which varies with u. Thus, $\langle Au, v \rangle = I(u)$ is a *linear functional* on H.

If there exists an operator A^* on a Hilbert space $(A^* : H \to H)$ such that

$$\langle Au, v \rangle = \langle u, A^*v \rangle \quad \text{for all } u, v \in H,$$

then A^* is called the *adjoint* of A. In general, $A \neq A^*$. If $A = A^*$, that is, $\langle Au, v \rangle = \langle u, Av \rangle$ for all u, v in H, then A is called *self-adjoint*.

It is important to note that any bounded operator T on a real Hilbert space $(T : H \to H)$ of the form $T = A^*A$ is self-adjoint. This follows from the fact that

$$\langle Tu, v \rangle = \langle A^*Au, v \rangle = \langle Au, Av \rangle = \langle u, A^*Au \rangle = \langle u, Tv \rangle.$$

A self-adjoint operator A on a Hilbert space H is said to be *positive* if $\langle Au, u \rangle \geq 0$ for all u in H, where equality implies that $u = 0$ in H. Further, if there exists a positive constant k such that $\langle Au, u \rangle \geq k \langle u, u \rangle$ for all u in H, then A is called *positive definite* in H.

The rest of this section is essentially concerned with linear operators in a real Hilbert space, which means that the associated scalar field involved is real. Some specific inner products which will be used in the subsequent sections include

$$\langle u, v \rangle = \int_a^b u(x) v(x) \, dx, \qquad \langle u, v \rangle = \iint_D u(x, y) v(x, y) \, dx \, dy,$$

where $D \subset R^2$.

Example 14.6.4. Determine whether the differentiable operators (i) $A = d/dx$, (ii) $A = d^2/dx^2$, and (iii) $A = \nabla^2 = (\partial^2/\partial x^2) + (\partial^2/\partial y^2)$ are self-adjoint for functions that are differentiable in $a \leq x \leq b$ or in $D \subset R^2$ and vanish on the boundary.

$$\text{(i)} \qquad \langle Au, v \rangle = \int_a^b \left(\frac{du}{dx} \right) v \, dx = \int_a^b u \left(-\frac{dv}{dx} \right) dx + [u, v]_a^b$$

$$= \langle u, A^* v \rangle \quad \text{where} \quad A^* = -\frac{d}{dx} \neq A.$$

Hence, A is not self-adjoint.

$$\text{(ii)} \qquad \langle Au, v \rangle = \int_a^b \left(\frac{d^2 u}{dx^2} \right) v \, dx = \int_a^b \left(\frac{du}{dx} \right) \left(-\frac{dv}{dx} \right) dx + \left[v \frac{du}{dx} \right]_a^b$$

$$= \int_a^b u \left(\frac{d^2 v}{dx^2} \right) dx + \left[v \frac{du}{dx} - u \frac{dv}{dx} \right]_a^b$$

$$= \int_a^b u \left(\frac{d^2 v}{dx^2} \right) dx = \langle u, Av \rangle.$$

Thus, A is self-adjoint.

$$\text{(iii)} \qquad \langle Au, v \rangle = \iint_D (\nabla^2 u) v \, dx \, dy = \iint_D [\nabla \cdot (\nabla u) v - \nabla u \cdot \nabla v] \, dx \, dy$$

$$= \int_{\partial D} (\widehat{\mathbf{n}} \cdot \nabla u) v \, dS - \iint_D \nabla u \cdot \nabla v \, dx \, dy$$

$$= -\iint_D (\nabla u \cdot \nabla v) \, dx \, dy,$$

where the divergence theorem is used with the unit outward normal vector $\widehat{\mathbf{n}}$.

Noting the symmetry of the right hand side in u and v, it follows that

$$\langle Au, v \rangle = \langle Av, u \rangle = \langle u, Av \rangle.$$

This means that the operator $A = \nabla^2$ is self-adjoint.

Example 14.6.5. Use the inner product $\langle \mathbf{u}, \mathbf{v} \rangle = \iiint_D (\mathbf{u} \cdot \mathbf{v}) \, dV$ and the operator $A = \text{grad}$ to show that $A^* = -\text{div}$ provided the functions vanish on the boundary surface ∂D of D.

We use the divergence theorem to obtain

$$\langle A\phi, \mathbf{v} \rangle = \iiint_D (\text{grad}\,\phi \cdot \mathbf{v}) \, dV = \iiint_D [\text{div}\,(\phi\mathbf{v}) - \phi\,\text{div}\,\mathbf{v}] \, dV,$$

$$= \iiint_D \phi\,(-\text{div}\,\mathbf{v}) \, dV + \iint_{\partial D} (\widehat{\mathbf{n}} \cdot \phi\mathbf{v}) \, dS,$$

$$= \iiint_D \phi\,(-\text{div}\,\mathbf{v}) \, dV = \langle \phi, -\text{div}\,\mathbf{v} \rangle = \langle \phi, A^*v \rangle.$$

In the theory of calculus of variations it is a common practice to use δu, $\delta^2 u$, etc. to denote the first and second variations of a function u. Thus, δ can be regarded as an operator that changes u into δu, u_x into $\delta\,(u_x)$, and u_{xx} into $\delta\,(u_{xx})$ with the meaning, $\delta u = \varepsilon v$, $\delta\,(u_x) = \varepsilon v_x$, $\delta\,(u_{xx}) = \varepsilon v_{xx}$, where ε is a small arbitrary real parameter. The operators δ, δ^2 are called the *first* and *second variational operators* respectively.

Some simple properties of the operator δ are given by

$$\frac{\partial}{\partial x}\,(\delta u) = \frac{\partial}{\partial x}\,(\varepsilon v) = \varepsilon\frac{\partial v}{\partial x} = \delta\left(\frac{\partial u}{\partial x}\right), \tag{14.6.1}$$

$$\delta\left[\int_a^b u\,dx\right] = \varepsilon\int_a^b v\,dx = \int_a^b \varepsilon\,v\,dx = \int_a^b (\delta u)\,dx. \tag{14.6.2}$$

The variational operator can be interchanged with the differential and integral operators, and proves to be very useful in the calculation of the variation of a functional.

The main task of the calculus of variations is concerned with the problem of minimizing or maximizing functionals involved in mathematical, physical and engineering problems. The variational principles have their origins in the simplest kind of variational problem, which was first considered by Euler in 1744 and Lagrange in 1760-61.

The classical Euler–Lagrange variational problem is to determine the extremum value of the functional

$$I\,(u) = \int_a^b F\,(x, u, u')\,dx, \qquad u' = \frac{du}{dx}, \tag{14.6.3}$$

with the boundary conditions

$$u\,(a) = \alpha \quad \text{and} \quad u\,(b) = \beta, \tag{14.6.4}$$

where u belongs to the class $C^2\,([a, b])$ of functions which have continuous derivatives up to second-order in $a \leq x \leq b$, and F has continuous second-order derivatives with respect to all of its arguments.

We assume that $I(u)$ has an extremum at some $u \in C^{(2)}([a,b])$. Then we consider the set of all variations $u + \epsilon v$ for fixed u where v is an arbitrary function belonging to $C^2([a,b])$ such that $v(a) = v(b) = 0$. We next consider the increment of the functional

$$\delta I = I(u + \epsilon v) - I(u) = \int_a^b [F(x, u + \epsilon v, u' + \epsilon v') - F(x, u, u')]\, dx.$$

$$(14.6.5)$$

From the Taylor series expansion

$$F(x, u + \epsilon v, u' + \epsilon v') = F(x, u, u') + \epsilon \left(v \frac{\partial F}{\partial u} + v' \frac{\partial F}{\partial u'} \right)$$
$$+ \frac{\epsilon^2}{2!} \left(v \frac{\partial F}{\partial u} + v' \frac{\partial F}{\partial u'} \right)^2 + \cdots,$$

it follows from (14.6.5) that

$$I(u + \epsilon v) = I(u) + \epsilon\, \delta I + \frac{\epsilon^2}{2!} \delta^2 I + \cdots, \qquad (14.6.6)$$

where the first and second variations of I are given by

$$\delta I = \int_a^b \left(v \frac{\partial F}{\partial u} + v' \frac{\partial F}{\partial u'} \right) dx, \qquad (14.6.7)$$

$$\delta^2 I = \int_a^b \left(v \frac{\partial F}{\partial u} + v' \frac{\partial F}{\partial u'} \right)^2 dx. \qquad (14.6.8)$$

The necessary condition for the functional $I(u)$ to have an extremum (that is, $I(u)$ is stationary at u) is that the first variation becomes zero at u so

$$0 = \delta I = \int_a^b \left(v \frac{\partial F}{\partial u} + v' \frac{\partial F}{\partial u'} \right) dx \qquad (14.6.9)$$

which is, by partial integration of the second integral,

$$= \int_a^b \left[\frac{\partial F}{\partial u} - \frac{d}{dx} \left(\frac{\partial F}{\partial u'} \right) \right] v\, dx + \left[v \frac{\partial F}{\partial u'} \right]_a^b.$$

Because $v(a) = v(b) = 0$, this means that

$$\int_a^b \left[\frac{\partial F}{\partial u} - \frac{d}{dx} \left(\frac{\partial F}{\partial u'} \right) \right] v\, dx = 0. \qquad (14.6.10)$$

Since v is arbitrary in $a \le x \le b$, it follows from (14.6.10) that

$$\frac{\partial F}{\partial u} - \frac{d}{dx} \left(\frac{\partial F}{\partial u'} \right) = 0. \qquad (14.6.11)$$

This is the famous *Euler–Lagrange equation*. We therefore can state:

Theorem 14.6.1. *A necessary condition for the functional* $I(u)$ *to be stationary at* u *is that* u *is a solution of the Euler–Lagrange equation*

$$\frac{\partial F}{\partial u} - \frac{d}{dx}\left(\frac{\partial F}{\partial u'}\right) = 0, \qquad a \le x \le b \qquad (14.6.12)$$

with

$$u(a) = \alpha, \qquad u(b) = \beta. \qquad (14.6.13)$$

This is called the Euler–Lagrange variational principle.

Note that, in general, equation (14.6.12) is a nonlinear second-order ordinary differential equations, and, although such an equation is very difficult to solve, still it seems to be more accessible analytically than the functional (14.6.3) from which it is derived.

The derivative $\frac{d}{dx}$ in (14.6.12) can be computed by recalling $u = u(x)$ and $u' = \frac{du}{dx}$, and equation (14.6.12) becomes

$$\frac{\partial F}{\partial u} - \frac{\partial^2 F}{\partial x \partial u'} - \frac{\partial^2 F}{\partial u \partial u'}\left(\frac{du}{dx}\right) - \frac{\partial^2 F}{\partial u'^2}\frac{d^2 u}{dx^2} = 0.$$

It is left to the reader to verify that the functional with one independent variable and nth-order derivatives in the form

$$I(u) = \int_a^b F\left(x, u, u_x, u_{xx}, \ldots, u_{x^n}, \ldots,\right) dx$$

admits the Euler–Lagrange equation

$$\frac{\partial F}{\partial u} - \frac{d}{dx}\left(\frac{\partial F}{\partial u_x}\right) + \frac{d^2}{dx^2}\left(\frac{\partial F}{\partial u_{xx}}\right) - \ldots (-1)^n \frac{d^n}{dx^n}\left(\frac{\partial^2 F}{\partial u_{x^n}}\right) = 0.$$

After we have determined the function u which makes $I(u)$ stationary, the question of the nature of the extremum arises, that is, its minimum, maximum, or saddle point properties. To answer this question, we look at the second variation defined in (14.6.8). If terms of $O(\varepsilon^3)$ can be neglected in (14.6.6), or if they vanish for the case of quadratic F, it follows from (14.6.6) that a necessary condition for the functional $I(u)$ to have a minimum $I(u) \ge I(u_0)$ at $u = u_0$ is that $\delta^2 I \ge 0$, for $I(u)$ to have a maximum $I(u) \le I(u_0)$ at $u = u_0$ is that $\delta^2 I \le 0$ at $u = u_0$ respectively for all admissible values of v. These results enable us to determine the upper or lower bounds for the stationary value $I(u_0)$ of the functional.

Example 14.6.6. Find out the shortest distance between given points A and B in the (x, y)-plane.

Suppose APB is any curve in the plane through A and B, and $s = \text{arc} AP$. The problem is to determine the curve for which the functional

$$I\left(y\right) = \int_A^B ds, \tag{14.6.14}$$

is a minimum.

Since $ds/dx = \left(1 + y'^2\right)^{\frac{1}{2}}$, functional (14.6.14) becomes

$$I\left(y\right) = \int_{x_1}^{x_2} \left(1 + y'^2\right)^{\frac{1}{2}} dx. \tag{14.6.15}$$

In this case, $F = \left(1 + y'^2\right)^{\frac{1}{2}}$ which depends on y' only, so $\partial F/\partial y = 0$. Hence, the Euler–Lagrange equation (14.6.12) becomes

$$\frac{d}{dx}\left(\frac{\partial F}{\partial y'}\right) = 0.$$

This gives the differential equation

$$y'' = 0. \tag{14.6.16}$$

This means that the curvature for all points on the curve AB is zero. Hence, the path AB is a straight line. It follows from the integration of (14.6.16) that $y = mx + c$ is a two-parameter family of straight lines.

Example 14.6.7. (*Fermat principle in optics*). In an optically homogeneous isotro-pic medium, light travels from one point A to another point B along the path for which the travel time is minimum.

The velocity of light v is the same at all points of the medium; hence, the minimum time is equivalent to the minimum path length. For simplicity, consider a path joining the two points A and B in the (x, y)-plane. The time to travel an elementary arc length ds is ds/v. Thus, the variational problem is to find the path for which

$$\int_A^B \frac{ds}{v} = \int_{x_1}^{x_2} \frac{\left(1 + y'^2\right)^{\frac{1}{2}} dx}{v} = \int_{x_1}^{x_2} F\left(y, y'\right) dx \tag{14.6.17}$$

is a minimum, where $y' = dy/dx$, and $v = v\left(y\right)$.

When F is a function of y and y', the Euler–Lagrange equation (14.6.12) becomes

$$\frac{d}{dx}\left(F - y' F_{y'}\right) = 0. \tag{14.6.18}$$

This follows from the result

$$\begin{aligned}
\frac{d}{dx}\left(F - y' F_{y'}\right) &= \frac{d}{dx} F\left(y, y'\right) - y'' F_{y'} - y' \frac{d}{dx}\left(F_{y'}\right) \\
&= y' F_y + y'' F_y - y'' F_y - y' \frac{d}{dx}\left(F_{y'}\right) \\
&= y' \left[F_y - \frac{d}{dx}\left(F_{y'}\right)\right] = 0, \quad \text{by} \quad (14.6.12).
\end{aligned}$$

Hence,

$$F - y'F_{y'} = \text{constant}, \tag{14.6.19}$$

or

$$\frac{\left(1 + y'^2\right)^{\frac{1}{2}}}{v} - \frac{y'^2}{v\left(1 + y'^2\right)^{\frac{1}{2}}} = \text{constant},$$

or

$$v^{-1}\left(1 + y'^2\right)^{-\frac{1}{2}} = \text{constant}. \tag{14.6.20}$$

In order to give a simple physical interpretation, we rewrite (14.6.20) in terms of the angle ϕ made by the tangent to the minimum path with the vertical y-axis so that

$$\sin\phi = \left(1 + y'^2\right)^{-\frac{1}{2}}.$$

Hence,

$$\frac{1}{v}\sin\phi = \text{constant} = K \tag{14.6.21}$$

for all points on the minimum curve. For a ray of light, $(1/v)$ must be directly proportional to the refractive index n of the medium through which light is travelling. Equation (14.6.21) is called the *Snell law of refraction of light*. Often this law is stated as

$$n\sin\phi = \text{constant}. \tag{14.6.22}$$

(A) Hamilton Principle

The difference between the kinetic energy T and the potential energy V of a dynamical system is denoted by $L = T - V$. The quantity L is called the *Lagrangian* of the system. The *Hamilton principle* states that the first variation of the time integral of L is zero, that is,

$$\delta \int_{t_1}^{t_2} L\,dt = \delta \int_{t_1}^{t_2} (T - V)\,dt = 0. \tag{14.6.23}$$

This result is supposed to be valid for all dynamical systems whether they are conservative or nonconservative.

For a conservative system the force field $\mathbf{F} = -\nabla V$ and $T + V = C$, where C is a constant, and so (14.6.23) gives the *principle of least action*

$$\delta A = 0, \qquad A = \int_{t_1}^{t_2} L\,dt, \tag{14.6.24}$$

where A is called the *action integral* or simply the *action* of the system.

Example 14.6.8. Derive the Newton second law of motion from the Hamilton principle.

Consider a particle of mass m at the position $\mathbf{r} = (x, y, z)$ which is moving under the action of a field of force $\mathbf{F}$. The kinetic energy of the particle is $T = \frac{1}{2} m \dot{\mathbf{r}}^2$, and the variation of work done is $\delta W = \mathbf{F} \cdot \delta \mathbf{r}$ and $\delta V = -\delta W$. Thus, the Hamilton principle for the system is

$$0 = \delta \int_{t_1}^{t_2} (T - V)\, dt = \int_{t_1}^{t_2} (\delta T - \delta V)\, dt = \int_{t_1}^{t_2} (m\dot{\mathbf{r}} \cdot \delta \dot{\mathbf{r}} + \mathbf{F} \cdot \delta \mathbf{r})\, dt.$$

Integrating this result by parts and noting that $\delta \mathbf{r}$ vanishes at $t = t_1$ and $t = t_2$, we obtain

$$\int_{t_1}^{t_2} (m\ddot{\mathbf{r}} - \mathbf{F}) \cdot \delta \mathbf{r}\, dt = 0.$$

This is true for every virtual displacement $\delta \mathbf{r}$, and hence, the integrand must vanish, that is,

$$m\ddot{\mathbf{r}} = \mathbf{F}. \tag{14.6.25}$$

This is the celebrated *Newton second law of motion*.

Example 14.6.9. Derive the equation for a simple harmonic oscillator in a non-resisting medium from the Hamilton principle.

For a simple harmonic oscillator, $T = \frac{1}{2} m \dot{x}^2$ and $V = \frac{1}{2} m \omega^2 x^2$. According to the Hamilton principle

$$\delta \int_{t_1}^{t_2} \left(\frac{1}{2} m \dot{x}^2 - \frac{1}{2} m \omega^2 x^2 \right) dt = \delta \int_{t_1}^{t_2} F(x, \dot{x})\, dt = 0.$$

This leads to the Euler–Lagrange equation

$$\frac{\partial F}{\partial x} - \frac{d}{dt}(m\dot{x}) = 0,$$

or

$$\ddot{x} + \omega^2 x = 0. \tag{14.6.26}$$

This is the equation for the simple harmonic oscillator.

Example 14.6.10. A straight uniform elastic beam of length l, line density ρ, cross-sectional moment of inertia I, and modulus of elasticity E is fixed at each end. The beam performs small transverse oscillations in the horizontal (x, y)-plane. Derive the equation of motion of the beam.

The potential energy of the elastic beam is

$$V = \frac{1}{2} \int_0^l \frac{M^2}{EI}\, dx = \frac{1}{2} \int_0^l EIy''^2 dx,$$

where the bending moment M is proportional to the curvature so that

$$M = EI\frac{y''}{(1+y'^2)^{\frac{1}{2}}} \sim EIy'' \quad \text{for small } y'.$$

The variational principle gives

$$\delta \int_{t_1}^{t_2} (T - V)\, dt = \delta \int_{t_1}^{t_2} F\left(y'', \dot{y}\right) dt = 0,$$

where

$$F\left(y'', \dot{y}\right) = \frac{1}{2} \int_0^l \left(\rho \dot{y}^2 - EIy''^2\right) dx.$$

This principle leads to the Euler–Lagrange equation

$$-\int_0^l \left(\rho \ddot{y} + EIy^{(iv)}\right) dx = 0,$$

or

$$\rho \ddot{y} + EIy^{(iv)} = 0. \tag{14.6.27}$$

This represents the partial differential equation of the transverse vibration of the beam.

(B) The Generalized Coordinates, Lagrange Equation, and Hamilton Equation

The Euler–Lagrange analysis of a dynamical system can be extended to more complex cases where the configuration of the system is described by *generalized coordinates* $q_1, q_2, \ldots, q_n$. Without loss of generality, we consider a system of three variables where the familiar Cartesian coordinates x, y, z can be expressed in terms of the generalized coordinates q_1, q_2, q_3 as

$$x = x\left(q_1, q_2, q_3\right), \qquad y = y\left(q_1, q_2, q_3\right), \qquad z = z\left(q_1, q_2, q_3\right). \tag{14.6.28}$$

For example, if (q_1, q_2, q_3) represents the cylindrical polar coordinates (r, θ, z), the above result becomes

$$x = r\cos\theta, \qquad y = r\sin\theta, \qquad z = z.$$

Since the coordinates are functions of time t, we obtain the following result by differentiation

$$\dot{x} = \frac{\partial x}{\partial q_1}\dot{q}_1 + \frac{\partial x}{\partial q_2}\dot{q}_2 + \frac{\partial x}{\partial q_3}\dot{q}_3 \tag{14.6.29}$$

with similar expressions for $\dot{y}$ and $\dot{z}$.

If these results are substituted into $T = \frac{1}{2}m\left(\dot{x}^2 + \dot{y}^2 + \dot{z}^2\right)$ and $V = V(x, y, z)$, then both T and V can be written in terms of the generalized coordinates q_i and the generalized velocities $\dot{q}_i$, as

$$T = T\left(q_1, q_2, q_3; \dot{q}_1, \dot{q}_2, \dot{q}_3\right), \qquad V = V\left(q_1, q_2, q_3\right), \qquad (14.6.30)$$

so that the Lagrangian has the form

$$L = T - V = L\left(q_i, \dot{q}_i\right). \qquad (14.6.31)$$

The Hamilton principle gives

$$\delta \int_{t_1}^{t_2} L\left(q_i, \dot{q}_i\right) dt = 0. \qquad (14.6.32)$$

The simple variation of this integral with fixed end points, the interchange of the variation operations and time derivatives for the variation of the generalized velocities, and then integration by parts yield

$$\int_{t_1}^{t_2} \left[\sum_{i=1}^{3}\left\{\frac{\partial L}{\partial q_i} - \frac{d}{dt}\left(\frac{\partial L}{\partial \dot{q}_i}\right)\right\}\delta q_i\right] dt = 0, \qquad (14.6.33)$$

where the integrated components vanish because of the conditions $\delta q_i = 0$ $(i = 1, 2, 3)$ at $t = t_1$ and $t = t_2$.

When the generalized coordinates are independent and the variations δq_i are independent for all t in (t_1, t_2), the coefficients of the variations δq_i vanish independently for arbitrary values of t_1 and t_2. This means that the integrand in (14.6.33) vanishes, that is,

$$\frac{d}{dt}\left(\frac{\partial L}{\partial \dot{q}_i}\right) - \frac{\partial L}{\partial q_i} = 0, \qquad i = 1, 2, 3. \qquad (14.6.34)$$

These are called the *Lagrange equations* of motion.

If a particle of mass m at position $r = (x_1, x_2, x_3)$ moves under the action of a conservative force field $F_i = -\partial V/\partial x_i$, the Lagrangian function is

$$L = T - V = \frac{1}{2}m\left(\dot{x}_1^2 + \dot{x}_2^2 + \dot{x}_3^2\right) - V\left(x_1, x_2, x_3\right). \qquad (14.6.35)$$

Consequently,

$$\frac{\partial L}{\partial \dot{x}_i} = m\dot{x}_i, \qquad \frac{\partial L}{\partial x_i} = -\frac{\partial V}{\partial x_i} = F_i. \qquad (14.6.36)$$

The former represents the momentum of the particle and the latter is the force acting on the particle. In view of (14.6.36), the Lagrange equation (14.6.34) gives the Newton second law of motion in the form

$$\frac{d}{dt}\left(m\dot{x}_i\right) = F_i. \qquad (14.6.37)$$

Example 14.6.11. Apply the Lagrange equations of motion to derive the equations of motion of a particle under the action of a central force, $-mF(r)$ where r is the distance of the particle of mass m from the center of force.

It is convenient to use the polar coordinates r and θ. In terms of the generalized coordinates $q_1 = r$ and $q_2 = \theta$, we write

$$x = r\cos\theta = q_1\cos q_2, \qquad y = r\sin\theta = q_1\sin q_2.$$

The kinetic energy T is

$$T = \frac{1}{2}m\left(\dot{x}^2 + \dot{y}^2\right) = \frac{1}{2}m\left(\dot{r}^2 + r^2\dot{\theta}^2\right) = \frac{1}{2}m\left(\dot{q}_1^2 + q_1^2\dot{q}_2^2\right). \quad (14.6.38)$$

Since $\mathbf{F} = \nabla V$, the potential is

$$V(r) = \int^r F(r)\,dr = \int^{q_1} F(q_1)\,dq_1. \qquad (14.6.39)$$

Then, the Lagrangian L is

$$L = T - V = \frac{1}{2}m\left[\left(\dot{q}_1^2 + q_1^2\dot{q}_2^2\right) - 2\int^{q_1} F(q_1)\,dq_1\right]. \quad (14.6.40)$$

Thus, the Lagrange equations (14.6.34) with $i = 1, 2, 3$ give the equations of motion

$$\ddot{q}_1 - q_1\dot{q}_2^2 + F(q_1) = 0, \qquad \frac{d}{dt}\left(q_1^2\dot{q}_2\right) = 0. \qquad (14.6.41)$$

In term of the polar coordinates, these equations become

$$\ddot{r} - r\dot{\theta}^2 = -F(r), \qquad \frac{d}{dt}\left(r^2\dot{\theta}\right) = 0. \qquad (14.6.42\text{ab})$$

Equation (14.6.42b) gives immediately

$$r^2\dot{\theta} = h, \qquad (14.6.43)$$

where h is a constant. In this case, $r\dot{\theta}$ represents the transverse velocity component, and $mr^2\dot{\theta} = mh$ is the constant angular momentum of the particle about the center of force.

Introducing $r = 1/u$, we obtain

$$\dot{r} = \frac{dr}{dt} = -\frac{1}{u^2}\frac{du}{dt} = -\frac{1}{u^2}\frac{du}{d\theta}\cdot\frac{d\theta}{dt} = -h\frac{du}{d\theta},$$

$$\ddot{r} = \frac{d^2r}{dt^2} = -h\frac{d}{dt}\left(\frac{du}{d\theta}\right) = -h\frac{d^2u}{d\theta^2}\frac{d\theta}{dt} = -h^2u^2\frac{d^2u}{d\theta^2}.$$

Substituting these into (14.6.42a) gives

$$-h^2 u^2 \frac{d^2 u}{d\theta^2} - h^2 u^3 = -F\left(\frac{1}{u}\right),$$

or

$$\frac{d^2 u}{d\theta^2} + u = \frac{1}{h^2 u^2} F\left(\frac{1}{u}\right). \tag{14.6.44}$$

This is the differential equation of the central orbit and can be solved by standard methods.

In particular, if the law of force is the attractive inverse square, $F(r) = \mu/r^2$ so that the potential $V(r) = -\mu/r$, the differential equation (14.6.44) becomes

$$\frac{d^2 u}{d\theta^2} + u = \frac{\mu}{h^2}, \tag{14.6.45}$$

if the particle is projected initially from distance a with velocity V at an angle β that the direction of motion makes with the outward radius vector. Thus, the constant h in (14.6.43) is $h = Va\sin\beta$. The angle ϕ between the tangent and radius vector of the orbit at any point is given by

$$\cot\phi = \frac{1}{r}\frac{dr}{d\theta} = u\frac{d}{d\theta}\left(\frac{1}{u}\right) = -\frac{1}{u}\frac{du}{d\theta}. \tag{14.6.46}$$

At $t = 0$, the initial conditions are

$$u = \frac{1}{a}, \quad \frac{du}{d\theta} = -\frac{1}{a}\cot\beta \quad \text{when } \theta = 0. \tag{14.6.47}$$

The general solution of equation (14.6.45) is

$$u = \frac{\mu}{h^2}\left[1 + e\cos(\theta + \alpha)\right], \tag{14.6.48}$$

where e and α are constants to be determined from the initial data.

Finally, the solution can be written as

$$\frac{l}{r} = 1 + e\cos(\theta + \alpha), \tag{14.6.49}$$

where

$$l = \frac{h^2}{\mu} = (Va\sin\beta)^2/\mu. \tag{14.6.50}$$

This represents a conic section of semi-latus rectum l and eccentricity e with its axis inclined at an angle α to the radius vector at the point of projection.

The initial conditions (14.6.47) lead to

$$\frac{l}{a} = 1 + e \cos \alpha, \qquad -\frac{l}{a} \cot \beta = -e \sin \alpha, \qquad (14.6.51)$$

which give

$$\tan \alpha = \frac{l \cot \beta}{l - a},$$

$$e^2 = \left(\frac{l}{a} - 1\right)^2 + \frac{l^2}{a^2} \cot^2 \beta = \frac{l^2}{a^2} \csc^2 \beta - \frac{2l}{a} + 1,$$

$$= 1 - \frac{2aV^2 \sin^2 \beta}{\mu} + \frac{a^2 V^4 \sin^2 \beta}{\mu^2}. \qquad (14.6.52)$$

Thus, the conic is an ellipse, parabola, or hyperbola accordingly as $e <=> 1$ that is, $V^2 <=> 2\mu/a$.

To derive the Hamilton equations, we introduce the concept of generalized momentum, p_i and generalized force, F_i as

$$p_i = \frac{\partial L}{\partial \dot{q}_i}, \qquad F_i = \frac{\partial L}{\partial q_i}. \qquad (14.6.53\text{ab})$$

Consequently, the Lagrange equations (14.6.34) become

$$\frac{\partial L}{\partial q_i} = \frac{d}{dt} p_i = \dot{p}_i. \qquad (14.6.54)$$

The Hamiltonian function H is defined by

$$H = \sum_{i=1}^{n} p_i \dot{q}_i - L. \qquad (14.6.55)$$

In general, $L = L(q_i, \dot{q}_i, t)$ is a function of q_i, $\dot{q}_i$ and t, where $\dot{q}_i$ enters through the kinetic energy as a quadratic term. Hence, equation (14.6.53a) will give p_i as a linear function of $\dot{q}_i$. This system of linear equations involving p_i and $\dot{q}_i$ can be solved to determine $\dot{q}_i$ in terms of p_i, and then, the $\dot{q}_i$ can, in principle, be eliminated from (14.6.55). This means that H can always be expressed as a function of p_i, q_i and t so that $H = H(p_i, q_i, t)$. Thus,

$$dH = \sum \frac{\partial H}{\partial p_i} dp_i + \sum \frac{\partial H}{\partial q_i} dq_i + \frac{\partial H}{\partial t} dt. \qquad (14.6.56)$$

On the other hand, differentiating H in (14.6.55) with respect to t gives

$$\frac{dH}{dt} = \sum p_i \frac{d}{dt} \dot{q}_i + \sum \dot{q}_i \frac{d}{dt} p_i - \sum \frac{\partial L}{\partial q_i} \frac{d}{dt} q_i - \sum \frac{\partial L}{\partial \dot{q}_i} \frac{d}{dt} \dot{q}_i - \frac{\partial L}{\partial t},$$

$$(14.6.57)$$

or

$$dH = \sum p_i \, d\dot{q}_i + \sum \dot{q}_i \, dp_i - \sum \frac{\partial L}{\partial q_i} \, dq_i - \sum \frac{\partial L}{\partial \dot{q}_i} \, d\dot{q}_i - \frac{\partial L}{\partial t} \, dt, \quad (14.6.58)$$

which becomes, in view of (14.6.53a),

$$dH = \sum \dot{q}_i \, dp_i - \sum \frac{\partial L}{\partial q_i} \, dq_i - \frac{\partial L}{\partial t} \, dt. \quad (14.6.59)$$

Evidently, two expressions of dH in (14.6.56) and (14.6.59) must be equal so that the coefficients of the corresponding differentials can be equated to obtain

$$\dot{q}_i = \frac{\partial H}{\partial p_i}, \qquad -\frac{\partial L}{\partial q_i} = \frac{\partial H}{\partial q_i}, \qquad -\frac{\partial L}{\partial t} = \frac{\partial H}{\partial t}. \quad (14.6.60\text{abc})$$

Using the Lagrange equation (14.6.54), the first two of the above equations become

$$\dot{q}_i = \frac{\partial H}{\partial p_i}, \qquad \dot{p}_i = -\frac{\partial H}{\partial q_i}. \quad (14.6.61\text{ab})$$

These are commonly known as the *Hamilton canonical equations* of motion. They play a fundamental role in advanced analytical dynamics.

Finally, the Lagrange–Hamilton theory can be used to derive the law of conservation of energy. In general, the Lagrangian L is independent of time t and hence, (14.6.60c) implies that $H = \text{constant}$. Again, T involved in $L = T - V$ is given by

$$T = \frac{1}{2} \sum_{i=1}^{n} \sum_{j=1}^{n} a_{ij} \, \dot{q}_i \, \dot{q}_j, \quad (14.6.62)$$

where the coefficients a_{ij} are symmetric functions of the generalized coordinates q_{ij}, that is, $a_{ij} = a_{ji}$.

On the other hand, V is, in general, independent of q_i and hence,

$$p_i = \frac{\partial L}{\partial \dot{q}_i} = \frac{\partial T}{\partial \dot{q}_i} = \sum_{j=1}^{n} a_{ij} \, \dot{q}_j. \quad (14.6.63)$$

Thus, the Hamiltonian H becomes

$$H = \sum_{i=1}^{n} p_i \dot{q}_i - L = \sum_{i=1}^{n} \left(\sum_{j=1}^{n} a_{ij} \, \dot{q}_j \right) \dot{q}_i - L = 2T - L = T + V. \quad (14.6.64)$$

Thus, H is equal to the total energy. It has already been observed that, if L does not contain t explicitly, H is a constant. This means that the sum of the potential and kinetic energies is constant. This is the *law of the conservation of energy*.

Example 14.6.12. Use the Hamiltonian equations to derive the equations of motion for the problem stated in Example 14.6.11.

The Lagrangian L for this problem is given by (14.6.40) with $q_1 = r$ and $q_2 = \theta$. It follows from the definition (14.6.53a) of the generalized momentum that

$$p_1 = m\dot{q}_1 = m\dot{r}, \qquad p_2 = mq_1^2\dot{q}_2 = mr^2\dot{\theta}. \qquad (14.6.65)$$

Expressing the results of the kinetic energy (14.6.38) and the potential energy (14.6.39) in terms of p_1 and p_2 the Hamiltonian $H = T + V$ can be written as

$$H = \frac{1}{2m}\left(p_1^2 + \frac{p_2^2}{q_1^2}\right) + m\int^{q_1} F(q_1)\,dq_1. \qquad (14.6.66)$$

Then, equations (14.6.65) and the Hamilton equation (14.6.61b) give

$$p_1 = m\dot{r}, \qquad p_2 = mr^2\dot{\theta}, \qquad (14.6.67)$$

$$\dot{p}_1 = \frac{1}{m}\frac{p_2^2}{q_1^3} + mF(q_1), \qquad \dot{p}_2 = 0. \qquad (14.6.68)$$

Clearly, these equations are identical with the equations of motion (14.6.42ab).

Example 14.6.13. Derive the equation of a simple pendulum by using (i) the Lagrange equations and (ii) the Hamilton equations.

We consider the motion of simple pendulum of mass m attached at the end of a rigid massless string of length l that pivots about a fixed point. We suppose that the pendulum makes an angle θ with its vertical position. The force F acting on the mass m is $F = -mg\sin\theta$, so that the potential V is obtained from $F = -\nabla V$ as $V = mgl\,(1 - \cos\theta)$. The kinetic energy $T = \frac{1}{2}ml^2\dot{\theta}^2$.

Thus the Lagrangian L is

$$L = T - V = \frac{1}{2}ml^2\dot{\theta}^2 - mgl\,(1 - \cos\theta) = L\left(\theta, \dot{\theta}\right). \qquad (14.6.69)$$

The Lagrange equation is

$$\frac{\partial L}{\partial \theta} - \frac{d}{dt}\left(\frac{\partial L}{\partial \dot{\theta}}\right) = 0, \qquad (14.6.70)$$

or

$$-mgl\sin\theta - \frac{d}{dt}\left(ml^2\dot{\theta}\right) = 0,$$

or

$$\ddot{\theta} + \omega^2\sin\theta = 0, \qquad \omega^2 = g/l. \qquad (14.6.71ab)$$

This is the equation of the simple pendulum.

To derive the same equation from the Hamilton equations, we choose $q_1 = l \, (\dot{q}_1 = 0)$ and $q_2 = \theta$ as the generalized (polar) coordinates. The kinetic and potential energies are

$$T = \frac{1}{2} m l^2 \dot{q}_2^2, \qquad V = mgl \, (1 - \cos q_2). \qquad (14.6.72\text{ab})$$

Thus, $H = T + V$ and $L = T - V$ are given by

$$(H, L) = \frac{1}{2} m l^2 \dot{q}_2^2 \pm mgl \, (1 - \cos q_2). \qquad (14.6.73\text{ab})$$

From the definition of the generalized momentum, we find that

$$p_2 = \frac{\partial L}{\partial \dot{q}_2} = m l^2 \dot{q}_2$$

so that the Hamiltonian H in terms of p_2 and q_2 is

$$H = \frac{1}{2} \frac{p_2^2}{m l^2} + mgl \, (1 - \cos q_2).$$

Thus, the Hamilton equation (14.6.61ab) gives

$$\ddot{\theta} + \omega^2 \sin \theta = 0, \qquad \omega^2 = \frac{g}{l}. \qquad (14.6.74)$$

The variational methods can be further extended for functionals depending on functions or more independent variables in the form

$$I \, [u \, (x, y)] = \iint_D F \, (x, y, u, u_x, u_y) \, dx \, dy \qquad (14.6.75)$$

where the values of the function $u \, (x, y)$ are prescribed on the boundary ∂D of a finite domain D in the (x, y)-plane. We assume that F is differentiable and the surface $u = u \, (x, y)$ giving an extremum is also continuously differentiable twice.

The first variation δI of I is defined by

$$\delta I \, [u, \varepsilon] = I \, (u + \varepsilon) - I \, (u) \qquad (14.6.76)$$

which is, by Taylor's expansion theorem

$$= \iint_D [\varepsilon F_u + \varepsilon_x F_p + \varepsilon_y F_q] \, dx \, dy \qquad (14.6.77)$$

where $\varepsilon \equiv \varepsilon \, (x, y)$ is small and $p = u_x$ and $q = u_y$. According to the variational principle, $\delta I = 0$ for all admissible values of ε. The partial integration of (14.6.77) combined with $\varepsilon = 0$ on ∂D gives

$$0 = \delta I = \iint_D \left[F_u - \frac{\partial}{\partial x} F_p - \frac{\partial}{\partial y} F_q \right] \varepsilon(x, y) \, dx \, dy. \quad (14.6.78)$$

This is true for all arbitrary ε, and hence, the integrand must vanish, that is

$$\frac{\partial}{\partial x} F_p + \frac{\partial}{\partial y} F_q - F_u = 0. \quad (14.6.79)$$

This is the *Euler–Lagrange equation* which is the second-order partial differential equation to be satisfied by the extremizing function $u(x, y)$.

Example 14.6.14. Derive the equation of motion for the free vibration of an elastic string of length l.

The potential energy V of the string is

$$V = \frac{1}{2} T^* \int_0^l u_x^2 \, dx \quad (14.6.80)$$

where $u = u(x, y)$ is the displacement of the string from its equilibrium position and T^* is the constant tension of the string.

The kinetic energy T is

$$T = \frac{1}{2} \int_0^l \rho u_t^2 \, dx \quad (14.6.81)$$

where ρ is the constant line-density of the string.

According to the Hamilton principle

$$\delta I = \delta \int_{t_1}^{t_2} (T - V) \, dt = \delta \int_{t_1}^{t_2} \int_0^l \frac{1}{2} \left(\rho u_t^2 - T^* u_x^2 \right) \, dx \, dt = 0 \quad (14.6.82)$$

which has the form

$$\delta \int_{t_1}^{t_2} \int_0^l L(u_t, u_x) = 0, \quad (14.6.83)$$

where

$$L = \frac{1}{2} \left(\rho u_t^2 - T^* u_x^2 \right).$$

Then the Euler–Lagrange equation is given by

$$\frac{\partial}{\partial t} (\rho u_t) - \frac{\partial}{\partial x} (T^* u_x) = 0, \quad (14.6.84)$$

or

$$u_{tt} - c^2 u_{xx} = 0, \qquad c^2 = T^* / \rho. \quad (14.6.85)$$

This is the wave equation of motion of the string.

Example 14.6.15. Derive the Laplace equation from the functional

$$I\left(u\right) = \iint_D \left(u_x^2 + u_y^2\right) dx\, dy$$

with a boundary condition $u = f\left(x, y\right)$ on ∂D.

The variational principle gives

$$\delta I = \delta \iint_D \left(u_x^2 + u_y^2\right) dx\, dy = 0.$$

This leads to the Euler–Lagrange equation

$$u_{xx} + u_{yy} = 0 \quad \text{in} \quad D.$$

Similarly, the functional

$$I\left[u\left(x, y, z\right)\right] = \iiint_D \left(u_x^2 + u_y^2 + u_z^2\right) dx\, dy\, dz$$

will lead to the three-dimensional Laplace equation

$$\nabla^2 u = u_{xx} + u_{yy} + u_{zz} = 0.$$

14.7 The Rayleigh–Ritz Approximation Method

We consider the boundary-value problem governed by the differential equation

$$Au = f \quad \text{in} \quad D \tag{14.7.1}$$

with the boundary condition

$$B\left(u\right) = 0 \quad \text{on} \quad \partial D \tag{14.7.2}$$

where A is a self-adjoint differential operator in a Hilbert space H and $f \in H$.

In general, the determination of the exact solution of the problem is often a difficult task. However, it can be shown that the solution of (14.7.1)–(14.7.2) is equivalent to finding the minimum of a functional $I\left(u\right)$ associated with the differential system. In other words, the solution can be characterized as the function which minimizes (or maximizes) the functional $I\left(u\right)$. A simple and efficient method for an approximate solution of the extremum problem was independently formulated by Lord Rayleigh and W. Ritz.

We next prove a fundamental result which states that the solution of the equation (14.7.1) is equivalent to finding the minimum of the quadratic functional

$$I(u) \equiv A \langle u, u \rangle - 2 \langle f, u \rangle. \qquad (14.7.3)$$

Suppose that $u = u_0$ is the solution of (14.7.1) so that $Au_0 = f$. Consequently,

$$I(u) \equiv A \langle u, u \rangle - 2 \langle Au_0, u \rangle = \langle A(u - u_0), u \rangle - \langle Au_0, u \rangle.$$

Since the inner product is symmetrical and $\langle Au_0, u \rangle = \langle u_0, Au \rangle = \langle Au, u_0 \rangle$, $I(u)$ can be written as

$$\begin{aligned}
I(u) &= \langle A(u - u_0), u \rangle - \langle Au, u_0 \rangle + \langle Au_0, u \rangle - \langle Au_0, u_0 \rangle, \\
&= \langle A(u - u_0), u - u_0 \rangle - \langle Au_0, u_0 \rangle, \\
&= \langle A(u - u_0), u - u_0 \rangle + I(u_0). \qquad (14.7.4)
\end{aligned}$$

Since A is a positive operator, $\langle A(u - u_0), u - u_0 \rangle \geq 0$ where equality holds if and only if $u - u_0 = 0$. It follows that

$$I(u) \geq I(u_0), \qquad (14.7.5)$$

where equality holds if and only if $u = u_0$. We conclude from this inequality that $I(u)$ assumes its minimum at the solution $u = u_0$ of equation (14.7.1).

Conversely, the function $u = u_0$ that minimizes $I(u)$ is a solution of equation (14.7.1). Clearly, $I(u) \geq I(u_0)$, that is, in particular, $I(u_0 + \alpha v) \geq I(u_0)$ for any real α and any function v. Explicitly,

$$\begin{aligned}
I(u_0 + \alpha v) &= \langle A(u_0 + \alpha v), u_0 + \alpha v \rangle - 2 \langle f, u_0 + \alpha v \rangle, \\
&= \langle Au_0, u_0 \rangle + 2\alpha \langle Au_0, v \rangle + \alpha^2 \langle Av, v \rangle - 2 \langle f, u_0 \rangle - 2\alpha \langle f, v \rangle.
\end{aligned}$$

This means that $I(u_0 + \alpha v)$ is a quadratic expression in α. Since $I(u)$ is minimum at $u = u_0$, then $\delta I(u_0, v) = 0$, that is,

$$\begin{aligned}
0 &= \left[\frac{d}{d\alpha} I(u_0 + \alpha v) \right]_{\alpha=0} \\
&= 2 \langle Au_0, v \rangle - 2 \langle f, v \rangle \\
&= 2 \langle Au_0 - f, v \rangle.
\end{aligned}$$

This is true for any arbitrary but fixed v. Hence, $Au_0 - f = 0$. This proves the assertion.

In the Rayleigh–Ritz method an approximate solution of (14.7.1)–(14.7.2) is sought in the form

$$u_n(\mathbf{x}) = \sum_{i=1}^{n} a_i \phi_i(\mathbf{x}), \qquad (14.7.6)$$

where $a_1, a_2, \ldots, a_n$ are n unknown coefficients to be determined so that $I(u_n)$ is minimum, and $\phi_1, \phi_2, \ldots, \phi_n$ represent a linearly independent and complete set of arbitrarily chosen functions that satisfy (14.7.2). This set of functions is often called a *trial set*. We substitute (14.7.6) into (14.7.3) to obtain

$$I(u_n) = \left\langle \sum_{i=1}^{n} a_i A(\phi_i), \sum_{j=1}^{n} a_j \phi_j \right\rangle - 2 \left\langle f, \sum_{i=1}^{n} a_i \phi_i \right\rangle.$$

Then the necessary condition for I to obtain a minimum (or maximum) is that

$$\frac{\partial I}{\partial a_j}(a_1, a_2, \ldots, a_n) = 0, \qquad j = 1, 2, \ldots, n, \tag{14.7.7}$$

or

$$\frac{\partial I}{\partial a_j} \left[\left\langle \sum_{i=1}^{n} a_i A(\phi_i), \sum_{j=1}^{n} a_j \phi_j \right\rangle - 2 \left\langle f, \sum_{i=1}^{n} a_i \phi_i \right\rangle \right] = 0,$$

or

$$\sum_{i=1}^{n} \langle A(\phi_i), \phi_j \rangle a_i + \sum_{i=1}^{n} \langle A(\phi_j), \phi_i \rangle a_j - 2 \langle f, \phi_j \rangle = 0,$$

or

$$2 \sum_{i=1}^{n} \langle A(\phi_i), \phi_j \rangle a_i = 2 \langle f, \phi_j \rangle.$$

Therefore,

$$\sum_{i=1}^{n} \langle A(\phi_i), \phi_j \rangle a_i = \langle f, \phi_j \rangle, \qquad j = 1, 2, \ldots, n. \tag{14.7.8}$$

This is a linear system of n equations for the n unknown coefficients a_j. Once $a_1, a_2, \ldots, a_n$ are determined, the approximate solution is given by (14.7.6).

In particular, when

$$\langle A(\phi_i), \phi_j \rangle = \begin{cases} 0, & i \neq j \\ 1, & i = j, \end{cases} \tag{14.7.9}$$

equation (14.7.8) gives a_j as

$$a_j = \langle \phi_j, f \rangle, \tag{14.7.10}$$

so that the Rayleigh–Ritz approximate series (14.7.6) becomes

$$u_n\left(\mathbf{x}\right) = \sum_{i=1}^{n} \langle \phi_i, f \rangle\, \phi_i\left(\mathbf{x}\right). \qquad (14.7.11)$$

This is similar to the Fourier series solution with known Fourier coefficients a_i.

In the limit $n \to \infty$, a limit function can be obtained from (14.7.6) as

$$u\left(\mathbf{x}\right) = \lim_{n\to\infty} \sum_{i=1}^{n} a_i\, \phi_i\left(\mathbf{x}\right) = \sum_{i=1}^{\infty} a_i\, \phi_i\left(\mathbf{x}\right), \qquad (14.7.12)$$

provided that the series converges. Under certain assumptions imposed on the functional $I\left(u\right)$ and the trial functions $\phi_1, \phi_2, \ldots, \phi_n$, the limit function $u\left(\mathbf{x}\right)$ represents an exact solution of the problem. In any event, (14.7.6) or (14.7.11) gives a reasonable approximate solution.

In the simplest case corresponding to $n = 1$, the Rayleigh–Ritz method gives a simple form of the functional

$$I\left(u_1\right) = I\left(a_1\phi_1\right) = a_1^2 \langle A\phi_1, \phi_1 \rangle - 2a_1 \langle f, \phi_1 \rangle,$$

where a_1 is readily determined from the necessary condition for extremum

$$0 = \frac{\partial}{\partial a_1} I\left(a_1\phi_1\right) = 2a_1 \langle A\phi_1, \phi_1 \rangle - 2 \langle f, \phi_1 \rangle,$$

or

$$a_1 = \frac{\langle f, \phi_1 \rangle}{\langle A\phi_1, \phi_1 \rangle}. \qquad (14.7.13)$$

The corresponding minimum value of the functional is given by

$$I\left(a_1\phi_1\right) = - \frac{\langle f, \phi_1 \rangle^2}{\langle A\phi_1, \phi_1 \rangle}. \qquad (14.7.14)$$

Thus, the essence of the Rayleigh–Ritz method is as follows. For a given boundary-value problem, an approximate series solution is sought so that the trial functions ϕ_i satisfy the boundary conditions. We solve the system of algebraic equations (14.7.7) to determine the coefficients a_i.

We now illustrate the method by several examples.

Example 14.7.1. Find an approximate solution of the Dirichlet problem

$$\nabla^2 u \equiv u_{xx} + u_{yy} = 0 \quad \text{in} \quad D$$
$$u = f \quad \text{on} \quad \partial D,$$

where $D \subset R^2$, and f is a given function.

This problem is equivalent to finding the minimum of the associated functional

$$I\left(u\right) = \iint_D \left(u_x^2 + u_y^2\right) dx\, dy.$$

We seek an approximate series solution in the form

$$u_2\left(x, y\right) = a_1\phi_1 + a_2\phi_2$$

with $a_1 = 1$ so that u_2 satisfies the given boundary conditions, that is, $\phi_1 = f$ and $\phi_2 = 0$ on ∂D. Substituting u_2 into the functional gives

$$I\left(u_2\right) = \iint_D \left[\left(\frac{\partial u_2}{\partial x}\right)^2 + \left(\frac{\partial u_2}{\partial y}\right)^2\right] dx\, dy$$

$$= \iint_D \left(\nabla\phi_1\right)^2 dx\, dy + 2a_2 \iint_D \left(\nabla\phi_1 \cdot \nabla\phi_2\right) dx\, dy$$

$$+ a_2^2 \iint_D |\nabla\phi_2|^2 \, dx\, dy.$$

The necessary condition for an extremum of $I\left(u_2\right)$ is

$$\frac{\partial I}{\partial a_2} = 0,$$

or

$$2\iint_D \left(\nabla\phi_1 \cdot \nabla\phi_2\right) dx\, dy + 2a_2 \iint_D |\nabla\phi_2|^2 \, dx\, dy = 0.$$

Therefore,

$$a_2 = -\frac{\iint \left(\nabla\phi_1 \cdot \nabla\phi_2\right) dx\, dy}{\iint |\nabla\phi_2|^2 \, dx\, dy}.$$

This a_2 minimizes the functional and the approximate solution is obtained.

However, this procedure can be generalized by seeking an approximate solution in the form

$$u_n = \sum_{i=1}^{n} a_i\phi_i \qquad (a_1 = 1)$$

so that $\phi_1 = f$ and $\phi_i = 0 \, (i = 2, 3, \ldots, n)$ on ∂D.

The coefficients a_i can be obtained by solving the system (14.7.7) with $j = 2, 3, \ldots, n$.

Example 14.7.2. A uniform elastic beam of length l carrying a uniform load W per unit length is freely hinged at $x = 0$ and $x = l$. Find the approximate solution of the boundary-value problem

$$EI_y^{(iv)}\left(x\right) = W,$$

$$y = y'' = 0 \quad \text{at} \quad x = 0 \quad \text{and} \quad x = l,$$

where $y = y(x)$ is the displacement function.

This problem is equivalent to finding a function $y(x)$ that minimizes the energy functional

$$I(y) = \int_0^l \left(Wy - \frac{EI}{2} y''^2 \right) dx.$$

We seek an approximate solution

$$y_n(x) = \sum_{r=1}^{n} a_r \sin\left(\frac{r\pi x}{l}\right)$$

which satisfies the boundary conditions.

Substitution of this solution into the energy functional gives

$$I(y_n) = \sum_{r=1}^{n} \left[\int_0^l W a_r \sin\left(\frac{r\pi x}{l}\right) dx - \frac{EI}{2} \int_0^l \frac{r^4 \pi^4}{l^4} a_r^2 \sin^2\left(\frac{r\pi x}{l}\right) dx \right]$$

$$= \frac{2Wl}{\pi} \sum_{r=1}^{n} \frac{a_r}{r} - \frac{EI\pi^4}{4l^3} \sum_{r=1}^{n} r^4 a_r^2.$$

The necessary conditions for extremum are

$$0 = \frac{\partial I}{\partial a_r} = \frac{2Wl}{r\pi} - \frac{EI\pi^4}{4l^3} 2a_r r^4, \qquad r = 1, 2, \ldots, n$$

which give a_r as

$$a_r = \frac{4Wl^4}{\pi^5 r^5 EI}, \qquad r = 1, 2, \ldots, n.$$

Thus, the approximate function $y(x)$ is

$$y_n(x) = \frac{4Wl^4}{\pi^5 EI} \sum_{r=1}^{n} \frac{1}{r^5} \sin\left(\frac{r\pi x}{l}\right).$$

The maximum deflection at $x = l/2$ is

$$y_{max} = \frac{4Wl^4}{\pi^5 EI} \left(1 - \frac{1}{3^5} + \frac{1}{5^5} - \cdots \right).$$

In this case, the first term of the series solution gives a reasonably good approximate solution as

$$y_1(x) \sim \frac{4Wl^4}{EI\pi^5} \sin\left(\frac{\pi x}{l}\right).$$

Example 14.7.3. Apply the Rayleigh–Ritz method to investigate the free vibration of a fixed elastic wedge of constant thickness governed by the energy functional

$$I\left(y\right) = \int_0^1 \left(\alpha x^3 y''^2 - \omega x y^2\right) dx, \qquad y\left(1\right) = y''\left(1\right) = 0,$$

where the free vibration is described by the function $u\left(x,t\right) = e^{i\omega t} y\left(x\right)$, ω is the frequency.

We seek an approximate solution in the form

$$y_n\left(x\right) = \sum_{r=1}^n a_r y_r\left(x\right) = \sum_{r=1}^n a_r \left(x-1\right)^2 x^{r-1}$$

which satisfies the given boundary conditions.

We take only the first two terms so that $y_2\left(x\right) = a_1 y_1 + a_2 y_2 = \left(x-1\right)^2 \left(a_1 + a_2 x\right)$. Substituting y_2 into the functional we obtain

$$I_2 = I\left(y_2\right) = \int_0^1 \left[\alpha x^3 \left(6a_2 x + 2a_1 - 4a_2\right)^2 - \omega x \left(x-1\right)^4 \left(a_1 + a_2 x\right)^2\right] dx$$

$$= \alpha \left[\left(a_1 - 2a_2\right)^2 + \frac{24}{5}\left(a_1 - 2a_2\right)a_2 + 6a_2^2\right] - \frac{\omega}{5}\left[\frac{a_1^2}{6} + \frac{2a_1 a_2}{21} + \frac{a_2^2}{56}\right].$$

The necessary conditions for an extremum are

$$\frac{\partial I_2}{\partial a_1} = 2a_1 \left(\alpha - \frac{\omega}{30}\right) + \frac{2}{5}a_2 \left(2\alpha - \frac{\omega}{21}\right) = 0,$$

$$\frac{\partial I_2}{\partial a_2} = \frac{2a_1}{5}\left(2\alpha - \frac{\omega}{21}\right) + \frac{2a_2}{5}\left(2\alpha - \frac{\omega}{56}\right) = 0.$$

For nontrivial solutions, the determinant of this algebraic system must be zero, that is,

$$\begin{vmatrix} \alpha - \frac{\omega}{30} & \frac{1}{5}\left(2\alpha - \frac{\omega}{21}\right) \\ 2\alpha - \frac{\omega}{21} & 2\alpha - \frac{\omega}{56} \end{vmatrix} = 0,$$

or

$$5\left(\alpha - \frac{\omega}{30}\right)\left(2\alpha - \frac{\omega}{56}\right) - \left(2\alpha - \frac{\omega}{21}\right)^2 = 0.$$

This represents the frequency equation of the vibration which has two roots ω_1 and ω_2. The smaller of these two frequencies gives an approximate value of the fundamental frequency of the vibration of the wedge.

Example 14.7.4. An elastic beam of length l, density ρ, cross-sectional area A, and modulus of elasticity E has its end $x = 0$ fixed and the other end connected to a rigid support through a linear elastic spring with spring constant k. Apply the Rayleigh–Ritz method to investigate the harmonic axial motion of the beam.

The kinetic energy and the potential energy associated with the axial motion of the beam are

$$T = \int_0^l \frac{\rho A}{2} U_t^2 \, dx, \qquad V = \int_0^l \frac{EA}{2} U_x^2 \, dx + \frac{k}{2} U^2 \left(l, t\right),$$

where $U\left(x, t\right)$ is the displacement function.

Since the axial motion is simple harmonic, $U\left(x, t\right) = u\left(x\right) e^{i\omega t}$, where ω is the frequency of vibration. Consequently, the expressions for T and V can be written in terms of $u\left(x\right)$. We then apply the Hamilton variational principle

$$\delta I \left(u\right) = \delta \left[\int_{t_1}^{t_2} \int_0^l \frac{1}{2} \left(\rho A \omega^2 u^2 - EA u_x^2\right) dx - \frac{k}{2} u^2 \left(l\right) \right] dt = 0.$$

The Euler–Lagrange equation for the variational principle is

$$\frac{d}{dx} \left(EA \frac{du}{dx}\right) + \rho A \omega^2 u = 0, \qquad 0 < x < l,$$

$$EA \frac{du}{dx} + ku = 0, \qquad \text{at} \quad x = l.$$

In terms of nondimensional variables $\left(x^*, u^*\right) = \left(1/l\right)\left(x, u\right)$ and parameters $\lambda = \left(\omega^2 \rho l^2 / E\right)$ and $\alpha = \left(kl/EA\right)$, this system becomes, dropping the asterisks,

$$u_{xx} + \lambda u = 0, \qquad 0 < x < 1,$$

$$u_x + \alpha u = 0, \qquad \text{at} \quad x = 1.$$

The associated functional for the system is

$$I\left(u\right) = \frac{1}{2} \int_0^1 \left(\lambda u^2 - u_x^2\right) dx - \frac{\alpha}{2} u^2 \left(1\right).$$

According to the Rayleigh–Ritz method, we seek approximate solution with $\alpha = 1$ in the form

$$u_2\left(x\right) = a_1 x + a_2 x^2$$

so that $I\left(u_2\right)$ is minimum.

We substitute $u_2\left(x\right)$ into the functional to obtain

$$I_2 = I\left(u_2\right) = \frac{1}{2} \int_0^l \left[\lambda \left(a_1 x + a_2 x^2\right)^2 - \left(a_1 + 2a_2 x\right)^2\right] dx - \frac{1}{2}\left(a_1 + a_2\right)^2.$$

The necessary conditions for extremum of the functional are

$$0 = \frac{\partial I_2}{\partial a_1} = a_1 \left(\frac{\lambda}{3} - 2 \right) + a_2 \left(\frac{\lambda}{4} - 2 \right),$$

$$0 = \frac{\partial I_2}{\partial a_2} = a_1 \left(\frac{\lambda}{4} - 2 \right) + a_2 \left(\frac{\lambda}{5} - \frac{7}{3} \right).$$

For nontrivial solutions, the determinant of system must be zero, that is,

$$\begin{vmatrix} \frac{\lambda}{3} - 2 & \frac{\lambda}{4} - 2 \\ \frac{\lambda}{4} - 2 & \frac{\lambda}{5} - \frac{7}{3} \end{vmatrix} = 0,$$

or

$$3\lambda^2 - 128\lambda + 480 = 0.$$

This quadratic equation gives two solutions:

$$\lambda_1 = 4.155, \qquad \lambda_2 = 38.512.$$

The corresponding values of the frequency are given by

$$\omega_1 = 2.038 \left(\frac{E}{\rho l^2} \right)^{\frac{1}{2}}, \qquad \omega_2 = 6.206 \left(\frac{E}{\rho l^2} \right)^{\frac{1}{2}}.$$

The exact solution is determined by the transcendental equation

$$\sqrt{\lambda} + \tan \sqrt{\lambda} = 0.$$

The first two roots of this equation can be obtained graphically as

$$\omega_{01} \sim 2.0288 \left(\frac{E}{\rho l^2} \right)^{\frac{1}{2}}, \qquad \omega_{02} \sim 4.9132 \left(\frac{E}{\rho l^2} \right)^{\frac{1}{2}}.$$

14.8 The Galerkin Approximation Method

As an extension of the Rayleigh–Ritz method, Galerkin formulated an ingenious approximation method which may be applied to a problem for which no simple variational principle exists. The differential operator A in equation (14.7.1) need not be linear for the solution of this equation. In order to solve the boundary-value problem (14.7.1)–(14.7.2), we construct an approximate solution $u(\mathbf{x})$ in the form

$$u_n(\mathbf{x}) = u_0(\mathbf{x}) + \sum_{i=1}^{n} a_i \phi_i(\mathbf{x}), \qquad (14.8.1)$$

where the $\phi_i(\mathbf{x})$ are known functions, u_0 is introduced to satisfy the boundary conditions, and the coefficients a_i are to be determined. Substituting (14.8.1) into (14.7.1) gives a non-zero *residual* R_n

$$R_n(a_1, a_2, \ldots, a_n, x, y) = A(u_n) = A(u_0) + \sum_{i=1}^{n} a_i A(\phi_i). \quad (14.8.2)$$

In this method the unknown coefficients a_i are determined by solving the following system of equations

$$\langle R_n, \phi_j \rangle = 0, \qquad j = 1, 2, \ldots, N. \quad (14.8.3)$$

Since A is linear, this can be written as

$$\sum_{i=1}^{n} a_i \langle A(\phi_i), \phi_j \rangle = - \langle A u_0, \phi_j \rangle, \quad (14.8.4)$$

which determines the a_j's. Substitution of the a_j's obtained from the solution of (14.8.4) into (14.8.1) gives the required approximate solution u_n.

We find an interesting connection between the Galerkin solution and the Fourier representation of the function u. We seek a Galerkin solution in the form

$$u_n(\mathbf{x}) = \sum_{i=1}^{n} a_i \phi_i(\mathbf{x}), \quad (14.8.5)$$

with a special restriction on the operator A which satisfies the condition

$$\langle A\phi_i, \phi_j \rangle = \begin{cases} 0, & i \neq j \\ 1, & i = j. \end{cases} \quad (14.8.6)$$

Thus, the application of the Galerkin method to (14.7.1) gives

$$\sum_{i=1}^{n} \langle A(\phi_i), \phi_j \rangle a_i = \langle f, \phi_i \rangle, \quad (14.8.7)$$

which is, by (14.8.6),

$$a_j = \langle f, \phi_j \rangle, \quad (14.8.8)$$

so the Galerkin solution (14.8.5) becomes

$$u_n(\mathbf{x}) = \sum_{i=1}^{n} \langle f, \phi_i \rangle \phi_i(\mathbf{x}). \quad (14.8.9)$$

Evidently, the Galerkin solution (14.8.5) is just the finite Fourier series solution.

Finally, we shall cite an example to show the equivalence of the Galerkin and Rayleigh–Ritz methods. We consider the Poisson equation

$$u_{xx} + u_{yy} = f(x, y) \quad \text{in} \quad D \subset R^2 \tag{14.8.10}$$

with a homogeneous boundary condition $u = 0$ on ∂D. The solution of this equation is equivalent to finding the minimum of the functional

$$I(u) = \iint_D \left(u_x^2 + u_y^2 + 2fu \right) dx\, dy. \tag{14.8.11}$$

According to the Rayleigh–Ritz method, we seek a trial solution in the form

$$u_n = \sum_{i=1}^n a_i \, \phi_i(x, y), \tag{14.8.12}$$

where the trial functions ϕ_i are chosen so that they satisfy the given boundary condition on ∂D.

We substitute the Rayleigh–Ritz solution u_n into $I(u)$ and then use $\partial I(u_n)/\partial a_k = 0$, for $k = 1, 2, \ldots, n$ to obtain

$$2 \iint_D \left(\frac{\partial u_n}{\partial x} \frac{\partial \phi_k}{\partial x} + \frac{\partial u_n}{\partial y} \frac{\partial \phi_k}{\partial y} + f\phi_k \right) dx\, dy = 0. \tag{14.8.13}$$

Application of Greens theorem with the homogeneous boundary condition leads to

$$\iint_D \left(\nabla^2 u_n - f \right) \phi_k \, dx\, dy = 0, \tag{14.8.14}$$

or

$$\langle R_n, \phi_k \rangle = 0, \qquad R_n \equiv \nabla^2 u_n - f. \tag{14.8.15}$$

This is the Galerkin equation for the undetermined coefficients a_k. This establishes the equivalence of the two methods.

Example 14.8.1. Use the Galerkin method to find an approximate solution of the Poisson equation

$$\nabla^2 u \equiv u_{xx} + u_{yy} = -1 \quad \text{in} \quad D = \{(x, y) : |x| < a, |y| < b\}$$

with the boundary conditions

$$u = 0 \quad \text{on} \quad \partial D = \{(x, y) : x = \pm a, \, y = \pm b\}.$$

We seek a trial solution in the form

$$u_N(x, y) = \sum_{m,n=1 \; 3,5,\dots}^{N} \sum^{N} a_{mn} \, \phi_{mn}(x, y),$$

where

$$\phi_{mn}(x, y) = \cos\left(\frac{m\pi x}{2a}\right) \cos\left(\frac{n\pi y}{2b}\right).$$

In this case $A = \nabla^2$ and the residual R_N is

$$R_N = Au_N + 1 = \nabla^2 u_N + 1$$

$$= -\left[\sum_{m=1}^{N}\sum_{n=1}^{N}\left(\frac{m^2\pi^2}{4a^2} + \frac{n^2\pi^2}{4b^2}\right) a_{mn}\,\phi_{mn}\right] + 1.$$

According to the Galerkin method

$$0 = \langle R_N, \phi_{kl}\rangle = \int_{-a}^{a}\int_{-b}^{b} R_N \cos\left(\frac{k\pi x}{2a}\right)\cos\left(\frac{l\pi y}{2b}\right) dx\, dy$$

$$= \frac{ab\pi^2}{4}\left(\frac{k^2}{a^2} + \frac{l^2}{b^2}\right) a_{kl} - \frac{16ab}{\pi^2 kl}(-1)^{\{(k+l)/2\}-1},$$

or

$$a_{kl} = \left(\frac{8ab}{\pi^2}\right)^2 \frac{(-1)^{\frac{1}{2}(k+l)-1}}{(b^2k^2 + a^2l^2)}.$$

Thus, the solution of the problem is

$$u_N(x, y) = \left(\frac{8ab}{\pi^2}\right)^2 \sum_{m,n=1\; 3,5,\dots}^{N}\sum^{N} \frac{(-1)^{\frac{1}{2}(m+n)-1}\,\phi_{mn}(x, y)}{(b^2m^2 + a^2n^2)}.$$

In particular, the solution for $u_N(x, y)$ can be derived for the square domain $D = \{(x, y) : |x| < 1, |y| < 1\}$. In the limit $N \to \infty$, these solutions are in perfect agreement with those obtained by the double Fourier series.

Example 14.8.2. Solve the problem in Example 14.8.1 by using algebraic polynomials as trial functions.

We seek an appropriate solution in the form

$$u_N(x, y) = \left(x^2 - a^2\right)\left(y^2 - b^2\right)\left(a_1 + a_2 x^2 + a_3 y^2 + a_4 x^4 y^4 + \dots\right).$$

Obviously, this satisfies the boundary conditions. In the first approximation, the solution assumes the form

$$u_1(x, y) \equiv a_1\phi_1 = a_1\left(x^2 - a^2\right)\left(y^2 - b^2\right),$$

where the coefficient a_1 is determined by the Galerkin integral

$$\int_{-a}^{a} \int_{-b}^{b} \left(\nabla^2 u_1 + 1\right) \phi_1 dx \, dy = 0,$$

or

$$\int_{-a}^{a} \int_{-b}^{b} \left[2a_1 \left(y^2 - b^2\right) + 2a_1 \left(x^2 - a^2\right) + 1\right] \left(x^2 - a^2\right) \left(y^2 - b^2\right) dx \, dy = 0.$$

A simple evaluation gives

$$a_1 = \frac{5}{4} \left(a^2 + b^2\right)^{-1},$$

and hence, the solution is

$$u_1 (x, y) = \frac{5}{4} \left(a^2 + b^2\right)^{-1} \left(x^2 - a^2\right) \left(y^2 - b^2\right).$$

14.9 The Kantorovich Method

In 1932, Kantorovich gave an interesting generalization of the Rayleigh–
Ritz method which leads from the solution of a partial differential equation
to the solution of a system of algebraic equations in terms of unknown
coefficients. The essence of the Kantorovich method is to reduce the problem
of the solution of partial differential equations to the solution of ordinary
differential equations in terms of undetermined functions.

Consider the boundary-value problem governed by (14.7.1)–(14.7.2). It
has been shown in Section 14.7 that the solution of the problem is equivalent
to finding the minimum of the quadratic functional $I(u)$ given by (14.7.3).

When the Rayleigh–Ritz method is applied to this problem, we seek
an approximate solution in the form (14.7.6) where the coefficients a_k are
constants. We then determine a_k so as to minimize $I(u_n)$.

In the Kantorovich method, we assume that a_k in (14.7.6) are no longer
constants but unknown functions of one of the independent variables x of
$\mathbf{x}$ so that the Kantorovich solution has the form

$$u_n (\mathbf{x}) = \sum_{k=1}^{n} a_k (x) \phi_k (\mathbf{x}), \tag{14.9.1}$$

where the products $a_k (x) \phi_k (\mathbf{x})$ satisfy the same boundary conditions as
u. Thus, the problem leads to minimizing the functional

$$I (u_n (\mathbf{x})) = I \left(\sum_{k=1}^{n} a_k (x) \phi_k (\mathbf{x}) \right). \tag{14.9.2}$$

Since $\phi_k(\mathbf{x})$ are known functions, we can perform integration with respect to all independent variables except x and obtain a functional $\bar{I}(a_1(x), a_2(x), \ldots, a_n(x))$ depending on n unknown functions $a_k(x)$ of one independent variable x. These functions must be so determined that they minimize the functional $\bar{I}(a_1, a_2, \ldots, a_n)$. Finally, under certain conditions, the solution $u_n(\mathbf{x})$ converges to the exact solution $u(\mathbf{x})$ as $n \to \infty$.

In order to describe the method more precisely, we consider the following example in two dimensions:

$$\nabla^2 u = f(x,y) \quad \text{in} \quad D, \qquad (14.9.3)$$
$$u(x,y) = 0 \qquad \text{on} \quad \partial D, \qquad (14.9.4)$$

where D is a closed domain bounded by the curves $y = \alpha(x)$, $y = \beta(x)$ and two vertical lines $x = a$ and $x = b$.

The solution of the problem is equivalent to finding the minimum of the functional

$$I(u) = \iint_D \left(u_x^2 + u_y^2 + 2fu \right) dx\, dy. \qquad (14.9.5)$$

We seek the solution in the form

$$u_n(x,y) = \sum_{k=1}^{n} a_k(x)\, \phi_k(x,y) \qquad (14.9.6)$$

which satisfies the given boundary condition, where $\phi_k(x,y)$ are known trial functions, and $a_k(x)$ are unknown functions to be determined so that they minimize $I(u_n)$. Substitution of u_n in the functional $I(u)$ gives

$$I(u_n) = \iint_D \left[\left(\frac{\partial u_n}{\partial x} \right)^2 + \left(\frac{\partial u_n}{\partial y} \right)^2 + 2f u_n \right] dx\, dy$$

$$= \int_a^b dx \int_{\alpha(x)}^{\beta(x)} \left\{ \left[\sum_{k=1}^n \left(\frac{\partial \phi_k}{\partial x} a_k - \phi_k a_k' \right) \right]^2 \right.$$

$$\left. + \left[\sum_{k=1}^n a_k \frac{\partial \phi_k}{\partial y} \right]^2 + 2f \sum_{k=1}^n a_k \phi_k \right\} dy \qquad (14.9.7)$$

$$= \int_a^b F(x, a_k, a_k')\, dx, \qquad (14.9.8)$$

where the integrand in (14.9.7) is a known function of y and the integration with respect to y is assumed to have been performed so that the result can be denoted by $F(x, a_k, a_k')$. Thus, the problem is reduced to determining the functions a_k so that they minimize $I(u_n)$. Hence, $a_k(x)$ can be found by solving the following system of linear Euler equations:

$$\frac{\partial F}{\partial a_k} - \frac{d}{dx}\left(\frac{\partial F}{\partial a'_k}\right) = 0, \qquad k = 1, 2, 3, \ldots, n. \qquad (14.9.9)$$

This system of ordinary differential equations for the functions a_k is to be solved with the boundary conditions $a_k(a) = a_k(b) = 0$, $k = 1, 2, \ldots, n$. Consequently, the required solution $u_n(x, y)$ is determined.

Example 14.9.1. Find a solution of the torsion problem governed by the Poisson equation $\nabla^2 u = -2$ in the rectangle $D = \{(x, y) : -a < x < a,$ $-b < y < b\}$ with the boundary condition $u = 0$ on $\partial D = \{x = \pm a, y = \pm b\}$.
   In the first approximation, we seek a solution in the form

$$u_1(x, y) = \left(y^2 - b^2\right) a_1(x)$$

which satisfies the boundary condition on $y = \pm b$. We next determine $a_1(x)$ so that $a_1(x) = a_1(-a) = 0$.
   The functional associated with the problem is

$$I(u) = \iint_D \left(u_x^2 + u_y^2 - 4u\right) dx\, dy.$$

Substituting u_1 into this functional yields

$$I(u_1) = \int_{-a}^{a} dx \int_{-b}^{b} \left[\left(y^2 - a^2\right)^2 a_1'^2 + 4y^2 a_1^2 - 4\left(y^2 - b^2\right) a_1\right] dy$$

$$= \int_{-a}^{a} \left[\frac{16}{15} b^5 a_1'^2 + \frac{8}{3} b^3 a_1^2 + \frac{16}{3} b^3 a_1\right] dx.$$

The Euler equation for the functional is

$$a_1'' - \frac{5}{2b^2} a_1 - \frac{5}{2b^2} = 0.$$

This is a linear ordinary differential equation for a_1 with constant coefficients, and has the general solution

$$a_1(x) = A \cosh kx + B \sinh kx - 1, \qquad k = \frac{1}{b}\sqrt{\frac{5}{2}},$$

where the constants A and B are determined by the boundary conditions $a_1(a) = a_1(-a) = 0$ so that $B = 0$ and

$$A = \frac{1}{\cosh ka}.$$

Thus, an approximate solution is

$$u_1(x, y) = \left(y^2 - b^2\right)\left(\frac{\cosh kx}{\cosh ka} - 1\right).$$

Finally, the torsional moment is given by

$$M = 2\mu\alpha \iint_D u\, dx\, dy \sim 2\mu\alpha \iint_D u_1 dx\, dy$$

$$= 2\mu\alpha \int_{-a}^{a} \left(\frac{\cosh kx}{\cosh ka} - 1 \right) dx \int_{-b}^{b} \left(y^2 - b^2 \right) dy$$

$$= \frac{16}{3}\mu\alpha b^3 a \left[1 - \frac{1}{ka} \tanh\left(ak \right) \right].$$

Example 14.9.2. Solve the Poisson equation $\nabla^2 u = -1$ in a triangular domain D bounded by $x = a$ and $y = \pm \left(x/\sqrt{3} \right)$ with $u = 0$ on ∂D.

The associated functional for the Poisson equation is

$$I\left(u \right) = \iint_D \left(u_x^2 + u_y^2 - 2u \right) dx\, dy.$$

We seek the Kantorovich solution in the first approximation

$$u\left(x, y \right) \sim \left(y^2 - \frac{x^2}{3} \right) u_1\left(x \right)$$

so that $u_1\left(a \right) = 0$.

Substituting the solution into $I\left(u \right)$ gives

$$I\left(u_1 \right) = \int_0^a dx \int_{-x/\sqrt{3}}^{x/\sqrt{3}} \left[\left\{ \left(y^2 - \frac{x^2}{3} \right) u_1' - \frac{2xu_1}{3} \right\}^2 + \right.$$

$$\left. + 4y^2 u_1^2\left(x \right) - 2\left(y^2 - \frac{x^2}{3} \right) u_1 \right] dy$$

$$= \frac{8}{135\sqrt{3}} \int_0^a \left(2x^5 u_1'^2 + 10x^4 u_1 u_1' + 30x^3 u_1^2 + 15x^3 u_1 \right) dx.$$

The Euler equation for this functional is

$$4\left(x^2 u_1'' + 5xu_1' - 5u_1 \right) = 15.$$

This is a nonhomogeneous ordinary differential equation of order two. We seek a solution of the corresponding homogeneous equation in the form x^r where r is determined by the equation $(r - 1)(r + 5) = 0$. The particular integral of the equation is $u_1 = -\frac{3}{4}$. Hence, the general solution is

$$u_1\left(x \right) = Ax + Bx^{-5} - \frac{3}{4},$$

where the constants A and B are to be determined by the boundary conditions. For the bounded solution, $B \equiv 0$. The condition $u_1\left(a \right) = 0$ implies $A = (3/4a)$. Therefore, the final solution is

$$u\left(x, y \right) = \frac{3}{4} \left(\frac{x}{a} - 1 \right) \left(y^2 - \frac{x^2}{3} \right).$$

14.10 The Finite Element Method

Many problems in mathematics, science and engineering are not simple and cannot be solved by exact closed-form analytical formulas. It is often necessary to obtain approximate numerical or asymptotic solutions rather than exact solutions. Many numerical methods that have evolved over the years reduce algebraic or differential equations to discrete form which can be solved easily by computer. However, if the numerical method is *not* carefully chosen, the numerically computed solution may *not* be anywhere close to the true solution. Another problem is that the computation for a difficult problem may take so long that it is impractical for a computer to carry out. The most commonly used numerical methods are finite differences that give *pointwise approximations* of the governing equations. These methods can be used successfully to solve many fairly difficult problems, but their major weakness is that they are not suitable for problems with irregular geometries, curved boundaries or unusual boundary conditions. For example, the finite difference methods are not particularly effective for a circular domain because a circle cannot be accurately partitioned into rectangles. However, there are other numerical methods including the finite element method and the boundary element method.

Unlike finite difference methods, the finite element method can be used effectively to determine fairly accurate approximate solutions to a wide variety of governing equations defined over irregular regions. The entire solution domain can be modeled analytically or approximated by replacing it with small, interconnected discrete finite elements (hence the name *finite element*). The solution is then approximated by extremely simple functions (linear functions) on these small elements such as triangles. These small elements are collected together and requirements of continuity and equilibrium are satisfied between neighboring elements.

In a nutshell, the basic idea of the finite element method (FEM) consists of decomposing a given domain into a set of finite elements of arbitrary shape and size. This decomposition is usually called a *mesh* or a *grid* with the restriction that elements cannot overlap nor leave any part of the domain uncovered. For each element, a certain number of points is introduced that can be located on the edges of the elements or inside. These points are called *nodes* that are usually vertices of triangles as shown in Figure 14.10.1. Finally, these nodes are used to approximate a function under consideration over the whole domain by *interpolation* in the finite elements.

Historically, the finite element method was developed originally to study stress fields in complicated aircraft structures in the early 1960s. Subsequently, it has been extended and widely applied to find approximate solutions to a wide variety of problems in mathematics, science, and engineering. It was Richard Courant (1888–1972) who first introduced piecewise continuous functions defined over triangular domains in 1943; he then used these triangular elements combined with the principle of minimum potential energy to study the St. Venant torsion problem in continuum mechanics. He also described element properties and finite element equations based on a variational principle. In 1965, the finite element method received an even broader interpretation when Zienkiewicz and Cheung (1965) suggested that it is applicable to all field problems that can be cast in variational form. During the late 1960s and early 1970s, considerable attention has been given to errors, bounds and convergence criteria for finite element approximations to solutions of various problems in continuum mechanics.

In order to develop the finite element method, we recall the celebrated Euler–Lagrange equation (14.6.12) with $u(a) = \alpha$ and $u(b) = \beta$. We divide the interval $a \leq x \leq b$ into n parts by the R_{n+1} set: $a = x_0 < x_1 < x_2 < \ldots < x_n = b$. Each such subinterval is called an *element*. In general, the length of the elements need not be equal, though for simplicity, we assume that they are equal in length so that $h = \frac{1}{n}(b - a)$. We set $u_k = u(x_k)$, $k = 0, 1, 2, \ldots, n$ so that $u_0(x_0) = \alpha$ and $u_n(x_n) = \beta$, while $u_1, u_2, \ldots, u_{n-1}$ are unknown quantities. We next rewrite the functional (14.6.3) as

$$I(u) = \int_{x_0}^{x_1} F(x, u, u')\, dx + \int_{x_1}^{x_2} F(x, u, u')\, dx + \ldots + \int_{x_{n-1}}^{x_n} F(x, u, u')\, dx.$$

$$(14.10.1)$$

We define a piecewise linear interpolating function $L(x)$ of u_i as the function which is continuous on $[a, b]$ and whose graph consists of straight line segments joining the consecutive pairs of points (x_k, u_k), (x_{k+1}, u_{k+1}) for $k = 0, 1, 2, \ldots, (n-1)$, that is,

$$L(x) = u_k + \frac{1}{h}(u_{k+1} - u_k)(x - x_k), \qquad x_k \leq x \leq x_{k+1}, \quad (14.10.2)$$

where $k = 0, 1, 2, \ldots, (n-1)$.

Substituting L for u and L' for u' in (14.10.1) and assuming that the integrals can be computed exactly yields

$$I_{n-1} = I_{n-1}(u_1, u_2, \ldots, u_{n-1}).$$

$$(14.10.3)$$

We next find the minimum of I_{n-1} by solving the system of equations

$$\frac{\partial I_{n-1}}{\partial u_k} = 0, \qquad k = 1, 2, \ldots, (n-1).$$

$$(14.10.4)$$

The solution of this system (14.10.4) is then substituted into (14.10.2) to obtain a continuous, piecewise linear approximation for the exact solution $u(x)$.

Example 14.10.1. (The Dirichlet problem for the Poisson equation in a plane).

We consider the problem

$$- \triangle u = f(x, y) \quad \text{in } D, \tag{14.10.5}$$

$$u = 0 \qquad \text{on the boundary } \partial D. \tag{14.10.6}$$

The region D is first triangulated so that it is approximated by a region D_n which is the union of a finite number of triangles as shown in Figures 14.10.1 (a) and 14.10.1 (b). We denote the interior vertices by V_1, V_2, ..., V_n.

We next choose n trial functions $v_1(x, y)$, $v_2(x, y)$, ..., $v_n(x, y)$, one for each interior vertex. Each trial function $v_m(x, y)$ is assumed to be equal to 1 at its vertices V_m and equal to zero at all other vertices as in shown in Figure 14.10.1 (c).

Each linear trial function $v_m(x, y) = ax + by + c$, where a, b, and c are different for each trial function and for each triangle. This requirement determines $v_m(x, y)$ uniquely. Indeed, its graph is simply a pyramid of unit height with its peak at V_m and it is zero on all the triangles which do not touch V_m.

We next approximate the solution $u(x, y)$ by a linear combination of the $v_m(x, y)$ so that

$$u_n(x, y) = a_1 v_1(x, y) + a_2 v_2(x, y) + \ldots + a_n v_n(x, y) = \sum_{m=1}^{n} a_m v_m(x, y),$$

$$\tag{14.10.7}$$

where the coefficients a_1, a_2, ..., a_n are to be determined.

We multiply the Poisson equation (14.10.5) by any function $v(x, y)$ which is zero on ∂D and next use Green's first identity to obtain

$$\iint_D \nabla u \cdot \nabla v \, dx \, dy = \iint_D f v \, dx \, dy. \tag{14.10.8}$$

We assume that (14.10.8) is valid only for the first n trial functions so that $v = v_k$ for $k = 1, 2, \ldots, n$. With $u(x, y) = u_n(x, y)$ and $v(x, y) = v_k(x, y)$, result (14.10.8) becomes

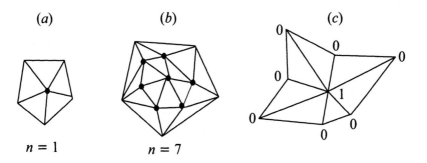

Figure 14.10.1 (a), (b), and (c). Triangular elements.

$$\sum_{m=1}^{n} a_m \left[\iint_D (\nabla u_m \cdot \nabla v_k) \, dx \, dy \right] = \iint_D f \, v_k \, dx \, dy. \quad (14.10.9)$$

This is a system of n linear equations, where $m = 1, 2, \ldots, n$ in the n unknown coefficients a_m, and can be rewritten in the form

$$\sum_{m=1}^{n} \alpha_{mk} \, a_m = f_k, \qquad k = 1, 2, \ldots, n, \qquad (14.10.10)$$

where

$$\alpha_{mk} = \iint_D (\nabla u_m \cdot \nabla v_k) \, dx \, dy, \qquad f_k = \iint_D f \, v_k \, dx \, dy. \quad (14.10.11)$$

Consequently, the finite element method leads to finding α_{mk} and f_k from (14.10.11) and then solving (14.10.10). Finally, the approximate value of the solution $u(x, y)$ is then given by (14.10.7).

Several comments are in order. First, the trial functions $v_m(x, y)$ depend on the geometry of the problem and are completely known. Second, the approximate solution $u_n(x, y)$ vanishes on the boundary ∂D_n. Third, at a vertex $V_i = (x_i, y_i)$,

$$u_n(x_i, y_i) = a_i v_i(x_i, y_i) + \ldots + a_n v_n(x_i, y_i) = \sum_{r=1}^{n} a_r v_r(x_i, y_i) = a_i$$

where

$$v_r(x_k, y_k) = \begin{cases} 0, & r \neq k \\ \\ 1, & r = k. \end{cases}$$

Fourth, the coefficients a_i are exactly the values of the approximate solution at the vertices $V_i = (x_i, y_i)$.

Example 14.10.2. We consider the variational problem of finding the extremes of the functional

$$I(u) = \int_0^6 \left(u'^2 + u^2 - 2u - 2xu \right) dx \qquad (14.10.12)$$

with the boundary conditions $u(0) = 1$ and $u(6) = 7$.

We divide $0 \leq x \leq 6$ into three equal parts of length $h = 2$ by $x_0 = 0$, $x_1 = 2$, $x_2 = 4$ and $x_3 = 6$. We set $u_k = u(x_k)$ so that $u_0 = u(0) = 1$ and $u_3 = u(6) = 7$, while u_1 and u_2 are unknown quantities. We have

$$F(x, u, u') = u'^2 + u^2 - 2u - 2xu \qquad (14.10.13)$$

so that

$$I = \int_0^2 F(x, u, u') \, dx + \int_2^4 F(x, u, u') \, dx + \int_4^6 F(x, u, u') \, dx. \quad (14.10.14)$$

We take

$$L(x) = \begin{cases} u_0 + \frac{1}{2}(u_1 - u_0)x, & 0 \le x \le 2 \\ u_1 + \frac{1}{2}(u_2 - u_1)(x - 2), & 2 \le x \le 4 \\ u_2 + \frac{1}{2}(u_3 - u_1)(x - 4), & 4 \le x \le 6, \end{cases} \quad (14.10.15)$$

so that its derivative is

$$L'(x) = \begin{cases} \frac{1}{2}(u_1 - u_0), & 0 \le x \le 2 \\ \frac{1}{2}(u_2 - u_1), & 2 < x \le 4 \\ \frac{1}{2}(u_3 - u_1), & 4 < x \le 6. \end{cases} \quad (14.10.16)$$

Substituting (14.10.15)–(14.10.16) into (14.10.14) and using (14.10.13) we get

$$I = \int_0^2 \left[\left(\frac{u_1 - u_0}{2} \right)^2 + \left\{ u_0 + \frac{1}{2}(u_1 - u_0)x \right\}^2 - 2 \left\{ u_0 + \frac{1}{2}(u_1 - u_0)x \right\} \right.$$

$$\left. -2x \left\{ u_0 + \frac{1}{2}(u_1 - u_0)x \right\} \right] dx$$

$$+ \int_2^4 \left[\left(\frac{u_2 - u_1}{2} \right)^2 + \left\{ u_1 + \frac{1}{2}(u_2 - u_1)(x - 2) \right\}^2 \right.$$

$$\left. -2 \left\{ u_1 + \frac{1}{2}(u_2 - u_1)(x - 2) \right\} - 2x \left\{ u_1 + \frac{1}{2}(u_2 - u_1)(x - 2) \right\} \right] dx$$

$$+ \int_4^6 \left[\left(\frac{u_3 - u_1}{2} \right)^2 + \left\{ u_2 + \frac{1}{2}(u_3 - u_1)(x - 4) \right\}^2 \right.$$

$$\left. -2 \left\{ u_2 + \frac{1}{2}(u_3 - u_1)(x - 4) \right\} - 2x \left\{ u_2 + \frac{1}{2}(u_3 - u_1)(x - 4) \right\} \right] dx.$$

Using the known values of u_0 and u_3 and integrating we obtain ($n = 3$)

$$I_2 = \frac{7}{3}(u_1^2 + u_2^2) - \frac{1}{3}u_1 u_2 - \frac{37}{3}u_1 - \frac{67}{3}u_2 - \frac{101}{3}.$$

Consequently, equation (14.10.4) gives two equations

$$\frac{\partial I_2}{\partial u_1} = \frac{14}{3}u_1 - \frac{1}{3}u_2 - \frac{37}{3} = 0,$$

$$\frac{\partial I_2}{\partial u_2} = -\frac{1}{3}u_1 + \frac{14}{3}u_2 - \frac{67}{3} = 0.$$

Thus, the solutions for u_1 and u_2 are $u_1 = 3$ and $u_2 = 5$.

Putting these values into (14.10.15) leads to the approximate solution

$$L(x) = 1 + x, \qquad 0 \le x \le 6. \tag{14.10.17}$$

In this problem, the Euler–Lagrange equation for (14.10.12) is given by

$$u'' - u = -(1 + x). \tag{14.10.18}$$

Solving this equation with $u(0) = 1$ and $u(6) = 7$ yields the exact solution

$$u(x) = 1 + x.$$

In this example, the exact and approximate solutions are identical due to the simplicity of the problem. In general, these solutions will be different.

We close this section by adding some comments on another numerical technique known as the *boundary element method* (*boundary integral equation method*). This method was widely used in early research in solid mechanics, fluid mechanics, potential theory and electromagnetic theory. However, the major breakthrough in the boundary integral equation method came in 1963 when two classic papers were published by Jaswon (1963) and Symm (1963). The boundary element method is based on the mathematical aspect of finding the Green's function solution of differential equations with prescribed boundary conditions. It also uses Green's theorem to reduce a volume problem to a surface problem, and a surface problem to a line problem. This technique is not only very useful but also very accurate for linear problems, especially for three dimensional problems with rapidly changing variables in fracture and contact problems in solid mechanics. However, this method is computationally less efficient than the finite element method, and is not widely used in industry. It is fairly popular for finding numerical solutions of acoustic problems. Since the early 1970s the boundary element method has continued to develop at a fast pace and has been extended to include a wide variety of linear and nonlinear problems in continuum mechanics.

14.11 Exercises

1. Obtain the explicit finite difference solution of the problem

$$u_{tt} - 4u_{xx} = 0, \qquad 0 < x < 1, \qquad t > 0,$$
$$u(0, t) = u(1, t) = 0, \qquad t \ge 0,$$
$$u(x, 0) = \sin 2\pi x, \qquad u_t(x, 0) = 0, \qquad 0 \le x \le 1.$$

Compare the numerical solution with the analytical solution

$$u(x, t) = \cos 4\pi t \sin 2\pi x$$

at several points.

2. (a) Calculate an explicit finite difference solution of the wave equation

$$u_{xx} - u_{tt} = 0, \qquad 0 < x < 1, \qquad t > 0,$$

satisfying the boundary conditions

$$u(0,t) = u(1,t) = 0, \qquad t \geq 0,$$

and the initial conditions

$$u(x,0) = \frac{1}{8} \sin \pi x, \qquad u_t(x,0) = 0, \qquad 0 \leq x \leq 1.$$

Show that the exact solution of the problem is

$$u(x,t) = \frac{1}{8} \cos \pi t \sin \pi x.$$

Compare the two solutions at several points.

(b) Solve the wave equation in (a) with the same boundary data and the initial data

$$u(x,0) = \sin \pi x, \qquad u_t(x,0) = 0, \qquad 0 \leq x \leq 1.$$

3. Use the Lax–Wendroff method to find a numerical solution of the problem

$$\begin{aligned}
u_x + u_t &= 0, & x > 0, \qquad t > 0, \\
u(x,0) &= 2 + x, & x > 0, \\
u(0,t) &= 2 - t, & t > 0.
\end{aligned}$$

Show that the exact solution of the problem is

$$u(x,t) = 2 + (x - t).$$

Compare the two solutions at various points.

4. Show that the finite difference approximation to the equation

$$au_t + bu_x = f(x,t)$$

is

$$u_{i,j+1} - \frac{1}{2}(u_{i+1,j} + u_{i-1,j}) + \left(\frac{\varepsilon b}{2a}\right)(u_{i+1,j} - u_{i-1,j}) - f_{i,j} = 0,$$

where a, b are constants and $\varepsilon = k/h$.

5. Obtain a finite difference solution of the heat conduction problem

$$u_t = \kappa\, u_{xx}, \qquad 0 < x < l, \quad t > 0,$$

with the boundary conditions

$$u(0,t) = u(l,t) = 0, \quad t > 0,$$

and the initial condition

$$u(x,0) = \frac{4x}{l}(l-x), \qquad 0 \le x \le l.$$

6. (a) Find an explicit finite difference solution of the parabolic system

$$u_t = u_{xx}, \qquad 0 < x < 1, \quad t > 0,$$
$$u(0,t) = u(1,t) = 0, \qquad\qquad t > 0,$$
$$u(0,t) = \sin x\pi \quad \text{on} \quad 0 \le x \le 1.$$

Compare the numerical results with the analytical solution

$$u(x,t) = e^{-\pi^2 t} \sin \pi x,$$

at $t = 0.5$ and $t = 0.05$.

(b) Prove that the Richardson finite difference scheme for problem 6(a) is

$$u_{i,j+1} = u_{i,j-1} + 2\varepsilon\, \delta_x^2\, u_{i,j}.$$

Hence, show that the exact solution of this equation is

$$u_{i,j} = \left(A_1 \alpha_1^j + A_2 \alpha_2^j\right) \sin \pi h i,$$

where α_1 and α_2 are the roots of the quadratic equation

$$x^2 + 8\varepsilon x \sin^2(\pi h/2) - 1 = 0.$$

7. Using four internal grid points, find the explicit finite difference solution of the Dirichlet problem

$$\nabla^2 u \equiv u_{xx} + u_{yy} = 0, \qquad 0 < x < 1, \qquad 0 < y < 1,$$
$$u(x,0) = x(1-x), \qquad\qquad u(x,1) = 0 \quad \text{on} \quad 0 \le x \le 1,$$
$$u(0,y) = u(1,y) = 0, \quad \text{on} \quad 0 \le y \le 1.$$

Compare the numerical solution with the exact analytical solution

$$u(x,y) = \frac{4}{\pi^3} \sum_{n=0}^{\infty} \frac{2}{(2n+1)^3} \frac{\sin n\pi x \, \sinh n\pi (1-y)}{\sinh n\pi}$$

at the point $(x,y) = \left(\frac{1}{3}, \frac{1}{3}\right)$.

8. Solve the Dirichlet problem by the explicit finite difference method

$$u_{xx} + u_{yy} = 0, \qquad 0 < x < 1, \qquad\qquad 0 < y < 1,$$
$$u(x,0) = \sin \pi x, \quad u(x,1) = 0 \qquad \text{on} \quad 0 \le x \le 1,$$
$$u(x,y) = 0, \quad \text{for} \quad x = 0, \quad x = 1 \quad \text{and} \quad 0 \le y \le 1.$$

9. Using a square grid system with $h = \frac{1}{2}$, find the finite difference solution of the Laplace equation on the quarter-disk given by

$$u_{xx} + u_{yy} = 0, \qquad x^2 + y^2 < 1, \quad y > 0,$$
$$u(x,0) = 0, \qquad\qquad -1 < x < 1,$$
$$u(x,y) = 10^2, \qquad\quad x^2 + y^2 = 1, \quad y > 0.$$

10. Find a finite difference solution of the wave problem

$$u_{tt} - u_{xx} = 0, \qquad\qquad 0 < x < 1, \qquad t > 0,$$
$$u(0,t) = u(1,t) = 0, \qquad\qquad\qquad\qquad t \ge 0,$$
$$u(x,0) = \frac{1}{2} x(1-x), \quad u_t(x,0) = 0, \quad 0 \le x \le 1.$$

Compare the numerical results with the exact analytical solution

$$u(x,t) = \frac{2}{\pi^3} \sum_{r=1}^{\infty} \frac{1}{r^3} \{1 - (-1)^r\} \cos \pi rt \sin \pi rx,$$

at various points.

11. Obtain a finite difference solution of the problem

$$u_{tt} = c^2 u_{xx}, \qquad\qquad 0 < x < 1, \quad t > 0,$$
$$u(0,t) = \sin \pi ct, \qquad u(1,t) = 0, \qquad t \ge 0,$$
$$u(x,0) = u_t(x,0) = 0, \qquad 0 \le x \le 1.$$

12. Show that the transformation $v = \log u$ transforms the nonlinear system

$$v_t = v_{xx} + v_x^2, \quad 0 < x < 1, \qquad t > 0,$$
$$v_x(0,t) = 1, \qquad v(1,t) = 0, \quad t \ge 0,$$
$$v(x,0) = 0, \qquad 0 \le x \le 1$$

into the linear system

$$u_t = u_{xx}, \qquad\qquad 0 < x < 1, \quad t > 0,$$
$$u_x(0,t) = u(0,t), \quad u(x,1) = 1, \quad t \ge 0,$$
$$u(x,0) = 1, \qquad\qquad 0 \le x \le 1.$$

Solve the linear system by the explicit finite difference method with the derivative boundary condition approximated by the central difference formula.

13. Solve the following parabolic system by the Crank–Nicolson method

$$u_t = u_{xx}, \qquad 0 < x < 1, \quad t > 0,$$
$$u(0,t) = u(1,t) = 0, \qquad\qquad t \geq 0,$$

with the initial condition

(a) $u(x,0) = 1, \qquad\qquad 0 \leq x \leq 1,$

(b) $u(x,0) = \sin \pi x, \qquad 0 \leq x \leq 1.$

(c) $u(x,0) = \sin \pi x, \qquad 0 \leq x \leq 1$
with $0 \leq t \leq 0.2$ and in formula (14.5.2) $\kappa = 1$, $k = h^2$.

14. Use the Crank–Nicolson implicit method with the central difference formula for the boundary conditions to find a numerical solution of the differential system

$$u_t = u_{xx}, \qquad 0 < x < 1, \quad t > 0,$$
$$u_x(0,t) = u_x(1,t) = -u, \qquad t \geq 0,$$
$$u(x,0) = 1, \qquad\qquad 0 \leq x \leq 1.$$

15. Find a numerical solution of the wave equation

$$u_{tt} = c^2 u_{xx}, \qquad 0 < x < l, \quad t > 0,$$

with the boundary and initial conditions

$$u = \frac{1}{20} u_x \quad \text{at} \quad x = 0 \quad \text{and} \quad x = l, \quad t > 0,$$
$$u(x,0) = 0, \quad u_t(x,0) = a \sin\left(\frac{\pi x}{l}\right) \qquad 0 \leq x \leq l.$$

16. Determine the function representing a curve which makes the following functional extremum:

(a) $I(y(x)) = \displaystyle\int_0^1 (y'^2 + 12xy)\,dx, \qquad y(0) = 0, \qquad y(1) = 1,$

(b) $I(y(x)) = \displaystyle\int_0^{\pi/2} (y'^2 - y^2)\,dx, \qquad y(0) = 0, \qquad y\left(\frac{\pi}{2}\right) = 1,$

(c) $I(y(x)) = \displaystyle\int_{x_0}^{x_1} \frac{1}{x}\left(1 + y'^2\right)^{\frac{1}{2}}\,dx.$

17. In the problem of tautochroneous motion, find the equation of the curve joining the origin O and a point A in the vertical (x,y)-plane so that a particle sliding freely from A to O under the action of gravity reaches the origin O in the shortest time, friction and resistance of the medium being neglected.

18. In the problem of minimum surface of revolution, determine a curve with given boundary points (x_0, y_0) and (x_1, y_1) such that rotation of the curve about the x-axis generates a surface of revolution of minimum area.

19. Show that the Euler equation of the variational principle

$$\delta I \left[u\left(x,y\right) \right] = \delta \iint_D F\left(x,y,u,p,q,l,m,n\right) dx \, dy = 0$$

is

$$F_u - \frac{\partial}{\partial x} F_p - \frac{\partial}{\partial y} F_q + \frac{\partial^2}{\partial x^2} F_l + \frac{\partial^2}{\partial x \partial y} F_m + \frac{\partial^2}{\partial y^2} F_n = 0,$$

where

$$p = u_x, \quad q = u_y, \quad l = u_{xx}, \quad m = u_{xy}, \quad n = u_{yy}.$$

20. Prove that the Euler–Lagrange equation for the functional

$$I = \iiint_R F\left(x,y,z,u,p,q,r,l,m,n,a,b,c\right) dx \, dy \, dz$$

is

$$F_u - \frac{\partial}{\partial x} F_p - \frac{\partial}{\partial y} F_q + \frac{\partial^2}{\partial z} F_r + \frac{\partial^2}{\partial x^2} F_l + \frac{\partial^2}{\partial y^2} F_m$$

$$+ \frac{\partial^2}{\partial z^2} F_n + \frac{\partial^2}{\partial x \partial y} F_a + \frac{\partial^2}{\partial y \partial z} F_b + \frac{\partial^2}{\partial z \partial x} F_c = 0,$$

where $(p,q,r) = (u_x, u_y, u_z)$, $(l,m,n) = (u_{xx}, u_{yy}, u_{zz})$, and $(a,b,c) = (u_{xy}, u_{yz}, u_{zx})$.

21. In each of the following cases apply the variational principle or its simple extension with appropriate boundary conditions to derive the corresponding equations:

(a) $F = u_x^2 + u_y^2 + 2u_{xy}^2$.

(b) $F = \frac{1}{2} \left[u_t^2 - \alpha \left(u_x^2 + u_y^2 \right) - \beta^2 u^2 \right]$,

(c) $F = \frac{1}{2} \left(u_t u_x + \alpha u_x^2 - \beta u_{xx}^2 \right)$,

(d) $F = \frac{1}{2} \left(u_t^2 - \alpha^2 u_{xx}^2 \right)$,

(e) $F = p\left(x\right) u'^2 + \frac{d}{dx} \left(q\left(x\right) u^2 \right) - \left[r\left(x\right) + \lambda s\left(x\right) \right] u^2$,

where p, q, r, and s are given functions of x, and α, β are constants.

22. Derive the Schrödinger equation from the variational principle

$$\delta \iiint_R \left[\frac{\hbar^2}{2m} \left(\psi_x^2 + \psi_y^2 + \psi_z^2 \right) + (V - E) \psi^2 \right] dx \, dy \, dz = 0,$$

where $h = 2\pi\hbar$ is the Planck constant, m is the mass of a particle moving under the action of a force field described by the potential $V(x, y, z)$ and E is the total energy of the particle.

23. Derive the Poisson equation $\nabla^2 u = F(x, y)$ from the variational principle with the functional

$$I(u) = \iint_D \left[u_x^2 + u_y^2 + 2uF(x, y) \right] dx \, dy,$$

where $u = u(x, y)$ is given on the boundary ∂D of D.

24. Derive the equation of motion of a vibrating string of length l under the action of an external force $F(x, t)$ from the variational principle

$$\delta \int_{t_1}^{t_2} \int_0^l \left[\left(\frac{1}{2} \rho u_t^2 - T^* u_x^2 \right) + \rho u F(x, t) \right] dx \, dt = 0,$$

where ρ is the line density and T^* is the constant tension of the string.

25. The kinetic and potential energies associated with the transverse vibration of a thin elastic plate of constant thickness h are

$$T = \frac{1}{2} \rho \iint_D \dot{u}^2 dx \, dy,$$

$$V = \frac{1}{2} \mu_0 \iint_D \left[(\nabla u)^2 - 2(1 - \sigma) \left(u_{xx} u_{yy} - u_{xy}^2 \right) \right] dx \, dy,$$

where ρ is the surface density and $\mu_0 = 2h^3 E / 3 \left(1 - \sigma^2 \right)$.

Use the variational principle

$$\delta \int_{t_1}^{t_2} \iint_D \left[(T - V) + fu \right] dx \, dy \, dt = 0$$

to derive the equation of motion of the plate

$$\rho \ddot{u} + \mu_0 \nabla^4 u = f(x, y, t),$$

where f is the transverse force per unit area acting on the plate.

26. The kinetic and potential energies associated with the wave motion in elastic solids are

$$T = \frac{1}{2} \iiint_D \rho \left(u_t^2 + v_t^2 + w_t^2 \right) dx \, dy \, dz$$

$$V = \frac{1}{2} \iiint_D \left[\lambda \left(u_x + v_y + w_z \right)^2 + 2\mu \left(u_x^2 + v_y^2 + w_z^2 \right) \right.$$
$$\left. + \mu \left\{ \left(v_x + u_y \right)^2 + \left(w_y + v_z \right)^2 + \left(u_z + w_x \right)^2 \right\} \right] dx \, dy \, dz.$$

Use the variational principle

$$\delta \int_{t_1}^{t_2} \iiint_D (T - V) \, dx \, dy \, dz = 0$$

to derive the equation of wave motion in an elastic medium

$$(\lambda + \mu) \operatorname{grad} \operatorname{div} \mathbf{u} + \mu \, \nabla^2 \mathbf{u} = \rho \, \mathbf{u}_{tt},$$

where $\mathbf{u} = (u, v, w)$ is the displacement vector.

27. From the variational principle

$$\delta \iint_D L \, d\mathbf{x} \, dt = 0 \quad \text{with} \quad L = -\rho \int_{-h}^{\eta} \left\{ \phi_t + \frac{1}{2} \left(\nabla \phi \right)^2 + gz \right\} dz$$

derive the basic equations of water waves

$$\nabla^2 \phi = 0, \qquad -h \, (x, y) < z < \eta \, (x, y, t), \quad t > 0,$$
$$\eta_t + \nabla \phi \cdot \nabla \eta - \phi_z = 0, \quad \text{on} \quad z = \eta,$$
$$\phi_z + \frac{1}{2} \left(\nabla \phi \right)^2 + gz = 0, \quad \text{on} \quad z = \eta,$$
$$\phi_z = 0, \quad \text{on} \quad z = -h,$$

where $\phi \, (x, y, z, t)$ is the velocity potential, and $\eta \, (x, y, t)$ is the free surface displacement function in a fluid of depth h.

28. Derive the Boussinesq equation for water waves

$$u_{tt} - c^2 u_{xx} - \mu \, u_{xxtt} = \frac{1}{2} \left(u^2 \right)_{xx}$$

from the variational principle

$$\delta \iint L \, dx \, dt = 0,$$

where $L \equiv \frac{1}{2} \phi_t^2 - \frac{1}{2} c^2 \phi_x^2 + \frac{1}{2} \mu \, \phi_{xt}^2 - \frac{1}{6} \phi_x^3$ and ϕ is the potential for $u \, (u = \phi_x)$.

29. Determine an approximate solution of the problem of finding an extremum of the functional

$$I \, (y \, (x)) = \int_0^1 \left(y'^2 - y^2 - 2xy \right) dx, \qquad y \, (0) = y \, (1) = 0.$$

30. Find an approximate solution of the torsion problem of a cylinder with an elliptic base; the domain of integration D is the interior of the ellipse with the major and minor axes $2a$ and $2b$ respectively. The associated functional is

$$I\left(u\left(x,y\right)\right) = \iint_D \left[\left(\frac{\partial u}{\partial x} - y\right)^2 u + \left(\frac{\partial u}{\partial y} + x\right)^2 u\right] dx\, dy.$$

31. Use the Rayleigh–Ritz method to find an approximate solution of the problem

$$\nabla^2 u = 0, \qquad 0 < x < 1, \quad 0 < y < 1,$$
$$u\left(0,y\right) = 0 = u\left(1,y\right),$$
$$u\left(x,0\right) = x\left(1-x\right).$$

32. Find an approximate solution of the boundary-value problem

$$\nabla^2 u = 0, \quad x > 0, \quad y > 0, \qquad x + 2y < 2,$$
$$u\left(0,y\right) = 0, \quad u\left(x,0\right) = x\left(2-x\right),$$
$$u\left(2-2y,y\right) = 0.$$

33. In the torsion problem in elasticity, the Prandtl stress function $\Psi\left(x,y\right) = \psi\left(x,y\right) - \frac{1}{2}\left(x^2 + y^2\right)$ satisfies the boundary value problem

$$\nabla^2 \Psi = -2 \quad \text{in} \quad D$$
$$\Psi = 0 \quad \text{on} \quad \partial D.$$

Use the Galerkin method to find an approximate solution of the problem in a rectangular domain $D = \{(x,y) : -a \le x \le a, -b \le y \le b\}$.

34. Apply the Galerkin approximation method to find the first eigenvalue of the problem of a circular membrane of radius a governed by the equation

$$\nabla^2 u \equiv \frac{d^2 u}{dr^2} + \frac{1}{r}\frac{du}{dr} = \lambda u \quad \text{in} \quad 0 < r < a$$
$$u = 0, \quad \text{on} \quad r = a.$$

35. Use the Rayleigh–Ritz method to find the solution in the first approximation of the problem of deformation of an elastic plate $(-a \le x \le a, -a \le y \le a)$ by a parabolic distribution of tensile forces over its opposite sides at $x = \pm a$. The problem is governed by the differential system

$$\nabla^4 u = \frac{2\alpha}{a^2} \quad \text{in} \quad |x| < a \quad \text{and} \quad |y| < a,$$
$$\left(u, \frac{\partial u}{\partial x}\right) = (0,0) \quad \text{on} \quad |x| = a,$$
$$\left(u, \frac{\partial u}{\partial y}\right) = (0,0) \quad \text{on} \quad |y| = a,$$

where $U = u_0 + u = \frac{1}{2}\alpha y^2 \left(1 - \frac{1}{6}y^2\right) + u$, and $\nabla^4 U = 0$.

36. Show that the Kantorovich solution of the torsion problem in exercise 33 is

$$\Psi_1(x,y) = \frac{1}{2}\left(b^2 - y^2\right)\left(1 - \frac{\cosh kx}{\cosh ka}\right), \qquad k = \frac{1}{b}\sqrt{\frac{5}{2}}.$$

37. (a) If the functional I in (14.6.3) depends on two functions u and v, that is,

$$I(u,v) = \int_a^b F\left(x,u,v,u',v'\right)dx,$$

show that there are two Euler–Lagrange equations for this functional

$$\frac{\partial F}{\partial u} - \frac{d}{dx}\left(\frac{\partial F}{\partial u'}\right) = 0, \qquad \text{and} \qquad \frac{\partial F}{\partial v} - \frac{d}{dx}\left(\frac{\partial F}{\partial v'}\right) = 0.$$

(b) Generalize the above result for the functional

$$I(\mathbf{u}) = \int_a^b F\left(x,\mathbf{u},\mathbf{u}'\right)dx,$$

where $\mathbf{u} = (u_1, u_2, \ldots, u_n)$, $u_i \in C^2\left([a,b]\right)$, and $u_i(a) = a_i$ and $u_i(b) = b_i$, $i = 1, 2, \ldots, n$.

38. Show that the Euler–Lagrange equation for the functional

$$I[u(x,y)] = \iint_D \left(1 + u_x^2 + u_y^2\right)^{\frac{1}{2}} dx\, dy$$

is

$$\left(1 + u_y^2\right)u_{xx} - 2u_x\,u_y\,u_{xy} + \left(1 + u_x^2\right)u_{yy} = 0.$$

39. Show that the Euler–Lagrange equation for the functional

$$I(u) = \iint_D F\left(x,y,u,u_x,u_y,u_{xx},u_{xy},u_{yy}\right)dx\,dy$$

is

$$\frac{\partial F}{\partial u} - \frac{\partial}{\partial x}\left(F_{u_x}\right) - \frac{\partial}{\partial y}\left(F_{u_y}\right) + \frac{\partial^2}{\partial x^2}\left(F_{u_{xx}}\right) + \frac{\partial^2}{\partial x \partial y}\left(F_{u_{xy}}\right)$$

$$+ \frac{\partial^2}{\partial y^2}\left(F_{u_{yy}}\right) = 0.$$

40. Derive the Euler–Lagrange equation for the functional

$$I\left[y\left(x\right)\right] = \int_a^b F\left(x, y, y'\right) dx$$

where

(a) $F\left(x, y, y'\right) = u\left(x, y\right)\sqrt{1 + y'^2}$,

(b) $F\left(x, y, y'\right) = \frac{1}{\sqrt{2g}}\left(\frac{1+y'^2}{y_1-y}\right)^{\frac{1}{2}}$

with $y\left(a\right) = y_1$ and $y\left(b\right) = y_2 < y_1$, (Brachistochrone problem).

(c) $F\left(x, y, y'\right) = y'^2\left(1 + y'^2\right)^2$,

(d) $F\left(x, y, y'\right) = \left(xy^3 - y'^2 + 3xyy'\right)$.

41. Show that there are an infinite number of continuous functions with piecewise continuous first derivatives that minimize the functional

$$I\left[y\left(x\right)\right] = \int_0^2 y'^2\left(1 + y'\right)^2 dx$$

with $y\left(0\right) = 1$ and $y\left(2\right) = 0$.

42. The torsion of a prismatic rod of rectangular cross section of length $2a$ and width $2b$ is governed by

$$\nabla^2 u = 2 \quad \text{in } R = \{(x, y) : -a < x < a, \quad -b < y < b\}$$
$$u = 0 \quad \text{on } \partial R.$$

(a) Find an approximate solution $u_1\left(x, y\right)$. Hence, calculate the torsional moment M for $a \neq b$ and for $a = b$.

(b) Find the exact classical solution for $u\left(x, y\right)$ and the torsional moment M.

43. Find an approximate solution of the biharmonic problem

$$\nabla^4 u = 0, \qquad R = \{(x, y) : -a < x < a, \quad -b < y < b\}$$

with the boundary conditions

$$u_{xy} = 0, \qquad u_{yy} = c\left(1 - \frac{y^2}{b^2}\right) \qquad \text{for} \quad x = \pm a,$$
$$u_{xy} = 0, \qquad u_{xx} = 0 \qquad \text{for} \quad y = \pm b,$$

where c is a constant.

44. Show that the Euler–Lagrange equation (14.6.12) can be written in the form

$$\frac{d}{dx}\left(F - u'\frac{\partial F}{\partial u'}\right) - \frac{\partial F}{\partial x} = 0.$$

45. Show that the extremals of the functional

$$I\left[y\left(x\right)\right] = \int_a^b \left[p\left(x\right)y'^2 - q\left(x\right)y^2\right]dx$$

subject to the constraint

$$J\left[y\left(x\right)\right] = \int_a^b r\left(x\right)y^2 dx = 1,$$

are solutions of the Sturm–Liouville equation

$$\frac{d}{dx}\left[p\left(x\right)\frac{dy}{dx}\right] + \left[q\left(x\right) + \lambda r\left(x\right)\right]y = 0.$$

46. Consider the finite element method for the wave equation

$$u_{tt} - u_{xx} = 0, \qquad 0 \le x \le l, \quad t > 0,$$
$$u\left(0\right) = u\left(l\right) = 0,$$

with given initial conditions.

(a) Show that an appropriate requirement is that

$$\sum_{i=1}^n A_i''\left(t\right)\int_0^l v_i\left(x\right)v_j\left(x\right)dx + \sum_{i=1}^n A_i\left(t\right)\int_0^l \frac{\partial v_i}{\partial x}\cdot\frac{\partial v_j}{\partial x}\,dx = 0,$$

where $j = 1, 2, \ldots, n$ and that the approximate solution is given by

$$u_n\left(x\right) = A_1\left(t\right)v_1\left(x\right) + \ldots + A_n\left(t\right)v_n\left(t\right) = \sum_{i=1}^n A_i\left(t\right)v_i\left(x\right).$$

(b) Show that the finite element method leads to a system of ordinary differential equations

$$B\frac{d^2 A}{dt^2} + C\,A\left(t\right) = 0, \qquad A\left(0\right) = D$$

where B and C are $n \times n$ matrices, $A\left(t\right)$ is a n-vector function and D is an n-vector.

15

Tables of Integral Transforms

In this chapter we provide a set of *short* tables of integral transforms of the functions that are either cited in the text or are in most common use in mathematical, physical, and engineering applications. For exhaustive lists of integral transforms, the reader is referred to Erdélyi et al. (1954), Campbell and Foster (1948), Ditkin and Prudnikov (1965), Doetsch (1970), Marichev (1983), Debnath (1995), and Oberhettinger (1972).

15.1 Fourier Transforms

$f(x)$	$F(k) = \frac{1}{\sqrt{2\pi}} \int_{-\infty}^{\infty} \exp(-ikx) f(x)\, dx$
1 $\exp(-a\,|x|)$, $\quad a > 0$	$\left(\sqrt{\frac{2}{\pi}}\right) a \left(a^2 + k^2\right)^{-1}$
2 $x\,\exp(-a\,|x|)$, $\quad a > 0$	$\left(\sqrt{\frac{2}{\pi}}\right)(-2aik)\left(a^2 + k^2\right)^{-2}$
3 $\exp\left(-ax^2\right)$, $\quad a > 0$	$\frac{1}{\sqrt{2a}} \exp\left(-\frac{k^2}{4a}\right)$
4 $\left(x^2 + a^2\right)^{-1}$, $\quad a > 0$	$\sqrt{\frac{\pi}{2}}\, \frac{\exp(-a|k|)}{a}$
5 $x\left(x^2 + a^2\right)^{-1}$	$\sqrt{\frac{\pi}{2}}\left(\frac{ik}{2a}\right)\exp\left(-a\,|k|\right)$

$f(x)$	$F(k) = \frac{1}{\sqrt{2\pi}} \int_{-\infty}^{\infty} \exp(-ikx) f(x)\, dx$
6 $\begin{cases} c, & a \le x \le b \\ 0, & \text{outside.} \end{cases}$	$\frac{ic}{\sqrt{2\pi}} \frac{1}{k} \left(e^{-ibk} - e^{-iak} \right)$
7 $\|x\| \exp(-a\|x\|), \qquad a > 0$	$\sqrt{\frac{2}{\pi}} \left(a^2 - k^2 \right) \left(a^2 + k^2 \right)^{-2}$
8 $\frac{\sin ax}{x}$	$\sqrt{\frac{\pi}{2}} H(a - \|k\|)$
9 $\exp\{-x(a - i\omega)\} H(x)$	$\frac{1}{\sqrt{2\pi}} \frac{i}{(\omega - k + ia)}$
10 $\left(a^2 - x^2 \right)^{-\frac{1}{2}} H(a - \|x\|)$	$\sqrt{\frac{\pi}{2}} J_0(ak)$
11 $\dfrac{\sin\left[b\left(x^2 + a^2 \right)^{\frac{1}{2}} \right]}{\left(x^2 + a^2 \right)^{\frac{1}{2}}}$	$\sqrt{\frac{\pi}{2}} J_0\left(a\sqrt{b^2 - k^2} \right) H(b - \|k\|)$
12 $\dfrac{\cos\left(b\sqrt{a^2 - x^2} \right)}{\left(a^2 - x^2 \right)^{\frac{1}{2}}} H(a - \|x\|)$	$\sqrt{\frac{\pi}{2}} J_0\left(a\sqrt{b^2 + k^2} \right)$
13 $e^{-ax} H(x), \qquad a > 0$	$\frac{1}{\sqrt{2\pi}} (a - ik) \left(a^2 + k^2 \right)^{-1}$
14 $\frac{1}{\sqrt{\|x\|}} \exp(-a\|x\|), \qquad a > 0$	$\left(a^2 + k^2 \right)^{-\frac{1}{2}} \left[a + \left(a^2 + k^2 \right)^{\frac{1}{2}} \right]^{\frac{1}{2}}$
15 $\delta^{(n)}(x - a), \; n = 0, 1, 2, \ldots$	$\frac{1}{\sqrt{2\pi}} (ik)^n \exp(-iak)$
16 $\exp(iax)$	$\sqrt{2\pi}\, \delta(k - a)$

15.2 Fourier Sine Transforms

	$f(x)$	$F_s(k) = \sqrt{\frac{2}{\pi}} \int_0^\infty \sin(kx) f(x)\,dx$
1	$\exp(-ax), \quad a > 0$	$\sqrt{\frac{2}{\pi}} k \left(a^2 + k^2\right)^{-1}$
2	$x\exp(-ax), \quad a > 0$	$\sqrt{\frac{2}{\pi}} (2ak) \left(a^2 + k^2\right)^{-2}$
3	$x^{\alpha-1}, \quad 0 < \alpha < 1$	$\sqrt{\frac{2}{\pi}} k^{-\alpha} \Gamma(\alpha) \sin\left(\frac{\pi\alpha}{2}\right)$
4	$\frac{1}{\sqrt{x}}$	$\frac{1}{\sqrt{k}}, \quad k > 0$
5	$x^{\alpha-1}e^{-ax}, \quad \alpha > -1,$ $a > 0$	$\sqrt{\frac{2}{\pi}} \Gamma(\alpha) r^{-\alpha} \sin(\alpha\theta)$, where $r = \left(a^2 + k^2\right)^{\frac{1}{2}}, \quad \theta = \tan^{-1}\left(\frac{k}{a}\right)$
6	$x^{-1}e^{-ax}, \quad a > 0$	$\sqrt{\frac{2}{\pi}} \tan^{-1}\left(\frac{k}{a}\right), \quad k > 0$
7	$x\exp\left(-a^2x^2\right)$	$2^{-3/2}\left(\frac{k}{a^3}\right)\exp\left(-\frac{k^2}{4a^2}\right)$
8	$\operatorname{erfc}(ax)$	$\sqrt{\frac{2}{\pi}}\frac{1}{k}\left[1 - \exp\left(-\frac{k^2}{4a^2}\right)\right]$
9	$x\left(a^2 + x^2\right)^{-1}$	$\sqrt{\frac{\pi}{2}}\exp(-ak), \quad a > 0$
10	$x\left(a^2 + x^2\right)^{-2}$	$\frac{1}{\sqrt{2\pi}}\left(\frac{k}{a}\right)\exp(-ak), \quad (a > 0)$

$f(x)$	$F_s(k) = \sqrt{\frac{2}{\pi}} \int_0^\infty \sin(kx) f(x)\, dx$
11 $H(a-x), \quad a>0$	$\sqrt{\frac{2}{\pi}} \frac{1}{k}(1-\cos ak)$
12 $x^{-1} J_0(ax)$	$\begin{cases} \sqrt{\frac{2}{\pi}} \sin^{-1}\left(\frac{k}{a}\right), & 0<k<a \\ \sqrt{\frac{\pi}{2}}, & a<k<\infty \end{cases}$
13 $x\left(a^2+x^2\right)^{-1} J_0(bx),$ $a>0,\ b>0$	$\sqrt{\frac{\pi}{2}}\, e^{-ak} I_0(ab), \quad a<k<\infty$
14 $J_0(a\sqrt{x}), \quad a>0$	$\sqrt{\frac{2}{\pi}} \frac{1}{k} \cos\left(\frac{a^2}{4k}\right)$
15 $\left(x^2-a^2\right)^{\nu-\frac{1}{2}} H(x-a),$ $\lvert\nu\rvert < \frac{1}{2}$	$2^{\nu-\frac{1}{2}} \left(\frac{a}{k}\right)^\nu \Gamma\left(\nu+\frac{1}{2}\right) J_{-\nu}(ak)$
16 $x^{1-\nu}\left(x^2+a^2\right)^{-1} J_\nu(ax),$ $\nu > -\frac{3}{2}, \quad a,b>0$	$\sqrt{\frac{\pi}{2}}\, a^{-\nu} \exp(-ak) I_\nu(ab),$ $a<k<\infty$
17 $x^{-\nu} J_{\nu+1}(ax), \quad \nu > -\frac{1}{2}$	$\dfrac{k\left(a^2-k^2\right)^{\nu-\frac{1}{2}}}{2^{\nu-\frac{1}{2}} a^{\nu+1} \Gamma\left(\nu+\frac{1}{2}\right)} H(a-k)$
18 $\operatorname{erfc}(ax)$	$\sqrt{\frac{2}{\pi}} \frac{1}{k}\left[1-\exp\left(-\frac{k^2}{4a^2}\right)\right]$
19 $x^{-\alpha}, \quad 0<\operatorname{Re}\alpha<2$	$\sqrt{\frac{2}{\pi}} \Gamma(1-\alpha) k^{\alpha-1} \cos\left(\frac{\alpha\pi}{2}\right)$
20 $\left(ax-x^2\right)^{\alpha-\frac{1}{2}} H(a-x),$ $\alpha > -\frac{1}{2}$	$\sqrt{2}\, \Gamma\left(\alpha+\frac{1}{2}\right) \left(\frac{a}{k}\right)^\alpha \sin\left(\frac{ak}{2}\right) J_\alpha\left(\frac{ak}{2}\right)$

15.3 Fourier Cosine Transforms

$f(x)$	$F_c(k) = \sqrt{\frac{2}{\pi}} \int_0^\infty \cos(kx) f(x) \, dx$
1 $\exp(-ax), \quad a > 0$	$\left(\sqrt{\frac{2}{\pi}}\right) a \left(a^2 + k^2\right)^{-1}$
2 $x \exp(-ax), \qquad a > 0$	$\left(\sqrt{\frac{2}{\pi}}\right) \left(a^2 - k^2\right) \left(a^2 + k^2\right)^{-2}$
3 $\exp\left(-a^2 x^2\right)$	$\frac{1}{a\sqrt{2}} \exp\left(-\frac{k^2}{4a^2}\right)$
4 $H(a - x)$	$\sqrt{\frac{2}{\pi}} \left(\frac{\sin ak}{k}\right)$
5 $x^{a-1}, \quad 0 < a < 1$	$\sqrt{\frac{2}{\pi}} \, \Gamma(a) \, k^{-a} \cos\left(\frac{a\pi}{2}\right)$
6 $\cos\left(ax^2\right), \quad a > 0$	$\frac{1}{2\sqrt{a}} \left[\cos\left(\frac{k^2}{4a}\right) + \sin\left(\frac{k^2}{4a}\right)\right]$
7 $\sin\left(ax^2\right), \quad a > 0$	$\frac{1}{2\sqrt{a}} \left[\cos\left(\frac{k^2}{4a}\right) - \sin\left(\frac{k^2}{4a}\right)\right]$
8 $\left(a^2 - x^2\right)^{\nu - \frac{1}{2}} H(a - x), \ \nu > -\frac{1}{2}$	$2^{\nu - \frac{1}{2}} \Gamma\left(\nu + \frac{1}{2}\right) \left(\frac{a}{k}\right)^\nu J_\nu(ak)$
9 $\left(a^2 + x^2\right)^{-1} J_0(bx), \ a, b > 0$	$\sqrt{\frac{\pi}{2}} \, a^{-1} \exp(-ak) I_0(ab),$ $b < k < \infty$
10 $x^{-\nu} J_\nu(ax), \ \nu > -\frac{1}{2}$	$\dfrac{\left(a^2 - k^2\right)^{\nu - \frac{1}{2}} H(a - k)}{2^{\nu - \frac{1}{2}} a^\nu \Gamma\left(\nu + \frac{1}{2}\right)}$

$f(x)$	$F_c(k) = \sqrt{\frac{2}{\pi}} \int_0^\infty \cos(kx) f(x) \, dx$
11 $\left(x^2 + a^2\right)^{-\frac{1}{2}} \exp\left[-b\left(x^2 + a^2\right)^{\frac{1}{2}}\right]$	$K_0\left[a\left(k^2 + b^2\right)^{\frac{1}{2}}\right]$, $a > 0$, $b > 0$
12 $x^{\nu-1} e^{-ax}$, $\nu > 0$, $a > 0$	$\sqrt{\frac{2}{\pi}}\, \Gamma(\nu)\, r^{-\nu} \cos n\theta$, where $r = \left(a^2 + k^2\right)^{\frac{1}{2}}$, $\theta = \tan^{-1}\left(\frac{k}{a}\right)$
13 $\frac{2}{x} e^{-x} \sin x$	$\sqrt{\frac{2}{\pi}} \tan^{-1}\left(\frac{2}{k^2}\right)$
14 $\sin\left[a\left(b^2 - x^2\right)^{\frac{1}{2}} H(b-x)\right]$	$\sqrt{\frac{\pi}{2}}\, (ab)\left(a^2 + k^2\right)^{-\frac{1}{2}}$ $\times J_1\left[b\left(a^2 + k^2\right)^{\frac{1}{2}}\right]$
15 $\frac{\left(1 - x^2\right)}{\left(1 + x^2\right)^2}$	$\sqrt{\frac{\pi}{2}}\, k \exp(-k)$
16 $x^{-\alpha}$, $0 < \alpha < 1$	$\sqrt{\frac{\pi}{2}}\, \frac{k^{\alpha-1}}{\Gamma(\alpha)} \sec\left(\frac{\pi\alpha}{2}\right)$
17 $\left(\frac{1}{a} + x\right) e^{-ax}$, $a > 0$	$\sqrt{\frac{\pi}{2}}\, \frac{2a^2}{\left(a^2 + k^2\right)^2}$
18 $\log\left(1 + \frac{a^2}{x^2}\right)$, $a > 0$	$\sqrt{2\pi}\, \frac{\left(1 - e^{-ak}\right)}{k}$
19 $\log\left(\frac{a^2 + x^2}{b^2 + x^2}\right)$, $a, b > 0$	$\sqrt{2\pi}\, \frac{\left(e^{-bk} - e^{-ak}\right)}{k}$
20 $a\left(x^2 + a^2\right)^{-1}$, $a > 0$	$\sqrt{\frac{\pi}{2}} \exp(-ak)$, $k > 0$

15.4 Laplace Transforms

	$f(t)$	$\overline{f}(s) = \int_0^\infty \exp(-st) f(t)\, dt$
1	$f^{(n)}(t)$	$s^n \overline{f}(s) - \displaystyle\sum_{r=0}^{n-1} s^{n-r-1} f^{(r)}(0)$
2	$\displaystyle\int_0^t f(t-\tau) g(\tau)\, d\tau$	$\overline{f}(s)\, \overline{g}(s)$
3	$t^n f(t)$	$(-1)^n \dfrac{d^n}{ds^n} \overline{f}(s)$
4	$f(t-a) H(t-a)$	$\exp(-as)\, \overline{f}(s)$
5	$t^n \qquad (n = 0,1,2,3,\ldots)$	$\dfrac{n!}{s^{n+1}}$
6	e^{at}	$\dfrac{1}{s-a}$
7	$t^n e^{-at}$	$\dfrac{\Gamma(n+1)}{(s+a)^{n+1}}$
8	$t^a \qquad (a > -1)$	$\dfrac{\Gamma(a+1)}{s^{a+1}}$
9	$e^{at} \cos bt$	$\dfrac{s-a}{(s-a)^2 + b^2}$
10	$e^{at} \sin bt$	$\dfrac{b}{(s-a)^2 + b^2}$
11	$\dfrac{1}{\sqrt{t}}$	$\sqrt{\dfrac{\pi}{s}}$
12	$2\sqrt{t}$	$\dfrac{1}{s}\sqrt{\dfrac{\pi}{s}}$

$f(t)$	$\overline{f}(s) = \int_0^\infty \exp(-st)\, f(t)\, dt$
13 $t^{-1/2} \exp\left(-\frac{a}{t}\right)$	$\sqrt{\frac{\pi}{s}} \exp\left(-2\sqrt{as}\right)$
14 $t^{-3/2} \exp\left(-\frac{a}{t}\right)$	$\sqrt{\frac{\pi}{a}} \exp\left(-2\sqrt{as}\right)$
15 $\frac{1}{\sqrt{\pi t}}(1 + 2at)\,e^{at}$	$\frac{s}{(s-a)\sqrt{s-a}}$
16 $\frac{1}{2\sqrt{\pi t^3}}\left(e^{bt} - e^{at}\right)$	$\sqrt{s-a} - \sqrt{s-b}$
17 $\exp\left(a^2 t\right)\operatorname{erf}\left(a\sqrt{t}\right)$	$\frac{a}{\sqrt{s}(s-a^2)}$
18 $\exp\left(a^2 t\right)\operatorname{erfc}\left(a\sqrt{t}\right)$	$\frac{1}{\sqrt{s}\left(\sqrt{s}+a\right)}$
19 $\frac{1}{\sqrt{\pi t}} + a\exp\left(a^2 t\right)\operatorname{erf}\left(a\sqrt{t}\right)$	$\frac{\sqrt{s}}{(s-a^2)}$
20 $\frac{1}{\sqrt{\pi t}} - a\exp\left(a^2 t\right)\operatorname{erfc}\left(a\sqrt{t}\right)$	$\frac{1}{\sqrt{s}+a}$
21 $\frac{\exp(-at)}{\sqrt{b-a}}\operatorname{erf}\left(\sqrt{(b-a)\,t}\right)$	$\frac{1}{(s+a)\sqrt{s+b}}$
22 $\frac{1}{2}e^{i\omega t}\left[\exp(-\lambda z)\operatorname{erfc}\left(\zeta - \sqrt{i\omega t}\right)\right.$ $\left. + \exp(\lambda z)\operatorname{erfc}\left(\zeta + \sqrt{i\omega t}\right)\right],$ where $\zeta = z/2\sqrt{\nu t}, \quad \lambda = \sqrt{\frac{i\omega}{\nu}}$	$(s - i\omega)^{-1}\exp\left(-z\sqrt{\frac{s}{\nu}}\right)$
23 $\frac{1}{2}\left[\exp(-ab)\operatorname{erfc}\left(\frac{b-2at}{2\sqrt{t}}\right)\right.$ $\left. + \exp(ab)\operatorname{erfc}\left(\frac{b+2at}{2\sqrt{t}}\right)\right]$	$\exp\left[-b\left(s + a^2\right)^{\frac{1}{2}}\right]$

	$f(t)$	$\overline{f}(s) = \int_0^\infty \exp(-st) f(t)\, dt$
24	$J_0(at)$	$\left(s^2 + a^2\right)^{-\frac{1}{2}}$
25	$I_0(at)$	$\left(s^2 - a^2\right)^{-\frac{1}{2}}$
26	$t^{\alpha-1} \exp(-at), \qquad \alpha > 0$	$\Gamma(\alpha)(s+a)^{-\alpha}$
27	$t^{-1} J_\nu(at)$	$\nu^{-1} a^\nu \left(\sqrt{s^2 + a^2} + s\right)^{-\nu},$ $\operatorname{Re}\nu > -\frac{1}{2}$
28	$J_0\left(a\sqrt{t}\right)$	$\frac{1}{s} \exp\left(-\frac{a^2}{4s}\right)$
29	$\left(\frac{2}{a}\right)^\nu t^{\nu/2} J_\nu\left(a\sqrt{t}\right)$	$s^{-(\nu+1)} \exp\left(-\frac{a^2}{4s}\right),$ $\operatorname{Re}\nu > -\frac{1}{2}$
30	$\frac{a}{2t\sqrt{\pi t}} \exp\left(-\frac{a^2}{4t}\right)$	$\exp\left(-a\sqrt{s}\right), \quad a > 0$
31	$\frac{1}{\sqrt{\pi t}} \exp\left(-\frac{a^2}{4t}\right)$	$\frac{1}{\sqrt{s}} \exp\left(-a\sqrt{s}\right), \quad a \geq 0$
32	$\exp\left(-\frac{a^2 t^2}{4}\right)$	$\frac{\sqrt{\pi}}{a} \exp\left(\frac{s^2}{a^2}\right) \operatorname{erfc}\left(\frac{s}{a}\right),$ $a \geq 0$
33	$\left(t^2 - a^2\right)^{-\frac{1}{2}} H(t-a)$	$K_0(as), \quad a > 0$
34	$\delta^{(n)}(t-a), \quad n = 0, 1, \ldots$	$s^n \exp(-as)$

$f(t)$	$\bar{f}(s) = \int_0^\infty \exp(-st) f(t)\, dt$
35 $t^{m\alpha+\beta-1} E_{\alpha,\beta}^{(m)}(\pm at)$, $m = 0, 1, 2, \ldots$	$\dfrac{m!\, s^{\alpha-\beta}}{\left(s^\alpha \mp a\right)^{m+1}}$
36 $\dfrac{\sqrt{\pi}}{\Gamma(\nu+\frac{1}{2})} \left(\dfrac{t}{2a}\right)^\nu J_\nu(at)$	$\left(s^2 + a^2\right)^{-\left(\nu+\frac{1}{2}\right)}$, $\operatorname{Re} \nu > -\frac{1}{2}$
37 $\frac{1}{2} e^{-ct} \left[\exp\left(-a\sqrt{b-c}\right) \right.$ $\times \operatorname{erfc}\left\{\frac{a}{\sqrt{4t}} - \sqrt{(b-c)t}\right\}$ $- \exp\left(a\sqrt{b-c}\right)$ $\left. \times \operatorname{erfc}\left\{\frac{a}{\sqrt{4t}} + \sqrt{(b-c)t}\right\}\right]$	$\dfrac{\exp\left(-a\sqrt{s+b}\right)}{(s+c)\sqrt{(s+b)}}$
38 $\frac{1}{2} e^{-ct} \left[\exp\left(-a\sqrt{b-c}\right) \right.$ $\times \operatorname{erfc}\left\{\frac{a}{\sqrt{4t}} - t\sqrt{b-c}\right\}$ $- \exp\left(a\sqrt{b-c}\right)$ $\left. \times \operatorname{erfc}\left\{\frac{a}{\sqrt{4t}} + t\sqrt{b-c}\right\}\right]$	$\dfrac{\exp\left(-a\sqrt{s+b}\right)}{(s+c)}$
39 $e^{-bt}\left[\sqrt{\frac{4t}{\pi}} \exp\left(-\frac{a^2}{4t}\right) \right.$ $\left. -a \operatorname{erfc}\left(\frac{a}{\sqrt{4t}}\right)\right]$	$\dfrac{\exp\left(-a\sqrt{s+b}\right)}{(s+b)^{3/2}}$
40 $e^{-bt}\left[\left(t + \frac{1}{2}a^2\right) \operatorname{erfc}\left(\frac{a}{\sqrt{4t}}\right) \right.$ $\left. -\sqrt{\frac{ta^2}{\pi}} \exp\left(-\frac{a^2}{4t}\right)\right]$	$\dfrac{\exp\left(-a\sqrt{s+b}\right)}{(s+b)^2}$

15.5 Hankel Transforms

	$f(r)$	order n	$\tilde{f}_n(k) = \int_0^\infty r\, J_n(kr)\, f(r)\, dr$
1	$H(a-r)$	0	$\frac{a}{k} J_1(ak)$
2	$\exp(-ar)$	0	$a\left(a^2+k^2\right)^{-\frac{3}{2}}$
3	$\frac{1}{r}\exp(-ar)$	0	$\left(a^2+k^2\right)^{-\frac{1}{2}}$
4	$\left(a^2-r^2\right)H(a-r)$	0	$\frac{4a}{k^3} J_1(ak) - \frac{2a^2}{k^2} J_0(ak)$
5	$a\left(a^2+r^2\right)^{-\frac{3}{2}}$	0	$\exp(-ak)$
6	$\frac{1}{r}\cos(ar)$	0	$\left(k^2-a^2\right)^{-\frac{1}{2}} H(k-a)$
7	$\frac{1}{r}\sin(ar)$	0	$\left(a^2-k^2\right)^{-\frac{1}{2}} H(a-k)$
8	$\frac{1}{r^2}\left(1-\cos ar\right)$	0	$\cosh^{-1}\left(\frac{a}{k}\right) H(a-k)$
9	$\frac{1}{r} J_1(ar)$	0	$\frac{1}{a} H(a-k), \quad a>0$
10	$Y_0(ar)$	0	$\left(\frac{2}{\pi}\right)\left(a^2-k^2\right)^{-1}$
11	$K_0(ar)$	0	$\left(a^2+k^2\right)^{-1}$

	$f(r)$	order n	$\tilde{f}_n(k) = \int_0^\infty r\, J_n(kr)\, f(r)\, dr$
12	$\frac{\delta(r)}{r}$	0	1
13	$\left(r^2 + b^2\right)^{-\frac{1}{2}}$ $\times \exp\left\{-a\left(r^2 + b^2\right)^{\frac{1}{2}}\right\}$	0	$\left(k^2 + a^2\right)^{-\frac{1}{2}} \exp\left\{-b\left(k^2 + a^2\right)^{\frac{1}{2}}\right\}$
14	$\left(r^2 + a^2\right)^{-\frac{1}{2}}$	0	$\frac{1}{k}\exp\left(-ak\right)$
15	$\exp\left(-ar\right)$	1	$k\left(a^2 + k^2\right)^{-3/2}$
16	$\frac{\sin ar}{r}$	1	$\frac{a\, H(k-a)}{k\left(k^2 - a^2\right)^{\frac{1}{2}}}$
17	$\frac{1}{r}\exp\left(-ar\right)$	1	$\frac{1}{k}\left[1 - \frac{a}{\left(k^2 + a^2\right)^{\frac{1}{2}}}\right]$
18	$\frac{1}{r^2}\exp\left(-ar\right)$	1	$\frac{1}{k}\left[\left(k^2 + a^2\right)^{\frac{1}{2}} - a\right]$
19	$r^n\, H(a - r)$	> -1	$\frac{1}{k}a^{n+1}\, J_{n+1}(ak)$
20	$r^n \exp\left(-ar\right), \ (\operatorname{Re} a > 0)$	> -1	$\frac{1}{\sqrt{\pi}}\, \frac{2^{n+1}\, \Gamma\left(n + \frac{3}{2}\right) a\, k^n}{\left(a^2 + k^2\right)^{n+\frac{3}{2}}}$
21	$r^n \exp\left(-ar^2\right)$	> -1	$\frac{k^n}{(2a)^{n+1}} \exp\left(-\frac{k^2}{4a}\right)$

	$f(r)$	order n	$\tilde{f}_n(k) = \int_0^\infty r\, J_n(kr)\, f(r)\, dr$
22	r^{a-1}	> -1	$\dfrac{2a\,\Gamma\left[\frac{1}{2}(a+n+1)\right]}{k^{a+1}\,\Gamma\left[\frac{1}{2}(1-a+n)\right]}$
23	$r^n\left(a^2-r^2\right)^{m-n-1}$ $\times H(a-r)$	> -1	$2^{m-n-1}\,\Gamma(m-n)\,a^m k^{n-m} J_m(ak)$
24	$r^m \exp\left(-r^2/a^2\right)$	> -1	$\dfrac{k^n\,a^{m+n+2}}{2^{n+1}\,\Gamma(n+1)}\,\Gamma\left(1+\frac{m}{2}+\frac{n}{2}\right)$ $\times\,_1F_1\left(1+\frac{m}{2}+\frac{n}{2};\,n+1;\,-\frac{1}{4}a^2k^2\right)$
25	$\frac{1}{r}J_{n+1}(ar)$	> -1	$k^n a^{-(n+1)} H(a-k),\ a > 0$
26	$r^n\left(a^2-r^2\right)^m H(a-r),$ $m > -1$	> -1	$2^m a^n \Gamma(m+1)\left(\frac{a}{k}\right)^{m+1}$ $\times\, J_{n+m+1}(ak)$
27	$\frac{1}{r^2}J_n(ar)$	$> \frac{1}{2}$	$\begin{cases} \frac{1}{2n}\left(\frac{k}{a}\right)^n,\ 0 < k \le a \\[6pt] \frac{1}{2n}\left(\frac{a}{k}\right)^n,\ a < k < \infty \end{cases}$
28	$\frac{r^n}{(a^2+r^2)^{m+1}},\ a > 0$	> -1	$\left(\frac{k}{2}\right)^m \frac{a^{n-m}}{\Gamma(m+1)} K_{n-m}(ak)$
29	$\exp\left(-p^2 r^2\right) J_n(ar)$	> -1	$\left(2p^2\right)^{-1}\exp\left(-\frac{a^2+k^2}{4p^2}\right) I_n\left(\frac{ak}{2p^2}\right)$
30	$\frac{1}{r}\exp(-ar)$	> -1	$\dfrac{\left[\left(k^2+a^2\right)^{\frac{1}{2}}-a\right]^n}{k^n\left(k^2+a^2\right)^{\frac{1}{2}}}$
31	$\frac{r^n}{(r^2+a^2)^{n+1}}$	> -1	$\left(\frac{k}{2}\right)^n \dfrac{K_0(ak)}{\Gamma(n+1)}$

	$f(r)$	order n	$\tilde{f}_n(k) = \int_0^\infty r\, J_n(kr)\, f(r)\, dr$
32	$\dfrac{r^n}{(a^2-r^2)^{n+\frac{1}{2}}} H(a-r)$	< 1	$\dfrac{1}{\sqrt{\pi}} \left(\dfrac{k}{2}\right)^n \Gamma\left(\dfrac{1}{2}-n\right) \left(\dfrac{\sin ak}{k}\right)$
33	$f(ar)$	n	$\dfrac{1}{a^2} \tilde{f}_n\left(\dfrac{k}{a}\right)$
34	$r^{-1} \exp\left(-ar^2\right)$	1	$\dfrac{1}{k}\left[1 - \exp\left(-\dfrac{k^2}{4a}\right)\right]$
35	$r^{-1} \sin\left(ar^2\right), \qquad a > 0$	1	$\dfrac{1}{k} \sin\left(\dfrac{k^2}{4a}\right)$
36	$r^{-1} \cos\left(ar^2\right), \qquad a > 0$	1	$1 - \cos\left(\dfrac{k^2}{4a}\right)$
37	$\exp\left(-ar\right), \qquad a > 0$	> -1	$\dfrac{\left(a+n\sqrt{k^2+a^2}\right)}{(k^2+a^2)^{3/2}} \left(\dfrac{k}{a+\sqrt{a^2+k^2}}\right)^n$
38	$\exp\left(-ar^2\right) J_0(br)$	0	$\dfrac{a}{2} \exp\left(-\dfrac{k^2-b^2}{4a}\right) I_0\left(\dfrac{bk}{2a}\right)$
39	$\dfrac{H(a-r)}{\sqrt{a^2-r^2}}$	0	$\sqrt{\dfrac{a\pi}{2k}} J_{\frac{1}{2}}(ak), \qquad a > 0$
40	$\dfrac{r^n H(a-r)}{\sqrt{a^2-r^2}}$	> -1	$\sqrt{\dfrac{\pi}{2k}} a^{n+\frac{1}{2}} J_{n+1}(ak), \qquad a > 0$
41	$r^{-2} \sin r$	0	$\sin^{-1}\left(\dfrac{1}{k}\right), \qquad (k > 1)$

15.6 Finite Hankel Transforms

	$f(r)$	order n	$\tilde{f}_n(k_i) = \int_0^a r J_n(r k_i) f(r) dr$
1	c, where c is a constant	0	$\left(\frac{ac}{k_i}\right) J_1(ak_i)$
2	$(a^2 - r^2)$	0	$\frac{4a}{k_i^3} J_1(ak_i)$
3	$(a^2 - r^2)^{-\frac{1}{2}}$	0	$k_i^{-1} \sin(ak_i)$
4	$\frac{J_0(\alpha r)}{J_0(\alpha a)}$	0	$-\frac{ak_i}{(\alpha^2 - k_i^2)} J_1(ak_i)$
5	$\frac{1}{r}$	1	$k_i^{-1}\{1 - J_0(ak_i)\}$
6	$r^{-1}(a^2 - r^2)^{-\frac{1}{2}}$	1	$\frac{(1 - \cos ak_i)}{(ak_i)}$
7	r^n	> -1	$\frac{a^{n+1}}{k_i} J_{n+1}(ak_i)$
8	$\frac{J_\nu(\alpha r)}{J_\nu(\alpha a)}$	> -1	$\frac{ak_i}{(\alpha^2 - k_i^2)} J_\nu'(ak_i)$
9	$r^{-n}(a^2 - r^2)^{-\frac{1}{2}}$	> -1	$\frac{\pi}{2}\left\{J_{\frac{n}{2}}\left(\frac{ak_i}{2}\right)\right\}^2$
10	$r^n(a^2 - r^2)^{-(n+\frac{1}{2})}$	$< \frac{1}{2}$	$\frac{\Gamma(\frac{1}{2}-n)}{\sqrt{\pi}\, 2^n} k_i^{n-1} \sin(ak_i)$
11	$r^{n-1}(a^2 - r^2)^{n-\frac{1}{2}}$	$> -\frac{1}{2}$	$\frac{\sqrt{\pi}}{2} \Gamma\left(n + \frac{1}{2}\right) \left(\frac{2}{k_i}\right)^n a^{2n} J_n^2\left(\frac{ak_i}{2}\right)$

Answers and Hints to Selected Exercises

1.6 Exercises

1. (a) Linear, nonhomogeneous, second-order; (b) quasi-linear, first-order; (c) nonlinear, first-order; (d) linear, homogeneous, fourth-order; (e) linear, nonhomogeneous, second-order; (f) quasi-linear, third-order; (g) nonlinear, second-order; and (h) nonlinear, homogeneous.

5. $u(x, y) = f(x) \cos y + g(x) \sin y$.

6. $u(x, y) = f(x) e^{-y} + g(y)$.

7. $u(x, y) = f(x + y) + g(3x + y)$.

8. $u(x, y) = f(y + x) + g(y - x)$.

11. $u_x = v_y \Rightarrow u_{xx} = v_{xy}, \ v_x = -u_y \Rightarrow v_{yx} = -u_{yy}$.

Thus, $u_{xx} + u_{yy} = 0$. Similarly, $v_{xx} + v_{yy} = 0$.

12. Since $u(x, y)$ is a homogeneous function of degree n, $u = x^n f\left(\frac{y}{x}\right)$.

$u_x = n x^{n-1} f\left(\frac{y}{x}\right) - x^{n-2} y f'\left(\frac{y}{x}\right)$, and $u_y = x^{n-1} f'\left(\frac{y}{x}\right)$.

Thus, $x u_x + y u_y = n x^n f\left(\frac{y}{x}\right) = n u$.

23. $u_x = -\frac{1}{b} \exp\left(-\frac{x}{b}\right) f(ax - by)$

$+ \exp\left(-\frac{x}{b}\right) \frac{d}{d(ax - by)} f(ax - by) \cdot \frac{d(ax - by)}{dx}$

$= -\frac{1}{b} \exp\left(-\frac{x}{b}\right) f + a \exp\left(-\frac{x}{b}\right) f'(ax - by)$

$u_y = (-b) \exp\left(-\frac{x}{b}\right) f'(ax - by)$. Thus, $b u_x + a u_y + u = 0$.

24. $V''(t) + 2bV'(t) + k^2c^2V(t) = 0$.

25. Differentiating with respect to r and t partially gives

$$V''(r) + n^2V(r) = 0.$$

2.8 Exercises

2. (a) $xp - yq = x - y$,   (d) $yp - xq = y^2 - x^2$.

3. (a) $u = f(y)$,   (b) $u = f(bx - ay)$,   (c) $u = f(ye^{-x})$,

   (d) $u = f(y - \tan^{-1}x)$,   (e) $u = f\left(\frac{x^2-y^2}{x}\right)$,

   (f) Hint: $\frac{dx}{y+u} = \frac{dy}{y} = \frac{du}{x-y} = \frac{d(x+u)}{x+u} = \frac{d(u+y)}{x}$, $x\,dx = (u+y)\,d(u+y) \Rightarrow$

   $(u+y)^2 - x^2 = c_1$. $\frac{d(u+x)}{u+x} = \frac{dy}{y} \Rightarrow \frac{u+x}{y} = c_2$,

   $f\left(\frac{u+x}{y}, (u+y)^2 - x^2\right) = 0$.

   (g) $\frac{dx}{y^2} = \frac{dy}{-xy} = \frac{du}{xu - 2xy} = \frac{d(u-y)}{x(u-y)}$.

   From the second and the fourth, $(u-y)y = c_1$ and $x^2 + y^2 = c_2$.

   Hence, $(u-y)y = f(x^2 + y^2)$. Thus, $u = y + y^{-1}f(x^2 + y^2)$.

   (h) $u + \log x = f(xy)$,   (i) $f(x^2 + u^2, y^3 + u^3) = 0$.

4. $u(x,y) = f(x^2 + y^{-1})$. Verify by differentiation that u satisfies the original equation.

5. (a) $u = \sin\left(x - \frac{3}{2}y\right)$,   (b) $u = \exp(x^2 - y^2)$,

   (c) $u = xy + f\left(\frac{y}{x}\right)$, $u = xy + 2 - \left(\frac{y}{x}\right)^3$,   (d) $u = \sin\left(y - \frac{1}{2}x^2\right)$,

   (e) $u = \begin{cases} \frac{1}{2}y^2 + \exp\left[-(x^2 - y^2)\right] & \text{for } x > y, \\ \frac{1}{2}x^2 + \exp\left[-(y^2 - x^2)\right] & \text{for } x < y. \end{cases}$

   (f)   Hint: $y = \frac{1}{2}x^2 + C_1$, $u = C_1^2\,x + C_2$,

   $u = x\left(y - \frac{1}{2}x^2\right)^2 + f\left(y - \frac{1}{2}x^2\right)$, $u = x\left(y - \frac{1}{2}x^2\right)^2 + \exp\left(y - \frac{1}{2}x^2\right)$.

   (g) $\frac{y}{x} = C_1$ and $\frac{u+1}{y} = C_2$, $C_2 = 1 + \frac{1}{C_1^2}$. Thus, $u = y + \frac{x^2}{y} - 1$,

   $y \neq 0$.

   (h)   Hint: $x + y = C_1$, $\frac{dy}{-u} = \frac{du}{u^2 + C_1^2}$, $u^2 + C_1^2 = C_2\exp(-2y)$.

From the Cauchy data, it follows that $1 + C_1^2 = C_2$, and hence,

$$u = \left[\left\{1 + (x+y)^2\right\}e^{-2y} - (x+y)^2\right]^{\frac{1}{2}}.$$

(i) $\frac{dy}{dx} - \frac{y}{x} = 1$, $\quad \frac{d}{dx}\left(\frac{y}{x}\right) = \frac{1}{x}$ which implies that $x = C_1 \exp\left(\frac{y}{x}\right)$.

$\frac{u+1}{x} = C_2$. Hence, $f\left(\frac{u+1}{x}, x\exp\left(-\frac{y}{x}\right)\right) = 0$.

Initial data imply $x = C_1$ and $\frac{x^2+1}{x} = C_2$. Hence $C_2 = C_1 + \frac{1}{C_1}$.

$\frac{u+1}{x} = x\exp\left(-\frac{y}{x}\right) + \frac{1}{x}\exp\left(\frac{y}{x}\right)$. Thus, $u = x^2\exp\left(-\frac{y}{x}\right) + \exp\left(\frac{y}{x}\right) - 1$.

(j) $\frac{dx}{\sqrt{x}} = \frac{dy}{u} = \frac{du}{-u^2}$. The second and the third give $y = -\log(Au)$ and

hence, $A = 1$ and $u = \exp(-y)$. The first and the third yield

$u^{-1} = 2\sqrt{x} - B$. At $(x_0, 0)$, $x_0 > 0$, $B = 2\sqrt{x_0} - 1$. Hence,

$u^{-1} = 2\left(\sqrt{x} - \sqrt{x_0}\right) + 1 = \frac{1}{y}$. The solution along the characteristic is

$u = \exp(-y)$ or $u^{-1} = 2\left(\sqrt{x} - \sqrt{x_0}\right) + 1$.

(k) $\frac{dx}{ux^2} = \frac{dy}{\exp(-y)} = \frac{du}{-u^2}$. The first and the third give $x^{-1} = \log u + A$

and hence, $A = \frac{1}{x_0}$, $x_0 > 0$. The second and third yield $u = \exp(-y)$.

Or, eliminating u gives $y = \left(x_0^{-1} - x^{-1}\right)$.

6. $u^2 - 2ut + 2x = 0$, and hence, $u = t \pm \sqrt{t^2 - 2x}$.

7. $u(x, y) = \exp\left(\frac{x}{x^2 - y^2}\right)$.

8. (a) $u = f\left(\frac{y}{x}, \frac{z}{x}\right)$ $\qquad$ (b) Hint: $u_1 = \frac{x-y}{xy} = C_1$,

$\frac{d(x-y)}{x^2 - y^2} = \frac{dz}{z(x+y)}$ gives $u = \frac{x-z}{z} = C_2$. Hence, $u = f\left(\frac{x-y}{xy}, \frac{x-y}{z}\right)$.

(c) $\phi = (x + y + z) = C_1$.

Hint: $\frac{\left(\frac{dx}{x}\right)}{y-z} = \frac{\left(\frac{dy}{y}\right)}{z-x} = \frac{\left(\frac{dz}{z}\right)}{x-y} = \frac{\frac{dx}{x} + \frac{dy}{y} + \frac{dz}{z}}{0} = \frac{d\log(xyz)}{0}$,

$\psi = xyz = C_2$, and hence, $u = f(x+y+z, xyz)$ is the general

solution.

(d) Hint: $x\,dx + y\,dy = 0$, $x^2 + y^2 = C_1$

$z\,dz = -\left(x^2 + y^2\right)y\,dy = -C_1 y\,dy$, $\quad z^2 + \left(x^2 + y^2\right)y^2 = C_2$,

$u = f\left(x^2 + y^2, z^2 + \left(x^2 + y^2\right)y^2\right)$.

(e) $\frac{x^{-1}dx}{y^2 - z^2} = \frac{y^{-1}dy}{z^2 - x^2} = \frac{z^{-1}dz}{y^2 - x^2} = \frac{d(\log xyz)}{0}$. $\quad u = f\left(x^2 + y^2 + z^2, xyz\right)$.

9. (a) Hint: $y - \frac{x^2}{2} = C_1$, $u = xy - \frac{x^3}{3} + C_2$, $\phi \left(u - xy + \frac{x^3}{3}, \; y - \frac{x^2}{2} \right) = 0$.

$u = xy - \frac{x^3}{3} + f \left(y - \frac{x^2}{2} \right)$, $\quad u = xy - \frac{x^3}{3} + \left(y - \frac{x^2}{2} \right)^2$.

(b) $u = xy - \frac{1}{3}x^3 + y - \frac{x^2}{2} + \frac{5}{6}$.

11. $\frac{x+u}{y} = C_1$, $u^2 - (x - y)^2 = C_2$, $u^2 - \frac{2u}{y} - (x - y)^2 - \frac{2}{y}(x - y) = 0$.

$u = \frac{2}{y} + (x - y)$, $y > 0$.

12. (a) $x = \frac{\tau^2}{2} + \tau s + s$, $y = \tau + 2s$, $u = \tau + s = \frac{(2x - 2y + y^2)}{2(y-1)}$

(b) $x = \frac{\tau^2}{2} + \tau s + s^2$, $y = \tau + 2s$, $u = \tau + s$,

$(y - s)^2 = 2x - s^2$, which is a set of parabolas.

(c) $x = \frac{1}{2}(\tau + s)^2$, $y = u = \tau + s$.

13. Hint: The initial curve is a characteristic, and hence, no solution exists.

14. (a) $u = \exp \left(\frac{xy}{x+y} \right)$, $\qquad$ (b) $u = \sin \left[\left(\frac{x^2 - y^2 + 1}{2} \right)^{\frac{1}{2}} \right]$,

(c) $u = 2 \left(\frac{xy}{3} \right)^{\frac{1}{2}} + \frac{1}{2} \log \left(\frac{y}{3x} \right)$, $\qquad$ (e) $u = \frac{1}{2}x^2 - \frac{1}{4}y^2 + \frac{1}{2}x^2y + \frac{1}{4}$.

(f) Hint: $\frac{dx}{1} = \frac{dy}{2} = \frac{du}{1+u}$, $y - 2x = c_1$ and $(1 + u)e^{-x} = c_2$,

$(1 + u)e^{-x} = f(y - 2x)$, $1 + u = \exp(3x - y + 1)[1 + \sin(y - 2x - 1)]$.

(g) Hint: $\frac{dx}{1} = \frac{dy}{2} = \frac{du}{u}$, $y - 2x = c_1$, and $u e^{-x} = c_2$, $u e^{-x} = f(y - 2x)$,

$u = \exp \left(\frac{y-x}{2} \right) \cos \left(\frac{y-3x}{2} \right)$.

(h) $\frac{dx}{1} = \frac{dy}{2x} = \frac{du}{2x\,u}$, $(y - x)^2 = c_1$, and $u e^{-x^2} = c_2$, $u e^{-x^2} = f(y - x^2)$, $u(x, y) = (x^2 - y)e^y$.

(i) $\frac{dx}{u} = \frac{dy}{1} = \frac{du}{u}$, $u - x = c_1$, and $u e^{-y} = c_2$, $f(u e^{-y}, u - x) = 0$,

$u e^y = g(u - x)$, $u = \frac{2x\,e^y}{2e^y - 1}$, $\frac{dx}{dy} = u$, $x = A(2e^y - 1)$ is the family of

characteristics.

(j) $\frac{dx}{1} = \frac{dy}{1} = \frac{du}{u^2}$, $y - x = c_1$, and $\frac{1}{u} + x = c_2$, $\frac{1}{u} + x = f(y - x)$,

$f(x) = - \left(\frac{1 - \tanh x}{\tanh x} \right)$, $\qquad u(x, y) = \frac{\tanh(x-y)}{1 - y\tanh(x-y)}$.

15. $3uy = u^2 + x^2 + y^2$. $\qquad$ Hint: $\frac{x\,dx + y\,dy + u\,du}{0}$, $x^2 + y^2 + u^2 = c_1$,

$\frac{dy}{y} = -\frac{du}{u}$ gives $uy = c_2$.

$x^2 + y^2 + u^2 = f(uy)$, and hence, $3u^2 = f(u^2)$.

16. (a) $x\,(s,\tau) = \tau$, $\ y\,(s,\tau) = \frac{\tau^2}{2} + a\tau s + s$, $\ u\,(s,\tau) = \tau + as$.

$\tau = x$, $\ s = (1+ax)^{-1}\left(y - \frac{1}{2}x^2\right)a$, and hence,

$u\,(x,y) = x + as = (1+ax)^{-1}\left\{x + a\left(y + \frac{1}{2}x^2\right)\right\}$, singular at $x = -\frac{1}{a}$.

(b) $y = \frac{u^2}{2} + f\,(u - x)$, $\quad 2y = u^2 + (u-x)^2$, $\quad u\,(0,y) = \sqrt{y}$.

17. (a) Hint: $\dfrac{d(x+y+u)}{2(x+y+u)} = \dfrac{d(y-u)}{-(y-u)} = \dfrac{d(u-x)}{-(u-x)}$

$(x+y+u)\,(y-u)^2 = c_1$ and $(x+y+u)\,(u-x)^2 = c_2$.

(b) Hint: $\dfrac{dx}{x} = \dfrac{dy}{-y}$. $\qquad\qquad$ Hence, $xy = a$.

$\dfrac{dx}{xu(u^2+a)} = \dfrac{du}{x^4}$. $\quad$ So, $\dfrac{dx}{du} = \dfrac{u(u^2+a)}{x^3}$ giving $x^4 = u^4 + 2au^2 + b$

and, thus, $x^4 - u^4 - 2u^2 xy = b$.

(c) $\dfrac{dx}{x+y} = \dfrac{dy}{x-y} = \dfrac{dy}{0}$ (exact equation). $\ u = f\left(x^2 - 2xy - y^2\right)$.

(d) $f\left(x^2 - y^2, \ u - \frac{1}{2}y^2\left(x^2 - y^2\right)\right) = 0$.

(e) $f\left(x^2 + y^2 + z^2, \ ax + by + cz\right) = 0$.

18. Hint: $\dfrac{dx}{x} = \dfrac{dy}{y} = \dfrac{dz}{z}$, and hence, $\dfrac{x}{z} = c$, $\quad \dfrac{y}{z} = d$.

$x^2 + y^2 = a^2$ and $z = \tan^{-1}\left(\frac{y}{x}\right)$ give $\left(c^2 + d^2\right)z^2 = a^2$

and $z = b\tan^{-1}\left(\frac{d}{c}\right)$.

$c = \left(\frac{a}{z}\right)\cos\theta$, $d = \left(\frac{a}{z}\right)\sin\theta$, and $z = b\tan^{-1}(\tan\theta) = b\theta$.

Thus, the curves are $x\,b\,\theta = az\cos\theta$ and $y\,b\,\theta = az\sin\theta$.

19. $F\left\{x + y + u, (x - 2y)^2 + 3u^2\right\} = 0$. Hint: $\dfrac{(dx - 2dy)}{9u} = \dfrac{du}{-3(x-2y)}$.

$(x - 2y)^2 + 3u^2 = (x + y + u)^2$.

20. $F\left(x^2 + y, \ yu\right) = 0$, $\quad \left(x^2 + y\right)^4 = yu$.

21. Hint: $x - y + z = c_1$, $\quad \dfrac{dz}{-(x+y+z)} = \dfrac{(dx+dy+dz)}{8z}$, and hence,

$8z^2 + (x + y + z)^2 = c_2$. $\quad F\left\{(x - y + z), \ 8z^2 + (x + y + z)^2\right\} = 0$.

$c_1^2 + c_2 = 2a^2$, or $(x - y + z)^2 + (x + y + z)^2 + 8z^2 = 2a^2$.

22. $F\left(x^2 + y^2 + z^2, \ y^2 - 2yz - z^2\right) = 0$.

(a) $y^2 - 2yz - z^2 = 0$, two planes $y = \left(1 \pm \sqrt{2}\right)z$.

(b) $x^2 + 2yz + 2z^2 = 0$, a quadric cone with vertex at the origin.

(c) $x^2 - 2yz + 2y^2 = 0$, a quadric cone with vertex at the origin.

23. Use the Hint of 17(c).

$\frac{dx}{dt} = x + y, \quad \frac{dy}{dt} = x - y, \quad \frac{d^2x}{dt^2} = 2x.$

$\left(\frac{dx}{dt}\right)^2 = 2x^2 + c.$ When $x = 0 = y, \quad \frac{dx}{dt} = \sqrt{2}\,x.$

$\sqrt{2}\,u = \ln x + x^2 - 2xy + 2y.$

24. (a) $a = f\left(x + \frac{3}{2}y\right).$

(b) $x = at + c_1, \quad y = bt, \quad u = c_2\,e^{ct}, \quad c_2 = f(c_1),$

$u(x,y) = f\left(x - \frac{a}{b}y\right)\exp\left(\frac{cy}{b}\right).$

(c) $u = f\left(\frac{x}{1-y}\right)(1-y)^c.$

(d) $x = \frac{1}{2}t^2 + ast + s, \quad y = t; \quad u = y + \frac{1}{2}\alpha\,(\alpha y + 1)^{-1}\left(2x - y^2\right).$

26. (a) Hint: $(f')^2 = 1 - (g')^2 = \lambda^2; \; f'(x) = \lambda$ and $g'(y) = \sqrt{1 - \lambda^2}.$

$f(x) = \lambda x + c_1$ and $g(y) = y\sqrt{1 - \lambda^2} + c_2.$

Hence, $u(x,y) = \lambda x + y\sqrt{1 - \lambda^2} + c.$

(b) Hint: $(f')^2 + (g')^2 = f(x) + g(y)$ or $(f')^2 - f(x) = g(y) - (g')^2 = \lambda.$

Hence, $(f')^2 = f(x) + \lambda \quad$ and $\quad g' = \sqrt{g(y) - \lambda}.$

Or, $\frac{df}{\sqrt{f+\lambda}} = dx \quad$ and $\quad \frac{dg}{\sqrt{g-\lambda}} = dy.$

$f(x) + \lambda = \left(\frac{x+c_1}{2}\right)^2$ and $g(y) - \lambda = \left(\frac{y+c_2}{2}\right)^2.$

$u(x,y) = \left(\frac{x+c_1}{2}\right)^2 + \left(\frac{y+c_2}{2}\right)^2.$

(c) Hint: $(f')^2 + x^2 = -g'(y) = \lambda^2.$

Or $f'(x) = \sqrt{\lambda^2 - x^2}, \quad$ and $\quad g(y) = -\lambda^2 y + c_2.$

Putting $x = \lambda\sin\theta$, we obtain

$f(x) = \frac{1}{2}\lambda^2\sin^{-1}\left(\frac{x}{\lambda}\right) + \frac{x}{2}\sqrt{\lambda^2 - x^2} + c_1,$

$u(x,y) = \frac{1}{2}\lambda^2\sin^{-1}\left(\frac{x}{\lambda}\right) + \frac{x}{2}\sqrt{\lambda^2 - x^2} - \lambda^2 y + (c_1 + c_2).$

(d) Hint: $x^2(f')^2 = \lambda^2$ and $1 - y^2(g')^2 = \lambda^2.$

Or, $f(x) = \lambda\ln x + c_1 \quad$ and $\quad g(y) = \sqrt{1 - \lambda^2}\,\ln y + c_2.$

27. (a) Hint: $v = \ln u$ gives $v_x = \frac{1}{u}\cdot u_x, \quad$ and $\quad v_y = \frac{1}{u}\cdot u_y.$

$x^2\left(\frac{u_x}{u}\right)^2 + y^2\left(\frac{u_y}{u}\right)^2 = 1.$

Or, $x^2 v_x^2 + y^2 v_y^2 = 1 \quad$ gives $\quad x^2(f')^2 + y^2(g')^2 = 1.$

$x^2 \{f'(x)\}^2 = 1 - y^2 (g')^2 = \lambda^2.$

Or, $f(x) = \lambda \ln x + c_1$ and $g(y) = \sqrt{1 - \lambda^2} (\ln y) + c_2.$

Thus, $v(x, y) = \lambda \ln x + \sqrt{1 - \lambda^2} (\ln y) + \ln c$, $(c_1 + c_2 = \ln c).$

$u(x, y) = c x^\lambda y^{\sqrt{1 - \lambda^2}}.$

(b) Hint: $v = u^2$ and $v(x, y) = f(x) + g(y)$ may not work.

Try $u = u(s)$, $s = \lambda x y$, so that $u_x = u'(y) \cdot (\lambda y)$ and $u_y = u'(s) \cdot (\lambda x).$

Consequently, $2\lambda^2 \left(\frac{1}{u} \frac{du}{ds}\right)^2 = 1.$ Or, $\frac{1}{u} \frac{du}{ds} = \frac{1}{\sqrt{2}} \frac{1}{\lambda}.$

Hence, $u(s) = c_1 \exp \left(\frac{s}{\lambda \sqrt{2}}\right).$ $\qquad u(x, y) = c_1 \exp \left(\frac{xy}{\sqrt{2}}\right).$

28. Hint: $v_x = \frac{1}{2} \frac{u_x}{\sqrt{u}}$, $v_y = \frac{1}{2} \frac{u_y}{\sqrt{u}}$. This gives $x^4 (f')^2 + y^2 (g')^2 = 1.$

Or, $x^4 (f')^2 = 1 - y^2 (g')^2 = \lambda^2.$

Or, $x^4 (f')^2 = \lambda^2$ and $y^2 (g')^2 = 1 - \lambda^2.$

Hence, $f(x) = -\frac{\lambda}{x} + c_1$ and $g(y) = \sqrt{1 - \lambda^2} \ln y + c_2$

$u(x, y) = \left(-\frac{\lambda}{x} + \sqrt{1 - \lambda^2} \ln y + c\right)^2.$

29. Hint: $v_x = \frac{u_x}{u}$, $v_y = \frac{u_y}{u}$. $\frac{v_x^2}{x^2} + \frac{v_y^2}{y^2} = 1$, and $v = f(x) + g(y).$

Or, $\frac{(f')^2}{x^2} = 1 - \frac{1}{y^2} (g')^2 = \lambda^2.$

$f'(x) = \lambda x$, and $g'(y) = \sqrt{1 - \lambda^2} y.$

Or, $f(x) = \frac{\lambda}{2} x^2 + c_1$, and $g(y) = \frac{1}{2} y^2 \sqrt{1 - \lambda^2} + c_2.$

$v(x, y) = \frac{\lambda}{2} x^2 + \frac{y^2}{2} \sqrt{1 - \lambda^2} + c = \ln u.$

$u(x, y) = c \exp \left(\frac{\lambda}{2} x^2 + \frac{y^2}{2} \sqrt{1 - \lambda^2}\right)$, $\quad c_1 + c_2 = \ln c.$

$e^{x^2} = u(x, 0) = c e^{\frac{\lambda}{2} x^2}$, which gives $c = 1$ and $\lambda = 2.$

30. (a) Hint: $\xi = x - y$, $\eta = y$; $\quad u(x, y) = e^y f(x - y),$

(b) $\xi = x$, $\eta = y - \frac{x^2}{2}$, $u_\xi = \eta + \frac{1}{2} \xi^2$, $u = \xi \eta + \frac{1}{6} \xi^3 + f(\eta).$

$u(x, y) = xy - \frac{1}{3} x^3 + f \left(y - \frac{x^2}{2}\right).$

(c) $\xi = y \exp(-x^2)$, $\eta = y$, and $e^{2u} f(\xi) = \eta$, $e^{2u} f \left(y e^{-x^2}\right) = y,$

(d) $\frac{dx}{1} = \frac{dy}{-y} = \frac{du}{1+u}$, $\qquad \xi = y e^x$, $\quad \eta = y.$

Thus, $(1 + u) f(\xi) = \frac{1}{\eta}.$ $\qquad$ Or, $(1 + u) f(y e^x) = y^{-1}.$

31. (c) $u(x, y) = \alpha \exp \left(\beta x - \frac{a}{b} \beta y\right).$

32. (a) $v(x,t) = x + ct,$ $u(x,t) = \frac{(6x+3ct^2+5ct^3)}{6(1+2t)}.$

   (b) $v(x,t) = x + ct,$ $u(x,t) = \frac{(6x+3ct^2+4ct^3)}{6(1+2t)}.$

33. (a) $v(x,t) = e^{x+at},$ $u - \frac{1}{a}e^{at} = c_1,$ and

   $u - \frac{1}{a}e^{at} = f\left(x - ut + \frac{t}{a}e^{at} - \frac{1}{a^2}e^{at}\right).$

   $u(x,t) = (1+t)^{-1}\left\{(x-ut) + \left(\frac{1}{a} + \frac{t}{a} - \frac{1}{a^2}\right)e^{at} + \left(\frac{1}{a^2} - \frac{1}{a}\right)\right\}.$

   (b) $v = x - ct,$ $u(x,t) = \frac{(6x-3ct^2+4ct^3)}{6(1-2t)}.$

34. $\frac{dt}{1} = \frac{dy}{-x} = \frac{du}{u},$ $t + \ln x = c_1,$ and $xu = c_2.$ $g(xu, t + \ln x) = 0.$

   Or, $u = \frac{1}{x}h(t + \ln x).$ $u(x,t) = e^t \ln(xe^t),$

   where g and h are arbitrary functions.

3.9 Exercises

11. Hint: Differentiate the first equation with respect to t to obtain $\rho_{tt} +$ $\rho_0 \text{div} u_t = 0.$ Take gradient of the last equation to get $\nabla\rho = -\left(\rho_0/c_0^2\right) u_t.$ We next combine these two equations to obtain $\rho_{tt} = c_0^2 \nabla^2\rho.$ Application of ∇^2 to $p - p_0 = c_0^2(\rho - \rho_0)$ leads to $\nabla^2 p = c_0^2 \nabla^2\rho.$ Also $p_{tt} = c_0^2 \rho_{tt} = c_0^4 \nabla^2\rho = c_0^2 \nabla^2 p.$

   Using $u = \nabla\phi$ in the first equation gives $\rho_t + \rho_0\nabla^2\phi = 0,$ and differenting the last equation with respect to t yields $\rho_t = -\left(\rho_0/c_0^2\right)\phi_{tt}.$ Combining these two equations produces the wave equation for $\phi.$ Finally, we take gradient of the first and the last equations to obtain $\nabla\rho_t + \rho_0\nabla^2 u = 0$ and $\nabla\rho = -\left(\rho_0/c_0^2\right)u_t$ that leads to the wave equation for $u_t.$

14. (a) Differentiate the first equation with respect to t and the second equation with respect to $x.$ Then eliminate V_{xt} and V_x to obtain the desired telegraph equation.

   (e) (i) $\frac{\partial^2}{\partial x^2}(I, V) = \frac{1}{c^2}\frac{\partial^2}{\partial t^2}(I, V),$ $c^2 = \frac{1}{LC}.$

(ii) $\frac{\partial}{\partial t}(I,V) = \kappa \frac{\partial^2}{\partial x^2}(I,V), \qquad \kappa = \frac{1}{RC}.$

(iii) $\left(\frac{\partial^2}{\partial t^2} + 2k\frac{\partial}{\partial t} + k^2\right)(I,V) = c^2 \frac{\partial^2}{\partial x^2}(I,V).$

17. (a) The two-dimensional unsteady Euler equations are

$$\frac{du}{dt} = -\frac{1}{\rho}\frac{\partial p}{\partial x}, \qquad \frac{dv}{dt} = -\frac{1}{\rho}\frac{\partial p}{\partial y},$$

where $\frac{d}{dt} = \frac{\partial}{\partial t} + \mathbf{u}\cdot\nabla = \frac{\partial}{\partial t} + u\frac{\partial}{\partial x} + v\frac{\partial}{\partial y}$, and $\mathbf{u} = (u,v)$.

(b) For two-dimensional steady flow, the Euler equations are

$$u\,u_x + v\,u_y = -\frac{1}{\rho}p_x, \qquad u\,v_x + v\,v_y = -\frac{1}{\rho}p_y.$$

Using $\frac{dp}{d\rho} = c^2$, these equations become

$$u\,u_x + v\,u_y = -c^2\,(\rho_x/\rho), \qquad u\,v_x + v\,v_y = -c^2\,(\rho_y/\rho).$$

Multiply the first equation by u and the second by v and add to obtain

$$u^2\,u_x + uv\,(u_y + v_x) + v^2 v_y = -\left(\frac{c^2}{\rho}\right)(u\rho_x + v\rho_y).$$

Using the continuity equation $(\rho u)_x + (\rho v)_y = 0$, the right hand of this

equation becomes $c^2\,(u_x + v_y)$. Hence is the desired equation.

(c) Using $\mathbf{u} = \nabla\phi = (\phi_x, \phi_y)$, the result follows.

(d) Substitute ρ_x and ρ_y from 17(b) into the continuity equation

$u\rho_x + v\rho_y + \rho\,(u_x + v_y) = 0$ to obtain

$$\left(c^2 - u^2\right)\phi_{xx} - 2uv\phi_{xy} + \left(c^2 - v^2\right)\phi_{yy} = 0.$$

Also

$$du = u_x\,dx + u_y\,dy = -\phi_{xx}dx - \phi_{xy}dy,$$

$$dv = v_x\,dx + v_y\,dy = -\phi_{xy}dx - \phi_{yy}dy.$$

Denoting D for the coefficient determinant of the above equations for

ϕ_{xx}, ϕ_{xy} and ϕ_{yy} gives the solutions

$$\phi_{xx} = -\frac{D_1}{D}, \qquad \phi_{xy} = \frac{D_2}{D}, \qquad \phi_{yy} = \frac{-D_3}{D}.$$

$D = 0$ gives a quadratic equation for the slope of the characteristic C,

that is,

$$\left(c^2 - u^2\right)\left(\frac{dy}{dx}\right)^2 + 2uv\left(\frac{dy}{dx}\right) + \left(c^2 - v^2\right) = 0.$$

Thus, directions are real and distinct provided

$$4u^2v^2 - 4\left(c^2 - u^2\right)\left(c^2 - v^2\right) > 0, \quad \text{or} \quad \left(u^2 + v^2\right) > c^2.$$

$$D_2 = 0 \text{ gives } -\frac{dy}{dx} = \frac{\left(c^2 - v^2\right)}{\left(c^2 - u^2\right)}\left(\frac{dv}{du}\right).$$

Substitute into the quadratic equation to obtain

$$\left(c^2 - v^2\right)\left(\frac{dv}{du}\right)^2 - 2uv\left(\frac{dv}{du}\right) + \left(c^2 - u^2\right) = 0.$$

Note that when $D_1 = D_2 = D_3 = D = 0$, any one of the second order ϕ derivatives can be discontinuous.

18. (a) Hint: Use $\nabla \times \frac{\partial \mathbf{u}}{\partial t} = \frac{\partial \omega}{\partial t}$, $\nabla \times (\mathbf{u} \cdot \nabla \mathbf{u}) \, \mathbf{u} \cdot \nabla \omega - \omega \nabla \mathbf{u}$, where we have used $\nabla \cdot \mathbf{u} = 0$ and $\nabla \cdot \omega = 0$. Since $\nabla \times \nabla f = \mathbf{0}$ for any scalar function f, these lead to the vorticity equation in this simplified model.

(b) The rate of change of vorticity as we follow the fluid is given by the term $\omega \cdot \nabla \mathbf{u}$.

(c) $\mathbf{u} = \mathbf{i}\,u\,(x,y) + \mathbf{j}\,v\,(x,y)$ and $\omega = w\,(x,y)\,\mathbf{k}$ and hence,

$\omega \cdot \nabla \mathbf{u} = w\,(x,y)\,\frac{\partial}{\partial z}\,[\mathbf{i}\,u\,(x,y) + \mathbf{j}\,v\,(x,y)] = 0$. This gives the result.

20. We differentiate the first equation partially with respect to t to find $\mathbf{E}_{tt} = c \operatorname{curl} \mathbf{H}_t$. We then substitute $\mathbf{H}_t$ from the second equation to obtain $\mathbf{E}_{tt} = -c^2 \operatorname{curl}(\operatorname{curl} \mathbf{E})$. Using the vector identity $\operatorname{curl}(\operatorname{curl} \mathbf{E}) = \operatorname{grad}(\operatorname{div} \mathbf{E}) - \nabla^2 \mathbf{E}$ with $\operatorname{div} \mathbf{E} = 0$ gives the desired equation. A similar procedure shows that $\mathbf{H}$ satisfies the same equation.

21. When Hooke's law is used to the rod of variable cross section, the tension at point P is given by $T_P = \lambda A\,(x)\,u_x$, where λ is a constant. A longitudinal vibration would displace each cross sectional element of the rod along the x-axis of the rod. An element PQ of length δx and mass $m = \rho A\,(x)\,\delta x$ will be displaced to $P'Q'$ with length $(\delta x + \delta u)$ with the same mass m. The acceleration of the element $P'Q'$ is u_{tt} so that the difference of the tensions at P' and Q' must be equal to the product $m\,u_{tt}$. Hence, $m\,u_{tt} = T_{Q'} - T_{P'} = \left(\frac{\partial}{\partial t}\,T_{P'}\right)\delta x = \frac{\partial}{\partial x}\left(\lambda A\,(x)\,u_x\right)\delta x$. This gives the equation.

4.6 Exercises

1. (a) $x < 0$, hyperbolic;

$$u_{\xi\eta} = \frac{1}{4}\left(\frac{\xi-\eta}{4}\right)^4 - \frac{1}{2}\left(\frac{1}{\xi-\eta}\right)(u_\xi - u_\eta),$$

$x = 0$, parabolic, the given equation is then in canonical form;

$x > 0$, elliptic and the canonical form is

$$u_{\alpha\alpha} + u_{\beta\beta} = \frac{1}{\beta}u_\beta + \frac{\beta^4}{16}.$$

(b) $y = 0$, parabolic; $y \neq 0$, elliptic, and hence,

$$u_{\alpha\alpha} + u_{\beta\beta} = u_\alpha + e^\alpha.$$

(d) Parabolic everywhere and hence,

$$u_{\eta\eta} = \frac{2\xi}{\eta^2}u_\xi + \frac{1}{\eta^2}e^{\xi/\eta}.$$

(f) Elliptic everywhere for finite values of x and y, then

$$u_{\alpha\alpha} + u_{\beta\beta} = u - \frac{1}{\alpha}u_\alpha - \frac{1}{\beta}u_\beta.$$

(g) Parabolic everywhere

$$u_{\eta\eta} = \frac{1}{1-e^{2(\eta-\xi)}}\left[\sin^{-1}\left(e^{\eta-\xi}\right) - u_\xi\right].$$

(h) $B^2 - 4AC = y - 4x$. Equation is hyperbolic if $y > 4x$, parabolic if $y = 4x$ and elliptic if $y < 4x$.

(i) $y = 0$, parabolic; and $y \neq 0$, hyperbolic,

$$u_{\xi\eta} = \frac{(1+\xi-\ln\eta)}{\eta}u_\xi + u_\eta + \frac{1}{\eta}u.$$

2. (i) $u(x,y) = f(y/x) + g(y/x)e^{-y}$,

(ii) Hint: $\varphi = ru$ and check the solution by substitution.

$u(r,t) = (1/r)f(r+ct) + (1/r)g(r-ct)$;

(iii) $A = 4$, $B = 12$, $C = 9$. Hence, $B^2 - 4AC = 0$. Parabolic at every point in (x,y)-plane. $\frac{dy}{dx} = \frac{3}{2}$ or $y = \frac{3}{2}x + c \Rightarrow 2y - 3x = c_1$, $\xi = 2y - 3x$, $\eta = y$. The canonical form is $u_{\eta\eta} - u = 1 \Rightarrow u(\xi,\eta) = f(\xi)\cosh\eta + g(\xi)\sinh\eta - 1$. Or, $u(x,y) = f(2y - 3x)\cosh y + g(2y - 3x)\sinh y - 1$.

(iv) Hyperbolic at all points in the (x,y)-plane. $\xi = y - 2x$, $\eta = y + x$. Thus, $u_{\xi\eta} + u_\eta = \xi$, $u(\xi,\eta) = \eta(\xi-1) + f(\xi)e^{-\xi} + g(\xi)$.

$u\left(x,y\right)=\left(x+y\right)\left(y-2x-1\right)+f\left(x+y\right)\exp\left(2x-y\right)+g\left(y-2x\right).$

(v) Hyperbolic. $\xi=y,\ \eta=y-3x.\ \xi u_{\xi\eta}+u_{\eta}=0,\ u\left(\xi,\eta\right)=\frac{1}{\xi}f\left(\eta\right)+g\left(\xi\right).$
$u\left(x,y\right)=\frac{1}{y}fu\left(y-3x\right)+g\left(y\right).$

(vi) $A=1,\ B=0,\ C=1,\ B^{2}-4AC=-4<0.$ So, this equation is elliptic. $\frac{dy}{dx}=\pm i$ or $\frac{dx}{dy}=\mp i$ or $\xi=x+iy=c_{1}$ and $\eta=x-iy=c_{2}.$

The general solution is $u=\phi\left(\xi\right)+\psi\left(\eta\right)=\phi\left(x+iy\right)+\psi\left(x-iy\right).$

(vii) $u=\phi\left(x+2iy\right)+\psi\left(x-2iy\right).$

(viii) $B^{2}-4AC=0.$ Equation is parabolic. The general solution is given by (4.3.16), where $\lambda=\left(\frac{B}{2A}\right)=-1$ and hence, the general solution becomes $u=\phi\left(y+x\right)+y\psi\left(y+x\right).$

(xi) $B^{2}-4AC=25>0.$ Hyperbolic. The general solution is
$$u=\phi\left(y-\tfrac{3}{2}x\right)+\psi\left(y-\tfrac{1}{6}x\right).$$

3. (a) $\xi=\left(y-x\right)+i\sqrt{2}\,x,\ \eta=\left(y-x\right)-i\sqrt{2}\,x,\qquad\alpha=y-x,\ \beta=\sqrt{2}\,x,$
$u_{\alpha\alpha}+u_{\beta\beta}=-\frac{1}{2}u_{\alpha}-2\sqrt{2}\,u_{\beta}-\frac{1}{2}u+\frac{1}{2}\exp\left(\beta/\sqrt{2}\right).$

(b) $\xi=y+x,\ \ \eta=y;\ \ u_{\eta\eta}=-\frac{3}{2}u.$

(c) $\xi=y-x,\ \ \eta=y-4x;\ \ u_{\xi\eta}=\frac{7}{9}\left(u_{\xi}+u_{\eta}\right)-\frac{1}{9}\sin\left[\left(\xi-\eta\right)/3\right].$

(d) $\xi=y+ix,\ \eta=y-ix.$ Thus, $\alpha=y,\ \ \beta=x.$

The given equation is already in canonical form.

(e) $\xi=x,\ \eta=x-\left(y/2\right);\ \ u_{\xi\eta}=18u_{\xi}+17u_{\eta}-4.$

(f) $\xi=y+\left(x/6\right),\qquad\eta=y;\ \ u_{\xi\eta}=6u-6\eta^{2}.$

(g) $\xi=x,\ \ \eta=y;$ the given equation is already in canonical form.

(h) $\xi=x,\ \ \eta=y;$ the given equation is already in canonical form.

(i) Hyperbolic in the $\left(x,y\right)$ plane except the axes $x=y=0.$ $\xi=xy,\ \eta=\left(y/x\right);\ y=\sqrt{\xi\eta},\ x=\sqrt{\xi/\eta};\ \ u_{\xi\eta}=\frac{1}{2}\left(1+\frac{1}{2}\sqrt{\frac{\eta}{\xi}}\right)u_{\eta}-\frac{1}{4}\frac{1}{\sqrt{\xi\eta}}-\frac{1}{4\xi\eta}-\frac{1}{2}.$

(j) Elliptic when $y>0;\ \ \frac{dy}{dx}=\pm i\sqrt{y},\ \ \alpha=2\sqrt{y}$ and $\beta=-x;$
$u_{\alpha\alpha}+u_{\beta\beta}=\alpha^{2}u_{\beta}.$ Parabolic when $y=0;\ \ u_{xx}+\frac{1}{2}u_{y}=0.$

Hyperbolic when $y < 0$; $\xi = x - 2\sqrt{-y}$, $\eta = -x - 2\sqrt{-y}$.

The canonical form is $u_{\xi\eta} = \frac{1}{16}(\xi + \eta)^2 (u_\eta - u_\xi)$.

(k) Parabolic, $\frac{dy}{dx} = (xy)^{-1}$.   Integrating gives $\frac{1}{2}y^2 = \ln x + \ln \xi$,   where

ξ is an integrating constant.   Hence, $\xi = \frac{1}{x} \exp\left(\frac{1}{2}y^2\right)$, $\eta = x$.

$u_{xx} = x^{-4} e^{y^2} u_{\xi\xi} - 2x^{-2} \exp\left(\frac{1}{2}y^2\right) u_{\xi\eta} + u_{\eta\eta} + 2x^{-2} \exp\left(\frac{1}{2}y^2\right) u_\xi$,

$u_{xy} = -yx^{-3} \exp\left(y^2\right) u_{\xi\xi} + yx^{-1} \exp\left(\frac{1}{2}y^2\right) u_{\xi\eta} - yx^{-2} \exp\left(\frac{1}{2}y^2\right) u_\xi$,

$u_{yy} = \left(y^2 x^{-2}\right) e^{y^2} u_{\xi\xi} + \left(y^2 x^{-2}\right) \exp\left(\frac{1}{2}y^2\right) u_\xi$. $u_{\eta\eta} + (\xi/\eta^2) u_\xi = 0$.

(l) Elliptic if $y > 0$, $\xi = x + 2i\sqrt{y}$, $\eta = x - 2i\sqrt{y}$,

$\alpha = \frac{1}{2}(\xi + \eta) = x$, $\beta = \frac{1}{2i}(\xi - \eta) = 2\sqrt{y}$; $u_{\alpha\alpha} + u_{\beta\beta} = \frac{1}{\beta} u_\beta$.

Hyperbolic $y < 0$, $\xi = x + 2i\sqrt{y}$, $\eta = x - 2i\sqrt{y}$, $\xi - \eta = 4i\sqrt{y}$;

$u_{\xi\eta} + \frac{1}{2}\left(\frac{u_\xi - u_\eta}{\xi - \eta}\right) = 0$. The equations of the characteristic curves are

$\frac{dy}{dx} = \pm i\sqrt{y}$ that gives $2\sqrt{y} = \pm i(x - c)$,   or $y = \mp \frac{1}{4}(x - c)^2$,

where c is an integrating constant. Two branches of parabolas with

positive or negative slopes.

4. (i) $u(x,y) = f(x + cy) + g(x - cy)$; (ii) $u(x,y) = f(x + iy) + f(x - iy)$;

  (iii) Use $z = x + iy$. Hence, $u(x,y) = (x - iy) f_1(x + iy) + f_2(x + iy)$

  $+ (x + iy) + f_3(x - iy) + f_4(x - iy)$

  (iv) $u(x,y) = f(y + x) + g(y + 2x)$;   (v) $u(x,y) = f(y) + g(y - x)$;

  (vi) $u(x,y) = (-y/128)(y - x)(y - 9x) + f(y - 9x) + g(y - x)$.

5. (i) $v_{\xi\eta} = -(1/16) v$,   (ii) $v_{\xi\eta} = (84/625) v$.

7. (ii) Use $\alpha = \frac{3y}{2}$, $\beta = -x^3/2$.

8. $x = r\cos\theta$, $y = r\sin\theta$; $r = \sqrt{x^2 + y^2}$, $\theta = \tan^{-1}\left(\frac{y}{x}\right)$.

$\frac{\partial u}{\partial x} = \frac{\partial u}{\partial r} \cdot \frac{\partial r}{\partial x} + \frac{\partial u}{\partial \theta} \cdot \frac{\partial \theta}{\partial x} = u_r \cdot \frac{x}{r} + u_\theta \cdot \left(-\frac{y}{r^2}\right)$.

$u_{xx} = (u_x)_x = (u_x)_r \cdot \frac{\partial r}{\partial x} + (u_x)_\theta \frac{\partial \theta}{\partial x}$

$= \left(\frac{x}{r} u_r - \frac{y}{r^2} u_\theta\right)_r \left(\frac{x}{r}\right) + \left(\frac{x}{r} u_r - \frac{y}{r^2} u_\theta\right)\left(-\frac{y}{r^2}\right)$

$= \left(\frac{x}{r} u_{rr} - \frac{x}{r^2} u_r + \frac{1}{r} u_r \frac{\partial x}{\partial r}\right)\frac{x}{r} - \left(\frac{y}{r^2} u_{r\theta} - \frac{2y}{r^3} u_\theta + \frac{1}{r^2} \frac{\partial y}{\partial r} u_\theta\right)\frac{x}{r}$

$+ \left(\frac{x}{r} u_{r\theta} + \frac{1}{r} u_r \cdot \frac{\partial x}{\partial \theta}\right)\left(-\frac{y}{r^2}\right) + \left(\frac{y}{r^2} u_{\theta\theta} + \frac{1}{r^2} u_\theta \cdot \frac{\partial y}{\partial \theta}\right)\frac{y}{r^2}$

$$= \tfrac{x^2}{r^2}\, u_{rr} - \tfrac{2xy}{r^3}\, u_{r\theta} + \tfrac{y^2}{r^4}\, u_{\theta\theta} + \tfrac{y^2}{r^3}\, u_r + \tfrac{2xy}{r^4}\, u_\theta.$$

Similarly,

$$u_{yy} = \tfrac{y^2}{r^2}\, u_{rr} + \left(\tfrac{2xy}{r^3}\right) u_{r\theta} + \tfrac{x^2}{r^4}\, u_{\theta\theta} + \tfrac{x^2}{r^3}\, u_r - \tfrac{2xy}{r^4}\, u_\theta.$$

Adding gives the result: $\nabla^2 u = u_{xx} + u_{yy} = u_{rr} + \tfrac{1}{r}\, u_r + \tfrac{1}{r^2} u_{\theta\theta} = 0.$

9. (c) Use Exercise 8.

10. (a) $u_x = u_\xi \xi_x + u_\eta \eta_x = a\, u_\xi + c\, u_\eta = \left(a\, \tfrac{\partial}{\partial \xi} + c\, \tfrac{\partial}{\partial \eta}\right) u,$

$u_y = u_\xi \xi_y + u_\eta \eta_y = b\, u_\xi + d\, u_\eta = \left(b\, \tfrac{\partial}{\partial \xi} + d\, \tfrac{\partial}{\partial \eta}\right) u.$

$u_{xx} = (u_x)_x = \left(a\, \tfrac{\partial}{\partial \xi} + c\, \tfrac{\partial}{\partial \eta}\right)\left(a\, \tfrac{\partial}{\partial \xi} + c\, \tfrac{\partial}{\partial \eta}\right) u$

$\qquad = \left(a^2 u_{\xi\xi} + 2ac\, u_{\xi\eta} + c^2 u_{\eta\eta}\right).$

$u_{yy} = (u_y)_y = \left(b\, \tfrac{\partial}{\partial \xi} + d\, \tfrac{\partial}{\partial \eta}\right)\left(b\, \tfrac{\partial}{\partial \xi} + d\, \tfrac{\partial}{\partial \eta}\right) u$

$\qquad = b^2 u_{\xi\xi} + 2bd\, u_{\xi\eta} + d^2 u_{\eta\eta}.$

$u_{xy} = (u_y)_x = \left(a\, \tfrac{\partial}{\partial \xi} + c\, \tfrac{\partial}{\partial \eta}\right)\left(b\, \tfrac{\partial}{\partial \xi} + d\, \tfrac{\partial}{\partial \eta}\right) u$

$\qquad = ab\, u_{\xi\xi} + (ad + bc)\, u_{\xi\eta} + cd\, u_{\eta\eta}.$

Consequently,

$$0 = A\, u_{xx} + 2B\, u_{xy} + C\, u_{yy}$$

$$= \left(A\, a^2 + 2Bab + C\, b^2\right) u_{\xi\xi} + 2\left[ac\, A + (ad + bc)\, B + bd\, C\right] u_{\xi\eta}$$

$$+ \left(A\, c^2 + 2Bcd + C\, d^2\right) u_{\eta\eta}.$$

Choose arbitrary constants a, b, c and d such that $a = c = 1$ and such that b and d are the two roots of the equation

$$C\lambda^2 + 2B\lambda + A = 0, \quad \text{and}$$

$$\lambda = \tfrac{-B \pm \sqrt{D}}{C} = b, d, \quad D = B^2 - AC.$$

Thus, the transformed equation with $a = c = 1$ is given by

$$[A + (b + d)\, B + bd\, C]\, u_{\xi\eta} = 0.$$

Or,

$$\left(\tfrac{2}{C}\right)\left(AC - B^2\right) u_{\xi\eta} = 0.$$

If $B^2 - AC > 0$, the equation is hyperbolic, and the equation $u_{\xi\eta} = 0$ is in the canonical form. The general solution of this canonical equation is $u = \phi(\xi) + \psi(\eta)$, where ϕ and ψ are arbitrary functions and the transformation becomes $\xi = x + by$ and $\eta = x + dy$, where b, d are real and distinct.

If $B^2 - AC < 0$, the equation is elliptic, and b and d are complex conjugate numbers $\left(d = \bar{b}\right)$. With $a = c = 1$, the transformation is given by $\xi = x + by$ and $\eta = x + \bar{b}y$. Then $\alpha = \frac{1}{2}(\xi + \eta)$ and $\beta = \frac{1}{2i}(\xi - \eta)$ can be used to transform the equation into the canonical form $u_{\alpha\alpha} + u_{\beta\beta} = 0$.

If $B^2 - AC = 0$, the equation is parabolic, here $b = -\frac{B}{C}$, a, c and d are arbitrary, but c and d are not both zero. Choose $a = c = 1$, $d = 0$ so that $\xi = x - \frac{B}{C}y$ and $\eta = y$ are used to transform the equation into the form $u_{\eta\eta} = 0$. The general solution is $u = \phi(\xi) + \eta\psi(\eta)$, where ϕ and ψ are arbitrary functions, and b is the double root of $C\lambda^2 + 2B\lambda + A = 0$, and $\xi = x + by$.

11. Seek a trial solution $u(x,y) = f(x + my)$ so that $u_{xx} = f''$, $u_{yy} = m^2 f''$. Substituting into the Laplace equation yields $(m^2 + 1) f'' = 0$ which gives that either $f'' = 0$ or $m^2 + 1 = 0$. Thus, $m = \pm i$. The general solution is $u(x,y) = F(x + iy) + G(x - iy)$. Identifying c with i gives the d'Alembert solution

$$u(x,y) = \frac{1}{2}[f(x + iy) + f(x - iy)] + \frac{1}{2i}\int_{x-iy}^{x+iy} g(\alpha)\, d\alpha.$$

12. (a) Hyperbolic. $(\xi, \eta) = \frac{2}{3}\left(y^{3/2} \pm x^{3/2}\right)$, $\quad 3\left(\xi^2 - \eta^2\right)u_{\xi\eta} = \eta u_\eta - \xi u_\eta$.

   (b) Elliptic. $\frac{dy}{dx} = \pm i\,\text{sech}^2 x$, $\quad \xi = y + i\tanh x$, $\quad \eta = y - i\tanh x$;
   $\alpha = y$, $\quad \beta = \tanh x$. $\quad$ Thus, $u_{\alpha\alpha} + u_{\beta\beta} = \frac{2\beta}{(1-\beta^2)}u_\beta$.

   (d) Hyperbolic. $\xi = y + \tanh x$, $\quad \eta = y - \tanh x$.
   $u_{\xi\eta} = \left[4 - (\xi - \eta)^2\right]^{-1}(\eta - \xi)(u_\xi - u_\eta)$ in the domain $(\eta - \xi)^2 < 4$.

   (e) Parabolic. $\xi = y - 3x$, $\eta = y$; $\quad u_{\eta\eta} = -\frac{\eta}{3}(u_\xi + u_\eta)$.

   (f) Elliptic. $\alpha = \frac{1}{2}(y^2 - x^2)$, $\beta = \frac{1}{2}x^2$. $\quad$ The canonical form is

$$u_{\alpha\alpha} + u_{\beta\beta} = [2\beta (\alpha + \beta)]^{-1} [\alpha u_\alpha - (\alpha + 2\beta) u_\beta].$$

(g) Elliptic. $\alpha = \sin x + y,$ $\beta = x,$ $u_{\alpha\alpha} + u_{\beta\beta} = (\sin \beta) u_\alpha - u.$

(h) Parabolic. $\xi = x + \cos y,$ $\eta = y.$ Thus, $u_{\eta\eta} = (\sin^2 \eta \cos \eta) u_\xi.$

13. The general solution is

$$u(x, y) = e^x \int_0^x e^{-\alpha} \cos(\alpha + y e^{\alpha - x}) \, d\alpha + e^x f(y e^{-x}) + g(x),$$

where f and g are arbitrary functions.

5.12 Exercises

1. (a) $u(x, t) = t,$ (b) $u(x, t) = \sin x \cos ct + x^2 t + \frac{1}{3}c^2 t^3,$

   (c) $u(x, t) = x^3 + 3c^2 x t^2 + xt,$ (d) $u(x, t) = \cos x \cos ct + (t/e),$

   (e) $u(x, t) = 2t + \frac{1}{2} \left[\log(1 + x^2 + 2cxt + c^2 t^2) \right.$

$$\left. + \log(1 + x^2 - 2cxt + c^2 t^2) \right],$$

   (f) $u(x, t) = x + (1/c) \sin x \sin ct.$

2. (a) $u(x, t) = 3t + \frac{1}{2}xt^2.$

   (c) $u(x, t) = 5 + x^2 t + \frac{1}{3}c^2 t^3 + (1/2c^2)(e^{x+ct} + e^{x-ct} - 2e^x),$

   (e) $u(x, t) = \sin x \cos ct + (e^t - 1)(xt + x) - x t e^t,$

   (f) $u(x, t) = x^2 + t^2(1 + c^2) + (1/c) \cos x \sin ct.$

$$3. \ s(r, t) = \begin{cases} 0, & 0 \le t < r - R \\ \frac{s_0(r-t)}{2r}, & r - R < t < r + R \\ 0, & r + R < t < \infty \end{cases}$$

4. $u(x, t) = \frac{1}{4}\sin(y + x) + \frac{3}{4}\sin(-y/3 + x) + y^2/3 + xy.$

5. $u(x, t) = \sin x \cos at + xt,$ where a is a physical constant.

6. (b) $u(x, t) = \frac{1}{2}f(xt) + \frac{1}{2}tf\left(\frac{x}{t}\right) + \frac{1}{4}\sqrt{xt} \int_{xt}^{x/t} \frac{f(\tau)}{\tau^{\frac{3}{2}}} d\tau - \frac{1}{2}\sqrt{xt} \int_{xt}^{x/t} \frac{g(\tau)}{\tau^{\frac{3}{2}}} d\tau.$

19. $u(x, t) = f\left(\frac{\sqrt{y^2 + x^2 - 8}}{2}\right) + g\left(\frac{\sqrt{y^2 - x^2 + 16}}{2}\right) - f(2).$

22. $u(x, t) = g\left(\frac{y - \frac{x^2}{2} + 4}{2}\right) + f\left(\frac{y + x^2/2}{2}\right) - f(2).$

28. The wave equation is

$$u_{tt} - c^2 u_{xx} = q(x, t), \quad c^2 = T/\rho.$$

Multiply the wave equation by u_t and rewrite to obtain

$$\frac{d}{dt}\left(\frac{1}{2}u_t^2 + \frac{1}{2}c^2 u_x^2\right) - c^2 \frac{\partial}{\partial x}(u_t u_x) = q(x, t) u_t.$$

Integrating this result gives the energy equation. In view of the fact that $u_t(0, t) = 0 = u_t(l, t)$, the energy equation with no external forces gives

$$\frac{dE}{dt} = 0 \Rightarrow E(t) = \text{constant}.$$

29. This problem is identically the same as that of (5.5.1). In this case $f(x) = 0 = g(x)$, and $U(t) = p(t)$. So the solution is given by (5.5.2) and (5.5.3). Consequently,

$$u(x, t) = \begin{cases} U\left(t - \frac{x}{c}\right), & x < ct, \\ 0, & x > ct. \end{cases}$$

30. (a) When $\omega \neq ck$,

$$u(x, t) = \frac{1}{(k^2 c^2 - \omega^2)} \sin(kx - \omega t) - \frac{(\omega - kc)}{2kc(\omega^2 - c^2 k^2)} \sin[k(x + ct)]$$
$$+ \frac{(\omega + kc)}{2kc(\omega^2 - c^2 k^2)} \sin[k(x - ct)].$$

This solution represents three harmonic waves which propagate with different amplitudes and with speeds $\pm c$ and the phase velocity (ω/k).

(b) When $\omega = ck$,

$$u(x, t) = \frac{1}{4}\sin(x - t) - \frac{1}{4}\sin(x + t) + \frac{1}{2}t\cos(x - t).$$

This solution represents two harmonic waves with constant amplitude and another harmonic wave whose amplitude grows linearly with time.

31. (a) $u(x, t) = \frac{1}{2}[\cos(x - 3t) + \cos(x + 3t)] + \frac{1}{6}\int_{x-3t}^{x+3t} \sin(2\alpha)\, d\alpha$
$$= \cos x \cos(3t) + \frac{1}{6}\sin(2x)\sin(6t).$$

(c) $u(x, t) = \cos(3x)\cos(21t) + tx.$

(e) $u(x, t) = x^3 + 27xt^2 + \frac{1}{6}[\cos(x + 3t) - \cos(x - 3t)]$
$$+ \frac{1}{6}[(x + 3t)\sin(x + 3t) - (x - 3t)\sin(x - 3t)].$$

(f) $u\left(x,t\right) = \frac{1}{2}\left[\cos\left(x - 4t\right) + \cos\left(x + 4t\right)\right]$

$$+\frac{1}{8}\,e^{-x}\left[\left(x + 1 - 4t\right)e^{4t} - \left(x + 1 + 4t\right)e^{-4t}\right].$$

32. Verify that

$$u\left(x,t\right) = \int_0^t v\left(x,t;\tau\right)d\tau$$

satisfies the Cauchy problem.

$$u_t\left(x,t\right) = v\left(x,t;t\right) + \int_0^t v_t\left(x,t;\tau\right)d\tau = \int_0^t v_t\left(x,t;\tau\right)d\tau$$

$$u_{tt}\left(x,t\right) = v_t\left(x,t;t\right) + \int_0^t v_{tt}\left(x,t;\tau\right)d\tau = p\left(x,t\right) + \int_0^t v_{tt}\left(x,t;\tau\right)d\tau$$

$$u_{xx}\left(x,t\right) = \int_0^t v_{xx}\left(x,t;\tau\right)d\tau.$$

Thus,

$$u_{tt} - c^2 u_{xx} = p\left(x,t\right) + \int_0^t \left(v_{tt} - c^2 v_{xx}\right)d\tau = p\left(x,t\right).$$

33. $u_t = v\left(x,t;t\right) + \int_0^t v_t\left(x,t;\tau\right)d\tau = p\left(x,t\right) + \int_0^t v_t\left(x,t;\tau\right)d\tau$

$$u_{xx} = \int_0^t v_{xx}\left(x,t;\tau\right)d\tau.$$

Hence,

$$u_t - \kappa\,u_{xx} = p\left(x,t\right) + \int_0^t \left(v_t - \kappa\,v_{xx}\right)d\tau = p\left(x,t\right).$$

34. According to the Duhamel principle

$$u\left(x,t\right) = \int_0^t v\left(x,t;\tau\right)d\tau$$

is the solution of the problem where $v\left(x,t;\tau\right)$ satisfies

$$v_t = \kappa\,v_{xx}, \qquad 0 \le x < 1, \quad t > 0,$$

$$v\left(0,t;\tau\right) = 0 = v\left(1,t;\tau\right),$$

$$v\left(x,\tau;\tau\right) = e^{-\tau}\sin\pi x, \qquad 0 \le x \le 1.$$

Using the separation of variables, the solution is given by

$v\left(x,t\right) = X\left(x\right)T\left(t\right)$ so that

$$X'' + \lambda^2 X = 0 \quad \text{and} \quad T' + \kappa\lambda^2 T = 0.$$

The solution is

$$v\left(x,t;\tau\right) = \sum_{n=1}^{\infty} a_n\left(\tau\right)e^{-\lambda_n^2\kappa t}\sin\lambda_n x,$$

when $\lambda_n = n\pi, \qquad n = 1, 2, 3, \ldots.$

Since $v\left(x, \tau; \tau\right) = e^{-\tau} \sin \pi x$,

$$e^{-\tau} \sin \pi x = \sum_{n=1}^{\infty} a_n\left(\tau\right) \exp\left(-n^2 \pi^2 \kappa \tau\right) \sin\left(\pi n x\right).$$

Equating the coefficients gives

$$e^{-\tau} = a_1\left(\tau\right) \exp\left(-\pi^2 \kappa \tau\right), \quad a_n\left(\tau\right) = 0, \; n = 2, 3, \ldots.$$

Consequently,

$$v\left(x, t; \tau\right) = \exp\left[\left(\pi^2 \kappa - 1\right)\tau\right] \exp\left(-\pi^2 \kappa t\right) \sin \pi x.$$

Thus,

$$u\left(x, t\right) = \exp\left(-\pi^2 \kappa t\right) \sin \pi x \int_0^t \exp\left[\left(\pi^2 \kappa - 1\right)\tau\right] d\tau$$

$$= \frac{e^{-t} - \exp\left(-\pi^2 \kappa t\right)}{\left(\pi^2 \kappa - 1\right)} \cdot \sin \pi x.$$

36. (a) The solution is $u\left(x, t\right) = \frac{1}{n}\left[e^{nx} \sin\left(2n^2 t + nx\right) + e^{-nx} \sin\left(2n^2 t - nx\right)\right]$,

and $u\left(x, t\right) \to \infty$ as $n \to \infty$ for certain values of x and t.

(b) $u_n\left(x, y\right) = \frac{1}{n} \exp\left(-\sqrt{n}\right) \sin nx \sinh ny$ is the solution. For $y \neq 0$,

$u_n\left(x, y\right) \to \infty$ as $n \to \infty$. But $\left(u_n\right)_y \left(x, 0\right) = \exp\left(-\sqrt{n}\right) \sin nx \to 0$ as

$n \to \infty$.

6.14 Exercises

1. (a) $f\left(x\right) = -\frac{\pi}{4} + \frac{h}{2} + \sum_{k=1}^{\infty} \left\{\frac{1}{\pi k^2}\left[1 + \left(-1\right)^{k+1}\right] \cos kx\right.$

$$\left. + \frac{1}{\pi k}\left[h + \left(h + \pi\right)\left(-1\right)^{k+1}\right] \sin kx\right\}.$$

(c) $f\left(x\right) = \sin x + \sum_{k=1}^{\infty} \frac{2\left(-1\right)^{k+1}}{k} \sin kx.$

(e) $f\left(x\right) = \frac{\sinh \pi}{\pi}\left[1 + \sum_{k=1}^{\infty} \frac{2\left(-1\right)^k}{1+k^2}\left(\cos kx - k \sin kx\right)\right].$

2. (a) $f\left(x\right) = \sum_{k=1}^{\infty} \frac{2}{k} \sin kx$

(b) $f\left(x\right) = \sum_{k=1}^{\infty} \left(\frac{2}{\pi k}\right)\left[1 - 2\left(-1\right)^k + \cos \frac{k\pi}{2}\right] \sin kx.$

(c) $f(x) = \sum_{k=1}^{\infty} \left[2(-1)^{k+1} \frac{\pi}{k} + \frac{4}{\pi k^3} \left((-1)^k - 1 \right) \right] \sin kx.$

(d) $f(x) = \sum_{k=2}^{\infty} \frac{2k}{\pi} \left[\frac{1+(-1)^k}{k^2-1} \right] \sin kx.$

3. (a) $f(x) = \frac{3}{2}\pi + \sum_{k=1}^{\infty} \frac{2}{\pi k^2} \left[(-1)^k - 1 \right] \cos kx.$

(b) $f(x) = \frac{\pi}{2} + \sum_{k=1}^{\infty} \frac{2}{\pi k^2} \left[(-1)^k - 1 \right] \cos kx.$

(c) $f(x) = \frac{\pi^2}{3} + \sum_{k=1}^{\infty} \frac{4(-1)^k}{k^2} \cos kx.$

(d) $f(x) = \frac{2}{3\pi} + \sum_{k=1,2,4,\dots}^{\infty} \frac{6}{\pi} \left[\frac{1+(-1)^k}{9-k^2} \right] \cos kx, \qquad k \neq 3.$

4. (b) $f(x) = \sum_{k=1}^{\infty} \left(\frac{2}{k\pi} \right) \sin \frac{k\pi}{2} \cos \left(\frac{k\pi x}{6} \right).$

(c) $f(x) = \frac{2}{\pi} + \sum_{k=2}^{\infty} \left(\frac{2}{k\pi} \right) \left[\frac{1+(-1)^k}{1-k^2} \right] \cos \left(\frac{k\pi x}{l} \right).$

(f) $f(x) = \sum_{k=1}^{\infty} \frac{k\pi}{1+k^2\pi^2} (-1)^{k+1} \left(e - e^{-1} \right) \sin (k\pi x).$

5. (a) $f(x) = \sum_{k=-\infty}^{\infty} \frac{1}{\pi} \left(\frac{2+ik}{4+k^2} \right) (-1)^k \sinh 2\pi \, e^{ikx}.$

(b) $f(x) = \sum_{k=-\infty}^{\infty} \frac{(-1)^k}{\pi(1+k^2)} \sinh \pi \, e^{ikx}.$

(d) $f(x) = \sum_{k=-\infty}^{\infty} (-1)^k \left(\frac{i}{k\pi} \right) e^{ik\pi x}.$

6. (a) $f(x) = \frac{\pi}{8} + \sum_{k=1}^{\infty} \left[\frac{1}{2\pi k^2} \left\{ (-1)^k - 1 \right\} \cos kx + \frac{(-1)^{k+1}}{2k} \sin kx \right].$

7. (a) $f(x) = \frac{l^2}{3} + \sum_{k=1}^{\infty} 4(-1)^k \left(\frac{1}{k\pi} \right)^2 \cos \left(\frac{k\pi x}{l} \right).$

8. (a) $\sin^2 x = \sum_{k=1,3,4,\dots}^{\infty} \frac{4(1-\cos k\pi)}{k\pi(4-k^2)} \sin kx.$

(b) $\cos^2 x = \sum_{k=1,3,4,\dots}^{\infty} \frac{2}{k\pi} \left(\frac{1-k^2}{4-k^2} \right) (1 - \cos k\pi) \sin kx.$

(d) $\sin x \cos x = \sum_{k=1,3,4,\dots}^{\infty} \frac{2}{\pi} \left(\frac{1-\cos k\pi}{4-k^2} \right) \cos kx.$

9. (a) $\frac{x^2}{4} = \frac{\pi^2}{12} - \sum_{k=1}^{\infty} \frac{(-1)^{k+1}}{k^2} \cos kx$.

(c) $\int_0^{\infty} \ln\left(2\cos\frac{x}{2}\right) dx = \sum_{k=1}^{\infty} (-1)^{k+1} \frac{\sin kx}{k^2}$.

(e) $\frac{\pi}{2} - \frac{4}{\pi} \sum_{k=1}^{\infty} \frac{\cos(2k-1)x}{(2k-1)^2} = \begin{cases} -x, & -\pi < x < 0 \\ x, & 0 < x < \pi. \end{cases}$

10. (a) $f(x,y) = \frac{16}{\pi^2} \sum_{m=1,3,\ldots} \sum_{n=1,3,\ldots} \left(\frac{1}{mn}\right) \sin mx \sin ny$

(c) $f(x,y) = \frac{\pi^4}{9} + \frac{1}{2}\sum_{m=1}^{\infty} \frac{8}{3}\pi^2 \frac{(-1)^m}{m^2} \cos mx + \frac{1}{2}\sum_{n=1}^{\infty} \frac{8}{3}\pi^2 \frac{(-1)^m}{n^2} \cos ny$

$+ \sum_{m=1}^{\infty} \sum_{n=1}^{\infty} \frac{16(-1)^{m+n}}{m^2 n^2} \cos mx \cos ny$.

(e) $f(x,y) = \sum_{m=1}^{\infty} \frac{2(-1)^{m+1}}{m} \sin mx \sin y$.

(g) $f(x,y) = \sum_{m=1}^{\infty} \sum_{n=1}^{\infty} d_{mn} \sin\left(\frac{m\pi x}{1}\right) \sin\left(\frac{n\pi y}{2}\right)$,

where

$$d_{mn} = \frac{4}{1.2} \int_0^2 \int_0^1 xy \sin(m\pi x) \sin\left(\frac{n\pi y}{2}\right) dx\, dy$$

$$= 2 \int_0^2 \left[\frac{\sin m\pi x}{m^2\pi^2} - \frac{x\cos m\pi x}{m\pi}\right]_0^1 y \sin\left(\frac{n\pi y}{2}\right) dy$$

$$= \frac{-2(-1)^m}{m\pi} \int_0^2 y\sin\left(\frac{n\pi y}{2}\right) dy = \frac{-2(-1)^m}{m\pi}\left(\frac{-4(-1)^n}{n\pi}\right)$$

$$= \frac{8(-1)^{m+n}}{\pi^2 mn}.$$

(h) $f(x,y) = \left(\frac{16}{\pi^2}\right) \sum_{m=1}^{\infty} \sum_{n=1}^{\infty} [(2m-1)(2n-1)]^{-1} \sin\left[\frac{(2m-1)\pi x}{a}\right]$

$\times \sin\left[\frac{(2n-1)\pi y}{b}\right]$.

(Double Fourier sine series).

(i) $f(x,y) = \left(\frac{\sin 2}{\pi}\right) \sum_{m=1}^{\infty} \frac{(-1)^{m+1}}{m} \sin \pi mx$

$$+ \left(\tfrac{8\sin 2}{\pi}\right) \sum_{m=1}^{\infty}\sum_{n=1}^{\infty} \left[\tfrac{(-1)^{m+n+1}}{m(4-\pi^2 n^2)}\right] \sin\left(m\pi x\right)\cos\left(\tfrac{\pi n y}{2}\right).$$

(j) $f\left(x,y\right) = \tfrac{2}{3}\pi^2 \sum_{m=1}^{\infty} \tfrac{(-1)^{m+1}}{m}\sin mx + \sum_{m=1}^{\infty}\sum_{n=1}^{\infty} \tfrac{8(-1)^{m+n+1}}{m\,n^2}\sin mx\cos ny.$

(k) $f\left(x,y\right) = \tfrac{\pi^4}{9} + \left(\tfrac{4\pi^2}{3}\right)\left[\sum_{m=1}^{\infty}\tfrac{(-1)^m}{m^2}\cos mx + \sum_{n=1}^{\infty}\tfrac{(-1)^n}{n^2}\cos ny\right]$

$$+16\sum_{m=1}^{\infty}\sum_{n=1}^{\infty}\tfrac{(-1)^{m+n}}{m^2 n^2}\cos mx\cos ny.$$

20. (a) $b_{2n} = 0$, $b_{2n+1} = \tfrac{8}{\pi(2n+1)^3}$.

   $f\left(x\right) = x\left(\pi - x\right) = \tfrac{8}{\pi}\left(\tfrac{\sin x}{1^3} + \tfrac{\sin 3x}{3^3} + \tfrac{\sin 5x}{5^3} + \ldots\right).$

   (b) Put $x = \tfrac{\pi}{2}$ and $x = \tfrac{\pi}{4}$ to find the sum of the series.

21. (a) $b_n = \tfrac{2}{\pi}\displaystyle\int_0^{\pi} f\left(x\right)\sin nx\,dx = \left(\tfrac{8}{n^2\pi^2}\right)\sin\left(\tfrac{n\pi}{2}\right), \qquad n = 0,1,2,\ldots.$

   $b_{2n} = 0$ and $b_{2n+1} = \tfrac{8(-1)^n}{\pi^2(2n+1)^2}, \qquad n = 0,1,2,\ldots.$

   (b) Put $x = \tfrac{\pi}{2}$.

22. (a) $f\left(x\right) = \displaystyle\sum_{n=1}^{\infty} b_n\sin\left(\tfrac{n\pi x}{a}\right),$

$$b_n = \tfrac{2}{a}\int_0^a \sin\left(\tfrac{n\pi x}{a}\right)dx = \tfrac{2}{n\pi}\left(1 - \cos n\pi\right)$$

$$= \tfrac{2}{n\pi}\left[1 - (-1)^n\right] = \begin{cases} \tfrac{4}{n\pi}, & \text{for odd } n, \\[2mm] 0, & \text{for even } n. \end{cases}$$

   $f\left(x\right) \sim \tfrac{4}{\pi}\left[\sin\left(\tfrac{\pi x}{a}\right) + \tfrac{1}{3}\sin\left(\tfrac{3\pi x}{a}\right) + \tfrac{1}{5}\sin\left(\tfrac{5\pi x}{a}\right) + \ldots\right]$

   $f\left(x\right) \sim \tfrac{1}{2}a_0 + \displaystyle\sum_{n=1}^{\infty} a_n\cos\left(\tfrac{n\pi x}{a}\right), \qquad a_0 = 2,$

   $a_n = \tfrac{2}{n\pi}\left(\sin n\pi - 0\right) = 0, \qquad n \neq 0,$

   $1 = 1 + 0\cdot\cos\left(\tfrac{n\pi}{a}\right) + 0\cdot\cos\left(\tfrac{2n\pi}{a}\right) + \ldots$

   (b) $f\left(x\right) = \displaystyle\sum_{n=1}^{\infty}(-1)^{n+1}\left(\tfrac{2a}{n\pi}\right)\sin\left(\tfrac{n\pi x}{a}\right)$

   $f\left(x\right) \sim \tfrac{1}{2}a_0 + \displaystyle\sum_{n=1}^{\infty} a_n\cos\left(\tfrac{n\pi x}{a}\right), \quad \text{where} \quad a_0 = \tfrac{2}{a}\int_0^a x\,dx = a,$

   $a_n = \tfrac{2}{a}\displaystyle\int_0^a x\cos\left(\tfrac{n\pi x}{a}\right)dx = \tfrac{2a}{n^2\pi^2}\left((-1)^n - 1\right)$

   $= 0$, for even n, and $-\tfrac{4a}{n^2\pi^2}$, for odd n.

23. (a) $a_0 = \frac{1}{a} \int_{-a}^{a} x \, dx = 0,$

$$a_n = \frac{1}{a} \int_{-a}^{a} x \cos\left(\frac{n\pi x}{a}\right) dx = \left[\frac{x}{n\pi} \sin\left(\frac{n\pi x}{a}\right) + \frac{a}{\pi^2 n^2} \cos\left(\frac{n\pi x}{a}\right)\right]_{-a}^{a}$$

$$= \frac{a}{n^2\pi^2}\left(\cos n\pi - \cos\left(-n\pi\right)\right) = 0$$

$$b_n = \frac{1}{a} \int_0^a x \sin\left(\frac{n\pi x}{a}\right) dx$$

$$= \frac{-x}{n\pi} \cos\left(\frac{n\pi x}{a}\right) + \frac{a}{n^2\pi^2} \sin\left(\frac{n\pi x}{a}\right)\Big|_{-a}^{a}$$

$$= \frac{-a}{n\pi} \cos n\pi + \frac{-a}{n\pi} \cos\left(-n\pi\right) = (-1)^{n+1}\left(\frac{2a}{n\pi}\right).$$

(c) $a_0 = \frac{1}{2\pi} \int_0^{2\pi} f(x) \, dx = \frac{1}{2\pi} \int_{\pi}^{2\pi} dx = \frac{1}{2}$

$$a_n = \frac{1}{\pi} \int_0^{2\pi} f(x) \cos nx \, dx = \frac{1}{\pi} \int_{\pi}^{2\pi} \cos nx \, dx = 0 \quad \text{for all } n.$$

$$b_n = \frac{1}{\pi} \int_0^{2\pi} f(x) \sin nx \, dx = \frac{1}{\pi} \int_{\pi}^{2\pi} \sin nx \, dx$$

$$= \frac{1}{\pi}\left[-1 + (-1)^n\right] = \begin{cases} 0, & n \text{ is even} \\ -\frac{2}{n\pi}, & n \text{ is odd.} \end{cases}$$

$$f(x) = \frac{1}{2} - \frac{2}{\pi} \sum_{k=0}^{\infty} \frac{\sin(2k+1)x}{(2k+1)}.$$

24. $f(x) = \left(\frac{1}{\pi} + \frac{1}{2} \cos x\right) + \frac{2}{\pi} \sum_{n=1}^{\infty} \frac{(-1)^{n+1} \cos 2nx}{(4n^2 - 1)}.$

26. Hint: An argument similar to that used in Section 6.5 can be employed to prove this general Parseval relation. More precisely, use (6.5.10) for $(f \pm g)$ to obtain

$$\frac{1}{\pi} \int_{-\pi}^{\pi} (f \pm g)^2 \, dx = \frac{1}{2}(a_0 \pm \alpha_0)^2 + \sum_{k=1}^{\infty}\left[(a_k \pm \alpha_k)^2 + (b_k \pm \beta_k)^2\right].$$

Subtracting the later equality from the former gives

$$\frac{1}{\pi} \int_{-\pi}^{\pi} f(x) g(x) \, dx = \frac{a_0 \alpha_0}{2} + \sum_{k=1}^{\infty}(a_k \alpha_k + b_k \beta_k).$$

27. (a) $f(x) = \frac{2}{\pi} \int_0^{\infty} \frac{1}{\alpha} \cos \alpha x \sin a\alpha \, d\alpha,$ (b) $f(x) = \frac{2}{\pi} \int_0^{\infty} \frac{\sin \pi\alpha \sin \alpha x}{(1 - \alpha^2)} d\alpha.$

33. $$c_k = \frac{1}{2\pi} \int_{-\pi}^{\pi} x e^{-ikx} dx = \frac{1}{2\pi} \left[\frac{x e^{-ikx}}{-ik} \bigg|_{-\pi}^{\pi} + \frac{1}{ik} \int_{-\pi}^{\pi} e^{-ikx} dx \right]$$

$$= \frac{1}{2\pi} \left[\frac{\pi e^{-ik\pi}}{-ik} + \frac{\pi e^{ik\pi}}{(-ik)} + \frac{e^{-ik\pi}}{k^2} \right]$$

$$= \frac{\pi}{2\pi} \left[\frac{(e^{ik\pi} + e^{-ik\pi})}{(-ik)} \right] = \frac{i \cos k\pi}{k} = (-1)^k \frac{i}{k}.$$

$$c_0 = \frac{1}{2\pi} \int_{-\pi}^{\pi} x \, dx = 0.$$

35. (a) $f(x) = \displaystyle\sum_{k=-\infty}^{\infty} c_k e^{ikx}$,

$$c_k = \frac{1}{2} (a_k - ib_k) = \frac{1}{2\pi} \int_{-\pi}^{\pi} f(x) e^{-ikx} dx,$$

$$= \frac{1}{4\pi} \int_{-\pi}^{\pi} \left[e^{i(a-k)x} + e^{-i(a+k)x} \right] dx,$$

$$= \frac{1}{4\pi i (a-k)} \left[e^{i(a-k)x} \right]_{-\pi}^{\pi} - \frac{1}{4\pi i (a+k)} \left[e^{-i(a+k)x} \right]_{-\pi}^{\pi}$$

$$= \frac{1}{2\pi} \left[\frac{1}{a-k} \sin(a-k)\pi + \frac{1}{a+k} \sin(a+k)\pi \right].$$

This is a real quantity and hence, $b_k = 0$ for $k = 1, 2, 3, \ldots$, and

$$a_k = \frac{2(-1)^k a \sin(\pi a)}{\pi (a^2 - k^2)}, \qquad k = 0, 1, 2, \ldots.$$

Thus,

$$\cos(ax) = \frac{2a \sin a\pi}{\pi} \left(\frac{1}{2a^2} - \frac{\cos x}{k^2 - 1^2} + \frac{\cos 2x}{k^2 - 2^2} - \cdots \right).$$

Since $\cos ax$ is even, the above series is continuous, even at $x = \pm n\pi$.

(b) Putting π for x and treating $a = x$ is a variable

$$\cot \pi x = \frac{2x}{\pi} \left(\frac{1}{2x^2} + \frac{1}{x^2 - 1^2} + \frac{1}{x^2 - 2^2} + \cdots \right)$$

or,

$$\cot \pi x - \frac{1}{\pi x} = -\frac{2x}{\pi} \sum_{n=1}^{\infty} \frac{1}{(n^2 - x^2)}.$$

(c) Since the convergence is uniform in any interval of the x-axis that does not contain any integers, term-by-term integration in $0 < a < x < 1$, gives

$$\pi \int_a^x \left(\cos \pi t - \frac{1}{\pi t} \right) dt = \ln \left(\frac{\sin \pi x}{\pi x} \right) - \ln \left(\frac{\sin \pi a}{\pi a} \right)$$

$$= \sum_{n=1}^{\infty} \ln \left(\frac{n^2 - x^2}{n^2 - a^2} \right).$$

In the limit as $a \to 0$, we find

$$\ln \left(\frac{\sin \pi x}{\pi x} \right) = \sum_{n=1}^{\infty} \ln \left(1 - \frac{x^2}{n^2} \right),$$

or

$$\sin \pi x = \pi x \prod_{n=1}^{\infty} \left(1 - \frac{x^2}{n^2} \right).$$

This is the product representation of $\sin \pi x$.

(d) Putting $x = \frac{1}{2}$, we obtain the Wallis formula for $\frac{\pi}{2}$ as an infinite product

$$\frac{\pi}{2} = \prod_{n=1}^{\infty} \frac{(2n)^2}{(2n-1)(2n+1)} = \frac{2}{1} \cdot \frac{2}{3} \cdot \frac{4}{3} \cdot \frac{4}{5} \cdot \frac{6}{5} \cdot \frac{6}{7} \cdot \frac{8}{7} \cdots.$$

36. (a) $f(x + 2\pi) = f(x)$.

$$a_k = \frac{1}{\pi} \int_0^{2\pi} e^x \cos kx \, dx, \qquad b_k = \frac{1}{\pi} \int_0^{2\pi} e^x \sin kx \, dx.$$

Evaluating these integrals gives

$$f(x) \sim \frac{1}{\pi} \left(e^{2\pi} - 1 \right) \left[\frac{1}{2} + \sum_{k=1}^{\infty} \frac{1}{(k^2 + 1)} \left(\cos kx - k \sin kx \right) \right].$$

(b) In this case, $f(-x) = -f(x)$ and $f(x)$ is odd, and periodic of period 2π.

Hence, $a_k = 0$, and b_k is given by

$$b_k = \frac{2}{\pi} \int_0^{\pi/2} \sin kx \, dx - \frac{2}{\pi} \int_{\frac{\pi}{2}}^{\pi} \sin kx \, dx.$$

Thus,

$$f(x) = \frac{4}{\pi} \left(\sin 2x + \frac{1}{3} \sin 6x + \frac{1}{5} \sin 10x + \ldots \right).$$

(c) $f(x)$ is periodic with period 1. Hence,

$$a_k = 2 \int_0^1 f(x) \cos(2\pi kx)\, dx, \qquad b_k = 2 \int_0^1 f(x) \sin(2\pi kx)\, dx,$$

$$f(x) \sim \frac{1}{2} - \frac{1}{\pi} \sum_{k=1}^{\infty} \frac{\sin(2\pi kx)}{k}, \qquad x \neq k.$$

$f(x) = 0$ for all $x = k$, and the corresponding infinite series does not represent the value 0 of $f(x)$ for $x = 0, \pm 1, \pm 2, \ldots$. Thus,

$$\frac{1}{2}[f(n+) + f(n-)] = \frac{1}{2}.$$

37. (a) This represents a square wave function.

$$a_0 = \frac{1}{l} \int_{-l}^{l} f(x)\, dx = \int_0^l dx = l,$$

$$a_k = \frac{1}{l} \int_{-l}^{l} f(x) \cos\left(\frac{\pi kx}{l}\right) dx$$

$$= \int_0^l \cos\left(\frac{\pi kx}{l}\right) dx = 0, \quad k \neq 0.$$

$$b_k = \frac{1}{l} \int_{-l}^{l} f(x) \sin\left(\frac{\pi kx}{l}\right) dx = \int_0^l \sin\left(\frac{\pi kx}{l}\right) dx$$

$$= \left(\frac{l}{\pi k}\right)(1 - \cos \pi k) = \begin{cases} 0, & k \text{ is even}, \\ \frac{2l}{\pi k}, & k \text{ is odd}. \end{cases}$$

$$f(x) = \frac{l}{2} + \left(\frac{2l}{\pi}\right)\left[\sin\left(\frac{\pi x}{l}\right) + \frac{1}{3}\sin\left(\frac{3\pi x}{l}\right) + \frac{1}{5}\sin\left(\frac{5\pi x}{l}\right) + \ldots\right]$$

$$= \frac{l}{2} + \left(\frac{2l}{\pi}\right) \sum_{k=1}^{\infty} \frac{\sin\left[(2k-1)\left(\frac{\pi x}{l}\right)\right]}{(2k-1)}.$$

At $x = 0, \pm nl$, f is not continuous, all terms in the above series after the first vanish and the sum is $(l/2)$. The graphs of the partial sums

$$s_n(x) = \frac{l}{2} + \left(\frac{2l}{\pi}\right)\left[\sin\left(\frac{\pi x}{l}\right) + \ldots + \frac{1}{(2n-1)}\sin\left\{(2n-1)\left(\frac{\pi x}{l}\right)\right\}\right]$$

can be drawn. These graphs show how the Fourier series converges at the points of continuity of $f(x)$, $s_n(x) \to f(x)$ as $n \to \infty$. However, at points of discontinuity at $x = 0$ and $x = \pm l$, $s_n(x)$ does not converge to the mean value. Just beyond the discontinuities at $x = 0$ and $x = \pm l$, the partial sums, $s_n(x)$ overshoot the value $|l|$. This behavior of the Fourier series at points of discontinuity is known as the Gibbs phenomenon.

(b) $a_0 = l$, $a_k = \dfrac{2l}{(\pi k)^2}\left(\cos k\pi - 1\right)$, $b_k = 0$, $k = 1, 2, \ldots$.

$$f(x) = \frac{l}{2} - \left(\frac{4l}{\pi^2}\right)\left[\cos\left(\frac{\pi x}{l}\right) + \frac{1}{3^2}\cos\left(\frac{3\pi x}{l}\right)\right.$$
$$\left. + \frac{1}{5^2}\cos\left(\frac{5\pi x}{l}\right) + \cdots\right]$$
$$= \frac{l}{2} - \left(\frac{4l}{\pi^2}\right)\sum_{k=1}^{\infty}\frac{1}{(2k-1)^2}\cos\left\{(2k-1)\frac{\pi x}{l}\right\}.$$

(c) $f(x) = \dfrac{l}{2} + \left(\dfrac{4l}{\pi^2}\right)\displaystyle\sum_{k=1}^{\infty}\dfrac{1}{(2k-1)^2}\cos\left[(2k-1)\dfrac{\pi x}{l}\right]$.

(d) $f(x) = \dfrac{a_0}{2} + \displaystyle\sum_{k=1}^{\infty}\left[a_k\cos(\pi kx) + b_k\sin(\pi kx)\right]$,

$$a_0 = \frac{l}{3}, \quad a_k = \frac{2(-1)^k}{\pi^2 k^2}, \quad b_k = \begin{cases} -\frac{1}{\pi k}, & k \text{ is even,} \\ \left(\frac{1}{\pi k} - \frac{4}{\pi^3 k^3}\right), & k \text{ is odd.} \end{cases}$$

38. (a) We have

$$\cos kx \sin\frac{1}{2}x = \frac{1}{2}\left[\sin\left(k + \frac{1}{2}\right)x - \sin\left(k - \frac{1}{2}\right)x\right].$$

Summing from $k = 1$ to $k = n$ gives

$$\left(\sin\frac{1}{2}x\right)\sum_{k=1}^{n}\cos kx = \frac{1}{2}\sum_{k=1}^{n}\left[\sin\left(k + \frac{1}{2}\right)x - \sin\left(k - \frac{1}{2}\right)x\right]$$
$$= \frac{1}{2}\left[\sin\left(n + \frac{1}{2}\right)x - \sin\frac{1}{2}x\right]$$

Dividing this result by $\sin\frac{1}{2}x$ and adding $\frac{1}{2}$ to both sides gives the result.

(b) $s_n(x) = \dfrac{a_0}{2} + \displaystyle\sum_{k=1}^{n}(a_k\cos kx + b_k\sin kx)$,   where

$(a_k\cos kx + b_k\sin kx)$
$$= \left[\frac{1}{\pi}\int_{-\pi}^{\pi}f(t)\cos kt\,dt\right]\cos kx$$
$$+ \left[\frac{1}{\pi}\int_{-\pi}^{\pi}f(t)\sin kt\,dt\right]\sin kx = \frac{1}{\pi}\int_{-\pi}^{\pi}f(t)\cos k(t - x)\,dt.$$

Thus,

$$\frac{a_0}{2} = \frac{1}{2\pi} \int_{-\pi}^{\pi} f(t)\, dt \quad \text{and hence,}$$

$$s_n(x) = \frac{a_0}{2} + \sum_{k=1}^{n} (a_k \cos kx + b_k \sin kx)$$

$$= \frac{1}{2\pi} \int_{-\pi}^{\pi} f(t)\, dt + \frac{1}{\pi} \sum_{k=1}^{n} \int_{-\pi}^{\pi} f(t) \cos k\,(t-x)\, dt$$

$$= \frac{1}{\pi} \int_{-\pi}^{\pi} f(t) \left[\frac{1}{2} + \sum_{k=1}^{n} \cos k\,(t-x) \right] dt$$

$$= \frac{1}{\pi} \int_{-\pi}^{\pi} f(t)\, \frac{\sin\left\{ \left(n+\frac{1}{2}\right)(t-x) \right\}}{2 \sin \frac{1}{2}(t-x)}\, dt \quad \text{from (a)}$$

which is, from $t - x = \xi$,

$$= \frac{1}{\pi} \int_{-\pi-x}^{\pi-x} f(x+\xi)\, \frac{\sin\left(n+\frac{1}{2}\right)\xi}{2 \sin \frac{1}{2}\xi}\, d\xi,$$

since the integrand has a period of 2π, we can replace the interval $-\pi-x$, $\pi-x$ by any interval of length 2π, that is, $(-\pi, \pi)$. This gives the result.

7.9 Exercises

1. (a) $u(x,t) = \displaystyle\sum_{n=1}^{\infty} \frac{4}{(n\pi)^3} \left[1 - (-1)^n\right] \cos(n\pi ct) \sin(n\pi x)$.

   (b) $u(x,t) = 3 \cos ct \sin x$.

2. (a) $u(x,t) = \displaystyle\sum_{n=1,3,4,\ldots}^{\infty} \frac{32[(-1)^n - 1]}{\pi c n^2 (n^2 - 4)} \sin(nct) \sin(nx)$.

5. $u(x,t) = \displaystyle\sum_{n=1}^{\infty} a_n T_n(t) \sin\left(\frac{n\pi x}{l}\right)$, where $a_n = \frac{2}{l} \int_0^l f(x) \sin\left(\frac{n\pi x}{l}\right) dx$, and

$$T_n(t) = \begin{cases} e^{-at/2} \left(\cosh \alpha t + \frac{a}{2\alpha} \sinh \alpha t\right), & \text{for} \quad \alpha^2 > 0 \\[2mm] e^{-at/2} \left(1 + \frac{at}{2}\right), & \text{for} \quad \alpha = 0 \\[2mm] e^{-at/2} \left(\cos \beta t + \frac{a}{2\beta} \sin \beta t\right), & \text{for} \quad \alpha^2 < 0, \end{cases}$$

in which

$$\alpha = \frac{1}{2} \left[a^2 - 4\left(b + \frac{n^2 \pi^2 c^2}{l^2}\right) \right]^{\frac{1}{2}}, \qquad \beta = \frac{1}{2} \left[4\left(b + \frac{n^2 \pi^2 c^2}{l^2}\right) - a^2 \right]^{\frac{1}{2}}.$$

6. $u(x,t) = \sum\limits_{n=1}^{\infty} a_n T_n(t) \sin\left(\frac{n\pi x}{l}\right),$ $a_n = \frac{2}{l}\int_0^l g(x)\sin\frac{n\pi x}{l}dx,$ and

$$T_n(t) = \begin{cases} \frac{2e^{-at/2}}{\sqrt{(a^2-\alpha)}}\sinh\left(\frac{\sqrt{(a^2-\alpha)}}{2}t\right), & \text{for } a^2 > \alpha, \\ te^{-at/2}, & \text{for } a^2 = \alpha, \\ \frac{2e^{-at/2}}{\sqrt{(\alpha-a^2)}}\sin\left(\frac{\sqrt{(\alpha-a^2)}}{2}t\right), & \text{for } a^2 < \alpha. \end{cases}$$

7. $\theta(x,t) = \sum\limits_{n=1}^{\infty} a_n \cos(a\alpha_n t)\sin(\alpha_n x + \phi_n),$ where

$$a_n = \frac{2(\alpha_n^2+h^2)}{2h+(\alpha_n^2+h^2)l}\int_0^l f(x)\sin(\alpha_n x + \phi_n)dx$$

and

$$\phi_n = \tan^{-1}\left(\frac{\alpha_n}{h}\right); \alpha_n \text{ are the roots of the equation } \tan\alpha l = \left(\frac{2h\alpha}{\alpha^2-h^2}\right).$$

11. $u(x,t) = v(x,t) + U(x),$ where

$$v(x,t) = \sum\limits_{n=1}^{\infty}\left[-\left(\frac{2}{l}\right)\int_0^l U(\tau)\sin\left(\frac{n\pi\tau}{l}\right)d\tau\right]\cos\left(\frac{n\pi ct}{l}\right)\sin\left(\frac{n\pi x}{l}\right)$$

and

$$U(x) = -\frac{A}{c^2}\sinh x + \frac{A}{c^2}\sinh(l+k-h)\frac{x}{l} + h.$$

12. $u(x,t) = \frac{A}{6c^2}x^2(1-x) + \sum\limits_{n=1}^{\infty}\frac{12}{(n\pi)^3}(-1)^n\cos(n\pi ct)\sin(n\pi x).$

14. (a) $u(x,t) = -\frac{hx^2}{2k} + \left(2u_0 + \frac{h}{2k}\right)x - \frac{4h}{k\pi}e^{-k\pi^2 t}\sin(\pi x)$
$$+ \sum\limits_{n=2}^{\infty} a_n e^{-kn^2\pi^2 t}\sin(n\pi x),$$

where

$$a_n = \frac{2u_0}{n\pi}[1+(-1)^n] + \frac{2u_0 n}{(n^2-1)\pi}[1+(-1)^n] + \frac{2h}{k\pi^3 n^3}[(-1)^n - 1].$$

(b) Hint: $v(x,t) = e^{-ht}u(x,t).$
$$u(x,t) = e^{-ht}\left[\frac{1}{2}a_0 + \sum\limits_{n=1}^{\infty}a_n\cos\left(\frac{n\pi x}{l}\right)\exp\left(-n^2\pi^2 kt/l^2\right)\right], \quad \text{where}$$

$$a_n = \frac{2}{l}\int_0^l f(\xi)\cos\left(\frac{n\pi\xi}{l}\right)d\xi.$$

15. (a) $u(x,t) = \sum\limits_{n=1}^{\infty}\frac{4}{n^3\pi^3}\left[2(-1)^{n+1} - 1\right]e^{-4n^2\pi^2 t}\sin(n\pi x).$

(b) $u(x,t) = \sum\limits_{n=1,3,4,\ldots}^{\infty}[(-1)^n - 1]\left[\frac{n}{\pi(4-n^2)} - \frac{1}{n\pi}\right]e^{-n^2 kt}\sin(nx).$

16. $u\left(x,t\right)=\sum_{n=1}^{\infty}\frac{2l^{2}}{n^{3}\pi^{3}}\left[1-\left(-1\right)^{n}\right]e^{-\left(n\pi/l\right)^{2}t}\cos\left(\frac{n\pi x}{l}\right).$

18. $v\left(x,t\right)=Ct\left(1-\frac{x}{l}\right)-\frac{Cl^{2}}{6k}\left[\left(\frac{x}{l}\right)^{3}-3\left(\frac{x}{l}\right)^{2}+2\left(\frac{x}{l}\right)\right]$

$$+\left(\frac{2Cl^{2}}{\pi^{3}k}\right)\sum_{k=1}^{\infty}\frac{e^{-n^{2}\pi^{2}kt/l^{2}}}{n^{3}}\sin\left(\frac{n\pi x}{l}\right).$$

21. $u\left(x,t\right)=v\left(x,t\right)+w\left(x\right),$ where

$$v\left(x,t\right)=e^{-kt}\sin x+\sum_{n=1}^{\infty}a_{n}\,e^{-n^{2}kt}\sin\left(nx\right),\text{ and}$$

$$a_{n}=\frac{-n}{\left(n^{2}+a^{2}\right)}\left[\left(-1\right)e^{-n-ax}-1\right]+\frac{2A}{a^{2}k\pi}\left[\frac{1}{n}\left\{\left(-1\right)^{n}-1\right\}\right]$$

$$+\frac{\left(-1\right)^{n}}{n}\left[e^{-a\pi}-1\right]$$

$$w\left(x\right)=\frac{A}{a^{2}k}\left[1-e^{-ax}+\frac{x}{\pi}\left(e^{-a\pi}-1\right)\right].$$

36. **Hint:** Suppose R is the rectangle $0\leq x\leq a,\,0\leq y\leq b$ and ∂R is its boundary positively oriented. Suppose that u_{1} and u_{2} are solutions of the problem, and put $v=\left(u_{1}-u_{2}\right)$. Then v satisfies the Laplace equation with $v=\left(x,0\right)=0=v\left(x,b\right),\;v_{x}\left(0,y\right)=0=v_{x}\left(a,y\right).$

8.14 Exercises

1. (a) $\lambda_{n}=n^{2},\,\phi_{n}\left(x\right)=\sin nx$ for $n=1,2,3,\dots$

   (b) $\lambda_{n}=\left(\left(2n-1\right)/2\right)^{2},\,\phi_{n}\left(x\right)=\sin\left(\left(2n-1\right)/2\right)\pi x$ for $n=1,2,3,\dots$

   (c) $\lambda_{n}=n^{2},\,\phi_{n}\left(x\right)=\cos nx$ for $n=1,2,3,\dots$

2. (a) $\lambda_{n}=0,\,n^{2}\pi^{2},\,\phi_{n}\left(x\right)=1,\,\sin n\pi x,\,\cos n\pi x$ for $n=1,2,3,\dots$

   (b) $\lambda_{n}=0,\,n^{2},\,\phi_{n}\left(x\right)=1,\,\sin nx,\,\cos nx$ for $n=1,2,3,\dots$

   (c) $\lambda_{n}=0,\,4n^{2},\,\phi_{n}\left(x\right)=1,\,\sin 2nx,\,\cos 2nx$ for $n=1,2,3,\dots$

3. (a) $\lambda_{n}=-\left(3/4+n^{2}\pi^{2}\right),\,\phi_{n}\left(x\right)=e^{-x/2}\sin n\pi x,\,n=1,2,3,\dots$

4. (a) $\lambda_{n}=1+n^{2}\pi^{2},\,\phi_{n}\left(x\right)=\left(1/x\right)\sin\left(n\pi\ln x\right),\,n=1,2,3,\dots$

   (b) $\lambda_{n}=\frac{1}{4}+\left(n\pi/\ln 3\right)^{2},\,\phi_{n}\left(x\right)=\left[1/\left(x+2\right)^{\frac{1}{2}}\right]\sin\left[\left(n\pi/\ln 3\right)\ln\left(x+2\right)\right],$

   $n=1,2,3,\dots$

(c) $\lambda_n = \frac{1}{12}\left[1 + (2n\pi/\ln 2)^2\right]$,

$\phi_n(x) = \left[1/(1+x)^{\frac{1}{2}}\right] \sin\left[(n\pi/\ln 2)\ln(1+x)\right]$, $n = 1, 2, 3, \ldots$

5. (a) $\phi(x) = \sin\left(\sqrt{\lambda}\ln x\right)$, $\lambda > 0$.

   (b) $\phi(x) = \sin\left(\sqrt{\lambda}x\right)$, $\lambda > 0$.

7. $f(x) \sim \sum_{n=1}^{\infty} \frac{2}{\pi}\left[\frac{(-1)^n - 1}{n^2}\right] \cos nx$.

11. (a) $G(x, \xi) = \begin{cases} x, & x \le \xi \\ \\ \xi, & x > \xi. \end{cases}$

12. (a) $u(x) = -\cos x + \left(\frac{\cos 1 - 1}{\sin 1}\right)\sin x + 1$,

   (b) $u(x) = -\frac{2}{5}\cos 2x - \frac{1}{10}\left(\frac{1 + 2\sin 2}{\cos 2}\right)\sin 2x + \frac{1}{5}e^x$.

16. $G(x, \xi) = \begin{cases} (x^3\xi/2) + (x\xi^3/2) - (9x\xi/5) + x, & \text{for } 0 \le x < \xi \\ \\ (x^3\xi/2) + (x\xi^3/2) - (9x\xi/5) + \xi, & \text{for } \xi \le x \le 1. \end{cases}$

24. (a) Hint: Differentiate $\cot\theta = \frac{py'}{y}$ with respect to x to find

$$-\csc^2\theta\, \frac{d\theta}{dx} = \frac{1}{y}(py')' - \frac{1}{y^2}(py'^2) = -\left(\lambda r + q + \frac{1}{p}\cot^2\theta\right).$$

$$\frac{d\theta}{dx} = (q + \lambda r)\sin^2\theta + \frac{1}{p}\cos^2\theta, \quad \frac{dr}{dx} = \frac{r}{2}\left(\frac{1}{p} - q - \lambda r\right)\sin 2\theta.$$

(b) At $\theta = n\pi$, $\frac{d\theta}{dx} = \frac{1}{p}$, and at $\theta = \left(n + \frac{1}{2}\right)\pi$, $\frac{d\theta}{dx} = (q + \lambda p)$.

9.10 Exercises

8. (a) $u(r, \theta) = \frac{4}{3}\left(\frac{1}{r} - \frac{r}{4}\right)\sin\theta$.

   (c) $u(r, \theta) = \sum_{n=1}^{\infty} a_n \sinh\left[(n\pi/\ln 3)\left(\theta - \frac{\pi}{2}\right)\right]\sin\left[(n\pi/\ln 3)\ln r\right]$,

   where

$$a_n = \frac{2}{\ln 3 \sinh(n\pi^2/2 \ln 3)} \left\{ \frac{n\pi \ln 3}{n^2\pi^2 + 4(\ln 3)^2} \left[9(-1)^n - 1 \right] \right.$$

$$\left. - \frac{4n\pi \ln 3}{n^2\pi^2 + (\ln 3)^2} \left[3(-1)^n - 1 \right] + \frac{3 \ln 3}{n\pi} \left[(-1)^n - 1 \right] \right\}.$$

9. $u(r, \theta) = \sum_{n=1}^{\infty} a_n \left(r^{-n\pi/\alpha} - b^{-2n\pi/\alpha} r^{n\pi/\alpha} \right) \sin\left(\frac{n\pi\theta}{\alpha} \right)$

$$+ \sum_{n=1}^{\infty} b_n \sinh\left[\frac{n\pi}{\ln(b/a)} (\theta - \alpha) \right] \sin\left[\frac{n\pi}{\ln(b/a)} (\ln r - \ln a) \right],$$

where

$$a_n = 2 \left[\alpha \left(a^{-n\pi/\alpha} - b^{-2n\pi/\alpha} a^{n\pi/\alpha} \right) \right]^{-1} \int_0^\alpha f(\theta) \sin\left(\frac{n\pi\theta}{\alpha} \right) d\theta,$$

$$b_n = -2 \left[\ln\left(\frac{b}{a} \right) \sinh\left\{ \alpha n\pi / \ln(b/a) \right\} \right]^{-1}$$
$$+ \int_a^b f(r) \sin\left[(n\pi / \ln[b/a]) \ln(ra) \right] \frac{dr}{r}.$$

12. $u(r, \theta) = \sum_{n=1}^{\infty} \frac{2}{\alpha J_\nu(a)} \left[\int_0^\alpha f(\theta) \sin\left(\frac{n\pi\tau}{\alpha} \right) d\tau \right] J_\nu(r) \sin\left(\frac{n\pi\theta}{\alpha} \right), \quad \nu = n\pi/\alpha.$

13. $u(r, \theta) = \frac{1}{2} \left(a^2 - r^2 \right).$

14. (a) $u(r, \theta) = -\frac{1}{3} \left(r + \frac{4}{r} \right) \sin\theta + \text{constant}.$

16. $u(r, \theta) = \sum_{n=1}^{\infty} \frac{1}{nR^{n-1}} r^n \sin n\theta.$

18. $u(r, \theta) = \frac{a_0}{2} + \sum_{n=1}^{\infty} r^n \left(a_n \cos n\theta + b_n \sin n\theta \right),$

where

$$a_n = \frac{R^{1-n}}{(n+Rh)\pi} \int_0^{2\pi} f(\theta) \cos n\theta \, d\theta, \qquad n = 0, 1, 2, \ldots.$$

$$b_n = \frac{R^{1-n}}{(n+Rh)\pi} \int_0^{2\pi} f(\theta) \sin n\theta \, d\theta, \qquad n = 1, 2, 3, \ldots.$$

20. $u(r, \theta) = c - \frac{r^4}{12} \sin 2\theta + \frac{1}{6} \left(\frac{r_1^6 - r_2^6}{r_1^4 - r_2^4} \right) r^2 \sin 2\theta + \frac{1}{6} \left(\frac{r_1^2 - r_2^2}{r_1^4 - r_2^4} \right) r_1^4 r_2^4 r^{-2} \sin 2\theta.$

21. $u(r, \theta) = \sum_{n=1}^{\infty} a_n \left(r^{-n\pi/\alpha} - b^{-2n\pi/\alpha} r^{n\pi/\alpha} \right) \sin\left(\frac{n\pi\theta}{\alpha} \right)$

$$+ \sum_{n=1}^{\infty} b_n \sinh\left[\frac{n\pi}{\ln(b/a)} (\theta - \alpha) \right] \sin\left[\frac{n\pi}{\ln(b/a)} (\ln r - \ln a) \right].$$

22. (a) $u(x,y) = \sum_{n=1}^{\infty} \frac{4[1-(-1)^n]}{(n\pi)^3 \sinh n\pi} \sin n\pi x \sinh \{n\pi(y-1)\}$

(c) $u(x,y) = \sum_{n=1}^{\infty} a_n (\sinh n\pi x - \tanh n\pi \cosh n\pi x) \sin n\pi y$, where

$$a_n = \frac{1}{\tanh n\pi} \left[\frac{2n\pi^3}{n^2\pi^4-4} + \frac{1-(-1)^n}{n\pi} \right].$$

23. (a) $u(x,y) = c + \sum_{n=1}^{\infty} a_n (\cosh nx - \tanh n\pi \sinh nx) \cos ny$, where

$$a_n = 2[1-(-1)^n]/(n^3\pi \tanh n\pi).$$

(c) $u(x,y) = -\frac{1}{\tanh \pi} [\cosh y - \tanh \pi \sinh y] \cos x + C.$

25. $u(x,y) = xy(1-x) + \sum_{n=1}^{\infty} \frac{4(-1)^n}{(n\pi)^3 \sinh n\pi} \sin(n\pi x) \sinh(n\pi y).$

27. $u(x,y) = c + (x^2/2)\left(\frac{x^2}{3} - y^2\right) + \sum_{n=1}^{\infty} \frac{8a^4(-1)^{n+1}}{(n\pi)^3 \sinh n\pi} \cosh\left(\frac{n\pi x}{a}\right) \cos\left(\frac{n\pi y}{a}\right).$

29. $u(x,y) = x[(x/2) - \pi] + \sum_{n=1}^{\infty} a_n \sin\left\{\left(\frac{2n-1}{2}\right)x\right\} \cosh\left\{\left(\frac{2n-1}{2}\right)y\right\}$, where

$$a_n = \frac{2}{A\pi} \int_0^\pi \left[f(x) - h\left(\frac{x^2}{2} - \pi x\right) \right] \sin\left[\left(\frac{2n-1}{2}\right)x\right] dx$$

with

$$A = \left(\frac{2n-1}{2}\right) \sinh\left(\frac{2n-1}{2}\right)\pi + h\cosh\left(\frac{2n-1}{2}\right)\pi.$$

32. Hint: The solution is given by (9.5.3) and the boundary conditions require

$$\sin^2\theta = \tfrac{1}{2}a_0 + \sum_{n=1}^{\infty} [(a_n + b_n)\cos n\theta + (c_n + d_n)\sin n\theta],$$

$$0 = \tfrac{1}{4}b_0 + \sum_{n=1}^{\infty} n\left[(a_n 2^{n-1} - b_n 2^{-n-1})\cos n\theta \right.$$

$$\left. + (c_n 2^{n-1} - d_n 2^{-n-1})\sin n\theta\right].$$

Using $\sin^2\theta = \tfrac{1}{2}(1 - \cos 2\theta)$, we equate coefficients to obtain

$$a_0 = 1, \quad b_0 = 0; \quad a_2 + b_2 = -\tfrac{1}{2}, \quad 2a_2 - \tfrac{1}{8}b_2 = 0;$$

$$\left.\begin{array}{l} a_n + b_n = 0 \\ 2^{n-1} a_n - 2^{-n-1} b_n = 0 \end{array}\right\} n = 1, 3, 4, 5, \ldots.$$

$$c_n + d_n = 0$$

$$2^{n-1} c_n - 2^{-n-1} d_n = 0 \left.\right\} \quad n = 1, 2, 3, \dots.$$

Thus, $a_0 = 1$, $b_0 = 1$, $a_2 = -\frac{1}{34}$, $b_2 = -\frac{8}{17}$, and the remaining coefficients are zero; finally

$$u(r, \theta) = \tfrac{1}{2} - \tfrac{1}{34} \left(r^2 + \tfrac{16}{r^2} \right) \cos 2\theta.$$

33. (a) Hint: Seek a separable solution $u(r, z) = R(r) Z(z)$ so that

$r^2 R'' + r R' - \lambda r^2 R = 0$, and $Z'' + \lambda Z = 0$, with $Z(0) = 0 = Z(h)$.

The solution of this eigenvalue problem is

$$\lambda_n = \left(\tfrac{n\pi}{h} \right)^2, \quad Z_n(z) = \sin \left(\tfrac{n\pi z}{h} \right), \qquad n = 1, 2, 3, \dots.$$

The solution of the radial equation is

$$R_n(r) = a_n I_0 \left(\tfrac{n\pi r}{h} \right) + b_n K_0 \left(\tfrac{n\pi r}{h} \right),$$

where I_0 and K_0 are modified Bessel functions. Since K_0 is unbounded at $r = 0$, all $b_n \equiv 0$. Thus,

$$u(r, z) = \sum_{n=1}^{\infty} a_n I_0 \left(\tfrac{n\pi r}{h} \right) \sin \left(\tfrac{n\pi z}{h} \right).$$

$$f(z) = u(1, z) = \sum_{n=1}^{\infty} a_n I_0 \left(\tfrac{n\pi}{h} \right) \sin \left(\tfrac{n\pi z}{h} \right).$$

This is a Fourier sine series for $f(z)$ and hence,

$$a_n I_0 \left(\tfrac{n\pi}{h} \right) = \tfrac{2}{h} \int_0^h f(z) \sin \left(\tfrac{n\pi z}{h} \right) dz.$$

(d) $u(r, z) = a I_0 \left(\tfrac{3\pi r}{h} \right) \sin \left(\tfrac{3\pi z}{h} \right) / I_0 \left(\tfrac{3\pi}{h} \right)$.

35. (a) $u(r, z) = 8 \sum_{n=1}^{\infty} \dfrac{\sinh z k_n}{z k_n \sinh k_n} \dfrac{J_0(k_n r)}{J_0(k_n)}$.

(b) $u(r, z) = \left(\tfrac{4a}{\pi} \right) \sum_{n=1}^{\infty} \dfrac{1}{(2n-1)} \dfrac{I_0 \left[\frac{1}{2}(2n-1)r \right]}{I_0 \left[\frac{1}{2}(2n-1) \right]} \sin \left\{ \tfrac{1}{2} (2n-1) z \right\}$.

(c) $u(r, z) = \sum_{n=1}^{\infty} a_n I_0 \left(\tfrac{n\pi r}{h} \right) \sin \left(\tfrac{n\pi z}{h} \right)$, where

$$a_n I_0 \left(\tfrac{n\pi a}{h} \right) = \tfrac{2}{h} \int_0^h f(z) \sin \left(\tfrac{n\pi z}{h} \right) dz.$$

10.13 Exercises

1. $u(x, y, z) = \dfrac{\sinh\left[(\pi/b)^2 + (\pi/c)^2\right]^{\frac{1}{2}}(a-x)}{\sinh\left[(\pi/b)^2 + (\pi/c)^2\right]^{\frac{1}{2}}a} \sin\left(\frac{\pi y}{b}\right) \sin\left(\frac{\pi z}{c}\right).$

2. $u(x, y, z) = \left[\dfrac{\sinh(\sqrt{2}\,\pi z)}{\sqrt{2}\,\pi} - \dfrac{\cosh(\sqrt{2}\,\pi z)}{\sqrt{2}\,\pi \tanh\sqrt{2}\,\pi}\right] \cos\pi x \cos\pi y.$

4. (a) $u(r, \theta, z) = \displaystyle\sum_{m=0}^{\infty}\sum_{n=1}^{\infty} (a_{mn}\cos m\theta + b_{mn}\sin m\theta)\, J_{mn}(a_{mn}r/a)$

$$\times \frac{\sinh \alpha_{mn}(l-z)/a}{\sinh \alpha_{mn}l/a},$$

where

$$a_{mn} = \frac{2}{a^2\pi\varepsilon_n\left[J_{m+1}(\alpha_{mn})\right]^2} \int_0^{2\pi}\int_0^a f(r,\theta)\, J_m(\alpha_{mn}r/a)\cos m\theta\, r\, dr\, d\theta$$

$$b_{mn} = \frac{2}{a^2\pi\left[J_{m+1}(\alpha_{mn})\right]^2} \int_0^{2\pi}\int_0^a f(r,\theta)\, J_m(\alpha_{mn}r/a)\sin m\theta\, r\, dr\, d\theta$$

with

$$\varepsilon_n = \begin{cases} 1, & \text{for } m \neq 0 \\ 2, & \text{for } m = 0 \end{cases}$$

and α_{mn} is the nth root of the equation $J_m(\alpha_{mn}) = 0$.

5. $u(r, \theta) = \frac{1}{3} + \left(2/3a^2\right) r^2 P_2(\cos\theta).$

7. $u(r, z) = \displaystyle\sum_{n=1}^{\infty} \frac{a_n \sinh \alpha_n(l-z)/a}{\cosh \alpha_n l/a} J_0(\alpha_n r/a)$, where $a_n = \dfrac{2qu}{k\alpha_n^2 J_0(\alpha_n)}$ and α_n is the root of $J_0(\alpha_n) = 0$ and k is the coefficient of heat conduction.

8. $u(r, z) = \dfrac{4u_0}{\pi}\displaystyle\sum_{n=1}^{\infty} \frac{[I_0(2n+1)(\pi r/l)]}{[I_0(2n+1)(\pi a/l)]}\frac{\sin(2n+1)\pi z/l}{(2n+1)}.$

9. $u(r, \theta) = u_2 + \left(\frac{u_1-u_2}{2}\right)\displaystyle\sum_{n=1}^{\infty}\left(\frac{2n+1}{n+1}\right) P_{n-1}(0)\left(\frac{r}{a}\right)^n P_n(\cos\theta).$

11. $u(r, \theta, \phi) = C + \displaystyle\sum_{n=1}^{\infty}\sum_{m=0}^{\infty} r^n P_n^m(\cos\theta)[a_{nm}\cos m\phi + b_{nm}\sin n\phi],$

where

$$a_{nm} = \frac{(2n+1)(n-m)!}{2n\pi(n+m)!} \int_0^{2\pi} \int_0^{\pi} f(\theta, \varphi) P_n^m(\cos\theta) \cos m\varphi \sin\theta \, d\theta \, d\varphi$$

$$b_{nm} = \frac{(2n+1)(n-m)!}{2n\pi(n+m)!} \int_0^{2\pi} \int_0^{\pi} f(\theta, \varphi) P_n^m(\cos\theta) \sin m\varphi \sin\theta \, d\theta \, d\varphi$$

$$a_{n0} = \frac{(2n+1)}{4n\pi} \int_0^{2\pi} \int_0^{\pi} f(\theta, \varphi) P_n(\cos\theta) \sin\theta \, d\theta \, d\varphi.$$

12. $u(x,y,t) = \displaystyle\sum_{n=1,3,4,\dots}^{\infty} \left(-\frac{4}{\pi}\right) \frac{[1-(-1)^n]}{n(n^2-4)} \cos\left(\sqrt{(n^2+1)}\pi ct\right) (\sin n\pi x \sin n\pi y).$

13. $u(r,\theta,t) = \displaystyle\sum_{n=0}^{\infty}\sum_{m=1}^{\infty} J_n(\alpha_{mn}r/a) \cos(\alpha_{mn}ct/a) [a_{mn}\cos n\theta + b_{mn}\sin n\theta]$

$$+ \sum_{n=0}^{\infty}\sum_{m=1}^{\infty} J_n(\alpha_{mn}r/a) \sin(\alpha_{mn}ct/a) [c_{mn}\cos n\theta + d_{mn}\sin n\theta],$$

where

$$a_{mn} = \frac{2}{\pi a^2 \varepsilon_n [J_n'(\alpha_{mn})]^2} \int_0^{2\pi}\int_0^a f(r,\theta) J_n(\alpha_{mn}r/a) \cos n\theta \, r \, dr \, d\theta$$

$$b_{mn} = \frac{2}{\pi a^2 [J_n'(\alpha_{mn})]^2} \int_0^{2\pi}\int_0^a f(r,\theta) J_n(\alpha_{mn}r/a) \sin n\theta \, r \, dr \, d\theta$$

$$c_{mn} = \frac{2}{\pi a c \alpha_{mn} \varepsilon_n [J_n'(\alpha_{mn})]^2} \int_0^{2\pi}\int_0^a g(r,\theta) J_n(\alpha_{mn}r/a) \cos n\theta \, r \, dr \, d\theta$$

$$d_{mn} = \frac{2}{\pi a c \alpha_{mn} [J_n'(\alpha_{mn})]^2} \int_0^{2\pi}\int_0^a g(r,\theta) J_n(\alpha_{mn}r/a) \sin n\theta \, r \, dr \, d\theta$$

in which α_{mn} is the root of the equation $J_n(\alpha_{mn}) = 0$ and

$$\varepsilon_n = \begin{cases} 2 & n = 0 \\ 1 & n \neq 0 \end{cases}.$$

15. $u\left(r,\theta,t\right)=\sum\limits_{n=0}^{\infty}\sum\limits_{m=1}^{\infty}J_n\left(\alpha_{mn}r\right)\exp\left(-\alpha_{mn}kt\right)\left[a_{mn}\cos n\theta+b_{mn}\sin n\theta\right],$

where

$$a_{nm}=\frac{2}{\pi\varepsilon_n\left[J_n'\left(\alpha_{mn}\right)\right]^2}\int_0^{2\pi}\int_0^1 f\left(r,\theta\right)J_n\left(\alpha_{mn}r\right)\cos n\theta\,r\,dr\,d\theta,$$

$$b_{nm}=\frac{2}{\pi\left[J_n'\left(\alpha_{mn}\right)\right]^2}\int_0^{2\pi}\int_0^1 f\left(r,\theta\right)J_n\left(\alpha_{mn}r\right)\sin n\theta\,r\,dr\,d\theta,$$

where α_{mn} is the root of the equation $J_n\left(\alpha_{mn}\right)=0$ and

$$\varepsilon_n=\begin{cases}1 & \text{for}\quad n\neq 0\\[2mm]2 & \text{for}\quad n=0.\end{cases}$$

16. $u\left(x,y,z,t\right)=\sin\pi x\sin\pi y\sin\pi z\cos\left(\sqrt{3}\,\pi ct\right).$

18. $u\left(r,\theta,z,t\right)=\sum\limits_{n=0}^{\infty}\sum\limits_{m=1}^{\infty}\sum\limits_{l=1}^{\infty}J_n\left(\alpha_{mn}r/a\right)\sin\left(m\pi z/l\right)\cos\left(\omega ct\right)$

$$\times\left[a_{nml}\cos n\theta+b_{nml}\sin n\theta\right]$$

$$+\sum\limits_{n=0}^{\infty}\sum\limits_{m=1}^{\infty}\sum\limits_{l=1}^{\infty}J_n\left(\alpha_{mn}r/a\right)\sin\left(m\pi z/l\right)\sin\left(\omega ct\right)$$

$$\times\left[c_{nml}\cos n\theta+d_{nml}\sin n\theta\right],$$

where

$$a_{nml}=\frac{4}{\pi a^2 l\varepsilon_n\left[J_n'\left(\alpha_{mn}\right)\right]^2}\int_0^a\int_0^{2\pi}\int_0^l f\left(r,\theta,z\right)J_n\left(\alpha_{mn}r/a\right)$$

$$\times\sin\left(m\pi z/l\right)\cos n\theta\,r\,dr\,d\theta\,dz,$$

$$b_{nml}=\frac{4}{\pi a^2 l\left[J_n'\left(\alpha_{mn}\right)\right]^2}\int_0^a\int_0^{2\pi}\int_0^l f\left(r,\theta,z\right)J_n\left(\alpha_{mn}r/a\right)$$

$$\times\sin\left(m\pi z/l\right)\sin n\theta\,r\,dr\,d\theta\,dz,$$

$$c_{nml}=\frac{4\omega^{-1}}{\pi a^2 l\varepsilon_n\left[J_n'\left(\alpha_{mn}\right)\right]^2}\int_0^a\int_0^{2\pi}\int_0^l g\left(r,\theta,z\right)J_n\left(\alpha_{mn}r/a\right)$$

$$\times\sin\left(m\pi z/l\right)\cos n\theta\,r\,dr\,d\theta\,dz,$$

$$d_{nml} = \frac{4\,\omega^{-1}}{\pi a^2 l \left[J_n' \left(\alpha_{mn} \right) \right]^2} \int_0^a \int_0^{2\pi} \int_0^l g\left(r, \theta, z \right) J_n \left(\alpha_{mn} r/a \right)$$
$$\times \sin \left(m\pi z/l \right) \sin n\theta\, r\, dr\, d\theta\, dz,$$

where α_{mn} is the root of the equation $J_n \left(\alpha_{mn} \right) = 0$ and

$$\omega = \left[(m\pi/l)^2 + (\alpha_{mn}/a)^2 \right]^{\frac{1}{2}}, \qquad \varepsilon_n = \begin{cases} 1; & \text{for } n \neq 0 \\ 2; & \text{for } n = 0. \end{cases}$$

20. $u\left(r, \theta, z, t \right) = \displaystyle\sum_{n=0}^{\infty} \sum_{m=1}^{\infty} \sum_{p=1}^{\infty} \left(a_{nmp} \cos n\theta + b_{nmp} \sin n\theta \right)$
$$\times J_n \left(\alpha_{mn} r/a \right) \sin \left(p\pi z/l \right) e^{-\omega t},$$

where

$$a_{nmp} = \frac{4}{\pi a^2 l \varepsilon_n \left[J_n' \left(\alpha_{mn} \right) \right]^2} \int_0^a \int_0^{2\pi} \int_0^l f\left(r, \theta, z \right) J_n \left(\alpha_{mn} r/a \right)$$
$$\times \sin \left(p\pi z/l \right) \cos n\theta\, r\, dr\, d\theta\, dz$$

$$b_{nmp} = \frac{4}{\pi a^2 l \left[J_n' \left(\alpha_{mn} \right) \right]^2} \int_0^a \int_0^{2\pi} \int_0^l f\left(r, \theta, z \right) J_n \left(\alpha_{mn} r/a \right)$$
$$\times \sin \left(p\pi z/l \right) \sin n\theta\, r\, dr\, d\theta\, dz,$$

in which

$$\varepsilon_n = \begin{cases} 1 & \text{for } n \neq 0 \\ 2 & \text{for } n = 0 \end{cases} \qquad \text{and} \quad \omega = \left[(p\pi/l)^2 + (\alpha_{mn}/a)^2 \right].$$

23. $u\left(x, y, t \right) = \displaystyle\sum_{m=1}^{\infty} \sum_{n=1}^{\infty} u_{mn}\left(t \right) \sin mx \sin ny,$

where

$$u_{mn}(t) = \frac{4(-1)^{m+n+1}}{mn\,\alpha_{mn}c}\left[\sin\left(\alpha_{mn}ct\right)\left\{\frac{\cos\left(1-\alpha_{mn}c\right)t-1}{2\left(1-\alpha_{mn}c\right)}\right.\right.$$

$$\left.+\frac{\cos\left(1+\alpha_{mn}c\right)t-1}{2\left(1+\alpha_{mn}c\right)}\right\}\right]$$

$$+\cos\left(\alpha_{mn}ct\right)\left\{\frac{\sin\left(1-\alpha_{mn}c\right)t}{2\left(1-\alpha_{mn}c\right)}+\frac{\sin\left(1+\alpha_{mn}c\right)t}{2\left(1+\alpha_{mn}c\right)}\right\},$$

and $\quad \alpha_{mn} = \left(m^2+n^2\right)^{\frac{1}{2}}$.

25. $u\left(x,y,t\right) = \sum\limits_{n=1}^{\infty}\sum\limits_{m=1}^{\infty}\frac{4A}{mn\pi^2}\frac{[(-1)^n-1][(-1)^n-1]}{k(n^2+m^2)}\left[1-e^{-k\left(n^2+m^2\right)t}\right]$

$$\times \sin nx\left(\sin my - m\cos my\right).$$

27. $u\left(x,y,t\right) = x\left(x-\pi\right)\left(1-\frac{y}{\pi}\right)\sin t + \sum\limits_{n=1}^{\infty}\sum\limits_{m=1}^{\infty}v_{mn}\left(t\right)\sin nx\sin my$, where

$$\alpha_{mn}^2 = \left(m^2+n^2\right) \text{ and}$$

$$v_{mn}\left(t\right) = \frac{8\,\exp\left(-c^2\alpha^2 t\alpha_{mn}^2\right)\left[1-(-1)^n\right]}{\pi^2 mn\left(1+c^4\alpha_{mn}^4\right)}\left[\frac{c^2}{n^2}\left(\alpha_{mn}^2-n^2\right)\right.$$

$$\times\left\{\cos t\,\exp\left(-c^2\alpha^2 t\alpha_{mn}^2\right)-1\right\}$$

$$\left.+\left(\frac{1}{n^2}+c^4\alpha_{mn}^2\right)\sin t\,\exp\left(-c^2\alpha^2 t\alpha_{mn}^2\right)\right].$$

30. $u\left(x,y,t\right) = \left(\frac{4qb^4}{\pi^5 D}\right)\sum\limits_{n=1,3,\ldots}^{\infty}\frac{1}{n^5}\left[1-\frac{v_n(x)}{1+\cosh(n\pi a/b)}\right]\sin\left(n\pi y/b\right)$, where

$$v_n\left(x\right) = 2\cosh\left(\frac{n\pi a}{2b}\right)\cosh\left(\frac{n\pi x}{b}\right)+\left(\frac{n\pi a}{2b}\right)\sinh\left(\frac{n\pi a}{2b}\right)\cosh\left(\frac{n\pi x}{b}\right)$$

$$-\left(\frac{n\pi x}{b}\right)\sinh\left(\frac{n\pi x}{b}\right)\cosh\left(\frac{n\pi a}{2b}\right).$$

32. Hint: In region 1, $x \le -a$, the solution of the Schrödinger equation

$$\frac{d^2\psi}{dx^2} = \kappa^2\psi, \qquad \kappa^2 = \frac{2M}{\hbar^2}\left(V_0 - E\right),$$

is

$$\psi_1\left(x\right) = A\,e^{\kappa x} + B\,e^{-\kappa x},$$

where A and B are constants. For boundedness of the solution as

$x \to -\infty$, $B \equiv 0$, and hence, $\psi_1(x) = A e^{\kappa x}$.

In region 2, $x \geq a$, the solution of the Schrödinger equation is

$$\psi_2(x) = C e^{\kappa x} + D e^{-\kappa x}.$$

For boundedness as $x \to \infty$, $C \equiv 0$. The solution is $\psi_2(x) = D e^{-\kappa x}$.

In region 3, $-a \leq x \leq a$, the potential is zero and hence, the equation

takes the simple form $\psi_{xx} + k^2 \psi = 0$, where $k^2 = \left(\frac{2M}{\hbar^2}\right) E$. The solu-

tion is $\psi_3(x) = E \sin kx + F \cos kx$. For matching conditions at $x = a$,

$\psi_2(a) = \psi_3(a)$,

or, $\qquad\qquad\qquad De^{-a\kappa} = E \sin ka + F \cos ak.$ $\qquad\qquad$ (1)

Similarly, matching conditions at $x = -a$ gives $\psi_1(-a) = \psi_3(-a)$,

or $\qquad\qquad\qquad A e^{-a\kappa} = -E \sin ak + F \cos ak.$ $\qquad\qquad$ (2)

Further, matching the derivatives $\psi'(a)$, $\psi'(-a)$ gives $\psi_2'(a) = \psi_3'(a)$

and $\psi_1'(-a) = \psi_3'(-a)$,

or $\qquad\qquad -\kappa De^{-a\kappa} = k(E \cos ak - F \sin ak),$ $\qquad\qquad$ (3)

$\qquad\qquad\qquad \kappa Ae^{-\kappa a} = k(E \cos ak - F \sin ak).$ $\qquad\qquad$ (4)

Adding and subtracting (1) and (2) gives

$$2F \cos ak = (A + D) e^{-a\kappa}, \quad 2E \sin ak = -(A - D) e^{-a\kappa}.$$

Adding and subtracting (3) and (4) gives

$$2k E \cos ak = -\kappa (A - D) e^{-a\kappa}, \quad 2k F \sin ak = \kappa (A + D) e^{-a\kappa}.$$

Setting $A - D = -A_1$ and $A + D = A_2$, the last two sets of equations

can be combined and rewritten as

$$\left.\begin{aligned} 2E \sin ak - A_1 e^{-a\kappa} &= 0 \\ 2k E \cos ak + \kappa A_1 e^{-a\kappa} &= 0 \end{aligned}\right\} \qquad (5)$$

and

$$\left.\begin{aligned} 2F \cos ak - A_2 e^{-a\kappa} &= 0 \\ 2k F \sin ak - \kappa A_2 e^{-a\kappa} &= 0 \end{aligned}\right\}. \qquad (6)$$

The set (5) has nontrivial solutions for E and A_1 only if

$$\begin{vmatrix} 2\sin ak & -e^{-a\kappa} \\ \\ 2k\cos ak & \kappa e^{-a\kappa} \end{vmatrix} = 0 \quad \text{which gives} \quad k\cot ak = -\kappa.$$

Similarly, the set (6) has nontrivial solutions for F and A_2 only if

$$k\tan ak = \kappa.$$

Note that it is impossible to satisfy both $k\cot ak = -\kappa$ and $k\tan ak = \kappa$ simultaneously. Hence, there are two classes of solutions, and solution is possible in quantum mechanics only if the energy satisfies certain conditions.

1. Odd solutions: $k\cot ak = -\kappa$. In this case, $F = A_2 = 0$. In terms of dimensionless variables, $\xi = ak$ and $\eta = a\kappa$ with definitions of k and κ, it follows that

$$\xi^2 + \eta^2 = a^2\left(k^2 + \kappa^2\right) = a^2\left[\tfrac{2M}{\hbar^2}\left(V_0 - E\right) + \tfrac{2M}{\hbar^2}E\right] = \tfrac{2M\,V_0 a^2}{\hbar^2}. \qquad (7)$$

This represents a circle. In terms of ξ and η, we write $k\cot ak = -\kappa$ as

$$ak\cot ak = -a\kappa, \text{ or } \xi\cot\xi = -\eta. \qquad (8)$$

The simultaneous solutions of equations (7) and (8) can be determined from graphs of these functions at their point of intersection. It turns out that both ξ and η assume the positive values in the first quadrant only. Clearly, in the range $0 \le \alpha = \tfrac{2M\,V_0 a^2}{\hbar^2} < \tfrac{\pi^2}{4}$, there is no solution. For $\left(\tfrac{\pi}{2}\right)^2 \le \alpha \le \left(\tfrac{3\pi}{2}\right)^2$, there is one solution. Thus, the existence of solutions depends on the parameters, M, V_0 and the range of the potential. A simultaneous solution determines the allowed energy for which the quantum mechanical motion is described by an odd solution.

2. Even solutions: $k\tan ak = \kappa$. In this case, $E = A_1 = 0$, and (5) still holds. We can write the above condition in terms of nondimensional variables as

$$\xi \tan \xi = \eta. \tag{9}$$

The simultaneous solutions of (7) and (9) can be found graphically as before. It follows from the graphical representation that (7) and (9) intersect once if $0 \leq \alpha < \pi^2$ in the first quadrant. There are two points of intersection if $\pi^2 \leq \alpha < (2\pi)^2$. The number of intersections (solutions) increases with the value of the parameter α. For each such allowed value of the energy determined from the points of intersection, there is an even solution in the present case.

Note also that, for both even and odd solutions, $\psi(x)$ is nonzero outside the finite square well so that there exists a nonzero probability for finding the particle there. This result is different from what is expected in classical mechanics. Finally, if $V_0 \to \infty$, it is easy to see that the intersections occur at $\xi = n\pi$, $\left(n + \frac{1}{2}\right)\pi$ which are in agreement with the analysis of the infinite square well potential discussed in Example 10.10.1.

33. Hint: The boundary conditions at $x = -a$ yields the matching conditions

$$A e^{-ika} + B e^{ika} = C e^{a\kappa} + D e^{-a\kappa},$$
$$A e^{-ika} - B e^{ika} = \left(\frac{i\kappa}{k}\right)\left(C e^{a\kappa} - D e^{-a\kappa}\right).$$

These results give the desired solution.

Similarly, matching conditions at $x = a$ gives the desired answer.

Combining the matching relations leads to the final matrix equation.

11.11 Exercises

3. $u(\rho, \theta) = \frac{1}{2\pi} \int_0^{2\pi} \frac{(\rho^2 - 1)f(\beta)d\beta}{[1 - \rho^2 - 2\rho \cos(\beta - \theta)]}.$

7. $u(x, y) = -\left(\frac{2}{b}\right) \sum_{n=1}^{\infty} \frac{\sin(n\pi y/b)}{\sinh(n\pi a/b)} \left[\sinh\left\{\frac{n\pi}{b}(a - x)\right\} \int_0^x f(\xi) \sinh \frac{n\pi\xi}{b} d\xi \right.$

$$\left. + \sinh\left(\frac{n\pi x}{b}\right) \int_0^a f(\xi) \sinh\left\{\frac{n\pi}{b}(a - \xi)\right\} d\xi \right].$$

8. $u(r, \theta) = -\sum_{n=0}^{\infty} \sum_{k=1}^{\infty} (R/\alpha_{nk})^2 J_n(\alpha_{nk}r/R)(A_{nk}\cos n\theta + B_{nk}\sin n\theta),$

where

$$A_{0k} = \frac{1}{\pi R^2 J_1^2(\alpha_{0k})} \int_0^R \int_0^{2\pi} rf(r, \theta) J_0(\alpha_{0k}r/R) \, dr \, d\theta$$

$$A_{nk} = \frac{2}{\pi R^2 J_{n+1}^2(\alpha_{nk})} \int_0^R \int_0^{2\pi} rf(r, \theta) J_n(\alpha_{nk}r/R) \cos n\theta \, dr \, d\theta$$

$$B_{nk} = \frac{2}{\pi R^2 J_{n+1}^2(\alpha_{nk})} \int_0^R \int_0^{2\pi} rf(r, \theta) J_n(\alpha_{nk}r/R) \sin n\theta \, dr \, d\theta$$

$n = 1, 2, 3, \ldots;$ $k = 1, 2, 3, \ldots$ and α_{nk} are the roots of $J(\alpha_{nk}) = 0$.

9. $G(r, r') = \frac{e^{ik|r - r'|}}{|r - r'|} - \frac{e^{ik|\rho - r'|}}{|\rho - r'|},$

where $r = (\xi, \eta, \zeta)$, $r' = (x, y, z)$, and $\rho = (\xi, \eta, -\zeta)$.

10. $G(r, r') = \frac{e^{ik|r - r'|}}{|r - r'|} + \frac{e^{ik|\rho - r'|}}{|\rho - r'|}.$

14. $G = -\frac{4a}{\pi} \sum_{n=1}^{\infty} \int_0^{\infty} \frac{1}{(\alpha^2 a^2 + n^2 \pi^2)} \sin\left(\frac{n\pi x}{a}\right) \sin\left(\frac{n\pi\xi}{a}\right) \sin \alpha y \, \sin \alpha \eta \, d\alpha.$

16. $u(r, z) = \frac{2C}{\pi} \int_0^{\infty} \int_0^{\infty} \frac{1}{(\kappa^2 - \lambda^2 - \beta^2)} J_0(\beta r) J_1(\beta a) \cos \lambda z \, d\beta \, d\lambda.$

17. $u(r, \theta) = A r^{\frac{1}{2}} \sin(\theta/2).$

18. $G = -\frac{2}{a} \sum_{n=1}^{\infty} \frac{\sinh \sigma y' \, \sinh \sigma(y - b)}{\sigma \sinh \sigma b} \sin\left(\frac{n\pi x}{a}\right) \sin\left(\frac{m\pi x'}{a}\right),$

$\sigma = \sqrt{(\kappa^2 + (n^2 \pi^2)/a^2)}, \qquad 0 < x' < x < a, \qquad 0 < y' < y < b.$

12.18 Exercises

1. (a) $F(k) = \sqrt{(1/2a)}\, e^{-k^2/4a}$,　(b) $F(k) = \sqrt{\frac{2}{\pi}}\, \frac{a}{(a^2+k^2)}$.

2. $F_a(k) = \sqrt{(2/\pi)}\, (\sin ka)/k$. This function $F_a(k)$ is called the *band limited function*.

3. (a) $F_a(k) = 1/|k|$.　(b) $F(k) = \frac{1}{\sqrt{2\pi}} \int_{-a}^{a} e^{-i\omega x}\, dx = \sqrt{\frac{2}{\pi}} \left(\frac{\sin a\omega}{\omega} \right)$.

   (c) $F(k) = \frac{1}{\sqrt{2\pi}} \int_{-a}^{a} e^{-ikx} \left(1 - \frac{|x|}{a} \right) dx = \frac{2}{\sqrt{2\pi}} \int_{0}^{a} \left(1 - \frac{x}{a} \right) \cos kx\, dx$

   $= \frac{2a}{\sqrt{2\pi}} \int_{0}^{1} (1-x) \cos(akx)\, dx = \frac{2a}{\sqrt{2\pi}} \int_{0}^{1} (1-x) \frac{d}{dx} \left(\frac{\sin akx}{ak} \right) dx$

   $= \frac{2a}{\sqrt{2\pi}} \int_{0}^{1} \frac{\sin(akx)}{ak}\, dx = \frac{a}{\sqrt{2\pi}} \int_{0}^{1} \frac{d}{dx} \left[\frac{\sin^2\left(\frac{1}{2}akx\right)}{\left(\frac{1}{2}ak\right)^2} \right] dx$

   $= \frac{a}{\sqrt{2\pi}} \frac{\sin^2\left(\frac{1}{2}ak\right)}{\left(\frac{1}{2}ak\right)^2}$.

   (d) $F(k) = \sqrt{\frac{\pi}{2}}\, \frac{\exp(-a|k|)}{a}$.

4. (a) $F(k) = \sqrt{(1/2)} \sin\left(\frac{k^2}{4} + \frac{\pi}{4} \right)$,　(b) $F(k) = \sqrt{(1/2)} \cos\left(\frac{k^2}{4} - \frac{\pi}{4} \right)$.

6. Hint: $I(a,b) = \int_{0}^{\infty} e^{-a^2 x^2} \cos bx\, dx$.

   $\frac{\partial I}{\partial b} = -\frac{b}{2a^2} I \Rightarrow I = C \exp\left(-\frac{b^2}{4a^2} \right)$.

   Since $I(a,0) = C = \int_{0}^{\infty} e^{-a^2 x^2}\, dx = \frac{\sqrt{\pi}}{2a}$, $I(a,b) = \frac{\sqrt{\pi}}{2a} \exp\left(-b^2/4a^2 \right)$.

7. (a) $f(x) = \int_{-\infty}^{\infty} f(t)\, \delta(x-t)\, dt = \int_{-\infty}^{\infty} f(t)\, dt\, \frac{1}{2\pi} \int_{-\infty}^{\infty} e^{ik(x-t)}\, dk$

   $= \frac{1}{\sqrt{2\pi}} \int_{-\infty}^{\infty} e^{ikx}\, dk\, \frac{1}{\sqrt{2\pi}} \int_{-\infty}^{\infty} e^{-ikt} f(t)\, dt = \frac{1}{\sqrt{2\pi}} \int_{-\infty}^{\infty} e^{ikx} F(k)\, dk$.

8. (b) Omit the factor $\frac{1}{\sqrt{2\pi}}$ in the definition of convolution.

   $[f * (g * h)](x) = \int_{-\infty}^{\infty} f(x - \xi)(g * h)(\xi)\, d\xi$

   $= \int_{-\infty}^{\infty} f(x - \xi)\, d\xi \int_{-\infty}^{\infty} g(\xi - t)\, h(t)\, dt$

   $= \int_{-\infty}^{\infty} \left[\int_{-\infty}^{\infty} f(x - \xi) g(\xi - t)\, \right] h(t)\, dt \quad (\xi - t = \eta)$

$$= \int_{-\infty}^{\infty} \left[\int_{-\infty}^{\infty} f(x-t-\eta)\, g(\eta)\, d\eta \right] h(t)\, dt$$

$$= \int_{-\infty}^{\infty} (f*g)(x-t)\, h(t)\, dt = [(f*g)*h](t).$$

(g) $\mathcal{F}^{-1}\{(F*G)(k)\} = \frac{1}{\sqrt{2\pi}} \int_{-\infty}^{\infty} e^{ikx} [F(k)*G(k)]\, dk$

$$= \frac{1}{\sqrt{2\pi}} \int_{-\infty}^{\infty} e^{ikx}\, dk\, \frac{1}{\sqrt{2\pi}} \int_{-\infty}^{\infty} F(k-\xi)\, G(\xi)\, d\xi$$

$$= \frac{1}{\sqrt{2\pi}} \int_{-\infty}^{\infty} G(\xi)\, d\xi\, \frac{1}{\sqrt{2\pi}} \int_{-\infty}^{\infty} e^{ikx}\, F(k-\xi)\, dk$$

$$= \frac{1}{\sqrt{2\pi}} \int_{-\infty}^{\infty} G(\xi)\, d\xi\, \frac{1}{\sqrt{2\pi}} \int_{-\infty}^{\infty} e^{i(\xi+\eta)x}\, F(\eta)\, d\eta = f(x)\, g(x).$$

9. (a) $(f*g)'(x) = \frac{d}{dx} \int_{-\infty}^{\infty} f(x-\xi)\, g(\xi)\, d\xi = \int_{-\infty}^{\infty} \left[\frac{d}{dx} f(x-\xi)\right] g(\xi)\, d\xi$

$$= \int_{-\infty}^{\infty} f'(x-\xi)\, g(\xi)\, d\xi = (f'*g)(x).$$

(c) Apply the Fourier transform and then use the convolution theorem. The use of the inverse Fourier transform proves the result.

(d) $\int_{-\infty}^{\infty} (f*g)(x)\, dx = \int_{-\infty}^{\infty} \left[\int_{-\infty}^{\infty} f(x-\xi)\, g(\xi)\, d\xi \right] dx$

$$= \int_{-\infty}^{\infty} g(\xi) \left[\int_{-\infty}^{\infty} f(x-\xi)\, dx \right] d\xi$$

$$= \int_{-\infty}^{\infty} g(\xi) \left[\int_{-\infty}^{\infty} f(\eta)\, d\eta \right] d\xi$$

$$= \int_{-\infty}^{\infty} g(\xi)\, d\xi \int_{-\infty}^{\infty} f(\eta)\, d\eta.$$

(f) Apply the definition of the Fourier transform without the factor $\frac{1}{\sqrt{2\pi}}$. This means that $\mathcal{F}\{\exp(-ax^2)\} = \sqrt{\frac{\pi}{a}} \exp\left(-\frac{k^2}{4a}\right)$.

$$\mathcal{F}\{(G_t*G_s)(x)\} = \mathcal{F}\{G_t(x)\}\, \mathcal{F}\{G_s(x)\}$$

$$= \exp(-k^2\kappa\, t) \exp(-k^2\kappa\, s).$$

$$= \exp[-k^2(t+s)\kappa].$$

The inverse Fourier transform gives the result.

10. (b) L.H.S. $= \int_{-\infty}^{\infty} g(k) e^{ikx} dk \frac{1}{\sqrt{2\pi}} \int_{-\infty}^{\infty} e^{-iky} f(y) dy$

$= \int_{-\infty}^{\infty} f(y) dy \frac{1}{\sqrt{2\pi}} \int_{-\infty}^{\infty} e^{-ik(y-x)} g(k) dk$

$= \int_{-\infty}^{\infty} G(y-x) f(y) dy.$

(c) Putting $x = 0$ in the result of 10(b) gives the desired result.

(d) $\sin x * e^{-a|x|} = \frac{1}{\sqrt{2\pi}} \int_{-\infty}^{\infty} \sin(x-\xi) e^{-a|\xi|} d\xi$

$= \frac{1}{\sqrt{2\pi}} \left[\int_{0}^{\infty} \sin(x+\eta) e^{-a\eta} d\eta + \int_{0}^{\infty} \sin(x-\xi) e^{-a\xi} d\xi \right]$

$= \frac{1}{\sqrt{2\pi}} \int_{0}^{\infty} [\sin(x+\xi) + \sin(x-\xi)] e^{-a\xi} d\xi$

$= \frac{1}{\sqrt{2\pi}} 2\sin x \int_{0}^{\infty} \cos\xi\, e^{-a\xi} d\xi = \frac{1}{\sqrt{2\pi}} \sin x \left(\frac{a}{1+a^2} \right).$

(e) $e^{ax} * \chi_{[0,\infty)}(x) = \frac{1}{\sqrt{2\pi}} \int_{0}^{\infty} e^{a(x-\xi)} d\xi = \frac{1}{a} \frac{e^{ax}}{\sqrt{2\pi}}.$

(f) Apply the Fourier transform to the left hand side to obtain

$$\frac{1}{\sqrt{4ab}} \mathcal{F}\left\{ \exp\left(-\frac{x^2}{4a}\right) \right\} \mathcal{F}\left\{ \exp\left(-\frac{x^2}{4b}\right) \right\}$$
$$= \frac{1}{\sqrt{4ab}} \sqrt{2a}\, e^{-ak^2} \cdot \sqrt{2b}\, e^{-bk^2} = e^{-(a+b)k^2}$$

which can be inverted to find the result.

11. $u(x,t) = \frac{1}{2} [f(x+ct) + f(x-ct)] + \frac{1}{2c} \int_{x-ct}^{x+ct} g(\tau) d\tau.$

12. $u(x,t) = \frac{1}{2\sqrt{\pi t}} \int_{0}^{\infty} \left[e^{-(x-\xi)^2/4t} - e^{-(x+\xi)^2/4} \right] f(\xi) d\xi.$

13. $u(x,t) = \frac{1}{\sqrt{2\pi}} \int_{-\infty}^{\infty} F(\xi) \cos\left(c\xi^2 t\right) e^{-i\xi x} d\xi.$

14. $u(x,t) = \frac{1}{\sqrt{\pi}} \int_{x/\sqrt{2at}}^{\infty} g\left(\frac{t-x^2}{2a\xi^2}\right) \left(\sin\frac{\xi^2}{2} + \cos\frac{\xi^2}{2} \right) d\xi.$

15. $u(x,t) = \frac{\delta_0}{2\pi} \int_{-\infty}^{\infty} \frac{\sin a\xi}{\xi} \frac{e^{i\xi x}}{|\xi|} e^{-|\xi|y} d\xi.$

16. $\phi(x,t) = \sqrt{4\pi t} \int_{-\infty}^{\infty} e^{-(x-\xi)^2/4t} f(\xi) d\xi.$

19. $u\left(x,y\right)=\frac{2y}{\pi}\int_{0}^{\infty}\frac{\left(x^{2}+\tau^{2}+y^{2}\right)f(\tau)d\tau}{\left[y^{2}+(x-\tau)^{2}\right]\left[y^{2}+(x+\tau)^{2}\right]}$

$$-\frac{1}{2\pi}\int_{0}^{\infty}g\left(\tau\right)\log\left(\frac{\left[x^{2}+(y+\tau)^{2}\right]}{\left[x^{2}+(y-\tau)^{2}\right]}\right)d\tau.$$

21. $u\left(x,y\right)=\frac{2}{\pi}\int_{0}^{\infty}\xi\sin x\xi\int_{0}^{t}e^{-\xi^{2}(t-\tau)}f\left(\tau\right)d\tau\,d\xi.$

22. $u\left(x,y\right)=\frac{2}{\pi}\int_{0}^{\infty}\frac{\sin\xi x\,\sin h\xi(1-y)}{\sin h\xi}\int_{0}^{\infty}f\left(\tau\right)\sin\xi\tau\,d\tau\,d\xi.$

24. (a) $\left(b^{2}-a^{2}\right)^{-1}\left(\cos at-\cos bt\right),$ (b) $\left(b^{2}-a^{2}\right)^{-1}\left(\frac{\sin at}{a}-\frac{\sin bt}{b}\right),$

     (c) $(a-b)^{-1}\left(e^{at}-e^{bt}\right),$ (d) $1-e^{-at}-ate^{-at},$

     (e) $\frac{1}{a}\left(1-e^{-at}\right),$ (f) $\frac{1}{a}\sin at-2\left(\sin at*\sin at\right).$

28. $u\left(x,t\right)=\begin{cases}\frac{1}{2}\left[f\left(ct+x\right)-f\left(ct-x\right)\right] & \text{for}\quad t>x/c\\[2mm]\frac{1}{2}\left[f\left(x+ct\right)+f\left(x-ct\right)\right] & \text{for}\quad t<x/c.\end{cases}$

29. $u\left(x,t\right)=\begin{cases}0, & \text{for}\quad t<x/c\\[2mm]f\left(t-x/c\right), & \text{for}\quad x/c<t\le\left(2-x\right)/c.\end{cases}$

30. $u\left(x,t\right)=f_{0}+\left(f_{1}-f_{0}\right)\operatorname{erfc}\left(\sqrt{\left(x^{2}/4\kappa t\right)}\right).$

31. $u\left(x,t\right)=x-x\operatorname{erfc}\left(x/\sqrt{4\kappa t}\right).$

32. $u\left(x,t\right)=2\int_{0}^{t}\int_{0}^{\eta}\operatorname{erfc}\left(x/\sqrt{4\kappa\xi}\right)d\eta\,d\xi.$

33. $u\left(x,t\right)=f_{0}\,e^{-ht}\left[1-\operatorname{erfc}\left(x/\sqrt{4\kappa t}\right)\right].$

34. $u\left(x,t\right)=f_{0}\operatorname{erfc}\left(x/\sqrt{4\kappa t}\right).$

35. $u\left(x,t\right)=\begin{cases}f_{0}t, & \text{for}\quad t<x/c\\[2mm]f_{0}x/c, & \text{for}\quad t>x/c.\end{cases}$

36. $u\left(x,t\right)=\begin{cases}\frac{1}{2}\left[f\left(x+ct\right)-f\left(ct-x\right)\right], & t<x/c\\[4mm]\frac{1}{2}\left[f\left(x+ct\right)+f\left(x-ct\right)\right], & t>x/c.\end{cases}$

37. $V(x,t) = V_0 \left(t - \frac{x}{c}\right) H \left(t - \frac{x}{c}\right)$.

    (i) $V = V_0 H \left(t - \frac{x}{c}\right)$,   (ii) $V = V_0 \cos\left\{w\left(t - \frac{x}{c}\right)\right\} H \left(t - \frac{x}{c}\right)$.

38. $u(z,t) = Ut \left[\left(1 + 2\zeta^2\right) \mathrm{erfc}\,(\zeta) - \frac{2\zeta}{\sqrt{\pi}} e^{-\zeta^2}\right]$,   where $\zeta = \frac{z}{2\sqrt{\nu t}}$.

41. $V(x,t) = V_0 \mathrm{erfc}\left(\frac{x}{2\sqrt{\kappa t}}\right)$.

42. $q(z,t) = \frac{a}{2} e^{iwt} \left[e^{-\lambda_1 z}\mathrm{erfc}\left\{\zeta - \left[it\left(2\Omega + w\right)\right]^{\frac{1}{2}}\right\}\right.$

$\qquad\qquad\qquad\qquad + e^{\lambda_1 z}\mathrm{erfc}\left\{\zeta + \left[it\left(2\Omega + w\right)\right]^{\frac{1}{2}}\right\}\Big]$

$\qquad\qquad\qquad + \frac{b}{2} e^{-iwt}\left[e^{-\lambda_2 z}\mathrm{erfc}\left\{\zeta - \left[it\left(2\Omega - w\right)\right]^{\frac{1}{2}}\right\}\right.$

$\qquad\qquad\qquad\qquad + e^{\lambda_2 z}\mathrm{erfc}\left\{\zeta + \left[it\left(2\Omega - w\right)\right]^{\frac{1}{2}}\right\}\Big]$,

where

$\lambda_{1,2} = \left\{\frac{i\left(2\Omega \pm w\right)}{\nu}\right\}$.

$q(z,t) \sim a\exp\left(iwt - \lambda_1 z\right) + b\exp\left(-iwt - \lambda_2 z\right)$, $\delta_{1,2} = \left\{\frac{\nu}{|2\Omega \pm w|}\right\}^{\frac{1}{2}}$.

43. $\left(\frac{\nu}{2\Omega}\right)^{\frac{1}{2}}$.

45. $f(t) = f(0) + \frac{1}{\Gamma(\alpha)\Gamma(1-\alpha)} \int_0^t g(x)(t-x)^{\alpha-1}\,dx$.

46. $x = a(\theta - \sin\theta)$, $y = a(1 - \cos\theta)$.

55. $u(x,t) = \sum_{n=1}^{\infty} \sin nx \int_0^t e^{-n^2(t-\tau)}a_n(\tau)\,d\tau + \sum_{n=1}^{\infty} b_n(0) \sin nx\, e^{-n^2 t}$,

where

$a_n(t) = \frac{2}{\pi}\int_0^\pi g(x,t)\sin nx\,dx$, $b_n(0) = \frac{2}{\pi}\int_0^\pi f(x)\sin nx\,dx$.

57. $u(x,t) = \sum_{n=1}^{\infty}\frac{2}{\pi}\sin\left\{\left(n - \frac{1}{2}\right)x\right\}\int_0^t\int_0^\pi e^{-(2n-1)^2(t-\tau)/4}$

$\qquad\qquad\qquad\qquad\qquad\qquad \times \sin\left\{\left(n - \frac{1}{2}\right)\xi\right\}g(\xi,\tau)\,d\xi\,d\tau$.

61. $u(x,t) = \frac{c}{2}\left[\frac{\sinh\left(x\sqrt{1/c}\right)}{\sinh\left(\pi\sqrt{1/c}\right)} - \frac{\sin\left(x\sqrt{1/c}\right)}{\sin\left(\pi\sqrt{1/c}\right)}\right]\sin t$

$\qquad\qquad\qquad + \left(\frac{2}{\pi c}\right)\sum_{n=1}^{\infty}(-1)^{n+1}\frac{n}{n^4 - (1/c)^2}\sin n^2 ct\,\sin nx$,

in which $\sqrt{1/c}$ is not an integer.

65. $f(x) = \int_0^\infty g(t) h(xt) dt,$ where $h(x) = \mathcal{M}^{-1} \left[\frac{1}{K(1-p)} \right].$

14.11 Exercises

13. (c) With $h = 0.2$, the initial values are $u_{i,0}(ih, 0) = \sin \pi (ih)$.

   $u_{1,0} = \sin 0.2\pi = 0.5878$, $u_{2,0} = \sin 0.4\pi = 0.9511$.

   Also, $u_{2,0} = u_{3,0}$ and $u_{1,0} = u_{4,0}$.

   In each time step, there are 4 internal mesh points. We have to solve 4 equations with 4 unknowns. However, the initial temperature distribution is symmetric about $x = 0.5$, and $u = 0$ at the endpoints for all time t. We have $u_{3,1} = u_{2,1}$ and $u_{4,1} = u_{1,1}$ in the first time row and similarly for the other time rows. This gives two equations with two unknowns.

16. (a) $y = x^3$,   (b) $y = \sin x$,   (c) $x^2 + (y - \beta)^2 = r^2$,

   where β and r are constants.

17. $x = a(\theta - \sin \theta)$, $y = a(1 - \cos \theta)$.

18. Hint: $I(y(x)) = 2\pi \int_{x_0}^{x_1} y \left(1 + y'^2\right)^{\frac{1}{2}} dx$.

   $x = c_1 t + c_2$, $y = c_1 \cosh t = c_1 \cosh \left(\frac{x - c_2}{c_1} \right)$.

   (A surface generated by rotation of a catenary is called a *catenoid*).

21. (a) $\nabla^4 u = 0$ (Biharmonic equation)

   (b) $u_{tt} - \alpha^2 \nabla^2 u + \beta^2 u = 0$ (Klein–Gordon equation)

   (c) $\phi_t + \alpha \phi_x + \beta \phi_{xxx} = 0$, $(\phi = u_x)$ (KdV equation)

   (d) $u_{tt} + \alpha^2 u_{xxxx} = 0$ (Elastic beam equation)

   (e) $\frac{d}{dx}(pu') + (r + \lambda s) u = 0$ (Sturm–Liouville equation).

22. $\frac{\hbar^2}{2m} \nabla^2 \psi + (E - V) \psi = 0$ (Schrödinger equation).

24. $u_{tt} - c^2 u_{xx} = F(x, t)$,   where $c^2 = T^*/\rho$.

29. Hint: $y_n = x\left(1-x\right)\sum_{r=1}^{n}a_r x^{r-1}$. Find the solution for $n=1$ and $n=2$.

$n=1:$ $a_1 = \frac{5}{18}$, $y_1 = a_1 x\left(1-x\right)$.

$n=2:$ $a_1 = \frac{71}{369}$, $a_2 = \frac{7}{41}$, $y_2 = x\left(1-x\right)\left(a_1+a_2 x\right)$.

30. Hint: $u_1\left(x,y\right) = a_1 xy$.

$I\left(u_1\right) = \frac{\pi ab}{4}\left[\left(a_1+1\right)^2 a^2 + \left(a_1-1\right)^2 b^2\right]$, $a_1 = \left(\frac{b^2-a^2}{b^2+a^2}\right)$.

31. Hint: $u_3 = x\left(1-x\right)\left(1-y\right) + x\left(1-x\right)y\left(1-y\right)\left(a_2+a_3 y\right)$.

32. Hint: $u_2 = x\left(2-x-2y\right) + a_2 xy\left(2-x-2y\right)$.

33. Hint: $\Psi_N = \sum_{m,n=1}^{N}a_{mn}\phi_{mn} = \sum_{m,n=1}^{N}a_{mn}\cos\left(\frac{m\pi x}{2a}\right)\cos\left(\frac{n\pi y}{2b}\right)$.

34. Hint: $\phi_n = \cos\left[\left(2n-1\right)\frac{\pi r}{2a}\right]$.

35. Hint: $I\left(u\right) = \int_{-a}^{a}\int_{-a}^{a}\left[\left(\nabla^2 u\right)^2 - \left(\frac{4\alpha}{a^2}\right)u\right]dx\,dy = \min$, and

$u_n = \left(x^2-a^2\right)^2\left(y^2-a^2\right)^2\left(a_1+a_2 x^2+a_3 y^3+\ldots\right)$.

36. Hint: $\Psi_1 = \left(b^2-y^2\right)U\left(x\right)$.

37. (a) Introduce two functions ϕ and ψ and two parameters α and β such that $U = u+\alpha\phi\left(x\right)$ and $V = v+\beta\psi\left(x\right)$. Then $\frac{\partial I}{\partial\alpha}=0$ and $\frac{\partial I}{\partial\beta}=0$.

40. (a) $u_y - u_x y' = \frac{uy''}{1+y'^2}$.

(b) $y'^2 = \frac{1-A^2(y_1-y)}{A^2(y_1-y)}$, (d) $2y''-3y+3xy^2 = 0$.

42. Seek an approximate solution

$u_n\left(x,y\right) = \left(a^2-x^2\right)\left(b^2-y^2\right)\left(a_1+a_2 x^2+a_3 y^2+\ldots+a_n x^{2r}y^{2s}\right)$.

For $n=1$, $f=2$

$0 = \iint_R\left(-u_{1xx}-u_{1yy}-2\right)\left(a^2-x^2\right)\left(b^2-y^2\right)dx\,dy$

$= 2\int_{-a}^{a}\int_{-b}^{b}\left[1-a_1\left(a^2-x^2\right)-a_1\left(b^2-y^2\right)\right]\left(a^2-x^2\right)\left(b^2-y^2\right)dx\,dy$

$= \frac{32}{9}a^2 b^3 - \frac{128}{45}\left(ab\right)^3\left(a^2+b^2\right)a_1$, $a_1 = \frac{5}{4}\left(a^2+b^2\right)^{-1}$

and $u_1 = \frac{5}{4}\frac{\left(a^2-x^2\right)\left(b^2-y^2\right)}{\left(a^2+b^2\right)}$.

The torsional moment

$$M = 2G\theta \int_{-a}^{a}\int_{-b}^{b} u_1 dx\, dy = \left(\tfrac{40}{9}\right)(G\theta)\left(\tfrac{a^3 b^3}{a^2+b^2}\right),$$

where G is the shear modulus and θ is the angle of twist per unit length.

When $a = b$, $M = \left(\tfrac{20}{9}\right)G\theta a^4 \sim 0.1388\,(2a)^4\, G\theta$. The tangential stresses

are

$$\tau_{zx} = G\theta\left(\tfrac{\partial u_1}{\partial y}\right), \qquad \tau_{zy} = G\theta\left(\tfrac{\partial u_1}{\partial x}\right).$$

(b) The exact solution is

$$u(x,y) = x(a-x) - \tfrac{8a^2}{\pi^3}\sum_{n=1}^{\infty} \frac{\cosh\left\{(2n-1)\tfrac{\pi y}{2a}\right\}\sin\left\{(2n-1)\tfrac{\pi x}{a}\right\}}{(2n-1)^3 \cosh\left\{(2n-1)\tfrac{\pi b}{2a}\right\}}.$$

$$M = 2G\theta\left[\tfrac{a^3 b}{6} - \tfrac{32a^4}{\pi^5}\sum_{n=1}^{\infty}\tfrac{1}{(2n-1)^5}\tanh\left\{(2n-1)\tfrac{\pi b}{2a}\right\}\right].$$

For $a = b$, $M = 0.1406\,(2a)^4\, G\theta$.

43. This problem deals with the expansion of a rectangular plate under tensile forces. Make the boundary conditions homogeneous. Integrating the boundary conditions gives

$$u_0 = \tfrac{1}{2}cy^2\left(1 - \tfrac{y^2}{6b^2}\right).$$

Set $u = u_0 + \tilde{u}$ so that $\nabla^4 \tilde{u} = \left(\tfrac{2c}{b^2}\right)$ and the boundary conditions become

$$\tilde{u}_{xy} = 0 = \tilde{u}_{yy} \quad \text{for} \quad x = \pm a, \qquad \tilde{u}_{xy} = 0 = \tilde{u}_{xx} \quad \text{for} \quad y = \pm b.$$

These boundary conditions hold if

$$\tilde{u} = 0, \qquad \tilde{u}_x = 0 \quad \text{for} \quad x = \pm a,$$
$$\tilde{u} = 0, \qquad \tilde{u}_y = 0 \quad \text{for} \quad y = \pm b.$$

By the Rayleigh–Ritz method

$$\iint_R \left(\nabla^4 u_n - f\right)\phi_k dx\, dy = 0, \qquad k = 1, 2, \ldots, n,$$

where the nth approximate solution $u_n(x,y)$ has the form

$$u_n(x,y) = \left(x^2 - a^2\right)^2\left(y^2 - b^2\right)^2\left(a_1 + a_2 x^2 + a_3 y^2 + \ldots\right).$$

For $n = 1$,

$$0 = \int_{-a}^{a}\int_{-b}^{b}\left[24a_1\left(y^2 - b^2\right)^2 + 16a_1\left(3x^2 - a^2\right)\left(3y^2 - b^2\right)\right.$$
$$\left. + 24a_1\left(x^2 - a^2\right)^2 - \left(\tfrac{2c}{b}\right)\right]\left(x^2 - a^2\right)^2\left(y^2 - b^2\right)^2 dx\, dy,$$

or

$$\left(\tfrac{54}{7} + \tfrac{256}{49}\tfrac{b^2}{a^2} + \tfrac{64}{7}\tfrac{b^4}{a^4}\right) a_1 = \tfrac{c}{a^6 b^2}.$$

When $a = b$, $a_1 = (0.04325)\tfrac{c}{a^6}$.

$$u_1 \sim u_0 + \tilde{u}_1 = \tfrac{1}{2}cy^2\left(1 - \tfrac{y^2}{6b^2}\right) + (0.04325)\left(ca^{-6}\right)$$
$$\times\left(x^2 - a^2\right)^2\left(y^2 - b^2\right)^2.$$

44. $\frac{dF}{dx} = \frac{\partial F}{\partial x} + \frac{\partial F}{\partial u}\frac{du}{dx} + \frac{\partial F}{\partial u'}\cdot\frac{du'}{dx} = \frac{\partial F}{\partial x} + u'\frac{\partial F}{\partial u} + u''\frac{\partial F}{\partial u'}$

$\frac{d}{dx}\left(u'\frac{\partial F}{\partial u'}\right) = u'\frac{d}{dx}\left(\frac{\partial F}{\partial u'}\right) + \frac{\partial F}{\partial u'}\cdot u''.$

Subtracting the latter from the former with (14.6.12) we obtain

$$\tfrac{d}{dx}\left(F - u'\tfrac{\partial F}{\partial u'}\right) = \tfrac{\partial F}{\partial x} + u'\left[\tfrac{\partial F}{\partial u} - \tfrac{d}{dx}\left(\tfrac{\partial F}{\partial u'}\right)\right] = \tfrac{\partial F}{\partial x}.$$

45. $$H = I - \lambda J = \int_a^b F\left(x, y, y'\right) dx$$
$$= \int_0^b \left[p\left(x\right)y'^2 - q\left(x\right)y^2 - \lambda\, r\left(x\right)y^2\right] dx.$$

The extremum of H leads to the Euler–Lagrange equation

$$\tfrac{d}{dx}\left(\tfrac{\partial F}{\partial y'}\right) - \tfrac{\partial F}{\partial y} = 0.$$

This leads to the answer.

46. (a) For simplicity, we assume that l is an integer and partition the interval into l equal subintervals. Each of the $l - 1 = n$ interior vertices has the trial function $v_j\left(x\right)$ defined by

$$v_j\left(x\right) = \begin{cases} 1 - j + x & \text{for } j - 1 \le x \le j, \\ 1 + j - x & \text{for } j \le x \le j + 1, \\ 0 & \text{otherwise.} \end{cases}$$

$v_j\left(x\right)$ is continuous and piecewise linear with $v_j\left(j\right) = 1$ and $v_j\left(k\right) = 0$ for all integers $k \ne j$.

Appendix: Some Special Functions and Their Properties

"One of the properties inherent in mathematics is that any real progress is accompanied by the discovery and development of new methods and simplifications of previous procedures ... The unified character of mathematics lies in its very nature; indeed, mathematics is the foundation of all exact natural sciences."

David Hilbert

This appendix is a short introduction to some special functions used in the book. These functions include gamma, beta, error, and Airy functions and their main properties. Also included are Hermite and Webber–Hermite functions and their properties. Our discussion is brief since we assume that the reader is already familiar with this material. For more details, the reader is referred to appropriate books listed in the bibliography.

A-1 Gamma, Beta, Error, and Airy Functions

The *Gamma function* (also called the *factorial function*) is defined by a definite integral in which a variable appears as a parameter

$$\Gamma(x) = \int_0^\infty e^{-t} t^{x-1} dt, \quad x > 0. \tag{A-1.1}$$

In view of the fact that the integral (A-1.1) is uniformly convergent for all x in $[a, b]$ where $0 < a \le b < \infty$, $\Gamma(x)$ is a continuous function for all $x > 0$.

Integrating (A-1.1) by parts, we obtain the fundamental property of $\Gamma(x)$

$$\Gamma(x) = \left[-e^{-t}t^{x-1}\right]_0^\infty + (x-1)\int_0^\infty e^{-t}t^{x-2}dt$$
$$= (x-1)\,\Gamma(x-1), \quad \text{for} \quad x-1 > 0.$$

Then we replace x by $x+1$ to obtain the fundamental result

$$\Gamma(x+1) = x\,\Gamma(x). \tag{A-1.2}$$

In particular, when $x = n$ is a positive integer, we make repeated use of (A-1.2) to obtain

$$\Gamma(n+1) = n\,\Gamma(n) = n(n-1)\,\Gamma(n-1) = \cdots$$
$$= n(n-1)(n-2)\cdots 3\cdot 2\cdot 1\,\Gamma(1) = n!, \tag{A-1.3}$$

where $\Gamma(1) = 1$.

We put $t = u^2$ in (A-1.1) to obtain

$$\Gamma(x) = 2\int_0^\infty \exp\left(-u^2\right) u^{2x-1}du, \quad x > 0. \tag{A-1.4}$$

Letting $x = \frac{1}{2}$, we find

$$\Gamma\left(\frac{1}{2}\right) = 2\int_0^\infty \exp\left(-u^2\right) du = 2\frac{\sqrt{\pi}}{2} = \sqrt{\pi}. \tag{A-1.5}$$

Using (A-1.2), we deduce

$$\Gamma\left(\frac{3}{2}\right) = \frac{1}{2}\,\Gamma\left(\frac{1}{2}\right) = \frac{\sqrt{\pi}}{2}. \tag{A-1.6}$$

Similarly, we can obtain the values of $\Gamma\left(\frac{5}{2}\right)$, $\Gamma\left(\frac{7}{2}\right), \ldots, \Gamma\left(\frac{2n+1}{2}\right)$.

The gamma function can also be defined for negative values of x by rewriting (A-1.2) as

$$\Gamma(x) = \frac{\Gamma(x+1)}{x}, \quad x \neq 0, -1, -2, \ldots \tag{A-1.7}$$

For example

$$\Gamma\left(-\frac{1}{2}\right) = \frac{\Gamma\left(\frac{1}{2}\right)}{-\frac{1}{2}} = -2\,\Gamma\left(\frac{1}{2}\right) = -2\sqrt{\pi}, \tag{A-1.8}$$

$$\Gamma\left(-\frac{3}{2}\right) = \frac{\Gamma\left(-\frac{1}{2}\right)}{-\frac{3}{2}} = \frac{4}{3}\sqrt{\pi}. \tag{A-1.9}$$

We differentiate (A-1.1) with respect to x to obtain

$$\Gamma(x)$$

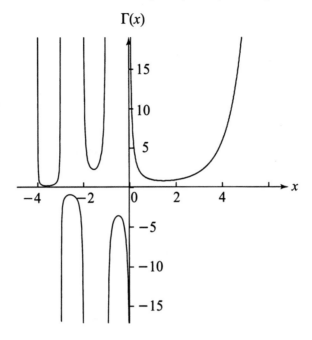

Figure A-1.1 The gamma function.

$$\frac{d}{dx}\Gamma\left(x\right)=\Gamma'\left(x\right)=\int_{0}^{\infty}\frac{d}{dx}\left(t^{x}\right)\frac{e^{-t}}{t}dt$$

$$=\int_{0}^{\infty}\frac{d}{dx}\left[\exp\left(x\log t\right)\right]\frac{e^{-t}}{t}dt$$

$$=\int_{0}^{\infty}t^{x-1}\left(\log t\right)e^{-t}dt. \tag{A-1.10}$$

At $x = 1$, this gives

$$\Gamma'\left(1\right)=\int_{0}^{\infty}e^{-t}\log t\,dt=-\gamma, \tag{A-1.11}$$

where γ is called the *Euler constant* and has the value 0.5772.

The graph of the gamma function is shown in Figure A-1.1.

Several useful properties of the gamma function are recorded below without proof for reference.

Legendre Duplication Formula

$$2^{2x-1}\,\Gamma\left(x\right)\Gamma\left(x+\frac{1}{2}\right)=\sqrt{\pi}\,\Gamma\left(2x\right), \tag{A-1.12}$$

In particular, when $x = n \; (n = 0, 1, 2, \ldots)$

$$\Gamma\left(n + \frac{1}{2}\right) = \frac{\sqrt{\pi} \; (2n)!}{2^{2n} \, n!}. \tag{A-1.13}$$

The following properties also hold for $\Gamma(x)$:

$$\Gamma(x) \, \Gamma(1 - x) = \pi \, \mathrm{cosec} \, \pi x, \qquad x \text{ is a noninteger}, \tag{A-1.14}$$

$$\Gamma(x) = p^x \int_0^\infty \exp(-pt) \, t^{x-1} dt, \tag{A-1.15}$$

$$\Gamma(x) = \int_{-\infty}^\infty \exp\left(xt - e^t\right) dt. \tag{A-1.16}$$

$$\Gamma(x + 1) \sim \sqrt{2\pi} \, \exp(-x) \, x^{x + \frac{1}{2}} \qquad \text{for large } x, \tag{A-1.17}$$

$$n! \sim \sqrt{2\pi} \, \exp(-n) \, x^{n + \frac{1}{2}} \qquad \text{for large } n. \tag{A-1.18}$$

The *incomplete gamma function*, $\gamma(x, a)$, is defined by the integral

$$\gamma(a, x) = \int_0^x e^{-t} t^{a-1} dt, \quad a > 0. \tag{A-1.19}$$

The *complementary incomplete gamma* function, $\Gamma(a, x)$, is defined by the integral

$$\Gamma(a, x) = \int_x^\infty e^{-t} \, t^{a-1} dt, \quad a > 0. \tag{A-1.20}$$

Thus, it follows that

$$\gamma(a, x) + \Gamma(a, x) = \Gamma(a). \tag{A-1.21}$$

The *beta function*, denoted by $B(x, y)$ is defined by the integral

$$B(x, y) = \int_0^t t^{x-1} (1 - t)^{y-1} \, dt, \quad x > 0, \quad y > 0. \tag{A-1.22}$$

The beta function $B(x, y)$ is *symmetric* with respect to its arguments x and y, that is,

$$B(x, y) = B(y, x). \tag{A-1.23}$$

This follows from (A-1.22) by the change of variable $1 - t = u$, that is,

$$B(x, y) = \int_0^1 u^{y-1} (1 - u)^{x-1} \, du = B(y, x).$$

If we make the change of variable $t = u / (1 + u)$ in (A-1.22), we obtain another integral representation of the beta function

$$B(x,y) = \int_0^\infty u^{x-1}(1+u)^{-(x+y)}\,du = \int_0^\infty u^{y-1}(1+u)^{-(x+y)}\,du,$$

$$\text{(A-1.24)}$$

Putting $t = \cos^2\theta$ in (A-1.22), we derive

$$B(x,y) = 2\int_0^{\pi/2} \cos^{2x-1}\theta\,\sin^{2y-1}\theta\,d\theta. \qquad \text{(A-1.25)}$$

Several important results are recorded below without proof for ready reference.

$$B(1,1) = 1, \qquad B\left(\frac{1}{2},\frac{1}{2}\right) = \pi, \qquad \text{(A-1.26)}$$

$$B(x,y) = \left(\frac{x-1}{x+y-1}\right)B(x-1,y), \qquad \text{(A-1.27)}$$

$$B(x,y) = \frac{\Gamma(x)\,\Gamma(y)}{\Gamma(x+y)}, \qquad \text{(A-1.28)}$$

$$B\left(\frac{1+x}{2},\frac{1-x}{2}\right) = \pi\sec\left(\frac{\pi x}{2}\right), \qquad 0 < x < 1. \qquad \text{(A-1.29)}$$

The *error function*, erf (x) is defined by the integral

$$\mathrm{erf}(x) = \frac{2}{\sqrt{\pi}}\int_0^x \exp\left(-t^2\right)dt, \qquad -\infty < x < \infty. \qquad \text{(A-1.30)}$$

Clearly, it follows from (A-1.30) that

$$\mathrm{erf}(-x) = -\mathrm{erf}(x), \qquad \text{(A-1.31)}$$

$$\frac{d}{dx}\left[\mathrm{erf}(x)\right] = \frac{2}{\sqrt{\pi}}\exp\left(-x^2\right), \qquad \text{(A-1.32)}$$

$$\mathrm{erf}(0) = 0, \quad \mathrm{erf}(\infty) = 1. \qquad \text{(A-1.33)}$$

The *complementary error function*, erfc (x) is defined by the integral

$$\mathrm{erfc}(x) = \frac{2}{\sqrt{\pi}}\int_x^\infty \exp\left(-t^2\right)dt. \qquad \text{(A-1.34)}$$

Clearly, it follows that

$$\mathrm{erfc}(x) = 1 - \mathrm{erf}(x), \qquad \text{(A-1.35)}$$

$$\mathrm{erfc}(0) = 1, \quad \mathrm{erfc}(\infty) = 0, \qquad \text{(A-1.36)}$$

$$\mathrm{erfc}(x) \sim \frac{1}{x\sqrt{\pi}}\exp\left(-x^2\right) \quad \text{for large } x. \qquad \text{(A-1.37)}$$

The graphs of erf (x) and erfc (x) are shown in Figure A-1.2.

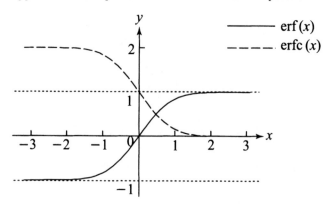

Figure A-1.2 The error function and the complementary error function.

Closely associated with the error function are the Fresnel integrals, which are defined by

$$C(x) = \int_0^x \cos\left(\frac{\pi t^2}{2}\right) dt \quad \text{and} \quad S(x) = \int_0^x \sin\left(\frac{\pi t^2}{2}\right) dt. \quad \text{(A-1.38)}$$

These integrals arise in diffraction problems in optics, in water waves, in elasticity, and elsewhere.

Clearly, it follows from (A-1.38) that

$$C(0) = 0 = S(0) \qquad\qquad\qquad\qquad \text{(A-1.39)}$$

$$C(\infty) = S(\infty) = \frac{\pi}{2}, \qquad\qquad\qquad \text{(A-1.40)}$$

$$\frac{d}{dx}C(x) = \cos\left(\frac{\pi x^2}{2}\right), \quad \frac{d}{dx}S(x) = \sin\left(\frac{\pi x^2}{2}\right). \quad \text{(A-1.41)}$$

It also follows from (A-1.38) that $C(x)$ has extrema at the points where $x^2 = (2n+1)$, $n = 0, 1, 2, 3, \ldots$, and $S(x)$ has extrema at the points where $x^2 = 2n$, $n = 1, 2, 3, \ldots$. The largest maxima occur first and are $C(1) = 0.7799$ and $S(\sqrt{2}) = 0.7139$. We also infer that both $C(x)$ and $S(x)$ are oscillatory about the line $y = 0.5$. The graphs of $C(x)$ and $S(x)$ for non-negative real x are shown in Figure A-1.3.

The Airy differential equation

$$\frac{d^2 y}{dx^2} - xy = 0 \qquad\qquad\qquad\qquad \text{(A-1.42)}$$

has solutions $y_1 = Ai(x)$ and $y_2 = Bi(x)$ which are called *Airy functions*. Using the transformation $y(x) = x^\alpha f(x^\beta)$, where α and β are constants, the Airy functions can be expressed in term of Bessel functions. Differentiating $y(x)$ with respect to x gives

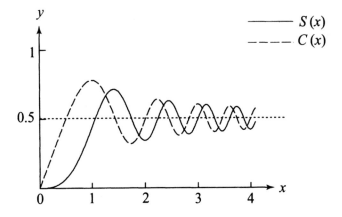

Figure A-1.3 The Fresnel integrals $C\left(x\right)$ and $S\left(x\right)$.

$$y'\left(x\right) = \alpha\, x^{\alpha-1}f + \beta\, x^{\alpha+\beta-1}f',$$
$$y''\left(x\right) = \alpha\left(\alpha-1\right)x^{\alpha-2}f + \left(2\alpha\beta + \beta^2 - \beta\right)x^{\alpha+\beta-2}f' + \beta^2 x^{\alpha+2\beta-2}f''.$$

Substituting y and y'' into the Airy equation (A-1.42) gives

$$\beta^2 x^{\alpha+2\beta-2}f'' + \left(2\alpha\beta + \beta^2 - \beta\right)x^{\alpha+\beta-2}f' + x^{\alpha-2}\left[\alpha\left(\alpha-1\right) - x^3\right]f = 0.$$

Multiplying this equation by $\left(x^{2-\alpha}\beta^{-2}\right)$ yields

$$x^{2\beta}f'' + \left(2\alpha + \beta - 1\right)\beta^{-1}x^\beta f' + \beta^{-2}\left[\alpha\left(\alpha-1\right) - x^3\right]f = 0. \quad \text{(A-1.43)}$$

Considering the coefficient of f, we require that x^3 be proportional to $x^{2\beta}$ so that $\beta = 3/2$. The coefficient of f' gives $\alpha = \frac{1}{2}$. Hence, $y\left(x\right) = x^{\frac{1}{2}}f\left(x^{3/2}\right)$. Consequently, equation (A-1.43) becomes

$$x^3 f'' + x^{3/2}f' - \left[\left(\frac{2}{3}x^{3/2}\right)^2 - \frac{1}{9}\right] = 0. \quad \text{(A-1.44)}$$

This equation admits solutions in terms of Bessel functions $K_{1/3}\left(\xi\right)$ and $I_{1/3}\left(\xi\right)$ where $\xi = \frac{2}{3}x^{3/2}$. The general solution of the Airy equation in terms of Bessel functions is given by

$$y\left(x\right) = \sqrt{x}\left[AK_{1/3}\left(\xi\right) + BI_{1/3}\left(\xi\right)\right]. \quad \text{(A-1.45)}$$

In fact, the Airy functions can be expressed as

$$Ai\left(x\right) = \frac{1}{\pi}\sqrt{\frac{x}{3}}\,K_{1/3}\left(\xi\right), \quad \text{(A-1.46)}$$

$$Bi\left(x\right) = \sqrt{\frac{x}{3}}\left[I_{1/3}\left(\xi\right) + I_{-1/3}\left(\xi\right)\right]. \quad \text{(A-1.47)}$$

The Airy equation (A-1.42) can also be solved by the method of Laplace, that is, by seeking a solution in integral form

$$y(x) = \int_C e^{xz} u(z)\, dz, \tag{A-1.48}$$

where the path of integration C is chosen such that u vanishes on the boundary. It follows that u satisfies the first-order differential equation

$$\frac{du}{dz} + z^2 u = 0.$$

The solution of this equation can be obtained in an integral form except for a normalization factor as

$$y(x) = \frac{1}{2\pi i} \int_{-i\infty}^{i\infty} \exp\left(xz - \frac{1}{3}z^3\right) dz = Ai(x), \tag{A-1.49}$$

or, equivalently,

$$Ai(x) = \frac{1}{2\pi} \int_{-\infty}^{\infty} \exp\left[i\left(xz + \frac{1}{3}z^3\right)\right] dz = \frac{1}{\pi} \int_0^{\infty} \cos\left(xz + \frac{1}{3}z^3\right) dz.$$
$$\tag{A-1.50}$$

More generally,

$$Ai(ax) = \frac{1}{2\pi a} \int_{-\infty}^{\infty} \exp\left[i\left(xz + \frac{z^3}{3a^3}\right)\right] dz$$
$$= \frac{1}{\pi a} \int_0^{\infty} \cos\left(xz + \frac{z^3}{3a^3}\right) dz. \tag{A-1.51}$$

The graph of $y = Ai(x)$ is shown in Figure A-1.4.

Finally, the method of power series can be used to solve the Airy equation to obtain the solution as

$$y(x) = a_0 \left[1 + \sum_{n=1}^{\infty} \frac{x^{3n}}{(3n)(3n-1)(3n-3)(3n-4)\ldots 3.2}\right]$$
$$+ a_1 \left[x + \sum_{n=1}^{\infty} \frac{x^{3n+1}}{(3n+1)(3n)(3n-2)(3n-3)\ldots 4.3}\right], \tag{A-1.52}$$

where a_0 and a_1 are constants.

The series solution (A-1.52) converges for all x rapidly due to the rapid decay of the coefficients as $n \to \infty$.

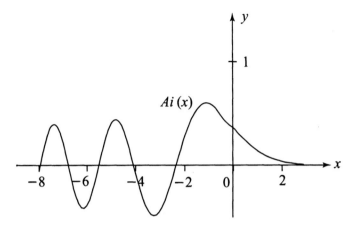

Figure A-1.4 Graph of the Airy function.

A-2 Hermite Polynomials and Weber–Hermite Functions

The Hermite polynomials $H_n(x)$ are defined by the Rodrigues formula

$$H_n(x) = (-1)^n \exp\left(x^2\right) \frac{d^n}{dx^n} \left[\exp\left(-x^2\right)\right],\qquad\text{(A-2.1)}$$

where $n = 0, 1, 2, 3, \ldots$.

The first seven Hermite polynomials are

$$H_0(x) = 1$$
$$H_1(x) = 2x$$
$$H_2(x) = 4x^2 - 2$$
$$H_3(x) = 8x^3 - 12x$$
$$H_4(x) = 16x^4 - 48x^2 + 12$$
$$H_5(x) = 32x^5 - 16x^3 + 120x$$
$$H_6(x) = 64x^6 - 480x^4 + 720x^2 - 120.$$

The *generating function of* $H_n(x)$ is

$$\exp\left(2xt - t^2\right) = \sum_{n=0}^{\infty} \frac{t^n}{n!} H_n(x).\qquad\text{(A-2.2)}$$

It follows from (A-2.2) that $H_n(x)$ satisfies the *parity relation*

$$H_n(-x) = (-1)^n H_n(x).\qquad\text{(A-2.3)}$$

Also, it follows from (A-2.2) that

$$H_{2n+1}(0) = 0, \quad H_{2n}(0) = (-1)^n \frac{(2n)!}{n!}. \qquad (A\text{-}2.4)$$

The *recurrence relations* for Hermite polynomials are

$$H_{n+1}(x) - 2x\, H_n(x) + 2n\, H_{n-1}(x) = 0, \qquad (A\text{-}2.5)$$
$$H_n'(x) = 2x\, H_{n-1}(x). \qquad (A\text{-}2.6)$$

The Hermite polynomials, $y = H_n(x)$, are solutions of the *Hermite differential equation*

$$y'' - 2xy' + 2ny = 0. \qquad (A\text{-}2.7)$$

The orthogonal property of Hermite polynomials is

$$\int_{-\infty}^{\infty} \exp\left(-x^2\right) H_n(x)\, H_m(x)\, dx = 2^n n! \sqrt{\pi}\, \delta_{mn}. \qquad (A\text{-}2.8)$$

With repeated use of integration by parts, it follows from (A-2.1) that

$$\int_{-\infty}^{\infty} \exp\left(-x^2\right) H_n(x)\, x^m\, dx = 0, \quad m = 0, 1, \ldots, (n-1), \qquad (A\text{-}2.9)$$
$$\int_{-\infty}^{\infty} \exp\left(-x^2\right) H_n(x)\, x^n\, dx = \sqrt{\pi}\, n!. \qquad (A\text{-}2.10)$$

The *Weber–Hermite* or, *simply the Hermite function*

$$y = h_n(x) = \exp\left(-\frac{x^2}{2}\right) H_n(x) \qquad (A\text{-}2.11)$$

satisfies the differential equation

$$h_n''(x) + \left(\lambda - x^2\right) h_n(x) = 0, \qquad (A\text{-}2.12)$$

where $\lambda = 2n + 1$. If $\lambda \neq 2n + 1$, then $h_n(x)$ is not finite as $|x| \to \infty$.

The Hermite functions $\{h_n(x)\}_0^{\infty}$ are an orthogonal basis for $L^2(\mathbb{R})$ with weight function one. They satisfy the following relations:

$$h_n'(x) = -x\, h_n(x) + 2n\, h_{n-1}(x), \qquad (A\text{-}2.13)$$
$$h_n'(x) = x\, h_n(x) - h_{n+1}(x), \qquad (A\text{-}2.14)$$
$$h_n''(x) - x^2\, h_n(x) + (2n+1)\, h_n(x) = 0. \qquad (A\text{-}2.15)$$

The normalized Weber–Hermite functions are given by

$$\psi_n(x) = \frac{h_n(x)}{(2^n n! \sqrt{\pi})} \exp\left(-\frac{x^2}{2}\right) H_n(x). \qquad (A\text{-}2.16)$$

The functions $\{\psi_n(x)\}$ form a orthornormal set in $(-\infty, \infty)$, that is,

$$\int_{-\infty}^{\infty} \psi_m(x)\,\psi_n(x)\,dx = \delta_{mn}. \qquad (A\text{-}2.17)$$

Physically, they represent quantum mechanical oscillator wave functions. Some graphs of these functions are shown in Figure A-2.1.

The Fourier transform of $h_n(x)$ is $(-i)^n h_n(x)$, that is,

$$\mathcal{F}\{h_n(x)\} = (-i)^n h_n(x). \qquad (A\text{-}2.18)$$

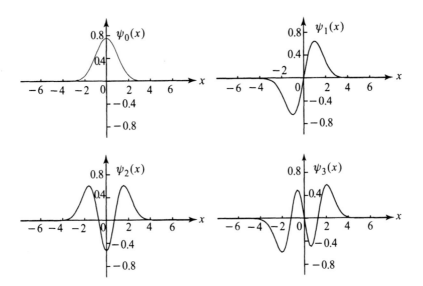

Figure A-2.1 The normalized Weber–Hermite functions.

Bibliography

The following bibliography is not, by any means, a complete one for the subject. For the most part, it consists of books and papers to which reference is made in text. Many other selected books and papers related to material in this book have been included so that they may serve to stimulate new interest in future advanced study and research.

[1] Ablowitz, M.J. and Segur, H., *Solitons and Inverse Scattering Transform,* SIAM, Philadelphia (1981).

[2] Arnold, V.I., *Ordinary Differential Equations,* Springer-Verlag, New York (1992).

[3] Bateman, H., *Partial Differential Equations of Mathematical Physics,* Cambridge University Press, Cambridge (1959).

[4] Becker, A.A., *The Boundary Element Method in Engineering,* Mc-Graw Hill, New York (1992).

[5] Benjamin, T.B., Bona, J.L., and Mahony, J.J., Model equations for long waves in nonlinear dispersive systems, *Phil. Trans. Roy. Soc.* **A272** (1972) 47–78.

[6] Berg, P. and McGregor, J., *Elementary Partial Differential Equations,* Holden-Day, New York (1966).

[7] Birkhoff, G. and Rota, G-C., *Ordinary Differential Equations* (Fourth Edition), John Wiley and Sons, New York (1989).

[8] Boas, M.L., *Mathematical Methods in the Physical Sciences,* John Wiley, New York (1966).

[9] Broman, A., *Introduction to Partial Differential Equations: from Fourier Series to Boundary-Value Problems,* Addison-Wesley, Reading, Massachusetts (1970).

[10] Brown, J.W. and Churchill, R.V., *Fourier Series and Boundary Value Problems* (Fifth Edition), McGraw-Hill, New York (1993).

[11] Burgers, J.M., A mathematical model illustrating the theory of turbulence, *Adv. Appl. Mech.* **1** (1948) 171–191.

[12] Byerly, W.E., *Fourier Series,* Dover Publications, New York (1959).

[13] Carleson, L., On the convergence and growth of partial sums of Fourier Series, *Acta Mathematica,* **116** (1966) 135–157.

[14] Carslaw, H.S., *Introduction to the Theory of Fourier Series and Integrals,* Dover Publications, New York (1950).

[15] Carslaw, H.S. and Jaeger, J.C., *Operational Methods in Applied Mathematics,* Oxford University Press, London (1949).

[16] Carslaw, H.S. and Jaeger, J.C., *Conduction of Heat in Solids* (Second Edition), Oxford University Press, Oxford (1986).

[17] Cartan, E., *Lecons sur les invariants intégraux,* Hermann, Paris (1922).

[18] Churchill, R.V., *Operational Mathematics* (Third Edition), McGraw-Hill, New York (1972).

[19] Clough, R.W., The finite element method after twenty five years; A personal view, *Computer Structures* **12** (1980) 361–370.

[20] Coddington, E.A. and Levinson, N., *Theory of Ordinary Differential Equations,* McGraw-Hill, New York (1955).

[21] Cole, R., *Theory of Ordinary Differential Equation,* Appleton-Century-Crofts (1968).

[22] Copson, E.T., *Asymptotic Expansions,* Cambridge University Press, Cambridge (1965).

[23] Coulson, C.A. and Jeffrey, A., *Waves: A Mathematical Approach to Common Types of Wave Motion* (Second Edition), Longman, London (1977).

[24] Courant, R., Variational methods for the solutions of problems of equilibrium and vibrations, *Bull. Amer. Math. Soc.* **49** (1943) 10–30.

[25] Courant, R. and Hilbert, D., *Methods of Mathematical Physics, Volume 1* (1953), *Volume 2* (1962), Interscience, New York.

[26] Courant, R., Friedrichs, K., and Lewy, H., Über die partiellen differenzengleichungen de mathematischen Physik, *Mathematische Annalen* **100** (1928) 32–74.

[27] Crank, J., *Mathematics of Diffusion* (Second Edition), Clarendon Press, Oxford (1975).

[28] Crank, J. and Nicholson, P., A practical method for numerical evaluation of solutions of partial differential equations of the heat conduction type, *Proc. Camb. Phil. Soc.* **43** (1947) 50–67.

[29] Davis, H.F., *Fourier Series and Orthogonal Functions,* Allyn and Bacon, New York (1963).

[30] Debnath, L., On asymptotic treatment of the transient development of surface waves, *Appl. Sci. Res.* **21** (1969) 24–36.

[31] Debnath, L., Unsteady axisymmetric capillary-gravity waves in a viscous liquid, *Indian J. Pure and Appl. Math.* **14** (1983) 540–553.

[32] Debnath, L., *Nonlinear Waves,* Cambridge University Press, Cambridge, England (1983).

[33] Debnath, L., Solitons and the inverse scattering transforms, *Bull. Cal. Math. Soc.* **84** (1992) 475–504.

[34] Debnath, L., *Nonlinear Water Waves,* Academic Press, Boston (1994).

[35] Debnath, L., *Integral Transforms and Their Applications*, CRC Press, Boca Raton, Florida (1995).

[36] Debnath, L., Some nonlinear evolution equations in water waves, *J. Math. Anal. Appl.* **251** (2000) 488–503.

[37] Debnath, L., Nonlinear dynamics of water waves and breaking phenomena, *Proc. Indian Nat. Sci. Acad.* Special Volume (2002) 683–753.

[38] Debnath, L., Fractional integral and fractional partial differential equation in fluid mechanics, *Fractional Calculus Appl. Analysis* **6** (2003) 119–155.

[39] Debnath, L., Recent applications of fractional calculus to science and engineering, *Internat. J. Math. & Math. Sci.* **2003, No. 54** (2003) 3413–3442.

[40] Debnath, L., *Nonlinear Partial Differential Equations for Scientists and Engineers* (Second Edition), Birkhauser Verlag, Boston (2005).

[41] Debnath, L. and Bhatta, D., On solutions to few linear fractional inhomogeneous partial differential equations in fluid mechanics, *Fractional Calculus Appl. Analysis* **7** (2004) 21–36.

[42] Debnath, L. and Bhatta, D., *Integral Transforms and Their Applications* (Second Edition), CRC Press, Boca Raton, Florida (2006).

[43] Debnath, L. and Mikusinski, P., *Introduction to Hilbert Spaces with Applications* (Third Edition), Elsevier Academic Press, London (2005).

[44] Debnath, L. and Rollins, D., The Cauchy–Poisson waves in an inviscid rotating stratified liquid, *Internat. J. Nonlinear Mech.* **27** (1992) 405–412.

[45] Debnath, L. and Shivamoggi, B.K., Three-dimensional nonlinear Schrodinger equation for finite-amplitude gravity waves in a fluid, *Nuovo Cimento* **94B** (1986) 140–148; **99B** (1987) 247.

[46] Debnath, L. and Shivamoggi, B.K., Pseudo-variational principles and nonlinear surface waves in a dissipative fluid, *Internat. J. Nonlinear Mech.* **25** (1990) 61–65.

[47] Dennemeyer, R., *Partial Differential Equations and Boundary Value Problems*, McGraw-Hill, New York (1968).

[48] Ditkin, V.A. and Prudnikov, A.P., *Integral transforms and Operational Calculus*, Pergamon Press, Oxford (1965).

[49] Doetsch, G., *Introduction to the Theory and Applications of the Laplace Transformation* (Translated by Walter Nader), Springer Verlag, New York (1970).

[50] Duff, G.F.D., *Partial Differential Equations*, University of Toronto, Toronto (1956).

[51] Duff, G.F.D. and Naylor, D., *Differential Equations of Applied Mathematics*, John Wiley, New York (1966).

[52] Dutta, M. and Debnath, L., *Elements of the Theory of Elliptic and Associated Functions with Applications*, World Press, Calcutta (1965).

[53] Epstein, B., *Partial Differential Equations*, McGraw-Hill, New York (1962).

[54] Erdélyi, A., Magnus, W., Oberhettinger, F. and Tricomi, F., *Tables of Integral Transforms* (Vols. I and II), McGraw-Hill, New York (1954).

[55] Folland, G.B., *Fourier Analysis and Its Applications*, Brooks/Cole Publishing Company, Pacific Grove, California (1992).

[56] Forsythe, G.E. and Wasow, W.R., *Finite-Difference Methods for Partial Differential Equations*, John Wiley, New York (1967).

[57] Fourier, J.B.J., *The Analytical Theory of Heat* (Translated by A. Freeman), Dover Publications, New York (1955).

[58] Fox, C., *An Introduction to the Calculus of Variations*, Oxford University Press, Oxford (1963).

[59] Garabedian, P.R., *Partial Differential Equations*, John Wiley, New York (1964).

[60] Gelfand, I.M. and Fomin, S.V., *Calculus of Variations* (Translated by R.A. Silverman), Prentice-Hall, Englewood Cliffs, New Jersey (1963).

[61] Ghosh, K.K. and Debnath, L., Some exact solutions of nonlinear shallow water equations, *Internal. J. Nonlinear Mech.* **31** (1997) 104–108.

[62] Greenberg, M., *Applications of Green's Functions in Science and Engineering*, Prentice-Hall, Englewood Cliffs, New Jersey (1971).

[63] Grunberg, G., A new method of solution of certain boundary problems for equations of mathematical physics permitting of a separation of variables, *J. Phys.* **10** (1946) 301–320.

[64] Haberman, R., *Mathematical Models: Mechanical Vibrations, Population Dynamics and Traffic Flow*, Prentice Hall, Englewood Cliffs, New Jersey (1977).

[65] Hadamard, J., *Lectures on Cauchy's Problem in Linear Partial Differential Equations*, Dover Publications, New York (1952).

[66] Haight, F.A., *Mathematical Theories of Traffic Flow*, Academic Press, New York (1963).

[67] Hellwig, G., *Partial Differential Equations*, Blaisdell, Waltham, Massachusetts (1964).

[68] Hildebrand, F., *Methods of Applied Mathematics*, Prentice-Hall, Englewood Cliffs, New Jersey (1965).

[69] Huebuer, K.H., Dewhirst, D.L., Smith, D.L., and Byrom, T.G., *The Finite Element Method for Engineers* (Fourth Edition), John Wiley & Sons, New York (2001).

[70] Irving, J. and Mullineux, N., *Mathematics in Physics and Engineering*, Academic, New York (1959).

[71]   Jackson, D., *Fourier Series and Orthogonal Polynomials*, Mathematical Association of America Monograph (1941).

[72]   Jaswon, M.A., Integral equation method in potential theory. I, *Proc. R. Soc. London, A* **275** (1963) 23–32.

[73]   Jeffrey, A., *Advanced Engineering Mathematics*, Academic Press, Boston (2002).

[74]   Jeffrey, A. and Engelbrecht, J., *Nonlinear Waves in Solids*, Springer-Verlag, New York (1994).

[75]   Jeffrey, A. and Kawahara, T., *Asymptotic Methods in Nonlinear Wave Theory*, Pitman Advanced Publishing Program, Boston (1982).

[76]   Jeffrey, A. and Taniuti, T., *Nonlinear Wave Propagation*, Academic Press, New York (1964).

[77]   Jeffreys, H., *Operational Methods in Mathematical Physics*, Cambridge University Press, New York (1931).

[78]   Jeffreys, J. and Jeffreys, B.S., *Methods of Mathematical Physics* (Third Edition), Cambridge University Press, Cambridge (1956).

[79]   John, F., *Partial Differential Equations* (Fourth Edition), Springer-Verlag, New York (1982).

[80]   Johnson, R.S., The KdV equation and related problems in water wave theory, in *Nonlinear Waves* (ed. L. Debnath), 25–43, Cambridge University Press, Cambridge (1983).

[81]   Johnson, R.S., *A Modern Introduction to the Mathematical Theory of Water Waves*, Cambridge University Press, Cambridge (1997).

[82]   Jones, D.S., *Generalized Functions*, Academic Press, New York (1966).

[83]   Kantorovich, L. and Krylov, V., *Approximate Methods in Higher Analysis,* Interscience, New York (1958).

[84]   Keener, J.P., *Principles of Applied Mathematics: Transformation and Approximation*, Addition-Wesley, Reading, Massachusetts (1988).

[85]   Keller, H., *Numerical Methods for Two Point Boundary Value Problems,* Blaisdell, Waltham, Massachusetts (1968).

[86]   Knobel, R., *An Introduction to the Mathematical Theory of Waves*, American Mathematical Society, Providence (1999).

[87]   Korteweg, D.J. and de Vries, G., On the change of form of long waves advancing in a rectangular canal, and on a new type of long stationary waves, *Phil. Mag.* **39** (1895) 422–443.

[88]   Koshlayokov, N.S., Smirnov, M.M. and Gliner, E.B., *Differential Equations of Mathematical Physics*, Holden-Day, New York (1964).

[89]   Kreider, D., Kuller, R., Ostberg, D., and Perkins, F., *An Introduction to Linear Analysis*, Addition-Wesley, Reading, Massachusetts (1966).

[90]   Kreyszig, E., *Advanced Engineering Mathematics* (Seventh Edition), John Wiley, New York (1993).

[91] Lamb, H., *Hydrodynamics* (Sixth Edition), Cambridge University Press, Cambridge, England (1932).

[92] Lax, P.D., Weak solutions of nonlinear hyperbolic equations and their numerical computation, *Comm. Pure Appl. Math.* **7** (1954) 159–193.

[93] Lax, P.D., The initial-value problem for nonlinear hyperbolic equations in two independent variables, *Ann. Math. Stud.* (Princeton) **33** (1954) 211–299.

[94] Lax, P.D., Hyperbolic systems of conservation law, II, *Comm. Pure Appl. Math.* **10** (1957) 537–566.

[95] Lax, P.D., *Partial Differential Equations*, Lectures on Hyperbolic Equations, Stanford University Press (1963).

[96] Lax, P.D., Development of singularities of solutions of nonlinear hyperbolic partial differential equations, *J. Math. Phys.* **5** (1964) 611–613.

[97] Lax, P.D., Integrals of nonlinear equations of evolution and solitary waves, *Comm. Pure Appl. Math.* **21** (1968) 467–490.

[98] Lax, P.D., The formation and decay of shock waves, *Amer. Math. Monthly* **79** (1972) 227–241.

[99] Lax, P.D., *Hyperbolic Systems of Conservation Lays and the Mathematical Theory of Shock Waves*, Society for Industrial and Applied Mathematics, Philadelphia (1973).

[100] Lax, P.D., Periodic solutions of the Korteweg–de Vries equation, *Comm. Pure Appl. Math.* **28** (1975) 141–188.

[101] Lax, P.D., A Hamiltonian approach to the KdV and other equations, in *Nonlinear Evolution Equations* (ed. M. Crandall), Academic Press, New York (1978) 207–224.

[102] Lax, P.D. and Wendroff, B., Systems of Conservation Laws, *Comm. Pure Appl. Math.* **13** (1960) 217–237.

[103] Leibovich, S. and Seebass, A.R., *Nonlinear Waves,* Cornell University Press, Ithaca and London (1972).

[104] Lighthill, M.J., *Introduction to Fourier Analysis and Generalized Functions,* Cambridge University Press, Cambridge (1964).

[105] Lighthill, M.J., *Waves in Fluids,* Cambridge University Press, Cambridge (1980).

[106] Logan, J.D., *Applied Mathematics* (Second Edition), John Wiley, New York (1997).

[107] Mainardi, F., On the initial-value problem for the fractional diffusion-wave equation, in *Waves and Stability in Continuous Media* (ed. S. Rionero and T. Ruggeri), World Scientific, Singapore (1994) 246–251.

[108] Mainardi, F., The time fractional diffusion-wave equation, *Radiofisika* **38** (1995) 20–36.

[109] Mainardi, F., Fractional relaxation-oscillation and fractional diffusion-wave phenomena, *Chaos, Solitons, Fractals* **7** (1996) 1461–1477.

[110] McLachlan, N.W., *Complex Variable and Operational Calculus with Technical Applications*, Cambridge University Press, London (1942).

[111] McLachlan, N.W., *Modern Operational Calculus*, Macmillan, London (1948).

[112] Mikhlin, S.G., *Variational Methods in Mathematical Physics*, Pergamon Press, Oxford (1964).

[113] Mikhlin, S.G., *Linear Equations of Mathematical Physics*, Holt, Rinehart and Winston, New York (1967).

[114] Miller, K.S., *Partial Differential Equations in Engineering Problems*, Prentice Hall, Englewood Cliffs, New Jersey (1953).

[115] Mitchell, A.R. and Griffiths, D.F., *The Finite Difference Method in Partial Differential Equations*, John Wiley, New York (1980).

[116] Morse, P.M. and Freshbach, H., *Methods of Theoretical Physics*, Volume 1 and 2, McGraw-Hill, New York (1953).

[117] Myint-U, T., *Ordinary Differential Equations*, Elsevier North Holland, Inc., New York (1978).

[118] Nigmatullin, R.R., The realization of the generalized transfer equation in a medium with fractal geometry, *Phys. Sta. Sol. (b)* **133** (1986) 425–430.

[119] Petrovsky, I., *Lectures on Partial Differential Equations*, Interscience, New York (1954).

[120] Picard, E., *Traité d'Analyse*, Gauthier-Villars, Paris (1896).

[121] Pinkus, A. and Zafrany, S., *Fourier Series and Integral Transforms*, Cambridge University Press, Cambridge (1997).

[122] Pipes, L.A., *Applied Mathematics for Engineers and Physicists*, McGraw-Hill, New York (1958).

[123] Qiao, Z., Generation of the hierarchies of solitons and generalized structure of the commutator representation, *Acta. Appl. Math. Sinica* **18** (1995) 287–301.

[124] Qiao, Z. and Strampp, W., Negative order MKdV hierarchy and a new integrable Neumann-like system, *Physica A* **313** (2002) 365–380.

[125] Raven, R.H., *Mathematics of Engineering Systems*, McGraw-Hill, New York (1966).

[126] Rayleigh, Lord, *The Theory of Sound*, Vol. I (1894), Vol. 2 (1896), Dover Publications, New York.

[127] Reif, F., *Fundamentals of Statistical and Thermal Physics*, McGraw Hill, New York (1965).

[128] Richtmyer, R.D. and Morton, K.W., *Difference Methods for initial-value problems*, Interscience, New York (1967).

[129] Roach, G.F., *Green's Functions* (Second Edition), Cambridge University Press, Cambridge (1982).

[130] Rogosinski, W.W., *Fourier Series*, Chelsea, London (1950).

[131] Sagan, H., *Boundary and Eigenvalue Problems in Mathematical Physics*, John Wiley, New York (1966).

[132] Sansone, G., *Orthogonal Functions*, Interscience, New York (1959).

[133] Schneider, W.R. and Wyss, W., Fractional diffusion and wave equations, *J. Math. Phys.* **30** (1989) 134–144.

[134] Schwartz, L., *Théorie des distributions*, Volume I (1950) and Volume II (1951), Herman and Cie, Paris.

[135] Scott, E.J., *Transform Calculus with an Introduction to Complex Variables*, Harper, New York (1955).

[136] Seeley, R., *An Introduction to Fourier Series and Integrals*, W.A. Benjamin, New York (1966).

[137] Smirnov, V.I., *Integral Equations and Partial Differential Equations*, Addison-Wesley, Reading, Massachusetts (1964).

[138] Smith, G.D., *Numerical Solution of the Partial Differential Equations* (Third Edition), Clarendon Press, Oxford (1985).

[139] Smith, M.G., *Introduction to the Theory of Partial Differential Equations*, Van Nostrand, Princeton, New Jersey (1967).

[140] Smoller, J., *Shock Waves and Reaction-Diffusion Equations* (Second Edition), Springer-Verlag, New York (1995).

[141] Sneddon, I.N., *Fourier Transforms*, McGraw-Hill, New York (1951).

[142] Sneddon, I.N., *Elements of Partial Differential Equations*, McGraw-Hill, New York (1957).

[143] Sneddon, I.N., *The Use of Integral Transforms*, McGraw-Hill, New York (1972).

[144] Sobolev, S.L., *Partial Differential Equations of Mathematical Physics*, Addison-Wesley, Reading, Massachusetts (1964).

[145] Soewono, E. and Debnath, L., Asymptotic stability of solutions of the generalized Burgers equation, *J. Math. Anal. Appl.* **153** (1990) 179–189.

[146] Soewono, E. and Debnath, L., Classification of self-similar solutions to a generalized Burgers equation, *J. Math. Anal. Appl.* **184** (1994) 389–398.

[147] Sokolnikoff, I.S., *Mathematical Theory of Elasticity*, McGraw-Hill, New York (1956).

[148] Sokolnikoff, I.S. and Redheffer, R.M., *Mathematics of Physics and Modern Engineering*, McGraw-Hill, New York (1966).

[149] Sommerfeld, A., *Partial Differential Equations in Physics*, Academic Press, New York (1964).

[150] Stakgold, I., *Green's Functions and Boundary Value Problems*, Wiley-Interscience, New York (1979).

[151] Stoker, J.J., *Water Waves*, Interscience, New York (1957).

[152] Stoker, G., On the theory of oscillatory waves, *Trans. Camb. Phil. Soc.* **8** (1847) 197–229.

[153] Strauss, W.A., *Partial Differential Equations: An Introduction*, John Wiley, New York (1992).

[154] Symm, G.T., Integral equation method in potential theory, *Proc. Roy Soc. London* **A275** (1963) 33–46.

[155] Tikhonov, A.N. and Samarskii, A.A., *Equations of Mathematical Physics*, Macmillan, New York (1963).

[156] Titchmarsh, E.C., *Eigenfunction Expansions Associated with Second-Order Differential Equations*, Oxford University Press, Oxford (1946).

[157] Titchmarsh, E.C., *An Introduction to the Theory of Fourier Integrals*, Oxford University Press, Oxford (1962).

[158] Toda, M., Vibration of a chain with nonlinear interaction, *J. Phys. Soc. Japan* **22** (1967) 431–436.

[159] Tolstov, G.P., *Fourier Series*, Prentice-Hall, Englewood Cliffs, New Jersey (1962).

[160] Tranter, C.J., *Integral Transforms in Mathematical Physics* (Third Edition), John Wiley, New York (1966).

[161] Tychonov, A.N. and Samarskii, A.A., *Partial Differential Equations of Mathematical Physics*, Holden-Day, New York (1964).

[162] Watson, G.N., *A Treatise on the Theory of Bessel Functions*, Cambridge University Press, Cambridge (1966).

[163] Webster, A.G., *Partial Differential Equations of Mathematical Physics*, Dover Publications, New York (1955).

[164] Weinberger, H., *A First Course in Partial Differential Equations*, Blaisdell, Waltham, Massachusetts (1965).

[165] Whitham, G.B., A general approach to linear and nonlinear dispersive waves using a Lagrangian, *J. Fluid Mech.* **22** (1965) 273–283.

[166] Whitham, G.B., A new approach to problems of shock dynamics. Part I. Two dimensional problems, *J. Fluid Mech.* **2** (1975) 146–171.

[167] Whitham, G.B., *Linear and Nonlinear Waves*, John Wiley, New York (1976).

[168] Whittaker, E.T. and Watson, G.N., *A Course on Modern Analysis*, Cambridge University Press, Cambridge (1952).

[169] Widder, D.V., *The Laplace Transform*, Princeton University Press, Princeton, New Jersey (1941).

[170] Zabusky, N.J. and Kruskal, M.D., Interaction of Solitons in a Collisionless Plasma and Recurrence of Initial States, *Phys. Rev. Lett.* **15** (1965) 240–243.

[171] Zakharov, V.E. and Shabat, A.B., Exact Theory of Two-Dimensional Self-Focusing and One-Dimensional Self Modulation of Waves in Nonlinear Media, *Sov. Phys. JETP* **34** (1972) 62–69.

[172] Zakharov, V.E. and Shabat, A.B., Interaction Between Solitons in a Stable Medium, *Sov. Phys. JETP* **37** (1974) 823–828.

[173] Zauderer, E., *Partial Differential Equations of Applied Mathematics* (Second Edition), John Wiley, New York (1989).

[174] Zienkiewicz, O.C. and Cheung, Y.K., Finite elements in the solutions of field problems, *Engineer.* **220** (1965) 370–507.

[175] Zienkiewicz, O.C. and Taylor, R.L., *The Finite Element Method*, McGraw-Hill, New York (1989).

Tables and Formulas

[176] Campbell, G.A. and Foster, R.M., *Fourier Integrals for Practical Applications*, Van Nostrand, New York (1948).

[177] Erdélyi, A., Magnus, W., Oberhettinger, F. and Tricomi, F. G., *Higher Transcendental Functions*, Volumes I, II, and III, McGraw-Hill, New York (1953).

[178] Jeffrey, A., *Handbook of Mathematical Formulas and Integrals* (Third Edition), Elsevier Academic Press, New York (2004).

[179] Jhnke, E., Emde, F., and Losch, F., *Tables of Higher Functions*, McGraw-Hill, New York (1960).

[180] Magnus, W., Oberhettinger, F., and Soni, R.P., *Formulas and Theorems for the Special Functions of Mathematical Physics* (Third Edition), Springer-Verlag, Berlin (1966).

[181] Oberhettinger, F., *Tables of Bessel Transforms*, Springer-Verlag, New York (1972).

[182] Marichev, O.I., *Handbook of Integral Transforms of Higher Transcendental Functions: Theory and Algorithmic Tables*, Ellis Horwood, Chichester, (1982).

[183] Zeidler, E., *Oxford User's Guide to Mathematics*, Oxford University Press, Oxford (2004).

Problem Books

[184] Budak, B.M., Samarskii, A.A., and Tikhonov, A.N., *A Collection of Problems on Mathematical Physics*, Macmillan, New York (1964).

[185] Lebedev, N.N., Skalskaya, I.P., and Uflyand, Y.S., *Problems of Mathematical Physics*, Prentice-Hall, Englewood Cliffs, New Jersey (1965).

[186] Smirnov, M.M., *Problems on the Equations of Mathematical Physics*, Noordhoff (1967).

Index

Printed in the United States of America

Lightning Source UK Ltd.
Milton Keynes UK
UKOW04n1312200514

231973UK00015B/492/P